Mathematical Physics Studies

More information about this series at http://www.springer.com/series/6316

Romeo Brunetti · Claudio Dappiaggi
Klaus Fredenhagen · Jakob Yngvason
Editors

Advances in Algebraic Quantum Field Theory

Springer

Editors
Romeo Brunetti
University of Trento
Trento
Italy

Klaus Fredenhagen
University of Hamburg
Hamburg
Germany

Claudio Dappiaggi
University of Pavia
Pavia
Italy

Jakob Yngvason
University of Vienna
Vienna
Austria

ISSN 0921-3767 ISSN 2352-3905 (electronic)
Mathematical Physics Studies
ISBN 978-3-319-21352-1 ISBN 978-3-319-21353-8 (eBook)
DOI 10.1007/978-3-319-21353-8

Library of Congress Control Number: 2015943794

Springer Cham Heidelberg New York Dordrecht London

Printed on acid-free paper

Springer International Publishing AG Switzerland is part of Springer Science+Business Media
(www.springer.com)

To the memory of our friend and colleague
John E. Roberts

Preface

Relativistic quantum field theory was conceived in the late 1920s as a framework unifying the two fundamental theories that revolutionized physics in the twentieth century: Quantum Mechanics and the Special Theory of Relativity. *Algebraic Quantum Field Theory* (AQFT) is relativistic quantum field theory regarded from a certain perspective, emphasizing localization of observables in space and time. It has its roots in the pioneering work of Rudolf Haag and Arthur Wightman from the 1950s and is a well-established branch of mathematical physics, distinguished by clear conceptual foundations and mathematically sound arguments.

AQFT is also called *Local Quantum Physics* which is the title of a monograph by Rudolf Haag[1] that summarizes the results and insights achieved by many researchers up to the mid-1990s. Since then a number of new developments have taken place. In May 19–23, 2014 a workshop with the title "Algebraic quantum field theory: Its status and its future" was held at the Erwin Schrödinger Institute for Mathematical Physics (ESI) in Vienna. The present volume reflects many of the themes discussed at this workshop but all contributions were written specially for this volume. The first contribution, by Klaus Fredenhagen, is a general introduction to the fundamentals of AQFT while other chapters focus on more specialized topics. The chapter by Fredenhagen and Katarzyna Rejzner in particular, shows how perturbative constructions of models with interactions fit elegantly into the formalism of AQFT.

One of the strengths of the algebraic approach is that it can be naturally applied when the underlying space-time manifold is curved, thus allowing to take important aspects of General Relativity into account. A large number of results on the characterization of physically important states, general covariance, perturbation theory, and renormalization, both on flat and curved space-times, have been obtained in the past 20 years and several of the contributions deal with these topics.

The introductory chapter by Klaus Fredenhagen has a section on AQFT on curved space times, the chapter by Marco Benini and Claudio Dappiaggi treats

[1]Rudolf Haag, Local Quantum Physics: Fields, Particles, Algebras, 2nd Ed., Springer 1996.

models of free quantum fields on such space-times, and that of Igor Khavkine and Valter Moretti is concerned with quasi-free Hadamard states. The contribution of Christopher Fewster and Rainer Verch deals with locally covariant AQFT on curved space-times employing the language of category theory, besides discussing selection criteria for physical states in terms of various stability conditions. The chapter by Thomas-Paul Hack and Nicola Pinamonti is concerned with applications of AQFT to cosmology. Here also, semiclassical back reaction of quantum fields on the metric of space-time is taken into account. The contribution by Dorothea Banhs, Sergio Doplicher, Gerardeo Morsella, and Gherardo Piacitelli goes further into the direction of quantum gravity by replacing classical space-time by a quantum space-time where the coordinates form a noncommutative algebra.

Conformal quantum field theory, that has several important applications in condensed matter physics, is a model example where algebraic structures and inequivalent Hilbert space representations that go far beyond standard Lagrangian field theory arise naturally. The chapter by Karl-Henning Rehren reviews this topic from a modern perspective. Pieter Naaijkens treats in his contribution Kitaev's quantum double model from a local quantum physics point of view. This is a further example where the algebraic and local way of thinking leads to important insights in a situation that is not directly connected with relativistic quantum theory but rather with condensed matter physics and quantum information theory.

The last chapter, by Gandalf Lechner, is concerned with new methods for constructing models of relativistic quantum fields. Here, a basic tool is the Tomita-Takesaki modular theory of von Neumann Algebras and the fact that the modular group and reflection of an algebra generated by a relativistic quantum field localized in a space-like wedge is explicitly known. Using these techniques, a large class of models in two space-time dimensions, that are so far inaccessible by other means, can be constructed. A further method described in this chapter is the use of deformations of the wedge algebras of known models to obtain new models.

Trento, Italy — Romeo Brunetti
Pavia, Italy — Claudio Dappiaggi
Hamburg, Germany — Klaus Fredenhagen
Vienna, Austria — Jakob Yngvason

Acknowledgments

The editors are grateful to the Erwin Schrödinger Institute for Mathematical Physics in Vienna for hosting and financing the workshop on AQFT in May 2014. Thanks are also due to Aldo Rampioni and to Kirsten Theunissen at Springer SBM NL for the fruitful and patient collaboration in the realization of this volume.

Contents

Chapter 1
An Introduction to Algebraic Quantum Field Theory

Klaus Fredenhagen

Abstract The algebraic approach to quantum field theory is reviewed, and its aims, successes and limitations are discussed.

1.1 Introduction

The algebraic properties of quantum observables, as discovered by Heisenberg, Born and Jordan and summarized in the canonical commutation relation between position q and momentum p,

$$[q, p] = i\hbar, \tag{1.1}$$

are crucial for the structure of quantum theory. A further crucial structure is the representability of observables as Hilbert space operators, as first discovered by Schrödinger and related to the probability interpretation of quantum mechanics by Born. It was later found that the existence of Hilbert space representations can be guaranteed if the algebra $\mathfrak{A}$ generated by observables is equipped with an involutive structure $A \mapsto A^*$, characterizing the real elements, and a norm which satisfies the C*-property

$$||A^*A|| = ||A||^2. \tag{1.2}$$

This involves the restriction to bounded observables, but this is neither from the point of view of physics (here it amounts to parametrize the possible results of a measurement by numbers in a finite interval) nor from the point of view of mathematics (there one exploits the spectral theorem for selfadjoint operators and considers bounded functions of the operator in question) a loss of generality. It may, however, lead to problems in calculations, since in concrete applications one often has to start from a formula involving unbounded operators. For our general considerations we may ignore these difficulties and consider the algebra of observables as a C*-algebra with unit, i.e. as an involutive unital algebra equipped with a C*-norm, which is

K. Fredenhagen (✉)
II. Institut für Theoretische Physik, 22761 Hamburg, Germany
e-mail: Klaus.Fredenhagen@desy.de

R. Brunetti et al. (eds.), *Advances in Algebraic Quantum Field Theory*,
Mathematical Physics Studies, DOI 10.1007/978-3-319-21353-8_1

complete as a normed space. Every such algebra arises as a norm closed algebra of Hilbert space operators, but may have in addition many other inequivalent representations by Hilbert space operators. Then the mathematical question arises which of the possible representations one should choose in order to describe a given physical situation.

In quantum mechanics with finitely many degrees of freedom a partial answer was given by von Neumann. He proved that the algebra generated by unitaries $e^{i\alpha q}$ and $e^{i\beta p}$ with $\alpha, \beta \in \mathbb{R}$ and satisfying Weyl's version of the canonical commutation relation,

$$e^{i\alpha q}e^{i\beta p} = e^{-i\alpha\beta}e^{i\beta p}e^{i\alpha q}, \tag{1.3}$$

has, up to unitary equivalence, only one regular irreducible representation, namely the representation found by Schrödinger. Here regular means that the maps $\alpha \mapsto e^{i\alpha q}$ and $\beta \mapsto e^{i\beta p}$ are strongly continuous.

In quantum field theory with its infinitely many degrees of freedom, the uniqueness result of von Neumann is no longer valid. Nevertheless, it took some time before the physical significance of this fact was realized. The crucial insight was due to Haag who found that theories with translation invariant interactions cannot have a ground state in the Hilbert space describing the vacuum of the free theory. It was then understood that this problem is generic, e.g. thermodynamical equilibrium states cannot be described in terms of density matrices in the vacuum Hilbert space.

Apart from these problems, there is a deeper reason why it is fortunate to separate the construction of observables from the construction of states. This is the apparent conflict between the principle of locality, which in particular governs classical field theory, and the existence of nonclassical correlations (*entanglement*) in quantum systems, often referred to as non-locality of quantum physics. As a matter of fact it turns out that the algebra of observables is completely compatible with the locality principle whereas the states typically exhibit nonlocal correlations. For this reason, *algebraic quantum field theory* is also called *local quantum physics* [63].

The concepts of algebraic quantum field theory were first introduced in a contribution of Haag to the Lille conference 1957 [61]. The main motivation at this time was an explanation why a quantum field theory yields a theory of interacting particles. The Haag-Ruelle scattering theory [62, 85] was a first success of these concepts.

It was then made mathematically precise by using the theory of operator algebras, mainly by Araki [5, 6]. The crucial step of considering the algebras of observables independently of their action on an underlying Hilbert space was performed in a programmatic paper by Haag and Kastler [64]. They formulated the following axioms:

- To each open bounded region $\mathcal{O}$ of Minkowski space $\mathbb{M}$ there is associated a unital C*-algebra $\mathfrak{A}(\mathcal{O})$, interpreted as the algebra of observables which can be measured within the spacetime region $\mathcal{O}$.
- If $\mathcal{O}_1 \subset \mathcal{O}_2$, then there exists an embedding (unital injective *-homomorphism)

$$i_{\mathcal{O}_2\mathcal{O}_1} : \mathfrak{A}(\mathcal{O}_1) \to \mathfrak{A}(\mathcal{O}_2). \tag{1.4}$$

- These embeddings satisfy the compatibility relation

$$i_{\mathcal{O}_3\mathcal{O}_2} \circ i_{\mathcal{O}_2\mathcal{O}_1} = i_{\mathcal{O}_3\mathcal{O}_1} \text{ if } \mathcal{O}_1 \subset \mathcal{O}_2 \subset \mathcal{O}_3. \quad (1.5)$$

- The identity component of of the Poincaré group is represented by automorphisms α_L with the properties

$$\alpha_L \circ \mathfrak{A} = \mathfrak{A} \circ L,$$
$$\alpha_L \circ i_{\mathcal{O}_1\mathcal{O}_2} = i_{L\mathcal{O}_1 L\mathcal{O}_2} \circ \alpha_L,$$
$$\alpha_{L_1} \circ \alpha_{L_2} = \alpha_{L_1 L_2}.$$

- If $\mathcal{O}_1$ is spacelike separated from $\mathcal{O}_2$ and $\mathcal{O} \supset \mathcal{O}_1 \cup \mathcal{O}_2$, then

$$[i_{\mathcal{O}\mathcal{O}_1}(A), i_{\mathcal{O}\mathcal{O}_2}(B)] = 0 \,\forall A \in \mathfrak{A}(\mathcal{O}_1),\, B \in \mathfrak{A}(\mathcal{O}_2).$$

These axioms do not involve any choice of Hilbert space representations. The system of algebras together with the embeddings and the Poincaré symmetries is called a *local net* (or a *Haag-Kastler net*). It contains the minimal requirements one may pose on observables localized in subregions of Minkowski space. Note, however, that fermionic fields which anticommute at spacelike separation are excluded from the local algebras. It was one of the main goals and finally successes of the algebraic approach that the occurrence of fermionic fields could be derived from the structure of local nets.

The axioms do not specify a dynamical law. The existence of a dynamical law, however, can be implied by a further axiom, namely the *time slice axiom* :

- If $\mathcal{O}_1 \subset \mathcal{O}_2$ contains a Cauchy surface of $\mathcal{O}_2$ (i.e. a hypersurface which is met exactly once by every non-extendible causal curve in $\mathcal{O}_2$), then the embedding $i_{\mathcal{O}_2\mathcal{O}_1}$ is an isomorphism.

The system of local algebras possesses an inductive limit, called the quasilocal algebra $\mathfrak{A}(\mathbb{M})$. It is the unique C*-algebra with embeddings $i_{\mathcal{O}} : \mathfrak{A}(\mathcal{O}) \to \mathfrak{A}(\mathbb{M})$ such that

$$i_{\mathcal{O}_2} \circ i_{\mathcal{O}_2\mathcal{O}_1} = i_{\mathcal{O}_1} \text{ if } \mathcal{O}_1 \subset \mathcal{O}_2,$$

which is generated by the local subalgebras $i_{\mathcal{O}}(\mathfrak{A}(\mathcal{O}))$. Since all embeddings $i_{\mathcal{O}}$ are injective we identify in the following $\mathfrak{A}(\mathcal{O})$ with its image under $i_{\mathcal{O}}$.

One might be worried by the task to specify algebras for all subregions of Minkowski space. But is suffices to associate algebras to so-called diamonds (or double cones). These regions are parametrized by a pair of points (x, y) with x in the chronological future of y and consist of all points z which are in the chronological past of x and the chronological future of y. We denote the set of double cones by $\mathcal{K}$. Algebras of other regions G are then defined as subalgebras of $\mathfrak{A}(\mathbb{M})$ which are generated by the algebras $\mathfrak{A}(\mathcal{O})$ with $G \supset \mathcal{O} \in \mathcal{K}$.

Once the local net is given, the space of states is obtained as a subset of the dual of $\mathfrak{A}(\mathbb{M})$ as a Banach space. Namely, states are just those linear functionals $\omega : \mathfrak{A}(\mathbb{M}) \to \mathbb{C}$ which satisfy the conditions

$$\omega(1) = 1, \tag{1.6}$$

$$\omega(A^*A) \geq 0 \; \forall A \in \mathfrak{A}(\mathbb{M}). \tag{1.7}$$

The values of a state are interpreted as expectation values, and the whole probability distribution $\mu_{\omega,A}$ of measured values $a \in \mathbb{R}$ of an observable $A = A^*$ is obtained from its moments, namely the expectation values of powers of A,

$$\int a^n d\mu_{\omega,A}(a) = \omega(A^n) \; , \; n \in \mathbb{N}. \tag{1.8}$$

Every state induces by the so-called GNS construction a representation π of the quasilocal algebra on some Hilbert space $\mathcal{H}$ together with a distinguished unit vector Ω such that

$$\omega(A) = \langle \Omega, \pi(A)\Omega \rangle,$$

and the set of vectors $\{\pi(A)\Omega | \; A \in \mathfrak{A}(\mathbb{M})\}$ is dense in $\mathcal{H}$.

The density matrices ρ in $\mathcal{H}$ form a family of states,

$$\omega_\rho(A) = \mathrm{Tr}\rho\pi(A),$$

the so-called folium of ω. The crucial mathematical fact is that infinitely dimensional C*-algebras in general have a huge number of disjoint folia. While this is welcome from the physics point of view in order to be able to describe macroscopically different situations, it makes the analysis of the structure of the state space difficult. Hence one of the main objectives in algebraic quantum field theory is to select a suitable subset of the state space which covers the situations of interest and admits a classification.

The most famous of these selection criteria is the DHR (for Doplicher, Haag and Roberts) criterion which is supposed to select the states of interest for elementary particle physics. These are states which differ from a distinguished state ω_0 only within some bounded region and its causal future and past.

The state ω_0 is interpreted as the vacuum. It is assumed to be a pure state (i.e. it cannot be decomposed into a convex combination of other states) and to be invariant under the Poincaré transformations α_L. Moreover, in its GNS-representation π_0, the associated unitary representation U of the identity component of the Poincaré group given by

$$U(L)\pi_0(A)\Omega = \pi_0(\alpha_L(A))\Omega,$$

which implements the automorphisms α_L,

$$U(L)\pi_0(A)U(L)^{-1} = \pi_0(\alpha_L(A)),$$

is strongly continuous and satisfies the relativistic spectrum condition

$$\mathrm{sp}(U_0) \subset \overline{V_+^*}.$$

Here U_0 is the restriction of U to the translation subgroup, V_+ is the cone of future directed timelike translations, V_+^* the dual cone in momentum space and $\overline{V_+^*}$ its closure. The representation U_0 is generated by mutually commuting selfadjoint operators P_μ, $\mu = 0, \ldots 3$ (the momentum operators)

$$U_0(x) = e^{iPx} \; , \; Px \equiv P_\mu x^\mu, \tag{1.9}$$

whose joint spectrum is by definition the spectrum of U_0.

In the presence of superselected charges there exist other representations π which are translation covariant, i.e. there exists a unitary strongly continuous representation of the translation group, which implements the translation automorphisms and fulfills the spectrum condition.

An important question is whether all observables can be expressed in terms of observables in arbitrarily small regions, as one might expect in models which are defined in terms of observable pointlike localized fields. A precise version of this property is additivity : a representation π satisfies additivity if for every covering of the open region $\mathcal{O}$ by double cones $\mathcal{O}_i$ one has

$$\pi(\mathfrak{A}(\mathcal{O})) \subset (\bigcup_i \pi(\mathfrak{A}(\mathcal{O}_i)))''. \tag{1.10}$$

Here $\mathcal{R}'$ denotes the commutant of a set of operators $\mathcal{R}$, i.e. the set of all bounded operators commuting with $\mathcal{R}$. The bicommutant of a selfadjoint set (a set of operators which is invariant under taking the adjoint) is a von Neumann algebra.[1]

The combination of the axiom of local commutativity with the spectrum condition imposes strong restrictions on the theory. The most remarkable one is the Reeh-Schlieder Theorem [63, 83].

Theorem 1.1.1 *Let (π, U_0) be a translation covariant representation satisfying additivity and the spectrum condition. Let $\Phi \in \mathcal{H}_\pi$ be a cyclic vector for $\mathfrak{A}(\mathbb{M})$, i.e.*

$$\pi(\mathfrak{A}(\mathbb{M}))\Phi \text{ is dense in } \mathcal{H}_\pi,$$

and let $\Psi = U_0(i\beta)\Phi$ with $\beta \in V_+$. Let $\mathcal{O}$ be a nonempty double cone in Minkowski space. Then the set $\pi(\mathfrak{A}(\mathcal{O}))\Psi$ is dense in $\mathcal{H}_\pi$, and Ψ is separating for $\pi(\mathfrak{A}(\mathcal{O}))'$, i.e. $A\Phi = 0$ implies $A = 0$ for $A \in \pi(\mathfrak{A}(\mathcal{O}))'$.

The interpretation of the Reeh-Schlieder Theorem induced an intense discussion in philosophy of science (see e.g. [67]). The theorem does not mean that there is an

---

[1] A von Neumann algebra is an algebra of Hilbert space operators which is invariant under involution, contains the unit operator and is closed with respect to the weak operator topology.

instantaneous effect of a local operation. A local operation would correspond to the application of a unitary $U \in \mathfrak{A}(\mathcal{O})$, and the set of vectors $\{\pi(U)\Phi|U \in \mathfrak{A}(\mathcal{O})\}$ is total, i.e. its finite linear combinations are dense in $\mathcal{H}_\pi$, but the set itself is not dense in the unit ball of $\mathcal{H}_\pi$. See [29] for a recent discussion.

An important generalization of the Reeh-Schlieder Theorem is the following theorem of Borchers [12]:

Theorem 1.1.2 *Let (π, U_0) as before and assume that the center of $\pi(\mathfrak{A}(\mathbb{M}))$ is trivial. Let $\mathcal{O}, \mathcal{O}_1$ be double cones with $\overline{\mathcal{O}} \subset \mathcal{O}_1$. Then to every nonzero projection $E \in \pi(\mathfrak{A}(\mathcal{O}))''$ there exists an isometry $V \in \pi(\mathfrak{A}(\mathcal{O}_1))''$ such that $VV^* = E$.*

This theorem shows that local algebras contain no nonzero finite dimensional projections. This property of relativistic quantum field theory is quite a surprise if confronted with structures known from nonrelativistic quantum mechanics.

The theorem reminds on the characterization of von Neumann algebras of type III. Namely, von Neumann algebras $\mathcal{N}$ with trivial centre (the so-called factors) can be classified in terms of the equivalence classes of their projections where 2 projections $E, F \in \mathcal{N}$ are called equivalent if there exists some $V \in \mathcal{N}$ such that

$$E = VV^* \text{ and } F = V^*V. \tag{1.11}$$

Type I factors are isomorphic to the algebra of all bounded operators on a Hilbert space $\mathcal{H}$. There the projections are equivalent if the subspaces $E\mathcal{H}$ and $F\mathcal{H}$ have the same dimension. In quantum field theory type I factors occur as the von Neumann algebras generated by $\pi(\mathfrak{A}(\mathbb{M}))$ in an irreducible representation π. Type II factors have projection classes which can be labeled by real numbers. They occur in physics only in extreme situations, for instance for a gas of fermions at infinite temperature. Type III factors are defined by the property that all nonzero projections are equivalent.

Borchers' Theorem above is somewhat weaker than the statement that the von Neumann algebras $\pi(\mathfrak{A}(\mathcal{O}))''$ are factors of type III. Actually, in many cases they are known to be of type III (see, e.g. [47]), and there are general reasons to expect that this is always the case [27].

As mentioned before, one of the early successes of the algebraic approach (in fact, the main motivation for its introduction) is the explanation of the particle structure observed in experiments. Namely, let $(\mathcal{H}, \pi_0, \Omega)$ be the GNS triple associated to the vacuum state ω_0, We assume that the spectrum of the translation group contains an isolated mass shell,

$$\mathrm{sp}(U_0) = \{0\} \cup H_m \cup H_{\geq M}\ ,\ 0 < m < M,$$

with $H_m = \{p|p_0 > 0, p^2 = m^2\}$ and $H_{\geq M} = \{p|p_0 > 0, p^2 \geq M^2\}$. The subspace $\mathcal{H}_1 \subset \mathcal{H}$ corresponding to H_m can be interpreted as 1-particle space. One then can show that the vacuum Hilbert space $\mathcal{H}$ contains also states which can be interpreted as multi-particle states at asymptotic times (Haag-Ruelle scattering theory). It is, however, an open question, whether all states in the vacuum representation admit a particle interpretation (problem of asymptotic completeness) (See [49] for recent progress).

One obvious obstruction to asymptotic completeness is that 1-particle states may exist which are disjoint from the states of the vacuum representation. This is true in particular for fermions, but also holds for bosonic particles which carry a superselected charge, e.g. the W-bosons of the electroweak theory.

In models these states are obtained by enlarging the algebra of observables to a so-called field algebra. In spite of the fact that the elements of the field algebra are not necessarily observable, one postulates that they satisfy local commutation, or, for fermionic fields, anticommutation relations. This allows to extend the construction of multi-particle states to charged particles which then satisfy Bose or Fermi statistics, respectively. A partial justification of this ansatz is the spin statistics theorem which states, under the hypothesis that either Bose or Fermi statistics holds, that only the observed connection between spin and statistics is possible. The ansatz is, however, ad hoc and is in fact not sufficiently general to cover all situations of interest, in particular in less than 4 dimensions when more general statistics are possible.

To incorporate these particles, Doplicher, Haag and Roberts formulated their selection criterion for "states of interest for particle physics". It incorporates the property of charge carrying fields that they satisfy local commutation relations with the observables but does not impose any commutation relations between the fields themselves. In a series of 4 seminal papers [40–43] they analyzed the arising representations and could show that

- The multi-particle structure previously seen in the vacuum sector extends to all DHR states.
- The particles satisfy Bose or Fermi statistics and may have additional degrees of freedom.
- Antiparticles exist and have the same mass and spin as the corresponding particle.

Moreover, Doplicher and Roberts succeeded in [46] to show that the internal degrees of freedom can be understood in terms of representations of a uniquely determined compact group. This led to an improved version of the Tannaka-Krein characterization of the dual of a compact group.

Interestingly, the DHR analysis yields a different structure if applied to quantum field theory in 2 spacetime dimensions [58]. This was used in particular in the operator algebraic approach to conformal field theory as explained in more detail in Rehren's contribution to this book.

The DHR criterion excludes states describing particles with an electric charge because of the associated electric flux which distinguishes these states from the vacuum even at arbitrarily large spacelike separation. This is a generic feature of charged particles in gauge theories [25]. If a particle is massive and if in its representation the energy-momentum spectrum has an upper mass gap, Buchholz and the author could show that the corresponding sector is related to a vacuum state such that the interpolating fields can be localized in an infinitely extended spacelike cone [26]. In 4 dimension this localization allows to apply the DHR analysis, but in 3 dimensions one observes a more general structure similar to the 2 dimensional situation for DHR states.

The massless situation which is relevant for quantum electrodynamics was recently solved by Buchholz and Roberts [32], based on an older proposal of Buchholz [24].

Algebraic quantum field theory, as may be seen from this overview, was very successful in analyzing the structure of quantum field theory, based on a few plausible axioms. It turned out, however, to be extremely difficult to construct concrete models which are physically interesting and satisfy these axioms. Actually, other approaches to quantum field theory suffer from the same problem, and sometimes these difficulties are taken as an indication for the need to go beyond quantum field theory, or even to give up the requirement of mathematical consistency. It is one aim of the present book to convince the reader that this is premature, and to indicate possible directions for improving the situation.

Let us summarize the known models of algebraic quantum field theory (see also [89] for an overview). There are first the models of free fields describing freely moving particles characterized by irreducible representations of the (covering of) the (connected component of) the Poincaré group. Traditionally these models are constructed in terms of annihilation and creation operators on the Fock space over the corresponding 1-particle space. The construction works for massive particles with spin $s \in \frac{1}{2}\mathbb{N}_0$ and for massless particles with helicities $h \in \frac{1}{2}\mathbb{Z}$. An elegant direct construction was performed in [20]. While physically of limited interest, they have a rich mathematical structure, moreover, they constitute a basis for the interpretation of scattering states, and they serve as a framework for the perturbative formulation of interacting models. Closely related are the generalized free fields, where the single particle space is replaced by a reducible representation of the Poincaré group. In case of a discrete mass spectrum they arise as tensor products of free theories. In the case of a continuous mass spectrum they are usually considered as unphysical, but recently, they were proposed under the name of unparticles [59].

Another class of models are the superrenormalizable models in 2 dimensions, in particular massive scalar fields with a polynomial interaction, and the Yukawa model [60, 87]. These models, unfortunately, seem to have no direct physical application.

A further class consists of conformally invariant theories in 2 dimensions (see the contribution of Rehren to this book). Here a huge class of models has been constructed (see e.g. [75]), and the results from the algebraic approach can be compared to results in other approaches, with mutual fertilization. These models are of high interest for mathematics, but they also have applications to physics, e.g. in critical phenomena of effectively low dimensional systems in condensed matter, and they appear naturally within string theory.

Another class are integrable models in 2 dimensions whose S-matrix satisfies the so-called factorization equations. In case the 2-particle scattering matrix is just multiplication by a pure phase depending on the incoming momenta, the corresponding local nets have been constructed by Lechner [78]. Constructions of more complicated cases have also been performed [1, 2]. In some cases, for example in the sinh-Gordon model with field equation

$$\Box\varphi + \lambda \sinh g\varphi = 0,$$

the algebraic construction can (and should) be compared with a direct construction. It is, however, remarkable that the algebraic construction works also in cases where no corresponding classical dynamics is known. See Chap. 10 for more details.

Motivated by attempts to replace spacetime by a noncommutative space where expected properties of quantum gravity are taken into account (see the contribution of Bahns et al. to this book), one can also construct models by deformation of a given net. This works even in higher dimension, but the models constructed so far satisfy only a restricted version of local commutativity, namely one obtains algebras associated to so-called wedges W, i.e. $W = L\{x \in \mathbb{M} | |x^0| < x^1\}$ for some Poincaré transformation L, and observables localized in spacelike separated wedges commute. This allows to construct 2-particle states and to compute the S-matrix for 2-particle scattering, but gives no hint whether a generalization to multi-particle states is possible [31].

A completely new elegant method of construction was recently performed by Barata, Jaekel and Mund (see a forthcoming publication). It starts from the construction of a Haag Kastler net on de Sitter space in terms of Tomita-Takesaki theory. The Minkowski space theory is then obtained by taking the limit of vanishing curvature in the spirit of the scaling limit of Buchholz and Verch [33].

The last class of models I want to mention is obtained by adopting the methods of renormalized perturbation theory to the algebraic framework. There one has to give up the condition that the local algebras are C*-algebras. Instead they are unital *-algebras, where the existence of Hilbert space representations is replaced by the existence of a representation on a space of formal power series $\mathcal{D}[[g]]$ with a dense subspace $\mathcal{D}$ of some Hilbert space where g is the coupling constant. Therefore all numerical predictions of the theory are formal power series in g, and a comparison with experiments requires a truncation of the series. As a matter of fact, in most cases only a few terms of the series can be computed by present techniques, and one finds in many cases an excellent agreement with measurements. Moreover, exploiting the concept of the renormalization group, one obtains different series labeled by a scale μ, and the series at different parameters μ_1, μ_2 are related by replacing the coupling constant by the so-called running coupling constant which typically is a formal power series in g and $\log \mu_2/\mu_1$. Observations at a given scale can then be compared with the truncated series at the same scale, where the running coupling constant is fitted with the data. This improves the agreement with the measured data considerably, and, moreover, confirms the predicted running of the coupling constant.

This weaker version of the Haag-Kastler axioms covers all models of quantum field theory which are relevant for elementary particle physics. Compared to the traditional way of treating perturbation theory, the algebraic approach allows a state independent renormalization and a construction of the local net without any infrared problems. Infrared problems can, in principle, reappear in the construction of states or in the computation of scattering cross sections. But there, the difficulties might have different reasons:

- The corresponding state may not exist, as e.g. the vacuum state for a massless scalar field in 2 dimensions, or a KMS-state of a scalar field with negative temperature.

- The states may belong to different superselection sectors.
- The usual method for construction does not work. This holds e.g. for KMS states of a massive scalar field, where Altherr [4] and Steinmann [88] observed infrared divergences at higher loop order. As was recently shown [57, 79], the problem disappears if the validity of the time slice axiom is exploited so that only the observables localized within a fixed time slice have to be taken into account.

The formalism of AQFT is motivated by the expected interplay between observables in quantum theory and the concept of locality arising from special relativity. In practical applications one often prefers to extend the framework by adding unobservable degrees of freedom. One example is the introduction of anticommuting fields in order to describe particles with Fermi statistics. As mentioned above, this extension of the framework can be intrinsically derived from the Haag-Kastler net of observables by the DHR theory. More generally this holds for all global gauge symmetries, where the compact group whose representations label the superselection sectors is shown to be constructible from the Haag-Kastler net.

A very important question is whether a corresponding statement can also be made for theories with local gauge symmetry, as QED or the standard model of elementary particle physics. In spite of quite a number of works dedicated to this question an answer is presently not known. Several strategies have been probed.

The most direct way is to mimick the traditional construction in terms of auxiliary fields (ghosts etc.). Here one necessarily gives up the framework of C*-algebras, but preserves locality. One obtains the algebra of observables as a resolution of a complex, in the sense of homological algebra [55]. An interesting alternative to this procedure is to enlarge the algebra of local observables by nonlocal quantities but to preserve the representability on a Hilbert space. See [81] for progress in this direction. In case of free abelian gauge theories as e.g. Maxwell's theory without charges, one can give a complete construction of the C*-algebraic Haag-Kastler net.

The most challenging open question is whether gravity can be included into the formalism of AQFT. Since the very concept of a region of spacetime has no intrinsic meaning in a generally covariant theory, a modification of the formalism would be necessary. But gravity is, at presently accessible observations, much weaker than all other known interactions, hence it is a good approximation to treat the spacetime of general relativity as a background which is not influenced by the other quantum fields. This is the motivation to study quantum field theory on generic spacetimes with a metric with Lorentzian signature.

1.2 AQFT on Curved Spacetime

As was first observed by Dimock [36] and Kay [73], AQFT is ideally suited for the treatment of QFT on curved backgrounds. The reason is that, contrary to the situation on Minkowski space, no intrinsic concept of a vacuum state nor of particles exist. The traditional approach to QFT therefore does not work, and attempts to use it lead to inconsistencies. As long as only free fields were discussed, a direct construction can be given on any globally hyperbolic spacetime, and an appropriate version of the Haag-Kastler axioms can be given. Even in free field theory, however, one is interested not only in the field itself, but also in nonlinear expressions in the field as e.g. the energy momentum tensor, in order to estimate the possible back reaction of the quantum fields on the spacetime geometry. It was observed that for free fields there exist states, the so-called Hadamard states, which have similar singularities as the vacuum state on Minkowski space. It was a major breakthrough when Radzikowski [82] observed that the class of Hadamard states can be completely characterized by the wave front set of the 2-point function. This allowed to extend the algebras from the linear fields (or, in order to work in the C*-framework, the exponentials of linear fields (Weyl operators)) to all local Wick polynomials [17]. Moreover, even renormalization could be performed in this extended framework [16].

But there remained an obstruction: namely, due to the absence of nontrivial isometries in generic spacetimes, the covariance axiom of Haag-Kastler is empty, in general. This leads within renormalization theory to the undesirable situation that the renormalization conditions at different points of spacetime cannot be compared with each other. The intuitive feeling is that the removal of divergences should be done in terms of prescriptions depending only on the local geometry. Actually, this idea was already used in Kay's treatment of the Casimir effect [74] and in Wald's discussion of the renormalization of the energy momentum tensor [92]. A precise version of this idea was given in the system of axioms for locally covariant quantum field theory [19] (see also [91] for a preliminary version with application to a spin statistics theorem on curved spacetimes as well as related work of Dimock [37]):

- To each time oriented globally hyperbolic spacetime M there is associated a unital C*-algebra $\mathfrak{A}(M)$.
- To each isometric, time orientation and causality preserving embedding $\chi : M \to N$ there is associated an embedding $\mathfrak{A}\chi : \mathfrak{A}(M) \to \mathfrak{A}(N)$.
- If $\chi : M \to N$ and $\psi : N \to L$ are embeddings of spacetimes as above, then $\mathfrak{A}\psi \circ \chi = \mathfrak{A}\psi \circ \mathfrak{A}\chi$.

These axioms tell that quantum field theory is a functor $\mathfrak{A}$ from the category of spacetimes, with embeddings as above as morphisms, to the category of unital C*-algebras, with injective unital *-homomorphisms as morphisms. One easily sees that, by restricting the spacetimes to subregions of a fixed spacetime, one obtains an associated Haag-Kastler net. The morphisms contain in particular the inclusions-this yields the isotony of the Haag-Kastler net- and the isometries -this yields the covariance of the Haag-Kastler net.

The remaining Haag-Kastler axioms (local commutativity and time slice axiom) can be easily formulated as additional properties of the functor.

Based on the concept of local covariance, the idea of allowing only operations, which are determined by the local geometry, can be made precise by requiring that they are natural transformations between the functor $\mathfrak{A}$ and other geometrically defined functors on the category of spacetimes. The program of renormalization on generic spacetimes was completed by Hollands and Wald [69, 70]. An important simplification of their construction was recently obtained by Khavkine and Moretti [77].

The axioms above were formulated for a scalar field. For other fields, in particular gauge fields, the relation to topological invariants of the spacetimes may become nontrivial, hence the admissible embeddings in the category of spacetimes have to preserve these structures. See e.g. [9] and references therein. More details on AQFT on curved spacetimes may be found in Chap. 4. See also the recent reviews by Benini et al. [10], by Hollands and Wald [71] as well as another review by Rejzner and the author [56].

One may even extend the given framework to the quantized gravitational field. There one has to split the metric into a background which defines a curved spacetime and a fluctuation which is treated as a quantum field. Since this split is arbitrary one has to prove that the theory does not depend on it. But there remains another difficulty, namely the absence of local observables due to the invariance of the theory under diffeomorphisms. In principle, this can be dealt with by introducing coordinates which are themselves dynamical fields. For a preliminary version of these ideas see [18] (compare also [76] for related ideas in the classical case and the discussion of relative observables within Loop Quantum Gravity [38, 84]).

1.3 Scattering Theory

Why do particles naturally occur in quantum field theory? After all, the theory is formulated without ever using the concept of particles, and as discussed before, no useful particle concept seems to exist for generic curved spacetime. Moreover, classical field theory typically does not admit, in general, an interpretation in terms of particles, only solitons which appear in some special cases remind on particles. Actually, the derivation of the particle structure in quantum field theory on Minkowski space proceeds in two quite different steps. The first is related to the phenomenon of discrete spectra in quantum physics. Therefore, in a translation covariant representation satisfying the spectrum condition, the mass operator

$$M = \sqrt{P_\mu P^\mu} \tag{1.12}$$

may have isolated eigenvalues. It is an old conjecture that the formation of mass eigenstates is related to an almost finite dimensionality of the state space corresponding to a localization within a bounded region with, at the same time, restricted total energy. This scarcity of the state space has been made precise in terms of compactness [65] or nuclearity [30] conditions. While this led to interesting results as split inclusion [39, 44] or the existence of KMS states [28], it was up to now not possible to derive the existence of mass eigenstates from this assumption (for a partial result in this direction see [50]). Therefore, one usually starts from the assumption that the mass operator has an isolated eigenvalue $m > 0$. The corresponding eigenstates are then interpreted as single particle states. We also assume here that the representation contains also a translationally invariant unit vector Ω, unique up to a phase, and call it the vacuum vector. The more general case was treated in [26].

The basic idea is that particle states can be generated from the vacuum vector by applying (almost) local operators, $\Psi = A(t)\Omega$. Since particle states satisfy the Klein Gordon equation

$$(\Box_x + m^2)U_0(x)\Psi = 0, \tag{1.13}$$

they describe freely moving particles, and the operators $A(t)$ generating them may be assumed to be essentially localized near to the actual position of the particle at time t. But then multi-particle states can be generated by products of (almost) local operators

$$\Psi_{1,\ldots,n}(t) = A_1(t)\ldots A_n(t)\Omega. \tag{1.14}$$

If the particles move with different velocities, the operators will, at large times, be localized in far spacelike separated regions and hence will commute, and thus the vector will depend only on the single particle vectors Ψ_j, $j = 1\ldots,n$. This rough idea does not work exactly, but approximately for large times, and one can show that it becomes exact in the limits $t \to \pm\infty$. Hence one obtains two in general different multi-particle states corresponding to each collection of single particle states, the outgoing ($t \to \infty$) and the incoming ($t \to -\infty$) state. The operator mapping the incoming state to the corresponding outgoing state is the scattering matrix, and it can be shown that it is always a partial isometry.

The rigorous argument proceeds at follows: given a single particle state Ψ with compact momentum support and a diamond $\mathcal{O} \in \mathcal{K}$, there is, by the Reeh-Schlieder theorem, for every $\varepsilon > 0$ an operator $A \in \mathfrak{A}(\mathcal{O})$ with $||\Psi - A\Omega|| < \varepsilon$. Let now f be a Schwartz function whose Fourier transform $\tilde{f}$ has a compact support within the forward lightcone, which intersects the energy momentum spectrum only on the mass shell and is equal to 1 on the momentum support of Ψ. Then $A(f) = \int dx\alpha_x(A)f(x)$ is almost local, in the sense that it can be fast approximated by local operators, and

$$||\Psi - A(f)\Omega|| = ||\tilde{f}(P)(\Psi - A\Omega)|| \leq \sup|\tilde{f}|\varepsilon. \tag{1.15}$$

We then consider the family of Schwartz functions $(f_t)_{t\in\mathbb{R}}$ with Fourier transforms

$$\tilde{f}_t(p) = e^{it(\sqrt{p^2}-m)}\tilde{f}(p) \tag{1.16}$$

and observe that the 1-particle vectors

$$\Psi(t) = A(t)\Omega \ \text{ with } A(t) = A(f_t) \tag{1.17}$$

do not depend on t.

The crucial fact is now that the functions f_t are essentially concentrated within the region tV_f, where V_f is the set of 4-velocities $p/\sqrt{p^2}$ with $p \in \operatorname{supp}\tilde{f}$. This follows by applying the argument of the stationary phase to the oscillating integral

$$f_t(x) = (2\pi)^{-4} \int d^4p \tilde{f}(p) e^{(it(\sqrt{p^2}-m)-ipx)}. \tag{1.18}$$

We now can realize the heuristic idea for the construction of multi-particle states by choosing single particle vectors $\Psi_1, \ldots \Psi_n$ with disjoint momentum support, the corresponding operators $A_i \in \mathfrak{A}(\mathcal{O})$ and the Schwartz functions f_i with Fourier transforms with mutually disjoint velocity supports V_{f_i}. We then find the following theorem:

Proposition 1.3.1 *The limits*

$$\lim_{t\to\pm\infty} A_1(t)\ldots A_n(t)\Omega \tag{1.19}$$

exist.

Proof The statement follows from the fact that the derivative with respect to t is integrable. Namely, using Leibniz' rule and the fact that the single particle vectors $A_i(t)\Omega$ are time independent, we can bound the derivative by

$$||\frac{d}{dt}A_1(t)\ldots A_n(t)\Omega|| \le \sum_{i<j} ||[\frac{d}{dt}A_i(t), A_j(t)]|| \prod_{k\neq i,j} ||A_k(t)||. \tag{1.20}$$

Since the commutators decay fast, due to the localization of the Schwartz functions $f_{i,t}$, and the norms of the operators $A_i(t)$ are polynomially bounded, the derivative decays fast and hence is integrable.

Clearly, the multi-particle vectors depend only on the corresponding single particle vectors, since the order of factors can be changed arbitrarily. Also their scalar products can be computed in terms of scalar products of single particle vectors:

Proposition 1.3.2 *Let $A_i, f_i, i = 1, \ldots, n$ and $B_j, g_j, j = 1, \ldots k$ be operators and Schwartz functions as above such that* $\mathrm{supp}\tilde{f}_i \cap \mathrm{supp}\tilde{g}_j = \emptyset$ *for $i \neq j$. Moreover, assume that the momentum support of each test function is so small that the set of differences $p - p'$, p, p' elements of the supports of the Fourier transforms of the test functions under consideration, intersects the energy momentum spectrum at most at $\{0\}$. Then*

$$\lim_{t\to\pm\infty} \langle A_1(t)\ldots A_n(t)\Omega, B_1(t)\ldots B_k(t)\Omega\rangle = \delta_{nk} \prod_{i=1}^{n} \langle A_i(0)\Omega, B_i(0)\Omega\rangle \tag{1.21}$$

Proof Due to the pairwise disjoint supports of the Fourier transforms of the test functions, the operators $A_i(t)^*$ commute with $B_j(t)$ for $i \neq j$ in the limits $t \to \pm\infty$. Hence the limit of the left hand side is equal to the limit of

$$\langle \prod_{j=k+1}^{n} A_j(t)\Omega, \prod_{i=1}^{k} A_i(t)^* B_i(t)\Omega\rangle \tag{1.22}$$

for $n \geq k$. (For $k \geq n$ we just have to exchange the roles of B's and A's.) But by the assumptions on the smallness of the momentum support of the test functions and on the uniqueness of the vacuum vector Ω, we have

$$A_i(t)^* B_i(t)\Omega = \Omega \langle A_i(0)\Omega, B_i(0)\Omega \rangle. \tag{1.23}$$

Since the momentum support of $\prod_{j=k+1}^{n} A_j(t)\Omega$ is contained in the forward light cone for $n > k$, the scalar product (1.22) vanishes in this case (as also in the case $n < k$), and we arrive at the formula in the proposition. □

The construction of scattering states can now easily be completed. Namely, let $\mathcal{H}_F$ denote the symmetric Fock space over the single particle space. The construction described above yields two densely defined isometries $W_\pm : \mathcal{H}_F \to \mathcal{H}$ with unique extensions to everywhere defined isometries (denoted by the same symbol). The images are the subspaces of incoming and outgoing scattering states, respectively, and the S-matrix is the partial isometry given by $S = W_+ W_-^*$.

Unfortunately, the construction breaks down in the presence of massless particles. But there, a different method was developed by Buchholz [23] which yields the scattering states of the massless particles. The question, however, about the particle structure for massive particles in the presence of massless particles is not yet completely clarified. Some progress was obtained by Dybalski [48].

1.4 Superselection Sectors

By definition, a superselection sector is an equivalence class of irreducible representations of the quasilocal algebra. We want to characterize representations by their relation to a distinguished representation π_0, the vacuum representation. Let for a double cone $\mathcal{O}$ of Minkowski space $\mathfrak{A}(\mathcal{O}')$ denote the C*-subalgebra of $\mathfrak{A}(\mathbb{M})$ which is generated by the algebras $\mathfrak{A}(\mathcal{O}_1)$ with double cones $\mathcal{O}_1 \subset \mathcal{O}'$ where $\mathcal{O}'$ denotes the spacelike complement of $\mathcal{O}$. The charge carrying fields are then characterized as partial intertwiners F between the vacuum representation π_0 and a representation π which contains charged states,

$$F : \mathcal{H}_{\pi_0} \to \mathcal{H}_\pi, \; F\pi_0(A) = \pi(A)F \; \forall \; A \in \mathfrak{A}(\mathcal{O}'). \tag{1.24}$$

Representations π satisfying the DHR selection criterion are those whose partial intertwiner spaces $\mathcal{F}_\pi(\mathcal{O})$ for every double cone $\mathcal{O}$ generate the representation space, i.e.

$$\mathcal{F}_\pi(\mathcal{O})\mathcal{H}_{\pi_0} \text{ is dense in } \mathcal{H}_\pi, \text{ and } \mathcal{F}_\pi(\mathcal{O})^*\mathcal{H}_\pi \text{ is dense in } \mathcal{H}_{\pi_0}. \tag{1.25}$$

This is equivalent to the statement that, after restriction to $\mathfrak{A}(\mathcal{O}')$, the state spaces of π and π_0 coincide (π and π_0 become quasiequivalent after restriction). The condition is usually motivated by the "particle behind the moon"-argument: the charge by which π and π_0 are distinguished might be compensated within an arbitrary region $\mathcal{O}$ which is inaccessible to observations. Note that the condition is satisfied if the charged states can be generated by local fields, but that it is violated if the charge can be measured at spacelike distances, as e.g. the electric charge by Gauss' law in electrodynamics.

The space of partial intertwiners between π_0 and π is a bimodule $\mathcal{F}_\pi(\mathcal{O})$ over $\mathfrak{A}(\mathcal{O})$ with the left and right action

$$A \cdot F \cdot B := \pi(A) F \pi_0(B). \tag{1.26}$$

This fact can be used to define products of fields in terms of tensor products of bimodules $\mathcal{F}_\pi(\mathcal{O})$ and $\mathcal{F}_{\pi'}(\mathcal{O})$. The elements of this tensor product shall then be interpreted as partial intertwiners between π_0 and a new representation $\pi \times \pi'$ corresponding to the composed charges. This product is called fusion in the framework of conformal field theory in 2 dimensions.

As a first step we observe that for $F, G \in \mathcal{F}_{\pi_0}(\mathcal{O})$, the operator F^*G commutes with $\pi_0(\mathfrak{A}(\mathcal{O}'))$. One now adds a crucial maximality condition to the algebra of bounded regions in the vacuum representation called Haag duality,

$$\pi_0(\mathfrak{A}(\mathcal{O}'))' = \pi_0(\mathfrak{A}(\mathcal{O}))''. \tag{1.27}$$

Haag duality implies that the operators F^*G as above are elements of $\pi_0(\mathfrak{A}(\mathcal{O}))''$. In the DHR analysis one further assumes that the partial intertwiner spaces even contain unitary elements, i.e. that the representations restricted to $\mathfrak{A}(\mathcal{O}')$ are even unitarily equivalent, whereas quasiequivalence means unitary equivalence of suitable multiples. We want to go the first steps without this assumption and use instead that, as a consequence of quasiequivalence and Haag duality, the localization regions of the partial intertwiners can be changed by local observables (charge transporter), namely if $\mathcal{O}_1, \mathcal{O}_2 \subset \mathcal{O}$ then $\mathcal{F}_\pi(\mathcal{O}_2) \subset \mathcal{F}_\pi(\mathcal{O}_1)\pi_0(\mathfrak{A}(\mathcal{O}))''$.

In the following we identify for all double cones $\mathcal{O}$ the local algebras $\mathfrak{A}(\mathcal{O})$ with the von Neumann algebra $\pi_0(\mathfrak{A}(\mathcal{O}))''$. Let $\mathfrak{A}_0$ denote the algebra of all local observables, $\mathfrak{A}_0 = \bigcup_\mathcal{K} \mathfrak{A}(\mathcal{O})$, and let $\mathcal{F}_\pi = \bigcup_\mathcal{K} \mathcal{F}_\pi(\mathcal{O})$ denote the vector space of all partial intertwiners between π_0 and π. Then $\mathcal{F}_\pi$ is a bimodule over $\mathfrak{A}_0$, and $\mathcal{H}_\pi$ is a left module. Moreover, due to Haag duality, $\mathcal{F}_\pi$ has an $\mathfrak{A}_0$ valued scalar product,

$$\langle F, G \rangle := F^*G, \tag{1.28}$$

and $\mathcal{H}_\pi$ is equal to the completion of the tensor product $\mathcal{F}_\pi \otimes_{\mathfrak{A}_0} \mathcal{H}_{\pi_0}$, equipped with the scalar product

$$\langle F \otimes \Phi, G \otimes \Psi \rangle = \langle \Phi, \langle F, G \rangle \Psi \rangle. \tag{1.29}$$

Also the tensor product of two DHR sectors, $\mathcal{F}_{\pi_1} \otimes_{\mathfrak{A}_0} \mathcal{F}_{\pi_2}$ carries an $\mathfrak{A}_0$ valued scalar product

$$\langle F_1 \otimes F_2, G_1 \otimes G_2 \rangle := F_2^* \pi_2(F_1^* G_1) G_2, \tag{1.30}$$

and a new Hilbert space representation $\pi_1 \times \pi_2$ is obtained by equipping the left module $\mathcal{F}_{\pi_1} \otimes_{\mathfrak{A}_0} \mathcal{F}_{\pi_2} \otimes_{\mathfrak{A}_0} \mathcal{H}_{\pi_0}$ with the scalar product

$$\langle F_1 \otimes F_2 \otimes \Phi, G_1 \otimes G_2 \otimes \Psi \rangle = \langle \Phi, \langle F_1 \otimes F_2, G_1 \otimes G_2 \rangle \Psi \rangle. \tag{1.31}$$

Intertwiners between DHR representations π and π', i.e. operators $T : \mathcal{H}_\pi \to \mathcal{H}_{\pi'}$ such that

$$T\pi(A) = \pi'(A)T \ \forall A \in \mathfrak{A}_0, \tag{1.32}$$

induce homomorphisms between the corresponding bimodules (denoted by the same symbol)

$$T : \mathcal{F}_\pi \to \mathcal{F}_{\pi'}, \ T : F \mapsto TF. \tag{1.33}$$

The structure we obtain is that of a tensor (or monoidal) C*-category with the DHR representations as objects and the intertwiners as morphisms where the tensor product is constructed in terms of the tensor products of the associated bimodules. This category has the following additional structures:

- It is additive, since direct sums of DHR representations are again DHR representations.
- It has subobjects, since any subrepresentation is again a DHR representation.

It is not a strict category, since the tensor product of bimodules is not strict. As usual we ignore this problem in the treatment of higher tensor products by using Mc Lane's coherence theorem which implies that different choices of brackets in tensor products are uniquely related by a natural isomorphism.

We now want to investigate the commutation relations between spacelike separated partial interwiners. Let $\mathcal{O}_1, \mathcal{O}_2 \in \mathcal{K}$ with $\mathcal{O}_1 \subset \mathcal{O}_2'$, and let π_1, π_2 be DHR representations. We define a right module homomorphism

$$\varepsilon(\pi_1, \pi_2) : \mathcal{F}_{\pi_1} \otimes_{\mathfrak{A}_0} \mathcal{F}_{\pi_2} \to \mathcal{F}_{\pi_2} \otimes_{\mathfrak{A}_0} \mathcal{F}_{\pi_1} \tag{1.34}$$

by

$$\varepsilon(\pi_1, \pi_2) F_1 \otimes F_2 = F_2 \otimes F_1 \tag{1.35}$$

with $F_i \in \mathcal{F}_{\pi_i}(\mathcal{O}_i)$, where we used that $\mathcal{F}_\pi = \mathcal{F}_\pi(\mathcal{O})\mathfrak{A}_0$, $\mathcal{O} \in \mathcal{K}$ for all DHR representations π. ε is called the statistics operator. We first check that it is well defined. Namely,

$$\sum_i F_{1,i} \otimes F_{2,i} \cdot A_i = 0 \text{ iff } \sum_{ij} A_i^* F_{2,i}^* \pi_2(F_{1,i}^* F_{1,j}) F_{2,j} A_j = 0, \tag{1.36}$$

with $F_{k,i} \in \mathcal{F}_{\pi_k}(\mathcal{O}_k)$ and $A_i \in \mathfrak{A}_0$. But $F_{k,i}^* F_{k,j} \in \mathfrak{A}(\mathcal{O}_k)$, hence

$$F_{2,i}^* \pi_2(F_{1,i}^* F_{1,j}) F_{2,j} = F_{2,i}^* F_{2,j} F_{1,i}^* F_{1,j} = F_{1,i}^* F_{1,j} F_{2,i}^* F_{2,j} = F_{1,i}^* \pi_1(F_{2,i}^* F_{2,j}) F_{1,j}. \tag{1.37}$$

We conclude that $\sum_i F_{1,i} \otimes F_{2,i} \cdot A_i = 0$ implies $\sum_i F_{2,i} \otimes F_{1,i} \cdot A_i = 0$. Moreover, we see that ε is norm preserving and that the induced operator on $\mathcal{H}_{\pi_1 \times \pi_2}$ is unitary.

The definition of ε depends on the choice of the double cones $\mathcal{O}_1$ and $\mathcal{O}_2$. But clearly ε does not change if we replace the double cones by smaller ones $\tilde{\mathcal{O}}_i \subset \mathcal{O}_i$, $i = 1, 2$. We now may deform the pair of spacelike separated double cones in the following way: Let $(\mathcal{O}_1^n, \mathcal{O}_2^n)_{n=1,\ldots,2N+1}$ be a sequence of pairs of spacelike separated double cones such that $\mathcal{O}_i^{2n\pm1} \subset \mathcal{O}_i^{2n}$, $n = 1, \ldots N$, $i = 1, 2$. Then ε defined for the pair $(\mathcal{O}_1^1, \mathcal{O}_2^1)$ coincides with ε defined for the pair $(\mathcal{O}_1^{2N+1}, \mathcal{O}_2^{2N+1})$. We conclude that in d dimensional Minkowski space with $d > 2$ the statistics operator is unique, whereas in 2 dimensions there are two choices. The crucial properties of the statistics operator are summarized in the following theorem:

Theorem 1.4.1 *1. ε is a bimodule homomorphism.*
2. In $d > 2$ dimensions we have

$$\varepsilon(\pi_1, \pi_2)\varepsilon(\pi_2, \pi_1) = 1. \tag{1.38}$$

3. Let π_1, π_2, π_3 be DHR representations. Then we have the identity

$$(\varepsilon(\pi_2, \pi_3) \otimes 1)(1 \otimes \varepsilon(\pi_1, \pi_3))(\varepsilon(\pi_1, \pi_2) \otimes 1) = (1 \otimes \varepsilon(\pi_1, \pi_2))(\varepsilon(\pi_1, \pi_3) \otimes 1)(1 \otimes \varepsilon(\pi_2, \pi_3)). \tag{1.39}$$

Proof 1. Let $A \in \mathfrak{A}(\mathcal{O})$ for some $\mathcal{O} \in \mathcal{K}$. Choose $\mathcal{O}_1, \mathcal{O}_2 \in \mathcal{K}$ such that the three double cones are pairwise spacelike separated. Let $F_i \in \mathcal{F}_{\pi_i}(\mathcal{O}_i)$, $i = 1, 2$. Then

$$A \cdot F_1 \otimes F_2 = F_1 \otimes F_2 \cdot A \text{ and } A \cdot F_2 \otimes F_1 = F_2 \otimes F_1 \cdot A, \tag{1.40}$$

hence

$$\varepsilon(\pi_1, \pi_2)(A \cdot F_1 \otimes F_2) = \varepsilon(\pi_1, \pi_2)(F_1 \otimes F_2 \cdot A) =$$
$$F_2 \otimes F_1 \cdot A = A \cdot F_2 \otimes F_1 = A \cdot \varepsilon(\pi_1, \pi_2)(F_1 \otimes F_2).$$

2. This follows from the uniqueness of ε.
3. This relation can be easily checked by application to $F_1 \otimes F_2 \otimes F_3$ with $F_i \in \mathcal{F}_{\pi_i}(\mathcal{O}_i)$ with pairwise spacelike separated double cones $\mathcal{O}_i$, $i = 1, 2, 3$. □

The statistics operators with the properties listed above equip the tensor category with a braiding or even a symmetry in $d > 2$. Physically it means that we can derive the commutation relations of fields and don't have to rely on an a priori choice.

An important question is whether the category is rigid, i.e. whether every π has a conjugate $\overline{\pi}$, unique up to equivalence. A DHR representation $\overline{\pi}$ is conjugate to π if there is an isometric intertwiner R such that

$$\overline{\pi} \times \pi(A)R = R\pi_0(A)\ \forall A \in \mathfrak{A}_0. \tag{1.41}$$

If π describes states of a particle we expect that its conjugate describes states of the corresponding antiparticle. Actually, in case the mass spectrum in π has an isolated eigenvalue, the existence of $\overline{\pi}$ could be proven [26, 52].

Theorem 1.4.2 *Let π be a translation covariant irreducible DHR representation of $\mathfrak{A}_0$ which satisfies the spectrum condition and whose mass spectrum contains an isolated eigenvalue. Then there exists a conjugate representation, which is irreducible and translation covariant and has the same mass spectrum as π.*

Actually, the theorem derived in [26, 52] is much stronger. In particular one does not need to know in advance that the representation π satisfies the DHR criterion. Instead one derives the existence of a vacuum representation π_0 such that π satisfies a weakened form of the DHR criterion where double cones have to be replaced by so-called spacelike cones. These are sets of the form

$$S = x + \bigcup_{\lambda>0} \lambda\mathcal{O} \tag{1.42}$$

with a double cone $\mathcal{O}$ whose closure is spacelike to the origin. If π_0 satisfies Haag duality for spacelike cones, one can redo the DHR analysis and one can directly construct the conjugate representation. There is, however, an important difference in the analysis of the statistics. Namely, due to the different localization one finds symmetry only in dimension $d > 3$. This allows the existence of particles with braid group statistics (plektons, anyons) in 3 dimensions.

We now restrict ourselves to DHR representations π with unitary partial intertwiners $V \in \mathcal{F}_\pi(\mathcal{O})$ for all $\mathcal{O} \in \mathcal{K}$, and we use the fact that the vacuum representation satisfies the conditions of Borchers theorem. Hence products, finite direct sums and subrepresentations also have unitary partial intertwiners. It is an interesting question how far the analysis of superselection sectors can be carried through for the general case. For progress in this direction see e.g. [86].

Let $V \in \mathcal{F}_\pi(\mathcal{O})$ be unitary. Then one can replace π by a unitarily equivalent representation ρ on the vacuum Hilbert space $\mathcal{H}_{\pi_0}$ by

$$\rho(A) = V^*\pi(A)V. \tag{1.43}$$

ρ is actually an endomorphism of $\mathfrak{A}_0$ by Haag duality, and it is localized in $\mathcal{O}$ in the sense that it acts trivially on all algebras $\mathfrak{A}(\mathcal{O}_1)$ with $\mathcal{K} \ni \mathcal{O}_1 \subset \mathcal{O}'$. Moreover, by the assumed existence of unitary elements in the partial intertwiner spaces $\mathcal{F}_\pi(\mathcal{O}_1)$ for all $\mathcal{O}_1 \in \mathcal{K}$ one finds endomorphisms ρ_1 localized in $\mathcal{O}_1$ which are equivalent as representations. But, again by Haag duality, the unitary operator in $\mathcal{H}_{\pi_0}$ which implements the equivalence is an element of $\mathfrak{A}(\mathcal{O}_2)$ for $\mathcal{K} \ni \mathcal{O}_2 \supset \mathcal{O} \cup \mathcal{O}_1$, hence ρ and ρ_1 are conjugate by inner automorphisms of $\mathfrak{A}_0$. Products of representations π_i now correspond to compositions of endomorphisms ρ_i in opposite order, $\pi_1 \times \pi_2 \simeq \rho_2 \circ \rho_1$.

The partial intertwiners $F \in \mathcal{F}_\rho(\mathcal{O})$ are identified with elements (ρ, F) of the so-called field bundle where $F \in \mathcal{B}(\mathcal{H}_{\pi_0}))$ with $\rho(A)F = FA$ for $A \in \mathfrak{A}(\mathcal{O}')$. If

$U \in \mathfrak{A}_0$ such that $\mathrm{Ad}U \circ \rho$ is localized in $\mathcal{O}$, then $\rho(A)U^* = U^*A$ for $A \in \mathfrak{A}(\mathcal{O}')$, hence $UF \in \mathfrak{A}(\mathcal{O}')' = \mathfrak{A}(\mathcal{O})$ and $F \in \mathfrak{A}_0$. The tensor product of the bimodules $\mathcal{F}_{\rho_i}$ is the field bundle product of [42],

$$(\rho_1, F_1) \otimes (\rho_2, F_2) = (\rho_2\rho_1, \rho_2(F_1)F_2). \tag{1.44}$$

Homomorphisms of bimodules are now given by operators $T \in \mathcal{B}(\mathcal{H}_{\pi_0})$ which satisfy the intertwiner relation

$$\sigma(A)T = T\rho(A),\ \ A \in \mathfrak{A}_0, \tag{1.45}$$

and where $T \in \mathfrak{A}(\mathcal{O})$ if σ and ρ are localized in $\mathcal{O}$. The statistics operator, in particular, can now be expressed in terms of endomorphisms and unitaries which move the localization regions of endomorphisms into spacelike separated regions.

Namely, let $(\rho_i, F_i) \in \mathcal{F}_{\rho_i}(\mathcal{O}_i), i = 1, 2$ with spacelike separated double cone $\mathcal{O}_1, \mathcal{O}_2$. Choose unitaries $U_i \in \mathfrak{A}_0$ such that $\rho_i' := \mathrm{Ad} \circ \rho_i$ is localized in $\mathcal{O}_i$. Then $F_i' := U_iF_i \in \mathfrak{A}(\mathcal{O}_i)$ and in (1.44) we obtain for the second entry on the right hand side

$$\rho_2(F_1)F_2 = \rho_2(U_1^*F_1')U_2^*F_2' = \rho_2(U_1^*)U_2^*\rho_2'(F_1')F_2' = \rho_2(U_1^*)U_2^*F_1'F_2'\,. \tag{1.46}$$

Inserting this into the formula for the definition of the statistics operator we obtain

$$\varepsilon(\rho_1, \rho_2) = \rho_1(U_2^*)U_1^*U_2\rho(U_1). \tag{1.47}$$

By passing from the representations to the endomorphisms we get a new category where the objects are the DHR endomorphisms and the arrows are intertwiners between these endomorphisms. This category is equivalent as a braided (and in $d > 2$ symmetric) tensor C*-category to the previous one (after adding the requirement of the existence of unitary partial intertwiners). In contrast to the previous one it is small (the objects form a set) and strict (the tensor product is now the composition of endomorphisms and hence associative).

Let us return to the question of rigidity (the existence and uniqueness of conjugates). This question is closely related to the classification of statistics. In order to characterize conjugates $\overline{\rho}$ of an DHR endomorphism ρ up to equivalence one requires in addition to the existence of an isometric intertwiner R from the vacuum id to the representation $\overline{\rho}\rho$ the existence of an isometry $\overline{R}$ from id to $\rho\overline{\rho}$ (this is automatically fulfilled by setting $\overline{R} = \varepsilon(\overline{\rho}, \rho)R$) and the equations

$$\overline{R}^*\rho(R) = R^*\overline{\rho}(\overline{R}) = \lambda 1,\ \lambda \neq 0. \tag{1.48}$$

(In category theory, it is common to normalize the intertwiners such that $\lambda = 1$. This, however, is in general not compatible with the required isometry.) We now compute λ in terms of the statistics operators. We find

$$\overline{R}^*\rho(R) = R^*\varepsilon(\overline{\rho}, \rho)^*\rho(R)\varepsilon(\mathrm{id}, \rho) = R^*\varepsilon(\overline{\rho}, \rho)^*\varepsilon(\overline{\rho}\rho, \rho)R = R^*\overline{\rho}(\varepsilon(\rho, \rho))R, \tag{1.49}$$

where we used $\varepsilon(\text{id}, \rho) = 1$, $\rho(R)\varepsilon(\text{id}, \rho) = \varepsilon(\overline{\rho}\rho, \rho)R$ and $\varepsilon(\overline{\rho}\rho, \rho) = \varepsilon(\overline{\rho}, \rho)$ $\overline{\rho}(\varepsilon(\rho, \rho))$. The completely positive map

$$\phi : \mathfrak{A}_0 \to \mathfrak{A}_0,\ \phi(A) = R^*\overline{\rho}(A)R$$

is a left inverse of ρ. In the case of a symmetry, it can be used for analysing the representation of the permutation group S_n generated by the operators $\rho^k(\varepsilon(\rho, \rho))$, $k = 0, \ldots, n-1$ in terms of the *statistics parameter* $\lambda = \phi(\varepsilon(\rho, \rho))$. One finds that, by positivity, the allowed values of λ are $\lambda \in \{\pm\frac{1}{d} | d \in \mathbb{N}\}$. The natural number d is called the statistical dimension within the DHR theory. As discovered by Longo [80], it coincides with the square root of the Jones index [72] of the inclusion $\rho(\mathfrak{A}(\mathcal{O})) \subset \rho(\mathfrak{A}(\mathcal{O}'))'$ which characterizes the degree of violation of Haag duality in the representation ρ (independently of the choice of $\mathcal{O} \in \mathcal{K}$). This relation remains valid also in the case where the symmetry has to be replaced by braiding. The statistical dimension coincides with the general notion of dimension of objects in symmetric tensor categories.

In case of a braiding, as it is the generic situation in 2 dimensions, one can approach a similar classification of the representation of the braid group B_n. This was successful under additional conditions on the spectrum of the statistics operators $\varepsilon(\rho, \rho)$. If e.g. it contains only two points as in the symmetric case, the representations can be classified in analogy to the representations of the symmetric group [58].

In the symmetric case, the DHR endomorphisms with finite statistical dimension form a rigid symmetric tensor C*-category. It was proven by Doplicher and Roberts that such a category is equivalent to a category of representations of a compact group G with a distinguished element k of order 2, where the representation spaces are graded by the eigenvalues of k and where the symmetry σ on the tensor product of representations is chosen such that for eigenvectors v, w

$$\sigma\ v \otimes w = \pm w \otimes v. \tag{1.50}$$

Here the minus sign appears only when both vectors have eigenvalue -1. Both, G and k are uniquely determined. The authors then exploit this equivalence and construct a net of von Neumann algebras $\mathfrak{F}$ on some Hilbert space $\mathcal{H}$ with the following properties:

- $\mathfrak{F}$ satisfies the condition of isotony.
- The group G (the gauge group) acts by automorphisms $g \mapsto \alpha_g$ such that $\alpha_g(\mathfrak{F}(\mathcal{O})) = \mathfrak{F}(\mathcal{O})$ for all $\mathcal{O}$.
- If $\mathcal{O}_1$ is spacelike to $\mathcal{O}_2$ and $F_i \in \mathfrak{F}(\mathcal{O}_i)$ with $\alpha_k(F_i) = +F_i$ (bosonic) or $-F_i$ (fermionic), then $F_1F_2 = \pm F_2F_1$, where the minus sign holds if both factors are fermionic.
- There is an action of the covering of the connected component of the Poincaré group $L \mapsto \alpha_L$ by automorphisms of $\mathfrak{F}$, such that

$$\alpha_L(\mathfrak{F}(\mathcal{O})) = \mathfrak{F}(L\mathcal{O}) \text{ and } \alpha_L\alpha_g = \alpha_g\alpha_L\ \forall g \in G. \tag{1.51}$$

- The subalgebras of fixed points under the gauge group,

$$\mathfrak{A}(\mathcal{O}) = \{A \in \mathfrak{F}(\mathcal{O}) | \alpha_g(A) = A \; \forall g \in G\}, \tag{1.52}$$

  form a net, which is equivalent to the original Haag-Kastler net and where the Poincaré symmetry of $\mathfrak{A}$ derives from the action $L \mapsto \alpha_L$ on $\mathfrak{F}$.
- Each irreducible DHR representation with finite statistics is equivalent to a subrepresentation of $\mathfrak{A}$ on $\mathcal{H}$.
- The net $\mathfrak{F}$ acts irreducibly on $\mathcal{H}$ in the following strong sense: Let $\mathcal{O}$ be a double cone. Then the only bounded operators which commute with $\mathfrak{F}(\mathcal{O})$ and $\mathfrak{A}(\mathbb{M})$ are the multiples of the identity.

For more details see the original papers [45, 46] and for a nice review [68].

The analysis described above concerns the sector structure on Minkowski space. In conformally invariant theories it is often useful to replace Minkowski space by its conformal completion. In such a spacetime the partially ordered set of double cones is no longer directed which leads to a modification of the DHR theory. More generally, in globally hyperbolic spacetimes, new phenomena might occur due to topological obstructions. See [21] and references therein for more information.

1.5 Structure of Local Algebras

The Haag-Kastler axioms, together with the existence of a vacuum representation, yield already quite a number of informations on the structure of local algebras. It was first observed on the example of the free scalar field that the local algebras are factors of type III [5, 6]. At that time, type III algebras were not well understood and were considered to be pathological. It was observed that the type III property is due to the sharp localization within a spacetime region, and Borchers conjectured that there exist intermediate type I factors between local algebras associated to regions $\mathcal{O}_1, \mathcal{O}_2$ with $\overline{\mathcal{O}_1} \subset \mathcal{O}_2$, a property which later was named the split property. Actually, Buchholz was able to prove this conjecture for the free scalar field [22].

A crucial progress in the structural analysis came with the advent of the Tomita-Takesaki theory. Namely, given a von Neumann algebra N with a cyclic and separating vector ξ, a situation present for local algebras due to the Reeh-Schlieder theorem, one can define an antilinear operator S with domain $N\xi$ by

$$SA\xi = A^*\xi, \; A \in N. \tag{1.53}$$

S is in general an unbounded operator. It is closable, and we denote the closure with the same symbol. The big surprise is that its polar decomposition

$$S = J\Delta^{1/2}, \; \Delta = S^*S \text{ (the } \textit{modular operator}) \tag{1.54}$$

has the following remarkable properties:

- J is an antiunitary involution (the *modular involution*), and $J\Delta = \Delta^{-1}J$.
- $JNJ = N'$.
- The unitaries Δ^{it} implement automorphisms of N (the *modular automorphisms*).
- The state induced by ξ satisfies the KMS condition with respect to the time evolution given by the modular automorphisms, with inverse temperature -1.

An immediate question is whether one can determine the modular structure for local algebras and their cyclic and separating vectors, in particular the vacuum, and whether these operations have a physical interpretation.

The first breakthrough was obtained by Bisognano and Wichmann [11]. They showed that for a generic Haag-Kastler net which is generated by Wightman fields in a specific way the problem could be solved for the algebra associated to the wedge $W = \{x \in \mathbb{M} | |x^0| < x^1\}$ and the vacuum. They proved that the modular automorphisms coincide with Lorentz boosts in the x^1-direction (hence the vacuum looks like a thermal state for a uniformly accelerated observer, an observation which was made independently at about the same time by Unruh [90] as an analog to Hawking radiation of black holes, see [34] for a discussion of the physical interpretation), and they showed that the modular involution coincides with the P_1CT-transformation where parity P is replaced by reflection P_1 of the x^1-coordinate only. As a consequence, Haag duality holds for wedges, since $P_1TW = W'$, a property called essential Haag duality, a concept introduced by Roberts.

There were also a few other cases where the modular structures could be uncovered, but for the generic case not much is known. There is, however, a very remarkable theorem of Borchers [13, 51] which can be seen as an abstract version of the Bisognano-Wichmann-Theorem. Namely, let M be a von Neumann algebras with a cyclic and separating vector Ω together with a strongly continuous 1-parameter unitary group $t \mapsto U(t)$ with the properties:

- The generator of U is positive.
- $U(t)\Omega = \Omega$.
- For $t > 0$, $\mathrm{Ad}U(t)$ maps M into itself.

Then, for the modular operator Δ and the modular involution J associated to the pair (N, Ω), the following relations hold:

$$\Delta^{it}U(s)\Delta^{-it} = U(e^{-2\pi t}s), \tag{1.55}$$

$$JU(s)J = U(-s). \tag{1.56}$$

These are the relations found by Bisognano and Wichmann for the wedge algebra on the vacuum where U describes the future directed lightlike translations within the wedge.

There is also a partial converse of Borchers' Theorem found by Wiesbrock [7, 93]. Namely, let N, M be von Neumann algebras acting on the same Hilbert space with a joint cyclic and separating vector Ω such that $N \subset M$ and that the modular automorphisms associated to M map (for $t > 0$) the algebra N into itself,

$$\mathrm{Ad}\Delta_M^{it}(N) \subset N, t > 0. \tag{1.57}$$

(This was termed *half sided modular inclusion*.) Then there exists a uniquely determined unitary 1-parameter group $a \mapsto U(a)$ with a positive generator which satisfies the relations above, such that

$$\mathrm{Ad}U(1)(N) = M. \tag{1.58}$$

It was shown by Borchers, Wiesbrock et al. that by using these theorems one can construct covariant Haag-Kastler nets from a finite family of half sided modular inclusions where the necessary number depends on the dimension of spacetime. Unfortunately, up to now, explicit examples of half sided modular inclusions have been found only within Haag-Kastler nets, so this method has not yet led to new examples of AQFT. See [14] for a very detailed overview which also contains many more references. See also [35] for an attempt to interpret the modular automorphisms as a state dependent intrinsic time evolution with applications to quantum gravity.

In spite of the fact that an explicit determination of the action of modular automorphisms on the local algebras was possible only in special cases, some general features could be established which finally led to the determination of the local von Neumann algebras up to isomorphy under plausible conditions. To explain this we first review the classification of type III algebras by Connes. Namely, Connes showed that the intersection of the spectra of all modular operators of a factor of type III with respect to its cyclic and separating vectors is one of the following possibilities:

- $\{0, 1\}$ (type III_0),
- $\{0, \lambda^n, n \in \mathbb{Z}\}$ for some $0 < \lambda < 1$ (type III_λ),
- $\mathbb{R}_+$ (type III_1).

One finds that in all known examples the algebras $\mathfrak{A}(\mathcal{O})$ are factors of type III_1. Moreover, under plausible assumptions, the type III_1 property is generic for theories generated by Wightman fields with a well behaved short distance scaling limit [53]. Again, this originally looked, from the mathematics perspective, as the least understood possibility. It soon turned out, however, that these factors have very nice properties. First of all, Haagerup [66] could show that there is up to isomorphy only one hyperfinite factor of type III_1, where hyperfinite means that the algebra is generated by an increasing sequence of full matrix algebras. It is exactly the factor already found for local algebras of free field theories. Hyperfiniteness of the local algebras is implied by the split property mentioned in the introduction. The split property, finally, can be derived from the so-called nuclearity condition [27, 30]. This condition may be understood as the condition that the partition function in a finite spatial volume is finite. Technically, it says that for each double cone $\mathcal{O}$ and each finite $\beta > 0$ the map

$$T_{\mathcal{O},\beta} : \mathfrak{A}(\mathcal{O}) \to \mathcal{H},\ A \mapsto e^{-\beta H} A\Omega \tag{1.59}$$

is nuclear. Here the algebra is represented on the vacuum Hilbert space $\mathcal{H}$, H is the Hamiltonian (the generator of time translations in a given Lorentz frame) and Ω is the unit vector inducing the vacuum state. A linear map f between Banach spaces X, Y is nuclear if it can be written in the form

$$f(x) = \sum y_n \langle l_n, x \rangle \tag{1.60}$$

with $y_n \in Y$, $l_n \in X^*$, the dual of X, and $\sum_n ||y_n|| \cdot ||l_n|| < \infty$. The infimum $\nu(f)$ over these sums for all such representations of f is called the nuclearity index. If $\nu(T_{\mathcal{O},\beta})$ behaves in the way expected from a partition function,

$$\nu(T_{\mathcal{O},\beta}) < e^{(\beta_0/\beta)^n}, \tag{1.61}$$

for some $\beta_0 > 0$ (depending on $\mathcal{O}$) and some $n \in \mathbb{N}$ (this was shown to be true for the free scalar field), then one finds that the split property holds and that the theory has KMS states for all $\beta > 0$ [28]. On the other hand, there are generalized free fields which violate the nuclearity condition as well as the split property.

Besides implying hyperfiniteness, the split property has many other nice consequences, which may be summarized in the universal localizing map [44]. Namely, let

$$\Lambda = (\mathfrak{A}(\mathcal{O}_1) \subset \mathfrak{A}(\mathcal{O}_2)) \tag{1.62}$$

be a split inclusion. Then there is a unit vector $\Omega_\Lambda \in \mathcal{H}$ which induces the product state

$$\omega_\Lambda(AB') = \omega_0(A)\omega_0(B'), \ A \in \mathfrak{A}(\mathcal{O}_1), B' \in \mathfrak{A}(\mathcal{O}_2'). \tag{1.63}$$

One then defines a unitary $U : \mathcal{H} \to \mathcal{H} \otimes \mathcal{H}$ by

$$UAB'\Omega_\Lambda = A\Omega \otimes B'\Omega. \tag{1.64}$$

The universal localizing map Φ_Λ now maps $\mathcal{B}(\mathcal{H})$ into the algebra $\mathfrak{A}(\mathcal{O}_2)$,

$$\Phi_\Lambda(A) = U^*(A \otimes 1)U, \tag{1.65}$$

in such a way that it acts trivially on $\mathfrak{A}(\mathcal{O}_1)$. We conclude that all operators on the Hilbert space are mapped by this procedure into the larger of the two local algebras. In particular, global charges and momentum operators can be localized in such a way that not only their action on the smaller algebra is preserved but also their spectrum. This is quite a surprise, since an integral over a charge density with a suitable test function would typically have a different spectrum than the global charge operator.

The split property allows to decouple the observables of the smaller region completely from the observables in the spacelike complement of the larger region, in the sense that normal states[2] exist (the product states) for which all correlations between

---

[2]states induced by density matrices.

these observables vanish. On the contrary, the fact that the local algebras are of type III implies that normal states always are correlated. They are even necessarily entangled, i.e. they cannot be uniformly approximated by convex combinations of product states. This mathematical fact is e.g. responsible for the Hawking effect; in any Hadamard state the algebra of observables outside the horizon is of type III and hence the state must be entangled with observables behind the horizon. (See [54] for a field theoretical derivation of the Hawking effect and, as an alternative, the recently postulated fire walls at the horizon which should deform the algebra into the type I case [3].)

Actually, far from being pathological, the type III_1-property of local algebras has many nice aspects. In particular, one can move by the adjoint action of unitary elements through the state space and finds almost transitivity, in the sense that each orbit is norm dense in the space of normal states. This is exploited in the Buchholz-Roberts analysis of superselection sectors for QED [32] and also in the discussion of transition probabilities in [29]. See also [94] and references therein for further information.

1.6 Conclusions

Algebraic Quantum Field Theory is an approach to quantum field theory which is in its aims essentially equivalent to other approaches, as e.g. the path integral approach or an approach based on canonical quantization of classical field theory. It offers some conceptual advantages compared with other approaches, in particular the separate discussion of observables and states which allows to incorporate the locality principle into the theory. Moreover, it is fully rigorous. In its formulation with C*- and von Neumann algebras the rich mathematical structure can be exploited and leads to an understanding of particle statistics and global gauge symmetries. Moreover, apparent contradictions between nonrelativistic quantum mechanics and relativistic quantum field theory find their natural explanation in the different structures of the occuring algebras of observables. The operator algebraic formulation is, on the other hand, rather rigid, which makes it difficult to deform a given model. Nevertheless, first examples have been obtained which satisfy a weakened form of the axioms. It is, however, also possible to relax the conditions on the operator algebras in order to make contact with the way QFT is treated in other approaches. The concepts from AQFT have turned out to be especially fruitful for the perturbative construction of interacting quantum field theories on curved spacetimes, a problem which could not been solved in other approaches.

References

1. Alazzawi, S.: Deformation of fermionic quantum field theories and integrable models. Lett. Math. Phys. **103**, 37–58 (2012)
2. Alazzawi, S: Deformation of quantum field theories and the construction of interacting models. PhD thesis, University of Vienna (2015)
3. Almheiri, A., Marolf, D., Polchinski, J., Sully, J.: Black holes: complementarity or firewalls? JHEP **1302**, 062 (2013)
4. Altherr, T.: Infrared problem in $g\phi^4$ theory at finite temperature. Phys. Lett. B **238**, 360 (1990)
5. Araki, H.: A lattice of von Neumann algebras associated with the quantum theory of a free Bose field. J. Math, Phys. **4**, 1343 (1963)
6. Araki, H.: Von Neumann algebras of local observables for free scalar field. J. Math, Phys. **5**, 1 (1964)
7. Araki, H., Zsid6, L.: Extension of the structure theorem of borchers and its application to half-sided modular inclusions. Rev. Math. Phys. **17**, 491 (2005)
8. Barata, J.C.A., Jäkel, C.D., Mund, J.: The $\mathcal{P}(\varphi)_2$ model on the de sitter space. arXiv:1311.2905 [math-ph]
9. Becker, C., Schenkel, A., Szabo, R.J.: Differential cohomology and locally covariant quantum field theory. arXiv:1406.1514 [hep-th]
10. Benini, M., Dappiaggi, C., Hack, T.-P.: Quantum field theory on curved backgrounds—a primer. Int. J. Modern Phys. **A 28**(17), 1330023 (2013)
11. Bisognano, J.J., Wichmann, E.H.: On the duality condition for a hermitian scalar field. J. Math. Phys. **16**, 985 (1975)
12. Borchers, H.J.: A remark on a theorem of B. Misra. Commun. Math. Phys. **4**, 315–323 (1967)
13. Borchers, H.J.: The CPT theorem in two-dimensional theories of local observables. Commun. Math. Phys. **143**, 315 (1992)
14. Borchers, H.J.: On revolutionizing quantum field theory with Tomita's modular theory. J. Math. Phys. **41**, 3604 (2000)
15. Brunetti, R., Duetsch, M., Fredenhagen, K.: Perturbative algebraic quantum field theory and the renormalization groups. Adv. Theor. Math. Phys. **13**, 1541 (2009)
16. Brunetti, R., Fredenhagen, K.: Microlocal analysis and interacting quantum field theories: renormalization on physical backgrounds. Commun. Math. Phys. **208**, 623 (2000)
17. Brunetti, R., Fredenhagen, K., Kohler, M.: The Microlocal spectrum condition and Wick polynomials of free fields on curved space-times. Commun. Math. Phys. **180**, 633 (1996)
18. Brunetti, R., Fredenhagen, K., Rejzner, K.: Quantum gravity from the point of view of locally covariant quantum field theory. arXiv:1306.1058 [math-ph]
19. Brunetti, R., Fredenhagen, K., Verch, R.: The generally covariant locality principle: a new paradigm for local quantum field theory. Commun. Math. Phys. **237**, 31 (2003)
20. Brunetti, R., Guido, D., Longo, R.: Modular localization and wigner particles. Rev. Math. Phys. **14**, 759 (2002)
21. Brunetti, R., Ruzzi, G.: Quantum charges and spacetime topology: the emergence of new superselection sectors. Commun. Math. Phys. **287**, 523 (2009)
22. Buchholz, D.: Product states for local algebras. Commun. Math. Phys. **36**, 287 (1974)
23. Buchholz, D.: Collision theory for massless bosons. Commun. Math. Phys. **52**, 147 (1977)
24. Buchholz, D.: The physical state space of quantum electrodynamics. Commun. Math. Phys. **85**, 49 (1982)
25. Buchholz, D.: Gauss' law and the infraparticle problem. Phys. Lett. B **174**, 331 (1986)
26. Buchholz, D., Fredenhagen, K.: Locality and the structure of particle states. Commun. Math. Phys. **84**, 1 (1982)
27. Buchholz, D., Fredenhagen, K., D'Antoni, C.: The universal structure of local algebras. Commun. Math. Phys. **111**, 123 (1987)
28. Buchholz, D., Junglas, P.: On the existence of equilibrium states in local quantum field theory. Commun. Math. Phys. **121**, 255 (1989)

29. Buchholz, D., Stœrmer, E.: Superposition, transition probabilities and primitive observables in infinite quantum systems. arXiv:1411.2100 [math-ph]
30. Buchholz, D., Wichmann, E.H.: Causal independence and the energy level density of states in local quantum field theory. Commun. Math. Phys. **106**, 321 (1986)
31. Buchholz, D., Lechner, G., Summers, S.J.: Warped convolutions, rieffel deformations and the construction of quantum field theories. Commun. Math. Phys. **304**, 95 (2011)
32. Buchholz, D., Roberts, J.E.: New light on infrared problems: sectors, statistics, symmetries and spectrum. Commun. Math. Phys. **330**, 935 (2014)
33. Buchholz, D., Verch, R.: Scaling algebras and renormalization group in algebraic quantum field theory. Rev. Math. Phys. **7**, 1195 (1995)
34. Buchholz, D., Verch, R.: Macroscopic aspects of the Unruh effect. arXiv:1412.5892 [gr-qc]
35. Connes, A., Rovelli, C.: Von neumann algebra automorphisms and time thermodynamics relation in general covariant quantum theories. Class. Quant. Grav. **11**, 2899 (1994)
36. Dimock, J.: Algebras of local observables on a manifold. Commun. Math. Phys. **77**, 219–228 (1980)
37. Dimock, J.: Dirac quantum fields on a manifold. Trans. Am. Math. Soc. **269**, 133 (1982)
38. Dittrich, B.: Partial and complete observables for canonical general relativity. Class. Quant. Grav. **23**, 6155 (2006)
39. Doplicher, S.: Local aspects of superselection rules. Commun. Math. Phys. **85**, 73 (1982)
40. Doplicher, S., Haag, R., Roberts, J.E.: Fields, observables and gauge transformations 1. Commun. Math. Phys. **13**, 1 (1969)
41. Doplicher, S., Haag, R., Roberts, J.E.: Fields, observables and gauge transformations 2. Commun. Math. Phys. **15**, 173 (1969)
42. Doplicher, S., Haag, R., Roberts, J.E.: Local observables and particle statistics 1. Commun. Math. Phys. **23**, 199 (1971)
43. Doplicher, S., Haag, R., Roberts, J.E.: Local observables and particle statistics 2. Commun. Math. Phys. **35**, 49 (1974)
44. Doplicher, S., Longo, R.: Local aspects of superselection rules II. Commun. Math. Phys. **88**, 399 (1983)
45. Doplicher, S., Roberts, J.E.: A new duality theory for compact groups. Inventiones Math. **98**, 157 (1989)
46. Doplicher, S., Roberts, J.E.: Why there is a field algebra with a compact gauge group describing the superselection structure in particle physics. Commun. Math. Phys. **131**, 51 (1990)
47. Driessler, W.: On the type of local algebras in quantum field theory. Commun. Math. Phys. **53**, 295 (1977)
48. Dybalski, W.: Haag-Ruelle scattering theory in presence of massless particles. Lett. Math. Phys. **72**, 27 (2005)
49. Dybalski, W., Gérard, C.: A criterion for asymptotic completeness in local relativistic QFT. Commun. Math. Phys. **332**(3), 1167 (2014)
50. Enss, V.: Characterization of particles by means of local observables in the relativistic quantum theory. Commun. Math. Phys. **45**, 35 (1975)
51. Florig, M.: On Borchers' theorem. Lett. Math. Phys. **46**, 289–293 (1998)
52. Fredenhagen, K.: On the existence of anti-particles. Commun. Math. Phys. **79**, 141 (1981)
53. Fredenhagen, K.: On the modular structure of local algebras of observables. Commun. Math. Phys. **97**, 79 (1985)
54. Fredenhagen, K., Haag, R.: On the derivation of hawking radiation associated with the formation of a black hole. Commun. Math. Phys. **127**, 273 (1990)
55. Fredenhagen, K., Rejzner, K.: Batalin-Vilkovisky formalism in perturbative algebraic quantum field theory. Commun. Math. Phys. **317**, 697 (2013)
56. Fredenhagen, K., Rejzner, K.: QFT on curved spacetimes: axiomatic framework and examples. arXiv:1412.5125 [math-ph]
57. Fredenhagen, K., Lindner, F.: Construction of KMS states in perturbative QFT and renormalized hamiltonian dynamics. Commun. Math. Phys. **332**(3), 895 (2014)

58. Fredenhagen, K., Rehren, K.H., Schroer, B.: Superselection sectors with braid group statistics and exchange algebras 1. Gen. Theor. Commun. Math. Phys. **125**, 201 (1989)
59. Georgi, H.: Unparticle physics. Phys. Rev. Lett. **98**, 221601 (2007)
60. Glimm, J., Jaffe, A.M.: Collected Papers: Constructive Quantum Field Theory. Selected Papers, vol. 2, 533 p. Birkhaeuser, Boston (1985) (Contemporary Physicists)
61. Haag, R.: Discussion des "axiomes" et des propriétés asymptotiques d'une théorie des champs locales avec particules composés. In: Les problémes mathématiques de la théorie quantique des champs. Colloques internationaux du CNRS, Paris. vol. 75, pp. 151-162. The english original appeared in R. Haag, Discussion of the 'axioms' and the asymptotic properties of a local field theory with composite particles, Eur. Phys. J. H **35**, 243 (2010)
62. Haag, R.: Quantum field theories with composite particles and asymptotic conditions. Phys. Rev. **112**, 669 (1958)
63. Haag, R: Local Quantum Physics: Fields, Particles, Algebras, 356 p. Springer, Berlin (1992) (Texts and monographs in physics)
64. Haag, R., Kastler, D.: An algebraic approach to quantum field theory. J. Math. Phys. **5**, 848 (1964)
65. Haag, R., Swieca, J.: When does a quantum field theory describe particles? Commun. Math. Phys. **1**, 308
66. Haagerup, U.: Connes bizentralizer problem and uniqueness of the injective factor of type III-1. Acta Math. **158**, 95–148 (1987)
67. Halvorson, H.: Reeh-Schlieder defeats Newton-Wigner: on alternative localization schemes in relativistic quantum field theory. Phil. Sci. **68**, 111 (2001) [quant-ph/0007060]
68. Halvorson, H., Muger, M.: Algebraic quantum field theory. [math-ph/0602036]
69. Hollands, S., Wald, R.M.: Local wick polynomials and time ordered products of quantum fields in curved space-time. Commun. Math. Phys. **223**, 289 (2001)
70. Hollands, S., Wald, R.M.: Existence of local covariant time ordered products of quantum fields in curved space-time. Commun. Math. Phys. **231**, 309 (2002)
71. Hollands, S., Wald, R.M.: Quantum fields in curved spacetime. arXiv:1401.2026 [gr-qc]
72. Jones, V.F.R.: Index for subfactors. Invent. Math. **72**, 1–25 (1983)
73. Kay, B.S.: Linear spin-zero quantum fields in external gravitational and scalar fields. Commun. Math. Phys. **62**, 55–70 (1978)
74. Kay, B.S.: The casimir effect without MAGIC. Phys. Rev. D **20**, 3052 (1979)
75. Kawahigashi, Y., Longo, R.: Classification of two-dimensional local conformal nets with c less than 1 and 2 cohomology vanishing for tensor categories. Commun. Math. Phys. **244**, 63 (2004)
76. Khavkine, I.: Local and gauge invariant observables in gravity. arXiv:1503.03754 [gr-qc]
77. Khavkine, I., Moretti, V.: Continuous and analytic dependence is an unnecessary requirement in renormalization of locally covariant QFT. arXiv:1411.1302 [gr-qc]
78. Lechner, G.: Construction of quantum field theories with factorizing s-matrices. Commun. Math. Phys. **277**, 821 (2008)
79. Lindner, F.: Perturbative algebraic quantum field theory at finite temperature. PhD thesis, University of Hamburg (2013)
80. Longo, R.: Index of subfactors and statistics of quantum fields I. Commun. Math. Phys. **126**, 217 (1989)
81. Mund, J., Schroer, B., Yngvason, J.: String-localized quantum fields and modular localization. Commun. Math. Phys. **268**, 621 (2006)
82. Radzikowski, M.J.: Micro-local approach to the Hadamard condition in quantum field theory on curved space-time. Commun. Math. Phys. **179**, 529 (1996)
83. Reeh, H., Schlieder, S.: Bemerkungen zur Unitäräquivalenz von Lorentzinvarianten Feldern. Nuovo Cimento **22**, 1051 (1961)
84. Rovelli, C.: Partial observables. Phys. Rev. D **65**, 124013 (2002)
85. Ruelle, D.: On the asymptotic condition in quantum field theory. Helv. Phys. Acta **35**, 147 (1962)

86. Ruzzi, G.: Essential properties of the vacuum sector for a theory of superselection sectors. Rev. Math. Phys. **15**, 1255-1283 (2003)
87. Schrader, R.: Yukawa quantum field theory in two space-time dimensions without cutoffs. Ann. Phys. **70**, 412 (1972)
88. Steinmann, O.: Perturbative quantum field theory at positive temperatures: an axiomatic approach. Commun. Math. Phys. **170**, 405 (1995)
89. Summers, S.J.: A perspective on constructive quantum field theory. arXiv:1203.3991 [math-ph]
90. Unruh, W.G.: Notes on black hole evaporation. Phys. Rev. D **14**, 870 (1976)
91. Verch, R.: A spin statistics theorem for quantum fields on curved space-time manifolds in a generally covariant framework. Commun. Math. Phys. **223**, 261 (2001)
92. Wald, R.M.: Quantum Field Theory in Curved Space-time and Black Hole Thermodynamics. University of Pr, Chicago (1994)
93. Wiesbrock, H.W.: Half sided modular inclusions of von Neumann algebras. Commun. Math. Phys. **157**, 83 (1993) [Erratum-ibid. **184**, 683 (1997)]
94. Yngvason, J.: Localization and entanglement in relativistic quantum physics. arXiv:1401.2652 [quant-ph]

Chapter 2
Perturbative Construction of Models of Algebraic Quantum Field Theory

Klaus Fredenhagen and Katarzyna Rejzner

Abstract The construction of models of algebraic quantum field theory by renormalized perturbation theory is reviewed.

2.1 Introduction

The axiomatic framework of AQFT allows for a qualitative description of a large class of phenomena occurring in particle physics and some parts of solid state physics. It does not, however, yield quantitative predictions, and there is a widespread impression that one has to abandon the formalism of AQFT if one wants to make real contact with experiments. Actually, as explained in Chap. 1, up to now no single model of an interacting AQFT in 4d Minkowski space has been constructed.

But what are the alternatives? Standard textbooks on QFT either start from canonical quantization of free field theory on Fock space and try to construct the interacting theory in the interaction picture, or they use the path integral formalism. The canonical approach ends up in the Gell-Mann Low formula for the vacuum expectation values of time ordered products of fields,

$$\omega_0(T\varphi(x_1)\dots\varphi(x_n)) = \frac{\langle\Omega, T\varphi_0(x_1)\dots\varphi_0(x_n)e^{\frac{i}{\hbar}\int\mathcal{L}_I(x)d^4x}\Omega\rangle}{\langle\Omega, Te^{\frac{i}{\hbar}\int\mathcal{L}_I(x)d^4x}\Omega\rangle}, \tag{2.1}$$

where φ_0 is the free field treated as an operator valued distribution on the Fock space, $\mathcal{L}_I$ is the interaction density treated as a Wick polynomial of φ_0 and Ω is the vacuum vector of the free theory. The time ordering symbol T means that the products have to be performed after ordering of the factors according to their time arguments.

K. Fredenhagen (✉)
II. Institut für Theoretische Physik, 22761 Hamburg, Germany
e-mail: klaus.fredenhagen@desy.de

K. Rejzner
University of York, Heslington, York YO10 5DD, UK
e-mail: kasia.rejzner@york.ac.uk

R. Brunetti et al. (eds.), *Advances in Algebraic Quantum Field Theory*,
Mathematical Physics Studies, DOI 10.1007/978-3-319-21353-8_2

The path integral approach reinterprets the Gell-Mann Low formula as an integral over all classical field configurations ϕ

$$\omega_0(T\varphi(x_1)\cdots\varphi(x_n)) = Z^{-1}\int \phi(x_1)\cdots\phi(x_n)e^{\frac{i}{\hbar}\int\mathcal{L}(x)d^4x}D\phi \tag{2.2}$$

where now $\mathcal{L}$ is the full classical Lagrangian, Z is a normalization factor, and $D\phi$ is thought of as the Lebesgue integral over field space.

Both versions are only heuristic, and it required the hard and ingenious work of several generations of physicists to turn these formal expressions into unambiguous computations. The state of the art is that one can create a formal power series in $\hbar$ where every term is well defined, up to some remaining infrared problems originating from the integral over Minkowski space in the exponent. The great success of QFT relies on the fact that already the first few terms of this series yield a good and often even excellent agreement with experimental data.

The path integral approach has the advantage that it is formally similar to probability theory. Actually, by passing to imaginary time (Wick rotation), one can interpret the vacuum expectation values of time ordered products of fields as correlation functions of a probability distribution (euclidean QFT). In particular, counter-intuitive properties of quantum physics as e.g., entanglement do not occur. Moreover, the momentum space integrals in the evaluation of Feynman diagrams have better convergence properties. Finally, due to the Osterwalder-Schrader theorem , a Wick rotation back to real time is possible under very general conditions.

The disadvantage of the path integral approach is that the noncommutative product of operators, which is crucial for the structure of quantum physics, appears only indirectly in terms of different boundary values of analytic functions. In the canonical approach, the operator product is given from the beginning, but there the definition of the time ordered product is problematic. First of all, it is not well defined as a product of operators, since, by the existence of a deterministic time evolution, fields at a given time can be expressed in terms of fields at an earlier time, and thus the time ordering prescription is ambiguous. One may instead define time ordered products $TA(t_1)\cdots A(t_n)$ of an operator valued function of time $t \mapsto A(t)$ as a symmetric operator valued function of n time variables such that

$$TA(t_1)\cdots A(t_n) = A(t_1)\cdots A(t_n) \text{ if } t_1 \geq \cdots \geq t_n. \tag{2.3}$$

This, however, does not work since the quantum fields are distributions, and the time ordering prescription would amount to multiplying them with a discontinuous function.

But there is a way out, as first observed by Stückelberg, further elaborated by Bogoliubov and collaborators and finally worked out by Epstein and Glaser (causal perturbation theory). Namely, one may define the time ordered product of n fields as an operator valued distribution which is already known for non-coinciding points. Due to the UV divergences of QFT, the extension to coinciding points is ambiguous, but the crucial observation is that this ambiguity is the same ambiguity which occurs in the removal of infinities in approaches where the theory is regularized by the introduction of a momentum cutoff, and where the theory without cutoff has to be fixed by renormalization conditions.

Originally, the insertion of a test function g into the interaction Lagrangian, instead of integrating it over all spacetime, was considered to be an intermediate step, and in the last step one aimed at the limit where g tends to 1 (adiabatic limit). In this limit one then finds vacuum expectation values of operator products of time ordered products of interacting fields, and using the Wightman reconstruction theorem , one obtains interacting fields as operator valued distributions on some "Hilbert space", of course only in the sense of formal power series. But as first observed in [32] and rediscovered in [8] the algebra of observables associated to some bounded region can be already constructed if one chooses a test function g which is equal to 1 on some slightly larger region. Actually, the full Haag-Kastler net of the interacting theory can be obtained in this way. Thus causal perturbation theory provides a direct way for a construction of the algebra of observables. Hence by replacing the condition that the local algebras have to be unital C*-algebras by the condition that they are isomorphic to unital *-algebras of formal power series of operators on a dense invariant subspace of some Hilbert space, one obtains a huge class of models, in particular the models used in elementary particle physics.

On this level, structural properties of the local net can be analyzed, but the powerful structural results on C*- and von Neumann algebras are not available. Nevertheless, one can derive interesting results, as e.g., the validity of the time-slice axiom [11], the existence of operator expansions [27] and an algebraic version of the Callan-Symanzik equation [9].

In order to reach numerical predictions, one needs in the next step a construction of states. States are here defined as linear maps from the algebra to the formal power series over $\mathbb{C}$, and the positivity condition on states now means that the expectation value of A^*A is the absolute square of another power series. The construction of states can be done via the adiabatic limit as described above; this is the way the vacuum state is constructed in [18]. As observed by Steinmann [43], this method does not work for the construction of KMS states. The reason is that the analog of the Gell-Mann Low formula does not hold at nonzero temperature, due to the different asymptotic time behavior of free and interacting systems at nonzero temperature. But here a structural result helps: namely, the time-slice axiom allows to treat only the theory within a short time interval, and the asymptotic behavior in time does not matter for the existence of states. What matters is the decay of correlations in spacelike directions which is exponentially fast for massive theories.

Up to now we considered the so-called on shell formalism , due to the fact that we constructed the operators on Fock space, thereby imposing the validity of the Klein Gordon equation for the free field. It turned out, however, to be more useful to replace Fock space operators by functionals of classical field configurations which are not restricted to those which satisfy the field equation. On the space of functionals one can then introduce several operations: the pointwise (classical) product, the involution by complex conjugation, the Peierls bracket (as a covariant version of a Poisson bracket on the space of functionals), the non-commutative, associative $\star$-product in the sense of deformation quantization and the time ordered product. It is the latter which is relevant for inducing the interaction and which requires renormalization. The other operations can be directly defined. This is trivial for the pointwise product and for the involution. The Peierls bracket is obtained by considering the linearized Euler-Lagrange operator, which e.g., for the φ^4-theory looks like

$$E'(\varphi) = \Box + m^2 + \frac{\lambda}{2}\varphi^2 \tag{2.4}$$

where the last term acts as a multiplication operator. We consider only theories where the linearized Euler-Lagrange operator is normally hyperbolic and hence has unique retarded and advanced propagators $\Delta^{R/A}(\varphi)$. The Peierls bracket is then defined by

$$\{F, G\}(\varphi) = \left\langle \frac{\delta F}{\delta \varphi}(\varphi), \Delta(\varphi)\frac{\delta G}{\delta \varphi}(\varphi) \right\rangle \tag{2.5}$$

where $\Delta = \Delta^R - \Delta^A$ and $\frac{\delta}{\delta\varphi}$ is the functional derivative (defined as the directional derivative). In free theories, E' and then also the propagators do not depend on φ. One then can define the $\star$-product (in the sense of formal power series in $\hbar$) by

$$(F \star G)(\varphi) = e^{\frac{i\hbar}{2}\left\langle \frac{\delta}{\delta\varphi}, \Delta \frac{\delta}{\delta\varphi'}\right\rangle} F(\varphi)G(\varphi')|_{\varphi'=\varphi}. \tag{2.6}$$

The time-ordered product is defined by a similar formula

$$(F \cdot_{\mathcal{T}} G) = e^{\hbar\left\langle \frac{\delta}{\delta\varphi}, \Delta^D \frac{\delta}{\delta\varphi'}\right\rangle} F(\varphi)G(\varphi')|_{\varphi'=\varphi} \tag{2.7}$$

with the Dirac propagator $\Delta^D = \frac{1}{2}(\Delta^R + \Delta^A)$.

Note that there is a crucial difference between the time ordered product and the other products. Namely, the ideal generated by the field equation with respect to the pointwise product is also an ideal with respect to the Poisson bracket and the $\star$-product, but not with respect to the time ordered product. This is actually a necessary condition which allows to use the time ordered product for introducing an interaction. Let V be the interaction. We then define the interacting observables by

$$R_V(F) = (e_{\mathcal{T}}^{V})^{\star -1} \star (e_{\mathcal{T}}^{V} \cdot_{\mathcal{T}} F). \tag{2.8}$$

Here $e_{\mathcal{T}}$ means the exponential series where powers are computed via the time ordered product. For an evaluation functional $\Phi_x(\varphi) = \varphi(x)$, the corresponding interacting field $x \to R_V(\Phi_x)$ satisfies the equation

$$E' R_V(\Phi_x) = E'\Phi_x + R_V\left(\frac{\delta V}{\delta \varphi(x)}\right) \tag{2.9}$$

which may be interpreted as the field equation with interaction V, when evaluated on some φ which satisfies the free field equation.

The rough description of the formalism has to be made precise in the following sense: One has to specify the functionals which are allowed, and one has to check whether this class contains the relevant ones. As it stands we need functionals whose functional derivatives are test functions in order that all operations are well defined. But we will see that by changing the products to equivalent ones, which corresponds to Wick ordering in the Fock space framework one can extend the Peierls bracket and the $\star$-product to a rather large class of functionals, which contains in particular the local functionals that appear as terms in the Lagrangian, and is stable under these operations. The definition of time ordered products is more involved and there are different possibilities, corresponding to the choice of renormalization conditions.

The plan of the paper is as follows: we will first outline the functional analytic tools which are needed for the operations. We then define the Peierls bracket and the $\star$-product on a class of functionals called microcausal. Thereafter we come to the problem to define the time ordered products. Here we first develop the general formalism and show that it leads to a construction of local nets. We review some structural properties of these nets, in particular their behavior under renormalization group transformations. Finally, we outline a possible construction of states.

2.2 Functional Derivatives, Wave Front Sets, and All That

Our approach to quantum field theory is based, in this respect similar to the path integral approach, on functionals of classical field configurations. But there, at least in its euclidean version, the measure theoretic aspects of the space of field configurations are of central importance; in our case, due to the frequent use of functional derivatives, the properties of the space of field configurations as a differential manifold are crucial.

For definiteness we concentrate on the case of a scalar field and fix an oriented, time-oriented globally hyperbolic spacetime M. There we consider the space of real valued smooth functions as the space of field configurations,

$$\mathcal{E} = C^{\infty}(M, \mathbb{R}). \tag{2.10}$$

We model it as a differentiable manifold over the space of compactly supported smooth functions

$$\mathcal{D} = C_c^{\infty}(M, \mathbb{R}) \tag{2.11}$$

where charts are defined as maps

$$\varphi + \mathcal{D} \to \mathcal{D},\ \varphi + \vec{\phi} \mapsto \vec{\phi}\,, \text{with } \varphi \in \mathcal{E}. \tag{2.12}$$

Clearly, $\mathcal{E}$ has the structure of an affine manifold. A similar affine structure can be introduced also for other fields, as e.g., gauge theories or gravity.

We will model observables as functionals on $\mathcal{E}$ and we allow only functionals which depend on the field configuration inside some compact region. This includes in particular the polynomial functionals

$$F(\varphi) = \sum_{k=1}^{n} \int \varphi(x_1) \cdots \varphi(x_k) f_k(x_1, \ldots, x_k) \tag{2.13}$$

with symmetrical distributional densities f_k with compact support. More generally, we consider functionals, for which all functional derivatives $F^{(n)}$ exist and are continuous. We recall after [24] (see [37] for a review) that a functional derivative of a functional is defined as

$$\langle F^{(1)}(\varphi), \vec{\phi} \rangle := \frac{d}{d\lambda} F(\varphi + \lambda \vec{\phi})|_{\lambda=0} \tag{2.14}$$

and a functional is differentiable if the derivative exists for all $\varphi \in \mathcal{E}$. It is continuously differentiable if the map $\mathcal{E} \times \mathcal{D} \to \mathbb{C}$, $(\varphi, \vec{\phi}) \mapsto \langle F^{(1)}(\varphi), \vec{\phi} \rangle$ is continuous. If $\mathcal{E}$ is taken with its natural Fréchet topology, this implies that $F^{(1)}(\varphi)$ is a compactly supported distributional density. Higher derivatives are obtained by iterating this definition, i.e.

$$F^{(n)}(\varphi)(\vec{\phi}_1, \ldots, \vec{\phi}_n) := \frac{d}{d\lambda} F^{(n-1)}(\varphi + \lambda \vec{\phi}_n)(\vec{\phi}_1, \ldots, \vec{\phi}_{n-1})|_{\lambda=0}, \tag{2.15}$$

and we find that the $F^{(n)}(\varphi)$'s are symmetric compactly supported distributional densities with support contained in K^n, for some compact set $K \subset M$. There remains, however, the problem that the propagators have singularities, and therefore the contractions with the distributional densities occurring as functional derivatives are not always well defined. The restriction to functionals whose functional derivatives are smooth densities, on the other side, would exclude almost all local functionals, i.e., functionals of the form

$$F(\varphi) = \int f(j_x(\varphi)), \tag{2.16}$$

where $j_x(\varphi) = (x, \varphi(x), \partial\varphi(x), \ldots)$ is the jet prolongation of φ and f is a density-valued function on the jet bundle. For these functionals, the derivatives are supported on the thin diagonal

$$D_n = \{(x_1, \ldots, x_n) \in M^n, x_1 = \cdots = x_n\} \tag{2.17}$$

and thus smooth for $n > 1$ only when they vanish.

The singularities of distributions can be analyzed using the concept of the wave front set. On the Minkowski spacetime, this concept arises in the study of the decay properties of the Fourier transform of the given distribution multiplied by some test function. A pair of a spacetime point x and a nonzero momentum k is an element of the wave front set of a given distribution t if, for any test function f with $f(x) \neq 0$ and some open cone around k, the Fourier transform of ft does not decay fast (i.e., faster than any power) inside the cone. The notion of the WF set can be generalized to an arbitrary smooth manifold M, and it is defined as a subset of T^*M.

As the first example we will consider the Dirac δ distribution on $\mathbf{R}$. Since $\langle f\delta, e^{ik\bullet}\rangle = f(0)$, where $e^{ik\bullet}(x) = e^{ikx}$, it follows that for any choice of f with $f(x) \neq 0$ the Fourier transform of ft does not decay fast in any direction and hence the wave front set of δ is

$$\mathrm{WF}(\delta) = \{(0,k), k \neq 0\}. \tag{2.18}$$

Another important example, also on $\mathbf{R}$, is the distribution $f \mapsto \lim_{\varepsilon\downarrow 0} \int \frac{f(x)}{x+i\varepsilon}dx$. Note that its Fourier transform is

$$\lim_{\varepsilon\downarrow 0} \int \frac{f(x)}{x+i\varepsilon} e^{ikx} dx = -i \int_k^\infty \hat{f}(k')dk'. \tag{2.19}$$

and $\int_k^\infty \hat{f}(k')dk'$ decays strongly as $k \to \infty$, while for $k \to -\infty$ we obtain

$$\lim_{k\to-\infty} \int_k^\infty \hat{f}(k')dk' = 2\pi f(0). \tag{2.20}$$

We can now conclude that

$$\mathrm{WF}(\lim_{\varepsilon\downarrow 0}(x+i\varepsilon)^{-1}) = \{(0,k), k < 0\}. \tag{2.21}$$

For more information on wave front sets see [26] or Chap. 4 of [3]. Using WF sets we can formulate a sufficient condition for a pointwise product of distributions to be well defined. Let t and s be distributions on M. The Whitney sum (i.e., pointwise sum) of their wave front sets is defined by

$$\mathrm{WF}(t) + \mathrm{WF}(s) = \{(x, k+k')|(x,k) \in \mathrm{WF}(t), (x,k') \in \mathrm{WF}(s)\} \tag{2.22}$$

If this set does not intersect the zero section of $T^*\mathbb{M}$, then we can define the pointwise product ts as

$$\langle ts, fg\rangle = \frac{1}{(2\pi)^n} \int \widehat{tf}(k)\widehat{sg}(-k)dk, \tag{2.23}$$

where $f, g \in \mathcal{D}$ are chosen with sufficiently small support. To see that the integral above converges, note that if $k \neq 0$, then either $\widehat{tf}$ is fast-decaying in a conical neighborhood around k or $\widehat{sg}$ is fast-decaying in a conical neighborhood around $-k$, while the other factor is polynomially bounded.

Beside the criterion for multiplying distributions, WF sets provide also a characterization of the propagation of singularities. Let P be a partial differential operator and σ_P its principal symbol. We can interpret σ_P as a function on the cotangent bundle T^*M, which carries a structure of a symplectic manifold. With the use of the canonical symplectic form, 1-forms on T^*M can be canonically identified with vector fields. Let X_P be the vector field (called the Hamiltonian vector field) corresponding to the 1-form $d\sigma_P$. In coordinates it is given by

$$X_P = \sum_{i=1}^{n} \frac{\partial \sigma_P}{\partial k_j}\frac{\partial}{\partial x_j} - \frac{\partial \sigma_P}{\partial x_j}\frac{\partial}{\partial k_j}.$$

Let $(x_j(t), k_j(t))$ be a curve that fulfills the system of equations (Hamilton's equations):

$$\frac{dx_j}{dt} = \frac{\partial \sigma_P}{\partial k_j},$$
$$\frac{dk_j}{dt} = -\frac{\partial \sigma_P}{\partial x_j}.$$

We call a solution $(x_j(t), k_j(t))$ of the above equations an *integral curve* of X_P and the bicharacteristic flow is defined as the set of all such solutions. Along this flow $\frac{d\sigma_P}{dt} = X_P(\sigma_P) = 0$, so σ_P is conserved under the bicharacteristic flow. We are now ready to state *the theorem on the propagation of singularities*: the wave front set of a solution u of the equation $Pu = f$ with f smooth is a union of orbits of the Hamiltonian flow X_P on the set of characteristics $\mathrm{char}P = \{(x,k) \in T^*M | \sigma(P)(x,k) = 0\}$ of P.

For hyperbolic differential operators on globally hyperbolic spacetimes, the characteristics is the light cone, and the principal symbol is the metric on the cotangent bundle. For such operators the wave front set of solutions is therefore a union of null geodesics γ together with their cotangent vectors $k = g(\dot{\gamma}, \cdot)$.

2.3 The Peierls Bracket and the ⋆-product

As outlined in the Introduction, we start our construction of a pAQFT model from the classical theory. To this end, we equip the space of functionals on the configuration space with a Poisson structure provided with the so called Peierls bracket. This bracket, introduced in [38], is the off-shell extension of the canonical bracket of classical mechanics, which is defined only on the space $\mathcal{E}_S$ of solutions to the equations of motion. To see how this works, we will start in a setting which resembles closely classical mechanics and then show the relation with the Peierls method on a concrete example.

2.3.1 Canonical Formalism and the Approach of Peierls

Let us start with the free scalar field with the field equation

$$P\varphi = 0, \tag{2.24}$$

where $P = \Box + m^2$ is the Klein-Gordon operator . For this equation the retarded and advanced Green's functions exist. We also know that for every $f \in \mathcal{E}$ whose support is past and future compact, Δf is a solution to (2.24). Conversely, every smooth solution of the Klein Gordon equation is of the form Δf for some $f \in \mathcal{E}$ with future and past compact support.

Without loss of generality, the spacetime can be assumed to be of the form $M = \mathbb{R} \times \Sigma$ with Cauchy surfaces $\{t\} \times \Sigma$, $t \in \mathbb{R}$. The space of Cauchy data $\Sigma \ni \mathbf{x} \mapsto (\varphi(t, \mathbf{x}), \dot{\varphi}(t, \mathbf{x}))$ on the surface $\{t\} \times \Sigma$ is

$$\mathcal{C} = \{(\phi, \psi) \in \mathcal{E}(\Sigma) \times \mathcal{E}(\Sigma)\},$$

where $\mathcal{E}(\Sigma) \doteq \mathcal{C}^\infty(\Sigma, \mathbb{R})$. This space is isomorphic to $\mathcal{E}_S$, the space of smooth solutions to (2.24).

As in classical mechanics, equations of motion can be derived from the least action principle. Elements of $\mathcal{C}$ play the role of generalized coordinates and generalized velocities, while a smooth trajectory $t \mapsto \phi(t)$, $t \in \mathbb{R}$ is a function which assigns to an instant of time t a function $\phi(t) \in \mathcal{E}(\Sigma)$ such that trajectories ϕ are in one to one correspondence with field configurations $\varphi : (t, \mathbf{x}) \to \phi(t)(\mathbf{x})$, i.e., elements of $\mathcal{E}$.

The Lagrangian L associates to every compact region $K \subset \Sigma$ a functional L_K on $\mathcal{C}$, typically given in terms of a Lagrangian density $\mathcal{L}$,

$$L_K(\phi, \psi) = \int_K \mathcal{L}(\phi(\mathbf{x}), \nabla\phi(\mathbf{x}), \psi(\mathbf{x}))d\sigma(\mathbf{x}),$$

and the action is, for every compact $K \subset \Sigma$ and every finite time interval I, a function on the space of trajectories defined by

$$S_{I\times K}(\phi) = \int_I L_K(\phi(t), \dot{\phi}(t))dt = \int_I \left(\int_K \mathcal{L}(\varphi(t, \mathbf{x}), \nabla_{\mathbf{x}}\varphi(t, \mathbf{x}), \dot{\varphi}(t, \mathbf{x}))d\sigma_t(\mathbf{x})\right) dt. \tag{2.25}$$

Solutions are configurations for which, for all compact K and I, $S_{I\times K}$ is stationary under variations $\delta\phi$ with support in the interior of $I \times K$. If e.g., $\mathcal{L}$ is the Lagrangian density of the free scalar field, then the least action principle yields (2.24) as the equation of motion.

Now let F, G be two functions on the space of trajectories which depend only on the restriction of the trajectory to $[t_1, t_2] \times K$ for some compact $K \subset \Sigma$ and $t_1 < t_2$. Let $\mathcal{E}_S$ be the space of solutions for an action S, and let $r_{\lambda G} : \mathcal{E}_S \to \mathcal{E}_{S+\lambda G}$ be the map which associates to a solution for S a solution for $S + \lambda G$ such that both solutions coincide for $t < t_1$ ($r_{\lambda G}$ is called the retarded Møller map) . Following the idea of Peierls, we consider the change of F under the change of the action and set, for a solution $\phi \in \mathcal{E}_S$,

$$\mathrm{D}_G F(\phi) = \frac{d}{d\lambda}|_{\lambda=0} F(r_{\lambda G}(\phi)).$$

Similarly, we introduce the advanced Møller map $a_{\lambda F} : \mathcal{E}_S \to \mathcal{E}_{S+\lambda F}$ where the solutions coincide for $t > t_2$, and set

$$\mathrm{Cl}_F G(\phi) = \frac{d}{d\lambda}|_{\lambda=0} G(a_{\lambda F}(\phi)).$$

The Peierls bracket of G and F is now defined by

$$\{G, F\}_{\mathrm{Pei}} \doteq \mathrm{D}_G F - \mathrm{Cl}_F G. \tag{2.26}$$

The advantage of the Peierls bracket is the fact that it is defined covariantly, directly in the Lagrangian formalism. As it stands, the Peierls bracket of two functionals is only defined on solutions, and one has to prove that it depends only on the restriction of the functionals to the space of solutions. In order to show that it satisfies the Jacobi identity, one has to extend it to a neighborhood of the space of solutions. It is, however, possible and also convenient to extend it to a Poisson bracket on functions of arbitrary configurations $\mathcal{E}$ (not only of solutions $\mathcal{E}_S$). We will now derive another formula for the Peierls bracket (formula (2.5) from the Introduction), which makes use of retarded and advanced Green's functions for normally hyperbolic operators. Next we will show that, restricted to the solution space, (2.5) is equivalent to the canonical bracket.

2.3.2 The Generalized Lagrangian Formalism

Before we continue, there is one small modification to the classical Lagrangian formalism, which we have to perform in order to make the quantization simpler. In formula (2.25), we have smeared the Lagrangian density $\mathcal{L}(\mathbf{x}, t)$ with a characteristic function of a certain compact region. Such sharp cut-offs would introduce additional divergences in the quantum theory, which we wish to avoid. Therefore, we replace the characteristic function by a smooth function that is equal to 1 on a sufficiently large region. Actually, it is convenient to consider *all* possible cutoffs and define the *generalized Lagrangian* as a map L from $\mathcal{D}$ to the space $\mathcal{F}_{\mathrm{loc}}$ of local functionals on $\mathcal{E}$. We require that

$$L(f + g + h) = L(f + g) - L(g) + L(g + h),$$

for $f, g, h \in \mathcal{D}$ and $\mathrm{supp}\, f \cap \mathrm{supp}\, h = \varnothing$. We also want

$$\mathrm{supp}\,(L(f)) \subseteq \mathrm{supp}\,(f),$$

where the support of a smooth functional $F \in \mathcal{C}^\infty(\mathcal{E}, \mathbb{C})$ is defined as

$$\begin{aligned}\mathrm{supp}\ F \doteq \{x \in M | \forall \text{ neighborhoods } U \text{ of } x\ \exists \varphi, \psi \in \mathcal{E}, \mathrm{supp}\ \psi \subset U, \\ \text{such that } F(\varphi + \psi) \neq F(\varphi)\}.\end{aligned} \tag{2.27}$$

The action is an equivalence class of Lagrangians, where $L_1 \sim L_2$ if

$$\mathrm{supp}\,(L_1 - L_2)(f) \subset \mathrm{supp}\ df. \tag{2.28}$$

The Euler-Lagrange derivative is a map $S' : \mathcal{E} \to \mathcal{D}'$ defined as

$$\langle S'(\varphi), h\rangle = \langle L(f)^{(1)}(\varphi), h\rangle, \tag{2.29}$$

with $f \equiv 1$ on supp$h$. Note that $S' \in \Gamma(T^*\mathcal{E})$. The field equation is now the condition that

$$S'(\varphi) = 0, \tag{2.30}$$

which coincides with the condition obtained from the variation of (2.25). We model observables as multilocal functionals on $\mathcal{E}$ (i.e., products of local functionals). The maps F, G considered in the previous section are examples of such functionals. The space of multilocal functionals on the space of solutions to (2.30) is given by the quotient $\mathcal{F}/\mathcal{F}_0$, where $\mathcal{F}_0$ denotes the space of multilocal functionals that vanish on $\mathcal{E}_S$.

The second variational derivative of the action is defined by

$$\langle S''(\varphi), h_1 \otimes h_2\rangle \doteq \langle L^{(2)}(f)(\varphi), h_1 \otimes h_2\rangle,$$

where $f \equiv 1$ on supp h_1 and supp h_2. S'' defined in such a way is symmetric two tensor on the affine manifold $\mathcal{E}$ (equipped with the smooth structure induced by τ_W) and, for each φ, it induces an operator from $\mathcal{D}$ to $\mathcal{D}'$. Moreover, since $L(f)$ is local, the second derivative has support on the diagonal, so $S''(\varphi)$ can be evaluated on smooth functions h_1, h_2, where only one of them is required to be compactly supported. This way we obtain an operator (the so called linearized Euler-Lagrange operator) $E'[S](\varphi) : \mathcal{E} \to \mathcal{D}'$.

We want to show now that the original formula of Peierls (2.26) is equivalent to (2.5), if $E'[S](\varphi)$ is a normally hyperbolic operator. Let $G \in \mathcal{F}_{\mathrm{loc}}$ be a local functional. We are interested in the flow (Φ_λ) on $\mathcal{E}$ which deforms solutions of the original field equation $S'(\varphi) = 0$ to those of the perturbed equation $S'(\varphi) + \lambda G^{(1)}(\varphi) = 0$. Let $\Phi_0(\varphi) = \varphi$ and

$$\frac{d}{d\lambda}\left(S'(\Phi_\lambda(\varphi)) + G^{(1)}(\Phi_\lambda(\varphi))\right)\Big|_{\lambda=0} = 0. \tag{2.31}$$

The vector field $\varphi \mapsto X(\varphi) = \frac{d}{d\lambda}\Phi_\lambda(\varphi)|_{\lambda=0}$ satisfies the equation

$$\langle E'[S](\varphi), X(\varphi)\rangle + G^{(1)}(\varphi) = 0. \tag{2.32}$$

Let $\Delta_S^{R/A}(\varphi)$ be the retarded/advanced Green's function of the normally hyperbolic operator $E'[S](\varphi)$ and let $\Delta_S(\varphi) = \Delta_S^R(\varphi) - \Delta_S^A(\varphi)$ be the causal propagator. We obtain now two distinguished solutions to the Eq. (2.32),

$$X^{R/A}(\varphi) = \langle \Delta_S^{R/A}(\varphi), G^{(1)}(\varphi)\rangle. \tag{2.33}$$

Note that $X^R(\varphi) = (\mathrm{D}_G\Phi)(\varphi)$, where Φ is the evaluation functional $\Phi_x(\varphi) \doteq \varphi(x)$. The difference $X = X^R - X^A$ defines a vector field $X \in \Gamma(T\mathcal{E})$ and it follows that

$$\{G, F\}_{\mathrm{Pei}}(\varphi) \doteq \mathrm{D}_G F(\varphi) - \mathrm{Cl}_F G(\varphi) = \langle F^{(1)}(\varphi), \Delta_S^{R/A}(\varphi) G^{(1)}(\varphi)\rangle.$$

Now we prove the equivalence between (2.5) and the canonical bracket. We fix a Cauchy surface $\{t\} \times \Sigma$. Note that, given Cauchy data $(\phi, \psi) \in \mathcal{C}$, we can write the unique solution φ corresponding to these Cauchy data as

$$\varphi(x) = \beta(\phi, \psi)(x) \equiv \int_\Sigma \left(\Delta_S(x; t, \mathbf{y})\psi(\mathbf{y}) - \frac{\partial}{\partial t}\Delta_S(x; t, \mathbf{y})\phi(\mathbf{y})\right) d\sigma_t(\mathbf{y}). \tag{2.34}$$

Canonical momenta are obtained as distributional densities by

$$\langle \pi(\phi, \psi), h\rangle \doteq \frac{d}{d\lambda} L_K(\phi, \psi + \lambda h)|_{\lambda=0}$$

We assume that for the Lagrangians of interest π is always smooth. The phase space is then

$$\mathcal{P} = \mathcal{E}(\Sigma) \times \mathcal{E}_d(\Sigma), \tag{2.35}$$

where $\mathcal{E}_d(\Sigma)$ is the space of smooth densities. The tangent space $T_{(\phi,\psi)}\mathcal{P}$ of $\mathcal{P}$ at some point (ϕ, ψ) consists of the compactly supported elements $(f, g) \in \mathcal{P}$. The phase space has the canonical symplectic form

$$\sigma_{(\phi,\psi)}((f_1, f_2), (g_1, g_2)) = \int_\Sigma (f_1 g_2 - f_2 g_1).$$

Note that $\mathcal{E}(\Sigma) \times \mathcal{E}_d(\Sigma) \subset \mathcal{E}(\Sigma) \times \mathcal{D}'(\Sigma) \cong T^*(\mathcal{E}(\Sigma))$, so $(\mathcal{P}, \sigma)$ is indeed the analog of the phase space in classical mechanics.

For simplicity we consider an action S induced by a Lagrangian L which depends on $\dot{\phi}$ only through the kinetic term $\frac{1}{2}\dot{\phi}^2$, hence $\pi(\mathbf{y}) \doteq \dot{\phi}(\mathbf{y}) d\sigma_t(\mathbf{y})$. Let $\alpha : (\phi, \pi) \mapsto (\phi, \dot{\phi})$ and $\tilde{\beta} \doteq \beta \circ \alpha : \mathcal{P} \to \mathcal{E}_S$. We can now prove the equivalence of the canonical and the Peierls bracket. Let $F, G \in \mathcal{F}$. Using (2.34) we obtain

$$\{F \circ \tilde{\beta}, G \circ \tilde{\beta}\}_{\mathrm{can}} = \int_\Sigma \left(\left\langle \frac{\delta F}{\delta \varphi} \circ \tilde{\beta}, \frac{\delta \tilde{\beta}}{\delta \phi(\mathbf{x})} \right\rangle \left\langle \frac{\delta G}{\delta \varphi} \circ \tilde{\beta}, \frac{\delta \tilde{\beta}}{\delta \pi(\mathbf{x})} \right\rangle - \left\langle \frac{\delta F}{\delta \varphi} \circ \tilde{\beta}, \frac{\delta \tilde{\beta}}{\delta \pi(\mathbf{x})} \right\rangle \left\langle \frac{\delta G}{\delta \varphi} \circ \tilde{\beta}, \frac{\delta \tilde{\beta}}{\delta \phi(\mathbf{x})} \right\rangle \right)$$
$$= \left\langle \Theta, F^{(1)} \circ \tilde{\beta} \otimes G^{(1)} \circ \tilde{\beta} \right\rangle,$$

where Θ is given by

$$\Theta(z', z) = \int_\Sigma \left(\dot{\Delta}_S(z'; t, \mathbf{x}) \Delta_S(z; t, \mathbf{x}) - \dot{\Delta}_S(z; t, \mathbf{x}) \Delta_S(z'; t, \mathbf{x}) \right) d\sigma(\mathbf{x}).$$

From general properties of the causal propagator Δ_S (the generalization of (2.34) to distributional Cauchy data) it follows that the convolution Θ above is equal to Δ_S. Hence, on the solution space $\mathcal{E}_S$,

$$\{F \circ \tilde{\beta}, G \circ \tilde{\beta}\}_{\mathrm{can}} = \{F, G\}_{\mathrm{Pei}} \circ \tilde{\beta}.$$

2.3.3 Example: The Poisson Bracket of the φ^4 Interaction

In this section, following [21], we give another argument for the equivalence of the Peierls and the canonical bracket on the example of the φ^4 interaction. Consider the generalized Lagrangian

$$L(f)(\varphi) = \int_M \left(\frac{1}{2} \nabla_\mu \varphi \nabla^\mu \varphi - \frac{m^2}{2} \varphi^2 - \frac{\lambda}{4!} \varphi^4 \right) f d\mu,$$

where $d\mu$ is the invariant measure on M, induced by the metric. Then $S'(\varphi) = -\left((\Box + m^2)\varphi + \frac{\lambda}{3!} \varphi^3 \right)$ and $E'[S](\varphi)$ is the linear operator

$$-\left(\Box + m^2 + \frac{\lambda}{2} \varphi^2 \right) \tag{2.36}$$

(the last term is to be understood as a multiplication operator). The Peierls bracket is

$$\{\Phi_x, \Phi_y\}_{\mathrm{Pei}} = \Delta_S(\Phi)(x, y), \tag{2.37}$$

where Φ_x, Φ_y are evaluation functionals on $\mathcal{E}$ and $x \mapsto \Delta_S(\varphi)(x, y)$ is a solution (at φ) of the linearized equation of motion with the initial conditions

$$\Delta_S(\varphi)(y^0, \mathbf{x}; y^0, \mathbf{y}) = 0, \quad \frac{\partial}{\partial x^0} \Delta_S(\varphi)(y^0, \mathbf{x}; y) = \delta(\mathbf{x}, \mathbf{y}). \tag{2.38}$$

This coincides with the Poisson bracket in the canonical formalism. Namely, let $\varphi \in \mathcal{E}_S$, then

$$0 = \left\{ (\Box + m^2)\Phi_x + \frac{\lambda}{3!}\Phi_x^3, \Phi_y \right\}_{\text{can}} = (\Box + m^2 + \frac{\lambda}{2}\Phi_x^2)\left\{ \Phi_x, \Phi_y \right\}_{\text{can}}. \quad (2.39)$$

In the first step we used the fact that φ is a solution of the equations of motion and in the second step we used the fact that the canonical bracket is a derivation in both arguments. We can see from the equation above that the canonical Poisson bracket satisfies the linearized field equation with the same initial conditions as the Peierls bracket. The uniqueness of solutions to these linearized equations implies that in fact $\{\Phi_x, \Phi_y\}_{\text{can}} = \{\Phi_x, \Phi_y\}_{\text{Pei}}$, on $\mathcal{E}_S$. This clearly extends to general functionals since $\{F, G\}_{\text{Pei}} = \langle \{\Phi_x, \Phi_y\}_{\text{Pei}}, F^{(1)} \otimes G^{(1)} \rangle$, and similarly for $\{.,.\}_{\text{can}}$.

2.3.4 Geometrical Structures in Classical Theory

The Peierls bracket, from now on denoted by $\{.,.\}_S$, introduces a symplectic structure on the space $\mathcal{F}/\mathcal{F}_0$ of on-shell multilocal functionals on $\mathcal{E}_S$. We can find a nice geometrical interpretation for this space using some basic notions of symplectic geometry. Let us assume that S is quadratic, so the equations of motion are of the form $S'(\varphi) = P\varphi = 0$ for some normally hyperbolic differential operator P. The space of solutions $\mathcal{E}_S$ is a vector space and hence an infinite dimensional manifold[1] with a tangent space $T\mathcal{E}_S = \mathcal{E}_S \times \mathcal{E}_S$ and cotangent space $T^*\mathcal{E}_S = \mathcal{E}_S \times \mathcal{E}'_S$. If F, G are multilocal functionals on $\mathcal{E}_S$, then their first functional derivatives are smooth, so

$$F^{(1)}(\varphi), G^{(1)}(\varphi) \in \mathcal{E}'_S \cap (\mathcal{D}/\{u \in \mathcal{D} | \langle u, \varphi \rangle = 0, \forall \varphi \in \mathcal{E}_S\})$$

for all $\varphi \in \mathcal{E}_S$, i.e. $\varphi \in \ker P$. Since P is a normally hyperbolic operator, one can show (see for example [21] for the proof in a more general setting) that $\{u \in \mathcal{D} | \langle u, \varphi \rangle = 0, \forall \varphi \in \mathcal{E}_S\} \cong P\mathcal{D}$, so functional derivatives of multilocal functionals are one forms in $\Gamma(\mathcal{E}_S \times \mathcal{D}/P\mathcal{D}) \subset \Gamma(T^*\mathcal{E}_S)$.

The causal propagator Δ_S induces a Poisson structure on $\mathcal{F}$, which is also well defined on the quotient $\mathcal{F}/\mathcal{F}_0$, as $\mathcal{F}_0$ is a Poisson ideal with respect to this structure. We can also use Δ_S to map one-forms in $\Gamma(\mathcal{E}_S \times \mathcal{D}/P\mathcal{D}) \subset \Gamma(T^*\mathcal{E}_S)$ to one-vectors in $\Gamma(\mathcal{E}_S \times \mathcal{E}_{S,\text{sc}}) \subset \Gamma(T\mathcal{E}_S)$, where "sc" indicates spacelike-compact support. To see how it works, note that Δ_S induces an operator $\mathcal{D} \to \mathcal{E}$ and ker $\Delta_S = P\mathcal{D}$, so Δ_S is well defined on equivalence classes in $\mathcal{D}/P\mathcal{D}$. To show that Δ_S is invertible on this space, it remains to show that it is surjective. We recall here the standard argument, which can also be found in [21]. Let f be a solution with a spacelike-compact support, $\chi \in \mathcal{E}$, and Σ_1, Σ_2 be Cauchy surfaces such that $\Sigma_1 \cap J_+(\Sigma_2) = \emptyset$.

[1] There is another natural way to introduce a smooth manifold structure on $\mathcal{E}_S$. We define the atlas where charts are given by maps $\varphi + \mathcal{D} \to \mathcal{E}_{S,sc}$, $\varphi + \vec{\phi} \mapsto \Delta_S \vec{\phi}$, with $\varphi \in \mathcal{E}_S$, where $\mathcal{E}_{S,sc}$ is the space of solutions with compactly supported Cauchy data. We have $\Delta_S : \mathcal{D} \to \mathcal{E}_{S,sc}$ and we equip $\mathcal{E}_{S,sc}$ with the final topology with respect to all curves of the form $\lambda \mapsto \varphi + \Delta_S(\vec{\phi}(\lambda))$, where $\lambda \mapsto \vec{\phi}(\lambda)$ is a smooth curve in $\mathcal{D}$. This gives $\mathcal{E}_S$ the structure of an affine manifold in the sense of convenient calculus [35].

Assume $\chi(x) = 0$ for $x \in J_-(\Sigma_1)$ and $\chi(x) = 1$ for $x \in J_+(\Sigma_2)$. Then $P\chi f = 0$ outside of the time slice bounded by Σ_1 and Σ_2 (χ = const. there) which implies that $P\chi f$ has compact support. Hence,

$$\Delta_S P\chi f = \Delta_S^R P\chi f + \Delta_S^A P(1-\chi)f = f.$$

We can now assign to a form $F^{(1)}$, the vector $\langle \Delta_S F^{(1)}, .\rangle$. On $\mathcal{E}_{S,\mathrm{sc}}$ we have the natural symplectic structure σ_1:

$$\sigma_1(f,g) = \int_\Sigma (f \wedge *dg - *df \wedge g) = \int_\Sigma (f(\partial_n g) - (\partial_n f)g) dvol_\Sigma,$$

where ∂_n is the normal derivative on Σ ($\partial_n f = n^\mu \partial_\mu f$, $n^\mu \xi_\mu = 0$ for $\xi \in T\Sigma$, $n^\mu n_\mu = 1$). Obviously, σ_1 extends to a constant 2-form on $\mathcal{E}_S$. The relation between the 2-form σ_1 and the "bi-vector field"[2] Δ_S is given by (see for example [46] for a proof based on the ideas of [12])

$$\sigma_1(\Delta_S F^{(1)}, \xi) = \langle F^{(1)}, \xi \rangle,$$

where $\xi \in \Gamma(\mathcal{E}_S \times \mathcal{E}_{S,\mathrm{sc}})$. In this sense we can think of Δ_S as the "inverse" of the symplectic structure σ_1. Setting $\xi = \Delta_S G^{(1)}$ for $G \in \mathcal{F}$, we obtain

$$\sigma_1(\Delta_S F^{(1)}, \Delta_S G^{(1)}) = \langle F^{(1)}, \Delta_S G^{(1)} \rangle.$$

If S is not quadratic, the situation is more complicated, since S' induces non-linear equations of motion. It turns out that for many classes of physically interesting systems solutions of $S'(\varphi) = 0$ develop singularities after a finite time, despite starting from smooth Cauchy data. Therefore the space of globally smooth solutions $\mathcal{E}_S$ might be very small and it does not necessarily capture all the interesting features of the theory. Moreover, in general it is not clear if $\mathcal{E}_S$ can be equipped with a manifold structure in the sense of infinite dimensional differential geometry (see for example [2] for the results on the space of solutions of Einstein's equations). A more general structure like a stratified space might be necessary.

For non-linear equations of motion it is therefore more convenient to replace the space of functionals on the space of solutions with the quotient $\mathcal{F}_S := \mathcal{F}/\{\langle S', X\rangle, X \in \mathcal{V}\}$, where $\mathcal{V} \subset \Gamma(T\mathcal{E})$ is the space of vector fields that are derivations of $\mathcal{F}$ and the duality denoted by $\langle ., . \rangle$ is the contraction of a 1-form $S' \in \Gamma(T^*\mathcal{E}(\mathcal{M}))$ with a vector field X. We say that we take the quotient of $\mathcal{F}$ by the *ideal generated by the equations of motion*. If the equations of motion are linear and normally hyperbolic, this ideal coincides with $\mathcal{F}_0$, so $\mathcal{F}_S$ is exactly the space of multilocal functionals on $\mathcal{E}_S$. In general, our point of view is more in line with the quantum theory and it avoids

[2]Since Δ_S is a bi-distribution rather than a smooth function, the map $\varphi \mapsto \Delta_S$ doesn't induce an actual bi-vector field on $\mathcal{E}$, but belongs to a suitable completion of $\Gamma(\Lambda^2 T\mathcal{E})$.

complications related to characterization of the geometrical structure of $\mathcal{E}_S$. It is also close in spirit to the way one studies varieties in algebraic geometry.

2.3.5 Deformation Quantization

Deformation quantization is a method to construct quantum theories from the classical ones by deforming the commutative product on the space $\mathcal{F}$ of functionals to a non-commutative product $\star$ on $\mathcal{F}[[\hbar]]$ (the space of formal power series in $\hbar$.

$$F \star G = \sum_{n=0}^{\infty} \hbar^n B_n(F, G), \tag{2.40}$$

and we require that

$$B_0(F, G) = F \cdot G,$$
$$B_1(F, G) - B_1(G, F) = i\hbar\{F, G\},$$

where $(F \cdot G)(\varphi) = F(\varphi)G(\varphi)$ is the pointwise product of functionals, $\{.,.\}$ is the Peierls bracket and the second condition is a realization of the idea that in the quantum theory one "replaces canonical brackets with commutators". The existence of higher order terms is necessary to avoid the Groenewald-van Hove no-go theorem. This result, established first for finite dimensional phase spaces, states that a Dirac type quantization prescription is not possible in the strict sense [23, 31]. More concretely (see [45]), consider the Lie algebra $\mathfrak{h}$ spanned by the canonical coordinate and momenta functions $q^1, \ldots, q^N, p_1, \ldots, p_N$ and 1, equipped with the canonical Poisson bracket $\{.,.\}_{\mathrm{can}}$. This is a Lie subalgebra of $\mathfrak{g} \doteq (\mathrm{Pol}(T^*\mathbb{R}^N), \{.,.\}_{\mathrm{can}})$ (polynomials on the phase space). The Groenewald-van Hove Theorem states that there exists no faithful irreducible representation of $\mathfrak{h}$ by operators on a dense domain of some Hilbert space which can be extended to a representation of $\mathfrak{g}$, so there is no quantization map Q from $\mathfrak{g}$ to the space of operators on some Hilbert space $\mathcal{H}$, such that

$$[Q(f), Q(g)] = i\hbar Q(\{f, g\}).$$

Deformation quantization [4, 5] provides a way out since it weakens the above condition to

$$[Q(f), Q(g)] = Q([f, g]_\star) = i\hbar Q(\{f, g\}) + \mathcal{O}(\hbar^2).$$

In field theory, as we have seen in the previous section, the space of functions on the N-dimensional phase space is replaced by $\mathcal{F}$, the space of multilocal functionals on $\mathcal{E}$, which is now infinite dimensional. The Poisson structure is provided by the Peierls bracket, defined with the use of the causal propagator Δ_S. In the simplest case, when S is quadratic, one can construct the $\star$-product using a Moyal-type formula. To avoid the functional analytic problems, for the moment we consider only regular functionals $\mathcal{F}_{\mathrm{reg}}$, i.e., those for which $F^{(n)}(\varphi)$ is a smooth compactly supported section for all $n \in \mathbb{N}$, $\varphi \in \mathcal{E}$. On $\mathcal{F}_{\mathrm{reg}}[[\hbar]]$ we can now define

$$(F \star G)(\varphi) \doteq \sum_{n=0}^{\infty} \frac{\hbar^n}{n!} \langle F^{(n)}(\varphi), \left(\tfrac{i}{2}\Delta_S\right)^{\otimes n} G^{(n)}(\varphi)\rangle, \tag{2.41}$$

which can be formally written as $e^{\frac{i\hbar}{2}\left\langle \Delta_S, \frac{\delta^2}{\delta\varphi\delta\varphi'}\right\rangle} F(\varphi)G(\varphi')|_{\varphi'=\varphi}$.

Let us consider the example of a free scalar field and regular functionals of the form

$$F_f(\varphi) = \int_{\mathrm{M}} f(x)\varphi(x)d\mu(x) \equiv \int f\varphi d\mu, \qquad \text{where } f \in \mathcal{D}.$$

We can now define $\mathcal{W}(f) \doteq \exp(iF_f)$ and check that

$$\begin{aligned}\langle(\mathcal{W}(f))^{(1)}(\varphi), h\rangle &= \frac{d}{d\lambda}(\mathcal{W}(f)(\varphi+\lambda h))\,|_{\lambda=0} = \frac{d}{d\lambda} e^{i\int f(\varphi+\lambda h)d\mu}\Big|_{\lambda=0} \\ &= \left(i\int fh\,d\mu_g\right)\mathcal{W}(f)(\varphi).\end{aligned}$$

and thus

$$\langle(\mathcal{W}(f))^{(n)}(\varphi), h^{\otimes n}\rangle = \left(i\int fh\,d\mu_g\right)^n \mathcal{W}(f)(\varphi).$$

Inserting this into the $\star$-product formula, we find,

$$\begin{aligned}\mathcal{W}(f)\star\mathcal{W}(\tilde{f}) &= \sum_{n=0}^{\infty}\left(\frac{i\hbar}{2}\right)^n \frac{(-1)^n}{n!}\left(\int \Delta_S(x,y)\tilde{f}(y)f(x)d\mu_g(x)d\mu_g(y)\right)^n \mathcal{W}(f+\tilde{f}) \\ &= e^{-\frac{i\hbar}{2}\Delta_S(f,\tilde{f})}\mathcal{W}(f+\tilde{f}),\end{aligned} \tag{2.42}$$

which reproduces the Weyl relations.

Having the interacting theory in mind, we will need to extend the star product to functionals more singular than the elements of $\mathcal{F}_{\mathrm{reg}}$, including, in particular, the non-linear local functionals. To understand possible obstructions to this extension we have to analyze the singularity structure of Δ_S. Using the theorem of propagation of singularities (see Sect. 2.2), we find that [39]

$$\mathrm{WF}(\Delta_S) = \{(x,k;x',-k') \in \dot{T}^*M^2 | (x,k) \sim (x',k')\},$$

where $\dot{T}$ denotes the tangent bundle minus the zero section and $(x,k) \sim (x',k')$ means that there exists a lighlike geodesic connecting x and x', to which k is co-tangent and k' is a parallel transport of k. We observe that the WF set of Δ_S is composed of two parts: one with $k \in (\overline{V}_+)_x$ and another with $k \in (\overline{V}_-)_x$, where $\overline{V}_\pm$ is (the dual of) the closed future/past lightcone This observation allows one to decompose Δ_S into two distributions with WF sets corresponding to these two components. Such a decomposition is a local version of the decomposition according to positive and negative energies [39]. definition of this space, let us give some motivation first. Therefore we can split Δ_S into

$$\tfrac{i}{2}\Delta_S = \Delta_S^+ - H,$$

where the WF set of Δ_S^+ is

$$\mathrm{WF}\,(\Delta_S^+) = \{(x,k;x',-k') \in \dot{T}^*M^2 | (x,k) \sim (x',k'), k \in (\overline{V}_+)_x\}, \tag{2.43}$$

and we also require that $\Delta_S = 2\mathrm{Im}(\Delta_S^+)$ and that Δ_S^+ is a distributional bisolution to the field equation and is of positive type (i.e. $\langle \Delta_S^+, \bar{f} \otimes f\rangle \geq 0$). On Minkowski space one could choose Δ_S^+ as the Wightman 2-point-function. On general globally hyperbolic spacetimes such a decomposition always exists but is not unique. If H and H' correspond to two such choices of decomposition, then $H - H'$ is a smooth symmetric bisolution to the field equations.

We can now replace $\frac{i}{2}\Delta_S$ with Δ_S^+ in (2.41) and the new product, denoted by $\star_H$ can be extended from $\mathcal{F}_{\mathrm{reg}}$ to $\mathcal{F}_{\mu c}$ defined as the space of functionals with functional derivatives satisfying

$$\mathrm{WF}\,(F^{(n)}(\varphi)) \subset \Xi_n, \quad \forall n \in \mathbb{N},\ \forall \varphi \in \mathcal{E}, \tag{2.44}$$

where Ξ_n is an open cone defined as

$$\Xi_n \doteq T^*\mathrm{M}^n \setminus \{(x_1,\dots,x_n;k_1,\dots,k_n) | (k_1,\dots,k_n) \in (\overline{V}_+^n \cup \overline{V}_-^n)_{(x_1,\dots,x_n)}\}, \tag{2.45}$$

where $(\overline{V}_\pm)_x$ is the closed future/past lightcone understood as a conic subset of $T_x^*\mathrm{M}$.

On $\mathcal{F}_{\mathrm{reg}}$ the two star products $\star$ and $\star_H$ are isomorphic structures and the intertwining map is given by

$$\alpha_H \doteq e^{\frac{\hbar}{2}\langle H, \frac{\delta^2}{\delta\varphi^2}\rangle}, \tag{2.46}$$

so that

$$F \star_H G = \alpha_H\left((\alpha_H^{-1}F) \star (\alpha_H^{-1}G)\right), \qquad F, G \in \mathcal{F}_{\mathrm{reg}}. \tag{2.47}$$

In the language of formal deformation quantization one says that products $\star$ and $\star_H$ are related by a gauge transformation, so they provide the same deformation quantization. In general a gauge transformation between star products is given by $F \mapsto F + \sum_{\hbar \geq 1} \hbar^n D_n(f)$, where each D_n is a differential operator. In our case, $D_n = \frac{1}{n!}\left\langle \frac{1}{2}(H - H'), \frac{\delta^2}{\delta\varphi^2} \right\rangle^n$.

Physically, the transition between $\star$ and $\star_H$ corresponds to normal ordering, so introducing the $\star_H$-product is just an algebraic version of Wick's theorem. As stated before, the codomain of $\alpha_H : \mathcal{F}_{\mathrm{reg}} \to \mathcal{F}_{\mathrm{reg}}$ can be "completed" (with the use of the Hörmander topology [15, 26]) to the larger space $\mathcal{F}_{\mu c}$ and we can also build a corresponding (sequential) completion $\alpha_H^{-1}(\mathcal{F}_{\mu c})$ of the domain. This amounts to extending $\mathcal{F}_{\mathrm{reg}}$ with all elements of the form $\lim_{n\to\infty} \alpha_H^{-1}(F_n)$, where (F_n) is a convergent sequence in $\mathcal{F}_{\mu c}$. The quantum algebra $\mathfrak{A}$ of the free theory is defined as the space of families F_H, labeled by possible choices of H, where $F_H \in \mathfrak{A}_H \doteq (\mathcal{F}_{\mu c}[[\hbar]], \star_H)$ fulfill the relations

$$F_{H'} = \alpha_{H'-H} F_H,$$

and the product is

$$(F \star G)_H = F_H \star_H G_H.$$

We can summarize the relations between the algebraic structures we have introduced so far by means of the following diagram:

$$\begin{array}{ccc}
(\mathcal{F}_{\mathrm{reg}}, \star) & \xrightarrow{\alpha_H} & (\mathcal{F}_{\mathrm{reg}}, \star_H) \\
\text{dense}\big\downarrow\cap & & \text{dense}\big\downarrow\cap \\
\mathfrak{A} & \xleftarrow{\alpha_H^{-1}} & (\mathcal{F}_{\mu c}, \star_H)
\end{array}$$

A family of coherent states on $\mathfrak{A}$ is obtained by the prescription

$$\omega_{H,\varphi}(F) \doteq \alpha_H(F)(\varphi) = F_H(\varphi),$$

where $\varphi \in \mathcal{E}_S$. This makes sense since F_H is a functional in $\mathcal{F}_{\mu c}$, so evaluation at a field configuration φ is well defined.

As an example we can consider the free scalar field with the generalized Lagrangian

$$L_0(f)(\varphi) = \frac{1}{2} \int_{\mathbb{M}} (\partial_\mu \varphi \partial^\mu \varphi - m^2 \varphi^2) f \, d\mu. \tag{2.48}$$

Let us define $\tilde{\mathfrak{A}}$ as the subalgebra of $\mathfrak{A}$ generated by the Weyl generators $\mathcal{W}(f) \doteq \exp(i F_f)$. Since

$$\left\langle H, \frac{\delta^2}{\delta\varphi^2} \right\rangle \left(i \int f\varphi d\mu \right)^n = -\frac{n!}{(n-2)!} H(f,f) \left(i \int f\varphi d\mu \right)^{n-2},$$

we conclude that

$$\left\langle H, \frac{\delta^2}{\delta\varphi^2} \right\rangle (\mathcal{W}(f)) = -H(f, f)\mathcal{W}(f)$$

so

$$\alpha_H (\mathcal{W}(f)) = e^{-\frac{\hbar}{2}H(f,f)}\mathcal{W}(f).$$

We can now consider a state obtained be evaluation at $\varphi = 0$. We see that

$$\omega_{H,0} (\mathcal{W}(f)) = e^{-\frac{\hbar}{2}H(f,f)},$$

so H plays the role of the covariance of the state $\omega_{H,0}$.

Going on-shell corresponds to taking the quotient of $\tilde{\mathfrak{A}}$ by the ideal $\tilde{\mathfrak{A}}_0$ generated by the elements

$$\mathcal{W}((\Box + m^2)f) - 1, \; f \in \mathcal{D}. \tag{2.49}$$

Note that $S_0'(\varphi) = (\Box + m^2)\varphi$, and using partial integration, we can conclude that $F_{(\Box+m^2)f}(\varphi) = \int S_0'(\varphi) f d\mu = \langle S_0', f\rangle$, so taking the quotient by $\tilde{\mathfrak{A}}_0$ implements the free field dynamics. We denote $\tilde{\mathfrak{A}}/\tilde{\mathfrak{A}}_0$ by $\tilde{\mathfrak{A}}_{S_0}$, and we see that $\omega_{H,0}$ is well defined on $\tilde{\mathfrak{A}}_{S_0}$, as H is a bisolution for the operator $P = \Box + m^2$.

2.3.6 Interpretation in Terms of Kähler Geometry

There is an elegant geometrical interpretation of the structures introduced in the previous section. Analogous to Kähler geometry, H plays the role of the Riemannian metric on $\mathcal{Y} \equiv \mathcal{D}/P\mathcal{D}$ and the 2-point function $\Delta_S^+ = \frac{i}{2}\Delta_S + H$ is a Hermitian 2-form on $\mathcal{Y}$.

The pair (H, Δ_S) induces an anti-involution J on $\mathcal{Y}$ (i.e. $J^2 = -1$) and if Δ_S^+ is a 2-point function of a quasi-free pure Hadamard state, the triple (H, Δ_S, J) is a Kähler structure on $\mathcal{Y}$. To see how this come about, let us recall some well known results (see for example [1, 13] for proofs). Let $\mathcal{Y}^{\mathbb{C}}$ denote the complexification of $\mathcal{Y}$. If $\Delta_S^{\mathbb{C}}$ and $H^{\mathbb{C}}$ are canonical extensions of Δ_S and H to $\mathcal{Y}^{\mathbb{C}}$, then the following are equivalent:

1. $H^{\mathbb{C}} + \frac{i}{2}\Delta_S^{\mathbb{C}} \geq 0$ on $\mathcal{Y}^{\mathbb{C}}$,
2. $|\langle f_1, \Delta_S f_2\rangle| \leq 2\langle f_1, Hf_1\rangle^{1/2}\langle f_2, Hf_2\rangle^{1/2}$, $f_1, f_2 \in \mathcal{Y}$.

We can complete $\mathcal{Y}$ with the product $(.,.)_H \doteq \langle ., H.\rangle$ to a real Hilbert space $\mathcal{H}$ and the inequality 2 implies that Δ_S is a bilinear form on $\mathcal{H}$ with norm less or equal 2. Therefore, there exists an operator $A \in \mathcal{B}(\mathcal{H})$ with $||A|| \leq 1$ such that

$$\langle f_1, \Delta_S f_2\rangle = 2(f_1, Af_2)_H$$

If A has a trivial kernel, then we can just construct the polar decomposition $A = -J|A|$ and J satisfies $J^2 = -1$, so we can use it to equip $\mathcal{Y}$ with an almost-complex structure. More generally, following the proof of theorem 17.12 of [13], we can define $\mathcal{Y}_{\text{sg}} \doteq \ker A$ and $\mathcal{Y}_{\text{reg}} \doteq \mathcal{Y}_{\text{sg}}^{\perp}$. We set $A_{\text{reg}} \doteq A \restriction_{\mathcal{Y}_{\text{reg}}}$ and construct the polar decomposition $A_{\text{reg}} = -J_{\text{reg}}|A_{\text{reg}}|$. If the dimension of $\mathcal{Y}_{\text{sg}}$ is even or infinite (which is the case in the situation we are interested in), then there exist an orthogonal anti-involution J_{sg} on $\mathcal{Y}_{\text{sg}}$ and we set $J = J_{\text{reg}} \oplus J_{\text{sg}}$.

Note that J induces also an almost complex structure on $\mathcal{E}_{S,\text{sc}}$ if we set $j\Delta_S f \doteq \Delta_S J f$, where $f \in \mathcal{Y}$. We define the holomorphic and anti-holomorphic subspaces of $\mathcal{Y}^{\mathbb{C}}$ as

$$\mathcal{Z} \doteq \{(f - iJf) | f \in \mathcal{Y}\},$$
$$\overline{\mathcal{Z}} \doteq \{(f + iJf) | f \in \mathcal{Y}\},$$

respectively. Projections onto these subspaces are given by $\mathbb{1}_{\mathcal{Z}} = \frac{1}{2}(\mathbb{1} - iJ^{\mathbb{C}})$ and $\mathbb{1}_{\overline{\mathcal{Z}}} = \frac{1}{2}(\mathbb{1} + iJ^{\mathbb{C}})$

The CCR algebra corresponding to $(\mathcal{Y}, \Delta_S)$ is just the algebra $\tilde{\mathfrak{A}}_{S_0}$ introduced at the end of the previous section. Note that $\omega_{H,0}$ is a state on $\tilde{\mathfrak{A}}_S$ with covariance H. This state is pure if and only if the triple (H, Δ_S, J) is a Kähler structure on $\mathcal{Y}$, i.e. all three structures are compatible and $\Delta_S \circ J = 2H$. We can now decompose Δ_S^+ in the holomorphic basis. A straightforward computation shows that

$$\langle \mathbb{1}_{\overline{\mathcal{Z}}} f_1, \Delta_S^+ (\mathbb{1}_{\mathcal{Z}} f_2) \rangle = \langle f_1, \Delta_S^+ f_2 \rangle,$$

where $f_1, f_2 \in \mathcal{Y}^{\mathbb{C}}$ and remaining components vanish, so in the holomorphic basis Δ_S^+ is represented by

$$\begin{pmatrix} 0 & 0 \\ \Delta_S^+ & 0 \end{pmatrix},$$

so it acts only on the holomorphic part of the first argument and the anti-holomorphic part of the second argument.

2.4 Time Ordered Products, and the Perturbative Construction of Local Nets

In the previous section we were concerned only with the quantization of free theories (quadratic actions). Given an arbitrary action S we first split $S = S_0 + S_I$, where S_0 is quadratic. We already know how to quantize the classical model defined by S_0, so now is the time to introduce the interaction. We will do it in this section, following the ideas of [6, 7, 18, 25, 41, 44], but before we start, we give a heuristic argument justifying our construction. The idea is to use the analogy with the interaction picture of quantum mechanics. Let H_0 be the Hamiltonian operator of the free theory and let $H_{t,I} = -\int_K :\mathcal{L}_I(0,\mathbf{x}): d\sigma_t$ be the interaction Hamiltonian, where $:\mathcal{L}_I:$ is the normal-ordered Lagrangian density, constructed from the classical quantity $\mathcal{L}_I$ and K is some compact subset of Σ (as explained in Sect. 2.3.1). The rigorous "smoothed-out" version of the Hamiltonian quantization will be given in Sect. 2.6.

We would like to use the Dyson formula and define the time evolution operator as a time ordered exponential, i.e.

$$U(t,s) = e^{itH_0} e^{-i(t-s)(H_0+H_I)} e^{-isH_0}$$

$$= 1 + \sum_{n=1}^{\infty} \frac{i^n}{n!} \int_{([s,t]\times\mathbb{R}^3)^n} T(:\mathcal{L}_I(x_1):\ldots:\mathcal{L}_I(x_n):) d^{4n}x,$$

where

$$x \mapsto \mathcal{L}_I(x) = e^{iH_0x^0} :\mathcal{L}_I(0,\mathbf{x}): e^{-iH_0x^0}$$

is an operator-valued function and T denotes time-ordering. Heuristically, one could use the unitary map defined above to obtain interacting fields as

$$\varphi_I(x) = U(x^0,s)^{-1}\varphi(x)U(x^0,s) = U(t,s)^{-1}U(t,x^0)\varphi(x)U(x^0,s), \tag{2.50}$$

where $s < x^0 < t$.

There are, however, serious problems with this heuristic formula. Firstly, typical Lagrangian densities, e.g. $:\mathcal{L}_I(x): = :\varphi(x)^4:$, can not be restricted to Σ_0 as operator valued distributions. This is the source of the so called UV problem. Moreover, as mentioned before, having the sharp cutoff function in the Lagrangian and Hamiltonian (like in (2.25)) leads to additional divergences (Stückelberg divergences). Finally there is the adiabatic limit problem related to the fact that the integral over $\mathbf{x}$ does not exist. Last but not least, the overall sum might not converge.

2.4.1 Causal Perturbation Theory

Some of these problems mentioned in the introduction can be easily dealt with by a slight modification of the above ansatz. For example, we avoid the Stückelberg divergences by replacing the sharp cutoffs with smooth test functions. The UV problem is solved by using causal perturbation theory in the sense of Epstein and Glaser [18]. In this method one switches the interaction on only in a compact region of spacetime and then takes the adiabatic limit (understood as a certain inductive limit) on the level of interacting observable algebras. These modifications of the Dyson formula ansatz lead to the definition of the *formal S-matrix*:

$$S(g) = 1 + \sum_{n=1}^{\infty} \frac{i^n}{n!} \int g(x_1) \cdots g(x_n) T(:\mathcal{L}_I(x_1):\ldots:\mathcal{L}_I(x_n):),$$

where g is a test density. In order to make this formula well defined, we need to make sense of the time-ordered products of $:\mathcal{L}_I(x_i):$. This will be done by Epstein-Glaser renormalization. Finally, the formula (2.50) has to be reinterpreted as a definition of a distribution, rather than a function. Hence, for a test density f we obtain

$$\int f(x)\varphi_I(x) = S(g)^{-1} \sum_{n=0}^{\infty} \frac{i^n}{n!} \int f(x) g(x_1) \cdots g(x_n) T\varphi(x)\mathcal{L}_I(x_1) \cdots \mathcal{L}_I(x_n)$$
$$= \frac{d}{d\lambda}|_{\lambda=0} S(g)^{-1} S(g, \lambda f),$$

where $S(g, f)$ is the formal S-matrix with the Lagrangian density $g\mathcal{L}_I + f\varphi$. This is the so called *Bogoliubov's formula* [7].

We are now left with the problem of defining the time-ordered products on Wick-ordered quantities $:\mathcal{L}_I(x):$. We have already mentioned in Sect. 2.3.5 that the normal ordering corresponds to passing between the star product $\star_H$ on $\mathfrak{A}_H$ and $\star$ on $\mathfrak{A}$. Note that elements of $\mathfrak{A}_H$ are functionals on $\mathcal{E}_S$, so we can identify classical quantities in $\mathcal{F}_{\text{loc}}$ with quantum ones by means of $\mathcal{T}_1^H : \mathcal{F}_{\text{loc}} \to \mathfrak{A}_H$ defined by $\mathcal{T}_1^H = \text{id}$. Composing with α_H^{-1} we obtain a map $\mathcal{T}_1 : \mathcal{F} \to \mathfrak{A}$, $\mathcal{T}_1 \doteq \alpha_H^{-1} \circ \mathcal{T}_1^H$ which maps "classical" to "quantum". This map is interpreted as the normal ordering and we can now make an identification

$$:F: \doteq \mathcal{T}_1 F, \qquad F \in \mathcal{F}_{\text{loc}}\,.$$

In the context of local covariance—see Chap. 4, this choice of normal ordering is not the most optimal one. This is because a family of Hadamard states cannot be chosen in a covariant way (i.e. compatible with embedding of globally hyperbolic spectimes), but a family of Hadamard parametrices can. The latter are bi-solutions of the linearized equations of motion only up to smooth terms. It is, therefore, more appropriate to define the normal ordering by a prescription where only the singular part of H is subtracted from the correlation function of two fields, as opposed to the prescription where one subtracts the full H. Concretely, we set $\mathcal{T}_1^H = \alpha_w$ so $\mathcal{T}_1 = \alpha_{H-w}^{-1}$, where w is the smooth part of the Hadamard 2-point function (see [21, 34] for a recent review). More precisely, this has to be understood as $\lim_{N\to\infty} \alpha_{H-w_N}^{-1} F$ for $F \in \mathcal{F}_{\text{loc}}$ and this limit makes sense, because the series converges after finitely many steps. The function w_N appearing in this prescription is $2N + 1$ times continuously differentiable and it appears in the 2-point function as $\Delta_S^+ = W_N^{sing} + w_N$. The singular part W_N^{sing} is of the form "$\frac{u}{\sigma} + v \ln \sigma$", with $\sigma(x, y)$ denoting the square of the length of the geodesic connecting x and y and with geometrical determined smooth functions u and v. For a more precise definition of what is the Hadamard form for of a 2-point function, see for example [34] or a recent review [21].

More concretely, for a density of the form

$$\Phi^{A,n}(f)(\varphi) \doteq \int f(x)\frac{d^n}{d\lambda^n}A(x)(\varphi)\big|_{\lambda=0}d^4x,$$

where $A(x)(\varphi) = e^{\lambda p(\nabla)\varphi(x)}$ (here p is a polynomial in covariant derivatives) we define

$$:\Phi^{A,n}:(f) \equiv \mathcal{T}_1(\Phi^{A,n}(f)) \doteq \alpha_H^{-1}\int f\frac{d^n}{d\lambda^n}A_H\big|_{\lambda=0}d^4x,$$

where

$$A_H(x) = e^{\frac{1}{2}p(\nabla)\otimes p(\nabla)w_N(x,x)}A(x),$$

Unfortunately, the modifications which we have done so far do not render the time-ordered products well defined. Heuristically, we would like the time-ordered product of two functionals to be (2.7), i.e.

$$(F\cdot_{\mathcal{T}}G) = e^{\hbar\left\langle \frac{\delta}{\delta\varphi},\Delta_{S_0}^D\frac{\delta}{\delta\varphi'}\right\rangle}F(\varphi)G(\varphi')|_{\varphi'=\varphi},$$

where $\Delta_{S_0}^D \doteq \frac{1}{2}(\Delta_{S_0}^R+\Delta_{S_0}^A)$ is the Dirac propagator. This makes sense if both F and G are regular functionals (i.e. elements of $\mathcal{F}_{\mathrm{reg}}$). This indeed provides the correct notion of time-ordering, since

$$F\cdot_{\mathcal{T}}G = \begin{cases} F\star G \text{ if } \operatorname{supp} G \prec \operatorname{supp} F,\\ G\star F \text{ if } \operatorname{supp} F \prec \operatorname{supp} G,\end{cases} \tag{2.51}$$

where the relation "$\prec$" means "not later than" i.e. there exists a Cauchy surface which separates $\operatorname{supp} G$ and $\operatorname{supp} F$ and in the first case $\operatorname{supp} F$ is in the future of this surface and in the second case it's in the past.

The time ordered product defined by (2.7) is associative, commutative and isomorphic to the point-wise product by means of

$$F\cdot_{\mathcal{T}}G = \mathcal{T}\left(\mathcal{T}^{-1}F\cdot\mathcal{T}^{-1}G\right), \tag{2.52}$$

where

$$\mathcal{T} = e^{i\hbar\langle\Delta_{S_0}^D,\frac{\delta^2}{\delta\varphi^2}\rangle} \tag{2.53}$$

or more precisely

$$(\mathcal{T}F)(\varphi) \doteq \sum_{n=0}^{\infty}\frac{\hbar^n}{n!}\left\langle (i\Delta_{S_0}^D)^{\otimes n}, F^{(2n)}(\varphi)\right\rangle.$$

The linear operator $\mathcal{T}$ defined above is sometimes called the "time-ordering operator" and it is interpreted as a map which goes from the "classical" to the "quantum", i.e.

$$\underset{\text{classical}}{(\mathcal{F}_{\text{reg}}, \cdot)} \xrightarrow{\mathcal{T}} \underset{\text{quantum}}{(\mathfrak{A}_{\text{reg}}, \star, \cdot_{\mathcal{T}})},$$

where $\mathfrak{A}_{\text{reg}} \subset \mathfrak{A}$ is the range of $\mathcal{T}$. Note that on the quantum side we have *two* products. Using the time-ordered product we can express the formal S-matrix $\mathcal{S}$: $\mathcal{F}_{\text{reg}}[[\hbar]] \to \mathcal{F}_{\text{reg}}[[\hbar]]$ as the time ordered exponential:

$$\mathcal{S}(V) \doteq e_{\mathcal{T}}^{iV/\hbar} = \mathcal{T}\big(e^{\mathcal{T}^{-1}iV/\hbar}\big). \tag{2.54}$$

According to our interpretation of $\mathcal{T}$, $\mathcal{S}$ is a map on the "quantum" algebra $\mathfrak{A}$ to itself. Interacting fields are obtained by means of the Bogoliubov formula, which reads

$$\begin{aligned} R_V(F) &= -i\hbar\frac{d}{d\lambda}\left(\mathcal{S}(V)^{\star-1} \star \mathcal{S}(V+\lambda F)\right)\big|_{\lambda=0} \\ &= \left(e_{\mathcal{T}}^{iV/\hbar}\right)^{\star-1} \star \left(e_{\mathcal{T}}^{iV/\hbar} \cdot_{\mathcal{T}} F\right). \end{aligned} \tag{2.55}$$

We interpret $R_V(F)$ as the interacting quantity corresponding to F. We can also define the interacting star product as

$$F \star_V G \doteq R_V^{-1}(R_V F \star R_V G).$$

The interacting theory is given in terms of the algebra $(\mathcal{F}_{\text{reg}}, \star_V)$ and R_V acts as the intertwining map between the free quantum theory and the interacting quantum theory, i.e.

$$\underset{\text{classical}}{(\mathcal{F}_{\text{reg}}, \cdot)} \xrightarrow{\mathcal{T}} \underset{\substack{\text{free}\\ \text{quantum}}}{(\mathfrak{A}_{\text{reg}}, \star, \cdot_{\mathcal{T}})} \xrightarrow{R_V^{-1}} \underset{\substack{\text{interacting}\\ \text{quantum}}}{(\mathfrak{A}_{\text{reg}}, \star_V)}. \tag{2.56}$$

All these formulas make sense if we restrict ourselves to regular functionals. This is, however, not satisfactory for our purposes, since typical interactions are local and non-linear, hence not regular. In the first attempt we could try to pass to a different star product, which amounts to replacing Δ_{S_0} by $\Delta_{S_0}^{+}$ and $\Delta_{S_0}^{D}$ by the Feynman propagator $\Delta_{S_0}^{F} = i\Delta_{S_0}^{D} + H$, so our diagram gets modified to

$$\underset{\text{classical}}{(\mathcal{F}_{\text{reg}}, \cdot)} \xrightarrow{\mathcal{T}^H} (\mathfrak{A}_{\text{reg}}, \star_H, \cdot_{\mathcal{T}^H}) \xrightarrow{\alpha_H^{-1}} \underset{\substack{\text{free}\\ \text{quantum}}}{(\mathfrak{A}_{\text{reg}}, \star, \cdot_{\mathcal{T}})} \xrightarrow{R_V^{-1}} \underset{\substack{\text{interacting}\\ \text{quantum}}}{(\mathfrak{A}_{\text{reg}}, \star_V)},$$

where $\mathcal{T}^H \doteq e^{i\hbar\langle \Delta_{S_0}^{F}, \frac{\delta^2}{\delta\varphi^2}\rangle}$, so $\mathcal{T} = \alpha_H^{-1} \circ \mathcal{T}^H$. This modification of the formalism, however, doesn't solve the problem yet. To extend our formalism to arbitrary local functionals, we need to perform the renormalization. The difficulty which we have to face is the fact that the WF set of $\Delta^F_{S_0}$ at 0 is like the WF set of the Dirac delta and therefore the tensor powers of $\Delta^F_{S_0}$ cannot be contracted with derivatives of local functionals.

However, there is a way to extend $\mathcal{T}^H$ to *local* functionals. First we extent $\mathcal{T}^H$ to $\mathcal{F}_{\mathrm{loc}}$ by setting $\mathcal{T}^H = \mathcal{T}^H_1$. We discuss here only the Minkowski spacetime situation, so we can set $\mathcal{T}^H_1 = \mathrm{id}$. The subspace $\mathcal{T}^H(\mathcal{F}_{\mathrm{loc}}) \subset \mathfrak{A}$ will be denoted by $\mathfrak{A}^H_{\mathrm{loc}}$. Let us define the n-th order time-ordered product as

$$\mathcal{T}^H_n(F_1, \dots, F_n) \doteq F_1 \cdot_{\mathcal{T}_H} \cdots \cdot_{\mathcal{T}_H} F_n,$$

whenever it exists. It is well defined for $F_1, \dots, F_n \in \mathcal{F}_{\mathrm{loc}}$ with pairwise disjoint supports and we will denote this domain of definition by $(\mathcal{F}_{\mathrm{loc}})^{\otimes n}_{\mathrm{pds}}$. Moreover

$$\mathcal{T}^H_n(F_1, \dots, F_n) = \mathcal{T}^H_k(F_1, \dots, F_k) \star_H \mathcal{T}^H_{n-k}(F_{k+1}, \dots, F_n), \tag{T1}$$

if the supports $\mathrm{supp}\, F_i$, $i = 1, \dots, k$ of the first k entries do not intersect the past of the supports $\mathrm{supp}\, F_j$, $j = k+1, \dots, n$ of the last $n - k$ entries. This property is called the *causal factorisation property* . We will take it as an axiom that we want to impose while extending time-ordered products to arbitrary local arguments. The other axioms include

(T 2) **Starting element**: $\mathcal{T}^H_0 = 1$, $\mathcal{T}^H_1 = \mathrm{id}$,
(T 3) **Symmetry**: Each $\mathcal{T}^H_n$ is symmetric (graded symmetric if Fermions are present).
(T 4) **φ-Locality** : $\mathcal{T}^H_n(F_1, \dots, F_n)$, as a functional on $\mathcal{E}$, depends on φ only via the functional derivatives of $F_1, \dots, F_n$.

In the seminal paper [18], Epstein and Glaser have shown that such a family of maps exists and non-uniqueness in defining $\mathcal{T}^H_n$'s is fully absorbed into adding multilinear maps $Z_n : \mathfrak{A}^{\otimes n}_{\mathrm{loc}} \to \mathfrak{A}_{\mathrm{loc}}$, i.e.

$$\widetilde{\mathcal{T}^H}_n(F_1, \dots, F_n) = \mathcal{T}^H_n(F_1, \dots, F_n) + \mathcal{Z}_n(F_1, \dots, F_n),$$

where $\{\mathcal{T}^H_n\}_{n\in\mathbb{N}}$ and $\{\widetilde{\mathcal{T}^H}_n\}_{n\in\mathbb{N}}$ are two choice of time-ordered products that coincide up to order $n - 1$. The renormalized S-matrix is now defined by

$$\mathcal{S}(V) = \sum_{n=0}^{\infty} \tfrac{1}{n!}\mathcal{T}_n(V, \dots, V) = \sum_{n=0}^{\infty} \tfrac{1}{n!}\alpha_H^{-1} \circ \mathcal{T}^H_n(\alpha_H V, \dots, \alpha_H V)\,.$$

The causal factorisation property for time ordered products implies that the S-matrix satisfies Bogoliubov's factorization relation

$$\mathcal{S}(V_1 + V_2 + V_3) = \mathcal{S}(V_1 + V_2)\mathcal{S}(V_2)^{-1}\mathcal{S}(V_2 + V_3) \tag{2.57}$$

if the support of V_1 does not intersect the past of the support of V_3.

We can also define the renormalized map $\mathcal{T}:\mathcal{F}\to\mathfrak{A}$ by $\mathcal{T}\doteq\bigoplus_n \alpha_H^{-1}\circ\mathcal{T}_n^H\circ m^{-1}$, where $m^{-1}:\mathcal{F}\to S^\bullet\mathcal{F}_{\mathrm{loc}}^{(0)}$ is the inverse of the multiplication, as defined in [20] and $\mathcal{F}_{\mathrm{loc}}^{(0)}$ is the space of local functionals that vanish at 0. The renormalized time ordered product $\cdot_\mathcal{T}$ is now a binary operation defined on the domain $D_\mathcal{T}\doteq\mathcal{T}(\mathcal{F})$. Analogously to the diagram (2.56), we obtain now

$$\underset{\text{classical}}{(\mathcal{F},\cdot)}\ \xrightarrow{\mathcal{T}}\ \underset{\substack{\text{free}\\ \text{quantum}}}{(\mathfrak{A},\star,\cdot_\mathcal{T})}\ \xrightarrow{R_V^{-1}}\ \underset{\substack{\text{interacting}\\ \text{quantum}}}{(\mathfrak{A},\star_V)}, \tag{2.58}$$

with the caveat that $\cdot_\mathcal{T}$ is well defined on $D_\mathcal{T}\subset\mathfrak{A}$.

We will now discuss in detail the ambiguity arising in defining $\mathcal{T}_n$'s. In physics this is known as the *renormalization ambiguity*. To understand it better and to relate it with the notion of the *renormalization group*, we first define a map $\mathcal{Z}:\mathfrak{A}_{\mathrm{loc}}[[\hbar]]\to\mathfrak{A}_{\mathrm{loc}}[[\hbar]]$ by summing up all the $\mathcal{Z}_n$'s relating two chosen prescriptions to define the time-ordered products. For any two choices of $\mathcal{T}_n$'s the corresponding map $\mathcal{Z}$ has the following properties:

(Z 1) $\mathcal{Z}(0)=0$,
(Z 2) $\mathcal{Z}^{(1)}(0)=\mathrm{id}$,
(Z 3) $\mathcal{Z}=\mathrm{id}+\mathcal{O}(\hbar)$,
(Z 4) $\mathcal{Z}(F+G+H)=\mathcal{Z}(F+G)+\mathcal{Z}(G+H)-\mathcal{Z}(G)$, if supp $F\cap$ supp G,
(Z 5) $\frac{\delta\mathcal{Z}}{\delta\varphi}=0$.

The group of formal diffeomorphisms of $\mathfrak{A}_{\mathrm{loc}}[[\hbar]]$ that fulfill **(Z 1)**–**(Z 5)** is called the *Stückelberg-Petermann renormalization group* $\mathcal{R}$. There is a relation between the formal S-matrices and elements of $\mathcal{R}$ provided by the main theorem of renormalization [15, 16]. It states that for two S-matrices $\mathcal{S}$ and $\hat{\mathcal{S}}$, built from time ordered products satisfying the axioms **(T 1)**–**(T 3)**, there exists $\mathcal{Z}\in\mathcal{R}$ such that

$$\hat{\mathcal{S}}=\mathcal{S}\circ\mathcal{Z}, \tag{2.59}$$

where $\mathcal{Z}\in\mathcal{R}$ and conversely, if $\mathcal{S}$ is an S-matrix satisfying the axioms **(T 1)**–**(T 4)** and $\mathcal{Z}\in\mathcal{R}$ then also $\hat{\mathcal{S}}$ fulfills the axioms.

2.4.2 *Methods for Explicit Construction of Time-Ordered Products*

The proof of existence of time-ordered products with properties **(T 1)**–**(T 4)** given in [18] is rather abstract and relies on an inductive argument. For practical purposes an existence result is not sufficient and one would like to obtain some explicit formulas for $\mathcal{T}_n$'s. In this section we will review results which show that the problem of constructing time-ordered products reduces to extending certain distributions.

Subsequently, we will give some concrete computational prescriptions for constructing such extensions.

We start with an example. Let $F = \frac{1}{2}\int \varphi^2 f d\mu$, $G = \frac{1}{2}\int \varphi^2 g d\mu$, $f, g \in \mathcal{D}$. If $\operatorname{supp} g \cap \operatorname{supp} f = \varnothing$, then the time ordered product $\cdot_{\mathcal{T}}$ of F and G is given by

$$\begin{aligned} &\mathcal{T}_2(F,G)(\varphi) \\ &= (F \cdot_{\mathcal{T}} G)(\varphi) = F(\varphi)G(\varphi) + i\hbar \int \varphi(x)\varphi(y) f(x) g(y) \Delta^F_{S_0}(x,y) d\mu(x) d\mu(y) \\ &- \frac{\hbar^2}{2}\int \Delta^F_{S_0}(x,y)^2 f(x) g(y) d\mu(x) d\mu(y). \end{aligned}$$

In the least term of the expression above we have a pointwise product of a distribution with itself. This could potentially cause problems. If $x \neq y$, then if (x, k) and $(y, -k')$ belong to the wave front set , then $k, -k'$ are cotangent to a null geodesics connecting x and y. Moreover, k is future directed if x is in the future of y and past directed otherwise, so the sum of two such covectors doesn't vanish. Hence, the condition on the multiplicability of distributions presented in Sect. 2.2 implies that $(\Delta^F_{S_0})^2$ as a distribution is well defined on the complement of the diagonal $\{(x, x)|x \in M\}$. Let us now consider what happens on the diagonal. There, the only restriction is $k = -k'$, hence the sum of WF $(\Delta^F_{S_0})$ with itself contains the zero section of the cotangent bundle at the diagonal. The problem of defining $\mathcal{T}^H_2(F, G)$ reduces now to the problem of extension of $\Delta^F_{S_0}$ to a distribution defined everywhere.

This generalizes, and the construction of $\mathcal{T}^H_n$'s reduces to extending numerical distributions defined everywhere outside certain subdiagonals in M^n. The construction proceeds recursively and, having constructed the time-ordered products of order $k < n$, at order n one is left with the problem of extending a distribution defined everywhere outside the thin diagonal of M^n. On Minkowski spacetime, exploiting the translational symmetry of $\mathbb{M}$, this reduces to extending a numerical distribution defined everywhere outside 0. One way of constructing explicitly such distributional extensions relies on the so called splitting method (see for example [40]). Here we will take a different approach, based on the notion of Steinmann's scaling degree [41]. Here is the definition:

Definition 2.4.1 Let $U \subset \mathbb{R}^n$ be a scale invariant open subset (i.e. $\lambda U = U$ for $\lambda > 0$), and let $t \in \mathcal{D}'(U)$ be a distribution on U. Let $t_\lambda(x) = t(\lambda x)$ be the scaled distribution. The scaling degree sd of t is

$$\operatorname{sd} t = \inf\{\delta \in \mathbb{R} | \lim_{\lambda \to 0} \lambda^\delta t_\lambda = 0\}. \tag{2.60}$$

The degree of divergence, another important concept used often in the literature, is defined as:

$$\operatorname{div}(t) \doteq \operatorname{sd}(t) - n\,.$$

The crucial result which allows us to construct time-ordered products is stated in the following theorem:

Theorem 2.4.2 *Let $t \in \mathcal{D}(\mathbb{R}^n \setminus \{0\})$ with scaling degree $\mathrm{sd}\, t < \infty$. Then there exists an extension of t to an everywhere defined distribution with the same scaling degree. The extension is unique up to the addition of a derivative $P(\partial)\delta$ of the delta function, where P is a polynomial with degree bounded by $\mathrm{div}(t)$ (hence vanishes for $\mathrm{sd}\, t < n$).*

In the example presented at the beginning of this subsection, the scaling degree of $(\Delta^F_{S_0})^2$ in 4 dimensions is 4, so the extension exists and is unique up to the addition of a multiple of the delta function.

The result above allows in principle to extend all the numerical distributions we need for the construction of time-ordered products. However, the computations can in general get very complicated, so it is convenient to formulate the combinatorics underlying our construction in terms of Feynman graphs. In the pAQFT framework, these are not fundamental objects, but instead they are derived (together with the corresponding Feynman rules) from time-ordered products.

Time-ordered products $\mathcal{T}^H_n$ should be maps from $\mathcal{F}_{\mathrm{loc}}^{\otimes n}$ to $\mathcal{F}_{\mu c}[[\hbar]]$ and, as indicated in the previous section, they are obtained by extending non-renormalized expressions that are originally defined only on $(\mathcal{F}_{\mathrm{loc}})^{\otimes n}_{\mathrm{pds}}$. Let us consider $F \equiv F_1 \otimes \cdots \otimes F_n \in (\mathcal{F}_{\mathrm{loc}})^{\otimes n}_{\mathrm{pds}}$ with the corresponding Wick-ordered quantities are elements of $\mathfrak{A}_{\mathrm{loc}}$ given by $A_1 \doteq \mathcal{T} F_1, \ldots, A_n \doteq \mathcal{T} F_n \in \mathcal{F}_{\mathrm{loc}}$. Note that F induces a map from $\mathcal{E}^n$ to $\mathbb{R}$ by $F(\varphi_1, \ldots, \varphi_2) = F_1(\varphi_1) \cdots F_n(\varphi_n)$. When we talk about functionals on $\mathcal{E}$ we will denote the variable by φ and for functionals on $\mathcal{E}^n$ we take an n-tuple $(\varphi_1, \ldots, \varphi_n)$.

Let us denote $D_{ij} \doteq i\hbar\langle \Delta^F_{S_0}, \frac{\delta^2}{\delta\varphi_i \delta\varphi_j}\rangle$ and $D \doteq i\hbar\langle \Delta^F_{S_0}, \frac{\delta^2}{\delta\varphi^2}\rangle$. The Leibniz rule for differentiation can be formulated as

$$\frac{\delta}{\delta\varphi} \circ m_n = m_n \circ (\sum_{i=1}^{n} \frac{\delta}{\delta\varphi_i}), \tag{2.61}$$

where m_n is the pointwise multiplication of n arguments, or in other words, a pullback through the diagonal map $\mathcal{E} \to \mathcal{E}^n$, $\varphi \mapsto (\varphi, \ldots, \varphi)$. The Leibniz rule implies that the non-renormalized expression for $\mathcal{T}^H$ satisfies

$$\mathcal{T}^H \circ m_n = e^{\frac{D}{2}} \circ m_n = m_n \circ e^{\sum_{i<j} D_{ij} + \sum_i \frac{1}{2} D_{ii}},$$

Hence

$$\begin{aligned} F_1 \cdot_{\mathcal{T}^H} \cdots \cdot_{\mathcal{T}^H} F_n &= e^{\frac{D}{2}} \circ m_n(e^{-\frac{1}{2}D_{11}} F_1, \ldots, e^{-\frac{1}{2}D_{nn}} F_n) \\ &= m_n \circ e^{\sum_{i<j} D_{ij}} (F_1, \ldots F_n) \equiv m_n \circ T_n(F_1, \ldots F_n) . \end{aligned}$$

We can now use an identity

$$e^{\sum_{i<j} D_{ij}} = \prod_{i<j} \sum_{l_{ij}=0}^{\infty} \frac{D_{ij}^{l_{ij}}}{l_{ij}!} \tag{2.62}$$

to express time ordered products in terms of graphs. Let $\mathcal{G}_n$ be the set of all graphs with vertex set $V(\Gamma) = \{1, \dots n\}$ and l_{ij} the number of lines $e \in E(\Gamma)$ connecting the vertices i and j. We set $l_{ij} = l_{ji}$ for $i > j$ and $l_{ii} = 0$. If e connects i and j we set $\partial e := \{i, j\}$. Then

$$T_n = \sum_{\Gamma \in \mathcal{G}_n} T_\Gamma, \tag{2.63}$$

where

$$T_\Gamma = \frac{1}{\mathrm{Sym}(\Gamma)} \langle t_\Gamma, \delta_\Gamma \rangle, \tag{2.64}$$

with

$$\delta_\Gamma = \frac{\delta^{2\,|E(\Gamma)|}}{\prod_{i \in V(\Gamma)} \prod_{e: i \in \partial e} \delta\varphi_i(x_{e,i})}$$

and

$$t_\Gamma = \prod_{e \in E(\Gamma)} \hbar \Delta_F(x_{e,i}, i \in \partial e) \tag{2.65}$$

The, so called, symmetry factor Sym is the number of possible permutations of lines joining the same two vertices, $\mathrm{Sym}(\Gamma) = \prod_{i<j} l_{ij}!$. Note that T_Γ is a map from $(\mathcal{F}_{\mathrm{loc}})_{\mathrm{pds}}^{\otimes V}$ to $\mathcal{C}^\infty(\mathcal{E}^{|V|}, \mathbb{R})[[\hbar]]$, where $\otimes V$ means that the factors in the tensor product are numbered by vertices and to a vertex $v \in V(\Gamma)$ we assign the variable φ_v. The renormalization problem is now the problem to extend T_n's to maps on $(\mathcal{F}_{\mathrm{loc}})^{\otimes n}$ and this can be achieved by extending all the maps T_Γ and using formula (2.63).

First we note that functional derivatives of local functionals are of the form

$$F^{(l)}(\varphi)(x_1, \dots, x_l) = \int \sum_{j=1}^{N} g_j[\varphi](y) p_j(\partial_{x_1}, \dots, \partial_{x_l}) \prod_{i=1}^{l} \delta(y - x_i) d\mu(y), \tag{2.66}$$

where $N \in \mathbb{N}$, p_j's are polynomials in partial derivatives and $g_j[\varphi]$ are φ-dependent test functions. The representation above is not unique, since some of the partial derivatives ∂_{x_i} can be replaced with ∂_y and applied to $g_j[\varphi]$. Another representation of $F^{(l)}(\varphi)$ is obtained by performing the integral above and using the centre of mass and relative coordinates:

$$F^{(l)}(\varphi)(x_1, \dots, x_l) = \sum_\beta f_\beta[\varphi](z) \partial^\beta \delta(x^{\mathrm{rel}}) \tag{2.67}$$

where $\beta \in \mathbb{N}_0^{4(l-1)}$, test functions $f_\beta[\varphi](x) \in \mathcal{D}$ are now φ-dependent functions of the center of mass coordinate $z = (x_1 + \cdots + x_k)/k$ and $x^{\mathrm{rel}} = (x_1 - z, \ldots, x_k - z)$ denotes the relative coordinates.

Using (2.66) we see that the functional differential operator δ_Γ applied to $F \in \mathcal{F}_{\mathrm{loc}}^{\otimes n}$ yields, at any n-tuple of field configurations $(\varphi_1, \ldots, \varphi_n)$, a compactly supported distribution in the variables $x_{e,i}$, $i \in \partial e$, $e \in E(\Gamma)$ with support on the partial diagonal $\Delta_\Gamma = \{x_{e,i} = x_{f,i}, i \in \partial e \cap \partial f, e, f \in E(\Gamma)\} \subset \mathbb{M}^{2|E(\Gamma)|}$ and with a wavefront set perpendicular to $T\Delta_\Gamma$. Note that the partial diagonal Δ_Γ can be parametrized using the center of mass coordinates

$$z_v \doteq \frac{1}{\mathrm{valence}(v)} \sum_{e|v\in\partial e} x_{e,v},$$

assigned to each vertex. The remaining relative coordinates are $x_{e,v}^{\mathrm{rel}} = x_{e,v} - z_v$, where $v \in V(\Gamma)$, $e \in E(\Gamma)$ and $v \in \partial e$. Obviously, we have $\sum_{e|v\in\partial e} x_{e,v}^{\mathrm{rel}} = 0$ for all $v \in V(\Gamma)$. In this parametrization $\delta_\Gamma F$ can be written as a finite sum

$$\delta_\Gamma F = \sum_{\mathrm{finite}} f^\beta \partial_\beta \delta_{\mathrm{rel}},$$

where $\beta \in \mathbb{N}_0^{4|V(\Gamma)|}$, each $f^\beta(\varphi_1, ..., \varphi_n)$ is a test function on Δ_Γ and δ_{rel} is the Dirac delta distribution in relative coordinates, i.e. $\delta_{\mathrm{rel}}(g) = g(0, \ldots, 0)$, where g is a function of $(x_{e,v}^{\mathrm{rel}}, v \in V(\Gamma), e \in E(\Gamma))$.

We can simplify our notation even further. Let Y_Γ denote the vector space spanned by derivatives of the Dirac delta distributions $\partial_\beta \delta_{\mathrm{rel}}$, where $\beta \in \mathbb{N}_0^{4|V(\Gamma)|}$. Obviously, Y_Γ is graded by $|\beta|$. Let $\mathcal{D}(\Delta_\Gamma, Y_\Gamma)$ denote the graded space of test functions on Δ_Γ with values in Y_Γ. With this notation we have $\delta_\Gamma F \in \mathcal{D}(\Delta_\Gamma, Y_\Gamma)$ and if $F \in (\mathcal{F}_{\mathrm{loc}})_{\mathrm{pds}}^{\otimes n}$, then $\delta_\Gamma F$ is supported on $\Delta_\Gamma \setminus \mathrm{DIAG}$, where DIAG is the large diagonal:

$$\mathrm{DIAG} = \{z \in \Delta_\Gamma | \exists v, w \in V(\Gamma), v \neq w : z_v = z_w\} .$$

We can now write (2.64) in the form

$$\frac{1}{\mathrm{Sym}(\Gamma)} \langle t_\Gamma, \delta_\Gamma \rangle = \sum_{\mathrm{finite}} \langle f^\beta \partial_\beta \delta_{\mathrm{rel}}, t_\Gamma \rangle$$

where t_Γ is now written in terms of centre of mass and relative coordinates. To see that this expression is well defined, note that we can move all the partial derivatives ∂_β to t_Γ by formal partial integration. Then the contraction with δ_{rel} is just the pullback through the diagonal map map $\rho_\Gamma : \Delta_\Gamma \to \mathbb{M}^{2|E(\Gamma)|}$ by

$$(\rho_\Gamma(z))_{e,v} = z_v \quad \text{if } v \in \partial e\,.$$

From the wavefront set properties of $\Delta^F_{S_0}$, we deduce that the pullback ρ^*_Γ of each $t^\beta_\Gamma \doteq \partial_\beta t_\Gamma$ is a well defined distribution on $\Delta_\Gamma\backslash\mathrm{DIAG}$, so (2.64) makes sense if $F \in (\mathcal{F}_{\mathrm{loc}})^{\otimes n}_{\mathrm{pds}}$, as expected. We conclude that $t_\Gamma \in \mathcal{D}'(\Delta_\Gamma\backslash\mathrm{DIAG}, Y_\Gamma)$, where the duality between t_Γ and a test function $f = \sum_{\mathrm{finite}} f^\beta \partial_\beta \delta$ is given by

$$\langle t_\Gamma, f\rangle \doteq \sum_\beta \langle t^\beta_\Gamma, f_\beta\rangle\,.$$

The renormalization problem now reduces to finding the extensions of t^β_Γ, so that t^β_Γ gets extended to an element of $\mathcal{D}'(\Delta_\Gamma, Y_\Gamma)$. The solution to this problem is obtained by using the inductive procedure of Epstein and Glaser. The induction step works as follows: if $t_{\Gamma'}$ is known for all graphs Γ' with fewer vertices than Γ, then t_Γ can be uniquely defined for all *disconnected*, all *connected one particle reducible* and all *one particle irreducible one vertex reducible graphs*. Graphs which are irreducible and do not contain any non-trivial irreducible subgraphs are called *EG-primitive*. For the remaining graphs , called *EG-irreducible*, t_Γ is defined uniquely on all $f \in \mathcal{D}(\Delta_\Gamma, Y_\Gamma)$ of the form above where f_β vanishes together with all its derivatives of order $\leq \omega_\Gamma + |\beta|$ on the thin diagonal of Δ_Γ. Here

$$\omega_\Gamma = (d-2)|E(\Gamma)| - d(|V(\Gamma)| - 1)$$

is the degree of divergence of the graph Γ. We denote this subspace by $\mathcal{D}_{\omega_\Gamma}(\Delta_\Gamma, Y_\Gamma)$. Graphs which are irreducible and do not contain any non-trivial irreducible subgraphs are called *EG-primitive*. Renormalization amounts to project a generic f to this subspace by a translation invariant projection $W_\Gamma : \mathcal{D}(\Delta_\Gamma, Y_\Gamma) \to \mathcal{D}_{\omega_\Gamma}(\Delta_\Gamma, Y_\Gamma)$. Different renormalization schemes differ by different choices of the projections W_Γ (see [17] for details).

On Minkowski spacetime we have further simplifications. By exploiting the translation invariance we find that, at each step of the recursive construction of time-ordered products, the renormalization problem reduces to the problem of extension of some distribution defined everywhere outside the origin, so this is what we will focus on now.

For concrete computations it is convenient to construct these extensions with the use of *regularization*. Let us first define the notion of a *regularization of a distribution*. Let $\tilde{t} \in \mathcal{D}'(\mathbb{R}^d \setminus \{0\})$, $d \in \mathbb{N}$, be a distribution with degree of divergence ω, and by $\bar{t} \in \mathcal{D}'_\omega(\mathbb{R}^d)$ we denote the unique extension of $\tilde{t}$ with the same degree of divergence. A family of distributions $\{t^\zeta\}_{\zeta\in\Omega\setminus\{0\}}$, $t^\zeta \in \mathcal{D}'(\mathbb{R}^d)$, with $\Omega \subset \mathbb{C}$ a neighborhood of the origin, is called a regularization of $\tilde{t}$, if

$$\forall g \in \mathcal{D}_\lambda(\mathbb{R}^d): \quad \lim_{\zeta\to 0}\langle t^\zeta, g\rangle = \langle \bar{t}, g\rangle\,. \tag{2.68}$$

We say that the regularization $\{t^\zeta\}$ is called analytic, if for all functions $f \in \mathcal{D}(\mathbb{R}^n)$ the map

$$\Omega \setminus \{0\} \ni \zeta \mapsto \langle t^\zeta, f\rangle \tag{2.69}$$

is analytic with a pole of finite order at the origin. The regularization $\{t^\zeta\}$ is called finite, if the limit $\lim_{\zeta\to 0}\langle t^\zeta, f\rangle \in \mathbb{C}$ exists $\forall f \in \mathcal{D}(\mathbb{R}^d)$.

For a finite regularization the limit $\lim_{\zeta\to 0} t^\zeta$ is, as expected, a solution t of the extension (renormalization) problem. Given a regularization $\{t^\zeta\}$ of t, it follows from (2.68) that for any projection $W : \mathcal{D} \to \mathcal{D}_\omega$

$$\langle \bar{t}, Wf\rangle = \lim_{\zeta\to 0}\langle t^\zeta, Wf\rangle \quad \forall f \in \mathcal{D}(\mathbb{R}^n)\,. \tag{2.70}$$

It was shown in [16] that any extension $t \in \mathcal{D}'(\mathbb{R}^d)$ of $\tilde{t}$ with the same scaling degree is of the form $\langle t, f\rangle = \langle \bar{t}, Wf\rangle$ with some W-projection of the form

$$Wf := f - \sum_{|\alpha|\leq\lambda} f^{(\alpha)}(0)\, w_\alpha, \tag{2.71}$$

where $w_\alpha \in \mathcal{D}(\mathbb{R}^d)$ such that for all multiindices $\beta \in \mathbb{N}_0^d$ with $|\beta| \leq \omega$ we have $\partial^\beta w_\alpha(0) = \delta_\alpha^\beta,\ |\alpha| \leq \omega$ Hence

$$\langle \bar{t}, Wf\rangle = \lim_{\zeta\to 0}\left[\langle t^\zeta, f\rangle - \sum_{|\alpha|\leq \mathrm{sd}(t)-n}\langle t^\zeta, w_\alpha\rangle\, f^{(\alpha)}(0)\right]. \tag{2.72}$$

In general, we cannot split the limit on the right hand side into two well defined terms. However, if the regularization $\{t^\zeta, \zeta \in \Omega \setminus \{0\}\}$ is analytic, then we can expand each term into a Laurent series around $\zeta = 0$, and because the overall limit is finite, the principal parts (pp) of these two Laurent series must be the same. This means that the principal part of any analytic regularization $\{t^\zeta\}$ of a distribution $t \in \mathcal{D}'(\mathbb{R}^d \setminus \{0\})$ is a local distribution of order $\mathrm{sd}(t) - d$. Following [17], we can now give a definition of the minimal subtraction in the EG framework.

Definition 2.4.3 (*Minimal Subtraction*) The regular part ($\mathrm{rp} = 1 - \mathrm{pp}$) of any analytic regularization $\{t^\zeta\}$ of a distribution $\tilde{t} \in \mathcal{D}'(\mathbb{R}^d \setminus \{0\})$ defines by

$$\langle t^{\mathrm{MS}}, f\rangle := \lim_{\zeta\to 0}\mathrm{rp}(\langle t^\zeta, f\rangle) \tag{2.73}$$

an extension of $\tilde{t}$ with the same scaling degree, $\mathrm{sd}(t^{\mathrm{MS}}) = \mathrm{sd}(\tilde{t})$. The extension t^{MS} defined by (2.73) is called the "minimal subtraction".

2.4.3 Interacting Theories

Let us now discuss the problem of constructing interacting nets of observables. We start from a space $\mathcal{D}^n$ of functions $f : \mathbb{M} \to \mathbb{R}^n$ with compact support. We assume that we have unitaries $S(f)$, $f \in \mathcal{D}^n$ with $S(0) = 1$, which generate a *-subalgebra $\tilde{\mathfrak{A}}$ of $\mathfrak{A}$ and satisfy for $f, g, h \in \mathcal{D}$ Bogoliubov's factorization relation

$$S(f+g+h) = S(f+g)S(g)^{-1}S(g+h)$$

if the past J_- of $\operatorname{supp} h$ does not intersect $\operatorname{supp} f$ (or, equivalently, if the future J_+ of $\operatorname{supp} f$ does not intersect $\operatorname{supp} h$). We can obtain these as formal S-matrices $S(f) \doteq \mathcal{S}(V(f))$, discussed in the previous section (see property (2.57)), for a generalized Lagrangian $V(f) = \alpha_H\left(\sum_{j=1}^n \int A_j(x) f^j(x) d\mu(x)\right)$, where $f \in \mathcal{D}^n$ and each $A_j(x)$ is a local function $\varphi \in \mathcal{E}$. Typically A'_j are polynomial and they represent Lagrangian densities of various interaction terms that one can add to the free action S_0.

We also assume that the translation group of Minkowski space acts by automorphisms α_x on $\tilde{\mathfrak{A}}$ such that

$$\alpha_x(S(f)) = S(f_x),\ \ f_x(y) = f(y-x).$$

Obviously, this is also satisfied for the S-matrices discussed so far. Under these general assumptions, we define local algebras $\mathfrak{A}(\mathcal{O})$, $\mathcal{O} \subset \mathbb{M}$, as the *-subalgebras of $\mathfrak{A}$ generated by $S(f)$, $\operatorname{supp} f \subset \mathcal{O}$ and obtain a translation covariant Haag-Kastler net on Minkowski space. To justify this claim, we will now check that all the axioms are satisfied.

Isotony and *Covariance* are obvious, and *Locality* follows from the fact that for functions f, g with spacelike separated supports

$$\operatorname{supp} f \cap J_\pm(\operatorname{supp} g) = \emptyset \tag{2.74}$$

and hence

$$S(f)S(g) = S(f+g) = S(g)S(f). \tag{2.75}$$

The crucial observation is now that the map $f \mapsto S(f)$ induces a large family of objects that satisfy Bogoliubov's factorisation relation, which are labeled by test functions $g \in \mathcal{D}^n$, namely the *relative S-matrices*

$$f \mapsto S_g(f) = S(g)^{-1}S(g+f).$$

We can choose $A_0(x) = \mathcal{L}_I(x)$ to be the Lagrangian density of the interaction term. Then, for $g = (g_0, 0, \ldots, 0)$, we obtain $V(g) = \int \mathcal{L}_I g_0 d\mu \equiv L_I(g_0)$, where $g_0 \in \mathcal{D}$. Note that $S(g + \lambda f) = \mathcal{S}(\alpha_H(\mathcal{L}_I(g_0) + \lambda \sum_j \int A_j f_j d\mu))$, so the derivative of S with respect to λ is just the retarded field $R_{\mathcal{L}_I(g_0)}(V(f))$. Let us now prove that the causal factorisation property indeed holds for $S_g(f)$. Let $f, h \in \mathcal{D}^n$ such that $\operatorname{supp} f$ does not intersect $J_-(\operatorname{supp} h)$. Let $g, g' \in \mathcal{D}^n$. Then

$$\begin{aligned} S_g(f+g'+h) &= S(g)^{-1}S(f+(g+g')+h) \\ &= S(g)^{-1}S(f+(g+g'))S(g+g')^{-1}S((g+g')+h) \\ &= S_g(f+g')S_g(g')^{-1}\underbrace{S(g)^{-1}S(g)}_{=1}S_g(g'+h). \quad \square \end{aligned}$$

We consider $S_g(f)$ as the retarded observable $S(f)$ under the influence of the interaction $L_I(g_0)$. The Haag-Kastler net $\mathfrak{A}_g$ of the interacting theory is then defined by the local algebras $\mathfrak{A}_g(\mathcal{O})$ which are generated by the relative S-matrices $S_g(f)$, $\operatorname{supp} f \subset \mathcal{O}$. These can indeed be interpreted as retarded observables, as $S_g(f)$ depends only on the behavior of g in the past of $\operatorname{supp} f$. More precisely, $\operatorname{supp}(g-g') \cap J_-(\operatorname{supp} f) = \emptyset$ implies

$$S_g(f) = S(g)^{-1}S((g-g')+g'+f)$$

$$= S(g)^{-1}S((g-g')+g')S(g')^{-1}S(g'+f) = S_{g'}(f).$$

The second observation is that $S_g(f)$ depends on the behavior of g outside of the future of $\operatorname{supp} f$ via a (formal) unitary transformation which does not depend on f. Namely, $\operatorname{supp}(g-g') \cap J_+(\operatorname{supp} f) = \emptyset$ implies

$$S_g(f) = S(g)^{-1}S(f+g'+(g-g'))$$

$$= S(g)^{-1}S(f+g')S(g')^{-1}S(g'+(g-g'))$$

$$= S(g)^{-1}S(g')S(g')^{-1}S(f+g')S_{g'}(g-g')$$

$$= \operatorname{Ad} S_{g'}(g-g')^{-1}(S_{g'}(f)).$$

Hence the structure of local algebras depends only locally on the interaction. This allows to perform the adiabatic limit directly on the level of local algebras.

In the next step we want to remove the restriction to interactions with compact support. Let $G : \mathbb{M} \to \mathbb{R}^n$ be smooth and $\mathcal{O}$ be bounded. Set

$$[G]_{\mathcal{O}} = \{g \in \mathcal{D}^n | g \equiv G \text{ on a neighborhood of } J_+(\mathcal{O}) \cap J_-(\mathcal{O})\}.$$

We consider the $\tilde{\mathfrak{A}}$-valued maps

$$S_{G,\mathcal{O}}(f) : [G]_{\mathcal{O}} \ni g \mapsto S_g(f) \in \tilde{\mathfrak{A}}.$$

The local algebra $\mathfrak{A}_G(\mathcal{O})$ is defined to be the algebra generated by $S_{G,\mathcal{O}}(f)$, $\operatorname{supp} f \subset \mathcal{O}$. Note that the evaluation maps

$$\gamma_{gG} : S_{G,\mathcal{O}}(f) \to S_g(f)$$

extend to isomorphisms of $\mathfrak{A}_G(\mathcal{O})$ and $\mathfrak{A}_g(\mathcal{O})$ for every $g \in [G]_{\mathcal{O}}$.

The local net is now defined by the embeddings $i_{\mathcal{O}_2\mathcal{O}_1}$ for $\mathcal{O}_1 \subset \mathcal{O}_2$

$$i_{\mathcal{O}_2\mathcal{O}_1} : S_{G,\mathcal{O}_1}(f) \mapsto S_{G,\mathcal{O}_2}(f)$$

for $f \in \mathcal{D}^n$ with supp $f \subset \mathcal{O}_1$. Let $\mathfrak{A}_G$ be the inductive limit with embeddings

$$i_{\mathcal{O}} : \mathfrak{A}_G(\mathcal{O}) \to \mathfrak{A}_G$$

and we set

$$S_G(f) = i_{\mathcal{O}}(S_{G,\mathcal{O}}(f)).$$

We are now ready to prove a crucial theorem about the net $\mathcal{O} \mapsto \mathfrak{A}_G(\mathcal{O})$.

Theorem 2.4.4 *Let G be translation invariant. Then the net becomes translation covariant by setting*

$$\alpha_x^G(S_G(f)) = S_G(f_x).$$

Proof We have to prove that α_x^G extends to an isomorphism from $\mathfrak{A}_G(\mathcal{O}) \to \mathfrak{A}_G(\mathcal{O}+x)$. Let $\mathcal{O}_1 \supset \mathcal{O} \cup \mathcal{O} - x$ and $g \in [G]_{\mathcal{O}_1}$. Then $g, g_x \in [G]_{\mathcal{O}}$ and $g_x = g + h_+^x + h_-^x$ with supp $h_\pm^x \cap J_\mp(\mathcal{O}) = \emptyset$. By causal factorization

$$\alpha_x^G = \gamma_{gG}^{-1} \circ \mathrm{Ad}U_g(x) \circ \alpha_x \circ \gamma_{gG}$$

with $U_g(x) = S_g(h_-^x)$. □

2.5 Time-Slice Axiom, Operator Product Expansions, and the Renormalization Group

We have seen that, starting from a free QFT and a definition of a time ordered product satisfying the axioms of Sect. 2.4.3 we can construct a local net (in the sense of formal power series) satisfying the Haag-Kastler axioms of Isotony, Locality and Covariance. In this section we want to analyze the net in more detail.

First we investigate whether the net satisfies the time-slice axiom . This can be done for the case that the net is defined on a generic Lorentzian globally hyperbolic spacetime M. It is known since a long time [22] that the free theory generated by linear functionals, modulo the ideal of the free field equation, satisfies this axiom, and by using the techniques of microlocal analysis, this result can be extended to the net $\mathfrak{A}$ generated by elements of the form $\alpha_H^{-1} F$, where $F \in \mathcal{F}_{\mu c}$ is a microcausal functional [11, 29]. In Chilian and Fredenhagen [11] it was shown that this implies that also the net $\mathfrak{A}_G$ introduced in the previous section satisfies the axiom. The argument relies on the fact that the algebra of the interacting theory associated to some bounded region can be constructed as a subalgebra of the free theory for a slightly larger region, and vice versa.

The problem is that these subalgebras are fixed only up to unitary equivalence, so one has in addition to show that these unitary transformations can be appropriately fixed. We use the fact that the relative S-matrices $S_g(f)$ are well defined also for test functions g with non-compact support provided the support is past compact, i.e. $\operatorname{supp} g \cap J_-(x)$ is compact for all $x \in M$.

Let Σ be a Cauchy surface of M and N a neighborhood of Σ. Let $\mathcal{O} \subset M$ be relatively compact. We choose a Cauchy surface Σ_- such that $\mathcal{O} \cup N \subset J_+(\Sigma_-)$ and a smooth function χ with past compact support such that $\operatorname{supp}(1-\chi) \subset J_-(\Sigma_-)$. We want to prove that $\mathfrak{A}_{G\chi}(\mathcal{O}) \subset \mathfrak{A}_{G\chi}(N)$.

By construction of the interacting theory we see immediately that $\mathfrak{A}_{G\chi}(\mathcal{O}) \subset \mathfrak{A}(M)$ holds. Due to the time slice property of the free theory, $\mathfrak{A}(M) = \mathfrak{A}(N')$ for each neighborhood N' of Σ. We now construct within the algebra $\mathfrak{A}_{G\chi}(N)$ an algebra which is isomorphic to $\mathfrak{A}(N')$ for a sufficiently small $\Sigma \subset N' \subset N$. For this purpose we choose another smooth function χ' with support contained in $J_+(N)$ and with $\operatorname{supp}(1-\chi') \subset J_-(N')$. Let now $\operatorname{supp} f \subset N'$. Then the unitaries

$$S_{G(\chi-\chi')}(f) = S_{G\chi}(g')^{-1} S_{G\chi}(g'+f), \text{ with } g' \equiv G\chi' \text{ on } J_-(\operatorname{supp} f), \operatorname{supp} g' \subset N, \tag{2.76}$$

generate an algebra isomorphic to $\mathfrak{A}(N')$ within $\mathfrak{A}_{G\chi}(N)$. The map

$$\alpha : S(f) \to S_{G(\chi-\chi')}(f) = \operatorname{Ad}(S(g-g'))^{-1}(S(f)) \tag{2.77}$$

with $g \equiv G\chi$ on $J_-(\operatorname{supp} f \cup \operatorname{supp} g')$ extends to an injective homomorphism from $\mathfrak{A}(M)$ into $\mathfrak{A}_{G\chi}(N)$. Since $\mathfrak{A}_{G\chi}(N) \subset \mathfrak{A}(M)$, α is an endomorphism of $\mathfrak{A}(M)$. We show that it is even an automorphism. For this purpose we construct the inverse of α. By exploiting the time slice property of the free theory, we can restrict ourselves to elements $S(f)$ with $\operatorname{supp} f \subset J_-(\Sigma_0)$. On these elements we have

$$\alpha^{-1}(S(f)) = \operatorname{Ad}(S(g-g'))(S(f)) = S(g-g'+f)S(g-g')^{-1} \tag{2.78}$$

where $g - g' \equiv G(\chi - \chi')$ on $J_+(\operatorname{supp} f)$. We conclude that $\mathfrak{A}_{G\chi}(N) = \mathfrak{A}(M)$. This proves the claim.

Another general property of the interacting net is the existence of an operator product expansion [27]. In the case of the product of two fields A and B it is an expansion

$$A(x)B(y) \sim \sum_k C^k_{AB}(x,y)\varphi_k(x) \tag{2.79}$$

with distributions C^k_{AB} and a basis of local fields φ_k, ordered with respect to the scaling dimension. This is an asymptotic expansion in the sense that after evaluation in a state coming from a Hadamard state of the free field, the difference between the right hand side of the relation and the left hand side, truncated at some k, tends to zero as $x \to y$, with an order depending on k.

The third property we look at is the behavior of the theory at different scales. In the standard formalism of QFT, one formulates this as a property of vacuum expectation values of products or time ordered products of fields, or one uses the concept of the so-called effective action. In this formulation one has to have control over the existence and uniqueness of the vacuum state. In the algebraic approach one can instead derive a relation between local nets. Namely given a local net $\mathcal{O} \mapsto \mathfrak{A}_1(\mathcal{O})$ one obtains another net by scaling the regions,

$$\mathfrak{A}_\lambda(\mathcal{O}) = \mathfrak{A}_1(\lambda\mathcal{O}). \tag{2.80}$$

If the net depends on some parameters (m, g), one can compensate the scaling by changing the parameters. One obtains the algebraic Callan-Symanzik equation [15]

$$\mathfrak{A}_\lambda^{m,g} \cong \mathfrak{A}_1^{m(\lambda),g(\lambda)} \tag{2.81}$$

The "running" of the parameters is as usual determined by the renormalization group equation which follows from the behavior of the time ordered product under scaling.

2.6 Hamiltonian Formalism for Quantum Field Theory, and the Construction of States

Up to now we remained in the realm of algebras. There we could study several structural properties of the theory. In order to get more detailed predictions of the theory one has to evaluate the algebra in specific states. A class of states on the local algebras can be obtained in terms of the states of the free theory by embedding the interacting theory into the free one, but this is highly ambiguous and gives no direct interpretation of the states. Conceptually, one does not need more, since the interpretation can be done in terms of the expectation values of observables. In practice, however, one would prefer to have states with an a priori interpretation as e.g. the vacuum state. The standard way to compute it is the evaluation of the product or the time ordered product of interacting fields with an interaction $L_I(g_0) = \int \mathcal{L}_I(x)g_0(x)d\mu(x)$ in the vacuum state of the free theory and performing the adiabatic limit $g_0 \to 1$. This limit is well behaved in massive theories, but exists also for a suitable sequence $(g_0)_n \to 1$ in certain massless theories such as massless φ^4 or QED. In the case of time ordered products one just reproduces the standard formulas in terms of Feynman such graphs; in the case of operator products one has to use Steinmann's sector graphs [42]. The adiabatic limit in this form, however, does not always exist, in particular not for states with nonzero temperature.

A more direct way of constructing states with specific properties could be imagined in a Hamiltonian formalism, as well known from nonrelativistic quantum mechanics. The difficulty is that the interaction Hamiltonian for a local QFT is very singular so that perturbation theory for selfadjoint operators cannot be used. There are two independent reasons for the singular character of perturbations in QFT. The first is translation symmetry. In Minkowski space this leads to Haag's theorem, which states that the ground state of the interacting theory cannot be represented by a vector in the Fock space of the free theory. If one takes this into account by restricting the interaction to a finitely extended spatial region, one can indeed apply the perturbation theory of selfadjoint operators in certain superrenormalizable models in 2 dimensions. One can then construct ground states and consider their limit if the cutoff is removed. In 4 dimensions, however, the local interaction densities are too singular, so that also the spatially restricted interaction is not an operator.

The Hamiltonian formalism relies on a split of spacetime into the product of a Cauchy surface and the time axis, and all the observables of the theory are constructed in terms of their initial values on this surface, which are supposed to be independent of the interaction. But from renormalization theory it is well known that in general one has to expect modifications of the canonical structure; moreover, even for free fields, the restriction to a Cauchy surface is singular for all nonlinear local fields.

Instead we use the fact that for generic perturbative QFT's the time-slice axiom holds. Moreover, as we saw from the discussion of the proof of this fact, the free and the interacting algebra of a time slice can be identified. This suggests to compare their time evolutions. Both are automorphism groups acting on the same algebra, and they differ by a cocycle. In case of a spatial cutoff of the interaction, the cocycle is implemented by a unitary cocycle within the algebra, whose generator is an integral over an operator valued function which may be interpreted as a regularized interaction Hamiltonian density $\mathcal{H}_I(\mathbf{x})$.

As in Sect. 2.4.3 we consider the space $\mathcal{D}^n$ of test functions and the algebra generated by $S(f) = \mathcal{S}(\alpha_H^{-1}(\sum_i \int A_i f^i d\mu))$. We also assume that $A_0 = \mathcal{L}_I$ is the interaction Lagrangian density. The time slice property proven in Sect. 2.5 induces isomorphisms between the free and the interacting algebras. Let χ be a smooth function of time t with $\chi(t) = 1$ for $t > -\varepsilon$ and $\chi(t) = 0$ for $t \leq -2\varepsilon$. Then $\mathrm{supp}\,((t, \mathbf{x}) \mapsto G(t, \mathbf{x})\chi(t))$ is past compact. We now define a map from $\mathfrak{A}_G$ to $\mathfrak{A}$ by

$$\gamma_\chi(S_G(f)) = S_{G\chi}(f), \ \ \mathrm{supp}\, f \subset (-\varepsilon, \varepsilon) \times \mathbb{R}^3.$$

Due to the time slice property this map extends to an isomorphism. Moreover, it only slightly changes the kinematical localization at $t = 0$. Let $\mathcal{O}_r = \{(t, \mathbf{x}) | |t| + |\mathbf{x}| < r\}$. Then

$$\gamma_\chi(\mathfrak{A}_G(\mathcal{O}_r)) \subset \mathfrak{A}(\mathcal{O}_{r+4\varepsilon}) \subset \gamma_\chi(\mathfrak{A}_G(\mathcal{O}_{r+8\varepsilon})).$$

Let G be constant, let $\alpha_x^{G,\chi} = \gamma_\chi \circ \alpha_x^G \circ \gamma_{\chi^{-1}}$ be the translations of the interacting theory mapped to the free theory, and consider the cocycle $\beta_x^{G,\chi} = \alpha_x^{G,\chi} \circ \alpha_{-x}$. We find $\beta_{(0,\mathbf{x})}^{G,\chi} = \mathrm{id}$ and, for f with $\mathrm{supp}\, f \subset \mathcal{O}_r$ and small t,

$$\beta_{(t,0)}^{G,\chi}(S(f)) = \mathrm{Ad} S_{h\chi}(h(\chi_t - \chi))(S(f))$$

where h is time independent, has compact spatial support and $h \equiv G$ on $\mathcal{O}_{r+4\varepsilon}$.

Proposition 2.6.1 *The unitaries* $U_t^{h\chi} = S_{h\chi}(h(\chi_t - \chi))$ *fulfill the cocycle equation*

$$U_{t+s}^{h\chi} = U_t^{h\chi}\alpha_t(U_s^{h\chi})$$

Proof For sufficiently large u (depending on s, t) we have

$$\begin{aligned}
&S_{h\chi}(h(\chi_t - \chi))\alpha_t(S_{h\chi}(h(\chi_s - \chi))) \\
&= S_{h(\chi-\chi_u)}(h(\chi_t - \chi))\alpha_t(S_{h(\chi-\chi_{u-t})}(h(\chi_s - \chi)) \\
&= S_{h(\chi-\chi_u)}(h(\chi_t - \chi))S_{h(\chi_t-\chi_u)}(h(\chi_{t+s} - \chi_t)) \\
&= S(h(\chi - \chi_u))^{-1}S(h(\chi_t - \chi_u))S(h(\chi_t - \chi_u))^{-1}S(h(\chi_{t+s} - \chi_u)) \\
&= S_{h(\chi-\chi_u)}(h(\chi_{t+s} - \chi)) = S_{h\chi}(h(\chi_{t+s} - \chi)).
\end{aligned}$$

□

We conclude that the unitary cocycle $U_t^{h\chi}$ describes the interacting time evolution (with spatial cutoff h) in the interaction picture. Due to the finite speed of propagation, it coincides with the full time evolution for small t.

We now consider a time translation covariant representation $(\mathcal{H}, \pi, U_0)$ and assume that the map $\mathcal{D} \ni f \to \pi(S(f))$ is strongly continuous. Then the cocycle $U_t^{h\chi}$ is strongly continuous, and

$$t \mapsto U_{h\chi}(t) = U_t^{h\chi}U_0(t) \tag{2.82}$$

is a strongly continuous 1-parameter group with selfadjoint generater $H_{h\chi}$ which describes the dynamics of the interacting system with spatial cutoff.

In case π is irreducible, one may now determine the spectrum of $H_{h\chi}$ and interpret it as the energy spectrum of the interacting theory with spatial cutoff (up to an additive constant). One may also look for a ground state and consider the limit of removal of the cutoff.

If π is a representation induced by a KMS state, and Ω_0 is the corresponding cyclic vector in the representation space, one knows by Connes' cocycle theorem that there exists a weight whose modular automorphims are the time translations of the interacting theory. If Ω_0 is in the domain of $e^{-\frac{\beta}{2}H_{h\chi}}$, then this weight is bounded and induced by the vector

$$\Omega_{h\chi} = e^{-\frac{\beta}{2}H_{h\chi}}\Omega_0. \tag{2.83}$$

If the cocycle is strongly differentiable on a dense domain, the interaction Hamiltonian can be defined as the generator of the cocycle. We obtain [19]

$$H_I^{h\chi} = \hbar\frac{d}{idt}U_t^{h\chi} = \hbar\, S(h\chi)^{-1}\frac{d}{idt}S(h\chi_t) = R_{V(h\chi)}(V(h\dot{\chi})),$$

where in the last step we have used Bogoliubov's formula (2.55) for interacting fields. In the limit $\varepsilon \to 0$, $\dot{\chi}$ tends to the δ-function and we obtain the usual interaction picture.

We illustrate the method on the example of an interaction with external sources. We start with the CCR algebra $\tilde{\mathfrak{A}}_{S_0}$ of the free scalar field introduced at the end of Sect. 2.3.5. The formal S-matrix is

$$S(f) = \mathcal{S}(F_f) = e^{iF_f/\hbar} e^{-\frac{i}{2\hbar}\langle f, \Delta^D f\rangle},$$

where $F_f(\varphi) = \int \varphi f d\mu$. One can verify it by direct computation (using the forumlas for time-ordered product given in Sect. 2.4.2) or, indirectly, by the verification of the causal factorization property (**T1**). Namely, we have

$$S(f+g)^{-1} S(f+g+h) \tag{2.84}$$

$$= e^{iF_h/\hbar} \exp \frac{i}{\hbar}(\langle f+g, \Delta^D(f+g)\rangle - \langle f+g+h, \Delta^D(f+g+h)\rangle + \langle f+g, \Delta h\rangle)$$

$$= e^{i\varphi(h)} \exp \frac{i}{2\hbar}(-\langle h, \Delta^D h\rangle - \langle f+g, \underbrace{(2\Delta^D - \Delta)}_{=\Delta^A} h\rangle),$$

$$= S(g)^{-1} S(g+h) e^{\frac{i}{2\hbar}\langle f, \Delta^A h\rangle}$$

hence if $\operatorname{supp} f \cap J_-(\operatorname{supp} h) = \emptyset$ then by the support property of the advanced propagator

$$\langle f, \Delta^A h\rangle = 0$$

and the factorization holds.

We find the interaction Hamiltonian (h time independent)

$$H_I^{h\chi} = -\varphi(h\dot{\chi}) - \text{const.}$$

Due to the smearing in time, this operator remains meaningful also for a pointlike source ($h \sim \delta(\mathbf{x})$).

In general, for the free theory we obtain the usual Fock space Hamiltonian H_0, and the Hamiltonian of the interacting theory with spatial cutoff is the sum of the free Hamiltonian and the interaction term,

$$H = H_0 + \int h(x)\mathcal{H}_I(\mathbf{x}) d^3\mathbf{x}. \tag{2.85}$$

In this framework, one can now apply the standard perturbative constructions of ground states and KMS states. In [19, 36] it was shown that in massive theories ground states and KMS states for positive temperatures exist. Some aspects of this formalism involving thermal mass were further developped in [14] in conjunction with the principle of perturbative agreement [30]. It is hoped that this regularized

Hamiltonian picture will allow to close the conceptual gap between the standard formalism in nonrelativistic quantum mechanics and quantum statistical mechanics and the formalism of relativistic QFT.

2.7 Conclusions

We have seen that the concepts of AQFT can be used in renormalized perturbative QFT and yield Haag-Kastler nets (in the sense of algebras of formal power series) for generic models of QFT. Due to its axiomatic formulation all possible renormalization methods are covered, and one has an a priori characterization of the class of renormalized theories associated to a classical Lagrangian, independent of any regularization scheme. For practical purposes, it is nevertheless often appropriate to introduce a regularization, and in particular analytic regularization schemes such as dimensional or analytic renormalization are useful, for computation but also for specifying a theory in its class (e.g. by minimal subtraction), see [17]. One may also incorporate the ideas of the renormalization flow equation in the sense of Polchinski and made rigorous in [33] . This is exposed in [15]. In these notes we restricted ourselves to scalar field theories. The generalization to other types of field theory have been discussed in several papers; fermionic theories can be treated essentially in the same way, and gauge theories can be treated after adding auxiliary fields (ghosts etc.) and constructing the time ordered products such that BRST symmetry is respected [15, 20, 27]. Even gravity can be included where however the concept of local algebras of observables has to be properly adapted [10].

References

1. Araki, H., Shiraishi, M.: On quasifree states of the canonical commutation relations (I). Publ. Res. Inst. Math. Sci. **7**, 105–120 (1971)
2. Arms, J.M., Marsden, J.E., Moncrief, V.: The structure of the space of solutions of Einstein's equations II: several Killing fields and the Einstein-Yang-Mills equations. Ann. Phys. **144**(1), 81–106 (1982)
3. Fredenhagen, K., Bär, C.H.: Quantum field theory on curved spacetimes, Lecture Notes in Physics, vol. 786. Springer, Berlin (2009)
4. Bayen, F., Flato, M., Fronsdal, C., Lichnerowicz, A., Sternheimer, D.: Deformation theory and quantization. I. Deformations of symplectic structures. Ann. Phys. **111**, 61–110 (1978)
5. Bayen, F., Flato, M., Fronsdal, C., Lichnerowicz, A., Sternheimer, D.: Deformation theory and quantization. II. Physical applications. Ann. Phys. **111**, 111–151 (1978)
6. Bogoliubov, N.N., Parasiuk, O.S.: Über die Multiplikation der Kausalfunktionen in der Quantentheorie der Felder. Acta Math. **97**, 227–266 (1957)
7. Bogoliubov, N.N., Shirkov, D.V.: Introduction to the Theory of Quantized Fields. Interscience Publishers (1959)
8. Brunetti, R., Fredenhagen, K.: Microlocal analysis and interacting quantum field theories: renormalization on physical backgrounds. Commun. Math. Phys. **208**, 623 (2000)

9. Brunetti, R., Duetsch, M., Fredenhagen, K.: Perturbative algebraic quantum field theory and the renormalization groups. Adv. Theor. Math. Phys. **13**, 1541 (2009)
10. Brunetti, R., Fredenhagen, K., Rejzner, K.: Quantum gravity from the point of view of locally covariant quantum field theory. arXiv:1306.1058 [math-ph]
11. Chilian, B., Fredenhagen, K.: The Time slice axiom in perturbative quantum field theory on globally hyperbolic spacetimes. Commun. Math. Phys. **287**, 513 (2009)
12. Crnkovic, C, Witten, E.: Covariant description of canonical formalism in geometrical theories. In: Three hundred years of gravitation, pp. 676–684 (1987)
13. Dereziński, J., Gérard, C.: Mathematics of Quantization and Quantum Fields. Cambridge University Press, Cambridge (2013)
14. Drago, N., Hack, T.-P., Pinamonti, N.: The generalised principle of perturbative agreement and the thermal mass. arXiv:1502.02705 [math-ph]
15. Duetsch, M., Fredenhagen, K.: A local (perturbative) construction of observables in gauge theories: the example of QED. Commun. Math. Phys. **203**, 71 (1999)
16. Dütsch, M., Fredenhagen, K.: Causal perturbation theory in terms of retarded products, and a proof of the action ward identity. Rev. Math. Phys. **16**(10), 1291–1348 (2004). hep-th/0403213
17. Duetsch, M., Fredenhagen, K., Keller, K.J., Rejzner, K.: Dimensional regularization in position space, and a forest formula for Epstein-Glaser renormalization. J. Math. Phys. **55**, 122303 (2014)
18. Epstein, H., Glaser, V.: The role of locality in perturbation theory. Ann. Poincare Phys. Theor. A **19**, 211 (1973)
19. Fredenhagen, K, Lindner, F.: Construction of KMS states in perturbative QFT and renormalized Hamiltonian dynamics. Commun. Math. Phys. **332**(3), 895 (2014)
20. Fredenhagen, K., Rejzner, K.: Batalin-Vilkovisky formalism in perturbative algebraic quantum field theory. Commun. Math. Phys. **317**, 697 (2013)
21. Fredenhagen, K., Rejzner, K.: QFT on curved spacetimes: axiomatic framework and examples. arXiv:1412.5125 [math-ph]
22. Fulling, S.A., Narcowich, F.J., Wald, R.M.: Singularity structure of the two-point function in quantum field theory in curved spacetime, ii. Ann. Phys. **136**, 243–272 (1981)
23. Groenewold, H.J.: On the principles of elementary quantum mechanics. Physica **12**, 405–460
24. Hamilton, R.S.: The inverse function theorem of Nash and Moser. Bull. Amer. Math. Soc. (N.S.) **7**(1)
25. Hepp, K.: Proof of the Bogoliubov-Parasiuk theorem on renormalization. Commun. Math. Phys. **2**, 301–326 (1966)
26. Hörmander, L.: The analysis of linear partial differential operators I: distribution theory and fourier analysis. Springer (2003)
27. Hollands, S.: The operator product expansion for perturbative quantum field theory in curved spacetime. Commun. Math. Phys. **273**, 1 (2007)
28. Hollands, S.: Renormalized quantum Yang-Mills fields in curved spacetime. Rev. Math. Phys. **20**, 1033 (2008)
29. Hollands, S., Wald, R.M.: Local wick polynomials and time ordered products of quantum fields in curved space-time. Commun. Math. Phys. **223**, 289 (2001)
30. Hollands, S., Wald, R.M.: Conservation of the stress tensor in perturbative interacting quantum field theory in curved spacetimes. Rev. Math. Phys. **17**, 227–311 (2007)
31. van Hove, L.: Sur certaines représentations unitaires d'un groupe infini de transformations. Proc. Roy. Acad. Sci. Belgium **26**, 1–102
32. Il'in, V.A., Slavnov, D.S.: Observable algebras in the S-matrix approach. Theor. Math. Phys. **36**, 32 (1978)
33. Keller, G., Kopper, C., Salmhofer, M.: Perturbative renormalization and effective Lagrangians in phi**4 in four-dimensions. Helv. Phys. Acta **65**, 32 (1992)
34. Kay, B.S., Wald, R.M.: Theorems on the uniqueness and thermal properties of stationary, nonsingular, quasifree states on spacetimes with a bifurcate killing horizon. Phys. Rep. **207**, 49–136 (1991)

35. Kriegl, A., Michor, P.: Convenient setting of global analysis, Mathematical Surveys and Monographs, vol. 53. American Mathematical Society, Providence (1997). http://www.ams.org/online_bks/surv53/
36. Lindner, F.: Perturbative Algebraic Quantum Field Theory at Finite Temperature. Ph.D. thesis, University of Hamburg (2013)
37. Neeb, K.-H.: Monastir lecture notes on infinite-dimensional lie groups. http://www.math.uni-hamburg.de/home/wockel/data/monastir.pdf
38. Peierls, R.E.: The commutation laws of relativistic field theory. Proc. Roy. Soc. **A214**, 143–157 (1952). London
39. Radzikowski, M.J.: Micro-local approach to the Hadamard condition in quantum field theory on curved space-time. Commun. Math. Phys. **179**, 529–553 (1996)
40. Scharf, G.: Finite QED: The Causal Approach. Springer, Verlag (1995)
41. Steinmann, O.: Perturbation Expansions in Axiomatic Field Theory. Lecture Notes in Physics 11. Springer (1971)
42. Steinmann, O.: Perturbation theory of wigthman functions. Commun. Math. Phys. **152**, 627 (1993)
43. Steinmann, O.: Perturbative quantum field theory at positive temperatures: an axiomatic approach. Commun. Math. Phys. **170**, 405 (1995)
44. Stückelberg, E.C.G., Rivier, D.: A propos des divergences en théorie des champs quantifiés. Helv. Phys. Acta **23**, 236–239 (1950)
45. Waldmann, S.: Poisson-Geometrie und Deformationsquantisierung: Eine Einführung. Springer, Verlag (2007)
46. Weise, J.C.: On the algebraic formulation of classical general relativity. Diploma thesis, Hamburg (2011). http://www.desy.de/uni-th/theses/Dipl_Weise.pdf

Chapter 3
Models of Free Quantum Field Theories on Curved Backgrounds

Marco Benini and Claudio Dappiaggi

Abstract Free quantum field theories on curved backgrounds are discussed via three explicit examples: the real scalar field, the Dirac field and the Proca field. The first step consists of outlining the main properties of globally hyperbolic spacetimes, that is the class of manifolds on which the classical dynamics of all physically relevant free fields can be written in terms of a Cauchy problem. The set of all smooth solutions of the latter encompasses the dynamically allowed configurations which are used to identify via a suitable pairing a collection of classical observables. As a last step we use such collection to construct a $*$-algebra which encodes the information on the dynamics and the canonical commutation or anti-commutation relations depending on the Bosonic or Fermionic nature of the underlying field.

3.1 Geometric Data

Goal of this section is to introduce all geometric concepts and tools which are necessary to discuss both the classical dynamics and the quantization of a free quantum field on a curved background. We assume that the reader is familiar with the basic notions of differential geometry and, to a minor extent, of general relativity. Therefore we will only sketch a few concepts and formulas, which we will use throughout this chapter; a reader interested to more details should refer to [6, 7, 52], yet paying attention to the conventions used here, which differ from time to time from those in the cited references.

Our starting point is $\mathcal{M}$, a smooth manifold which is endowed with a (smooth) Lorentzian metric g of signature $(+, -, \dots, -)$. Furthermore, although the standard

M. Benini
Department of Mathematics, Heriot-Watt University, Colin Maclaurin Building, Riccarton, Edinburgh EH14 4AS, UK
e-mail: mbenini87@gmail.com

C. Dappiaggi (✉)
Dipartimento di Fisica, Università Degli Studi di Pavia and INFN—Sezione di Pavia, Via Bassi, 6, I-27100 Pavia, Italy
e-mail: claudio.dappiaggi@unipv.it

R. Brunetti et al. (eds.), *Advances in Algebraic Quantum Field Theory*, Mathematical Physics Studies, DOI 10.1007/978-3-319-21353-8_3 generalizations to curved backgrounds of the field theories on Minkowski spacetime, on which the current models of particle physics are based, entail that $\mathcal{M}$ ought to be four dimensional, in this chapter we shall avoid this assumption. The only exception will be Sect. 3.3.2, where we will describe Dirac spinors in four dimensions only, for the sake of simplicity. Especially in order to make contact with the other chapters of this book, we introduce a few auxiliary, notable tensors. We employ an abstract index notation[1] and we stress that our conventions might differ from those of many textbooks, e.g., [52]. As a starting point, we introduce the Riemann tensor $\mathrm{Riem} : T\mathcal{M} \otimes T\mathcal{M} \to \mathrm{End}(T\mathcal{M})$, defined using the abstract index notation by the formula $(\nabla_a \nabla_b - \nabla_b \nabla_a)v^d = R_{abc}{}^d v^c$, where v is an arbitrary vector field and ∇ is the covariant derivative. The Ricci tensor is instead $\mathrm{Ric} : T\mathcal{M}^{\otimes 2} \to \mathbb{R}$ and its components are $R_{ab} = R^d{}_{adb}$, while the scalar curvature is simply $R \doteq g^{ab} R_{ab}$.

For later convenience we impose a few additional technical constraints on the structure of the admissible manifolds, which we recollect in the following definition:

Definition 3.1.1 For $n \geq 2$, we call the pair $(\mathcal{M}, g)$ a *Lorentzian manifold* if $\mathcal{M}$ is an n-dimensional, Hausdorff, second countable, connected, orientable, smooth manifold $\mathcal{M}$ endowed with a Lorentzian metric g.

Notice that henceforth we shall always assume that an orientation $\mathfrak{o}$ for $\mathcal{M}$ has been chosen. We could allow in principle more than one connected component, but, at least in this chapter, it would lead to no further insight and, thus, we avoid it for the sake of simplicity. The Lorentzian character of g plays a distinguished role since it entails that all spacetimes come endowed with a causal structure, which lies at the heart of several structural properties of a free quantum field theory. More precisely, let us start from Minkowski spacetime, $\mathcal{M} \equiv \mathbb{R}^4$, endowed with the standard Cartesian coordinates in which the metric tensor reads $\eta_{\mu\nu} = \mathrm{diag}(1, -1, -1, -1)$. Let $p \in \mathbb{R}^4$ be arbitrary. With respect to it, we can split the set of all other points of $\mathbb{R}^4$ in three separate categories: We say that $q \in \mathbb{R}^4$ is

- *timelike* separated from p if the connecting vector v_{pq} is such that $\eta(v_{pq}, v_{pq}) > 0$.
- *lightlike* separated if $\eta(v_{pq}, v_{pq}) = 0$.
- *spacelike* separated if $\eta(v_{pq}, v_{pq}) < 0$.

If we add to this information the possibility of saying that a point p lies in the future (*resp.* in the past) of q if $x^0(p) > x^0(q)$ (*resp.* $x^0(p) < x^0(q)$), x^0 being the Cartesian time coordinate, we can introduce $I^+_{\mathbb{R}^4}(p)$ and $I^-_{\mathbb{R}^4}(p)$, the chronological future ($+$) and past ($-$) of p, as the collection of all points which are timelike related to p and lie in its future ($+$) or in its past ($-$). Similarly, we define $J^\pm_{\mathbb{R}^4}(p)$ as the causal future ($+$) and past ($-$) of p adding also the points which are lightlike related to p. Notice that, per convention, p itself is included in both $J^+_{\mathbb{R}^4}(p)$ and $J^-_{\mathbb{R}^4}(p)$. In a language more commonly used in theoretical physics, $J^+_{\mathbb{R}^4}(p)$ and $J^-_{\mathbb{R}^4}(p)$ are the future and

[1]Notice that, in this chapter, we employ the following convention for the tensor components: Latin indices, $a, b, c, \ldots$, are used for abstract tensor indices, Greek ones, $\mu, \nu, \ldots$ for coordinates, while i, j, k are used for spatial components or coordinates.

the past light cones stemming from p. By extension, if Ω is an open subregion of $\mathbb{R}^4$ we introduce $J^{\pm}_{\mathbb{R}^4}(\Omega) \doteq \bigcup_{p\in\Omega} J^{\pm}_{\mathbb{R}^4}(p)$. Similarly we define $I^{\pm}_{\mathbb{R}^4}(\Omega)$.

On a generic background, the above structures cannot be transported slavishly. First of all, contrary to $\mathbb{R}^4$, a manifold M does not have to look like a Euclidean space globally. In order to circumvent this obstruction, one can start from a generic point $p \in \mathcal{M}$ and consider the tangent space $T_p\mathcal{M}$. Using the metric g, one can label a tangent vector $v \in T_p\mathcal{M}$ according to the value of $g(v, v)$. Specifically v is timelike if $g(v, v) > 0$, lightlike if $g(v, v) = 0$ and spacelike if $g(v, v) < 0$. Hence, we can associate to the vector space $T_p\mathcal{M}$ a two-folded light cone stemming from $0 \in T_p\mathcal{M}$ and we have the freedom to set one of the folds as the collection of future-directed vectors. If such choice can be made consistently in a smooth way for all points of the manifold, we say that $\mathcal{M}$ is *time orientable*. In a geometric language this is tantamount to requiring the existence of a global vector field on M which is timelike at each point. Henceforth we assume that this is indeed the case and that a time orientation $\mathfrak{t}$ has been fixed. Notice that, as a consequence, every background we consider is completely specified by a quadruple:

Definition 3.1.2 A *spacetime* $\boldsymbol{M}$ is a quadruple $(\mathcal{M}, g, \mathfrak{o}, \mathfrak{t})$, where $(\mathcal{M}, g)$ is a time-orientable n-dimensional Lorentzian manifold ($n \geq 2$), $\mathfrak{o}$ is a choice of orientation on $\mathcal{M}$ and $\mathfrak{t}$ is a choice of time-orientation.

The next step in the definition of a causal structure for a Lorentzian manifold consists of considering a piecewise smooth curve $\gamma : I \to \mathcal{M}$, $I = [0, 1]$. We say that γ is timelike (*resp.* lightlike, spacelike) if such is the vector tangent to the curve at each point. We say that γ is causal if the tangent vector is nowhere spacelike and that it is *future (past) directed* if each tangent vector to the curve is future (or past) directed. Taking into account these structures, we can define on an arbitrary spacetime $\boldsymbol{M}$ the chronological future and past of a point p as $I^{\pm}_{\boldsymbol{M}}(p)$, the collection of all points $q \in \mathcal{M}$ such that there exists a future- (past-)directed timelike curve $\gamma : I \to \mathcal{M}$ for which $\gamma(0) = p$ and $\gamma(1) = q$. In complete analogy we can define the causal future and past $J^{\pm}_{\boldsymbol{M}}(p)$ as well as, for any open subset $\Omega \subset \mathcal{M}$, $J^{\pm}_{\boldsymbol{M}}(\Omega) = \bigcup_{p\in\Omega} J^{\pm}_{\boldsymbol{M}}(p)$. Similarly we define $I^{\pm}_{\boldsymbol{M}}(\Omega)$. We will also denote the union of the causal future $J^{+}_{\boldsymbol{M}}(\Omega)$ and the causal past $J^{-}_{\boldsymbol{M}}(\Omega)$ of Ω with $J_{\boldsymbol{M}}(\Omega)$.

The identification of a causal structure is not only an interesting fingerprint of a spacetime, but it has also far-reaching physical consequences, as it suggests us that not all time-oriented spacetimes should be thought as admissible. As a matter of fact one can incur in pathological situations such as closed timelike and causal curves, which are often pictorially associated to evocative phenomena such as time travel. There are plenty of examples available in the literature ranging from the so-called Gödel Universe—see for example [43]—to the Anti-de Sitter spacetime (AdS), which plays nowadays a prominent role in many applications to high energy physics and string theory. Let us briefly sketch the structure of the latter in arbitrary n-dimensions, $n > 2$—see [32]. AdS$_n$ is a maximally symmetric solution to the Einstein's equations with a negative cosmological constant Λ. In other words it is a manifold of constant curvature $R = \frac{2n}{n-2}\Lambda$ with the topology $\mathbb{S}^1 \times \mathbb{R}^{n-1}$. It can be realized in the $(n+1)$-dimensional spacetime $\mathbb{R}^{n+1}$, endowed with the Cartesian coordinates x^μ, $\mu = 0, \ldots, n$, and with the metric $\widetilde{g} = \mathrm{diag}(1, 1, -1, \ldots, -1)$ as the hyperboloid

$$\tilde{g}(x, x) = R^2$$

If we consider the locus $x^i = 0, i = 2, \ldots, n$ we obtain the circle $(x^0)^2 + (x^1)^2 = R^2$ together with the induced line element $(dx^0)^2 + (dx^1)^2$. In other words we have found a closed curve in AdS_n whose tangent vector is everywhere timelike.

Even without making any contact with field theory, it is clear that scenarios similar to the one depicted are problematic as soon as one is concerned with the notion of causality. Therefore it is often customary to restrict the attention to a class of spacetimes which avoids such quandary, while being at the same time sufficiently large to include almost all interesting cases. These are the so-called *globally hyperbolic spacetimes*. We characterize them following [52, Chap. 8]. As a starting point, we consider a spacetime $\boldsymbol{M}$ and we introduce two additional notions:

1. A subset $\Sigma \subset \mathcal{M}$ is called *achronal* if each timelike curve in $\boldsymbol{M}$ intersects Σ at most once;
2. For any subset $\Sigma \subset \boldsymbol{M}$, we call future (+), respectively past (−) *domain of dependence* $D^{\pm}_{\boldsymbol{M}}(\Sigma)$, the collection of all points $q \in \mathcal{M}$ such that every past (+), respectively future (−) inextensible causal curve passing through q intersects Σ. With $D_{\boldsymbol{M}}(\Sigma) \doteq D^{+}_{\boldsymbol{M}}(\Sigma) \cup D^{-}_{\boldsymbol{M}}(\Sigma)$ we indicate simply the domain of dependence.

We state the following:

Definition 3.1.3 We say that $\boldsymbol{M}$ is *globally hyperbolic* if and only if there exists a Cauchy surface Σ, that is a closed achronal subset of $\boldsymbol{M}$ such that $D_{\boldsymbol{M}}(\Sigma) = \mathcal{M}$.

Notice that, as a by-product of this definition, one can conclude, not only that no closed causal curve exists in $\boldsymbol{M}$, but also that $\mathcal{M}$ is homeomorphic to $\mathbb{R} \times \Sigma$, while Σ is a C^0, $(n-1)$-dimensional submanifold of $\boldsymbol{M}$, *cf.* [7, Theorem 3.17] for the case $n = 4$. It is worth mentioning two apparently unrelated points: (1) in the past, it has been often assumed that the Cauchy surface could be taken as smooth and (2) Definition 3.1.3 does not provide any concrete mean to verify in explicit examples whether a spacetime $\boldsymbol{M}$ is globally hyperbolic or not. Alternative characterizations of global hyperbolicity, such as that M is strongly causal and $J^{+}_{\boldsymbol{M}}(p) \cap J^{-}_{\boldsymbol{M}}(q)$ is either empty or compact for all $p, q \in \mathcal{M}$, did not help in this respect. A key step forward was made a decade ago by the work of Bernal and Sanchez, see [12, 13]. Their main result is here stated following the formulation of [6, Sect. 1.3]:

Theorem 3.1.4 *Let $\boldsymbol{M}$ be given. The following statements are equivalent:*

1. *$\boldsymbol{M}$ is globally hyperbolic.*
2. *There exists no closed causal curve in $\boldsymbol{M}$ and $J^{+}_{\boldsymbol{M}}(p) \cap J^{-}_{\boldsymbol{M}}(q)$ is either compact or empty for all $p, q \in \mathcal{M}$.*
3. *$(\mathcal{M}, g)$ is isometric to $\mathbb{R} \times \Sigma$ endowed with the line element $ds^2 = \beta dt^2 - h_t$, where $t : \mathbb{R} \times \Sigma \to \mathbb{R}$ is the projection on the first factor, β is a smooth and*

strictly positive function on $\mathbb{R} \times \Sigma$ and $t \mapsto h_t$, $t \in \mathbb{R}$, yields a smooth one-parameter family of Riemannian metrics. Furthermore, for all $t \in \mathbb{R}$, $\{t\} \times \Sigma$ is an $(n-1)$-dimensional, spacelike, smooth Cauchy surface in $\boldsymbol{M}$.

The main advantage of this last theorem is to provide an easier criterion to verify explicitly whether a given time-oriented spacetime is globally hyperbolic. In order to convince the reader that this class of manifolds includes most of the physically interesting examples, we list a collection of the globally hyperbolic spacetimes which are often used in the framework of quantum field theory on curved backgrounds:

- We say that a spacetime $\boldsymbol{M}$ is *ultrastatic* if $(\mathcal{M}, g)$ is isometric to $\mathbb{R} \times \Sigma$ with line element $ds^2 = dt^2 - \pi^* h$, where $\pi^* h$ is the pullback along the projection $\pi : \mathbb{R} \times \Sigma \to \Sigma$ of a metric h on Σ. $\boldsymbol{M}$ is globally hyperbolic if and only if it is geodesically complete, that is every maximal geodesic is defined on the whole real line—see [27]. Minkowski spacetime falls in this category.
- All Friedmann-Robertson-Walker (FRW) spacetimes are four-dimensional homogeneous and isotropic manifolds diffeomorphic to $\mathbb{R} \times \Sigma$ with

$$ds^2 = dt^2 - a^2(t)\left(\frac{dr^2}{1-kr^2} + r^2 d\mathbb{S}^2(\theta, \varphi)\right),$$

  where $d\mathbb{S}^2(\theta, \varphi)$ is the standard line element of the unit 2-sphere, while $a(t)$ is a smooth and strictly positive function depending only on time. Furthermore k is a constant which, up to a normalization, can be set to 0, 1, -1 and, depending on this choice, Σ is a three-dimensional spacelike manifold whose model space is either $\mathbb{R}^3$, the 3-sphere $\mathbb{S}^3$ or the three dimensional hyperboloid $\mathbb{H}^3$. The remaining coordinate r has a domain of definition which runs over the whole positive real line if $k = 0, -1$, while $r \in (0, 1)$ if $k = 1$. On account of [7, Theorem 3.68], we can conclude that every Friedmann-Robertson-Walker spacetime is globally hyperbolic.[2] Notice that, in many concrete physical applications, the coordinate t runs only on an open interval of $\mathbb{R}$, but, as proven in [7, Theorem 3.69], it does not spoil the property of being globally hyperbolic. Following a similar argument one can draw similar conclusions when working with time oriented, homogeneous spacetimes, which are also referred to as Bianchi spacetimes.
- A noteworthy collection of solutions of the vacuum Einstein's equations consists of the Kerr family which describes a rotating, uncharged, axially symmetric, four-dimensional black hole [50]. In the so-called Boyer-Linquist chart, the underlying line element reads as

$$\begin{aligned} ds^2 =& \frac{\Delta - a^2 \sin^2\theta}{\Pi} dt^2 + \frac{4Mar \sin^2\theta}{\Pi} dt d\varphi - \frac{\Pi}{\Delta} dr^2 \\ & - \Pi d\theta^2 - \frac{(r^2+a^2)^2 - \Delta a^2 \sin^2\theta}{\Pi} d\varphi^2, \end{aligned}$$

[2] We are grateful to Zhirayr Avetisyan for pointing us out Theorem 3.68 in [7].

where $\Delta = r^2 - 2Mr + a^2$ and $\Pi = r^2 + a^2\cos^2\theta$, while M and $J = Ma$ are two real parameters which are interpreted respectively as the mass and total angular momentum of the black hole. Notice that t runs along the whole real line, θ, φ are the standard coordinates over the unit 2-sphere, while r plays the role of a radial-like coordinate. A generic Kerr spacetime possesses coordinate horizons at $r_\pm = M \pm \sqrt{M^2 - a^2}$ provided that $M^2 \geq a^2$ and the region for which $r \in (r_+, \infty)$, often also known as exterior region to the black hole, is actually a globally hyperbolic spacetime. If we set $a = 0$, that is the black hole does not rotate, we recover the spherically symmetric Schwarzschild spacetime and, consistently, the static region outside the event horizon located at $r = 2M$ is itself globally hyperbolic.

- Another spacetime, which is often used as a working example in quantum field theory on curved spacetime is *de Sitter* (dS_n), the maximally symmetric solution to the Einstein's equations with a positive cosmological constant Λ and $n > 2$. In $\mathbb{R}^{n+1}$ endowed both with the standard Cartesian coordinates x^μ, $\mu = 0, \dots, n$, and with the metric $\tilde{g} = \mathrm{diag}(1, -1, \dots, -1)$, dS_n can be realized as the hyperboloid $\tilde{g}(x, x) = -R^2$, where $R^2 = \frac{(n-1)(n-2)}{2\Lambda}$. After the change of coordinates $x^0 = R\sinh(t/R)$ and $x^i = R\cosh(t/R)e^i$, $i = 1, \dots, n$, where $\sum_{i=1}^n (e^i)^2 = 1$, the line element of dS_n reads

$$ds^2 = dt^2 - R^2\cosh^2(t/R)\, d\mathbb{S}^2(e^1, \dots, e^n),$$

where $d\mathbb{S}^2(e^1, \dots, e^n)$ represents the standard line element of the unit $(n-1)$-sphere and t runs along the whole real line. Per direct inspection we see that dS_n is diffeomorphic to an n-dimensional version of a Friedmann-Robertson-Walker spacetime with compact spatial sections and it is globally hyperbolic.

Since in the next section we will be interested in functions from a globally hyperbolic spacetime to a suitable target vector space and in their support properties, we conclude the section with a useful definition:

Definition 3.1.5 Let $\boldsymbol{M}$ be a globally hyperbolic spacetime and V a finite dimensional vector space. We call

(0) $C_0^\infty(\mathcal{M}; V)$ the space of smooth and *compactly supported* V-valued functions on $\mathcal{M}$,

(sc) $C_{sc}^\infty(\mathcal{M}; V)$ the space of smooth and *spacelike compact* V-valued functions on $\mathcal{M}$, that is $f \in C_{sc}^\infty(\mathcal{M}; V)$ if there exists a compact subset $K \subset \mathcal{M}$ such that $\mathrm{supp}\, f \subset J_{\boldsymbol{M}}(K)$,

(fc/pc) $C_{fc}^\infty(\mathcal{M}; V)$ the space of smooth and *future compact* V-valued functions on $\mathcal{M}$, that is $f \in C_{fc}^\infty(\mathcal{M}; V)$ if $\mathrm{supp}\, f \cap J_{\boldsymbol{M}}^+(p)$ is compact for all $p \in \mathcal{M}$. Mutatis mutandis, we shall also consider $C_{pc}^\infty(\mathcal{M}; V)$, namely the space of smooth and *past compact* V-valued functions on $\mathcal{M}$,

(tc) $C_{tc}^\infty(\mathcal{M}; V) \doteq C_{fc}^\infty(\mathcal{M}; V) \cap C_{pc}^\infty(\mathcal{M}; V)$ the space of smooth and *timelike compact* V-valued functions on $\mathcal{M}$.

3.2 On Green Hyperbolic Operators

Globally hyperbolic spacetimes play a pivotal role, not only because they do not allow for pathological situations, such as closed causal curves, but also because they are the natural playground for classical and quantum fields on curved backgrounds. More precisely, the dynamics of most (if not all) models, we are interested in, is either ruled by or closely related to wave-like equations. Also motivated by physics, we want to construct the associated space of solutions by solving an initial value problem. To this end we need to be able to select both an hypersurface on which to assign initial data and to identify an evolution direction. In view of Theorem 3.1.4, globally hyperbolic spacetimes appear to be indeed a natural choice. Goal of this section will be to summarize the main definitions and the key properties of the class of partial differential equations, useful to discuss the models that we shall introduce in the next sections. Since this is an overkilled topic, we do not wish to make any claim of being complete and we recommend to an interested reader to consult more specialized books and papers for more details. We suggest for example [33–36], the more recent [54] and also [6], on which most of this section is based; moreover, notice that, several examples of Green hyperbolic operators can be found in [4, 5], while a remarkable extension of their domain of definition, which we shall implicitly assume, is available in [2].

As a starting point we introduce the building block of any classical and quantum field theory:

Definition 3.2.1 A *vector bundle* of rank $k < \infty$ over an n-dimensional smooth manifold $\mathcal{M}$ (base space) is specified by $F \equiv F(\mathcal{M}, \pi, V)$, where F, (the total space), is a smooth manifold of dimension $n + k$, V (the typical fiber) is a k-dimensional vector space and $\pi : F \to \mathcal{M}$ is a smooth surjective map. Furthermore we require that:

1. There exists a vector space isomorphism between V and each fiber $F_p \doteq \pi^{-1}(p)$, $p \in \mathcal{M}$,
2. For each $p \in \mathcal{M}$ there exists an open neighbourhood $U \ni p$ and a diffeomorphism $\Psi : \pi^{-1}(U) \to U \times V$ such that $\pi_1 \circ \Psi = \pi$ on $\pi^{-1}(U)$, where $\pi_1 : U \times V \to U$ is the projection on the first factor of the Cartesian product;
3. The restriction of Ψ to each fiber is an isomorphism of vector spaces.

The pair (U, Ψ) fulfilling these conditions is called a *local trivialization* of E.

Notice that throughout the text we shall use the word *vector bundle atlas*, meaning a collection of local trivializations of F covering $\mathcal{M}$. We will not discuss the theory of vector bundles and, for more details, refer to [37]. The only exceptions are the following two definitions:

Definition 3.2.2 Let $F = F(\mathcal{M}, \pi, V)$ be a vector bundle and let N be a submanifold of $\mathcal{M}$. We call *restriction* of F to N the vector bundle $F|_N \equiv \widetilde{F} = \widetilde{F}(N, \pi', V)$, where $\widetilde{F} = \pi^{-1}(N)$ and $\pi' : \widetilde{F} \to N$ is defined by $\pi'(f) = \pi(f)$ for all $f \in \widetilde{F}$.

Definition 3.2.3 Let $F = F(\mathcal{M}, \pi, V)$ be a vector bundle. We call *dual bundle* F^* the vector bundle over $\mathcal{M}$ whose fiber over $p \in \mathcal{M}$ is $(F^*)_p = (F_p)^*$, the dual vector space to F_p.

We say that a vector bundle F is *(globally) trivial* if there exists a fiber preserving diffeomorphism from F to the Cartesian product $\mathcal{M} \times V$ restricting to a vector space isomorphism on each fiber. In practice, this corresponds to a trivialization of F which is defined everywhere, to be compared with the notion of a local trivialization as per Definition 3.2.1. Most of the examples we shall consider in this chapter come from globally trivial vector bundles. Bear in mind, however, that one of the canonical examples of a vector bundle, namely the tangent bundle $T\mathcal{M}$ to a manifold $\mathcal{M}$, is not trivial in general, e.g., when $\mathcal{M} = \mathbb{S}^2$. It is also noteworthy that, given any two vector bundles $F = F(\mathcal{M}, \pi, V)$ and $F' = F'(\mathcal{M}, \pi', V')$, we can construct naturally a third vector bundle, the *bundle of homomorphisms* $\mathrm{Hom}(F, F')$ over the base space $\mathcal{M}$. Its fiber over a base point $p \in \mathcal{M}$ is $\mathrm{Hom}(F_p, F'_p)$, which is a vector space isomorphic to the vector space $\mathrm{Hom}(V, V')$ of homomorphism from V to V'. If $F' = F$, then we shall write $\mathrm{End}(F)$ for $\mathrm{Hom}(F, F')$ and call it *bundle of endomorphisms*, whose typical fiber is $\mathrm{End}(V)$.

Another structure which plays a distinguished role is the following:

Definition 3.2.4 Given a vector bundle F, we call *space of smooth sections* $\Gamma(F) = \{\sigma \in C^\infty(\mathcal{M}; F) \mid \pi \circ \sigma = \mathrm{id}_\mathcal{M}\}$, where $\mathrm{id}_\mathcal{M} : \mathcal{M} \to \mathcal{M}$ is the identity map. By generalizing Definition 3.1.5, the subscripts 0, sc, fc/pc and tc shall refer to those sections whose support is respectively compact, spacelike compact, future or past compact and timelike compact.

Notice that $\Gamma(F)$ is an infinite-dimensional vector space and, whenever F is trivial, it is isomorphic to $C^\infty(\mathcal{M}; V)$. The next structure, which we introduce, will play an important role in the construction of an algebra of observables for a free quantum field on a curved background:

Definition 3.2.5 Let F be a vector bundle on a manifold $\mathcal{M}$. Denote with $F \times_\mathcal{M} F$ the fibered product of bundles obtained taking the Cartesian product fiberwise. A non-degenerate inner product on F is a smooth map $\cdot : F \times_\mathcal{M} F \to \mathbb{R}$ such that

1. the restriction of $\cdot$ to $F_p \times F_p$ is a bilinear form for each $p \in \mathcal{M}$,
2. $v \in F_p$ vanishes if $v \cdot w = 0$ for all $w \in F_p$.

Furthermore, if $\cdot$ is symmetric on each fiber, we call it *Bosonic*, if it is antisymmetric, we call it *Fermionic*.

Since we consider only spacetimes $\boldsymbol{M} = (\mathcal{M}, g, \mathfrak{o}, \mathfrak{t})$, the orientation of $\mathcal{M}$ is fixed by $\mathfrak{o}$ and therefore we can introduce the metric-induced volume form $\mathrm{dvol}_{\boldsymbol{M}}$ on $\mathcal{M}$. Then, any inner product as in Definition 3.2.5 induces a non-degenerate pairing between smooth sections and compactly supported smooth sections of F:

$$(\cdot,\cdot) : \Gamma_0(F) \times \Gamma(F) \to \mathbb{R}, \qquad (\sigma, \tau) \mapsto \int_\mathcal{M} (\sigma \cdot \tau)\, \mathrm{dvol}_{\boldsymbol{M}}. \tag{3.1}$$

Since we will make use of it later in this chapter, notice that, (3.1) is still meaningful if we consider $\tau, \sigma \in \Gamma(F)$ with $\text{supp}(\tau) \cap \text{supp}(\sigma)$ compact.

We have all ingredients to start addressing the main point of this section, namely partial differential equations. The building block is the following:

Definition 3.2.6 Let $F = F(\mathcal{M}, \pi, V)$ and $F' = F'(\mathcal{M}, \pi', V')$ be two vector bundles of rank k and k' respectively over the same manifold $\mathcal{M}$. A *linear partial differential operator* of order at most $s \in \mathbb{N}_0$ is a linear map $L : \Gamma(F) \to \Gamma(F')$ such that, for all $p \in \mathcal{M}$, there exist both a coordinate neighborhood (U, Φ) centered at p, local trivializations (U, Ψ) and (U, Ψ') respectively of F and of F', as well as a collection of smooth maps $A_\alpha : U \to \text{Hom}(V; V')$ labeled by multi-indices for which, given any $\sigma \in \Gamma(F)$, on U one has

$$L\sigma = \sum_{|\alpha| \leq s} A_\alpha \, \partial^\alpha \sigma.$$

Notice that here we are implicitly using both the coordinate chart Φ and the trivializations Ψ and Ψ'; moreover, the sum runs over all multi-indices $\alpha = (\alpha_0, \ldots, \alpha_{n-1}) \in \mathbb{N}_0^n$ such that $|\alpha| \doteq \sum_{\mu=0}^{n-1} \alpha_\mu \leq s$ and $\partial^\alpha = \prod_{\mu=0}^{n-1} \partial_\mu^{\alpha_\mu}$, where $\partial_0, \ldots, \partial_{n-1}$ are the partial derivatives with respect to the coordinates $x^0, \ldots, x^{n-1}$ coming from the chart (U, Φ). Furthermore, L is of order $s \in \mathbb{N}_0$ when it is of order at most s, but not of order at most $s - 1$.

Notice that linear partial differential operators cannot enlarge the support of a section, a property which will be often used in the rest of this chapter. Another related and useful concept intertwines linear partial differential operators with the pairing (3.1):

Definition 3.2.7 Consider a spacetime $\boldsymbol{M} = (\mathcal{M}, g, \mathfrak{o}, \mathfrak{t})$. Let $F = E(\mathcal{M}, \pi, V)$ and let $F' = F'(\mathcal{M}, \pi', V')$ be two vector bundles over the manifold $\mathcal{M}$, both endowed with a non-degenerate inner product. Denote the pairings defined in (3.1) for F and for F' respectively with $(\cdot, \cdot)_F$ and $(\cdot, \cdot)_{F'}$. Let $L : \Gamma(F) \to \Gamma(F')$ be a linear partial differential operator. We call *formal adjoint* of L the linear partial differential operator $L^* : \Gamma(F') \to \Gamma(F)$ such that, for all $\sigma \in \Gamma(F)$ and $\tau \in \Gamma(F')$ with supports having compact overlap, the following identity holds:

$$(L^*\tau, \sigma)_{F'} = (\tau, L\sigma)_F.$$

If $F = F'$, we say that L is *formally self-adjoint* whenever L^* coincides with L.

Existence of L^* is a consequence of Stokes theorem and uniqueness is instead due to the non-degeneracy of the pairing $(\cdot, \cdot)_{F'}$.

Definition 3.2.6 accounts for a large class of operators, most of which are not typically used in the framework of field theory, especially because they cannot be associated to an initial value problem. In order to select a relevant class for our purposes, we introduce a useful concept—see also [6, 33]:

Definition 3.2.8 Let $F = F(\mathcal{M}, \pi, V)$ and $F' = F'(\mathcal{M}, \pi', V')$ be two vector bundles over the same manifold $\mathcal{M}$ and let $L : \Gamma(F) \to \Gamma(F')$ be any linear partial differential operator of order s as per Definition 3.2.6. We call *principal symbol* of L the map $\sigma_L : T^*M \to \mathrm{Hom}(F, F')$ locally defined as follows: For $p \in \mathcal{M}$, mimicking Definition 3.2.6, consider a coordinate chart around $p \in \mathcal{M}$ and local trivializations of F and of F' and, for all $\zeta \in T_p^*\mathcal{M}$, set

$$\sigma_L(\zeta) = \sum_{|\alpha|=s} A_\alpha(p)\zeta^\alpha,$$

where $\zeta^\alpha = \prod_{\mu=0}^{n-1} \zeta_\mu^{\alpha_\mu}$ and ζ_μ are the components of ζ with respect to the chosen chart. Furthermore, given a Lorentzian manifold $(\mathcal{M}, g)$, we call a second order linear partial differential operator $P : \Gamma(F) \to \Gamma(F)$ *normally hyperbolic* if $\sigma_P(\zeta) = g(\zeta, \zeta)\,\mathrm{id}_{F_p}$ for all $p \in \mathcal{M}$ and all $\zeta \in T_p^*\boldsymbol{M}$.

In order to better grasp the structure of a normally hyperbolic operator P, we can write it in a local coordinate frame following both Definitions 3.2.6 and 3.2.8. Let thus p be in $\mathcal{M}$ and (U, Φ) be a chart centered at p where the vector bundle F is trivial. There exist both A and A^μ, $\mu = 0, \ldots, n-1$, smooth maps from U to $\mathrm{End}(V)$ such that, for any $\sigma \in \Gamma(F)$, on U one has

$$P\sigma = g^{\mu\nu}\mathrm{id}_V\, \partial_\mu \partial_\nu \sigma + A^\mu\, \partial_\mu \sigma + A\, \sigma,$$

where both the chart and the vector bundle trivializations are understood. One immediately notices that locally this expression agrees up to terms of lower order in the derivatives with the one for the d'Alembert operator acting on sections of F constructed out of a covariant derivative ∇ on F, that is the operator $\Box_\nabla = g^{\mu\nu}\nabla_\mu\nabla_\nu : \Gamma(F) \to \Gamma(F)$. Therefore, one realizes that normally hyperbolic operators provide the natural generalization of the usual d'Alembert operator. Besides this remark, Definition 3.2.8 becomes even more important if we assume, moreover, that the underlying background is globally hyperbolic. In fact, in this case we can associate to each normally hyperbolic operator P an initial value problem (also known as Cauchy problem). As a matter of fact, in view both of Definition 3.1.3 and of Theorem 3.1.4, initial data can be assigned on each Cauchy surface and the following proposition shows the well-posedness of the construction:

Proposition 3.2.9 *Let $\boldsymbol{M} = (\mathcal{M}, g, \mathfrak{o}, \mathfrak{t})$ be a globally hyperbolic spacetime and $\Sigma \subset \mathcal{M}$ any of its spacelike, smooth Cauchy surfaces together with its future-pointing unit normal vector field $\boldsymbol{n}$. Consider any vector bundle F over $\mathcal{M}$, a normally hyperbolic operator $P : \Gamma(F) \to \Gamma(F)$, and a P-compatible*[3] *covariant derivative ∇ on F. Let $F|_\Sigma$ be the restriction of F to Σ as per Definition 3.2.2. Then, for any $J \in \Gamma(F)$ and for any $u_0, u_1 \in \Gamma(F|_\Sigma)$, the following initial value problem admits a unique solution $u \in \Gamma(F)$:*

[3] A covariant derivative ∇ on F is P-compatible if there exists a section $A \in \Gamma(\mathrm{End}(F))$ such that $\Box_\nabla + A = P$.

$$\begin{aligned} Pu &= J && \text{on } \mathcal{M}, \\ u &= u_0 && \text{on } \Sigma, \\ \nabla_n u &= u_1 && \text{on } \Sigma. \end{aligned} \tag{3.2}$$

Furthermore, if we set $\Omega = \operatorname{supp} u_0 \cup \operatorname{supp} u_1 \cup \operatorname{supp} J$, *then* $\operatorname{supp} u \subset J_{\boldsymbol{M}}(\Omega)$.

The proof of this proposition has been given in different forms in several books, e.g. [6, 26] and [3, Sect. 3.5.3]. Notice that the Cauchy problem (3.2) is not linear since we allow for a non-vanishing source term. For pedagogical reasons, we shall henceforth consider only the case $J = 0$ although the reader should keep in mind that such constraint is not really needed and a treatment especially of quantization in this scenario has been given in [11] and further refined in [24].

The characterization of all smooth solutions of the equation $Pu = 0$ represents the first step in outlining a quantization scheme for a free field theory. To this end one does not consider directly the Cauchy problem (3.2), but rather exploits a notable property of normally hyperbolic operators, namely the fact that on any globally hyperbolic spacetime they come together with Green operators. Here we introduce them paying particular attention to the domain where they are defined. The reader should keep in mind that our presentation slightly differs in comparison for example to [6] and we make use of results which are presented in [2, 26, 48]:

Definition 3.2.10 Let $\boldsymbol{M} = (\mathcal{M}, g, \mathfrak{o}, \mathfrak{t})$ be a globally hyperbolic spacetime and consider a vector bundle F over $\mathcal{M}$. Furthermore, let $L : \Gamma(F) \to \Gamma(F)$ be a linear partial differential operator. We call *retarded* $(+)$ and *advanced* $(-)$ *Green operators* two linear maps

$$E^+ : \Gamma_{pc}(F) \to \Gamma(F), \qquad E^- : \Gamma_{fc}(F) \to \Gamma(F), \tag{3.3}$$

fulfilling the properties listed below:

1. For any $f \in \Gamma_{pc}(F)$, it holds $LE^+ f = f = E^+ L f$ and $\operatorname{supp}(E^+ f) \subset J^+_{\boldsymbol{M}}(\operatorname{supp} f)$,
2. For any $f \in \Gamma_{fc}(F)$, it holds $LE^- f = f = E^- L f$ and $\operatorname{supp}(E^- f) \subset J^-_{\boldsymbol{M}}(\operatorname{supp} f)$.

The operator $E \doteq E^- - E^+ : \Gamma_{tc}(F) \to \Gamma(F)$ will be referred to as *advanced-minus-retarded operator*. A linear partial differential operator admitting both E^+ and E^- will be called *Green hyperbolic*.

Notice that in the literature the symbols $E^\pm$ are often written as $G^\pm$, while the operator E is also called *causal propagator*. We avoid this nomenclature since, from time to time, it is also used for completely different objects and we wish to avoid a potential source of confusion for the reader. In view of the application of this material to some specific field theoretical models, see Sect. 3.3, we introduce now the canonical integral pairing $\langle \lambda \cdot, \cdot \rangle$ between the sections of a vector bundle F and those of its dual F^*. This is defined by integrating over the base manifold the fiberwise pairing between F^* and F:

$$\langle f', f\rangle \doteq \int_{\mathcal{M}} f'(f)\,\mathrm{dvol}_{\boldsymbol{M}}, \tag{3.4}$$

where $f \in \Gamma(F)$ and $f' \in \Gamma(F^*)$ have supports with compact overlap. Notice that the formula above provides non-degenerate bilinear pairings between $\Gamma_0(F^*)$ and $\Gamma(F)$ and between $\Gamma(F^*)$ and $\Gamma_0(F)$. In turn, this pairing allows for the following definition:

Definition 3.2.11 Let $\boldsymbol{M} = (\mathcal{M}, g, \mathfrak{o}, \mathfrak{t})$ be a globally hyperbolic spacetime and consider a vector bundle F over $\mathcal{M}$ and its dual F^*. Furthermore, let $L : \Gamma(F) \to \Gamma(F)$ be a partial differential operator. We call *formal dual* the linear partial differential operator $L^\star : \Gamma(F^*) \to \Gamma(F^*)$ defined through (3.4) via

$$\langle L^\star f', f\rangle = \langle f', Lf\rangle, \tag{3.5}$$

where $f \in \Gamma(F)$ and $f' \in \Gamma(F^*)$ have supports with compact overlap.

Notice that, if F is endowed with a non-degenerate inner product as per Definition 3.2.5, then we can identify F with F^*. Via this identification (3.4) becomes (3.1).

Proposition 3.2.12 *Let $\boldsymbol{M} = (\mathcal{M}, g, \mathfrak{o}, \mathfrak{t})$ be a globally hyperbolic spacetime and consider a vector bundle F over $\mathcal{M}$ and its dual F^*. Furthermore, let both $L : \Gamma(F) \to \Gamma(F)$ and its formal dual $L^\star : \Gamma(F^*) \to \Gamma(F^*)$ be Green hyperbolic operators, whose retarded and advanced Green operators are $E^\pm$ and $E^{\star\pm}$ respectively. Then, for all $f \in \Gamma_0(F)$ and $f' \in \Gamma_0(F^*)$, it holds that*

$$\langle E^{\star\mp} f', f\rangle = \langle f', E^\pm f\rangle,$$

where $\langle \cdot, \cdot\rangle$ is the pairing defined in (3.4).

Proof The statement is a consequence of the following chain of equalities, which holds true for arbitrary $f' \in \Gamma_0(F^\star)$ and $f \in \Gamma_0(F)$:

$$\langle E^{\star\mp} f', f\rangle = \langle E^{\star\mp} f', LE^\pm f\rangle = \langle L^\star E^{\star\mp} f', E^\pm f\rangle = \langle f', E^\pm f\rangle.$$

From the definition, L admits left-inverses E^+ and E^- on sections with past (respectively future) compact support. In other words L is injective thereon. As a consequence, E^+ and E^- are uniquely specified by their support properties and by the condition of being also right-inverses of L on $\Gamma_{pc}(F)$ and respectively on $\Gamma_{fc}(F)$.

Lemma 3.2.13 *Let $\boldsymbol{M} = (\mathcal{M}, g, \mathfrak{o}, \mathfrak{t})$ be a globally hyperbolic spacetime. Consider a vector bundle F over $\mathcal{M}$ endowed with a non-degenerate inner product as per Definition 3.2.5. Furthermore, let both $L : \Gamma(F) \to \Gamma(F)$ and its formal adjoint $L^* : \Gamma(F) \to \Gamma(F)$ be Green hyperbolic operators. Then, calling $E^\pm$ and $E^{*\pm}$ their respective retarded and advanced Green operators, the identity*

$$(E^{*\mp} f', f) = (f', E^\pm f)$$

holds for all $f' \in \Gamma_0(F)$ and $f \in \Gamma_0(F)$.

Proof Since F is endowed with a non-degenerate inner product, (3.4) reduces to (3.1) upon identification of F^* with F. Furthermore the formal dual of L coincides with its formal adjoint under this identification, i.e. $L^\star = L^*$. Hence we are falling in the hypotheses of Proposition 3.2.12, from which the sought result follows.

Notice that all normally hyperbolic operators are Green hyperbolic. This result follows from [2, 6]. In the latter reference the retarded and advanced Green operators for a normally hyperbolic operator are shown to exist, although with a smaller domain compared to the one we consider here, while in the first one the domains are uniquely extended, thus fulfilling the requirement of our Definition 3.2.10. Let us stress that there are physically interesting partial differential operators which are Green hyperbolic, but not normally hyperbolic. The most notable example is the Dirac operator which will be discussed in Sect. 3.3.2. The reader should also keep in mind that some authors are calling Green hyperbolic an operator which fulfills the hypotheses of Proposition 3.2.12 or of Lemma 3.2.13—see for example [2].

The usefulness of both retarded and advanced Green operators becomes manifest as soon as one notices that, to every timelike compact section f of a vector bundle F we can associate a solution of the linear equation $Lu = 0$ as $u = Ef = E^- f - E^+ f$. Yet, before concluding that we have given a characterization of all solutions, we need a few additional data:

Lemma 3.2.14 *Let $\boldsymbol{M} = (\mathcal{M}, g, \mathfrak{o}, \mathfrak{t})$ be a globally hyperbolic spacetime. Consider a vector bundle F over $\mathcal{M}$ and a Green hyperbolic operator $L : \Gamma(F) \to \Gamma(F)$. Let $E^\pm$ be the retarded and advanced Green operators for L and denote with E the corresponding advanced-minus-retarded operator. Then $f \in \Gamma_{tc}(F)$ is such that $Ef = 0$ if and only if $f = Lh$ for some $h \in \Gamma_{tc}(F)$. Furthermore, $f \in \Gamma_{tc}(F)$ is such that $Lf = 0$ if and only if $f = 0$ and, moreover, for any $f \in \Gamma(F)$ there exists $h \in \Gamma(F)$ such that $Lh = f$.*

Proof On account of Definition 3.2.10, it holds that $ELh = 0$ for all $h \in \Gamma_{tc}(F)$, thus we need only to show that, given $f \in \Gamma_{tc}(F)$ such that $Ef = 0$, then there exists $h \in \Gamma_{tc}(F)$ such that $f = Lh$. Taking any such f, $Ef = 0$ implies that $E^- f = E^+ f$. The support properties of the retarded and advanced Green operators entail that $\mathrm{supp}\,(E^- f) \subset J^+(\mathrm{supp}\, f) \cap J^-(\mathrm{supp}\, f)$. In other words $h = E^- f \in \Gamma_{tc}(F)$. If we apply the operator L, it holds $Lh = LE^- f = f$.

Suppose now that there exists $f \in \Gamma_{tc}(F)$ such that $Lf = 0$. By applying either the retarded or the advanced Green operators we obtain, $f = E^\pm Lf = 0$.

To conclude the proof, consider $f \in \Gamma(F)$. Taking a partition of unity $\{\chi_+, \chi_-\}$ on $\boldsymbol{M}$ such that $\chi_\pm = 1$ on a past/future compact region,[4] one can introduce $h = E^+(\chi_+ f) + E^-(\chi_- f) \in \Gamma(F)$. Since $\chi_+ + \chi_- = 1$ everywhere, $Lh = f$ as claimed.

---

[4]A partition of unity such as the one described exists on account of Theorem 3.1.4. In fact, after splitting the globally hyperbolic spacetime $\boldsymbol{M}$ in the Cartesian product of $\mathbb{R}$ and a spacelike Cauchy surface Σ and for any choice of $t_\pm \in \mathbb{R}$ with $t_- < t_+$, one can introduce a partition of unity $\{\chi_+, \chi_-\}$ on $\mathbb{R}$ such that $\chi_\pm(t) = 1$ for $\pm t \geq \pm t_\pm$. Pulling this partition of unity back to $\mathcal{M}$ along the projection on the time factor $t : \mathcal{M} \to \mathbb{R}$, one obtains a partition of unity on $\mathcal{M}$ of the sought type.

In view of this last result, we can finally characterize the space of solutions of $Lu = 0$ via the advanced-minus-retarded operator—see also [53]:

Theorem 3.2.15 *Let $M = (\mathcal{M}, g, \mathfrak{o}, \mathfrak{t})$ be a globally hyperbolic spacetime. Consider a vector bundle F over $\mathcal{M}$ and let $L : \Gamma(F) \to \Gamma(F)$ be a Green hyperbolic operator. Let $E^{\pm}$ be the retarded and advanced Green operators for L and denote with E the corresponding advanced-minus-retarded operator. The map presented below is a vector space isomorphism between* Sol, *the vector space of smooth solutions of the linear partial differential equation $Lu = 0$, $u \in \Gamma(F)$, and the quotient of $\Gamma_{tc}(F)$ by the image of L acting on $\Gamma_{tc}(F)$:*

$$\frac{\Gamma_{tc}(F)}{L(\Gamma_{tc}(F))} \to \mathsf{Sol}, \qquad [f] \mapsto Ef, \tag{3.6}$$

where $f \in \Gamma_{tc}(F)$ is any representative of the equivalence class $[f]$ in the quotient.

Proof Let us notice that the advanced-minus-retarded operator $E : \Gamma_{tc}(F) \to \Gamma(F)$ induces the sought map from $\Gamma_{tc}(F)/L(\Gamma_{tc}(F))$ to Sol. On account of Lemma 3.2.14 the image does not depend on the representative of $[f]$ and $u = Ef$ is a solution of $Lu = 0$. This map is injective since, given $f, f' \in \Gamma_{tc}(E)$ such that $Ef = Ef'$, per linearity of E and applying Lemma 3.2.14, one finds $h \in \Gamma_{tc}(E)$ such that $Lh = f - f'$. In other words f and f' are two representatives of the same equivalence class in $\Gamma_{tc}(F)/L(\Gamma_{tc}(F))$, which entails injectivity. Only surjectivity is still to be proven. Given $u \in \mathsf{Sol}$ and taking into account a partition of unity $\{\chi_+, \chi_-\}$ on $\mathcal{M}$ such that $\chi_\pm = 1$ in a past/future compact region, one finds $L(\chi_+ u + \chi_- u) = Lu = 0$, therefore $h = L(\chi_- u) = -L(\chi_+ u)$ is timelike compact. Exploiting the properties of retarded and advanced Green operators, one concludes the proof:

$$Eh = E^- L(\chi_- u) + E^+ L(\chi_+ u) = \chi_- u + \chi_+ u = u.$$

It might be useful to summarize the content of Lemma 3.2.14 and of Theorem 3.2.15 with the following exact sequence:

$$0 \longrightarrow \Gamma_{tc}(F) \xrightarrow{L} \Gamma_{tc}(F) \xrightarrow{E} \Gamma(F) \xrightarrow{L} \Gamma(F) \longrightarrow 0.$$

We remind the reader that this is simply a symbolic way of stating that the kernel of each of the arrows depicted above coincides with the image of the preceding one.

Although the last theorem provides a complete characterization of the solutions of the partial differential equation associated to a Green hyperbolic operator, we need to introduce and to study a vector subspace of Sol which will play a distinguished role in the analysis of explicit models.

Proposition 3.2.16 *Let $M = (\mathcal{M}, g, \mathfrak{o}, \mathfrak{t})$ be a globally hyperbolic spacetime. Consider a vector bundle F over $\mathcal{M}$ and let $L : \Gamma(F) \to \Gamma(F)$ be a Green hyperbolic operator. Let $E^{\pm}$ be the retarded and advanced Green operators for L and denote* *with E the corresponding advanced-minus-retarded operator. Then the following statements hold true:*

1. *If $f \in \Gamma_0(F)$ is such that $Lf = 0$, then $f = 0$;*
2. *If $f \in \Gamma_0(F)$ is such that $Ef = 0$, then there exists $h \in \Gamma_0(F)$ such that $Lh = f$;*
3. *For each $h \in \Gamma_{sc}(F)$ there exists $f \in \Gamma_{sc}(F)$ such that $Lf = h$.*

Furthermore, let $\mathsf{Sol}_{sc} \subset \mathsf{Sol}$ be the vector subspace whose elements are smooth and spacelike compact solutions of $Lu = 0$. Then the map presented below is an isomorphism between Sol_{sc} and the quotient of $\Gamma_0(F)$ by the image of L acting on $\Gamma_0(F)$:

$$\frac{\Gamma_0(F)}{L(\Gamma_0(F))} \to \mathsf{Sol}_{sc}, \qquad [f] \mapsto Ef, \tag{3.7}$$

where $f \in \Gamma_0(F)$ is any representative of the equivalence class $[f]$ in the quotient.

Proof The proof follows slavishly those of Lemma 3.2.14 and of Theorem 3.2.15 and therefore we shall not repeat it in details. One has only to keep in mind that E maps sections with compact support to sections with spacelike compact support and that the intersection between a spacelike compact region and a timelike compact one is compact.

In terms of an exact sequence, this last proposition translates to

$$0 \longrightarrow \Gamma_0(F) \xrightarrow{L} \Gamma_0(F) \xrightarrow{E} \Gamma_{sc}(F) \xrightarrow{L} \Gamma_{sc}(F) \longrightarrow 0. \tag{3.8}$$

Spacelike compact solutions to a linear partial differential equation are also noteworthy since, under certain additional assumptions, they can be naturally endowed with an additional structure which plays a key role in the construction of the algebra of observables for a Bosonic/Fermionic free quantum field theory—see also [6, 31, 40, 41, 53].

Proposition 3.2.17 *Let $\boldsymbol{M} = (\mathcal{M}, g, \mathfrak{o}, \mathfrak{t})$ be a globally hyperbolic spacetime. Consider a vector bundle F over $\mathcal{M}$ endowed with a non-degenerate inner product as per Definition 3.2.5. Let $L : \Gamma(F) \to \Gamma(F)$ be a formally self-adjoint Green hyperbolic operator and denote with $E^{\pm}$ the corresponding retarded and advanced Green operators and with E the associated advanced-minus-retarded operator. Then the map presented below defines a non-degenerate bilinear form on $\Gamma_0(F)$:*

$$\tau : \frac{\Gamma_0(F)}{L(\Gamma_0(F))} \times \frac{\Gamma_0(F)}{L(\Gamma_0(F))} \to \mathbb{R}, \qquad ([f],[f']) \mapsto (f, Ef') \tag{3.9}$$

where $(\cdot,\cdot)$ is the pairing defined in (3.1), while $f \in [f]$ and $f' \in [f']$ are two arbitrary representatives. Furthermore, τ is a symplectic form in the Bosonic case, namely when the inner product on F is symmetric, while it is a scalar product in the Fermionic case, namely when the inner product on F is anti-symmetric.

Proof Notice that the definition of the map τ is well-posed since it does not depend on the choice of representatives. In fact, on the one hand, $ELh = 0$ for all $h \in \Gamma_0(F)$ and, on the other hand, L is formally self-adjoint. From its definition, it immediately follows that τ is bilinear. Let us show that it is also non-degenerate. Suppose $f \in \Gamma_0(F)$ is such that $\tau([f],[f']) = 0$ for all $f' \in \Gamma_0(F)$. This means that $-(Ef, f') = (f, Ef') = 0$ for all $f' \in \Gamma_0(F)$. Here we exploited the fact that L is formally self-adjoint, therefore Lemma 3.2.13 holds with $L^* = L$. Since the pairing $(\cdot,\cdot)$ between $\Gamma(F)$ and $\Gamma_0(F)$ is non-degenerate, one deduces that $Ef = 0$. Recalling also (3.8), it follows that f lies in $L(\Gamma_0(F))$, meaning that $[f] = 0$. Similarly, one can show non-degeneracy in the other argument too. To conclude the proof, suppose that we are in the Bosonic (Fermionic) case, namely we have $a \cdot b = (-) b \cdot a$ for all $p \in \mathcal{M}$ and all $a, b \in F_p$. Therefore, τ_M is anti-symmetric (symmetric) as the following chain of identities shows:

$$-\tau([f],[f']) = -(f, Ef') = (Ef, f') = (-)(f', Ef) = (-)\tau([f'],[f]). \tag{3.10}$$

Note that we exploited the formal self-adjointness of L in the first place and then also the symmetry (anti-symmetry) of $\cdot$.

Notice that, in the literature it is also customary to denote $\tau([f],[f'])$ with $E(f, f')$.

3.3 Classical and Quantum Field Theory

In this section we shall construct the classical field theories and their quantum counterparts for three models, namely the real scalar field, the Proca field and the Dirac field. We shall consider an arbitrary, but fixed, globally hyperbolic spacetime $M = (\mathcal{M}, g, \mathfrak{o}, \mathfrak{t})$ as the background for the dynamical evolution of the fields under analysis. For each model, we shall introduce a suitable class of sufficiently well-behaved functionals defined on the space of classical field configurations. The goal is to find functionals which can be thought of as *classical observables* in the sense that one can extract any information about a given field configuration by means of these functionals and, moreover, each of them provides some information which cannot be detected by any other functional. Note that the approach we adopt allows for extensions in several directions. In fact, it has been followed both in the context of affine field theories [11] as well as for gauge field theories [8–10]. Even when a space of classical observables complying with these requirements has been found, a symplectic structure is still needed in order to have the full data describing our classical field theory. This structure will be induced in a natural way by the partial differential equation ruling the dynamics. The reasons for the need of a symplectic structure are manifold. Conceptually, the analogy with classical mechanics motivates chiefly this requirement; at a practical level, instead, this is a bit of information which is needed to step-up a quantization scheme for the models, we are interested in.

3.3.1 The Real Scalar Field

As mentioned above, let us fix once and for all a globally hyperbolic spacetime $\boldsymbol{M} = (\mathcal{M}, g, \mathfrak{o}, \mathfrak{t})$, which provides the background where to specify the field equation.

3.3.1.1 Classical Field Theory

For the real scalar case, the *off-shell* field configurations are real-valued smooth functions on $\mathcal{M}$. This means that, before imposing the field equation, the relevant space of configurations is $C^\infty(\mathcal{M})$. As a starting point, we introduce linear functionals on $C^\infty(\mathcal{M})$ as follows: Given $f \in C_0^\infty(\mathcal{M})$, we denote by $F_f : C^\infty(\mathcal{M}, \mathbb{R}) \to \mathbb{R}$ the map defined below:

$$F_f(\phi) = \int_{\mathcal{M}} f\,\phi\,\mathrm{dvol}_{\boldsymbol{M}}, \tag{3.11}$$

where $\mathrm{dvol}_{\boldsymbol{M}}$ is the standard volume form on $\boldsymbol{M} = (\mathcal{M}, g, \mathfrak{o}, \mathfrak{t})$ defined out of its orientation $\mathfrak{o}$ and of its metric g. Notice that the above definition of a functional makes use of the usual non-degenerate bilinear pairing between $C_0^\infty(\mathcal{M})$ and $C^\infty(\mathcal{M})$. It is a well-known result of functional analysis that this pairing is non-degenerate. This has two important consequences: First, the map $f \in C_0^\infty(\mathcal{M}) \mapsto F_f$ implicitly defined by (3.11) is injective, thus allowing us to identify the space of functionals $\{F_f : f \in C_0^\infty(\mathcal{M})\}$ with $C_0^\infty(\mathcal{M})$. Second, the class of functionals considered so far is rich enough to separate off-shell configurations, namely, given two different configurations $\phi, \psi \in C^\infty(\mathcal{M})$, there exists always $f \in C_0^\infty(M)$ such that $F_f(\phi) \neq F_f(\psi)$. In fact, this is equivalent to the following statement, which follows from the non-degeneracy of the bilinear pairing between $C_0^\infty(\mathcal{M})$ and $C^\infty(\mathcal{M})$: If $\phi \in C^\infty(\mathcal{M})$ is such that $F_f(\phi) = 0$ for all $f \in C_0^\infty(\mathcal{M})$, then $\phi = 0$. Phrased differently, we are saying that, off-shell functionals of the form F_f, *cf.* (3.11), are faithfully labeled by $f \in C_0^\infty(\mathcal{M})$. Later we shall restrict functionals to dynamically allowed field configurations. This restriction will break the one-to-one correspondence between functionals F_f and $f \in C_0^\infty(\mathcal{M})$.

So far, we did not take into account the dynamics of the real scalar field. This is specified by the following partial differential equation:

$$\Box_{\boldsymbol{M}}\phi + (m^2 + \xi R)\phi = 0, \tag{3.12}$$

where $\xi \in \mathbb{R}$, R stands for the scalar curvature built out of g, m^2 is a real number while $\Box_{\boldsymbol{M}} = g^{ab}\nabla_a\nabla_b : C^\infty(\mathcal{M}) \to C^\infty(\mathcal{M})$ is the d'Alembert operator on $\boldsymbol{M}$ defined out of the metric g via the associated Levi-Civita connection ∇. Notice that in this chapter we are not imposing any constraint on the sign of the mass term since it plays no role. Yet, in the other chapters, one might force additional conditions motivated especially by the physical interpretation of the model. When $\phi \in C^\infty(\mathcal{M})$ is a solution of equation (3.12), we say that ϕ is an *on-shell* field configuration. For convenience, we introduce the differential operator $P = \Box_M + m^2 + \xi R$, so that (3.12) reduces to $P\phi = 0$. We collect all on-shell field configurations in a vector space:

$$\mathsf{Sol} = \{\phi \in C^\infty(\mathcal{M}) : P\phi = 0\} \subset C^\infty(\mathcal{M}).$$

It is important to mention that the second order linear differential operator P is formally self-adjoint, *cf.* Definition 3.2.7, meaning that, for each $\phi, \psi \in C^\infty(\mathcal{M})$ with supports having compact intersection, one has

$$\int_{\mathcal{M}} P\phi\ \psi\ \mathrm{dvol}_{\boldsymbol{M}} = \int_{\mathcal{M}} \phi\ P\psi\ \mathrm{dvol}_{\boldsymbol{M}}. \tag{3.13}$$

This identity follows from a double integration by parts. Furthermore, P is normally hyperbolic. This entails that P admits unique retarded and advanced Green operators E^+ and E^-, see [2, 6, 54]. In particular, it is a Green hyperbolic operator as per Definition 3.2.10.

Since the functionals F_f are sufficiently many to separate points in $C^\infty(\mathcal{M})$, this is also the case for Sol, the latter being a subspace of $C^\infty(\mathcal{M})$. We already achieved our first requirement to define classical observables. In fact, the functionals F_f can detect any information about on-shell field configurations. Specifically, two on-shell configurations $\phi, \psi \in \mathsf{Sol}$ coincide if and only if the outcome of their evaluation is the same on all functionals, namely $F_f(\phi) = F_f(\psi)$ for all $f \in C_0^\infty(\mathcal{M})$. Yet, it is the case that some of the functionals considered give no information when evaluated on Sol in the sense that their evaluation on an on-shell configuration always vanishes. Here is an explicit example.

Example 3.3.1 Consider $f \in C_0^\infty(\mathcal{M})$. Clearly Pf is a smooth function with compact support, therefore it makes sense to consider the linear functional $F_{Pf} : C^\infty(\mathcal{M}) \to \mathbb{R}$. Just reading out (3.13), one gets $F_{Pf}(\phi) = F_f(P\phi)$ for all $\phi \in C^\infty(\mathcal{M})$. In particular, it follows that $F_{Pf}(\phi) = 0$ for all $\phi \in \mathsf{Sol}$, thus F_{Pf} vanishes of Sol.

The example above shows that functionals of the form F_f, after the restriction to Sol, are no longer faithfully labeled by $f \in C_0^\infty(\mathcal{M})$. In fact, there are indeed redundant functions in $C_0^\infty(\mathcal{M})$, which provide the same functional on Sol, an example being provided by 0 and Pf, for any $f \in C_0^\infty(\mathcal{M})$. According to our second requirement, to identify a suitable space of classical observables, one has to get rid of such redundancies. Therefore, one identifies two functions f and h in $C_0^\infty(\mathcal{M})$ if $F_f(\phi) = F_h(\phi)$ for all $\phi \in \mathsf{Sol}$, thus restoring a faithful labeling for functionals on solutions. This result can be easily achieved as follows. First, one introduces the subspace of those smooth functions with compact support providing functionals on $C^\infty(\mathcal{M})$ whose restriction to Sol vanishes:

$$N = \{f \in C_0^\infty(\mathcal{M}) : F_f(\phi) = 0,\ \forall \phi \in \mathsf{Sol}\} \subset C_0^\infty(\mathcal{M}). \tag{3.14}$$

Notice that, according to Example 3.3.1, $P(C_0^\infty(\mathcal{M}))$, the image of $C_0^\infty(\mathcal{M})$ via P, is a subspace of N.[5] Therefore, we can take the quotient of $C_0^\infty(\mathcal{M})$ by N, resulting in a new vector space:

$$\mathcal{E} = C_0^\infty(M)/N. \tag{3.15}$$

An equivalence class $[f] \in \mathcal{E}$ yields a functional $F_{[f]} : \mathsf{Sol} \to \mathbb{R}$ specified by $F_{[f]}(\phi) = F_f(\phi)$ for any on-shell configuration $\phi \in \mathsf{Sol}$ and for any choice of a representative f of the class $[f]$. $F_{[f]}$ is well-defined on account of the definition of N and of the fact that it is evaluated on solutions only. Furthermore, by construction, these functionals are in one-to-one correspondence with points in $\mathcal{E}$, therefore equivalence classes $[f] \in \mathcal{E}$ faithfully label functionals of the type $F_{[f]} : \mathsf{Sol} \to \mathbb{R}$. Since the quotient by N does not affect the property of separating points in Sol, we conclude that $\mathcal{E}$ has the properties required to be interpreted as a space of classical observables, namely by evaluation it can distinguish different on-shell configurations and, moreover, there are no redundancies since different points of $\mathcal{E}$ provide different functionals on Sol. This motivates the fact that we shall refer to $[f] \in \mathcal{E}$ as a *classical observable* for the real scalar field.

Remark 3.3.2 Besides implementing non-redundancy (essentially by definition), the quotient by N, corresponds to go on-shell at the level of functionals. Contrary to the functionals in (3.11), which were defined not only on the subspace Sol of on-shell configurations, but also on all of off-shell fields, an equivalence class $[f] \in \mathcal{E}$ provides a functional $F_{[f]}$, which is well-defined only on Sol. In fact, for $\phi \in C^\infty(\mathcal{M}) \setminus \mathsf{Sol}$, $F_f(\phi)$ depends on the choice of the representative $f \in [f]$. Thus $\mathcal{E}$ defines functionals on $\mathsf{Sol} \subset C^\infty(\mathcal{M})$ only. In this sense, the quotient by N implements the on-shell condition at the level of functionals.

Remark 3.3.3 Before dealing with the problem of endowing $\mathcal{E}$ with a suitable symplectic structure, we would like to point out that $N = P(C_0^\infty(\mathcal{M}))$. In fact, take $f \in C_0^\infty(\mathcal{M})$ such that $F_f(\phi) = 0$ for all $\phi \in \mathsf{Sol}$. According to Theorem 3.2.15, the advanced-minus-retarded operator $E : C_{tc}^\infty(\mathcal{M}) \to C^\infty(\mathcal{M})$ associated to P maps surjectively onto Sol. Therefore, we can rephrase our condition on f as $F_f(Eh) = 0$ for all $h \in C_{tc}^\infty(\mathcal{M})$. Exploiting (3.11) and recalling Lemma 3.2.13, one reads

$$F_f(Eh) = \int_{\mathcal{M}} f\ Eh\, \mathrm{dvol}_{\boldsymbol{M}} = -\int_{\mathcal{M}} Ef\ h\, \mathrm{dvol}_{\boldsymbol{M}}.$$

According to the hypothesis, the integral on the right-hand-side has to vanish for all $h \in C_{tc}^\infty(\mathcal{M})$, hence $Ef = 0$. In fact, it would be enough to consider $h \in C_0^\infty(\mathcal{M})$ to come to this conclusion. Recalling the properties of E again, one finds $f' \in C_0^\infty(\mathcal{M})$ such that $Pf' = f$, thus showing that $f \in N$ implies $f \in P(C_0^\infty(\mathcal{M}))$. Since the inclusion $P(C_0^\infty(\mathcal{M})) \subset N$ follows from Example 3.3.1, we conclude that $N = P(C_0^\infty(\mathcal{M}))$ as claimed.

---

[5] In Remark 3.3.3, we shall show that, in the case of the real scalar field, $N = P(C_0^\infty(\mathcal{M}))$. More generally, using the same argument, one can prove an analogous result for any field whose dynamics is ruled by a Green hyperbolic operator.

Up to now, a vector space $\mathcal{E}$ providing functionals on Sol has been determined such that points in Sol can be distinguished by evaluation on these functionals and, moreover, $\mathcal{E}$ does not contain redundancies, meaning that the map which assigns to $[f] \in \mathcal{E}$ the functional $F_f : \mathsf{Sol} \to \mathbb{R}$ is injective. Yet, to get the classical field theory of the scalar field, a symplectic structure[6] on $\mathcal{E}$ naturally induced by the field equation is still needed. For the following construction we shall need the tools developed in Sect. 3.2. Let E^+ and E^- denote the retarded and advanced Green operators associated to $P = \Box_{\boldsymbol{M}} + m^2 + \xi R$ and consider the corresponding advanced-minus-retarded operator $E = E^- - E^+$. On account of Remark 3.3.3, we have that $\mathcal{E} = C_0^\infty(\mathcal{M})/P(C_0^\infty(\mathcal{M}))$. Therefore, applying Proposition 3.2.17, we obtain a symplectic structure on $\mathcal{E}$:

$$\tau : \mathcal{E} \times \mathcal{E} \to \mathbb{R}, \qquad ([f],[h]) \mapsto F_f(Eh) = \int_{\boldsymbol{M}} f\, Eh \,\mathrm{dvol}_{\boldsymbol{M}}, \tag{3.16}$$

where f and h are arbitrary representatives of the equivalence classes $[f]$ and respectively $[h]$ in $\mathcal{E}$. The pair $(\mathcal{E}, \tau)$ is the symplectic space of observables describing the classical theory of the real scalar field on the globally hyperbolic spacetime $\boldsymbol{M}$ and it is the starting point for the quantization scheme that we shall discuss in the next section. As a preliminary step we discuss some relevant properties.

Theorem 3.3.1 *Consider a globally hyperbolic spacetime $\boldsymbol{M} = (\mathcal{M}, g, \mathfrak{o}, \mathfrak{t})$ and let $(\mathcal{E}, \tau)$ be the symplectic space of classical observables defined above for the real scalar field. The following properties hold:*

Causality *The symplectic structure vanishes on pairs of observables localized in causally disjoint regions. More precisely, let $f, h \in C_0^\infty(\mathcal{M})$ be such that $\operatorname{supp} f \cap J_{\boldsymbol{M}}(\operatorname{supp} h) = \emptyset$. Then $\tau([f],[h]) = 0$.*

Time-slice axiom *Let $\mathcal{O} \subset \mathcal{M}$ be a globally hyperbolic open neighborhood of a spacelike Cauchy surface Σ for $\boldsymbol{M}$, namely $\mathcal{O}$ is an open neighborhood of Σ in $\mathcal{M}$ containing all causal curves for $\boldsymbol{M}$ whose endpoints lie in $\mathcal{O}$. In particular, the restriction of $\boldsymbol{M}$ to $\mathcal{O}$ provides a globally hyperbolic spacetime $\boldsymbol{O} = (\mathcal{O}, g|_{\mathcal{O}}, \mathfrak{o}|_{\mathcal{O}}, \mathfrak{t}|_{\mathcal{O}})$. Denote with $(\mathcal{E}_{\boldsymbol{M}}, \tau_{\boldsymbol{M}})$ and with $(\mathcal{E}_{\boldsymbol{O}}, \tau_{\boldsymbol{O}})$ the symplectic spaces of observables for the real scalar field respectively over $\boldsymbol{M}$ and over $\boldsymbol{O}$. Then the map $L : \mathcal{E}_{\boldsymbol{O}} \to \mathcal{E}_{\boldsymbol{M}}$ defined by $L[f] = [f]$ for all $f \in C_0^\infty(\mathcal{O})$ is an isomorphism of symplectic spaces.*[7]

Proof Let us start from the causality property and take $f, h \in C_0^\infty(\mathcal{M})$ such that their supports are causally disjoint. Recalling the definition of τ given in (3.16), one has $\tau([f],[h]) = F_f(Eh)$. Taking into account the support properties of the advanced-minus-retarded operator E, one deduces that $\operatorname{supp}(Eh)$ is included in $J_{\boldsymbol{M}}(\operatorname{supp} h)$,

[6]Even though the term "symplectic structure" is mathematically correct, it would be more appropriate to refer to this as a constant Poisson structure. Yet, we shall adhere to the common nomenclature of quantum field theory on curved spacetimes.

[7]The function in the right-hand-side of the equation which defines L is the extension by zero to the whole spacetime of the function appearing in the left-hand-side.

which does not intersect the support of f per assumption. Since $F_f(Eh)$ is the integral of the pointwise product of f with Eh, see (3.11), $\tau([f],[h]) = F_f(Eh) = 0$ as claimed.

For the time-slice axiom, consider a globally hyperbolic open neighborhood $\mathcal{O} \subset \mathcal{M}$ of a spacelike Cauchy surface for $\boldsymbol{M}$ and consider the globally hyperbolic spacetime $\boldsymbol{O} = (\mathcal{O}, g|_{\mathcal{O}}, \mathfrak{o}|_{\mathcal{O}}, \mathfrak{t}|_{\mathcal{O}})$. The same construction applied to $\boldsymbol{M}$ and to $\boldsymbol{O}$ provides the symplectic spaces $(\mathcal{E}_{\boldsymbol{M}}, \tau_{\boldsymbol{M}})$ and respectively $(\mathcal{E}_{\boldsymbol{O}}, \tau_{\boldsymbol{O}})$. The function $f \in C_0^\infty(\mathcal{O})$ can be extended by zero to the whole $\mathcal{M}$ and we denote it still by f with a slight abuse of notation; moreover, for each $h \in C_0^\infty(\mathcal{O})$, the extension of $Ph = \Box_{\boldsymbol{O}} h + m^2 h + \xi R$ is of the form $Ph = \Box_{\boldsymbol{M}} h + m^2 h + \xi R$, where now $h \in C_0^\infty(\mathcal{M})$ denotes the extension of the original $h \in C_0^\infty(\mathcal{O})$. These observations entail that the map $L : \mathcal{E}_{\boldsymbol{O}} \to \mathcal{E}_{\boldsymbol{M}}$ specified by $L[f] = [f]$ for all $f \in C_0^\infty(\mathcal{O})$ is well-defined. Note that L is linear and that it preserves the symplectic form. In fact, given $[f], [h] \in \mathcal{E}_{\boldsymbol{O}}$, one has

$$\tau_{\boldsymbol{M}}(L[f], L[h]) = \int_{\mathcal{M}} f\ Eh \, \mathrm{dvol}_{\boldsymbol{M}} = \int_{\mathcal{O}} f\ Eh \, \mathrm{dvol}_{\boldsymbol{O}} = \tau_{\boldsymbol{O}}([f],[h]),$$

where the restriction from $\mathcal{M}$ to $\mathcal{O}$ in the domain of integration is motivated by the fact that, per construction, $f = 0$ outside $\mathcal{O}$. Being a symplectic map, L is automatically injective. In fact, given $[f] \in \mathcal{E}_{\boldsymbol{O}}$ such that $L[f] = 0$, one has $\tau_{\boldsymbol{O}}([f],[h]) = \tau_{\boldsymbol{M}}(L[f], L[h]) = 0$ for all $[h] \in \mathcal{E}_{\boldsymbol{O}}$ and the non-degeneracy of $\tau_{\boldsymbol{O}}$ entails that $[f] = 0$. It remains only to check that L is surjective. To this end, starting from any $f \in C_0^\infty(\mathcal{M})$, we look for $f' \in C_0^\infty(\mathcal{M})$ with support inside $\mathcal{O}$ such that $[f'] = [f]$ in $\mathcal{E}_{\boldsymbol{M}}$. Recalling that $\mathcal{O}$ is an open neighborhood of the spacelike Cauchy surface Σ and exploiting the usual space-time decomposition of $\boldsymbol{M}$, see Theorem 3.1.4, one finds two spacelike Cauchy surfaces Σ_+, Σ_- for $\boldsymbol{M}$ included in $\mathcal{O}$ lying respectively in the future and in the past of Σ. Let $\{\chi^+, \chi^-\}$ be a partition of unity subordinate to the open cover $\{I^+_{\boldsymbol{M}}(\Sigma_-), I^-_{\boldsymbol{M}}(\Sigma_+)\}$ of $\mathcal{M}$. By construction the intersection of the supports of χ^+ and of χ^- is a timelike compact region both of $\mathcal{O}$ and of $\mathcal{M}$. Since $PEf = 0$, $\chi^+ + \chi^- = 1$ on $\mathcal{M}$ and recalling the support properties of E, it follows that $f' = P(\chi^- Ef) = -P(\chi^+ Ef)$ is a smooth function with compact support inside $\mathcal{O}$. Furthermore, recalling also the identity $PE^- f = f$, one finds

$$\begin{aligned} f' - f &= P(\chi^- E^- f) - P(\chi^- E^+ f) - P(\chi^+ E^- f) - P(\chi^- E^- f) \\ &= P(-\chi^- E^+ f - \chi^+ E^- f). \end{aligned}$$

The support properties of both the retarded and advanced Green operators E^+, E^- entail that $-\chi^- E^+ f - \chi^+ E^- f$ is a smooth function with compact support on $\mathcal{M}$. In fact $\operatorname{supp} \chi^\mp \cap \operatorname{supp}(E^\pm f)$ is a closed subset of $J^\mp_{\boldsymbol{M}}(\Sigma_\pm) \cap J^\pm_{\boldsymbol{M}}(\operatorname{supp} f)$, which is compact. This shows that $f' - f \in P(C_0^\infty(\mathcal{M})) \subset N$, see also Example 3.3.1. Therefore we found $[f'|_{\mathcal{O}}] \in \mathcal{E}_{\boldsymbol{O}}$ such that $L[f'|_{\mathcal{O}}] = [f]$ showing that, besides being injective, the symplectic map L is also surjective and hence an isomorphism of symplectic spaces.

Remark 3.3.4 We comment briefly on the apparently different approach, which is often presented in the literature. In fact, in place of the pair $(\mathcal{E}, \tau)$, it is quite common to consider Sol_{sc}, the space of solutions with spacelike compact support, endowed with the following symplectic structure:

$$\sigma : \mathsf{Sol}_{sc} \times \mathsf{Sol}_{sc} \to \mathbb{R}, \qquad (\phi, \psi) \mapsto \int_\Sigma (\phi \nabla_{\boldsymbol{n}} \psi - \psi \nabla_{\boldsymbol{n}} \phi)\, d\Sigma, \tag{3.17}$$

where Σ is a spacelike Cauchy surface for the globally hyperbolic spacetime $\boldsymbol{M} = (\mathcal{M}, g, \mathfrak{o}, \mathfrak{t})$, $\boldsymbol{n}$ is the future-pointing unit normal vector field on Σ, and $d\Sigma$ is the induced volume form on Σ.[8] Notice that the integrand in (3.17) is implicitly meant to be restricted to Σ. Exploiting the fact that only solutions of the field equation are considered, one can prove that σ does not depend on the choice of the spacelike Cauchy surface Σ. The restriction to the subspace of solutions with spacelike compact support guarantees that the argument of the integral in (3.17) is an integrable function. We outline below an isomorphism of symplectic spaces between $(\mathcal{E}, \tau)$ and $(\mathsf{Sol}_{sc}, \sigma)$:

$$I : \mathcal{E} \to \mathsf{Sol}_{sc}, \qquad [f] \mapsto Ef, \tag{3.18}$$

where $f \in C_0^\infty(\mathcal{M})$ is any representative of $[f] \in \mathcal{E}$ and E denotes the advanced-minus-retarded operator associated to the differential operator $P = \Box_{\boldsymbol{M}} + m^2 + \xi R$, which rules the dynamics of the real scalar field. The map I is a by-product of Proposition 3.2.16 as soon as we remind that in (3.15) $N = P(C_0^\infty(\mathcal{M}))$, as shown in Remark 3.3.3. It remains only to check that $\sigma(Ef, Eh) = \tau([f], [h])$ for all $f, h \in C_0^\infty(\mathcal{M})$. Recalling that $\phi = Ef$ and $\psi = Eh$ are both solutions of the field equation, namely $P\phi = 0$ and $P\psi = 0$, by means of a double integration by parts, one gets the following:

$$\begin{aligned}
\int_{\mathcal{M}} f\, Eh\, \mathrm{dvol}_{\boldsymbol{M}} &= \int_{J^+_{\boldsymbol{M}}(\Sigma)} f\, \psi\, \mathrm{dvol}_{\boldsymbol{M}} + \int_{J^-_{\boldsymbol{M}}(\Sigma)} f\, \psi\, \mathrm{dvol}_{\boldsymbol{M}} \\
&= \int_{J^+_{\boldsymbol{M}}(\Sigma)} (PE^- f)\psi\, \mathrm{dvol}_{\boldsymbol{M}} + \int_{J^-_{\boldsymbol{M}}(\Sigma)} (PE^+ f)\psi\, \mathrm{dvol}_{\boldsymbol{M}} \\
&= -\int_\Sigma (\nabla_{\boldsymbol{n}}(E^- f))\psi\, d\Sigma + \int_\Sigma (E^- f)\nabla_{\boldsymbol{n}}\psi\, d\Sigma \\
&\quad + \int_\Sigma (\nabla_{\boldsymbol{n}}(E^+ f))\psi\, d\Sigma - \int_\Sigma (E^+ f)\nabla_{\boldsymbol{n}}\psi\, d\Sigma \\
&= \int_\Sigma (\phi \nabla_{\boldsymbol{n}} \psi - \psi \nabla_{\boldsymbol{n}} \phi)\, d\Sigma.
\end{aligned} \tag{3.19}$$

[8] The volume form $d\Sigma$ on Σ is defined out of the structure induced on Σ itself as a submanifold of the globally hyperbolic spacetime $\boldsymbol{M}$. More explicitly, on Σ we take the Riemannian metric $g|_\Sigma$ and the orientation specified by the orientation and time-orientation of $\boldsymbol{M}$. Then $d\Sigma$ is the natural volume form defined out of these data.

In the first step, we decomposed the integral by splitting the domain of integration into two subsets whose intersection has zero measure. The second step consisted of exploiting the properties of the retarded and advanced Green operators E^+ and E^- for P. Using $E^\mp$ inside the integral over $J^\pm_M(\Sigma)$ allows us to integrate by parts twice. For each integral, this operation produces two boundary terms and an integral which vanishes since the integrand contains $P\psi = 0$. Adding together the four boundary terms, one concludes that $\sigma(I[f], I[h]) = \tau([f], [h])$ as expected.

3.3.1.2 Quantum Field Theory

The next step consists of constructing a quantum field theory for the real scalar field out of the classical one, whose content is encoded in the symplectic space $(\mathcal{E}, \tau)$. This result is obtained by means of a construction that can be traced back to [14, 29, 51], while the generalization to curved backgrounds has been discussed from an axiomatic point of view first in [19]. The so-called algebraic approach can be seen as a two-step quantization scheme: First, one identifies a suitable unital $*$-algebra encoding the structural relations between the observables, such as causality and locality; second, one selects a state, that is a positive, normalized, linear functional on the algebra, which allows us to recover the standard probabilistic interpretation of quantum theories via the GNS theorem. We will focus only on the first step for three different models of free fields, while the second will be at the heart of Chap. 5. We consider the unital $*$-algebra $\mathcal{A}$ generated over $\mathbb{C}$ by the symbols $\mathbf{1}$ and $\Phi([f])$ for all $[f] \in \mathcal{E}$ and satisfying the following relations for all $[f], [g] \in \mathcal{E}$ and for all $a, b \in \mathbb{R}$:

$$\Phi(a[f] + b[g]) = a\Phi([f]) + b\Phi([g]), \tag{3.20}$$

$$\Phi([f])^* = \Phi([f]), \tag{3.21}$$

$$\Phi([f]) \cdot \Phi([h]) - \Phi([h]) \cdot \Phi([f]) = i\tau([f], [h])\mathbf{1}. \tag{3.22}$$

More concretely, one can start introducing an algebra A consisting of the vector space $\bigoplus_{k\in\mathbb{N}_0} \mathcal{E}_{\mathbb{C}}^{\otimes k}$ obtained as the direct sum of all the tensor powers of the complexification $\mathcal{E}_{\mathbb{C}}$ of the vector space $\mathcal{E}$, where we have set $\mathcal{E}_{\mathbb{C}}^{\otimes 0} = \mathbb{C}$. Therefore, elements of A can be seen as sequences $\{v_k \in \mathcal{E}_{\mathbb{C}}^{\otimes k}\}_{k\in\mathbb{N}_0}$ with only finitely many non-zero terms. Each v_k in the sequence is a finite linear combination with $\mathbb{C}$-coefficients of terms of the form $[f_1] \otimes \cdots \otimes [f_k]$ for $[f_1], \ldots, [f_k] \in \mathcal{E}$. A is endowed with the product $\cdot : A \times A \to A$ specified by

$$\{u_k\} \cdot \{v_k\} = \{w_k\}, \qquad w_k = \sum_{i+j=k} u_i \otimes v_j. \tag{3.23}$$

So far, A is an algebra whose generators satisfy (3.20). We specify an involution $* : A \to A$ by setting

$$\{\underbrace{0,\ldots,0}_{k\text{ times}},[f_1]\otimes[f_2]\otimes\cdots\otimes[f_k],0,\ldots\}^* = \{\underbrace{0,\ldots,0}_{k\text{ times}},[f_k]\otimes[f_{k-1}]\otimes\cdots\otimes[f_1],0,\ldots\},$$

for all $[f_1],\ldots,[f_k] \in \mathcal{E}$, and extending it by antilinearity to the whole of A. Therefore, A is now a $*$-algebra implementing the relation (3.21) too. It is straightforward to realize that the identity of the $*$-algebra A is $\mathbf{1} = \{1, 0, \ldots\}$, hence A is also unital. Note that an arbitrary element of A can be obtained as a finite $\mathbb{C}$-linear combination of $\mathbf{1}$ and of finite products of elements of the form $\{0, [f], 0, \ldots\} \in A$, which are in one-to-one correspondence with elements of $\mathcal{E}$. To match the notation used in the more abstract setting, let us introduce the map $\Phi : \mathcal{E} \to A$, $[f] \mapsto \{0, [f], 0, \ldots\}$, which embeds $\mathcal{E}$ into A. The $*$-algebra A already "knows" of the dynamics of the real scalar field since this is already encoded in $\mathcal{E}$, however, the canonical commutation relations (CCR) (3.22) are still missing. Therefore, using the symplectic structure τ on $\mathcal{E}$, we introduce the two-sided $*$-ideal I of A generated by terms of the form

$$\Phi([f])\cdot\Phi([h]) - \Phi([h])\cdot\Phi([f]) - i\tau([f],[h])\mathbf{1},$$

for all $[f],[h] \in \mathcal{E}$. Taking the quotient of A by I, one obtains the unital $*$-algebra $\mathcal{A} = A/I$ implementing the canonical commutation relations for the real scalar field. Note that, with a slight abuse of notation, we shall denote with $\Phi([f])$ also the equivalence class in $\mathcal{A}$ of any generator $\Phi([f])$ of A, thus completely matching the notation used in the abstract definition of $\mathcal{A}$ as the unital $*$-algebra generated by $\mathcal{E}$ over $\mathbb{C}$ implementing the relations (3.20)–(3.22). Note in particular that (3.22) is the smeared version of the usual commutation relations. This motivates our interpretation of $\mathcal{A}$ as the quantum field theory for the real scalar field on $\boldsymbol{M}$.

Remark 3.3.5 Before proceeding with the analysis of the properties of the quantum field theory for the real scalar field, we would like to emphasize that, under suitable conditions, our quantization procedure perfectly agrees with the standard textbook quantization involving creation and annihilation operators. In fact, assuming that $\boldsymbol{M}$ is Minkowski spacetime, one can relate directly our algebraic approach to the one more commonly used by means of an expansion in Fourier modes of the fundamental quantum fields $\Phi([f])$, which generate the algebra $\mathcal{A}$. In particular, one recovers the usual commutation relations between creation and annihilation operators out of the canonical commutation relations specified in (3.22)—see for example [53]. This argument should convince the reader that the approach presented above is a very effective extension to arbitrary globally hyperbolic spacetimes of the usual quantization procedure for Minkowski spacetime.

The properties of the classical field theory presented in Theorem 3.3.1 have counterparts at the quantum level as shown by the following theorem.

Theorem 3.3.2 *Consider a globally hyperbolic spacetime* $\boldsymbol{M} = (\mathcal{M}, g, \mathfrak{o}, \mathfrak{t})$ *and let* $\mathcal{A}$ *be the unital* $*$*-algebra of observables for the real scalar field introduced above. The following properties hold:*

Causality *Elements of the algebra $\mathcal{A}$ localized in causally disjoint regions commute. More precisely, let $f, h \in C_0^\infty(\mathcal{M})$ be such that supp $f \cap J_M(\text{supp}\, h) = \emptyset$. Then $\Phi([f]) \cdot \Phi([h]) = \Phi([h]) \cdot \Phi([f])$.*

Time-slice axiom *Let $\mathcal{O} \subset \mathcal{M}$ be a globally hyperbolic open neighborhood of a spacelike Cauchy surface Σ for $\boldsymbol{M}$, namely $\mathcal{O}$ is an open neighborhood of Σ in $\mathcal{M}$ containing all causal curves for $\boldsymbol{M}$ whose endpoints lie in $\mathcal{O}$. In particular, the restriction of $\boldsymbol{M}$ to $\mathcal{O}$ provides a globally hyperbolic spacetime $\boldsymbol{O} = (\mathcal{O}, g|_{\mathcal{O}}, \mathfrak{o}|_{\mathcal{O}}, \mathfrak{t}|_{\mathcal{O}})$. Denote with $\mathcal{A}_{\boldsymbol{M}}$ and with $\mathcal{A}_{\boldsymbol{O}}$ the unital $*$-algebras of observables for the real scalar field respectively over $\boldsymbol{M}$ and over $\boldsymbol{O}$. Then the unit-preserving $*$-homomorphism $\Phi(L) : \mathcal{A}_{\boldsymbol{O}} \to \mathcal{A}_{\boldsymbol{M}}$, $\Phi([f]) \mapsto \Phi(L[f])$ is an isomorphism of $*$-algebras, where L denotes the symplectic isomorphism introduced in Theorem 3.3.1.*

Proof The quantum version of the causality property follows directly from the classical version and the canonical commutation relations. In fact, taking $f, h \in C_0^\infty(\mathcal{M})$ with causally disjoint supports, one has $\tau([f], [h]) = 0$ due to Theorem 3.3.1. Recalling the canonical commutation relations (3.22), one has $\Phi([f]) \cdot \Phi([h]) - \Phi([h]) \cdot \Phi([f]) = i\tau([f], [h])\mathbf{1} = 0$ as claimed.

Also the time-slice axiom follows directly from its classical counterpart. In fact, setting $\Phi(L)\Phi([f]) = \Phi(L[f])$ for each generator $\Phi([f])$ of the unital $*$-algebra $\mathcal{A}_{\boldsymbol{O}}$ uniquely defines a unital $*$-homomorphism $\Phi(L) : \mathcal{A}_{\boldsymbol{O}} \to \mathcal{A}_{\boldsymbol{M}}$. Consider the inverse of L, which exists since the classical time-slice axiom states that L is a symplectic isomorphism, see Theorem 3.3.1. The same construction applied to L^{-1} provides the unital $*$-homomorphism $\Phi(L^{-1}) : \mathcal{A}_{\boldsymbol{M}} \to \mathcal{A}_{\boldsymbol{O}}$. If $\Phi(L^{-1})$ inverts $\Phi(L)$ on all generators $\Phi([f])$ of $\mathcal{A}$, then $\Phi(L^{-1})$ is the inverse of $\Phi(L)$ and thus $\Phi(L)$ is a $*$-isomorphism. Therefore, for all generators $\Phi([f])$ of $\mathcal{A}$, we have to check the identities $\Phi(L)\Phi(L^{-1})\Phi([f]) = \Phi([f])$ and $\Phi(L^{-1})\Phi(L)\Phi([f]) = \Phi([f])$. But these are obvious consequences of the definitions of $\Phi(L)$ and of $\Phi(L^{-1})$. Therefore $\Phi(L)$ is a $*$-isomorphism.

3.3.2 The Dirac Field

In this section we present the classical and quantum theory of the Dirac field on a globally hyperbolic spacetime. In analogy with the scalar case, we shall discuss first the classical model and later develop the corresponding quantum field theory implementing canonical anti-commutation relations. Note that, unlike the scalar case, to implement anti-commutation relations, we shall need a Hermitian structure in place of a symplectic one. Contrary to the real scalar field, the geometry of the space where the Dirac field takes its values requires much more attention. This will be the first topic of our presentation, providing the framework to write down the Dirac equation. Afterwards, we shall construct a suitable space of classical observables for on-shell configurations of the Dirac field. As in the previous case, we look for a space of sections that, by means of integration, provides functionals on on-shell configurations. Again, we will be guided by the requirement that the functionals obtained must be able to detect any on-shell configuration (separability). A quotient will remove all redundancies which might be present in the chosen space of sections. Separability and non-redundancy motivate our interpretation of this quotient as providing a space of classical observables for the Dirac field. To complete the classical part, we shall endow our space of observables with a Hermitian structure, which will be used to quantize the classical model, eventually leading to an algebra of observables implementing the usual anti-commutation relations for the Dirac field on a globally hyperbolic spacetime. Some references discussing the quantum Dirac field on globally hyperbolic spacetimes are [17, 20, 25, 47, 55].

3.3.2.1 Kinematics and Dynamics

Contrary to the case of the real scalar field, to specify the natural environment for the Dirac field, it is not enough to consider a globally hyperbolic spacetime $\boldsymbol{M} = (\mathcal{M}, g, \mathfrak{o}, \mathfrak{t})$. In fact, to introduce the kinematics of the Dirac field, one needs more data, namely a spin structure on $\boldsymbol{M}$.

Definition 3.3.3 Let $\boldsymbol{M} = (\mathcal{M}, g, \mathfrak{o}, \mathfrak{t})$ be an n-dimensional globally hyperbolic spacetime. Denote with $F\boldsymbol{M}$ the principal $\mathrm{SO}_0(1, n-1)$-bundle of oriented and time-oriented frames on $\boldsymbol{M}$, where $\mathrm{SO}_0(1, n-1)$ denotes the component connected to the identity of the proper Lorentz group $\mathrm{SO}(1, n-1)$ in n dimensions. Furthermore, consider the spin group $\mathrm{Spin}(1, n-1)$, namely the double cover of $\mathrm{SO}(1, n-1)$ and denote with $\Lambda : \mathrm{Spin}(1, n-1) \to \mathrm{SO}(1, n-1)$ the covering group homomorphism. Let us also indicate the component connected to the identity of $\mathrm{Spin}(1, n-1)$ with $\mathrm{Spin}_0(1, n-1)$. A *spin structure* on $\boldsymbol{M}$ consists of a pair $(S\boldsymbol{M}, \pi)$, where the *spin bundle* $S\boldsymbol{M}$ is a principal $\mathrm{Spin}_0(1, n-1)$-bundle, and the *spin frame projection* $\pi : S\boldsymbol{M} \to F\boldsymbol{M}$ is a bundle map covering the identity on the base and intertwining the right group actions of $\mathrm{Spin}_0(1, n-1)$ on $S\boldsymbol{M}$ and of $\mathrm{SO}_0(1, n-1)$ on $F\boldsymbol{M}$, namely such that $\pi(p\, S) = \pi(p)\, \Lambda(S)$ for all $p \in S\boldsymbol{M}$ and for all $S \in \mathrm{Spin}_0(1, n-1)$, where the group actions are denoted by juxtaposition.

Unfortunately, a spin structure does not always exist on globally hyperbolic spacetimes of arbitrary dimension and, even if it does, it might be non-unique. In fact, in general, there are topological obstructions both to existence and to uniqueness [42, Sect. 2.2]. Yet, four-dimensional globally hyperbolic spacetimes, the most relevant case to physics, always admit a spin structure, even though this need not be unique. First, all spin bundles over a four-dimensional globally hyperbolic spacetime are trivial on account of [38, Sect. 3]. Second, all orientable three-manifolds are parallelizable, see [44]. Since any four-dimensional globally hyperbolic spacetime $\boldsymbol{M}$ can be presented as the product of a real line (time) and an oriented 3-manifold (spatial Cauchy surface), see Theorem 3.1.4, it follows that it is parallelizable. In particular, there exists a global section ε of the principal bundle $F\boldsymbol{M}$, which consists of an ordered quadruple (ε_μ) of no-where vanishing orthonormal vector fields on

$\mathcal{M}$ whose orientation is chosen in order to agree with the orientation and the time-orientation of $\boldsymbol{M}$. In particular, this entails that $F\boldsymbol{M}$ is trivial, a trivialization being specified by the global frame ε itself. In fact, $F\boldsymbol{M} \simeq \mathcal{M} \times \mathrm{SO}_0(1,3)$ via the principal bundle map $(x, \lambda) \in \mathcal{M} \times \mathrm{SO}_0(1,3) \mapsto (\varepsilon(x), \lambda) \in F\boldsymbol{M}$. Since both $S\boldsymbol{M}$ and $F\boldsymbol{M}$ are trivial for all four-dimensional globally hyperbolic spacetimes $\boldsymbol{M}$, it follows that the freedom in the choice of the spin structure actually resides only in that of the spin frame projection $\pi : S\boldsymbol{M} \to F\boldsymbol{M}$, which, in turn, reduces to choosing a smooth $\mathrm{SO}_0(1,3)$-valued function over $\mathcal{M}$. In fact, all possible spin projections between the trivial principal bundles $S\boldsymbol{M}$ and $F\boldsymbol{M}$ are of the form

$$\begin{aligned} \pi : S\boldsymbol{M} \simeq \mathcal{M} \times \mathrm{Spin}_0(1,3) &\to F\boldsymbol{M} \simeq \mathcal{M} \times \mathrm{SO}_0(1,3), \\ (x, S) &\mapsto (x, f(x)\,\Lambda(S)), \end{aligned}$$

for some $f \in C^\infty(\mathcal{M}, \mathrm{SO}_0(1,3))$.

Once a spin structure $(S\boldsymbol{M}, \pi)$ has been chosen on the four-dimensional globally hyperbolic spacetime $\boldsymbol{M}$, at a kinematic level, a Dirac field is defined to be a section of the vector (Dirac) bundle $D\boldsymbol{M} = S\boldsymbol{M} \times_T \mathbb{C}^4$ with typical fiber $\mathbb{C}^4$ associated to the principal $\mathrm{Spin}_0(1,3)$-bundle $S\boldsymbol{M}$ via the Dirac representation $T = D^{\frac{1}{2},0} \oplus D^{0,\frac{1}{2}}$ of $\mathrm{Spin}_0(1,3)$ on $\mathbb{C}^4$.[9] Since $S\boldsymbol{M}$ is trivial, all its associated bundles are such, thus motivating the more direct definition of the spinor and cospinor bundles given below.

Definition 3.3.4 Let $\boldsymbol{M} = (\mathcal{M}, g, \mathfrak{o}, \mathfrak{t})$ be a four-dimensional and globally hyperbolic spacetime. We define the spinor bundle $D\boldsymbol{M}$ as the trivial vector bundle $M \times \mathbb{C}^4$, while the cospinor bundle $D^*\boldsymbol{M}$ is its dual $M \times (\mathbb{C}^4)^*$.

At this stage, one can talk about spinors and cospinors as sections of $D\boldsymbol{M}$ and respectively of $D^*\boldsymbol{M}$. In fact, both bundles being trivial, spinors and cospinors are just smooth functions on $\mathcal{M}$ taking values in either $\mathbb{C}^4$ or $(\mathbb{C}^4)^*$. Yet, to construct physical quantities out of spinors, such as scalars or currents, and to write down the Dirac equation, one still needs γ-matrices.

Definition 3.3.5 Consider the four-dimensional Minkowski space $\mathbb{M}^4 = (\mathbb{R}^4, \eta)$. The Dirac algebra $\mathcal{D}$ is the unital algebra generated over $\mathbb{R}$ by an orthonormal basis $\{\ell_\mu\}_{\mu=0,\dots,3}$ of $\mathbb{M}^4$ and satisfying the relation $\ell_\mu\,\ell_\nu + \ell_\nu\,\ell_\mu = 2\eta_{\mu\nu}\mathbf{1}$.

A choice of the γ-matrices amounts to fixing an irreducible complex representation of the Dirac algebra $\mathcal{D}$ on the algebra $\mathrm{M}(4,\mathbb{C})$ of four-by-four complex matrices. In fact, any choice of $\gamma_0, \dots, \gamma_3 \in \mathrm{M}(4,\mathbb{C})$ satisfying $\gamma_\mu\gamma_\nu + \gamma_\nu\gamma_\mu = 2\eta_{\mu\nu}1_4$ for all $\mu, \nu = 0,1,2,3$ induces an irreducible representation $\rho : \mathcal{D} \to \mathrm{M}(4,\mathbb{C})$ defined by $\rho(\ell_\mu) = \gamma_\mu$. Here 1_4 denotes the four-by-four identity matrix. Note that different choices of the γ-matrices induce equivalent representations [45], hence the same physical description. Yet, to be concrete, we shall consider a specific representation, namely the chiral one. Therefore we consider the following family of γ-matrices:

[9] Note that the Dirac representation T is usually regarded as a unitary representation of $\mathrm{SL}(2,\mathbb{C})$ on $\mathbb{C}^4$, yet $\mathrm{Spin}(1,3)$ is isomorphic to $\mathrm{SL}(2,\mathbb{C})$ as a Lie group.

$$\gamma_0 = \begin{pmatrix} 0_2 & 1_2 \\ 1_2 & 0_2 \end{pmatrix}, \qquad \gamma_i = \begin{pmatrix} 0_2 & \sigma_i \\ \sigma_i & 0_2 \end{pmatrix}, i = 1, 2, 3. \tag{3.24}$$

where 0_2, 1_2 and $\{\sigma_i\}_{i=1,2,3}$ respectively denote the zero matrix, the identity matrix and the Pauli matrices in $\mathrm{M}(2, \mathbb{C})$. As one can directly check, the γ-matrices of our choice satisfy the following relations:

$$\begin{aligned} &\gamma_\mu\gamma_\nu + \gamma_\nu\gamma_\mu = 2\eta_{\mu\nu}1_4, \quad \mu, \nu = 0, \dots, 3, \\ &\gamma_0^\dagger = \gamma_0, \qquad\qquad \gamma_i^\dagger = -\gamma_i, \quad i = 1, 2, 3, \\ &\overline{\gamma_\mu} = -\gamma_2\gamma_\mu\gamma_2^{-1} \quad \gamma_0\rho(n) > 0, \end{aligned} \tag{3.25}$$

n is any future pointing timelike vector in $\mathbb{M}^4$, $\overline{(\cdot)}$ denotes the complex conjugation of each entry, $(\cdot)^T$ is the transpose of a matrix and $(\cdot)^\dagger = \overline{(\cdot)}^T$. Since we defined the spinor bundle DM as a trivial bundle over $\mathcal{M}$ with fiber $\mathbb{C}^4$, we can interpret the γ-matrices as endomorphisms of this bundle:

$$\gamma_\mu : DM \to DM, \qquad\qquad (x, \sigma) \mapsto (x, \gamma_\mu\sigma). \tag{3.26}$$

Note that the action of the γ-matrices on cospinors is obtained by composing $\gamma_\mu : DM \to DM$ on the right. In fact, one can read the cospinors bundle D^*M as a bundle whose fibers are $\mathbb{C}$-linear functionals on the corresponding fiber of DM. Therefore it is natural to express the action of γ_μ on D^*M as $(x, \omega) \in D^*M \mapsto (x, \omega) \circ \gamma_\mu = (x, \omega \circ \gamma_\mu) \in D^*M$. For simplicity, in the following the composition will be left understood. Let us also mention that, the γ-matrices being invertible, see (3.25), the induced vector bundle maps are actually isomorphisms. In particular, by means of the γ-matrices, one can introduce complex anti-linear vector bundle isomorphisms covering the identity which implement adjunction and charge conjugation:

$$A : DM \to D^*M, \qquad\qquad (x, \sigma) \mapsto (x, \sigma^\dagger\gamma_0), \tag{3.27}$$

$$C_s : DM \to DM, \qquad\qquad (x, \sigma) \mapsto (x, \overline{\gamma_2\sigma}), \tag{3.28}$$

$$C_c : D^*M \to D^*M, \qquad\qquad (x, \omega) \mapsto (x, \overline{\omega}\gamma_2), \tag{3.29}$$

From (3.25) one can show that A intertwines C_s and C_c up to a sign, namely $A \circ C_s = -C_c \circ A$.

Furthermore, let us fix an oriented, orthochronous, orthonormal co-frame $e = (e^\mu)_{\mu=0,\dots,3}$ on M once and for all. The e^μ's are no-where vanishing one-forms on $\mathcal{M}$ which allow to completely reconstruct the structure of the globally hyperbolic spacetime $M = (\mathcal{M}, g, \mathfrak{o}, \mathfrak{t})$: $g = \eta_{\mu\nu}e^\mu \otimes e^\nu$, $\mathfrak{o} = [e^0 \wedge \dots \wedge e^3]$ and $\mathfrak{t} = [e^0]$, where the square brackets are used to indicate the (time-)orientation induced by the enclosed form. Fixing e is completely equivalent to the choice of a frame $\varepsilon = (\varepsilon_\mu)$, namely a section of the frame bundle FM. In fact, ε can be obtained from e setting $\varepsilon_\mu = \eta_{\mu\nu}(e^\nu)^\flat$ and similarly e can be obtained from ε as $e^\mu = \eta^{\mu\nu}\varepsilon_\nu^\sharp$, where $(\cdot)^\flat$ and $(\cdot)^\sharp$ are the canonical g-induced isomorphism which lower and raise the indices of tensors on $\mathcal{M}$, while η denotes the metric of Minkowski space $\mathbb{M}^4$. Using the fixed co-frame e of $\boldsymbol{M}$, one can specify a one-form γ over $\mathcal{M}$ taking values in the bundle of endomorphisms of the spinor bundle $D\boldsymbol{M}$:

$$\gamma : T\mathcal{M} \to \mathrm{End}(D\boldsymbol{M}), \qquad v \mapsto e^\mu(v)\gamma_\mu. \tag{3.30}$$

Note that $\{e^\mu(v) \in \mathbb{R}\}_{\mu=0,\dots,3}$ are the components of $v \in T_x M$ with respect to the frame ε obtained raising the indices of the fixed co-frame e, namely $v = e^\mu(v)\varepsilon_\mu$.

To write down the Dirac equation, the last necessary ingredient is a suitable covariant derivative on the spinor and cospinor bundles. Abstractly, one could start from the Levi-Civita connection, which is a principal bundle connection on the frame bundle $F\boldsymbol{M}$. Exploiting the spin structure $(S\boldsymbol{M}, \pi)$, one can pull-back the Levi-Civita connection form along π and then lift it along the double cover $\Lambda : \mathrm{Spin}_0(1,3) \to \mathrm{SO}_0(1,3)$ to obtain a 1-form on $S\boldsymbol{M}$ taking values in $\mathfrak{so}(1,3)$ the Lie algebra of both $\mathrm{SO}_0(1,3)$ and $\mathrm{Spin}_0(1,3)$. This procedure actually provides a principal bundle connection on $S\boldsymbol{M}$. Thinking of the spinor bundle as a vector bundle associated to $S\boldsymbol{M}$, a covariant derivative is naturally induced from the connection on $S\boldsymbol{M}$. This covariant derivative is the one relevant to the Dirac field. Yet, motivated by the fact that $S\boldsymbol{M}$ is trivial (and hence so is any associated bundle), we preferred to define directly the spinor bundle as a suitable trivial vector bundle. Following this approach, it seems more appropriate to define the covariant derivative on $D\boldsymbol{M}$ explicitly:

$$\nabla : \Gamma(T\mathcal{M}) \otimes \Gamma(D\boldsymbol{M}) \to \Gamma(D\boldsymbol{M}), \quad (X, \sigma) \mapsto \nabla_X \sigma = \partial_X \sigma + \frac{1}{4} X^\mu \Gamma^\rho_{\mu\nu} \gamma_\rho \gamma^\nu \sigma, \tag{3.31}$$

where σ is regarded as a smooth $\mathbb{C}^4$-valued function on $\mathcal{M}$, $X^\mu = e^\mu(X)$ are the components of X in the fixed frame and $\Gamma^\rho_{\mu\nu} = e^\rho(\nabla_{\varepsilon_\mu}\varepsilon_\nu)$ are the Christoffel symbols of the Levi-Civita connection with respect to the given frame. The covariant derivative is naturally extended to cospinors by imposing the identity

$$\partial_X(\omega(\sigma)) = (\nabla_X \omega)(\sigma) + \omega(\nabla_X \sigma),$$

for each vector field $X \in \Gamma(T\mathcal{M})$, for each spinor field $\sigma \in \Gamma(D\boldsymbol{M})$ and for each cospinor $\omega \in \Gamma(D^*\boldsymbol{M})$. We extend further ∇ to mixed spinor-tensor fields via the Leibniz rule. As an example, we show that $\nabla\gamma = 0$, the computation being carried out using frame components:

$$\begin{aligned}
\nabla_{\varepsilon_\mu}\gamma &= -\Gamma^\rho_{\mu\nu} e^\nu \otimes \gamma_\rho + \frac{1}{4}\Gamma^\sigma_{\mu\nu} e^\rho \otimes [\gamma_\sigma \gamma^\nu, \gamma_\rho] \\
&= -\Gamma^\rho_{\mu\nu} e^\nu \otimes \gamma_\rho + \frac{1}{4}\Gamma^\sigma_{\mu\nu} e^\rho \otimes (\gamma_\sigma\{\gamma^\nu, \gamma_\rho\} - \{\gamma_\sigma, \gamma_\rho\}\gamma^\nu) \\
&= -\frac{1}{2}\Gamma^\rho_{\mu\nu} e^\nu \otimes \gamma_\rho - \frac{1}{2}\Gamma^\sigma_{\mu\tau}\eta_{\sigma\nu}\eta^{\tau\rho} e^\nu \otimes \gamma_\rho = 0.
\end{aligned} \tag{3.32}$$

To conclude, we exploited the identity $\Gamma^{\sigma}_{\mu\nu}\eta_{\rho\sigma} + \Gamma^{\sigma}_{\mu\rho}\eta_{\nu\sigma} = 0$, which follows from $\nabla g = 0$ written in frame components. Notice that $[\cdot,\cdot]$ and $\{\cdot,\cdot\}$ are used here to denote respectively the commutator and the anti-commutator of matrices.

Using the covariant derivatives both for spinors and for cospinors, together with our choice of the γ-matrices, we can introduce the first order linear differential operators $\nabla\!\!\!\!/_{\mathrm{s}} : \Gamma(\boldsymbol{DM}) \to \Gamma(\boldsymbol{DM})$ and $\nabla\!\!\!\!/_{\mathrm{c}} : \Gamma(\boldsymbol{D^*M}) \to \Gamma(\boldsymbol{D^*M})$ defined according to

$$\nabla\!\!\!\!/_{\mathrm{s}}\sigma = \mathrm{Tr}_g(\gamma\,\nabla\sigma), \qquad \forall\sigma \in \Gamma(\boldsymbol{DM}), \tag{3.33}$$

$$\nabla\!\!\!\!/_{\mathrm{c}}\omega = \mathrm{Tr}_g(\nabla\omega\,\gamma), \qquad \forall\omega \in \Gamma(\boldsymbol{D^*M}), \tag{3.34}$$

where Tr_g denotes the metric-contraction of the covariant two-tensor $\gamma\,\nabla\sigma$ taking values in $\boldsymbol{DM}$ and similarly for $\nabla\omega\,\gamma \in \Gamma(T^*\mathcal{M} \otimes T^*\mathcal{M} \otimes \boldsymbol{D^*M})$. With respect to the fixed frame $(\varepsilon_\mu)_{\mu=0,\dots,3}$ (3.33) reads $\nabla\!\!\!\!/_{\mathrm{s}}\sigma = \eta^{\mu\nu}\gamma_\mu\nabla_{\varepsilon_\nu}$ while, using the abstract tensor notation, one has $\nabla\!\!\!\!/_{\mathrm{s}}\sigma = g^{ab}\gamma_a\nabla_b\sigma$. Similar considerations apply to $\nabla\!\!\!\!/_{\mathrm{c}}$.

We can now write down the Dirac equation both for spinors and for cospinors in the usual form:

$$i\nabla\!\!\!\!/_{\mathrm{s}}\sigma - m\sigma = 0, \qquad\qquad -i\nabla\!\!\!\!/_{\mathrm{c}}\omega - m\omega = 0. \tag{3.35}$$

For convenience, we introduce the differential operators $P_{\mathrm{s}} = i\nabla\!\!\!\!/_{\mathrm{s}} - m\,\mathrm{id}_{\Gamma(\boldsymbol{DM})}$ for spinors and $P_{\mathrm{c}} = -i\nabla\!\!\!\!/_{\mathrm{s}} - m\,\mathrm{id}_{\Gamma(\boldsymbol{D^*M})}$ for cospinors. Exploiting the properties of the γ-matrices listed in (3.25), and taking into account the action of the adjunction (3.27) and of the charge conjugations (3.28) and (3.29) on sections, one can easily prove that $A \circ P_{\mathrm{s}} = P_{\mathrm{c}} \circ A$, $C_{\mathrm{s}} \circ P_{\mathrm{s}} = P_{\mathrm{s}} \circ C_{\mathrm{s}}$ and $C_{\mathrm{c}} \circ P_{\mathrm{c}} = P_{\mathrm{c}} \circ C_{\mathrm{c}}$.

To investigate the properties of the Dirac equation, we introduce an integral pairing between sections of the spinor bundle $\boldsymbol{DM}$ and sections of its dual $\boldsymbol{D^*M}$. For each pair of sections $\omega \in \Gamma(\boldsymbol{D^*M})$ and $\sigma \in \Gamma(\boldsymbol{DM})$ such that $\mathrm{supp}\,\omega \cap \mathrm{supp}\,\sigma$ is compact, we define

$$\langle\omega,\sigma\rangle = \int_{\mathcal{M}} \omega(\sigma)\,\mathrm{dvol}_{\boldsymbol{M}}.$$

This pairing is per construction linear in both arguments. Since the adjunction A maps spinors to cospinors, we can use it to form integral pairings between spinors and between cospinors:

$$(\sigma,\tau)_{\mathrm{s}} = \langle A\sigma,\tau\rangle, \qquad (\omega,\zeta)_{\mathrm{c}} = \langle\zeta, A^{-1}\omega\rangle, \tag{3.36}$$

where $\sigma,\tau \in \Gamma(\boldsymbol{DM})$ are such that the intersection of their supports is compact and $\omega,\zeta \in \Gamma(\boldsymbol{D^*M})$ satisfy the same condition. Notice that, due to the anti-linearity of A, both pairings defined above are linear in the second argument and anti-linear in the first. Furthermore, it is easy to check that $(\cdot,\cdot)_{\mathrm{s}}$ induces a Hermitian form on $\Gamma_0(\boldsymbol{DM})$. In fact, given $\sigma,\tau \in \Gamma(\boldsymbol{DM})$ such that their supports have compact overlap, one has $\sigma^\dagger(\gamma_0\tau) = \tau^T(\gamma_0^T\overline{\sigma})$, hence, using also the identity $\gamma_0^\dagger = \gamma_0$, one deduces that

$$\overline{(\sigma,\tau)_s} = \overline{\int_{\mathcal{M}} \sigma^\dagger(\gamma_0\tau)\,\mathrm{dvol}_{\boldsymbol{M}}} = \int_{\mathcal{M}} \tau^\dagger(\gamma_0\sigma)\,\mathrm{dvol}_{\boldsymbol{M}} = (\tau,\sigma)_s. \tag{3.37}$$

Similarly, $(\cdot,\cdot)_c$ induces a Hermitian form on $\Gamma_0(D^*\boldsymbol{M})$.

Using the properties (3.25), one realizes that $\gamma : \Gamma(D\boldsymbol{M}) \to \Gamma(D\boldsymbol{M})$ is formally self-adjoint with respect to $(\cdot,\cdot)_s$. Similarly, $\gamma : \Gamma(D^*\boldsymbol{M}) \to \Gamma(D^*\boldsymbol{M})$ has the same property with respect to $(\cdot,\cdot)_c$. Furthermore, both $\nabla\!\!\!/_s$ and $\nabla\!\!\!/_c$ coincide with their formal adjoints with respect to $(\cdot,\cdot)_s$ and respectively to $(\cdot,\cdot)_c$ up to the sign and, moreover, still up to the sign, they are formal duals of each other with respect to $\langle\cdot,\cdot\rangle$. Specifically, consider any pair of spinors $\sigma,\tau \in \Gamma(D\boldsymbol{M})$, any pair of cospinors $\omega,\zeta \in \Gamma(D^*\boldsymbol{M})$ and any pair formed by a spinor $\upsilon \in \Gamma(D\boldsymbol{M})$ and a cospinor $\varpi \in \Gamma(D^*\boldsymbol{M})$. Assume that the supports of the sections in each pair have compact overlap. Then the following identities hold:

$$(\nabla\!\!\!/_s\sigma,\tau)_s = -(\sigma,\nabla\!\!\!/_s\tau)_s, \qquad (\nabla\!\!\!/_c\omega,\zeta)_c = -(\omega,\nabla\!\!\!/_c\zeta)_c, \qquad \langle\nabla\!\!\!/_c\varpi,\upsilon\rangle = -\langle\varpi,\nabla\!\!\!/_s\upsilon\rangle. \tag{3.38}$$

For the sake of clarity, below we prove the first identity. The proof of the others is analogous.

$$\begin{aligned}(\nabla\!\!\!/_s\sigma,\tau)_s + (\sigma,\nabla\!\!\!/_s\tau)_s &= \int_{\mathcal{M}} \mathrm{Tr}_g\Big(\big(A(\gamma\,\nabla\sigma)\big)(\tau) + (A\sigma)(\gamma\,\nabla\tau)\Big)\,\mathrm{dvol}_{\boldsymbol{M}}\\ &= \int_{\mathcal{M}} \mathrm{Tr}_g\Big(\big(\nabla(A\sigma)\big)(\gamma\tau) + (A\sigma)\big(\nabla(\gamma\tau)\big)\Big)\,\mathrm{dvol}_{\boldsymbol{M}}\\ &= \int_{\mathcal{M}} \mathrm{Tr}_g\Big(\nabla\big((A\sigma)(\gamma\tau)\big)\Big)\,\mathrm{dvol}_{\boldsymbol{M}}\\ &= \int_{\mathcal{M}} \mathrm{d}*\big((A\sigma)(\gamma\tau)\big) = 0,\end{aligned} \tag{3.39}$$

where we used (3.25), the Leibniz rule, the identity $\nabla\gamma = 0$ proved in (3.32) and Stokes' theorem. We remind the reader that d is the exterior derivative for differential forms over $\mathcal{M}$, while $*$ denotes the Hodge star operator defined out of the metric g and out of the orientation $\mathfrak{o}$ of the globally hyperbolic spacetime $\boldsymbol{M}$. From the identities in (3.38) it follows that both P_s and P_c are formally self-adjoint differential operators. Therefore, it is enough to exhibit retarded and advanced Green operators for each of them to conclude that those are unique, see Lemma 3.2.13, and that P_s and P_c are Green hyperbolic, see Definition 3.2.10. Furthermore, Proposition 3.2.12 entails that the retarded/advanced Green operator for P_c is the formal dual of the advanced/retarded Green operator for P_s. To construct the Green operators we are interested in, we observe that $\nabla\!\!\!/_s^2 = \nabla\!\!\!/_s \circ \nabla\!\!\!/_s$ and $\nabla\!\!\!/_c^2 = \nabla\!\!\!/_c \circ \nabla\!\!\!/_c$ are both normally hyperbolic operators. Consider for example $\nabla\!\!\!/_s^2$. For each $\sigma \in \Gamma(D\boldsymbol{M})$, one has the following:

$$
\begin{aligned}
\nabla\!\!\!\!/_{\rm s}^2\sigma &= \gamma^\mu\nabla_{\varepsilon_\mu}(\gamma^\nu\nabla_{\varepsilon_\nu}\sigma) = \gamma^\mu\gamma^\nu\nabla_{\varepsilon_\mu}\nabla_{\varepsilon_\nu}\sigma \\
&= -\gamma^\nu\gamma^\mu\nabla_{\varepsilon_\mu}\nabla_{\varepsilon_\nu}\sigma + 2\eta^{\mu\nu}\nabla_{\varepsilon_\mu}\nabla_{\varepsilon_\nu}\sigma \\
&= -\gamma^\nu\gamma^\mu(\nabla_{\varepsilon_\nu}\nabla_{\varepsilon_\mu}\sigma + \nabla_{\varepsilon_\mu}\nabla_{\varepsilon_\nu}\sigma - \nabla_{\varepsilon_\nu}\nabla_{\varepsilon_\mu}\sigma) + 2\eta^{\mu\nu}\nabla_{\varepsilon_\mu}\nabla_{\varepsilon_\nu}\sigma \\
&= -\nabla\!\!\!\!/_{\rm s}^2\sigma + 2\Box_\nabla\sigma + \frac{1}{2}R\sigma.
\end{aligned} \tag{3.40}
$$

Notice that the last computation has been performed with respect to the chosen co-frame $e = (e^\mu)_{\mu=0,\dots,3}$. $\Box_\nabla : \Gamma(DM) \to \Gamma(DM)$ denotes the d'Alembert operator constructed out of the connection ∇, while R is the scalar curvature. Let us also mention that, for the last equality in the computation above, we used the first Bianchi identity and the anti-commutation relations between the γ-matrices. Therefore, one concludes that $\nabla\!\!\!\!/_{\rm s}^2 = \Box_\nabla + R/4$, hence it is normally hyperbolic, *cf.* Definition 3.2.8.

Proposition 3.3.6 *Let $\boldsymbol{M} = (\mathcal{M}, g, \mathfrak{o}, \mathfrak{t})$ be a four-dimensional globally hyperbolic spacetime together with a co-frame $e = (e^\mu)_{\mu=0,\dots,3}$. The first order linear differential operators $P_{\rm s} : \Gamma(D\boldsymbol{M}) \to \Gamma(D\boldsymbol{M})$ and $P_{\rm c} : \Gamma(D^*\boldsymbol{M}) \to \Gamma(D^*\boldsymbol{M})$, which rule the dynamics of spinors and respectively of cospinors, are formally self-adjoint with respect to $(\cdot,\cdot)_{\rm s}$ and respectively to $(\cdot,\cdot)_{\rm c}$. Furthermore, they are both Green hyperbolic. In particular, their retarded and advanced Green operators are given by*

$$
E_{\rm s}^\pm = P_{\rm s}F_{\rm s}^\pm, \qquad\qquad E_{\rm c}^\pm = P_{\rm c}\mathcal{F}^\pm, \tag{3.41}
$$

where $F_{\rm s}^+$ and $F_{\rm s}^-$ denote the retarded and advanced Green operators for the Green hyperbolic operator $P_{\rm s}^2 = P_{\rm s}P_{\rm s} : \Gamma(D\boldsymbol{M}) \to \Gamma(D\boldsymbol{M})$, while $\mathcal{F}^+$ and $\mathcal{F}^-$ denote those corresponding to the Green hyperbolic operator $P_{\rm c}^2 = P_{\rm c}P_{\rm c} : \Gamma(D^*\boldsymbol{M}) \to \Gamma(D^*\boldsymbol{M})$.

Proof Formal self-adjointness of $P_{\rm s}$ follows directly from (3.38). In fact, the minus sign which appears while integrating by parts $\nabla\!\!\!\!/_{\rm s}$ is reabsorbed by the imaginary unit while passing from one argument of $(\cdot,\cdot)_{\rm s}$ to the other due to anti-linearity in the first argument of the pairing. A similar argument shows that also $P_{\rm c}$ is formally self-adjoint with respect to $(\cdot,\cdot)_{\rm c}$.

It is enough to exhibit retarded and advanced Green operators to conclude that both $P_{\rm s}$ and $P_{\rm c}$ are Green hyperbolic, *cf.* Definition 3.2.10. Specifically, in the following we shall prove that the operators introduced in (3.41) are actually the sought Green operators. We focus on the case of spinors, the other being completely analogous. First of all, we prove that the formally self-adjoint operator $P_{\rm s}^2$ is Green hyperbolic as claimed. In fact, on account of the identity $\nabla\!\!\!\!/_{\rm s}^2 = \Box_\nabla + R/4$, which is a consequence of (3.40), one concludes that $P_{\rm s}^2 = -\nabla\!\!\!\!/_{\rm s}^2 - 2im\nabla\!\!\!\!/_{\rm s} + m^2$. Therefore, according to Definition 3.2.8, $-P_{\rm s}^2$ is normally hyperbolic, hence it admits retarded and advanced Green operators, see [2] and [6, Chap. 3]. Reversing the sign, one gets retarded and advanced Green operators $F_{\rm s}^+$ and $F_{\rm s}^-$ for $P_{\rm s}^2$, thus showing that $P_{\rm s}^2$ is Green hyperbolic. To conclude the proof, we show that $E_{\rm s}^+ = P_{\rm s}F_{\rm s}^+$ and $E_{\rm s}^- = P_{\rm s}F_{\rm s}^-$ are retarded and advanced Green operators for $P_{\rm s}$. The support properties of retarded and advanced Green operators are satisfied since F_s^+ and F_s^- are retarded and advanced Green operators and, moreover, being a differential operator, P_s does not enlarge supports. Indeed, for each $\sigma \in \Gamma_{pc/fc}(DM)$, one has $P_s E_s^{\pm}\sigma = P_s^2 F_s^{\pm}\sigma = \sigma$. It remains only to check that $E_s^{\pm} P_s \sigma = \sigma$. Let us take $\tau \in \Gamma_0(DM)$ and consider $(E_s^{\pm} P_s \sigma, \tau)_s$:

$$
\begin{aligned}
(E_s^{\pm} P_s \sigma, \tau)_s &= (E_s^{\pm} P_s \sigma, P_s E_s^{\mp} \tau)_s = (P_s E_s^{\pm} P_s \sigma, E_s^{\mp} \tau)_s \\
&= (P_s \sigma, E_s^{\mp} \tau)_s = (\sigma, P_s E_s^{\mp} \tau)_s = (\sigma, \tau)_s.
\end{aligned}
$$

In the last chain of identities we exploited repeatedly the formal self-adjointness of P_s and the identity $E_s^{\pm} P_s \upsilon = \upsilon$, which holds true for all $\upsilon \in \Gamma_{pc/fc}(DM)$. Since $(\cdot, \cdot)_s$ provides a non-degenerate pairing between $\Gamma(DM)$ and $\Gamma_0(DM)$, we deduce that $E_s^{\pm} P_s \sigma = \sigma$, thus completing the proof.

Indeed, the fact that P_s and P_c are formally self-adjoint with respect to $(\cdot, \cdot)_s$ and to $(\cdot, \cdot)_c$ has a counterpart involving the corresponding retarded and advanced Green operators on account of Lemma 3.2.13. A similar argument applies to the fact P_c is the formal dual of P_s with respect to $\langle \cdot, \cdot \rangle$, see (3.38) and Proposition 3.2.12. Summing up, one has the following identities for all $\sigma, \tau \in \Gamma_0(DM)$ and for all $\omega, \zeta \in \Gamma_0(D^*M)$:

$$
(E_s^{\pm}\sigma, \tau)_s = (\sigma, E_s^{\mp}\tau)_s, \quad (E_c^{\pm}\omega, \zeta)_c = (\omega, E_c^{\mp}\zeta)_c, \quad \langle E_c^{\pm}\omega, \sigma \rangle = \langle \omega, E_s^{\mp}\sigma \rangle. \tag{3.42}
$$

Proposition 3.3.6 concludes our discussion about the dynamics of the Dirac field. In fact, introducing the advanced-minus-retarded operators $E_s = E_s^- - E_s^+$ and $E_c = E_c^- - E_c^+$ corresponding to P_s and respectively to P_c, one can easily represent all on-shell spinors and cospinors over the four-dimensional globally hyperbolic spacetime M, see Theorem 3.2.15.

3.3.2.2 Classical Observables

From the previous section we know that the on-shell configurations of the Dirac field are either spinors or cospinors, namely sections of either DM or D^*M, satisfying the Dirac equation (3.35). We shall consider now a class of functionals on these field configurations. As further properties, we shall require that this class is large enough to separate different on-shell configurations and that its elements are represented faithfully by some vector space, to be endowed later with the Hermitian structure canonically induced by the Dirac Lagrangian. Let us start with $\tau \in \Gamma_0(DM)$ and $\zeta \in \Gamma_0(D^*M)$ to introduce the functional S_τ for spinors and the functional C_ω for cospinors:

$$
\begin{aligned}
S_\tau : \Gamma(DM) \to \mathbb{C}, \quad & \sigma \mapsto (\tau, \sigma)_s, \\
C_\zeta : \Gamma(D^*M) \to \mathbb{C}, \quad & \omega \mapsto (\zeta, \omega)_c.
\end{aligned} \tag{3.43}
$$

Since both $(\cdot,\cdot)_s$ and $(\cdot,\cdot)_c$ induce non-degenerate bilinear pairings on $\Gamma_0(DM)\times\Gamma(DM)$ and respectively on $\Gamma_0(D^*M)\times\Gamma(DM)$, one can identify the vector spaces of functionals $\{S_\tau : \tau\in\Gamma_0(DM)\}$ and $\{C_\zeta : \zeta\in\Gamma_0(D^*M)\}$ with $\Gamma_0(DM)$ and respectively with $\Gamma_0(D^*M)$. These identifications are implemented via the anti-linear maps $\tau\in\Gamma_0(DM)\mapsto S_\tau$ and $\zeta\in\Gamma_0(D^*M)\mapsto C_\zeta$. Let us stress one fact, which follows from non-degeneracy of the pairings $(\cdot,\cdot)_s$ and $(\cdot,\cdot)_c$. The functionals $\{S_\tau : \tau\in\Gamma_0(DM)\}$ on spinors and the functionals $\{C_\zeta : \zeta\in\Gamma_0(D^*M)\}$ on cospinors are sufficiently many to separate different off-shell field configurations, hence on-shell ones in particular. Therefore, our separability requirement is already achieved.

The functionals introduced above do not take into account the dynamics for Dirac fields. We can easily overcome this hurdle restricting the domains to on-shell configurations. Let us introduce the spaces of on-shell spinors and of on-shell cospinors:

$$\mathsf{Sol}^s=\{\sigma\in\Gamma(DM):\ P_s\sigma=0\},\quad \mathsf{Sol}^c=\{\omega\in\Gamma(D^*M):\ P_c\omega=0\}. \tag{3.44}$$

Given $\tau\in\Gamma_0(DM)$ and $\zeta\in\Gamma_0(D^*M)$, with a slight abuse of notation, we denote the restrictions $S_\tau:\mathsf{Sol}^s\to\mathbb{C}$ and $C_\zeta:\mathsf{Sol}^c\to\mathbb{C}$ of the original functionals introduced in (3.43) by the same symbols. This restriction causes some redundancies in the spaces $\Gamma_0(DM)$ and $\Gamma_0(D^*M)$, which do not faithfully represent the functionals after the restriction to on-shell configurations. This fact is explicitly shown in the next example.

Example 3.3.6 Let us consider $\tau\in\Gamma_0(DM)$. Since P_s does not enlarge supports, $P_s\tau$ is still a compactly supported section of DM. Hence, we can consider the functional $S_{P_s\tau}$. We show that this functional vanishes when restricted to Sol^s. In fact, according to Proposition 3.3.6, one deduces that P_s is formally self-adjoint with respect to $(\cdot,\cdot)_s$, which entails that $S_{P_s\tau}(\sigma)=S_\tau(P_s\sigma)=0$ for all $\sigma\in\mathsf{Sol}^s$. In full analogy, $C_{P_c\zeta}$ vanishes on Sol^c for all $\zeta\in\Gamma_0(D^*M)$ since P_c is formally self-adjoint with respect to $(\cdot,\cdot)_c$. Summing up, the elements of $P_s(\Gamma_0(DM))$ and of $P_c(\Gamma_0(D^*M))$ are redundant since they provide only trivial functionals respectively on Sol^s and on Sol^c.

To implement our second requirement for classical observables, namely that the space representing functionals should be free of redundancies (or, equivalently, functionals should be represented faithfully by this space), we simply take a quotient by the subspace of (co)spinors inducing functionals which vanish on-shell:

$$N^s=\{\tau\in\Gamma_0(DM):\ S_\tau(\sigma)=0,\ \forall\sigma\in\mathsf{Sol}^s\} \tag{3.45}$$
$$N^c=\{\zeta\in\Gamma_0(D^*M):\ C_\zeta(\omega)=0,\ \forall\omega\in\mathsf{Sol}^c\}. \tag{3.46}$$

As anticipated, we introduce the quotient spaces

$$\mathcal{E}^s=\Gamma_0(DM)/N^s,\qquad \mathcal{E}^c=\Gamma_0(D^*M)/N^c. \tag{3.47}$$

$\mathcal{E}^s$ and $\mathcal{E}^c$ are regarded as the spaces of linear classical observables respectively for spinors and for cospinors. In fact, these spaces faithfully represent the restrictions to on-shell configurations of the functionals defined in (3.43), which are sufficiently many to distinguish between different on-shell configurations. For example, consider the case of spinors. It is clear that the equivalence class $[\tau] \in \mathcal{E}^{\mathrm{s}}$ induces a unique functional $S_\tau : \mathsf{Sol}^{\mathrm{s}} \to \mathbb{C}$, independent of the choice of $\tau \in [\tau]$. Indeed different representatives induce different functionals on $\Gamma(DM)$ (off-shell), but, per definition of N^{s}, all these functionals have the same restriction to $\mathsf{Sol}^{\mathrm{s}}$ (on-shell). Therefore one has an anti-linear map $[\tau] \in \mathcal{E}^{\mathrm{s}} \mapsto S_\tau$. Again per definition of N^{s}, this map is injective, thus providing a faithful way to represent by means of $\mathcal{E}^{\mathrm{s}}$ the restrictions to on-shell configurations of the functionals in (3.43).

Remark 3.3.7 Using our knowledge about the dynamics of the Dirac field, *cf.* Sect. 3.3.2.1, we can prove that

$$N^{\mathrm{s}} = P_{\mathrm{s}}(\Gamma_0(D\boldsymbol{M})), \qquad N^{\mathrm{c}} = P_{\mathrm{c}}(\Gamma_0(D^*\boldsymbol{M})), \tag{3.48}$$

meaning that all redundant functionals are of the form presented in Example 3.3.6. As always, we focus our attention to the case of spinors only, the argument being basically the same in the case of cospinors too. On account of Example 3.3.6, one already has the inclusion $P_{\mathrm{s}}\Gamma_0(D\boldsymbol{M}) \subset N^{\mathrm{s}}$. For the converse inclusion, take $\tau \in N^{\mathrm{s}}$ and notice that $(\tau, E_{\mathrm{s}}\sigma)_{\mathrm{s}} = S_\tau(E_{\mathrm{s}}\sigma) = 0$ for all $\sigma \in \Gamma_0(D\boldsymbol{M})$ since E_{s} is the advanced-minus-retarded operator for P_{s}. P_{s} is formally self-adjoint with respect to $(\cdot,\cdot)_{\mathrm{s}}$ as shown in Proposition 3.3.6. Therefore, recalling the properties of retarded and advanced Green operators, we have $(E_{\mathrm{s}}\tau, \sigma)_{\mathrm{s}} = -(\tau, E_{\mathrm{s}}\sigma)_{\mathrm{s}} = 0$ for all $\sigma \in \Gamma_0(D\boldsymbol{M})$, hence $E_{\mathrm{s}}\tau = 0$ due to the non-degeneracy of $(\cdot,\cdot)_{\mathrm{s}}$. Recalling (3.8), one finds $\upsilon \in \Gamma_0(D\boldsymbol{M})$ such that $P_{\mathrm{s}}\upsilon = \tau$, thus showing that $N^{\mathrm{s}} \subset P_{\mathrm{s}}(\Gamma_0(D\boldsymbol{M}))$.

So far, we determined the spaces $\mathcal{E}^{\mathrm{s}}$ and $\mathcal{E}^{\mathrm{c}}$ of classical observables for Dirac spinors and cospinors. Yet, to formulate the corresponding quantum field theory, one needs suitable Hermitian structures in order to write down the usual anti-commutation relations for Dirac fields. This is the purpose of the next proposition.

Proposition 3.3.7 *Consider a four-dimensional globally hyperbolic spacetime* $\boldsymbol{M} = (\mathcal{M}, g, \mathfrak{o}, \mathfrak{t})$ *and take a co-frame* $e = (e^\mu)_{\mu=0,\dots,3}$ *on it. Let* P_{s} *and* P_{c} *denote the differential operators ruling the dynamics of spinors and respectively of cospinors. Introduce the corresponding advanced-minus-retarded operators* E_{s} *and* E_{c}*. Those defined below are non-degenerate Hermitian forms on* $\mathcal{E}^{\mathrm{s}}$ *and respectively on* $\mathcal{E}^{\mathrm{c}}$*:*

$$\begin{aligned} \mathrm{h}^{\mathrm{s}} &: \mathcal{E}^{\mathrm{s}} \times \mathcal{E}^{\mathrm{s}} \to \mathbb{C}, \quad ([\sigma],[\tau]) \mapsto -i(\sigma, E_{\mathrm{s}}\tau)_{\mathrm{s}}, \\ \mathrm{h}^{\mathrm{c}} &: \mathcal{E}^{\mathrm{c}} \times \mathcal{E}^{\mathrm{s}} \to \mathbb{C}, \quad ([\omega],[\zeta]) \mapsto i(\omega, E_{\mathrm{c}}\zeta)_{\mathrm{c}}, \end{aligned} \tag{3.49}$$

where the representatives $\sigma \in [\sigma]$, $\tau \in [\tau]$, $\zeta \in [\zeta]$ *and* $\omega \in [\omega]$ *are chosen arbitrarily. Furthermore, following* (3.27)*, the antilinear isomorphism*

$$A : \mathcal{E}^{\mathrm{s}} \to \mathcal{E}^{\mathrm{c}}, \qquad [\tau] \mapsto [A\tau] \tag{3.50}$$

relates h^{s} *to* h^{c}*, namely one has* $\mathrm{h}^{\mathrm{c}}(A[\sigma], A[\tau]) = \mathrm{h}^{\mathrm{s}}([\tau],[\sigma])$.

Proof We shall discuss explicitly the spinor case. The argument in the case of cospinors is very similar. First of all, let us show that h^{s} is a well-defined non-degenerate Hermitian form. As a starting point, consider the map

$$(\cdot, E_{\mathrm{s}}\cdot)_{\mathrm{s}} : \Gamma_0(DM) \times \Gamma_0(DM) \to \mathbb{C}. \tag{3.51}$$

Since E_{s} is linear, this map is sesquilinear as $(\cdot,\cdot)_{\mathrm{s}}$ is. Furthermore, P_{s} is formally self-adjoint with respect to $(\cdot,\cdot)_{\mathrm{s}}$ and $P_{\mathrm{s}} \circ E_{\mathrm{s}} = 0 = E_{\mathrm{s}} \circ P_{\mathrm{s}}$ on $\Gamma_0(DM)$. This entails that $(\cdot, E_{\mathrm{s}}\cdot)_{\mathrm{s}}$ vanishes whenever one of its arguments is of the form $P_{\mathrm{s}}\tau$ for any $\tau \in \Gamma_0(DM)$. It follows that the form defined in (3.51) descends to the quotient $\Gamma_0(D^*M)/P_{\mathrm{s}}(\Gamma_0(D^*M))$. On account of (3.47) and of (3.48), the space of classical observables $\mathcal{E}^{\mathrm{s}}$ for spinors is exactly of this form, hence h^{s} is a well-defined sesquilinear form on $\mathcal{E}^{\mathrm{s}}$, namely it is anti-linear in the first argument and linear in the second.

The second part of the proof is devoted to showing that h^{s} is Hermitian. This follows from the fact that $(\cdot,\cdot)_{\mathrm{s}}$ provides a Hermitian form on $\Gamma_0(DM)$, cf. (3.37). Specifically, given $\sigma, \tau \in \Gamma_0(DM)$, the following holds:

$$\overline{\mathrm{h}^{\mathrm{s}}([\sigma],[\tau])} = \overline{-i(\sigma, E_{\mathrm{s}}\tau)_{\mathrm{s}}} = i(E_{\mathrm{s}}\tau,\sigma)_{\mathrm{s}} = -i(\tau, E_{\mathrm{s}}\sigma)_{\mathrm{s}} = \mathrm{h}^{\mathrm{s}}([\tau],[\sigma]). \tag{3.52}$$

Notice that in the third equality we exploited the formal self-adjointness of P_{s} with respect to $(\cdot,\cdot)_{\mathrm{s}}$, which entails that $(E_{\mathrm{s}}\sigma,\tau)_{\mathrm{s}} = -(\sigma, E_{\mathrm{s}}\tau)_{\mathrm{s}}$ for all $\sigma, \tau \in \Gamma_0(DM)$.

We still have to prove that h^{s} is non-degenerate. To this aim, consider $[\tau] \in \mathcal{E}^{\mathrm{s}}$ such that $\mathrm{h}^{\mathrm{s}}([\sigma],[\tau]) = 0$ for all $[\sigma] \in \mathcal{E}^{\mathrm{s}}$. Our goal is to show that this condition implies $[\tau] = 0$. In fact, one deduces that $(\sigma, E_{\mathrm{s}}\tau)_{\mathrm{s}}$ has to vanish for all $\sigma \in \Gamma_0(DM)$. Therefore, by non-degeneracy of the pairing $(\cdot,\cdot)_{\mathrm{s}}$ between $\Gamma_0(DM)$ and $\Gamma(DM)$, it descends that $E_{\mathrm{s}}\tau = 0$, hence there exists $\upsilon \in \Gamma_0(DM)$ such that $P_{\mathrm{s}}\upsilon = \tau$. This proves that $[\tau] = 0$.

The last part of the proof focuses on the relation between h^{s} and h^{c}. Since $A \circ P_{\mathrm{s}} = P_{\mathrm{c}} \circ A$, it follows that $A : \mathcal{E}^{\mathrm{s}} \to \mathcal{E}^{\mathrm{c}}$ is well-defined. Furthermore, this is an anti-linear isomorphism of vector spaces since $A : DM \to D^*M$ is an anti-linear vector bundle isomorphism. For each $\sigma, \tau \in \Gamma_0(DM)$, one has the following chain of equalities:

$$\begin{aligned}\mathrm{h}^{\mathrm{c}}(A[\sigma], A[\tau]) &= i(A\sigma, E_{\mathrm{c}}A\tau)_{\mathrm{c}} = i(A\sigma, AE_{\mathrm{s}}\tau)_{\mathrm{c}}\\ &= i(E_{\mathrm{s}}\tau,\sigma)_{\mathrm{s}} = -i(\tau, E_{\mathrm{s}}\sigma)_{\mathrm{s}} = \mathrm{h}^{\mathrm{s}}([\tau],[\sigma]).\end{aligned}$$

For the second equality, we used the identity $E_{\mathrm{c}} \circ A = A \circ E_{\mathrm{s}}$ on $\Gamma_0(DM)$, which follows from $A \circ P_{\mathrm{s}} = P_{\mathrm{c}} \circ A$ on $\Gamma(DM)$.

Remark 3.3.8 The general theory of Green hyperbolic operators provides an isomorphism between the space $\mathcal{E}^{\mathrm{s}}$ of classical observables for spinors and the space $\mathsf{Sol}^{\mathrm{s}}_{sc}$ of on-shell spinors with spacelike compact support, see Proposition 3.2.16. This isomorphism is realized by the advanced-minus-retarded operator E_{s} for P_{s}, which is the differential operator ruling the dynamics for Dirac spinors. Similarly, E_{c}, the advanced-minus-retarded operator corresponding to P_c, provides an isomorphism between $\mathcal{E}^c$ and Sol^c_{sc}:

$$I_s : \mathcal{E}^s \to \mathsf{Sol}^s_{sc}, \quad [\tau] \to E_s\tau,$$
$$I_c : \mathcal{E}^c \to \mathsf{Sol}^c_{sc}, \quad [\zeta] \to E_c\zeta.$$

I^s and I^c become isomorphisms of Hermitian spaces as soon as Sol^s_{sc} and Sol^c_{sc} are endowed with the usual Hermitian structures for Dirac fields written in terms of the initial data on a spacelike Cauchy surface Σ for the four-dimensional globally hyperbolic spacetime $\boldsymbol{M} = (\mathcal{M}, g, \mathfrak{o}, \mathfrak{t})$. Denoting with $\boldsymbol{n}$ the future-pointing unit normal vector field on Σ and with $d\Sigma$ the volume form naturally induced on Σ, one introduces the following non-degenerate Hermitian forms on Sol^s_{sc} and on Sol^c_{sc}:

$$\mathrm{H}^s : \mathsf{Sol}^s_{sc} \times \mathsf{Sol}^s_{sc} \to \mathbb{C}, \qquad \mathrm{H}^s(\sigma, \tau) = \int_\Sigma (A\sigma)(\not{n}\tau)\, d\Sigma, \tag{3.53}$$

$$\mathrm{H}^c : \mathsf{Sol}^c_{sc} \times \mathsf{Sol}^c_{sc} \to \mathbb{C}, \qquad \mathrm{H}^c(\omega, \zeta) = \int_\Sigma \zeta(\not{n}A^{-1}\omega)\, d\Sigma, \tag{3.54}$$

where $\not{n} = \gamma(\boldsymbol{n})$ denotes the section over Σ of $\mathrm{End}(D\boldsymbol{M})$ obtained evaluating the $\mathrm{End}(D\boldsymbol{M})$-valued one-form γ on the vector field $\boldsymbol{n}$ at each point of Σ. One can prove that I_s preserves the Hermitian structures by mimicking the strategy used in (3.19) for the real scalar field and by relying on the calculation presented in (3.39). More explicitly, given $\sigma, \tau \in \Gamma_0(D\boldsymbol{M})$, one finds the following:

$$\begin{aligned}
\mathrm{h}^s([\sigma], [\tau]) &= -i \int_{J^-_M(\Sigma)} (A(P_s E^+_s\sigma))(E_s\tau)\, \mathrm{dvol}_M - i \int_{J^+_M(\Sigma)} (A(P_s E^-_s\sigma))(E_s\tau)\, \mathrm{dvol}_M \\
&= - \int_{J^-_M(\Sigma)} \mathrm{d} * \Big((A(E^+_s\sigma))(\gamma(E_s\tau))\Big) - \int_{J^+_M(\Sigma)} \mathrm{d} * \Big((A(E^-_s\sigma))(\gamma(E_s\tau))\Big) \\
&= - \int_\Sigma (A(E^+_s\sigma))(\not{n}(E_s\tau))\, d\Sigma + \int_\Sigma (A(E^-_s\sigma))(\not{n}(E_s\tau))\, d\Sigma \\
&= \mathrm{H}^s(E_s\sigma, E_s\tau).
\end{aligned}$$

Notice that, after the integration by parts of the terms involving $\not{\nabla}_s$ has been performed, only boundary terms are left due to the fact that $P_s E_s \tau = 0$. The case of cospinors follows suit.

At this stage we have the Hermitian forms h^s and h^c defined on the spaces of classical observables $\mathcal{E}^s$ and $\mathcal{E}^c$ respectively for spinors and for cospinors. Therefore, we can consider the Hermitian spaces $(\mathcal{E}^s, \mathrm{h}^s)$ and $(\mathcal{E}^c, \mathrm{h}^c)$. Furthermore, according to Proposition 3.3.7, $A : \mathcal{E}^s \to \mathcal{E}^c$ establishes a strict relation between the two Hermitian structures h^s and h^c. These Hermitian spaces and their relation are exactly the data needed in order to pass from the classical Dirac field to the corresponding quantum counterpart. However, before turning our attention to the quantum case, we would like to investigate some properties of the Hermitian spaces $(\mathcal{E}^{\mathrm{s}}, \mathrm{h}^{\mathrm{s}})$ and $(\mathcal{E}^{\mathrm{c}}, \mathrm{h}^{\mathrm{c}})$.

Theorem 3.3.8 *Let* $\boldsymbol{M} = (\mathcal{M}, g, \mathfrak{o}, \mathfrak{t})$ *be a four-dimensional globally hyperbolic spacetime and take a co-frame* $e = (e^{\mu})_{\mu=0,\dots,3}$ *on it. Let* $(\mathcal{E}^{\mathrm{s}}, \mathrm{h}^{\mathrm{s}})$ *and* $(\mathcal{E}^{\mathrm{c}}, \mathrm{h}^{\mathrm{c}})$ *be the Hermitian spaces of classical observables defined above respectively for spinors and for cospinors. The following properties hold:*

Causality *The Hermitian structures vanish on pairs of observables localized in causally disjoint regions. More precisely, let* $\sigma, \tau \in \Gamma_0(D\boldsymbol{M})$ *be such that* $\mathrm{supp}\,\sigma \cap J(\mathrm{supp}\,\tau) = \emptyset$. *Then* $\mathrm{h}^{\mathrm{s}}([\sigma], [\tau]) = 0$. *Similarly, taking* $\omega, \zeta \in \Gamma_0(D^*\boldsymbol{M})$ *such that* $\mathrm{supp}\,\omega \cap J_{\boldsymbol{M}}(\mathrm{supp}\,\zeta) = \emptyset$, *one has* $\mathrm{h}^{\mathrm{c}}([\omega], [\zeta]) = 0$.

Time-slice axiom *Let* $\mathcal{O} \subset \mathcal{M}$ *be a globally hyperbolic open neighborhood of a spacelike Cauchy surface* Σ *for* $\boldsymbol{M}$, *namely* $\mathcal{O}$ *is an open neighborhood of* Σ *in* $\mathcal{M}$ *containing all causal curves for* $\boldsymbol{M}$ *whose endpoints lie in* $\mathcal{O}$. *In particular, the restriction of* $\boldsymbol{M}$ *to* $\mathcal{O}$ *provides a globally hyperbolic spacetime* $\boldsymbol{O} = (\mathcal{O}, g|_{\mathcal{O}}, \mathfrak{o}|_{\mathcal{O}}, \mathfrak{t}|_{\mathcal{O}})$. *Furthermore, as a co-frame on* $\boldsymbol{O}$, *we consider the restriction of the co-frame e on* $\boldsymbol{M}$. *Denote with* $(\mathcal{E}^{\mathrm{s}}_{\boldsymbol{M}}, \mathrm{h}^{\mathrm{s}}_{\boldsymbol{M}})$ *and with* $(\mathcal{E}^{\mathrm{s}}_{\boldsymbol{O}}, \mathrm{h}^{\mathrm{s}}_{\boldsymbol{O}})$ *the Hermitian spaces of observables for spinors respectively over* $\boldsymbol{M}$ *and over* $\boldsymbol{O}$. *Similarly, let* $(\mathcal{E}^{\mathrm{c}}_{\boldsymbol{M}}, \mathrm{h}^{\mathrm{c}}_{\boldsymbol{M}})$ *and* $(\mathcal{E}^{\mathrm{c}}_{\boldsymbol{O}}, \mathrm{h}^{\mathrm{c}}_{\boldsymbol{O}})$ *denote the Hermitian spaces of observables for cospinors respectively over* $\boldsymbol{M}$ *and over* $\boldsymbol{O}$. *Then the maps* $L_{\mathrm{s}} : \mathcal{E}^{\mathrm{s}}_{\boldsymbol{O}} \to \mathcal{E}^{\mathrm{s}}_{\boldsymbol{M}}$ *and* $L_{\mathrm{c}} : \mathcal{E}^{\mathrm{c}}_{\boldsymbol{O}} \to \mathcal{E}^{\mathrm{c}}_{\boldsymbol{M}}$, *defined by* $L_{\mathrm{s}}[\tau] = [\tau]$ *for all* $\tau \in \Gamma_0(D\boldsymbol{O})$ *and by* $L_{\mathrm{c}}[\zeta] = [\zeta]$ *for all* $\zeta \in \Gamma_0(D^*\boldsymbol{O})$, *are isomorphisms of Hermitian spaces.*[10]

Proof The proof of this theorem follows slavishly that of Theorem 3.3.1 for the real scalar case. In fact, the only difference is that we are replacing symplectic structures with Hermitian ones. Apart from that, the proof presented there holds in this case as well. In fact, the argument relies only on the Green hyperbolicity of the differential operators ruling the dynamics and indeed both P_{s} and P_{c} have this property according to Proposition 3.3.6.

3.3.2.3 Quantum Field Theory

Given a four-dimensional globally hyperbolic spacetime $\boldsymbol{M} = (\mathcal{M}, g, \mathfrak{o}, \mathfrak{t})$ and choosing a co-frame $e = (e^{\mu})_{\mu=0,\dots,3}$ on it, in the previous section we were able to construct all the kinematical and dynamical objects related to the classical theory of the Dirac field. In particular, we obtained two Hermitian spaces of classical observables for the Dirac field on $\boldsymbol{M}$. The first one, $(\mathcal{E}^{\mathrm{s}}, \mathrm{h}^{\mathrm{s}})$, is used to test on-shell spinors, while the second one, $(\mathcal{E}^{\mathrm{c}}, \mathrm{h}^{\mathrm{c}})$, pertains to cospinors. Furthermore, the two Hermitian spaces are related by an anti-linear isomorphism $A : \mathcal{E}^{\mathrm{s}} \to \mathcal{E}^{\mathrm{c}}$, which satisfies $\mathrm{h}^{\mathrm{c}}(A[\sigma], A[\tau]) = \mathrm{h}^{\mathrm{s}}([\tau], [\sigma])$ for all $[\sigma], [\tau] \in \mathcal{E}^{\mathrm{s}}$. Let us stress that these spaces faithfully represent a class of linear functionals defined on on-shell Dirac

[10]The sections on the right-hand-side in the definitions of L_{s} and of L_{c} are the extensions by zero to the whole spacetime of the sections which appear on the left-hand-side.

fields, which is rich enough to distinguish between different field configurations. These properties motivate our interpretation of $\mathcal{E}^{\mathrm{s}}$ and $\mathcal{E}^{\mathrm{c}}$ as spaces of classical observables for the Dirac field.

Now we want to switch from the classical field theoretical description to its quantum counterpart. As for the scalar case, we shall only construct a suitable algebra of observables, omitting any discussion concerning algebraic states, a topic which will be addressed in Chap. 5. This result is achieved by considering the unital $*$-algebra $\mathcal{A}$ defined as follows. Starting from the unital $*$-algebra freely generated over $\mathbb{C}$ by the symbols $\mathbf{1}$, $\Phi([\tau])$ and $\Psi([\zeta])$ for all $[\tau] \in \mathcal{E}^{\mathrm{s}}$ and for all $[\zeta] \in \mathcal{E}^{\mathrm{c}}$, we impose the relations listed below, thus obtaining the sought unital $*$-algebra $\mathcal{A}$:

$$\Phi(a[\sigma] + b[\tau]) = a\Phi([\sigma]) + b\Phi([\tau]), \tag{3.55}$$

$$\Phi([\sigma])^* = \Psi(A[\sigma]), \tag{3.56}$$

$$\Phi([\sigma]) \cdot \Phi([\tau]) + \Phi([\tau]) \cdot \Phi([\sigma]) = 0, \tag{3.57}$$

$$\Psi([\omega]) \cdot \Psi([\zeta]) + \Psi([\zeta]) \cdot \Psi([\omega]) = 0, \tag{3.58}$$

$$\Psi([\zeta]) \cdot \Phi([\tau]) + \Phi([\tau]) \cdot \Psi([\zeta]) = \mathrm{h}^{\mathrm{c}}(A[\tau], [\zeta])\mathbf{1}. \tag{3.59}$$

These relations must hold for all $a, b \in \mathbb{C}$, for all $[\sigma], [\tau] \in \mathcal{E}^{\mathrm{s}}$ and for all $[\omega], [\zeta] \in \mathcal{E}^{\mathrm{c}}$. In view of (3.55) the map $\Phi : \mathcal{E}^{\mathrm{s}} \to \mathcal{A}$, $[\tau] \mapsto \Phi([\tau])$ is linear. On account of (3.56) and $A : \mathcal{E}^{\mathrm{s}} \to \mathcal{E}^{\mathrm{c}}$ being anti-linear, the map $\Psi : \mathcal{E}^{\mathrm{c}} \to \mathcal{A}$, $[\zeta] \mapsto \Psi([\zeta])$, is linear too. To conclude, (3.57)–(3.59) provide the canonical anti-commutation relations (CAR) for the Dirac field. A more concrete construction can be obtained mimicking the one for the real scalar field, see Sect. 3.3.1.2 and [1]. Specifically, we consider the vector space $A = \bigoplus_{k \in \mathbb{N}_0} (\mathcal{E}^{\mathrm{s}} \oplus \mathcal{E}^{\mathrm{c}})^{\otimes k}$.[11] This is endowed with the product specified $\cdot : A \times A \to A$ defined by

$$\{u_k\} \cdot \{v_k\} = \{w_k\}, \qquad w_k = \sum_{i+j=k} u_i \otimes v_j. \tag{3.60}$$

Clearly, endowing A with $\cdot$ provides a unital algebra, whose unit is given by $\mathbf{1} = \{1, 0, \dots\}$. The generators of this algebra are

$$\Phi([\tau]) = \left\{0, \begin{pmatrix} [\tau] \\ 0 \end{pmatrix}, 0, \dots\right\}, \qquad \Psi([\zeta]) = \left\{0, \begin{pmatrix} 0 \\ [\zeta] \end{pmatrix}, 0, \dots\right\}, \tag{3.61}$$

for all $[\tau] \in \mathcal{E}^{\mathrm{s}}$ and for all $[\zeta] \in \mathcal{E}^{\mathrm{c}}$. So far, the construction is almost identical to the one for the real scalar field. The only difference is that we replaced the complexification of the space of classical observables for the scalar field with the direct sum of the spaces of classical observables for spinors and for cospinors. The involution $* : B \to B$ is implemented by means of the anti-linear isomorphism $A : \mathcal{E}^{\mathrm{s}} \to \mathcal{E}^{\mathrm{c}}$:

[11] As usual, the component of the direct sum corresponding to the degree $k = 0$ is simply $\mathbb{C}$.

$$\left\{0,\dots,0,\begin{pmatrix}[\tau_1]\\ [\zeta_1]\end{pmatrix}\otimes\cdots\otimes\begin{pmatrix}[\tau_k]\\ [\zeta_k]\end{pmatrix},0,\dots\right\}^*$$
$$=\left\{0,\dots,0,\begin{pmatrix}A^{-1}[\zeta_k]\\ A[\tau_k]\end{pmatrix}\otimes\cdots\otimes\begin{pmatrix}A^{-1}[\zeta_1]\\ A[\tau_1]\end{pmatrix},0,\dots\right\},$$

for all $k \in \mathbb{N}_0$, for all $[\tau_1],\dots,[\tau_k] \in \mathcal{E}^{\mathrm{s}}$ and for all $[\zeta_1],\dots,[\zeta_k] \in \mathcal{E}^{\mathrm{c}}$. As always, $*$ is extended to all elements of B by anti-linearity, thus turning B into a unital $*$-algebra. The canonical anti-commutation relations are implemented taking the quotient of B by the two-sided $*$-ideal I of B generated by the elements listed below:

$$\Phi([\sigma])\cdot\Phi([\tau])+\Phi([\tau])\cdot\Phi([\sigma]), \tag{3.62}$$
$$\Psi([\omega])\cdot\Psi([\zeta])+\Psi([\zeta])\cdot\Psi([\omega]), \tag{3.63}$$
$$\Psi([\zeta])\cdot\Phi([\tau])+\Phi([\tau])\cdot\Psi([\zeta])-\mathrm{h}^{\mathrm{c}}(A[\tau],[\zeta])\mathbf{1}, \tag{3.64}$$

for all $[\sigma],[\tau] \in \mathcal{E}^{\mathrm{s}}$ and for all $[\omega],[\zeta] \in \mathcal{E}^{\mathrm{c}}$. The unital $*$-algebra $\mathcal{A} = B/I$ resulting from the quotient is a concrete realization of the one presented in the first part of the present section.

Having established the algebra $\mathcal{A}$ describing the quantum theory of the free Dirac field on the four-dimensional globally hyperbolic spacetime $\boldsymbol{M}$, we would like to investigate some of its properties, as well as its relation to the traditional presentation of the quantum Dirac field.

Remark 3.3.9 Let us mention that, similarly to the scalar case, see Remark 3.3.5, the Dirac quantum field theory presented above reduces to the one usually found in any undergraduate textbook on quantum field theory as soon as $\boldsymbol{M}$ is Minkowski spacetime. This can be seen by means of a suitable Fourier expansion of the solutions to the field equations.

The properties of the classical theory of the Dirac field, which were investigated in Theorem 3.3.8, have counterparts at the quantum level. We conclude this section analyzing this aspect.

Theorem 3.3.9 *Let $\boldsymbol{M} = (\mathcal{M}, g, \mathfrak{o}, \mathfrak{t})$ be a four-dimensional globally hyperbolic spacetime and take a co-frame $e = (e^\mu)_{\mu=0,\dots,3}$ on it. Let $\mathcal{A}$ be the unital $*$-algebra of quantum observables for the Dirac field on $\boldsymbol{M}$. The following properties hold:*

Causality *The elements of $\mathcal{A}$ localized in causally disjoint regions anti-commute. To wit, let $\zeta \in \Gamma_0(D^*\boldsymbol{M})$ and $\tau \in \Gamma_0(D\boldsymbol{M})$ be such that $\operatorname{supp}\zeta \cap J_{\boldsymbol{M}}(\operatorname{supp}\tau) = \emptyset$. Then $\Psi([\zeta])\cdot\Phi([\tau])+\Phi([\tau])\cdot\Psi([\zeta]) = 0$. In particular, the even subalgebra $\mathcal{A}^{\mathrm{even}}$ of $\mathcal{A}$, whose elements are finite linear combinations of products of an even number of generators of $\mathcal{A}$, fulfills the bosonic version of causality, namely the elements of $\mathcal{A}^{\mathrm{even}}$ localized in causally disjoint regions commute.*

Time-slice axiom *Let $\mathcal{O} \subset \mathcal{M}$ be a globally hyperbolic open neighborhood of a spacelike Cauchy surface Σ for $\boldsymbol{M}$, namely $\mathcal{O}$ is an open neighborhood of Σ in $\mathcal{M}$ containing all causal curves for $\boldsymbol{M}$ whose endpoints lie in $\mathcal{O}$. In* *particular, the restriction of* $\boldsymbol{M}$ *to* $\mathcal{O}$ *provides a globally hyperbolic spacetime* $\boldsymbol{O} = (\mathcal{O}, g|_\mathcal{O}, \mathfrak{o}|_\mathcal{O}, \mathfrak{t}|_\mathcal{O})$*. Furthermore, as a co-frame on* $\boldsymbol{O}$*, we consider the restriction of the co-frame e on* $\boldsymbol{M}$*. Denote with* $\mathcal{A}_{\boldsymbol{M}}$ *and with* $\mathcal{A}_{\boldsymbol{O}}$ *the unital* $*$*-algebras of observables for the Dirac field respectively over* $\boldsymbol{M}$ *and over* $\boldsymbol{O}$*. Then the map* $I : \mathcal{A}_{\boldsymbol{O}} \to \mathcal{A}_{\boldsymbol{M}}$*, defined on generators by* $I(\Phi([\tau])) = \Phi(L_{\mathrm{s}}[\tau])$ *for all* $[\tau] \in \mathcal{E}^{\mathrm{s}}_{\boldsymbol{O}}$ *and by* $I(\Psi([\zeta])) = \Psi(L_{\mathrm{c}}[\zeta])$ *for all* $[\zeta] \in \mathcal{E}^{\mathrm{c}}_{\boldsymbol{O}}$*, is an isomorphism of unital* $*$*-algebras. Recall that the Hermitian isomorphisms* $L_{\mathrm{s}} : \mathcal{E}^{\mathrm{s}}_{\boldsymbol{O}} \to \mathcal{E}^{\mathrm{s}}_{\boldsymbol{M}}$ *and* $L_{\mathrm{c}} : \mathcal{E}^{\mathrm{c}}_{\boldsymbol{O}} \to \mathcal{E}^{\mathrm{c}}_{\boldsymbol{M}}$ *were introduced in Theorem 3.3.8.*

Proof Given $\zeta \in \Gamma_0(D^*\boldsymbol{M})$ and $\tau \in \Gamma_0(D\boldsymbol{M})$ such that $\operatorname{supp}\zeta \cap J(\operatorname{supp}\tau) = \emptyset$, from Theorem 3.3.8, one deduces that $\mathrm{h}^{\mathrm{c}}(A[\tau], [\zeta]) = 0$. Therefore, recalling (3.59), one concludes that $\Psi([\zeta]) \cdot \Phi([\tau]) + \Phi([\tau]) \cdot \Psi([\zeta]) = 0$. Let us now consider three generators G_1, G_2, G_3 of $\mathcal{A}$ (they can be either of the form $\Phi([\tau])$ for $[\tau] \in \mathcal{E}^{\mathrm{s}}$ or of the from $\Psi([\zeta])$ for $[\zeta] \in \mathcal{E}^{\mathrm{c}}$). We assume that G_1 and G_2 are localized in a region which is causally disjoint from the one where G_3 is localized. On account of the first part of this theorem, we deduce that $G_i \cdot G_3 + G_3 \cdot G_i = 0$ for $i = 1, 2$. The following chain of equalities follows from the last identity:

$$(G_1 \cdot G_2) \cdot G_3 - G_3 \cdot (G_1 \cdot G_2) = G_1 \cdot G_2 \cdot G_3 + G_1 \cdot G_3 \cdot G_2 \\ - G_1 \cdot G_3 \cdot G_2 - G_3 \cdot G_1 \cdot G_2 = 0.$$

This already entails that all elements of $\mathcal{A}^{\mathrm{even}}$ commute with all elements of $\mathcal{A}$ provided that they are localized in causally disjoint regions. The claim follows as a special case.

The quantum time-slice axiom follows directly from its classical counterpart. The procedure is very similar to the scalar case, see Theorem 3.3.2. In fact, $I : \mathcal{A}_{\boldsymbol{O}} \to \mathcal{A}_{\boldsymbol{M}}$ is a homomorphism of unital $*$-algebras by definition and, moreover, we can introduce an inverse $I^{-1} : \mathcal{A}_{\boldsymbol{M}} \to \mathcal{A}_{\boldsymbol{O}}$ of I simply setting $I^{-1}\Phi([\tau]) = \Phi(L_{\mathrm{s}}^{-1}[\tau])$ for all $[\tau] \in \mathcal{E}^{\mathrm{s}}$ and $I^{-1}\Psi([\zeta]) = \Psi(L_{\mathrm{c}}^{-1}[\zeta])$ for all $[\zeta] \in \mathcal{E}^{\mathrm{c}}$. It is straightforward to check that $I^{-1} \circ I = \mathrm{id}_{\mathcal{A}_{\boldsymbol{O}}}$ and $I \circ I^{-1} = \mathrm{id}_{\mathcal{A}_{\boldsymbol{M}}}$, so that I^{-1} is actually the inverse of I and then I is an isomorphism of unital $*$-algebras as claimed.

3.3.3 The Proca Field

The last example we shall analyze is the Proca field over globally hyperbolic spacetimes. We shall adopt the same approach used in the previous cases. Specifically, we shall start investigating the properties of the differential operator which rules the dynamics of the Proca field. After that, we shall introduce a suitable space of classical observables. In particular, we want a space of sections which can be used to define linear functionals on on-shell Proca fields. As usual, we shall require that the functionals obtained are sufficiently many to distinguish between different on-shell configurations. Furthermore, we want to get rid of the redundancies which might be contained in the space of sections we use to build functionals. As soon as these requirements are achieved, we shall interpret the result as a space of classical observables for the Proca field. In fact, this space faithfully represents a class of functionals defined on-shell, which is rich enough to detect any field configuration. Then we shall endow this space with a symplectic structure, which will play a central role in the prescription to quantize the classical Proca field. This topic will be addressed in the last part of this section. Before starting our analysis, let us mention some references where the Proca field has been studied using the language of algebraic quantum field theory. These are [16, 23, 28].

3.3.3.1 Dynamics and Classical Observables

Let us consider an n-dimensional globally hyperbolic spacetime $\boldsymbol{M} = (\mathcal{M}, g, \mathfrak{o}, \mathfrak{t})$. Unless stated otherwise, from now on, $\boldsymbol{M}$ shall be kept fixed. The off-shell configurations for the Proca field are sections of the cotangent bundle $T^*\mathcal{M}$, namely one-forms over $\mathcal{M}$. Adopting a standard convention, we shall denote the space of k-forms by $\Omega^k(\mathcal{M})$. Let us remind the reader that differential forms constitute a graded algebra with respect to the standard wedge product $\wedge : \Omega^k(\mathcal{M}) \times \Omega^{k'}(\mathcal{M}) \to \Omega^{k+k'}(\mathcal{M})$. To introduce the dynamics, we need two operations on forms, namely the differential $\mathrm{d} : \Omega^k(\mathcal{M}) \to \Omega^{k+1}(\mathcal{M})$ and the Hodge dual $* : \Omega^k(\mathcal{M}) \to \Omega^{n-k}(\mathcal{M})$. d is defined out of the differentiable structure on $\mathcal{M}$, see [15, Sect. 1.1], while $*$ depends also on the metric g and the orientation $\mathfrak{o}$, see [39, Sect. 3.3]. For our purposes it is enough to mention that d is a graded derivative with respect to the wedge product $\wedge$, that $\mathrm{dd} = 0$ and that $*$ is an isomorphism, hence $*^{-1}$ is well-defined. For further details on the theory of differential forms, see [15]. d and $*$ enable us to introduce the codifferential:

$$\delta : \Omega^k(\mathcal{M}) \to \Omega^{k-1}(\mathcal{M}), \qquad \delta = (-1)^k *^{-1} \mathrm{d} * . \tag{3.65}$$

Notice that $\delta\delta = 0$ due to $\mathrm{dd} = 0$. Introducing the symmetric pairing $(\cdot, \cdot)$ between k-forms, defined by

$$(\alpha, \beta) = \int_{\mathcal{M}} \alpha \wedge *\beta,$$

where $\alpha, \beta \in \Omega^k(\mathcal{M})$ have supports with compact intersection, one can prove that δ is the formal adjoint of d with respect to $(\cdot, \cdot)$, meaning that $(\alpha, \delta\beta) = (\mathrm{d}\alpha, \beta)$ for all $\alpha \in \Omega^k(\mathcal{M})$ and $\beta \in \Omega^{k+1}(\mathcal{M})$ such that $\operatorname{supp}\alpha \cap \operatorname{supp}\beta$ is compact. In fact, applying Stokes' theorem, one finds

$$\begin{aligned}
(\mathrm{d}\alpha, \beta) - (\alpha, \delta\beta) &= \int_{\mathcal{M}} (\mathrm{d}\alpha \wedge *\beta - \alpha \wedge *\delta\beta) \\
&= \int_{\mathcal{M}} (\mathrm{d}\alpha \wedge *\beta + (-1)^k \alpha \wedge \mathrm{d} * \beta) \\
&= \int_{\mathcal{M}} \mathrm{d}(\alpha \wedge *\beta) = 0.
\end{aligned}$$

After these preliminaries, we are ready to introduce the Proca equation over $\boldsymbol{M}$ for a form $A \in \Omega^1(\mathcal{M})$:

$$-\delta \mathrm{d}A + m^2 A = 0$$

where $m^2 \in \mathbb{R} \setminus \{0\}$. As for the case of a real scalar field, all our results are valid for all possible values of the mass. Yet, in other chapters of this book, it might be necessary to put restrictions on the sign on the basis of a physical reasoning. Using the abstract index notation, the Proca equation reads

$$\nabla^a(\nabla_a A_b - \nabla_b A_a) + m^2 A_b = 0.$$

For convenience, we introduce the second-order linear differential operator $P = -\delta\mathrm{d} + m^2 \mathrm{id}_{\Omega^1(\mathcal{M})}$. With this definition, the Proca equation can be rewritten as $PA = 0$. Since δ is the formal adjoint of d with respect to $(\cdot, \cdot)$, it follows that $P : \Omega^1(\mathcal{M}) \to \Omega^1(\mathcal{M})$ is formally self-adjoint:

$$(P\alpha, \beta) = -(\mathrm{d}\alpha, \mathrm{d}\beta) + m^2(\alpha, \beta) = (\alpha, P\beta), \tag{3.66}$$

for all $\alpha, \beta \in \Omega^k(\mathcal{M})$ whose supports have compact overlap. In the next proposition we show that P is Green hyperbolic by exhibiting its retarded and advanced Green operators, see Definition 3.2.10.

Proposition 3.3.10 *Let $\boldsymbol{M} = (\mathcal{M}, g, \mathfrak{o}, \mathfrak{t})$ be an n-dimensional globally hyperbolic spacetime and let $Q = -m^{-2}\mathrm{d}\delta + id_{\Omega^1(\mathcal{M})} : \Omega^1(\mathcal{M}) \to \Omega^1(\mathcal{M})$. The second order linear differential operator $P : \Omega^1(\mathcal{M}) \to \Omega^1(\mathcal{M})$, which rules the dynamics of the Proca field on $\boldsymbol{M}$, is formally self-adjoint with respect to $(\cdot, \cdot)$. Furthermore it is Green hyperbolic. In particular, its retarded and advanced Green operators are given by $E^{\pm} = QF^{\pm}$, where F^{+} and F^{-} denote the retarded and advanced Green operators for the normally hyperbolic operator $K = PQ : \Omega^1(\mathcal{M}) \to \Omega^1(\mathcal{M})$.*

Proof In (3.66) we have shown that P is formally self-adjoint. Let us consider $K = PQ$. Recalling that $\mathrm{dd} = 0$, one finds that $K = -\delta\mathrm{d} - \mathrm{d}\delta + m^2 \mathrm{id}_{\Omega^1(\mathcal{M})}$. Therefore K coincides with the Hodge-d'Alembert operator $-\delta\mathrm{d} - \mathrm{d}\delta$, up to a term of order zero in the derivatives. In particular, in local coordinates, the principal part of K is of the form $g^{\mu\nu}\partial_\mu\partial_\nu$, hence K is normally hyperbolic. On account of [2] and [6, Chap. 3], K admits unique retarded and advanced Green operators F^+ and F^-. To conclude the proof we have to show that $E^+ = QF^+$ and $E^- = QF^-$ are retarded and advanced Green operators for P. Since Q is a linear differential operator, it cannot enlarge the support. Therefore, $E^{\pm}$ inherits the correct support property for a retarded/advanced Green operator from $F^{\pm}$. Furthermore, for all $\alpha \in \Omega^1_{pc/fc}(\mathcal{M})$, one has $PE^{\pm}\alpha = KF^{\pm}\alpha = \alpha$. It remains only to check that $E^{\pm}P\alpha = \alpha$ for all $\alpha \in \Omega^1_{pc/fc}(\mathcal{M})$. Exploiting the formal self-adjointness of P and keeping in mind the first part of the proof, one gets the following chain of identities for all $\beta \in \Omega^1_0(\mathcal{M})$:

$$(\beta, E^{\pm} P\alpha) = (P E^{\mp}\beta, E^{\pm} P\alpha) = (E^{\mp}\beta, P E^{\pm} P\alpha)$$
$$= (E^{\mp}\beta, P\alpha) = (P E^{\mp}\beta, \alpha) = (\beta, \alpha).$$

Since $(\cdot,\cdot)$ provides a non-degenerate bilinear pairing between $\Omega^1_0(\mathcal{M})$ and $\Omega^1(\mathcal{M})$, we conclude that $E^{\pm} P\alpha = \alpha$ for all $\alpha \in \Omega^1_{pc/fc}(\mathcal{M})$, hence E^+ and E^- are retarded and advanced Green operators for P, which is consequently Green hyperbolic.

Due to Proposition 3.3.10, we can apply the general theory of Green hyperbolic operators presented in Sect. 3.2 to the operator P ruling the dynamics of the Proca field. In particular we find that $(E^{\pm}\alpha, \beta) = (\alpha, E^{\mp}\beta)$ for all $\alpha, \beta \in \Omega^1_0(\mathcal{M})$ and we can introduce the advanced-minus-retarded operator $E = E^- - E^+$, which enables us to represent any solution starting from a one-form with timelike compact support.

We have completed the analysis of the dynamics of the Proca field over the n-dimensional globally hyperbolic spacetime $\boldsymbol{M}$. In the following we shall focus on the construction of a suitable space of classical observables. Exploiting the non-degenerate bilinear pairing $(\cdot,\cdot)$ between $\Omega^1_0(\mathcal{M})$ and $\Omega^1(\mathcal{M})$, we can introduce a family of linear functionals on off-shell configurations. In fact, given $\alpha \in \Omega^1(\mathcal{M})$, we consider

$$F_\alpha : \Omega^1(\mathcal{M}) \to \mathbb{R}, \qquad A \mapsto (\alpha, A). \tag{3.67}$$

The fact that $(\cdot,\cdot)$ is non-degenerate has two consequences. The first one is that we can identify the vector space $\{F_\alpha : \alpha \in \Omega^1_0(\mathcal{M})\}$, formed by the functionals introduced in (3.67), with $\Omega^1_0(\mathcal{M})$. The second one is that the mentioned space of functionals is sufficiently rich to distinguish between different off-shell configurations. In particular, on-shell configurations can be separated as well, therefore our first requirement for the space of classical observables is achieved by $\Omega^1_0(\mathcal{M})$. Yet, as soon as we go on-shell, which corresponds to restricting the functionals defined above to field configurations $A \in \Omega^1(\mathcal{M})$ satisfying the equation of motion $PA = 0$, some of the functionals become trivial. Before presenting explicit examples of this kind of redundancy for certain elements of $\Omega^1_0(\mathcal{M})$, let us introduce:

$$\mathsf{Sol} = \{A \in \Omega^1(\mathcal{M}) :\ PA = 0\}.$$

Example 3.3.10 The situation here is basically the same as in Example 3.3.1 for the scalar field. In fact, formal self-adjointness of P with respect to $(\cdot,\cdot)$ entails that $F_{P\alpha}(A) = F_\alpha(PA)$ for all $\alpha \in \Omega^1_0(\mathcal{M})$ and for all $A \in \Omega^1(\mathcal{M})$. Therefore, we have $F_{P\alpha}(A) = 0$ for $A \in \mathsf{Sol}$, thus showing that one-forms in $P(\Omega^1_0(\mathcal{M})))$ are redundant in the sense that they provide functionals which always vanish on-shell.

As shown by the example above, $\Omega^1_0(\mathcal{M})$ does not provide a faithful way to represent the restrictions to on-shell configurations of the functionals defined in (3.67). Therefore $\Omega^1_0(\mathcal{M})$ does not meet our second requirement to be identified with the space of classical observables for the Proca field. In order to circumvent this issue, we proceed as in the previous cases. Introducing the subspace

$$N = \{\alpha \in \Omega_0^1(\mathcal{M}) : F_\alpha(A) = 0, \ \forall A \in \mathsf{Sol}\} \subset \Omega_0^1(\mathcal{M})$$

of those one-forms which produce functionals vanishing on-shell, we consider the quotient space

$$\mathcal{E} = \Omega_0^1(\mathcal{M})/N.$$

Per construction $\mathcal{E}$ has no redundancy left and therefore it represents faithfully the restrictions to Sol of the functionals in (3.67). Notice that this representation is realized by sending each equivalence class $[\alpha] \in \mathcal{E}$ to the functional $F_\alpha : \mathsf{Sol} \to \mathbb{R}$ defined by any representative $\alpha \in [\alpha]$. This assignment is well-defined because two representatives of $[\alpha]$ differ by a one-form which produces a functional vanishing on-shell. Since the original space $\Omega_0^1(\mathcal{M})$ is sufficient to separate solutions, this is the case for $\mathcal{E}$ too. These features motivate our interpretation of $\mathcal{E}$ as the space of classical observables for the Proca field on the globally hyperbolic spacetime $\boldsymbol{M}$.

To complete our analysis of the classical theory of the Proca field, we still have to endow $\mathcal{E}$ with a symplectic structure, which will eventually enable us to quantize the model by means of canonical commutation relations. We shall prove first that $N = P(\Omega_0^1(\mathcal{M}))$ and then Proposition 3.2.17 will provide the desired symplectic structure on $\mathcal{E}$. The situation is again basically the same as in the scalar case. In fact, Example 3.3.10 provides the inclusion $P(\Omega_0^1(\mathcal{M})) \subset N$ and we are left with the proof of the converse inclusion, which follows just from the Green hyperbolicity of P. Given $\alpha \in N$, $F_\alpha(E\beta) = 0$ for all $\beta \in \Omega_0^1(\mathcal{M})$ due to $E\beta$ being a solution. Yet, this means that $(E\alpha, \beta) = -(\alpha, E\beta) = 0$ for all $\beta \in \Omega_0^1(\mathcal{M})$, hence the non-degeneracy of $(\cdot, \cdot)$ entails that $E\alpha = 0$. Exploiting (3.8), we find $\gamma \in \Omega_0^1(\mathcal{M})$ such that $P\gamma = \alpha$, thus proving the desired inclusion $N \subset P(\Omega_0^1(\mathcal{M}))$. In particular, $\mathcal{E}$ is the same as the quotient $\Omega_0^1(\mathcal{M})/P(\Omega_0^1(\mathcal{M}))$. Therefore, recalling Proposition 3.2.17, we get a symplectic structure

$$\tau : \mathcal{E} \times \mathcal{E} \to \mathbb{R}, \qquad ([\alpha], [\beta]) \mapsto (\alpha, E\beta). \tag{3.68}$$

In particular, we can regard $(\mathcal{E}, \tau)$ as a symplectic space of classical observables for the Proca field over the globally hyperbolic spacetime $\boldsymbol{M}$.

Remark 3.3.11 It is often customary to present the symplectic form as an integral over a spacelike Cauchy surface Σ of the globally hyperbolic spacetime $\boldsymbol{M} = (\mathcal{M}, g, \mathfrak{o}, \mathfrak{t})$. The integrand is given in terms of those data on Σ which are needed to set up an initial value problem for the field equation of interest. Similarly to the scalar and Dirac cases, we show how to relate our approach to this one. Let us consider $\alpha, \beta \in \Omega_0^1(\mathcal{M})$ and note that $E\beta$ is a solution. We shall split the integral which defines $\tau([\alpha], [\beta])$ in two parts and we shall exploit the properties of the retarded and advanced Green operators to replace α with $PE^\pm\alpha$ in such a way that we are allowed to use Stokes' theorem:

$$\begin{aligned}
\tau([\alpha],[\beta]) &= \int_{J^-_{\mathcal{M}}(\Sigma)} (PE^+\alpha)\wedge *(E\beta) + \int_{J^+_{\mathcal{M}}(\Sigma)} (PE^-\alpha)\wedge *(E\beta) \\
&= -\int_{J^-_{\mathcal{M}}(\Sigma)} \mathrm{d}\big((E^+\alpha)\wedge *\mathrm{d}(E\beta) - (E\beta)\wedge *\mathrm{d}(E^+\alpha)\big) \\
&\quad - \int_{J^+_{\mathcal{M}}(\Sigma)} \mathrm{d}\big((E^-\alpha)\wedge *\mathrm{d}(E\beta) - (E\beta)\wedge *\mathrm{d}(E^-\alpha)\big) \\
&= \int_{\Sigma} \big((E\alpha)\wedge *\mathrm{d}(E\beta) - (E\beta)\wedge *\mathrm{d}(E\alpha)\big) \\
&= \int_{\Sigma} \Big(g\big(E\alpha, \iota_{\boldsymbol{n}}\mathrm{d}(E\beta)\big) - g\big(E\beta, \iota_{\boldsymbol{n}}\mathrm{d}(E\alpha)\big)\Big)\, d\Sigma,
\end{aligned} \tag{3.69}$$

where $d\Sigma$ is the naturally induced volume form on Σ, $\boldsymbol{n}$ denotes the future-pointing unit normal vector field on Σ and $\iota_{\boldsymbol{n}}$ is the operator which inserts the vector field $\boldsymbol{n}$ in the form to which it is applied. Notice that the integration by parts only gives boundary terms since $PE\beta = 0$. Due to (3.69), one can realize that our symplectic space $(\mathcal{E}, \tau)$ is isomorphic to the symplectic space $(\mathsf{Sol}_{sc}, \sigma)$ often considered in the literature, where Sol_{sc} denotes the space of on-shell configurations of the Proca field with spacelike compact support and $\sigma : \mathsf{Sol}_{sc} \times \mathsf{Sol}_{sc} \to \mathbb{R}$ is the symplectic form defined for all $A, B \in \mathsf{Sol}_{sc}$ by

$$\sigma(A, B) = \int_{\Sigma} \big(g(A, \iota_{\boldsymbol{n}}\mathrm{d}B) - g(B, \iota_{\boldsymbol{n}}\mathrm{d}A)\big)\, d\Sigma.$$

Before turning our attention to the quantum theory of the Proca field, we devote a few lines to examine some of the properties of the symplectic space $(\mathcal{E}, \tau)$ of classical observables for the Proca field over the globally hyperbolic spacetime $\boldsymbol{M}$. Notice that we shall not provide the details of the proof since this would be nothing more than a slavish copy of the proof of Theorem 3.3.1.

Theorem 3.3.11 *Let $\boldsymbol{M} = (\mathcal{M}, g, \mathfrak{o}, \mathfrak{t})$ be an n-dimensional globally hyperbolic spacetime and let $(\mathcal{E}, \tau)$ be the symplectic space of classical observables introduced above for the Proca field. The following properties hold:*

Causality *The symplectic structure vanishes on pairs of observables localized in causally disjoint regions. More precisely, let $\alpha, \beta \in \Omega^1_0(\mathcal{M})$ be such that $\mathrm{supp}\,\alpha \cap J_{\boldsymbol{M}}(\mathrm{supp}\,\beta) = \emptyset$. Then $\tau([\alpha],[\beta]) = 0$.*

Time-slice axiom *Let $\mathcal{O} \subset \mathcal{M}$ be a globally hyperbolic open neighborhood of a spacelike Cauchy surface Σ for $\boldsymbol{M}$, namely $\mathcal{O}$ is an open neighborhood of Σ in $\mathcal{M}$ containing all causal curves for $\boldsymbol{M}$ whose endpoints lie in $\mathcal{O}$. In particular, the restriction of $\boldsymbol{M}$ to $\mathcal{O}$ provides a globally hyperbolic spacetime $\boldsymbol{O} = (\mathcal{O}, g|_{\mathcal{O}}, \mathfrak{o}|_{\mathcal{O}}, \mathfrak{t}|_{\mathcal{O}})$. Denote with $(\mathcal{E}_{\boldsymbol{M}}, \tau_{\boldsymbol{M}})$ and with $(\mathcal{E}_{\boldsymbol{O}}, \tau_{\boldsymbol{O}})$ the symplectic spaces of observables for the Proca field respectively over $\boldsymbol{M}$ and over $\boldsymbol{O}$.*

Then the map $L : \mathcal{E}_O \to \mathcal{E}_M$ *defined by* $L[\alpha] = [\alpha]$ *for all* $\alpha \in \Omega^1_0(\mathcal{O})$ *is an isomorphism of symplectic spaces.*[12]

3.3.3.2 Quantum Field Theory

To complete our analysis of the Proca field, we present the quantization of the classical field theory developed in the previous section, which consists of a symplectic space $(\mathcal{E}, \tau)$ of classical observables for the Proca field over a globally hyperbolic spacetime $\boldsymbol{M} = (\mathcal{M}, g, \mathfrak{o}, \mathfrak{t})$. The quantization procedure is completely equivalent to the case of the real scalar field. For this reason we shall skip most of the details, referring the reader to Sect. 3.3.1.2. We introduce the quantum theory of the Proca field in terms of the unital $*$-algebra $\mathcal{A}$ generated over $\mathbb{C}$ by the symbols $\mathbf{1}$ and $\Phi([\alpha])$ for all classical observables $[\alpha] \in \mathcal{E}$ and satisfying the relations listed below:

$$\Phi(a[\alpha] + b[\beta]) = a\Phi([\alpha]) + b\Phi([\beta]), \tag{3.70}$$

$$\Phi([\alpha])^* = \Phi([\alpha]), \tag{3.71}$$

$$\Phi([\alpha]) \cdot \Phi([\beta]) - \Phi([\beta]) \cdot \Phi([\alpha]) = i\tau([\alpha], [\beta])\mathbf{1}, \tag{3.72}$$

for all $a, b \in \mathbb{C}$ and for all $[\alpha], [\beta] \in \mathcal{E}$. As usual, the first relation expresses the linearity of the quantum field, the second relation keeps track of the fact that classically the Proca field is a real field, therefore quantum Proca fields should be Hermitian, and finally the third relation implements the canonical commutation relations (CCR) for Bosonic field theories. We interpret $\mathcal{A}$ as the algebra of quantum observables for the Proca field over the globally hyperbolic spacetime $\boldsymbol{M}$.

We conclude analyzing certain properties of the quantum theory of the Proca field. Mimicking the proof of Theorem 3.3.2 for the real scalar field, and exploiting the properties of the classical theory of the Proca field, which have been developed in Theorem 3.3.11, one obtains the following result.

Theorem 3.3.12 *Let* $\boldsymbol{M} = (\mathcal{M}, g, \mathfrak{o}, \mathfrak{t})$ *be an* n*-dimensional globally hyperbolic spacetime and let* $\mathcal{A}$ *be the unital* $*$*-algebra of observables for the Proca field introduced above. The following properties hold:*

Causality *Elements of the algebra* $\mathcal{A}$ *localized in causally disjoint regions commute. More precisely, let* $\alpha, \beta \in \Omega^1_0(\mathcal{M})$ *be such that* $\operatorname{supp}\alpha \cap J_M(\operatorname{supp}\beta) = \emptyset$*. Then* $\Phi([\alpha]) \cdot \Phi([\beta]) = \Phi([\beta]) \cdot \Phi([\alpha])$.

Time-slice axiom *Let* $\mathcal{O} \subset \mathcal{M}$ *be a globally hyperbolic open neighborhood of a spacelike Cauchy surface* Σ *for* $\boldsymbol{M}$*, namely* $\mathcal{O}$ *is an open neighborhood of* Σ *in* $\mathcal{M}$ *containing all causal curves for* $\boldsymbol{M}$ *whose endpoints lie in* $\mathcal{O}$*. In particular, the restriction of* $\boldsymbol{M}$ *to* $\mathcal{O}$ *provides a globally hyperbolic spacetime* $\boldsymbol{O} = (\mathcal{O}, g|_\mathcal{O}, \mathfrak{o}|_\mathcal{O}, \mathfrak{t}|_\mathcal{O})$*. Denote with* $\mathcal{A}_M$ *and with* $\mathcal{A}_O$ *the unital* $*$*-algebras of*

[12]The differential form on the right-hand-side of the equation which defines L is the extension by zero to the whole spacetime of the differential form which appears on the left-hand-side.

observables for the Proca field respectively over $\boldsymbol{M}$ *and over* $\boldsymbol{O}$. *Then the unit-preserving* $*$*-homomorphism* $\Phi(L) : \mathcal{A}_O \to \mathcal{A}_M$, $\Phi([\alpha]) \mapsto \Phi(L[\alpha])$ *is an isomorphism of* $*$*-algebras, where L is the symplectic isomorphism introduced in Theorem 3.3.11.*

To conclude the chapter we would like to comment briefly on two aspects which have not been discussed. On the one hand we have only treated fields of spin 0, $1/2$ and 1, the latter under the assumption of a non-vanishing mass. This choice was made only for the sake of simplicity since all other cases would involve necessarily a discussion of local gauge invariance, a topic which is still under study and which would require a chapter on its own—for linear gauge theories refer to [31]. We mention a few references for an interested reader: for electromagnetism [8, 10, 18, 21, 23, 46, 49], for spin $3/2$ fields [30, 31], while for massless spin 2 fields and linearized gravity [9, 22]. Another important aspect, neglected in this chapter, concerns the discussion about the existence of a relation between the algebra of observables for a free field theory built on two globally hyperbolic spacetimes which can be related one to the other via an isometric embedding. The analysis of such aspect leads to the formulation of the so-called principle of general local covariance, one of the milestones of modern axiomatic quantum field theory. This principle, together with its consequences, is discussed in Chap. 4.

Acknowledgments The work of M.B. has been supported by a grant of the University of Pavia, which is gratefully acknowledged. C.D. is grateful to Zhirayr Avetisyan for the useful discussions.

References

1. Araki, H.: On Quasifree States of CAR and Bogoliubov Automorphisms. Publ. RIMS Kyoto Univ. **6**, 385 (1970/71)
2. Bär, C.: Green-hyperbolic operators on globally hyperbolic spacetimes. Commun. Math. Phys. **333**, 1585 (2015)
3. Bär, C., Fredenhagen, K. (eds.): Quantum Field Theory on Curved Spacetimes: Concepts and Mathematical Foundations, 1st edn. Springer, Berlin (2009)
4. Bär, C., Ginoux N.: CCR- versus CAR-quantization on curved spacetimes. In: Finster F. et al. (eds.) Quantum Field Theory and Gravity, pp. 183–206. Birkäuser, Basel (2012)
5. Bär, C., Ginoux, N.: Classical and quantum fields on Lorentzian manifolds. In: Bär, C., Lohkamp, J., Schwarz, M. (eds.) Global Differential Geometry, pp. 359–400. Springer, Berlin (2012)
6. Bär, C., Ginoux, N., Pfäffle, F.: Wave Equations on Lorentzian Manifolds and Quantization, 1st edn. Eur. Math. Soc, Zürich (2007)
7. Beem, J.K., Ehrlich, P.E., Easley, K.L.: Global Lorentzian Geometry. 2nd edn. CRC (1996)
8. Benini, M., Dappiaggi, C., Hack, T.-P., Schenkel, A.: A C^*-algebra for quantized principal $U(1)$-connections on globally hyperbolic Lorentzian manifolds. Commun. Math. Phys. **332**, 477 (2014)
9. Benini, M., Dappiaggi, C., Murro, S.: Radiative observables for linearized gravity on asymptotically flat spacetimes and their boundary induced states. J. Math. Phys. **55**, 082301 (2014)
10. Benini, M., Dappiaggi, C., Schenkel, A.: Quantized Abelian principal connections on Lorentzian manifolds. Commun. Math. Phys. **330**, 123 (2014)

11. Benini, M., Dappiaggi, C., Schenkel, A.: Quantum field theories on affine bundles. Ann. Henry Poincaré **15**, 171 (2014)
12. Bernal, A.N., Sanchez, M.: Smoothness of time functions and the metric splitting of globally hyperbolic spacetimes. Commun. Math. Phys. **257**, 43 (2005)
13. Bernal, A.N., Sanchez, M.: Further results on the smoothability of Cauchy hypersurfaces and Cauchy time functions. Lett. Math. Phys. **77**, 183 (2006)
14. Borchers, H.J.: On structure of the algebra of field operators. Il Nuovo Cimento **24**, 214 (1962)
15. Bott, R., Tu, L.W.: Differential forms in algebraic topology, 1st edn. Springer, New York (1982)
16. Dappiaggi, C.: Remarks on the Reeh-Schlieder property for higher spin free fields on curved spacetimes. Rev. Math. Phys. **23**, 1035 (2011)
17. Dappiaggi, C., Hack, T.-P., Pinamonti, N.: The extended algebra of observables for Dirac fields and the trace anomaly of their stress-energy tensor. Rev. Math. Phys. **21**, 1241 (2009)
18. Dappiaggi, C., Lang, B.: Quantization of Maxwell's equations on curved backgrounds and general local covariance. Lett. Math. Phys. **101**, 265 (2012)
19. Dimock, J.: Algebras of local observables on a manifold. Commun. Math. Phys. **77**, 219 (1980)
20. Dimock, J.: Dirac quantum fields on a manifold. Trans. Am. Math. Soc. **269**, 133 (1982)
21. Dimock, J.: Quantized electromagnetic field on a manifold. Rev. Math. Phys. **4**, 223 (1992)
22. Fewster, C.J., Hunt, D.S.: Quantization of linearized gravity in cosmological vacuum spacetimes. Rev. Math. Phys. **25**, 1330003 (2013)
23. Fewster, C.J., Pfenning, M.J.: A quantum weak energy inequality for spin one fields in curved space-time. J. Math. Phys. **44**, 4480 (2003)
24. Fewster, C.J., Schenkel, A.: Locally covariant quantum field theory with external sources. to appear in Ann. Henri Poincaré
25. Fewster, C.J., Verch, R.: A quantum weak energy inequality for Dirac fields in curved spacetime. Commun. Math. Phys. **225**, 331 (2002)
26. Friedlander, F.G.: The Wave Equation on a Curved Space-Time, 1st edn. Cambridge University Press, Cambridge (1975)
27. Fulling, S.A.: Aspects of Quantum Field Theory in Curved Space-Time. Cambridge University Press, Cambridge (1989)
28. Furlani, E.P.: Quantization of massive vector fields in curved space-time. J. Math. Phys. **40**, 2611 (1999)
29. Haag, R., Kastler, D.: An algebraic approach to quantum field theory. J. Math. Phys. **5**, 848 (1964)
30. Hack, T.-P., Makedonski, M.: A No-Go theorem for the consistent quantization of Spin 3/2 Fields on general curved spacetimes. Phys. Lett. B **718**, 1465 (2013)
31. Hack, T.-P., Schenkel, A.: Linear bosonic and fermionic quantum gauge theories on curved spacetimes. Gen. Rel. Grav. **45**, 877 (2013)
32. Hawking, S.W., Ellis, G.F.R.: The Large Scale Structure of Space-time. Cambridge University Press, Cambridge (1997)
33. Hörmander, L.: The Analysis of Linear Partial Differential Operators I, 2nd edn. Springer, Berlin (1990)
34. Hörmander, L.: The Analysis of Linear Partial Differential Operators II, 1st edn. Springer, Berlin (1983)
35. Hörmander, L.: The Analysis of Linear Partial Differential Operators III, 1st edn. Springer, Berlin (1985)
36. Hörmander, L.: The Analysis of Linear Partial Differential Operators IV, 1st edn. Springer, Berlin (1985)
37. Husemoller, D.: Fibre Bundles, 3rd edn. Springer, New York (2004)
38. Isham, C.J.: Spinor fields in four-dimensional space-time. Proc. Roy. Soc. Lond. A **364**, 591 (1978)
39. Jost, J.: Riemannian Geometry and Geometric Analysis, 6th edn. Springer, Berlin (2011)
40. Khavkine, I.: Covariant phase space, constraints, gauge and the Peierls formula. Int. J. Mod. Phys. A **29**(5), 1430009. arXiv:1402.1282 [math-ph] (2014)

41. Khavkine, I.: Characteristics, Conal Geometry and Causality in Locally Covariant Field Theory. arXiv:1211.1914 [gr-qc]
42. Lawson, H.B., Michelsohn, M.-L.: Spin Geometry, 1st edn. Princeton University Press, Princeton (1989)
43. Ozsváth, I., Schucking, E.: Gödel's trip. Am. J. Phys. **71**, 801 (2003)
44. Parker, P.E.: On some theorems of Geroch and Stiefel. J. Math. Phys. **25**, 597 (1984)
45. Pauli, W.: Contributions mathématiques à la théorie des matrices de Dirac. Ann. Inst. H. Poincaré **6**, 109 (1936)
46. Pfenning, M.J.: Quantization of the Maxwell field in curved spacetimes of arbitrary dimension. Class. Quant. Grav. **26**, 135017 (2009)
47. Sanders, K.: The locally covariant Dirac field. Rev. Math. Phys. **22**, 381 (2010)
48. Sanders, K.: A note on spacelike and timelike compactness. Class. Quant. Grav. **30**, 115014 (2013)
49. Sanders, K., Dappiaggi, C., Hack, T.-P.: Electromagnetism, local covariance, the Aharonov-Bohm effect and Gauss' law. Commun. Math. Phys. **328**, 625 (2014)
50. Townsend, P.K.: Black holes: lecture notes. gr-qc/9707012
51. Uhlmann, A.: Über die definition der quantenfelder nach Wightman und Haag. Wiss. Z. Karl-Marx-Univ. Leipzig. **11**, 213 (1962)
52. Wald, R.M.: General Relativity, 1st edn. The University of Chicago Press, Chicago (1984)
53. Wald, R.M.: Quantum Field Theory in Curved Spacetime and Black Hole Thermodynamics, 1st edn. The University of Chicago Press, Chicago (1994)
54. Waldmann, S.: Geometric wave equations. arXiv:1208.4706 [math.DG]
55. Zahn, J.: The renormalized locally covariant Dirac field. Rev. Math. Phys. **26**(1), 1330012 (2014)

Chapter 4
Algebraic Quantum Field Theory in Curved Spacetimes

Christopher J. Fewster and Rainer Verch

4.1 Introduction

There are many approaches to quantum field theory, almost all of which rely heavily, in one way or another, on concepts of symmetry. This refers, in particular, to the behaviour of a quantum field theory with respect to the symmetries of the spacetime (or space, for Euclidean formulations) on which it exists. For example, Poincaré covariance is one of the defining properties for a relativistic quantum field theory on Minkowski space, in conjunction with the concept of locality for observables [86, 143]. Any general account of quantum field theory on curved spacetimes faces the problem that reliance on symmetries is of little assistance except in special cases. For these reasons, most work in the area has, until recently, focussed on specific (usually non-self-interacting) models in specific spacetimes. Our purpose in this chapter is different: we will explore what can be said, in a model-independent way, about quantum fields in general curved spacetimes. This is motivated from several directions: (a) as a matter of general principle, one wishes to understand how far the ideas and methods of quantum field theory can be extended; (b) it recognizes that our knowledge of our own universe is limited in both scope and detail, and that its actual geometry (even setting aside questions of quantum geometric structure) is by no means that of a symmetric spacetime; (c) it allows for some macroscopic features (e.g., a collapsing star, or an experimental apparatus) to be treated as 'given' and not obtained from the microscopic theory; (d) it provides a framework in which controlled approximations of complex situations by simple ones can be discussed.

C.J. Fewster (✉)
Department of Mathematics, University of York, Heslington, York YO10 5DD, UK
e-mail: chris.fewster@york.ac.uk

R. Verch
Institut für Theoretische Physik, Universität Leipzig, Brüderstr. 16, 04103 Leipzig, Germany
e-mail: rainer.verch@uni-leipzig.de

R. Brunetti et al. (eds.), *Advances in Algebraic Quantum Field Theory*,
Mathematical Physics Studies, DOI 10.1007/978-3-319-21353-8_4

Time has proved that the mathematical framework of operator algebras permits a very clear and efficient way to precisely formulate the conceptual underpinnings of quantum field theory—locality and covariance—and to analyse the consequences [2, 6, 14, 86], hence the name "algebraic quantum field theory". Our presentation attempts to follow that line of thought for quantum field theory in curved spacetime. In place of symmetry, the concept of covariance or, more precisely, of local general covariance, is put at the centre of our approach, reflecting the considerable progress that has been made since this concept was given a concise formulation in the early years of this millennium [24, 156]. Early forms of the idea appear in work such as [41, 104, 163].

We should make clear that the main purpose of this contribution is to set out the conceptual and mathematical structure of algebraic quantum field theory in curved spacetime, not its application to concrete situations, such as the Hawking effect [80, 84, 85, 87, 107, 116] or cosmology [36, 40, 157] or the Casimir effect [52, 63, 114]. It is intended to be read alongside the other chapters, which provide the context and application for the structures discussed here. There are also a number of topics that have not been discussed owing to constraints of time, space and energy. Some of these will be listed below, after we indicate what is covered here.

We start by considering the quantized linear scalar Klein-Gordon field on globally hyperbolic spacetimes as a motivating example from which some basic concepts of locally covariant quantum field theory can be read off. Taking these as guidelines, the general concept of locally covariant quantum field theory will be formulated in terms of a functor between a category of spacetimes and a category of $*$-algebras. Further assumptions will be added and their consequences studied, so that the general structure of the theory, in a model-independent algebraic framework, begins to take shape. Central parts of that structure are played by Einstein causality, and the time-slice axiom, which, by interplay with local covariance, induces the notion of relative Cauchy evolution and provides the theory with a dynamical structure. Then the states and their Hilbert-space representations are discussed, as well as the concept of a state space for locally covariant quantum field theories. In that context, the microlocal spectrum condition makes its appearance as the most promising, and at the same time most general, selection criterion for a space of physical states.

A further step in developing the theory derives from the fact that globally hyperbolic spacetimes can be deformed to more symmetric spacetimes. Together with the time-slice axiom and local covariance, this allows one to transfer properties that hold on Minkowski spacetime to general spacetimes. This fact has been observed and exploited in the literature [54, 82, 135, 150, 151, 156], however here we systemize it as a "rigidity argument" for locally covariant quantum field theory, and this is a new and original ingredient of this contribution. We will show that Einstein causality, the Schlieder property and extended locality can all be extended to locally covariant QFTs in this way, and also that the Reeh-Schlieder and the split properties are consequences of closely related arguments.

We will then discuss the relation of several other selection criteria for state spaces of physical states in locally covariant quantum field theories and their relation to the microlocal spectrum condition, mainly quantum energy inequalities, and the existence of ground states (or more generally, passive states) in ultrastatic spacetimes. We go on to consider locally covariant quantum fields and present the spin and statistics relation in that framework. Furthermore, we discuss embeddings of locally covariant quantum field theories into each other, which leads to a locally covariant concept of internal symmetries. That provides a setting in which one can consider the question of what it means that quantum field theories on different spacetimes can be regarded as representing "the same physics". Another new feature of this contribution is the observation that models such as the Klein–Gordon theory can be given a "universal definition" at the functorial level, without direct reference to the theory on particular spacetimes. Finally, we present an argument showing that there is no locally covariant state under very general assumptions.

As mentioned above, there are various topics that we have not been able to include, and while the list of references is extensive, it is certainly not complete. Notable absences include discussion of gauge fields and charge superselection theory in curved spacetime [25, 26, 85, 131] as well as the perturbative construction of locally covariant interacting quantum fields in curved spacetime [20, 21, 81, 94–96]. The development of the latter has led to a formalization of operator product expansions that may be seen as a particular approach towards algebraic quantum field theory in curved spacetimes, mainly investigated by Hollands and Wald [93, 97] (besides other results, this has led to a version of a PCT-theorem in curved spacetimes [92]). We shall not discuss Haag duality [132] or situations of "geometric modular action" [28, 84, 85], nor the relation between Euclidean and Lorentzian quantum field theory for quantum fields in curved spacetimes [100, 101], nor any form of constructive quantum field theory (beyond free fields) in curved spacetimes, on this, cf. [5] and references cited there. A further omission concerns spacetimes that are not globally hyperbolic or have boundaries [105, 106, 110, 111, 128, 165].

Aside from a familiarity with quantum field theory on Minkowski spacetime and the standard terminology from general relativity e.g. at the level of [161], some basic knowledge of category theory is assumed on part of the reader (as regards concepts like category, functor, morphism, naturality, for which see e.g., [113]). Knowledge of typical mathematical concepts of functional analysis in Hilbert spaces are taken for granted, but we will summarize some of the relevant background on operator algebras as far as it is needed.

The abstract structures and arguments will be illustrated by the example of the free linear scalar field. This might give the impression that the theory only makes statements about linear quantum fields. We emphasize that this is not so and that locally covariant quantum field theories with self-interaction have been constructed perturbatively, and that there are also such interacting quantum field theories obeying the time-slice axiom [33, 94–96]; it is these assumptions on which our theoretical arguments mainly rest. The long-standing problem of establishing the existence of interacting quantum field theories beyond perturbation theory in physical spacetime dimension remains; yet we hope that the principle of local covariance will provide a new guideline in the attempts of their construction.

4.2 A Motivating Example

We begin with the simplest model of QFT in curved spacetime, the linear Klein–Gordon field with field equation $P_{\boldsymbol{M}}\phi := (\Box_{\boldsymbol{M}} + m^2 + \xi R_{\boldsymbol{M}})\phi = 0$ on spacetime $\boldsymbol{M}$,[1] where the mass $m \geq 0$ and coupling $\xi \in \mathbb{R}$ are fixed but arbitrary. Here, anticipating later developments, we have put subscripts on the d'Alembertian and scalar curvature, to indicate the spacetime under consideration. In each globally hyperbolic spacetime $\boldsymbol{M}$, the field algebra $\mathscr{A}(\boldsymbol{M})$ of this theory may be presented in terms of generators $\Phi_{\boldsymbol{M}}(f)$ labelled by complex-valued test functions $f \in C_0^\infty(\boldsymbol{M})$ and subject to relations

KG1 linearity of $f \mapsto \Phi_{\boldsymbol{M}}(f)$
KG2 hermiticity, $\Phi_{\boldsymbol{M}}(f)^* = \Phi_{\boldsymbol{M}}(\overline{f})$
KG3 the field equation, $\Phi_{\boldsymbol{M}}(P_{\boldsymbol{M}} f) = 0$
KG4 the canonical commutation relations, $[\Phi_{\boldsymbol{M}}(f), \Phi_{\boldsymbol{M}}(h)] = i E_{\boldsymbol{M}}(f, h) \mathbf{1}_{\mathscr{A}(\boldsymbol{M})}$

which hold for all $f, h \in C_0^\infty(\boldsymbol{M})$ (see, e.g., Chap. 5 modulo slight differences). Equivalently, one can define $\mathscr{A}(\boldsymbol{M})$ as the Borchers-Uhlmann algebra, i.e. the quotient of the tensor algebra over the test-function space $C_0^\infty(\boldsymbol{M})$ by the relations described in KG1 to KG4 [15, 149]. Thus defined, $\mathscr{A}(\boldsymbol{M})$ does not admit the structure of a C^*-algebra. This is not always required, but in some situations, it is useful to have $\mathscr{A}(\boldsymbol{M})$ as a C^*-algebra. The canonical way of reaching a C^*-algebraic description of the quantized linear Klein-Gordon field on $\boldsymbol{M}$ proceeds as follows: Define $\mathcal{K}(\boldsymbol{M}) = C_0^\infty(\boldsymbol{M}, \mathbb{R})/(P_{\boldsymbol{M}} C_0^\infty(\boldsymbol{M}, \mathbb{R}))$, and write $f_\sim = f + P_{\boldsymbol{M}} C_0^\infty(\boldsymbol{M}, \mathbb{R})$ for $f \in C_0^\infty(\boldsymbol{M}, \mathbb{R})$. Then define $\mathscr{A}(\boldsymbol{M})$ to be the *Weyl algebra* of the linear Klein-Gordon field on $\boldsymbol{M}$, which is defined as the (unique [18]) C^*-algebra generated by elements $W_{\boldsymbol{M}}(f_\sim)$, $f_\sim \in \mathcal{K}(\boldsymbol{M})$, and a unit element $\mathbf{1}$, subject to the relations $W_{\boldsymbol{M}}(f_\sim) W_{\boldsymbol{M}}(h_\sim) = \mathrm{e}^{i E_{\boldsymbol{M}}(f,h)/2} W_{\boldsymbol{M}}(f_\sim + h_\sim)$, $W_{\boldsymbol{M}}(-f_\sim) = W_{\boldsymbol{M}}(f_\sim)^*$ and $W_{\boldsymbol{M}}(0) = \mathbf{1}$.

The algebraic description of the theory on $\boldsymbol{M}$ is useful for many applications, when supplemented by a suitable class of states such as the Hadamard class described in Sect. 4.5 and Chap. 5. A rather richer structure is revealed, however, when one relates the algebras obtained on different, but suitably related, spacetimes.

Consider two spacetimes $\boldsymbol{M}$ and $\boldsymbol{N}$ and a smooth map $\psi : \boldsymbol{M} \to \boldsymbol{N}$. Our first aim is to understand what constraints $\boldsymbol{M}$, $\boldsymbol{N}$ and ψ should satisfy in order that there can be a meaningful relationship between $\mathscr{A}(\boldsymbol{M})$ and $\mathscr{A}(\boldsymbol{N})$. The sort of relationship we intend here is one in which the generating smeared fields are related directly to one another in the following way. Provided that ψ is smoothly invertible on its range, we may push forward test functions from $C_0^\infty(\boldsymbol{M})$ to $C_0^\infty(\boldsymbol{N})$ according to

$$(\psi_* f)(p) = \begin{cases} f(\psi^{-1}(p)) & p \in \psi(\boldsymbol{M}) \\ 0 & \text{otherwise.} \end{cases} \tag{4.1}$$

[1] We adopt signature convention $+ - \cdots -$ for the metric and in general adopt the conventions used in Chaps. 3, 5 and 6.

It is natural to use the push-forward to map smeared fields in $\boldsymbol{M}$ to smeared fields in $\boldsymbol{N}$, writing

$$\mathscr{A}(\psi)\Phi_{\boldsymbol{M}}(f) := \Phi_{\boldsymbol{N}}(\psi_* f). \tag{4.2}$$

(In the Weyl formulation, one uses $\mathscr{A}(\psi)W_{\boldsymbol{M}}(f_\sim) = W_{\boldsymbol{N}}((\psi_* f)_\sim)$ where the subscript $\sim$ refers to $\boldsymbol{M}$ and $\boldsymbol{N}$, respectively, and the following discussion would proceed completely analogously.) However, the assignment (4.2) is only well-defined if it is compatible with the algebraic relations holding in $\mathscr{A}(\boldsymbol{M})$ and $\mathscr{A}(\boldsymbol{N})$; in particular, we must have $\Phi_{\boldsymbol{N}}(\psi_* P_{\boldsymbol{M}} f) = 0$ for all test functions $f \in C_0^\infty(\boldsymbol{M})$. Further conditions arise if we wish to extend $\mathscr{A}(\psi)$ from the generators to the full algebra. Here, the simplest possibility is that $\mathscr{A}(\psi)$ should be a $*$-homomorphism that also preserves units. In that case, the commutation relations, together with (4.2), give

$$[\Phi_{\boldsymbol{N}}(\psi_* f), \Phi_{\boldsymbol{N}}(\psi_* h)] = \mathscr{A}(\psi)[\Phi_{\boldsymbol{M}}(f), \Phi_{\boldsymbol{M}}(h)] = iE_{\boldsymbol{M}}(f,h)\mathbf{1}_{\mathscr{A}(\boldsymbol{N})} \tag{4.3}$$

and hence $E_{\boldsymbol{N}}(\psi_* f, \psi_* h) = E_{\boldsymbol{M}}(f,h)$ for all $f, h \in C_0^\infty(\boldsymbol{M})$. We see that $E_{\boldsymbol{M}}$ is the pull-back, $E_{\boldsymbol{M}} = (\psi \times \psi)^* E_{\boldsymbol{N}}$, of $E_{\boldsymbol{N}}$ and its wave-front set [98] therefore obeys

$$\mathrm{WF}(E_{\boldsymbol{M}}) \subset (\psi \times \psi)^* \mathrm{WF}(E_{\boldsymbol{N}}).$$

Given the known structure of both sides, we deduce that ψ^* must map null-covectors on $\boldsymbol{N}$ to null covectors on $\boldsymbol{M}$, and preserve time-orientation. This already restricts ψ to be a conformal isometry, and in fact one may see that it must be an isometry unless $P_{\boldsymbol{M}}$ is conformally invariant. With an eye to other theories such as the pseudoscalar or Maxwell fields, we might reasonably require ψ to preserve not only the time-orientation, but also the spacetime orientation.

We have seen that the local structure of ψ is quite restricted if there is to be any hope of implementing (4.2). In fact, the condition $E_{\boldsymbol{M}} = (\psi \times \psi)^* E_{\boldsymbol{N}}$, of $E_{\boldsymbol{N}}$ also has global consequences: the image $\psi(\boldsymbol{M})$ must be a causally convex subset of $\boldsymbol{N}$, which requires that every causal curve in $\boldsymbol{N}$ whose endpoints lie in $\psi(\boldsymbol{M})$ should be contained entirely in $\psi(\boldsymbol{M})$. Examples showing the failure of this relation in the absence of causal convexity are to be found in [4, 105]; more generally, the conclusion follows from the fact that singularities propagate along null geodesics.

Our discussion has led us, with very little alternative, to a specification of those maps of interest: $\psi : \boldsymbol{M} \to \boldsymbol{N}$ is a smooth, isometric embedding, preserving orientation and time-orientation, and with causally convex image. For such ψ, we now have a unit-preserving $*$-homomorphism $\mathscr{A}(\psi) : \mathscr{A}(\boldsymbol{M}) \to \mathscr{A}(\boldsymbol{N})$ which turns out to be injective.[2] We may observe something more: if we also consider a map $\varphi : \boldsymbol{L} \to \boldsymbol{M}$ obeying these conditions, then the same is true of the composition $\psi \circ \varphi : \boldsymbol{L} \to \boldsymbol{N}$. It is clear from (4.2) that

$$\mathscr{A}(\psi \circ \varphi)\Phi_{\boldsymbol{L}}(f) = \Phi_{\boldsymbol{N}}((\psi \circ \varphi)_* f) = \Phi_{\boldsymbol{N}}(\psi_* \varphi_* f) = \mathscr{A}(\psi)(\mathscr{A}(\varphi)\Phi_{\boldsymbol{L}}(f)) \tag{4.4}$$

[2]The algebra $\mathscr{A}(\boldsymbol{M})$ is simple (and not the zero algebra!), so $\mathscr{A}(\psi)$ either has trivial kernel or full kernel; the latter case is excluded because $\mathscr{A}(\psi)\mathbf{1}_{\mathscr{A}(\boldsymbol{M})} = \mathbf{1}_{\mathscr{A}(\boldsymbol{N})} \neq 0$.

for all $f \in C_0^\infty(\boldsymbol{L})$, and we extend from the generators to obtain

$$\mathscr{A}(\psi \circ \varphi) = \mathscr{A}(\psi) \circ \mathscr{A}(\varphi). \tag{4.5}$$

In addition, it is clear that the identity map $\mathrm{id}_{\boldsymbol{M}}$ of $\boldsymbol{M}$ corresponds to $\mathscr{A}(\mathrm{id}_{\boldsymbol{M}}) = \mathrm{id}_{\mathscr{A}(\boldsymbol{M})}$. These observations may all be summarised in the single statement that *the theory is described by a covariant functor* $\mathscr{A}$ between two categories:

Loc the category whose objects are all globally hyperbolic spacetimes $\boldsymbol{M} = (\mathcal{M}, g, \mathfrak{o}, \mathfrak{t})$ of fixed dimension n with finitely many connected components, and whose morphisms are smooth isometric embeddings, preserving orientation and time-orientation, having causally convex image. Here $\mathcal{M}$ is the underlying manifold, with metric g, while the symbol $\mathfrak{o}$ stands for a choice of orientation, represented by one of the components of the set of nowhere-zero smooth n-forms on $\mathcal{M}$. Similarly, $\mathfrak{t}$ denotes the time-orientation, represented by one of the components of the set of nowhere-zero smooth g-timelike 1-forms on $\mathcal{M}$.

Alg the category of unital $*$-algebras *excluding the zero algebra*, with unit-preserving injective $*$-homomorphisms as morphisms.

Here, we have anticipated future developments by allowing for disconnected spacetimes when defining **Loc**. If $\boldsymbol{M}$ is disconnected, and $\psi : \boldsymbol{M} \to \boldsymbol{N}$ according to the definition above, then causal convexity of $\psi(\boldsymbol{M})$ forces its various components to be causally disjoint—no causal curve can join one component to another. Of course, one should be alive to the possibility that the example has features that might not be shared by all theories. In particular, **Loc** admits spacetime embeddings in which the components of the image have closures that are in causal contact, or allow for self-touchings at their boundaries. While the Klein–Gordon theory has well-defined morphisms corresponding to such embeddings, it is conceivable that there are reasonable theories that do not, and that a more conservative starting point should be found in due course. In the definition of **Alg**, we have excluded the zero unital algebra (consisting of a single element which is both the zero and unit) to avoid some pathologies and to ensure that **Alg** has an initial object, namely the algebra of complex numbers.[3] Accordingly, the unit is distinct from the zero element in every object of **Alg**.

Although we have reached this structure by means of an example, it has a clear physical interpretation and could be motivated in its own terms. Namely, the morphisms ψ specify embeddings in which all causal relations between points in the image $\psi(\boldsymbol{M})$ (with respect to $\boldsymbol{N}$) are already causal relations between the corresponding points in $\boldsymbol{M}$. Physics, by which we mean here degrees of freedom and laws of motion (without yet specifying boundary or initial conditions), in the image region would be expected to correspond to that in the domain spacetime: this is a version of the principle of locality. In particular, we expect the physics on the smaller

[3] An initial object in a category **C** is an object I with the property that there is, to each object C of **C**, exactly one morphism from I to C.

spacetime to be faithfully represented within that of the larger, and that there should be no distinction between physics in the embedded region $\psi(\boldsymbol{M})$ and in the spacetime $\boldsymbol{M}$. The functorial definition provides a consistency mechanism that protects the ignorance of an experimenter within $\psi(\boldsymbol{M})$ of the nature (or even existence) of the spacetime beyond the region under her control. Further discussion of these ideas can be found in [55].

Taking all of the above into account, we will adopt the general assumption that any theory that is both covariant and respects the principle of locality should be described by a covariant functor from **Loc** (or another suitable category of spacetimes) to a category of physical systems, in which morphisms represent embeddings of one system as a subsystem of another and are required to be monic. This is a working hypothesis for the development of a model-independent theory, but note that

1. the whole enterprise is questionable for spacetimes that are smaller in scale than the physical systems they support (e.g., measured by a Compton wavelength)
2. as mentioned, **Loc** may admit too wide a variety of morphisms for some theories
3. for gauge theories in particular the requirement of injectivity is sometimes in conflict with other desirable features of the theory, particularly in order to capture topological aspects. See, e.g., the discussion following Theorem 4.6.3.

4.3 General Assumptions and First Consequences

We begin a more formal development of the structure, which rests on a number of general assumptions. The first has already been motivated:

Assumption 4.3.1 (*Local covariance*) A locally covariant theory is a functor

$$\mathscr{A} : \mathsf{Loc} \to \mathsf{Alg}.$$

Depending on the application, one might wish to specify the target category more stringently, e.g., requiring that $\mathscr{A}$ takes values in the subcategory **C*-Alg** of **Alg**, consisting of unital C^*-algebras. One may also formulate locally covariant descriptions for theories other than QFT by allowing a more general category **Phys**. Here, however, we will remain in the algebraic description for the most part.

Given this starting point, we may define a net of local algebras in each spacetime $\boldsymbol{M} = (\mathcal{M}, g, \mathfrak{o}, \mathfrak{t})$. Let $\mathscr{O}(\boldsymbol{M})$ be the set of all open causally convex subsets of $\boldsymbol{M}$, with at most finitely many connected components (which are necessarily causally disjoint). Then, for each nonempty $O \in \mathscr{O}(\boldsymbol{M})$, we may define a new object $\boldsymbol{M}|_O = (O, g|_O, \mathfrak{o}|_O, \mathfrak{t}|_O)$ of **Loc**, which is simply the set O equipped with the causal structures induced from $\boldsymbol{M}$ and regarded as a spacetime in its own right. In addition, the subset embedding of O in $\mathcal{M}$ is evidently a smooth embedding which is an isometric (time)-orientation preserving map owing to the way we have defined $\boldsymbol{M}$. Thus it defines a morphism $\iota_{\boldsymbol{M};O} : \boldsymbol{M}|_O \to \boldsymbol{M}$ in **Loc**. The functor $\mathscr{A}$ therefore assigns both an algebra $\mathscr{A}(\boldsymbol{M}|_O)$ and a morphism $\mathscr{A}(\iota_{\boldsymbol{M};O})$ of $\mathscr{A}(\boldsymbol{M}|_O)$ into $\mathscr{A}(\boldsymbol{M})$. The image of this morphism,

$$\mathscr{A}^{\mathrm{kin}}(\boldsymbol{M};\, O) = \mathscr{A}(\iota_{\boldsymbol{M};O})(\mathscr{A}(\boldsymbol{M}_O)), \tag{4.6}$$

is called the *kinematic algebra* associated with region O, and gives a description of the physics of the theory within O.[4] The kinematic algebras have some immediate properties. First, suppose that $O_1 \subset O_2$. Then the factorisation $\iota_{\boldsymbol{M};O_1} = \iota_{\boldsymbol{M};O_2} \circ \iota_{\boldsymbol{M}|_{O_2};O_1}$ of the inclusion morphism implies that $\mathscr{A}(\iota_{\boldsymbol{M};O_1}) = \mathscr{A}(\iota_{\boldsymbol{M};O_2}) \circ \mathscr{A}(\iota_{\boldsymbol{M}|_{O_2};O_1})$ and hence that

$$\mathscr{A}^{\mathrm{kin}}(\boldsymbol{M};\, O_1) \subset \mathscr{A}^{\mathrm{kin}}(\boldsymbol{M};\, O_2). \tag{4.7}$$

In other words, the kinematic net is isotonous. Consequently, if $O_1, O_2 \in \mathscr{O}(\boldsymbol{M})$ have a nonempty intersection $O_1 \cap O_2$ (which is seen to be causally convex and therefore an element of $\mathscr{O}(\boldsymbol{M})$) then

$$\mathscr{A}^{\mathrm{kin}}(\boldsymbol{M};\, O_1 \cap O_2) \subset \mathscr{A}^{\mathrm{kin}}(\boldsymbol{M};\, O_1) \cap \mathscr{A}^{\mathrm{kin}}(\boldsymbol{M};\, O_2). \tag{4.8}$$

(One does not expect equality here.) On the other hand, if $O_1, O_2 \in \mathscr{O}(\boldsymbol{M})$ are nonempty and their union is causally convex then $\mathscr{A}^{\mathrm{kin}}(\boldsymbol{M};\, O_1 \cup O_2)$ contains both $\mathscr{A}^{\mathrm{kin}}(\boldsymbol{M};\, O_i)$ and therefore the algebra that they generate, so

$$\mathscr{A}^{\mathrm{kin}}(\boldsymbol{M};\, O_1) \vee \mathscr{A}^{\mathrm{kin}}(\boldsymbol{M};\, O_2) \subset \mathscr{A}^{\mathrm{kin}}(\boldsymbol{M};\, O_1 \cup O_2). \tag{4.9}$$

In the C^*-algebraic setting, where $\mathscr{A} : \mathsf{Loc} \to \mathsf{C^*\text{-}Alg}$, we may sharpen this result so that the left-hand side is the C^*-subalgebra of $\mathscr{A}^{\mathrm{kin}}(\boldsymbol{M};\, O_1 \cup O_2)$ generated by the $\mathscr{A}^{\mathrm{kin}}(\boldsymbol{M};\, O_i)$.[5] Both (4.8) and (4.9) extend to finitely many O_i in obvious ways; if equality holds in (4.9), the theory $\mathscr{A}$ will be described as *finitely additive*.

Next, consider a morphism $\psi : \boldsymbol{M} \to \boldsymbol{N}$. Then the spacetimes $\boldsymbol{M}|_O$ and $\boldsymbol{N}|_{\psi(O)}$ are isomorphic via the map $\hat{\psi}_O$ obtained as the restriction of ψ to O, obeying $\iota_{\boldsymbol{N};\psi(O)} \circ \hat{\psi}_O = \psi \circ \iota_{\boldsymbol{M};O}$. Applying the functor $\mathscr{A}$, and noting that $\mathscr{A}(\hat{\psi}|_O)$ is an isomorphism, we find

$$\mathscr{A}^{\mathrm{kin}}(\boldsymbol{N};\, \psi(O)) = \mathscr{A}(\psi)(\mathscr{A}^{\mathrm{kin}}(\boldsymbol{M};\, O)). \tag{4.10}$$

An important special case arises where $\psi : \boldsymbol{M} \to \boldsymbol{M}$, i.e., $\psi \in \mathrm{End}(\boldsymbol{M})$, in which $\mathscr{A}(\psi)$ defines an endomorphism of the kinematic net, or a net isomorphism in the case where ψ is an isomorphism, $\psi \in \mathrm{Aut}(\boldsymbol{M})$. In particular, we see that there is a homomorphism of monoids from $\mathrm{End}(\boldsymbol{M})$ to $\mathrm{End}(\mathscr{A}(\boldsymbol{M}))$, that restricts to a group homomorphism from $\mathrm{Aut}(\boldsymbol{M})$ to $\mathrm{Aut}(\mathscr{A}(\boldsymbol{M}))$.

[4] Alternatively, and perhaps more in the spirit of a categorical description, one might say that the morphism $\mathscr{A}(\iota_{\boldsymbol{M};O})$, regarded as defining a subobject of $\mathscr{A}(\boldsymbol{M})$, should be the focus here [69].

[5] In a general categorical setting, one would employ the *categorical union* of the $\mathscr{A}^{\mathrm{kin}}(\boldsymbol{M};\, O_i)$.

The properties (4.7), (4.8), (4.9) and (4.10) are direct generalisations of the properties of the nets of local algebras encountered in Minkowski space AQFT; see [86] and Chap. 1. It is remarkable that they all follow without further input from the single assumption that the theory is described functorially. As an application of (4.9) and (4.10), suppose that $\boldsymbol{M}$ has finitely many connected components $\mathcal{M}_i$ and $\psi : \boldsymbol{M} \to \boldsymbol{N}$. Then we have

$$\bigvee_i \mathscr{A}^{\mathrm{kin}}(\boldsymbol{N}; \psi(\mathcal{M}_i)) \subset \mathscr{A}^{\mathrm{kin}}(\boldsymbol{N}; \psi(\mathcal{M})). \tag{4.11}$$

If $\mathscr{A}$ is finitely additive, then equality holds in (4.11); this need not be true in general.

Before proceeding to the other standard assumptions, two further definitions are required. For a region $O \subset \boldsymbol{M}$, we write $O' := \boldsymbol{M} \setminus \overline{J_{\boldsymbol{M}}(O)}$ for its open causal complement; in addition, a morphism $\psi : \boldsymbol{M} \to \boldsymbol{N}$ will be described as *Cauchy* if $\psi(\boldsymbol{M})$ contains a Cauchy surface for $\boldsymbol{N}$ (equivalently, if every inextendible timelike curve in $\boldsymbol{N}$ intersects $\psi(\boldsymbol{M})$). The remaining assumptions are:

Assumption 4.3.2 (*Einstein Causality*) If $O_1, O_2 \in \mathcal{O}(\boldsymbol{M})$ are causally disjoint in the sense that $O_1 \subset O_2' := \boldsymbol{M} \setminus \overline{J_{\boldsymbol{M}}(O_2)}$, then

$$[\mathscr{A}^{\mathrm{kin}}(\boldsymbol{M}; O_1), \mathscr{A}^{\mathrm{kin}}(\boldsymbol{M}; O_2)] = \{0\}. \tag{4.12}$$

Assumption 4.3.3 (*Timeslice*) If $\psi : \boldsymbol{M} \to \boldsymbol{N}$ is Cauchy then $\mathscr{A}(\psi)$ is an isomorphism.

Unless otherwise specified, the term 'locally covariant QFT' will refer to a functor obeying Assumptions 4.3.1–4.3.3. In Sect. 4.6.2 it will be seen that Assumption 4.3.2 is partly redundant: it is enough that Einstein causality should hold for *one* pair of spacelike separated regions in *one* spacetime for it to hold for suitable spacelike separated regions in general spacetimes. Together with other assumptions, Einstein causality leads to an additional *monoidal structure* on the theory—see [22] and Sect. 4.6.2 for discussion. If the theory not only describes observables, but also smeared fermionic fields, for example, then a suitable graded commutator should be employed.

The timeslice assumption is one of the lynch-pins of the structure and encodes the idea that the theory *has* a dynamical law, although *what* it is is left unspecified. It has an immediate consequence: if $O \in \mathcal{O}(\boldsymbol{M})$ is nonempty, then O contains a Cauchy surface of the Cauchy development $D_{\boldsymbol{M}}(O)$—the set of all points p in $\boldsymbol{M}$ with the property that all inextendible piecewise-smooth causal curves through p intersect O, which is open, causally convex and therefore a member of $\mathcal{O}(\boldsymbol{M})$.[6] We already know that $\iota_{\boldsymbol{M};O}$ factors as $\iota_{\boldsymbol{M};O} = \iota_{\boldsymbol{M};D_{\boldsymbol{M}}(O)} \circ \iota_{\boldsymbol{M}|_{D_{\boldsymbol{M}}(O)};O}$. Applying the functor, the

[6]Some authors, notably Penrose [123] and Geroch [83], define the Cauchy development with timelike curves of various types. We follow [8, 90, 119, 161]. Many authors only define the Cauchy development for achronal sets. The fact that $D_{\boldsymbol{M}}(O)$ is open is most easily seen using limit curves cf. [8, Prop. 3.31] or [90, Lem. 6.2.1].

timeslice property entails that $\mathscr{A}(\iota_{M|_{D_M(O)};O})$ is an isomorphism and so

$$\mathscr{A}^{\rm kin}(\boldsymbol{M}; O) = \mathscr{A}^{\rm kin}(\boldsymbol{M}; D_{\boldsymbol{M}}(O)). \tag{4.13}$$

Hence we may immediately strengthen (4.8) to

$$\mathscr{A}^{\rm kin}(\boldsymbol{M}; D_{\boldsymbol{M}}(O_1) \cap D_{\boldsymbol{M}}(O_2)) \subset \mathscr{A}^{\rm kin}(\boldsymbol{M}; O_1) \cap \mathscr{A}^{\rm kin}(\boldsymbol{M}; O_2). \tag{4.14}$$

The timeslice property and its ramifications will be dominant themes in this chapter.

To conclude this section, we note that various models obeying the general assumptions listed have been constructed. The prototypical example is the free Klein–Gordon model described in Sect. 4.2. There, it was shown that the theory is given in terms of a functor $\mathscr{A}$: **Loc** $\to$ **Alg**, with algebras $\mathscr{A}(\boldsymbol{M})$ generated by 'smeared fields' $\Phi_{\boldsymbol{M}}(f)$ $(f \in C_0^\infty(\boldsymbol{M}))$ and subject to relations KG1–4. It is easily seen that, for non-empty $O \in \mathscr{O}(\boldsymbol{M})$, the kinematic algebra $\mathscr{A}^{\rm kin}(\boldsymbol{M}; O)$ is the subalgebra of $\mathscr{A}(\boldsymbol{M})$ generated by those $\Phi_{\boldsymbol{M}}(f)$ with $f \in C_0^\infty(O)$. Then (4.11) holds with equality and Einstein causality holds because $\operatorname{supp} E_{\boldsymbol{M}} f \subset J_{\boldsymbol{M}}(\operatorname{supp} f)$. The timeslice property can be shown by arguments used in Theorem 3 of Chap. 3 (which establishes the causality and timeslice properties at the level of symplectic spaces): if $\psi : \boldsymbol{M} \to \boldsymbol{N}$ is a Cauchy morphism, let $\chi \in C^\infty(\boldsymbol{N})$ be chosen so that $\chi \equiv 0$ to the future of Σ^+ and $\chi \equiv 1$ to the past of Σ^-, where $\Sigma^\pm$ are Cauchy surfaces in $\psi(\boldsymbol{M})$. If $f \in C_0^\infty(\boldsymbol{N})$ then $f' = P_{\boldsymbol{N}} \chi E_{\boldsymbol{N}} f$ may be shown to have compact support in $\psi(\boldsymbol{M})$ and to obey $f - f' \in P_{\boldsymbol{N}} C_0^\infty(\psi(\boldsymbol{M}))$. Then

$$\Phi_{\boldsymbol{N}}(f) = \Phi_{\boldsymbol{N}}(f') = \Phi_{\boldsymbol{N}}(\psi_* \psi^* f') = \mathscr{A}(\psi)\Phi_{\boldsymbol{M}}(\psi^* f')$$

by KG3, the support properties of f' and the definition of $\mathscr{A}(\psi)$. As every generator of $\mathscr{A}(\boldsymbol{N})$ lies in the image of the injective map $\mathscr{A}(\psi)$, it is an isomorphism.

Similarly, models such as the Proca and (with modifications) Dirac fields also fit into the framework [35, 135], as do the perturbatively constructed models of pAQFT—we refer to Chaps. 3 and 2 for details.

As a further model of interest, let us return to the Klein–Gordon theory $\mathscr{A}$. Each algebra $\mathscr{A}(\boldsymbol{M})$ contains a unital $*$-subalgebra $\mathscr{A}^{\rm ev}(\boldsymbol{M})$ of elements that generated by the unit together with bilinear elements $\Phi_{\boldsymbol{M}}(f)\Phi_{\boldsymbol{M}}(h)$ $(f, h \in \mathscr{D}(\boldsymbol{M}))$. It is easily seen that, for $\psi : \boldsymbol{M} \to \boldsymbol{N}$, the morphism $\mathscr{A}(\psi)$ restricts to a morphism $\mathscr{A}^{\rm ev}(\psi) : \mathscr{A}^{\rm ev}(\boldsymbol{M}) \to \mathscr{A}^{\rm ev}(\boldsymbol{N})$, which defines a new locally covariant theory $\mathscr{A}^{\rm ev}$: **Loc** $\to$ **Alg**. Like $\mathscr{A}$, this theory obeys Assumptions 4.3.1–4.3.3. However, relation (4.9) cannot be strengthened to an equality for this theory: consider space-like separated $O_i \in \mathscr{O}(\boldsymbol{M})$ and let $f_i \in C_0^\infty(O_i)$ $(i = 1, 2)$. Then the kinematic algebra $\mathscr{A}^{\rm ev,kin}(\boldsymbol{M}; O_1 \cup O_2)$ contains an element $\Phi_{\boldsymbol{M}}(f_1)\Phi_{\boldsymbol{M}}(f_2)$, which is not contained in $\mathscr{A}^{\rm ev,kin}(\boldsymbol{M}; O_1) \vee \mathscr{A}^{\rm ev,kin}(\boldsymbol{M}; O_2)$; in other words, $\mathscr{A}^{\rm ev}$ is not finitely additive.

One should also bear in mind that the general assumptions so far also allow for models that display unphysical properties. For example, define a theory by

$$\mathscr{B}(\boldsymbol{M}) = \begin{cases} \mathscr{A}(\boldsymbol{M}) & \text{if } \boldsymbol{M} \text{ has noncompact Cauchy surfaces} \\ \mathscr{A}(\boldsymbol{M}) \otimes \mathscr{A}(\boldsymbol{M}) & \text{if } \boldsymbol{M} \text{ has compact Cauchy surfaces} \end{cases} \tag{4.15}$$

and, for $\psi : \boldsymbol{M} \to \boldsymbol{N}$,

$$\mathscr{B}(\psi)A = \begin{cases} \mathscr{A}(\psi)A & \boldsymbol{N} \text{ has noncompact Cauchy surfaces} \\ (\mathscr{A}(\psi) \otimes \mathscr{A}(\psi))\, A & \boldsymbol{M} \text{ has compact Cauchy surfaces} \\ (\mathscr{A}(\psi))\, A \otimes \mathbf{1} & \text{otherwise} \end{cases} \tag{4.16}$$

(by a general result in Lorentzian geometry, these cases are disjoint and the only case that can arise under 'otherwise' is that of $\boldsymbol{N}$ having compact Cauchy surfaces and $\boldsymbol{M}$ having noncompact Cauchy surfaces; see [69, Prop. A.1], which is based on results in [32]).

The reader may verify that theory obeys Assumptions 4.3.1–4.3.3, while being a theory of a single scalar field in some spacetimes, and of two independent scalar fields in others. More examples in a similar vein can be found in [69]; indeed, one could employ the same construction using any locally covariant theory as a starting point. Therefore, the Assumptions 4.3.1–4.3.3 are not in themselves sufficient to guarantee that a theory represents the same physics in all spacetimes, an issue that will be studied further in Sect. 4.11.

4.4 Relative Cauchy Evolution

The locally covariant approach not only conveniently summarises many general facts about QFT in curved spacetimes, but has also led to new developments in the subject. One such is the idea of *relative Cauchy evolution*, introduced in [24] and further developed in [69], which allows for the comparison of the dynamics of a theory on different spacetimes, even when one cannot be embedded in the other.

Let $\boldsymbol{M} = (\mathcal{M}, g, \mathfrak{o}, \mathfrak{t})$ be a globally hyperbolic spacetime. If h is a smooth compactly supported rank-2 covariant tensor field that is 'not too big' then we can define a deformed spacetime $\boldsymbol{M}[h] = (\mathcal{M}, g + h, \mathfrak{o}, \mathfrak{t}[h])$ which is still globally hyperbolic, where $\mathfrak{t}[h]$ is the unique choice of time orientation agreeing with $\mathfrak{t}$ outside the support of h.[7] The set of all such metric perturbations will be denoted $H(\boldsymbol{M})$. The idea is now to select regions to the past and future of the metric perturbation that are common to $\boldsymbol{M}$ and $\boldsymbol{M}[h]$ and contain Cauchy surfaces thereof. This is achieved

[7]The orientation need not be changed when the metric changes; recall that $\mathfrak{o}$ is a component of the nonzero smooth n-forms, and not e.g., the volume form.

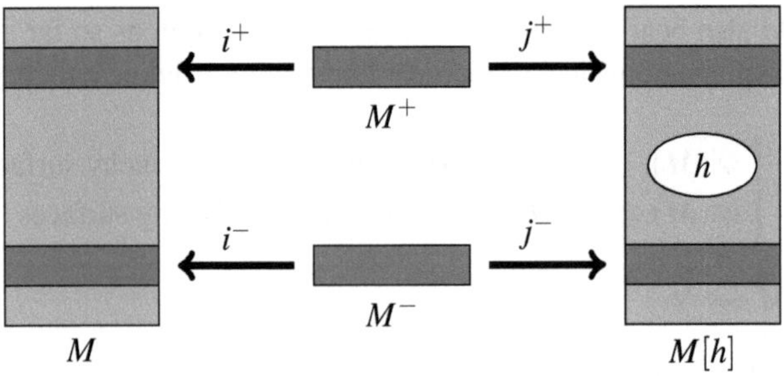

Fig. 4.1 Spacetimes involved in the construction of the relative Cauchy evolution

by choosing Cauchy morphisms $\iota^\pm : \boldsymbol{M}^\pm \to \boldsymbol{M}$ with images $\iota^\pm(\boldsymbol{M}^\pm)$ contained in $\mathcal{M} \setminus J^\mp_{\boldsymbol{M}}(\operatorname{supp} h)$, i.e., so that $\operatorname{supp} h$ has trivial intersection with the causal future of $\iota^+(\boldsymbol{M}^+)$ and the causal past of $\iota^-(\boldsymbol{M}^-)$. Given these choices, there are Cauchy morphisms $j^\pm : \boldsymbol{M}^\pm \to \boldsymbol{M}[h]$ with *the same underlying maps* as $\iota^\pm$.

The arrangement of spacetimes is displayed pictorially Fig. 4.1 and can be portrayed diagrammatically as

$$\begin{array}{ccccc} & \overset{\iota^+}{\swarrow} & \boldsymbol{M}^+ & \overset{j^+}{\searrow} & \\ \boldsymbol{M} & & & & \boldsymbol{M}[h] \\ & \underset{\iota^-}{\nwarrow} & \boldsymbol{M}^- & \underset{j^-}{\nearrow} & \end{array} \qquad \begin{array}{ccccc} & \overset{\mathscr{A}(\iota^+)}{\swarrow} & \mathscr{A}(\boldsymbol{M}^+) & \overset{\mathscr{A}(j^+)}{\searrow} & \\ \mathscr{A}(\boldsymbol{M}) & & & & \mathscr{A}(\boldsymbol{M}[h]) \\ & \underset{\mathscr{A}(\iota^-)}{\nwarrow} & \mathscr{A}(\boldsymbol{M}^-) & \underset{\mathscr{A}(j^-)}{\nearrow} & \end{array},$$

where the diagram on the right is obtained by applying the functor corresponding to a locally covariant theory $\mathscr{A}$. The important point is that the timeslice property ensures that all the morphisms in this second diagram are isomorphisms, and so can be inverted. This permits us to traverse the right-hand diagram clockwise, starting and ending at $\mathscr{A}(\boldsymbol{M})$, to obtain the relative Cauchy evolution

$$\mathrm{rce}_{\boldsymbol{M}}[h] = \mathscr{A}(\iota^-) \circ \mathscr{A}(j^-)^{-1} \circ \mathscr{A}(j^+) \circ \mathscr{A}(\iota^+)^{-1},$$

which encodes the response of the theory to the metric variation h as an automorphism of $\mathscr{A}(\boldsymbol{M})$. The relative Cauchy evolution is independent of the choices of Cauchy morphisms made—there is also a canonical choice in which $\boldsymbol{M}^\pm$ have underlying manifolds $\mathcal{M}^\pm = \mathcal{M} \setminus J^\mp_{\boldsymbol{M}}(\operatorname{supp} h)$ (see [69, §3.4]).

It is not hard to compute the relative Cauchy evolution for the real scalar field. Fix any $\boldsymbol{M} \in \mathsf{Loc}$ and any compactly supported metric perturbation $h \in H(\boldsymbol{M})$. As the relative Cauchy evolution is independent of the specific Cauchy morphisms used, convenient choices can be made. Choose Cauchy surfaces $\Sigma^\pm$ of $\boldsymbol{M}$ so that $\Sigma^\pm \subset I^\pm_{\boldsymbol{M}}(\Sigma^\mp)$ and with $\operatorname{supp} h \subset I^-_{\boldsymbol{M}}(\Sigma^+) \cap I^+_{\boldsymbol{M}}(\Sigma^-)$, i.e., h lies to the future of Σ^- and the past of Σ^+. Then let $\mathcal{M}^\pm = I^\pm_{\boldsymbol{M}}(\Sigma^\pm)$ and define $\boldsymbol{M}^\pm = \boldsymbol{M}|_{\mathcal{M}^\pm}$, letting $\iota^\pm$ and $j^\pm$ be the inclusion morphisms of $\mathcal{M}^\pm$ in $\boldsymbol{M}$ and $\boldsymbol{M}[h]$ respectively, which are necessarily Cauchy. It is enough to evaluate the action of the relative Cauchy evolution on the generators $\Phi_{\boldsymbol{M}}(f)$ of $\mathscr{A}(\boldsymbol{M})$—moreover, by the timeslice property (on $\boldsymbol{M}$) it is sufficient to restrict to test functions f supported in $\mathcal{M}^+$, for which

$$\mathscr{A}(j^+) \circ \mathscr{A}(\imath^+)^{-1} \Phi_{\boldsymbol{M}}(f) = \Phi_{\boldsymbol{M}[h]}(f).$$

Using the timeslice property on $\boldsymbol{M}[h]$ (cf. the discussion in Sect. 4.3) we may write $\Phi_{\boldsymbol{M}[h]}(f) = \Phi_{\boldsymbol{M}[h]}(P_{\boldsymbol{M}[h]}\chi E_{\boldsymbol{M}[h]} f)$, where $\chi \in C^\infty(\mathcal{M})$ has been chosen to vanish identically in $J^+_{\boldsymbol{M}[h]}(\Sigma^-)$, and to take the value 1 identically to the past of some other Cauchy surface in $\mathcal{M}^-$. In particular, χ vanishes on the support of h and also on $\mathcal{M}^+$. With these choices, $P_{\boldsymbol{M}[h]}\chi E_{\boldsymbol{M}[h]} f$ is supported in $\mathcal{M}^-$, whereupon

$$\begin{aligned} \mathrm{rce}_{\boldsymbol{M}}[h]\Phi_{\boldsymbol{M}}(f) &= \mathscr{A}(\imath^-) \circ \mathscr{A}(j^-)^{-1} \Phi_{\boldsymbol{M}[h]}(P_{\boldsymbol{M}[h]}\chi E_{\boldsymbol{M}[h]} f) \\ &= \Phi_{\boldsymbol{M}}(P_{\boldsymbol{M}[h]}\chi E_{\boldsymbol{M}[h]} f). \end{aligned}$$

This expression can be simplified so as to remove the dependence on χ. The support properties of χ entail that $\chi E_{\boldsymbol{M}[h]} f = \chi E^-_{\boldsymbol{M}[h]} f$, and so

$$\begin{aligned} P_{\boldsymbol{M}[h]}\chi E_{\boldsymbol{M}[h]} f &= f - P_{\boldsymbol{M}[h]}(1-\chi)E^-_{\boldsymbol{M}[h]} f \\ &= f - (P_{\boldsymbol{M}[h]} - P_{\boldsymbol{M}})(1-\chi)E^-_{\boldsymbol{M}[h]} f - P_{\boldsymbol{M}}(1-\chi)E^-_{\boldsymbol{M}[h]} f. \end{aligned}$$

Moreover, $P_{\boldsymbol{M}[h]}$ and $P_{\boldsymbol{M}}$ differ only where $\chi = 0$, and outside $J^+_{\boldsymbol{M}[h]}(\operatorname{supp} f)$, so

$$(P_{\boldsymbol{M}[h]} - P_{\boldsymbol{M}})(1-\chi)E^-_{\boldsymbol{M}[h]} f = (P_{\boldsymbol{M}[h]} - P_{\boldsymbol{M}})E^-_{\boldsymbol{M}[h]} f = (P_{\boldsymbol{M}[h]} - P_{\boldsymbol{M}})E_{\boldsymbol{M}[h]} f$$

which gives $P_{\boldsymbol{M}[h]}\chi E_{\boldsymbol{M}[h]} f = f - (P_{\boldsymbol{M}[h]} - P_{\boldsymbol{M}})E_{\boldsymbol{M}[h]} f - P_{\boldsymbol{M}}(1-\chi)E^-_{\boldsymbol{M}[h]} f$. Crucially, $(1-\chi)E^-_{\boldsymbol{M}[h]} f$ is compactly supported, and so the field equation axiom KG3 gives

$$\mathrm{rce}_{\boldsymbol{M}}[h]\Phi_{\boldsymbol{M}}(f) = \Phi_{\boldsymbol{M}}(f) - \Phi_{\boldsymbol{M}}((P_{\boldsymbol{M}[h]} - P_{\boldsymbol{M}})E_{\boldsymbol{M}[h]} f) \tag{4.17}$$

at least for those f supported in $\mathcal{M}^+$. As previously mentioned, this suffices to fix the action of $\mathrm{rce}_{\boldsymbol{M}}[h]$ on the whole of $\mathscr{A}(\boldsymbol{M})$.

Equation (4.17) clearly shows that the relative Cauchy evolution is trivial if f is supported within the causal complement of $\operatorname{supp} h$ (at least for our class of f). In fact, it is true for any locally covariant theory that $\mathrm{rce}_{\boldsymbol{M}}[h]$ acts trivially on $\mathscr{A}^{\mathrm{kin}}(\boldsymbol{M}; O)$ for any region $O \in \mathcal{O}(\boldsymbol{M})$ with $O \subset (\operatorname{supp} h)^\perp := \boldsymbol{M} \setminus \operatorname{supp} h$, which shows that the relative Cauchy evolution is local with respect to the metric perturbation [69, Prop. 3.5]. Moreover, it is also covariant (again, for any locally covariant theory): given $\boldsymbol{M} \to \boldsymbol{N}$ and $h \in H(\boldsymbol{M})$, it can be shown [69, Prop. 3.7] that $\psi_* h \in H(\boldsymbol{N})$ and

$$\mathrm{rce}_{\boldsymbol{N}}[\psi_* h] \circ \mathscr{A}(\psi) = \mathscr{A}(\psi) \circ \mathrm{rce}_{\boldsymbol{M}}[h].$$

A particular case of interest is where $\boldsymbol{N} = \boldsymbol{M}$ and $\psi \in \mathrm{Aut}(\boldsymbol{M})$ is a spacetime symmetry. Alternatively, if $K \subset \boldsymbol{M}$ is compact and contains the support of

$h \in H(\boldsymbol{M})$, then

$$\mathrm{rce}_{\boldsymbol{M}}[\psi_*(h+g)-g] = \mathrm{rce}_{\boldsymbol{M}}[h] \tag{4.18}$$

for any diffeomorphism ψ that acts trivially outside K [24, Prop. 4.1].

The interpretation of the relative Cauchy evolution is best understood by means of its functional derivatives, which turn out to be related to a stress-energy tensor. Let $s \mapsto h(s)$ be a smooth 1-parameter family of metric perturbations with $h(0) = 0$, so that $\boldsymbol{M}[h(s)]$ is a globally hyperbolic spacetime for all sufficiently small $|s|$. Assuming the relevant derivatives exist (in a suitable topology, which might, for instance, be a weak topology induced by a state space—see Sect. 4.5) we may define a derivation δ on $\mathscr{A}(\boldsymbol{M})$ by

$$\delta(A) = -2i\,\frac{d}{ds}\mathrm{rce}_{\boldsymbol{M}}[h(s)]A\bigg|_{s=0}$$

that depends linearly on $f = \dot{h}(0)$. It is convenient to denote this by $\delta(\cdot) = [\mathbf{T}_{\boldsymbol{M}}(f),\cdot]$, without any implication that $\mathbf{T}_{\boldsymbol{M}}(f)$ is an element of $\mathscr{A}(\boldsymbol{M})$. Moreover, the right-hand side can be written in functional derivative notation, leading to the suggestive equation

$$[\mathbf{T}_{\boldsymbol{M}}(f), A] = \frac{2}{i}\int_{\boldsymbol{M}} f_{\mu\nu}\frac{\delta \mathrm{rce}_{\boldsymbol{M}}}{\delta g_{\mu\nu}}(A).$$

Although $\mathbf{T}_{\boldsymbol{M}}(f)$ (or rather, the derivation it represents) has only been defined for symmetric test tensors f, we can extend it to arbitrary smearings by demanding that it vanish on antisymmetric f. Then (4.18) has an interesting consequence [24]. Let X^a be a smooth compactly supported vector field, and define $\psi_s = \exp(sX)$ be the 1-parameter family of diffeomorphisms it generates, which act trivially outside a fixed compact set (for $|s| < s_*$, say). This induces a 1-parameter family of metric perturbations $h(s) = \psi(s)_* g - g$, with

$$\dot{h}(0)_{ab} = (\pounds_X g)_{ab} = \nabla_a X_b + \nabla_b X_a.$$

By (4.18), $\mathrm{rce}_{\boldsymbol{M}}[h(s)] = \mathrm{rce}_{\boldsymbol{M}}[0] = \mathrm{id}_{\mathscr{A}(\boldsymbol{M})}$ for all s, so

$$[\mathbf{T}_{\boldsymbol{M}}(\pounds_X g), A] = -2i\,\frac{d}{ds}\mathrm{rce}_{\boldsymbol{M}}[h(s)]A\bigg|_{s=0} = 0,$$

which asserts that $\mathbf{T}_{\boldsymbol{M}}$ is conserved, when regarded as a symmetric derivation-valued tensor field.[8] If the derivation is inner, i.e., given by the commutator with elements $\mathbf{T}_{\boldsymbol{M}}(f) \in \mathscr{A}(\boldsymbol{M})$, then we deduce that $\nabla \cdot \mathbf{T}_{\boldsymbol{M}}$ belongs to the centre of $\mathscr{A}(\boldsymbol{M})$.

[8] It is natural to write $\mathbf{T}_{\boldsymbol{M}}(\pounds_X g) = -2(\nabla \cdot \mathbf{T}_{\boldsymbol{M}})(X)$, regarding the divergence in a weak sense.

To investigate the interpretation further, we return to the example of the scalar field. Starting from (4.17), it is clear that

$$\frac{d}{ds}\mathrm{rce}_{M}[sh]\Phi_{M}(f)\bigg|_{s=0} = \Phi_{M}(L_{M}[h]E_{M}f) \tag{4.19}$$

in any topology for which the derivative exists, where

$$L_{M}[h]\phi = -\frac{d}{ds}P_{M[sh]}\bigg|_{s=0}\phi = \nabla_a\left(h^{ab}\nabla_b\phi\right) - \frac{1}{2}\left(\nabla^a h^b{}_b\right)\nabla_a\phi.$$

Equation (4.19) is derived for f supported in $\mathcal{M}^+$, but is actually valid for all $f \in \mathscr{D}(\boldsymbol{M})$.[9] The link with the stress-energy tensor is obtained as follows. Working in 'unsmeared notation',

$$[\Phi_{M}(x)\Phi_{M}(x'), \Phi_{M}(f)] = i(E_{M}f)(x)\Phi_{M}(x') + i(E_{M}f)(x')\Phi_{M}(x)$$

and since renormalisation of the Wick square involves subtracting a C-number from the point-split square and then taking the points back together,

$$[\Phi^2_{M,\mathrm{ren}}(x), \Phi_{M}(f)] = 2i(E_{M}f)(x)\Phi_{M}(x)$$

or, smearing against $k \in C_0^\infty(\boldsymbol{M})$,

$$[\Phi^2_{M,\mathrm{ren}}(k), \Phi_{M}(f)] = 2i\Phi_{M}(kE_{M}f).$$

Similarly,

$$[((\nabla_a\Phi)(\nabla_b\Phi))_{M,\mathrm{ren}}(x), \Phi_{M}(f)] = 2(i\nabla_{(a}\Phi_{M}(x))(\nabla_{b)}E_{M}f|_x)$$

or, smearing against a symmetric tensor h,

$$[((\nabla\Phi)(\nabla\Phi))_{M,\mathrm{ren}}(h), \Phi_{M}(f)] = -2i\Phi_{M}(\nabla_a h^{ab}\nabla_b E_{M}f).$$

Applying these formulae to the stress-energy tensor,

$$\mathsf{T}_{ab} = (\nabla_a\phi)(\nabla_b\phi) - \frac{1}{2}g_{ab}g^{cd}(\nabla_c\phi)(\nabla_d\phi) + \frac{1}{2}m^2 g_{ab}\phi^2,$$

quantized by point-splitting, a short calculation using Leibniz' rule gives

$$[\mathsf{T}_{M,\mathrm{ren}}(h), \Phi_{M}(f)] = -2i\Phi_{M}(L_{M}[h]E_{M}f) + i\Phi_{M}(h^c{}_c(\Box + m^2)E_{M}f),$$

---

[9]Decompose $f = f_0 + P_{M}f_1$, where f_0 is supported in the image of $\mathcal{M}^+$, and $f_1 \in \mathscr{D}(\boldsymbol{M})$. As $\Phi_{M}(f) = \Phi_{M}(f_0)$, we may apply (4.19) to f_0 and then use the fact that $E_{M}f_0 = E_{M}f$.

of which the last term vanishes. Comparison with (4.19) yields the important formula

$$\frac{d}{ds}\mathrm{rce}_{\boldsymbol{M}}[sh]\Phi_{\boldsymbol{M}}(f)\bigg|_{s=0} = -\frac{1}{2i}[\mathbf{T}_{\boldsymbol{M},\mathrm{ren}}(h), \Phi_{\boldsymbol{M}}(f)],$$

where $\mathbf{T}_{\boldsymbol{M},\mathrm{ren}}$ is the renormalised stress-energy tensor of the theory.[10]

Similar computations for other models [9, 46, 59, 64, 136] support the view that, in general, the functional derivative of the relative Cauchy evolution may be interpreted as a stress-energy tensor. One is therefore led to regard the relative Cauchy evolution as a proxy for the action. This is quite remarkable, because we have not assumed that locally covariant theories are specified in terms of classical actions. It is a striking illustration of the power of the general framework.

4.5 States and State Spaces

4.5.1 States and Representations

While much of the structure of quantum field theory lies in the algebraic relations, particularly the concepts of locality and causality and their relation to covariance, an important role is played by the states (or, synonymously, expectation value functionals). In particular, states permit the comparison of the mathematical framework with experiment, and the interpretation of the formal framework in more concrete, physical terms, leading to an understanding of the meaning of certain observables, the charge structure, field content and the degrees of freedom of a quantum field theory. Moreover, certain aspects which distinguish quantized fields from classical fields, like entanglement, are best understood at the level of states.

The mathematical definition asserts only that a state on a $*$-algebra $\mathcal{A}$ is a positive, normalized and (suitably) continuous functional $\omega : \mathcal{A} \to \mathbb{C}$, interpreted as yielding the expectation values $\omega(A)$ of any observable A (an element $A \in \mathcal{A}$ with $A^* = A$). Here, positivity means $\omega(A^*A) \geq 0$ for all $A \in \mathcal{A}$, while normalization requires $\omega(\mathbf{1}) = 1$ for the unit element of $\mathcal{A}$ (which, by default, is assumed to exist if $\mathcal{A}$ is the algebra of observables of a physical system). Continuity may be an involved issue. If $\mathcal{A}$ is a C^*-algebra, however, norm continuity is already implied by positivity, and furthermore, the existence of a large set of states is warranted from the outset. This is one of the reasons why, from a mainly mathematical perspective, it is very convenient to treat observable algebras as C^*-algebras.

However, it is known by several examples that this very general mathematical description of a "state" allows for many which can hardly be interpreted as physically realistic configurations of a quantum field because their behaviour on certain

[10]This result differs by a sign from that in [24] (and repeated e.g., in [69, 70]). The source of the difference arises on p. 61 of [24], where the action of an '*advanced*' Green function is taken to have support in the *causal future* of the source. The sign error does not affect the results of [69, 70].

observables of interest (particularly those measuring local quantities of momentum and energy) is too singular [163]. Thus, suitable regularity properties must be imposed to select physically realistic states to which an interpretation of the observables can be tied and on which an identification of the field content can be built. In quantum field theory on Minkowski spacetime, Poincaré covariance is instrumental in specifying the vacuum state, starting from which one can proceed with a characterization (or at least selection) of physical states, but, in curved spacetimes, that is not at hand, and one must devise other criteria. We will address that issue in a while, but first, it is necessary to introduce some terminology.

We begin with the concept of a Hilbert space representation of a $*$-algebra $\mathcal{A}$, denoted by $(\mathcal{H}, \pi, \mathcal{D})$ and consisting of a Hilbert space $\mathcal{H}$ together with a dense subspace $\mathcal{D}$, and a representation π of $\mathcal{A}$ by closable operators defined on $\mathcal{D}$. Furthermore, it is required that $\pi(A)\mathcal{D} \subset \mathcal{D}$ $(A \in \mathcal{A})$ and that one has

$$\pi(AB) = \pi(A)\pi(B), \quad \pi(\alpha A + \beta B) = \alpha\pi(A) + \beta\pi(B) \quad \text{and} \quad \pi(A)^* = \pi(A^*)$$

holding on $\mathcal{D}$ for all $\alpha, \beta \in \mathbb{C}$, $A, B \in \mathcal{A}$. A further requirement is the continuity of $A \mapsto \pi(A)$ at least weakly with respect to $\mathcal{D}$ in the topology of $\mathcal{A}$. If $\mathcal{A}$ is a C^*-algebra, the $\pi(A)$ are necessarily bounded operators for $A \in \mathcal{A}$ and there is no restriction to assume $\mathcal{D} = \mathcal{H}$; hence we will simply write $(\mathcal{H}, \pi)$ for a Hilbert space representation of a C^*-algebra.

The *folium* of a representation $(\mathcal{H}, \pi, \mathcal{D})$ of $\mathcal{A}$, denoted by $\mathrm{Fol}(\pi)$, consists of all states on $\mathcal{A}$ which can be written as finite convex sums of states of the form $\omega_\xi(A) = (\xi, \pi(A)\xi)$ where ξ is any unit vector in $\mathcal{D}$. If $\mathcal{A}$ is a C^*-algebra, we can define the *W^*-folium* $\mathrm{Fol}^W(\pi)$ of $(\mathcal{H}, \pi)$ as the weak closure of $\mathrm{Fol}(\pi)$. The set of states $\mathrm{Fol}^W(\pi)$ is also called the set of *normal states* on $\mathcal{A}$ with respect to the representation $(\mathcal{H}, \pi)$.

Any state ω on a $*$-algebra $\mathcal{A}$, with suitable continuity properties, determines a unique (up to unitary equivalence) Hilbert space representation of $\mathcal{A}$, the *GNS representation*, denoted $(\mathcal{H}_\omega, \pi_\omega, \mathcal{D}_\omega, \Omega_\omega)$. Here, $(\mathcal{H}_\omega, \pi_\omega, \mathcal{D}_\omega)$ is a Hilbert space representation of $\mathcal{A}$ and Ω_ω is a unit vector in $\mathcal{D}_\omega$ such that $\omega(A) = (\Omega_\omega, \pi_\omega(A)\Omega_\omega)$ for all $A \in \mathcal{A}$ and $\mathcal{D}_\omega = \pi_\omega(A)\Omega_\omega$, implying that Ω_ω is a cyclic vector for the representation. If $\mathcal{A}$ is a C^*-algebra, one takes $\mathcal{D}_\omega = \mathcal{H}_\omega$ as before, denoting the GNS representation more simply by $(\mathcal{H}_\omega, \pi_\omega, \Omega_\omega)$. Furthermore one can assign a folium $\mathrm{Fol}^W(\omega) := \mathrm{Fol}^W(\pi_\omega)$ to any state ω, i.e. the folium of its GNS representation. Correspondingly, one calls any state in $\mathrm{Fol}^W(\pi_\omega)$ a *normal state* with respect to ω, or simply a state normal to ω.

Again for the case of a state ω on a C^*-algebra $\mathcal{A}$, we define the *induced von Neumann algebra*, $\mathcal{N}_\omega$, in the GNS-representation $(\mathcal{H}_\omega, \pi_\omega, \Omega_\omega)$ by

$$\mathcal{N}_\omega = \pi_\omega(\mathcal{A})'',$$

where the double prime denotes the bi-commutant and coincides with the weak closure of $\pi_\omega(\mathcal{A})$ by von Neumann's theorem.[11] This definition is mostly useful when applied to the local induced von Neumann algebras considered below.

Many quantum field theories are specified in terms of their fields and then, states usually arise from (and are defined by) their n-point functions. In some examples, e.g. if the algebras $\mathscr{A}(\boldsymbol{M})$ of observables assigned to a spacetime $\boldsymbol{M}$ are C^*-algebras, it may occur that the quantum field operators $\phi(f)$ are not contained in $\mathscr{A}(\boldsymbol{M})$, but arise as objects affiliated to the induced von Neumann algebra $\mathcal{N}_\omega(\boldsymbol{M}) = \pi_\omega(\mathscr{A}(\boldsymbol{M}))''$ in the GNS representations of physical states ω. Affiliation means that bounded functions of the operators U and $|\phi(f)|$ occurring in the polar decomposition $\phi(f) = U|\phi(f)|$ belong to the induced von Neumann algebra. As we wish to avoid a discussion of precise domain properties for quantum field operators, we will introduce n-point functions, or Wightman functions, in a manner that is quite close to the original idea. So let us suppose that $\boldsymbol{M}$ is an object of **Loc** and that $\mathscr{A}(\boldsymbol{M})$ is a $*$-algebra of observables assigned to spacetime $\boldsymbol{M}$ (it does not matter here if $\mathscr{A}(\boldsymbol{M})$ is part of a theory functor $\mathscr{A}$, or is a C^*-algebra). Then, we say that a state ω on $\mathscr{A}(\boldsymbol{M})$ *possesses affiliated n-point functions* if there are (1) a C^∞ complex vector bundle $\mathcal{V}_{\boldsymbol{M}}$ having $\boldsymbol{M}$ as base manifold together with a fibre-wise complex conjugation Γ on $\mathcal{V}_{\boldsymbol{M}}$ (2) a weakly dense (in the weak topology induced by the GNS-representation of ω) $*$-subalgebra $\mathscr{A}_0(\boldsymbol{M})$ and (3) a sequence w_n^ω ($n \in \mathbb{N}$) of distributions of positive type on $C_0^\infty(\mathcal{V}_{\boldsymbol{M}}^n)$ (the compactly supported C^∞ sections in $\mathcal{V}_{\boldsymbol{M}}^n$), such that for any $A \in \mathscr{A}_0(\boldsymbol{M})$ there is a sequence $F_A^{(n)} \in C_0^\infty(\mathcal{V}_{\boldsymbol{M}}^n)$ obeying

$$\sum_{|\boldsymbol{n}|\leq N} w_{|\boldsymbol{n}|}^\omega(F_{A_1}^{(n_1)} \otimes \cdots \otimes F_{A_k}^{(n_k)}) \underset{N\to\infty}{\longrightarrow} \omega(A_1 \cdots A_k)$$

for any finite collection of elements $A_1, \ldots, A_k \in \mathscr{A}_0(\boldsymbol{M})$. Here, $\boldsymbol{n} = (n_1, \ldots, n_k)$ is a multi-index and $|\boldsymbol{n}| = n_1 + \cdots + n_k$; the definition $F^{(0)} \in \mathbb{C}$ and $w_0^\omega(F^{(0)}) = F^{(0)}$ is adopted, and the positive type condition means that

$$\sum_{m,n=0}^{N} w_{n+m}^\omega(F^{(n)} \otimes \Gamma F^{(m)}) \geq 0$$

holds for any finite selection of $F^{(0)}, \ldots, F^{(N)}$, understanding that Γ acts on each of the m fibre factors, with $\Gamma F^{(0)} = \overline{F^{(0)}}$ (complex conjugation).

These definitions facilitate the introduction of conditions on the states ω via conditions on the wavefront sets [98] WF (w_n^ω) of their affiliated n-point functions w_n^ω. A generalized concept of wavefront set for states ω may be given without using affiliated n-point functions [155], but will not be pursued here.

[11] By default, all $*$-algebras here are unital, i.e. they have a unit element for the algebra product.

4.5.2 States in Locally Covariant Theories

Let $\mathscr{A}$ be a locally covariant theory, i.e. a functor $\mathscr{A} : \mathbf{Loc} \to \mathbf{Alg}$. By an $\mathscr{A}$-*state*, we mean a family $(\omega_M)_{M \in \mathbf{Loc}}$ indexed by the objects in **Loc**, where each ω_M is a state on $\mathscr{A}(M)$. This is a very general mathematical definition and does not involve, as it stands, any regularity criteria selecting physically realistic states. Furthermore, the definition does not relate the states ω_{M_1} and ω_{M_2} on spacetimes M_1 and M_2, even if (parts of) M_1 can be embedded into (parts of) M_2 by morphisms in **Loc**.

Given a globally hyperbolic spacetime M admitting a non-trivial group $\mathrm{Aut}(M)$ of spacetime isometries preserving orientation and time-orientation, it has been remarked before that $\mathscr{A}$ induces a group representation of $\mathrm{Aut}(M)$ by elements in $\mathrm{Aut}(\mathscr{A}(M))$. A state ω_M on $\mathscr{A}(M)$ is called $\mathrm{Aut}(M)$-*invariant* if

$$\omega_M \circ \mathscr{A}(\psi) = \omega_M \tag{4.20}$$

for all $\psi \in \mathrm{Aut}(M)$. To give a concrete example, suppose that M is Minkowski spacetime, then $\mathrm{Aut}(M)$ is the proper orthochronous Poincaré group, and invariance of a state ω_M is one of the important properties singling out a vacuum state. More generally, if a spacetime M is stationary, then there is a subgroup in $\mathrm{Aut}(\mathscr{A}(M))$ of time translations, and invariance of a state ω_M under this subgroup—i.e. time-translation invariance—is one of the required properties of a ground state or KMS state. As states of this type are commonly regarded as physical states of a quantum field, the invariance property of states is one of the features of states to look for. This prompts the question of whether the concept of invariant state can be generalized to the locally covariant setting. We refer to such a generalization as a *natural state*, and the definition is this: A state $(\omega_M)_{M \in \mathbf{Loc}}$ for a locally covariant theory $\mathscr{A}$ is called *natural* if $\omega_N \circ \mathscr{A}(\psi) = \omega_M$ whenever $M \xrightarrow{\psi} N$ is a morphism in **Loc**.

While this seems natural in the sense of reflecting the natural duality between algebras and states, it turns out to be asking too much: Under additional very mild regularity properties, which are expected to be general features of large sets of states in quantum field theories, and have been proved to hold in many examples and to be consequences of, e.g., the general Wightman framework on Minkowski spacetime, there are no natural states for locally covariant theories (Theorem 4.11.3).

The moral is that one should not expect physical states to be invariant under arbitrary spacetime embeddings in locally covariant quantum field theories. However, there may be sets of states assigned to spacetimes which behave invariantly under spacetime embeddings, and in fact, it is desirable to formulate selection criteria for sets of physical states in a way that such an invariance property is fulfilled.

4.5.3 State Spaces

Suppose that $\mathcal{A}$ is a $*$-algebra. Then we define a *state space* for $\mathcal{A}$ to be a set $\boldsymbol{S}$ of states on $\mathcal{A}$ having the property that $\boldsymbol{S}$ is invariant under operations in $\mathcal{A}$ and under forming finite convex sums, i.e., given any ω in $\boldsymbol{S}$, then the states

$$\omega'(A) = \sum_{i=1}^{N} \lambda_i \frac{\omega(B_i^* A B_i)}{\omega(B_i^* B_i)}$$

are also contained in $\boldsymbol{S}$, for any choice of finitely many $B_i \in \mathcal{A}$ (with $\omega(B_i^* B_i) > 0$) and $\lambda_i > 0$ with $\sum_i \lambda_i = 1$ ($i = 1, \ldots, N$, $N \in \mathbb{N}$).

Note that the folium of any Hilbert space representation of $\mathcal{A}$ is closed in the above sense. Thus one can introduce a category **Stsp** of state spaces which is "dual" to the category of algebras: The objects in **Stsp** are state spaces $\boldsymbol{S}$ of $*$-algebras $\mathcal{A}$; more precisely, they are pairs $(\boldsymbol{S}, \mathcal{A})$, where $\mathcal{A}$ indicates the $*$-algebra for which $\boldsymbol{S}$ is a state space. Then a morphism α^* in **Stsp** is the dual of a suitable morphism α in **Alg**. In more detail, if $(\boldsymbol{S}_1, \mathcal{A}_1)$ and $(\boldsymbol{S}_2, \mathcal{A}_2)$ are objects in **Stsp**, then any morphism $\alpha : \mathcal{A}_1 \to \mathcal{A}_2$ in **Alg** such that $\alpha^* \boldsymbol{S}_2 \subset \boldsymbol{S}_1$ defines a morphism $(\boldsymbol{S}_2, \mathcal{A}_2) \xrightarrow{\alpha^*} (\boldsymbol{S}_1, \mathcal{A}_1)$ in **Stsp**. Here, the dual action of α on states is given by $\alpha^* \omega_2 = \omega_2 \circ \alpha$ for $\omega_2 \in \boldsymbol{S}_2$. Thus (with some abuse of notation that is unlikely to give rise to ambiguities) $\alpha^*(\boldsymbol{S}_2, \mathcal{A}_2) = (\alpha^*(\boldsymbol{S}_2), \mathcal{A}_1)$. The morphism composition rule is defined as the composition rule of positive dual, convex maps between dual spaces of $*$-algebras.

Now let $\mathscr{A}$ be a locally covariant theory. Then we define a *state space for* $\mathscr{A}$ (henceforth, $\mathscr{A}$-state space) to be a contravariant functor $\mathscr{S}$ between **Loc** and **Stsp**, such that $\mathscr{S}(\boldsymbol{M})$ is a state space for $\mathscr{A}(\boldsymbol{M})$ for any object $\boldsymbol{M}$ of **Loc**, and such that, if ψ is a morphism in **Loc**, then $\mathscr{S}(\psi)$ is induced by $\mathscr{A}(\psi)^*$, the dual map of $\mathscr{A}(\psi)$.[12] The state space is said to obey the timeslice condition if $\mathscr{S}(\boldsymbol{M}) = \mathscr{S}(\psi)(\mathscr{S}(\boldsymbol{N}))$ for every Cauchy morphism $\psi : \boldsymbol{M} \to \boldsymbol{N}$.

4.5.4 Conditions on States

The definition of "state space" just given is very general, and to ensure that the states contained in the state space of a locally covariant theory is formed by states which can be given a reasonable and consistent physical interpretation, it needs to be supplemented with further conditions. The conditions are expressed as conditions which the states of a state space fulfil individually, or in relation to each other. Therefore we shall list some of them; they correspond to regularity properties one expects physical states to have for locally covariant theories and are motivated either by examples

[12] Note that $\mathscr{S}$, or its opposite covariant functor $\mathscr{S}^{\mathrm{op}}$: **Loc** $\to$ **Stsp**^{op}, contains all the information in $\mathscr{A}$, and could be used by itself to specify the theory in full—this is done e.g., in [48, 54].

featuring such properties (in curved or flat spacetime) or by structural arguments in favour of such properties in general quantum field theory (in flat spacetime). We refer in particular to [86] for discussion.

The $*$-algebra $\mathscr{A}(\boldsymbol{M})$ may, but need not, derive from a functor $\mathscr{A}$ from **Loc** to **Alg**. However, we assume that $\mathscr{A}(\boldsymbol{M})$ is generated by a system of $*$-subalgebras (C^*-subalgebras if $\mathscr{A}(\boldsymbol{M})$ is a C^*-algebra) $\mathscr{A}(\boldsymbol{M};\, O)$, $O \in \mathcal{O}(\boldsymbol{M})$ fulfilling isotony.

Reeh-Schlieder property. Let ω be a state on $\mathscr{A}(\boldsymbol{M})$ and let $O \in \mathcal{O}(\boldsymbol{M})$. Then one says that ω has the *Reeh-Schlieder property* with respect to O if, in the GNS-representation $(\mathcal{H}_\omega, \pi_\omega, \mathcal{D}_\omega, \Omega_\omega)$ of any $\omega \in \boldsymbol{S}(\boldsymbol{M})$, the set of vectors

$$\pi_\omega(\mathscr{A}(\boldsymbol{M};\, O))\Omega_\omega \quad \text{is dense in} \quad \mathcal{H}_\omega\,.$$

Split property. This property is conveniently formulated under the assumption that the $\mathscr{A}(\boldsymbol{M};\, O)$ are C^*-algebras, although a definition can also be given in more general cases. Let O_1 and O_2 be two spacetime regions in $\mathcal{O}(\boldsymbol{M})$ so that $\overline{O_1} \subset O_2$. A state ω on $\mathscr{A}(\boldsymbol{M})$ is said to fulfil the *split property* for the pair O_1 and O_2 of regions if there is some type I factor von Neumann subalgebra $\mathcal{N}$ of $\mathsf{B}(\mathcal{H}_\omega)$ so that

$$\mathcal{N}_\omega(O_1) \subset \mathcal{N} \subset \mathcal{N}_\omega(O_2)\,.$$

Here, $\mathcal{N}_\omega(\boldsymbol{M};\, O) = \pi_\omega(\mathscr{A}(\boldsymbol{M};\, O))''$ are the local von Neumann algebras induced in the GNS representations. At this point we recall that $\mathcal{N}$ is a factor if $\mathcal{N} \cap \mathcal{N}' = \mathbb{C}\mathbf{1}$, i.e. if only multiples of the identity operator are contained in both $\mathcal{N}$ and its commutant. A von Neumann algebra $\mathcal{N}$ is of type I if there is a von Neumann algebra isomorphism $\gamma : \mathcal{N} \to \mathsf{B}(\tilde{\mathcal{H}})$ where $\tilde{\mathcal{H}}$ is some (possibly inseparable) Hilbert space.

Intermediate factoriality. A state ω on $\mathscr{A}(\boldsymbol{M})$ will be defined as having the property of *intermediate factoriality* if for any $O \in \mathcal{O}(\boldsymbol{M})$ there are some $\tilde{O} \in \mathcal{O}(\boldsymbol{M})$ and some factor von Neumann subalgebra $\mathcal{N}$ of $\mathsf{B}(\mathcal{H})$ such that

$$\mathcal{N}_\omega(O) \subset \mathcal{N} \subset \mathcal{N}_\omega(\tilde{O})\,.$$

We note that, while this is technically reminiscent of the split property, the condition here is different, and it has a different purpose—as a consequence of intermediate factoriality, the GNS representations $\pi_{\omega|\mathscr{A}(\boldsymbol{M};O)}$ and $\pi_\omega|_{\mathscr{A}(\boldsymbol{M};O)}$ are quasiequivalent, i.e. the have the same folia. (For a fuller discussion and proofs, see [24]).

Primarity. A state ω fulfils the condition of *primarity* with respect to some region $O \in \mathcal{O}(\boldsymbol{M})$ if $\mathcal{N}_\omega(O)$ is a factor. An immediate consequence is this: If ω fulfils primarity for a subset of regions $O \in \mathcal{O}(\boldsymbol{M})$ such that any relatively compact subset of $\boldsymbol{M}$ is contained in some such O, then ω satisfies intermediate factoriality.

Duality. The condition of *duality* of ω with respect to some region $O \in \mathcal{O}(\boldsymbol{M})$ requires that

$$\mathcal{N}_\omega(O') = \mathcal{N}_\omega(O)'$$

where $O' = M \backslash \overline{J_M(O)}$ is the open causal complement of O in M and $\mathcal{N}_\omega(O')$ is the von Neumann algebra generated by all $\mathcal{N}_\omega(O_\times)$, for relatively compact $O_\times \subset O'$.

Local quasiequivalence. This condition is best formulated under the assumption that the $\mathscr{A}(M; O)$ are C^*-algebras. A set of states $S_0(M)$ is said to fulfil *local quasiequivalence* if for any pair of states $\omega_1, \omega_2 \in S_0(M)$ the equality

$$\mathrm{Fol}^W(\pi_{\omega_1}|_{\mathscr{A}(M;O)}) = \mathrm{Fol}^W(\pi_{\omega_2}|_{\mathscr{A}(M;O)}) \tag{4.21}$$

holds for all $O \in \mathcal{O}(M)$, i.e. if the GNS-representations of the states have the same folia when restricted to local algebras $\mathscr{A}(M; O)$.

Note that local quasiequivalence given in this form is equivalent to the condition

$$\mathrm{Fol}^W(\pi_{\omega_1|\mathscr{A}(M;O)}) = \mathrm{Fol}^W(\pi_{\omega_2|\mathscr{A}(M;O)}),$$

stating that the folia of the GNS-representations of states restricted to the local algebras coincide, once the states fulfil intermediate factoriality or the Reeh-Schlieder property. (Again, we refer to [24] for further discussion).

Triviality of local von Neumann algebras over points. Once more, this condition assumes that the $\mathscr{A}(M; O)$ are C^*-algebras. We say that a state ω is *point-trivial* if

$$\bigcap_{O \ni p} \mathcal{N}_\omega(M; O) = \mathbb{C}\mathbf{1}$$

for any $p \in M$. This says that the induced local von Neumann algebras induced by ω contain only multiples of the identity operator if their localization regions are shrunk to any point in spacetime.

Scaling limits. For simplicity of notation, we will introduce this concept for n-point functions w_n^ω of scalar type, i.e. $\mathcal{V}_M \cong \mathbb{C}$. The generalization to higher-dimensional vector bundles is not difficult (see, e.g., [134]). With this assumption, let ω be a state on $\mathscr{A}(M)$ with affiliated n-point functions w_n^ω, and let x be a point in M. With the help of the exponential map $\exp_x$ at x, T_xM can be identified with Minkowski spacetime (of suitable dimension). If $f \in C_0^\infty(T_xM)$, we define

$$f^{[\lambda]}(\exp_x(x')) = f(\lambda^{-1}x'), \quad \lambda > 0,$$

for x' in a neighbourhood of the origin in T_xM which is contained in the domain of the exponential map. This way, the functions $f^{[\lambda]}$ are defined and C_0^∞ on an open neighbourhood of x in M. Then one says that the state ω has a *regular scaling limit* at x if (i) the state is point-trivial (at x) and (ii) if there is a monotonous function $\nu(\lambda) > 0$ of the scaling parameter $\lambda > 0$ such that the limits

$$w_n^0(f_1 \otimes \cdots \otimes f_n) = \lim_{\lambda \to 0} \nu(\lambda)^n w_n^\omega(f_1^{[\lambda]} \otimes \cdots \otimes f_n^{[\lambda]})$$

exist for all $f_j \in C_0^\infty(T_x\boldsymbol{M})$ and if the n-point distributions thus obtained satisfy the Streater-Wightman axioms (in a nontrivial manner).

Wavefront set spectrum condition, or microlocal spectrum condition (μSC). If a state ω has affiliated n-point functions w_n^ω, a *microlocal spectrum condition*, abbreviated μSC, is a condition on the wavefront sets $\mathrm{WF}\,(w_n^\omega)$. We shall not pause here to give the definition of the wavefront sets of distributions defined on C_0^∞ sections in vector bundles as this is well-explained elsewhere [134], nor shall we record the precise form of the μSC which has been given in [21, 23]. The μSC can be seen as a microlocal remnant of the spectrum condition imposed on n-point functions in the Streater-Wightman approach to quantum field theory on Minkowski spacetime [143]. One of the main features of the μSC is that it is manifestly covariant (provided the vector bundle $\mathcal{V}_{\boldsymbol{M}}$ connects appropriately to the functorial structure of **Loc**) and this is at the heart of the considerable advances which quantum field theory in curved spacetimes has seen since the introduction of the μSC. A certain asymmetry under exchange of the order of entries in $\mathrm{WF}\,(w_n^\omega)$ is characteristic of the μSC. At the level of the 2-point function w_2^ω of a state, the μSC requires

$$\mathrm{WF}\,(w_2^\omega) \subset \{(x,\xi;x',\xi') \in T^*M \times T^*M : (x,\xi) \sim (x',-\xi'), \xi \rhd 0\}$$

where $(x,\xi) \sim (x',-\xi')$ means that the manifold base-points x and x' are connected by a lightlike geodesic and that ξ and $-\xi'$ are co-parallel to that geodesic, and $-\xi'$ is the parallel transport of ξ along the connecting geodesic. The relation $\xi \rhd 0$ means that ξ is future-pointing with respect to the time-orientation on $\boldsymbol{M}$.[13]

The conditions listed above are fulfilled in several models of quantum fields on curved spacetimes which fulfil linear hyperbolic field equations, and are quantized imposing canonical commutation relations (CCRs) in the case of integer spin fields, or canonical anti-commutation relations (CARs) in the case of half-integer spin fields, when choosing as set of states $\boldsymbol{S}_0(\boldsymbol{M})$ the set of quasifree Hadamard states. Hadamard states are specified by a particular form of the two-point function; on this matter, we refer to Chap. 5. The most complete investigation in this respect has been carried out for the minimally coupled Klein-Gordon field. Even though the conditions are partially inspired by the behaviour of linear quantum field models, many of them are viewed as being valid also for interacting quantum fields and required for a consistent interpretation of the theory. We collect results and references below.

As just mentioned, the Hadamard condition for linear quantum fields on curved spacetimes, which requires that the two-point function w_2^ω of a state ω takes the Hadamard form where the singular part of w_2^ω is determined by the spacetime metric and the field equation [107, 159], implies the microlocal spectrum condition. In fact, as was first shown in a seminal paper by Radzikowski for the quantized minimally coupled Klein-Gordon field, for quasifree states the Hadamard condition and the

[13] Our convention on Fourier transforms of compactly supported distributions is (in Minkowski space) $\hat{u}(k) = u(e_k)$, where $e_k(x) = e^{ik_\mu x^\mu}$; this is extended to manifolds using coordinate charts.

microlocal spectrum condition μSC are equivalent [127]. This was also shown to hold for other models of linear quantized fields on curved spacetimes [134].

When settling for some choice of sets of states $S_0(\boldsymbol{M})$ for any $\boldsymbol{M}$ in **Loc** for a given locally covariant theory $\mathscr{A}$, where $S_0(\boldsymbol{M})$ satisfies some, or even all, of the above stated conditions, one can obtain an $\mathscr{A}$-state space $\mathscr{S}$ in the following manner [24]: First, one must check if the $S_0(\boldsymbol{M})$, $\boldsymbol{M} \in$ **Loc**, transform contravariantly under the dualized morphisms of $\mathscr{A}$, which means $\mathscr{A}(\psi)^* S_0(\boldsymbol{N}) \subset S_0(\boldsymbol{M})$ whenever $\psi : \boldsymbol{M} \to \boldsymbol{N}$ is a morphism in **Loc**, with equality holding in case that $\psi(\boldsymbol{M}) = \boldsymbol{N}$. Next, one ought to check if the sets of states $S_0(\boldsymbol{M})$ fulfil the condition of local quasiequivalence for all $\boldsymbol{M} \in$ **Loc**, as well as intermediate factoriality—to facilitate the discussion, we will assume from now on that the $\mathscr{A}(\boldsymbol{M})$ are C^* algebras. If that is the case, one can augment the sets of states as

$$S(\boldsymbol{M}) = \{\omega \text{ is state on } \mathscr{A}(\boldsymbol{M}) : \omega|_{\mathscr{A}(\boldsymbol{M};O)} \in \mathrm{Fol}^W(\pi_{\omega_0}|_{\mathscr{A}(\boldsymbol{M};O)})\} \tag{4.22}$$

which is to hold for all $O \in \mathcal{O}(\boldsymbol{M})$ and any $\omega_0 \in S_0(\boldsymbol{M})$. Note that, in view of the assumed local quasiequivalence and intermediate factoriality of the $S_0(\boldsymbol{M})$, the definition of $S(\boldsymbol{M})$ is independent of the choice of $\omega_0 \in S_0(\boldsymbol{M})$ in (4.22).

Then, setting $\mathscr{S}(\boldsymbol{M}) = S(\boldsymbol{M})$ for objects $\boldsymbol{M}$ of **Loc** and $\mathscr{S}(\psi) = \mathscr{A}(\psi)^*$ for morphisms ψ of **Loc** yields a state space for $\mathscr{A}$ which inherits several of the properties featuring in the $S_0(\boldsymbol{M})$. More precisely, one finds the following statement, which, as mentioned, assumes that the $S_0(\boldsymbol{M})$ transform contravariantly under the dualized morphisms of $\mathscr{A}$.

Theorem 4.5.1 *Suppose that for any object $\boldsymbol{M}$ of* **Loc**, *the set of states $S_0(\boldsymbol{M})$ satisfies local quasiequivalence and intermediate factoriality. Then $\mathscr{S}$ is an $\mathscr{A}$-state space, thus any $\mathscr{S}(\boldsymbol{M})$ is closed under operations of $\mathscr{A}(\boldsymbol{M})$ and under forming convex sums, and consists, locally, of a single folium (so locally, i.e. in restriction to $\mathscr{A}(\boldsymbol{M}; O)$ for any $O \in \mathcal{O}(\boldsymbol{M})$, all states in $\mathscr{S}(\boldsymbol{M})$ are normal to any/all states in $S_0(\boldsymbol{M})$). Moreover, if the states in $S_0(\boldsymbol{M})$ are also point-trivial, the same holds for all the states in $\mathscr{S}(\boldsymbol{M})$, for any object $\boldsymbol{M}$ of* **Loc**.

The proof of this statement (with slight variations) can be found in [24], where it is referred to as *principle of local definiteness*, as put forward initially by Haag et al. [87]. We mention that it is also important that locally, the state space coincides with a single folium of states and therefore, is minimal, as this rules out the occurrence of local superselection rules, akin to charges which may sit somewhere locally, but cannot be moved around by any device. Thus, in this sense, the state space, at least formally, captures the idea that a replacement for a vacuum state should be a set of states which are in a formal sense vacuum-like, meaning that they have a low particle density, temperature, and stress-energy density. Of course, these properties are, on generic spacetimes, only approximately realized, and will in general only have a 'relative' meaning, e.g. compared to local curvature quantities. In the situation described here, where the $\mathscr{A}$-state space of a locally covariant quantum field theory $\mathscr{A}$ consists locally of a single folium, one has a situation very similar to quantum field theory on Minkowski spacetime where locally (in restriction to algebras $\mathscr{A}(\boldsymbol{M}; O)$ with $O \in \mathcal{O}(\boldsymbol{M})$), the state space consists of states which are normal to the vacuum state. This is the starting point for the theory of superselection charges, at least of localizable and transportable charges which are represented by (equivalence classes of) states which are normal to the vacuum state on algebras $\mathscr{A}(\boldsymbol{M}; O')$ if the spacetime region O' is the causal complement of a double cone (for Minkowski spacetime, O' is not relatively compact), but are not normal to the vacuum state on the full spacetime algebra $\mathscr{A}(\boldsymbol{M})$. See [6, 86] and literature cited there for an exposition of superselection theory, and background material. To some extent, the theory of localized, transportable superselection charges can be generalized to quantum field theory in curved spacetime [85], and also to locally covariant quantum field theory [25, 26, 131]. However, the case of non-localized superselection charges is more complicated in curved spacetimes since the topology of the Cauchy-surface of the spacetime under consideration may play an important role regarding the existence or non-existence of certain types of charges. Problems of that nature occur already when trying to obtain a locally covariant setting for free quantum electrodynamics at the field algebra level. However, we shall not pursue this circle of problems any further at this point, and instead refer to the literature [10, 64, 137].

There is an assertion about the type of the local von Neumann algebras $\mathcal{N}_\omega(O)$ which is implied if the state ω has a regular scaling limit at a point in the spacelike boundary of O (which means that O must have a non-trivial causal complement), together with causality of the local algebras. The statement has been given in the curved spacetime context in [6, 72, 153, 164]; it builds on a seminal paper by Fredenhagen [79]. We rephrase it here as follows.

Theorem 4.5.2 *Suppose that the state ω on $\mathscr{A}(\boldsymbol{M})$ possesses a regular scaling limit at some point x lying in the spacelike boundary of $O \in \mathcal{O}(\boldsymbol{M})$, that O has a non-trivial (open) spacelike complement O', and that $\mathcal{N}_\omega(O)$ and $\mathcal{N}_\omega(O')$ are pairwise commuting von Neumann algebras. Then $\mathcal{N}_\omega(O)$ is of type* III_1.

We remark that this result holds also if O is not relatively compact, provided the other conditions are met. The type III_1 property of $\mathcal{N}_\omega(O)$ means, roughly speaking, that $\mathcal{N}_\omega(O)$ contains no (non-zero) finite-dimensional projections. For the precise mathematical statement, see [13]. Suffice it to mention that the type III_1 property of local algebras of von Neumann algebras is a typical feature of local (von Neumann) algebras of observables in relativistic quantum field theory which does not appear in quantum mechanics, or quantum statistical mechanics. There are some interesting consequences—in particular, like the Reeh-Schlieder theorem to be discussed below, the type III_1 property of the local von Neumann algebras has as one of its consequences the ubiquity of states which are entangled across acausally separated spacetime regions. The reader is referred to the references [30, 89, 146, 158] for further material related to that theme.

Next, we shall compile what is known about states of linear quantum field models from the point of view of locally covariant quantum field theory, providing examples for the properties of states listed above, and for Theorems 4.5.1 and 4.5.2.

Proposition 4.5.3 *Assume that $\mathscr{A}$ is the C^*-algebra version of locally covariant quantum field theory of the quantized linear Klein-Gordon field with a general curvature coupling, corresponding to the field equation $(\Box_M + \xi R_M + m^2)\phi = 0$, where R_M is the scalar curvature of the underlying spacetime $\boldsymbol{M}$ of* **Loc**, *and $m^2 \geq 0$ and $\xi \geq 0$ are fixed constants (the same for all $\boldsymbol{M}$). (That is, $\mathscr{A}(\boldsymbol{M})$ is the Weyl algebra of the quantized Klein–Gordon field on each $\boldsymbol{M} \in$* **Loc**.*) Then the following hold:*

(1) Any quasifree Hadamard state on $\mathscr{A}(\boldsymbol{M})$ fulfils point-triviality, intermediate factoriality, existence of affiliated n-point functions, the μSC, and for a certain class of spacetime regions: Split-property, primarity, and Haag-duality [154].
(2) Any two quasifree Hadamard states ω_1 and ω_2 on $\mathscr{A}(\boldsymbol{M})$ are locally quasi-equivalent, i.e. the condition (4.21) *holds for any $O \in \mathcal{O}(\boldsymbol{M})$ [152].*
(3) A quasifree Hadamard state ω on $\mathscr{A}(\boldsymbol{M})$ fulfils the Reeh-Schlieder property with respect to any spacetime region $O \in \mathcal{O}(\boldsymbol{M})$ if the two-point function w_2^{ω} fulfils the analytic microlocal spectrum condition *[145]. Without assuming the analytic microlocal spectrum condition, there are also spacetime regions $O \in \mathcal{O}(\boldsymbol{M})$ and quasifree Hadamard states ω on $\mathscr{A}(\boldsymbol{M})$ such that ω has the Reeh-Schlieder property with respect to O [135, 144, 150].*
(4) Setting $\boldsymbol{S}_0(\boldsymbol{M})$ to coincide with the set of all quasifree Hadamard states in the case of the locally covariant quantized Klein-Gordon field, the assumptions stated for Theorems 4.5.1 are fulfilled [24].

We remark that similar results have also been obtained for the quantized Dirac, Proca, and (partially) electromagnetic fields, choosing in each case the set of quasifree Hadamard states as the set $\boldsymbol{S}_0(\boldsymbol{M})$ [34, 134].

4.6 Spacetime Deformation and the Rigidity Argument

Techniques based on deformations of globally hyperbolic spacetimes go back to the work of Fulling et al. [82] in which the existence of Hadamard states on ultrastatic spacetimes was used to deduce their existence on general globally hyperbolic spacetimes. As first recognised in [156], the same idea can be used to great effect in locally covariant QFT.

4.6.1 Spacetime Deformation

There are two basic components to the spacetime deformation construction: the existence of a standard form for globally hyperbolic spacetimes, and the actual deformation procedure. Consider any object $\boldsymbol{M} = (\mathcal{M}, g, \mathfrak{o}, \mathfrak{t})$ of **Loc**. An important property of globally hyperbolic spacetimes is that $\boldsymbol{M}$ admits foliations into smooth spacelike Cauchy surfaces. Moreover, every spacelike Cauchy surface Σ of $\boldsymbol{M} \in$ **Loc** also carries an orientation $\mathfrak{w}$ fixed by the requirement that $\mathfrak{t} \wedge \mathfrak{w}$ is the restriction of $\mathfrak{o}$ to Σ,[14] and all such oriented Cauchy surfaces are oriented-diffeomorphic (i.e., diffeomorphic via an orientation-preserving map). These facts may be used to prove the following structure theorem for **Loc** (see [69, §2.1]).

Proposition 4.6.1 *Supposing that $\boldsymbol{M} \in$ **Loc**, let Σ be a smooth spacelike Cauchy surface of $\boldsymbol{M}$ with induced orientation $\mathfrak{w}$, and let $t_* \in \mathbb{R}$. Then there is a* **Loc***-object $\boldsymbol{M}_{st} = (\mathbb{R} \times \Sigma, g, \mathfrak{t} \wedge \mathfrak{w}, \mathfrak{t})$ and an isomorphism $\rho : \boldsymbol{M}_{st} \to \boldsymbol{M}$ in* **Loc** *such that*

- *the metric g is of split form*

$$g = \beta dt \otimes dt - h_t, \tag{4.23}$$

  where t is the coordinate corresponding to the first factor of the Cartesian product $\mathbb{R} \times \Sigma$, the function $\beta \in C^\infty(\mathbb{R} \times \Sigma)$ is strictly positive and $t \mapsto h_t$ is a smooth choice of (smooth) Riemannian metrics on Σ;
- *each $\{t\} \times \Sigma$ is a smooth spacelike Cauchy surface, and $\rho(t_*, \cdot)$ is the inclusion of Σ in $\boldsymbol{M}$;*
- *the vector $\partial/\partial t$ is future-directed according to $\mathfrak{t}$.*

We refer to $\boldsymbol{M}_{st}$ as a standard form *for $\boldsymbol{M}$.*

This statement is a slight elaboration of results due to Bernal and Sánchez (see particularly, [12, Thm 1.2] and [11, Thm 2.4]), which were previously long-standing folk-theorems. The main deformation result can now be stated (see [69, Prop. 2.4]):

Proposition 4.6.2 *Spacetimes $\boldsymbol{M}$, $\boldsymbol{N}$ in* **Loc** *have oriented-diffeomorphic Cauchy surfaces if and only if there is a chain of Cauchy morphisms in* **Loc** *forming a diagram*

$$\boldsymbol{M} \xleftarrow{\alpha} \boldsymbol{P} \xrightarrow{\beta} \boldsymbol{I} \xleftarrow{\gamma} \boldsymbol{F} \xrightarrow{\delta} \boldsymbol{N}. \tag{4.24}$$

The importance of the deformation result is that a locally covariant theory $\mathscr{A}$ obeying the time-slice condition maps every Cauchy morphism of **Loc** to an isomorphism of **Alg**, so the chain of Cauchy morphisms in (4.24) induces an isomorphism

$$\mathscr{A}(\delta) \circ \mathscr{A}(\gamma)^{-1} \circ \mathscr{A}(\beta) \circ \mathscr{A}(\alpha)^{-1} : \mathscr{A}(\boldsymbol{M}) \to \mathscr{A}(\boldsymbol{N}). \tag{4.25}$$

This isomorphism is not canonical, owing to the many choices used to construct it. Nonetheless, we will see that it is possible to use results of this type to transfer information and structures between the instantiations of the theory on $\boldsymbol{M}$ and $\boldsymbol{N}$.

In the following, a few more definitions will be needed ([69, Def. 2.5]). A *Cauchy ball* in a Cauchy surface Σ of $\boldsymbol{M} \in$ **Loc** is a subset $B \subset \Sigma$ for which there is a chart (U, ϕ) of Σ such that $\phi(B)$ a nonempty open ball in $\mathbb{R}^{n-1}$ whose closure is contained in $\phi(U)$. A *diamond* in $\boldsymbol{M}$ is any open relatively compact subset of the

[14] Recall that $\mathfrak{t}$, $\mathfrak{o}$ and $\mathfrak{w}$ are all regarded as connected components of certain sets of nowhere zero forms; by $\mathfrak{t} \wedge \mathfrak{w}$ we denote the set of all possible exterior products from within $\mathfrak{t}$ and $\mathfrak{w}$.

form $D_{\boldsymbol{M}}(B)$, where B is a Cauchy ball. The diamond is said to have *base* B and be *based on* any Cauchy surface in which B is a Cauchy ball. A *multi-diamond* D is a union of finitely many causally disjoint diamonds; there exists a common Cauchy surface on which each component is based, and the intersection of D with an open causally convex neighbourhood of any such Cauchy surface is called a *truncated (multi)-diamond.*

4.6.2 The Rigidity Argument and Some Applications

It is often the case that if a locally covariant QFT has a property in one spacetime, then the same is true in all spacetimes, a phenomenon that we call *rigidity*. This testifies to the strength of the hypotheses given in Sect. 4.3, particularly the timeslice property. A simple example is provided by Einstein causality, which was originally included as Assumption 4.3.2 for a locally covariant quantum field theory. However, this assumption is largely redundant: Provided a theory is Einstein causal in one spacetime (e.g., Minkowski), it must be so in all spacetimes.

Let $\mathscr{A} : \mathsf{Loc} \to \mathsf{Alg}$ obey local covariance and the timeslice condition, but not necessarily Einstein causality, and for each $\boldsymbol{M} \in \mathsf{Loc}$ let $\mathcal{O}^{(2)}(\boldsymbol{M})$ denote the set of all ordered pairs $\langle O_1, O_2 \rangle \in \mathcal{O}(\boldsymbol{M}) \times \mathcal{O}(\boldsymbol{M})$ such that the O_i are nonempty and causally disjoint in the sense that $O_1 \subset O_2'$. For any such pair $\langle O_1, O_2 \rangle \in \mathcal{O}^{(2)}(\boldsymbol{M})$ let $P_{\boldsymbol{M}}(O_1, O_2)$ be the proposition that Einstein causality holds for O_1 and O_2, i.e., that $\mathscr{A}^{\text{kin}}(\boldsymbol{M}; O_1)$ and $\mathscr{A}^{\text{kin}}(\boldsymbol{M}; O_2)$ commute. These propositions have some simple properties:

R1 for all $\langle O_1, O_2 \rangle \in \mathcal{O}^{(2)}(\boldsymbol{M})$,

$$P_{\boldsymbol{M}}(O_1, O_2) \iff P_{\boldsymbol{M}}(D_{\boldsymbol{M}}(O_1), D_{\boldsymbol{M}}(O_2)).$$

R2 given $\psi : \boldsymbol{M} \to \boldsymbol{N}$ then, for all $\langle O_1, O_2 \rangle \in \mathcal{O}^{(2)}(\boldsymbol{M})$,

$$P_{\boldsymbol{M}}(O_1, O_2) \iff P_{\boldsymbol{N}}(\psi(O_1), \psi(O_2)).$$

R3 for all $\langle O_1, O_2 \rangle \in \mathcal{O}^{(2)}(\boldsymbol{M})$ and all nonempty $\widetilde{O}_i \in \mathcal{O}(\boldsymbol{M})$ with $\widetilde{O}_i \subset O_i$ $(i = 1, 2)$

$$P_{\boldsymbol{M}}(O_1, O_2) \implies P_{\boldsymbol{M}}(\widetilde{O}_1, \widetilde{O}_2).$$

Here, R1 holds trivially because $\mathscr{A}(\boldsymbol{M}; O) = \mathscr{A}(\boldsymbol{M}; D_{\boldsymbol{M}}(O))$, while to prove R2 we recall from (4.10) that $\mathscr{A}^{\text{kin}}(\boldsymbol{N}; \psi(O_i)) = \mathscr{A}(\psi)(\mathscr{A}^{\text{kin}}(\boldsymbol{M}; O_i))$, and use the equality

$$[\mathscr{A}(\boldsymbol{N}; \psi(O_1)), \mathscr{A}(\boldsymbol{N}; \psi(O_2))] = \mathscr{A}(\psi)([\mathscr{A}(\boldsymbol{M}; O_1), \mathscr{A}(\boldsymbol{M}; O_2)])$$

together with injectivity of $\mathscr{A}(\psi)$. R3 is also trivial as $\mathscr{A}^{\text{kin}}(\boldsymbol{M}; \widetilde{O}_i) \subset \mathscr{A}^{\text{kin}}(\boldsymbol{M}; O_i)$.

These facts allow us to prove the following.

Theorem 4.6.3 *Let the theory $\mathscr{A} : \mathsf{Loc} \to \mathsf{Alg}$ obey local covariance and time-slice (Assumptions 4.3.1 and 4.3.3) but not necessarily Einstein causality (Assumption 4.3.2). Suppose that, in the theory $\mathscr{A}$, Einstein causality holds for some pair of nonempty causally disjoint regions $O_1, O_2 \in \mathscr{O}(\boldsymbol{M})$ in some spacetime $\boldsymbol{M} \in$ Loc. Then Einstein causality holds in theory $\mathscr{A}$ for every pair of nonempty causally disjoint regions $\widetilde{O}_1, \widetilde{O}_2 \in \mathscr{O}(\widetilde{\boldsymbol{M}})$ in every spacetime $\widetilde{\boldsymbol{M}} \in \mathsf{Loc}$ for which either of the following hold:*

(a) the Cauchy surfaces of $\widetilde{O}_i$ are oriented diffeomorphic to those of O_i for $i = 1, 2$;
(b) each component of $\widetilde{O}_1 \cup \widetilde{O}_2$ has Cauchy surface topology $\mathbb{R}^3$ (e.g., if the $\widetilde{O}_i$ are truncated multi-diamonds.)

Remark 4.6.4 The regions O_i in the hypotheses might be a finite spacelike distance from one another. However, the regions $\widetilde{O}_i$ need not be separated in this way and could touch at their boundaries, or even link around each other if they have nontrivial topology. For example, consider a theory which obeys Einstein causality for a pair of causally disjoint diamonds based on the $t = 0$ hyperplane of Minkowski space $\boldsymbol{M}_0$. Within each diamond, choose a subregion $D_{\boldsymbol{M}_0}(T_i)$, where T_i is an open subset of the $t = 0$ hyperplane with topology $\mathbb{R}^{n-2} \times \mathbb{T}$ (e.g., a thickened closed curve). Einstein causality holds for these regions (by R3) and thus holds for every pair of causally disjoint regions $D_{\boldsymbol{M}_0}(\widetilde{T}_i)$, where the $\widetilde{T}_i$ have topology $\mathbb{R}^{n-2} \times \mathbb{T}$, even if they are linked through one another.

Proof (a) By Proposition 4.6.2 there is a chain of morphisms

$$\widetilde{\boldsymbol{M}} \xleftarrow{\widetilde{\iota}} \widetilde{\boldsymbol{M}}|_{\widetilde{O}_1\cup\widetilde{O}_2} \xleftarrow{\widetilde{\psi}} \widetilde{\boldsymbol{L}} \xrightarrow{\widetilde{\varphi}} \boldsymbol{I} \xleftarrow{\varphi} \boldsymbol{L} \xrightarrow{\psi} \boldsymbol{M}|_{O_1\cup O_2} \xrightarrow{\iota} \boldsymbol{M}$$

where $\psi, \widetilde{\psi}, \varphi, \widetilde{\varphi}$ are Cauchy morphisms and $\iota = \iota_{\boldsymbol{M};O_1\cup O_2}, \widetilde{\iota} = \iota_{\widetilde{\boldsymbol{M}};\widetilde{O}_1\cup\widetilde{O}_2}$. The spacetime $\boldsymbol{M}|_{O_1\cup O_2}$ has two connected components, which are just the subsets O_1 and O_2 (recall that the underlying manifold of $\boldsymbol{M}|_{O_1\cup O_2}$ is just $O_1 \cup O_2$ as a set); the same holds, *mutatis mutandis*, for $\widetilde{\boldsymbol{M}}|_{\widetilde{O}_1\cup\widetilde{O}_2}$. Each of the spacetimes $\boldsymbol{I}, \boldsymbol{L}, \widetilde{\boldsymbol{L}}$ has two connected components, which we label $I_i, L_i, \widetilde{L}_i$ respectively ($i = 1, 2$) so that

$$D_{\boldsymbol{M}|_{O_1\cup O_2}}(\psi(L_i)) = O_i, \qquad D_{\widetilde{\boldsymbol{M}}|_{\widetilde{O}_1\cup\widetilde{O}_2}}(\widetilde{\psi}(\widetilde{L}_i)) = \widetilde{O}_i,$$
$$D_{\boldsymbol{I}}(\varphi(L_i)) = I_i = D_{\boldsymbol{I}}(\widetilde{\varphi}(\widetilde{L}_i))$$

for $i = 1, 2$. Using properties R1 and R2 we may now argue

$$P_{\boldsymbol{M}}(O_1, O_2) \underset{\iota}{\overset{R2}{\iff}} P_{\boldsymbol{M}|_{O_1\cup O_2}}(O_1, O_2) \overset{R1}{\iff} P_{\boldsymbol{M}|_{O_1\cup O_2}}(\psi(L_1), \psi(L_2))$$
$$\underset{\psi}{\overset{R2}{\iff}} P_{\boldsymbol{L}}(L_1, L_2) \underset{\varphi}{\overset{R2}{\iff}} P_{\boldsymbol{I}}(\varphi(L_1), \varphi(L_2)) \overset{R1}{\iff} P_{\boldsymbol{I}}(I_1, I_2)$$

where we have indicated the morphism involved in each use of R2. By a similar chain of reasoning, $P_{\widetilde{\boldsymbol{M}}}(\widetilde{O}_1, \widetilde{O}_2) \iff P_{\boldsymbol{I}}(I_1, I_2)$. As $P_{\boldsymbol{M}}(O_1, O_2)$ holds by hypothesis, we deduce that $P_{\widetilde{\boldsymbol{M}}}(\widetilde{O}_1, \widetilde{O}_2)$ also holds.

For (b), we observe first that for $i = 1, 2$, O_i certainly contains a truncated multi-diamond D_i with the same number of components as $\widetilde{O}_i$. Then $P_{\boldsymbol{M}}(D_1, D_2)$ holds by R3 and so $P_{\widetilde{\boldsymbol{M}}}(\widetilde{O_1}, \widetilde{O_2})$ also holds by part (a). □

Note that this argument makes no specific reference to Einstein causality at all: it simply uses the *rigidity hypotheses* R1–3, and therefore allows a number of other results to be proved in a similar fashion.

Corollary 4.6.5 *Assume that, in addition to the hypotheses of Theorem 4.6.3, the theory $\mathscr{A}$ is* additive with respect to truncated multidiamonds, *i.e., each $\mathscr{A}(\boldsymbol{M})$ is generated by its subalgebras $\mathscr{A}^{\mathrm{kin}}(\boldsymbol{M}; D)$ as D runs over the truncated multidiamonds of $\boldsymbol{M}$. Then $\mathscr{A}$ obeys Einstein causality in full.*

Proof Let $\langle \widetilde{O}_1, \widetilde{O}_2 \rangle \in \mathscr{O}^{(2)}(\widetilde{\boldsymbol{M}})$ be chosen arbitrarily. It follows from the additional hypothesis that each $\mathscr{A}^{\mathrm{kin}}(\widetilde{\boldsymbol{M}}; \widetilde{O}_i)$ is generated by subalgebras of the form $\mathscr{A}^{\mathrm{kin}}(\widetilde{\boldsymbol{M}}; D)$ where D runs over the truncated multidiamonds in $\widetilde{O}_i$.[15] Applying Theorem 4.6.3(b), it follows that Einstein causality holds for $\widetilde{O}_1$ and $\widetilde{O}_2$. □

Remark 4.6.6 These results have an interesting consequence for free electromagnetism in $n = 4$ dimensions. Consider two observables, one of which is the magnetic flux Φ_1 through a 2-surface S_1 bounded by closed curve C_1, while the other is the electric flux Φ_2 through 2-surface S_2 bounded by C_2; we assume that these curves lie in the $t = 0$ hyperplane and have thickenings T_i that are causally disjoint.[16] Each observable can be written as a (gauge-invariant) line integral of suitable 1-form potentials around the relevant bounding curve and it would be natural to expect that $\Phi_i \in \mathscr{A}^{\mathrm{kin}}(\boldsymbol{M}_0; U_i)$, where $U_i = D_{\boldsymbol{M}_0}(T_i)$. But these two algebras commute, while the commutator $[\Phi_1, \Phi_2]$ is proportional to the linking number of C_1 and C_2 [129], so at least one of these natural expectations is incorrect. Indeed, if the theory respects electromagnetic duality as a local symmetry then neither Φ_1 nor Φ_2 can belong to the local algebra of the relevant thickened curve.[17] Provided that $\mathscr{A}$ maps spacetime embeddings to injective maps, the algebras $\mathscr{A}(\boldsymbol{M}_0|_{U_i})$ of the nonsimply connected spacetimes $\boldsymbol{M}_0|_{U_i}$ also fail to contain observables corresponding to the Φ_i.

Note that this conclusion required no discussion of how free electromagnetism should be formulated in spacetimes other than Minkowski, beyond the requirements of local covariance and the timeslice property. This explains why 'topological observables' are absent from quantizations of Maxwell theory obeying these properties [59]. To restore them, one must relax the assumption of local covariance to permit noninjective maps [7, 37, 59, 137].

As mentioned above, the rigidity argument can be used for other purposes. For instance, the *Schlieder property* [139] relates to the algebraic independence of local algebras of spacelike separated regions: specifically, it demands that the product of

[15]Here, we use the stability of (multi)-diamonds under **Loc** morphisms [25, Lem. 2.8].

[16]The same arguments could be applied to more general smearings of the field-strength.

[17]Of course, each Φ_i *is* contained in the local algebra for regions containing the 2-surfaces S_i.

elements taken from two such algebras can vanish only if at least one of the elements vanishes. In this case we will say that the Schlieder property holds for the given regions. Another example is *extended locality*, which requires that local algebras corresponding to spacelike separated regions intersect only in multiples of the unit operator. In its original formulation [108, 140] extended locality was established for the local von Neumann algebras of certain spacelike separated diamonds, under standard hypotheses of AQFT plus an additional condition on the absence of translationally invariant quasi-local observables; it is a necessary condition for the C^*-independence of the corresponding subalgebras [147, Def. 2.4]. Here we formulate extended locality in the category **Alg**.

Theorem 4.6.7 *Let $\mathscr{A}$: **Loc** → **Alg** obey local covariance and the timeslice condition. Then the statement of Theorem 4.6.3 holds with 'Einstein causality' replaced by (a) 'the Schlieder property', or (b) 'extended locality'.*

Proof Define $P_{\boldsymbol{M}}(O_1, O_2)$ to be the proposition that the Schlieder property (in case (a)) or extended locality (in case (b)) holds for the kinematic algebras associated with $\langle O_1, O_2\rangle \in \mathcal{O}^{(2)}(\boldsymbol{M})$. To apply the argument in the proof of Theorem 4.6.3 we need only check that the rigidity hypotheses R1–R3 hold. In each case, R1 and R3 hold by the reasoning used for Einstein causality. To see that R2 holds in case (a), consider $\psi : \boldsymbol{M} \to \boldsymbol{N}$ and suppose $A_i \in \mathscr{A}^{\mathrm{kin}}(\boldsymbol{N}; \psi(O_i))$ obey $A_1A_2 = 0$. By (4.10), there exist $B_i \in \mathscr{A}^{\mathrm{kin}}(\boldsymbol{M}; O_i)$ such that $A_i = \mathscr{A}(\psi)B_i$ for $B_i \in \mathscr{A}^{\mathrm{kin}}(\boldsymbol{M}; O_i)$, which necessarily obey $B_1B_2 = 0$ because $\mathscr{A}(\psi)$ is an injective algebra homomorphism. It follows that $P_{\boldsymbol{M}}(O_1, O_2) \implies P_{\boldsymbol{N}}(\psi(O_1), \psi(O_2))$. The converse is proved similarly.

In the case (b), injectivity of $\mathscr{A}(\psi)$ and the covariance property (4.10) give

$$\mathscr{A}^{\mathrm{kin}}(\boldsymbol{N}; \psi(O_1)) \cap \mathscr{A}(\boldsymbol{N}; \psi(O_2)) = \mathscr{A}(\psi)\left(\mathscr{A}^{\mathrm{kin}}(\boldsymbol{M}; O_1) \cap \mathscr{A}^{\mathrm{kin}}(\boldsymbol{M}; O_2)\right) \tag{4.26}$$

and R2 is immediate. □

If Einstein causality and the Schlieder property both hold for $\langle O_1, O_2\rangle \in \mathcal{O}^{(2)}(\boldsymbol{M})$, then there is an **Alg**-isomorphism

$$\begin{aligned}\mathscr{A}^{\mathrm{kin}}(\boldsymbol{M}; O_1) \odot \mathscr{A}^{\mathrm{kin}}(\boldsymbol{M}; O_2) &\longrightarrow \mathscr{A}^{\mathrm{kin}}(\boldsymbol{M}; O_1) \vee \mathscr{A}^{\mathrm{kin}}(\boldsymbol{M}; O_2)\\ \sum_i A_i \odot B_i &\longmapsto \sum_i A_i B_i\end{aligned} \tag{4.27}$$

as shown by Roos [130]. Here, $\odot$ denotes the algebraic tensor product. If $\mathscr{A}$ is finitely additive then the subalgebra on the right-hand side of (4.27) can be replaced by $\mathscr{A}^{\mathrm{kin}}(\boldsymbol{M}; O_1 \cup O_2)$, which is isomorphic to $\mathscr{A}(\boldsymbol{M}|_{O_1\cup O_2})$. Now the spacetime $\boldsymbol{M}_{O_1\cup O_2}$ is **Loc**-isomorphic to (but distinct from) the disjoint union $\boldsymbol{M}|_{O_1} \sqcup \boldsymbol{M}|_{O_2}$, so $\mathscr{A}(\boldsymbol{M}|_{O_1\cup O_2}) \cong \mathscr{A}(\boldsymbol{M}|_{O_1} \sqcup \boldsymbol{M}|_{O_2})$. The upshot is that there is an isomorphism

$$\mathscr{A}(\boldsymbol{M}|_{O_1}) \odot \mathscr{A}(\boldsymbol{M}|_{O_2}) \cong \mathscr{A}(\boldsymbol{M}|_{O_1} \sqcup \boldsymbol{M}|_{O_2}). \tag{4.28}$$

This idea may be extended to show that $\mathscr{A}$ is a monoidal functor between **Loc** (with $\sqcup$ as the monoidal product, and extended to include an empty spacetime as the monoidal unit) and **Alg** (with the algebraic tensor product); see [22], which, however, proceeds from different assumptions. Note that the monoidal property is not just a restatement of Einstein causality; as shown above, it involves additional properties, notably the Schlieder property (or, as in [22], a form of the split property).

In the C^*-algebraic setting, with $\mathscr{A}$: **Loc** $\to$ **C*-Alg**, the statements of Theorems 4.6.3 and 4.6.7 go through without change. However the isomorphism (4.27) remains at the algebraic level, and further conditions are needed to determine whether it can be extended to a **C*-Alg**-isomorphism between a C^*-tensor product and the C^*-algebra generated by the local algebras. In this context, it is most natural to employ the minimal C^*-tensor product—we refer to [22] for more discussion.

4.7 Analogues of the Reeh–Schlieder Theorem and Split Property

In this section, we discuss the (partial) Reeh–Schlieder and split properties described in Sect. 4.5 in greater detail. In particular, we show how spacetime deformation arguments can be used to deduce the existence of states with (partial) Reeh–Schlieder and split properties on a spacetime of interest, if such states exist on a spacetime to which it can be linked by Cauchy morphisms. The arguments are based on those of [151] (for split) and [150] (for Reeh–Schlieder) which applied to the Klein–Gordon theory. A general treatment of Reeh–Schlieder results for locally covariant quantum field theories was given by Sanders [135]. Our treatment follows [48], in which the geometrical underpinnings of these arguments were placed into a common streamlined form, yielding states that have both the split and (partial) Reeh–Schlieder properties, in general locally covariant theories. The Reeh-Schlieder theorem implies the ubiquity of long-range correlations in quantum field theory, which, among other things, lead generically to entanglement across acausally separated regions or spacetime horizons [86, 158, 162]. The split property implies, on the contrary, that it is also possible to fully isolate a local system in quantum field theory such that it has no correlations with its environment and that the states by which this can be achieved lie locally in the folium of physical states [86].

We will make use of some particular subsets of Cauchy surfaces.

Definition 4.7.1 Let $\boldsymbol{M} \in$ **Loc**. A *regular Cauchy pair* (S, T) in $\boldsymbol{M}$ is an ordered pair of subsets of $\boldsymbol{M}$, that are nonempty open, relatively compact subsets of a common smooth spacelike Cauchy surface in which $\overline{T}$ has nonempty complement, and so that $\overline{S} \subset T$. There is a preorder on regular Cauchy pairs so that $(S_1, T_1) \prec (S_2, T_2)$ if and only if $S_2 \subset D_{\boldsymbol{M}}(S_1)$ and $T_1 \subset D_{\boldsymbol{M}}(T_2)$ (see Fig. 4.2).[18]

[18]The preorder is not a partial order, because $(S_1, T_1) \prec (S_2, T_2) \prec (S_1, T_1)$ implies $D_{\boldsymbol{M}}(S_1) = D_{\boldsymbol{M}}(S_2)$ and $D_{\boldsymbol{M}}(T_1) = D_{\boldsymbol{M}}(T_2)$, but not necessarily $S_1 = S_2$ and $T_1 = T_2$.

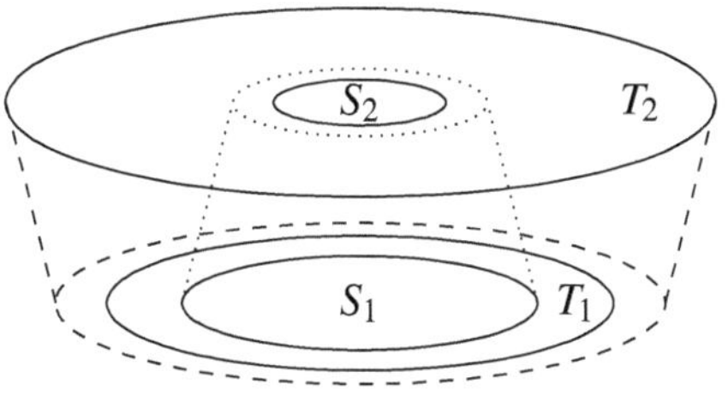

Fig. 4.2 Regular Cauchy pairs with $(S_1, T_1) \prec (S_2, T_2)$. *Dotted* (resp., *dashed*) *lines* indicate relevant portions of $D_M(S_1)$ (resp., $D_M(T_2)$).

These conditions ensure that $D_M(S)$ and $D_M(T)$ are open and casually convex, and hence elements of $\mathcal{O}(\boldsymbol{M})$. Moreover, if $\psi : \boldsymbol{M} \to \boldsymbol{N}$ is a Cauchy morphism, then a pair of subsets (S, T) of $\boldsymbol{M}$ is a regular Cauchy pair if and only if $(\psi(S), \psi(T))$ is a regular Cauchy pair for $\boldsymbol{N}$ with $\overline{\psi(T)} \subset \psi(\boldsymbol{M})$. The main property of the preorder that will be used is:

Lemma 4.7.2 *[48] Suppose that* $\boldsymbol{M}$ *takes standard form with underlying manifold* $\mathbb{R} \times \Sigma$*, and that* (S, T) *is a regular Cauchy pair in* $\boldsymbol{M}$*, lying in the surface* $\{t\} \times \Sigma$*. Then there exists an* $\varepsilon > 0$ *such that every Cauchy surface* $\{t'\} \times \Sigma$ *with* $|t' - t| < \varepsilon$ *contains a regular Cauchy pair preceding* (S, T) *and also a regular Cauchy pair preceded by* (S, T)*.*

Clearly, ε may be chosen uniformly for any finite collection of regular Cauchy pairs in the Cauchy surface $\{t\} \times \Sigma$.

In this section we consider a general locally covariant theory $\mathscr{A} : \mathsf{Loc} \to \mathsf{C^*\text{-}Alg}$, because the properties we describe are most naturally given in the C^*-context. We assume throughout that $\mathscr{A}$ has the timeslice property and obeys Einstein causality.

Definition 4.7.3 Let $\boldsymbol{M} \in \mathsf{Loc}$ and suppose that ω is a state of $\mathscr{A}(\boldsymbol{M})$ with GNS representation $(\mathscr{H}_\omega, \pi_\omega, \Omega_\omega)$. Then ω is said to have the *Reeh–Schlieder property* for a regular Cauchy pair (S, T) if the GNS vector Ω_ω is cyclic for $\mathcal{R}_S$ and separating for $\mathcal{R}_T$,[19] where $\mathcal{R}_U = \pi_\omega(\mathscr{A}^{\mathrm{kin}}(\boldsymbol{M}; D_{\boldsymbol{M}}(U)))''$ denotes the local von Neumann algebra corresponding to $U = S, T$. For brevity, we will sometimes say that *ω is Reeh–Schlieder for (S, T)*.

The state ω is said to have the *split property* for (S, T) (or to be 'split for' (S, T)) if there is a type-I factor $\mathcal{N}$ such that $\mathcal{R}_S \subset \mathcal{N} \subset \mathcal{R}_T$.

Remark 4.7.4 If a vector is separating for an algebra, it is separating for any subalgebra thereof; if it is cyclic for an algebra, it is cyclic for any algebra of which it is a subalgebra. Thus, if ω has the Reeh–Schlieder property for (S, T) then it does for every $(\tilde{S}, \tilde{T})$ with $(\tilde{S}, \tilde{T}) \prec (S, T)$. Similarly, if ω has the split property for (S, T) then it does for every $(\tilde{S}, \tilde{T})$ with $(S, T) \prec (\tilde{S}, \tilde{T})$, because $\mathcal{R}_{\tilde{S}} \subset \mathcal{R}_S \subset \mathcal{N} \subset \mathcal{R}_T \subset \mathcal{R}_{\tilde{T}}$.

[19] That is, we require $\mathcal{R}_S\Omega_\omega$ to be dense in $\mathscr{H}_\omega$ and $\mathcal{R}_T \ni A \mapsto A\Omega_\omega \in \mathscr{H}_\omega$ to be injective.

The proof of Theorem 4.7.6 below relies on a careful geometric construction together with the following result, which follows easily from the uniqueness of the GNS representation [48]:

Lemma 4.7.5 *Let (S, T) be a regular Cauchy pair in $\boldsymbol{M} \in \mathsf{Loc}$ and suppose $\psi : \boldsymbol{M} \to \boldsymbol{N}$ is Cauchy. (a) A state $\omega_{\boldsymbol{N}}$ on $\mathscr{A}(\boldsymbol{N})$ is Reeh–Schlieder for a regular Cauchy pair $(\psi(S), \psi(T))$ with $\overline{\psi(T)} \subset \psi(\boldsymbol{M})$ if and only if $\mathscr{A}(\psi)^*\omega_{\boldsymbol{N}}$ is Reeh–Schlieder for (S, T). As $\mathscr{A}(\psi)$ is an isomorphism, this implies that $\omega_{\boldsymbol{M}}$ is Reeh–Schlieder for (S, T) if and only if $(\mathscr{A}(\psi)^{-1})^*\omega_{\boldsymbol{M}}$ is Reeh–Schlieder for $(\psi(S), \psi(T))$. (b) The previous statement also holds if 'Reeh–Schlieder' is replaced by 'split'.*

The main result of this section is:

Theorem 4.7.6 *Let $\boldsymbol{M}, \boldsymbol{N} \in \mathsf{Loc}$ have oriented-diffeomorphic Cauchy surfaces and suppose $\omega_{\boldsymbol{N}}$ is a state on $\mathscr{A}(\boldsymbol{N})$ that has the Reeh–Schlieder and split properties for all regular Cauchy pairs. Given any regular Cauchy pair $(S_{\boldsymbol{M}}, T_{\boldsymbol{M}})$ in $\boldsymbol{M}$, there is a chain of Cauchy morphisms between $\boldsymbol{M}$ and $\boldsymbol{N}$ inducing an isomorphism $\nu : \mathscr{A}(\boldsymbol{M}) \to \mathscr{A}(\boldsymbol{N})$ such that $\omega_{\boldsymbol{M}} = \nu^*\omega_{\boldsymbol{N}}$ is Reeh–Schlieder and split for $(S_{\boldsymbol{M}}, T_{\boldsymbol{M}})$.*

Proof (Sketch) Assume, without loss, that $\boldsymbol{M}$ is in standard form $\boldsymbol{M} = (\mathbb{R} \times \Sigma, g_{\boldsymbol{M}}, \mathfrak{o}, \mathfrak{t}_{\boldsymbol{M}})$ so that $S_{\boldsymbol{M}}$ and $T_{\boldsymbol{M}}$ are contained in the Cauchy surface $\{t_{\boldsymbol{M}}\} \times \Sigma$ for some $t_{\boldsymbol{M}} \in \mathbb{R}$, and that $\boldsymbol{N}$ is also in standard form with $\boldsymbol{N} = (\mathbb{R} \times \Sigma, g_{\boldsymbol{N}}, \mathfrak{o}, \mathfrak{t}_{\boldsymbol{N}})$.

By Lemma 4.7.2 there exists $t_* > t_{\boldsymbol{M}}$ so that $\{t_*\} \times \Sigma$ contains regular Cauchy pairs $({}_*S, {}_*T)$ and (S_*, T_*) with

$$({}_*S, {}_*T) \prec_{\boldsymbol{M}} (S_{\boldsymbol{M}}, T_{\boldsymbol{M}}) \prec_{\boldsymbol{M}} (S_*, T_*), \tag{4.29}$$

where $\prec_{\boldsymbol{M}}$ indicates the preorder given by the causal structure of $\boldsymbol{M}$. As both $({}_*S, {}_*T)$ and (S_*, T_*) are regular Cauchy pairs in a common Cauchy surface of $\boldsymbol{N}$, we may also choose $t_{\boldsymbol{N}} > t_*$ so that $\{t_{\boldsymbol{N}}\} \times \Sigma$ contains regular Cauchy pairs $(S_{\boldsymbol{N}}, T_{\boldsymbol{N}})$ and $({}_{\boldsymbol{N}}S, {}_{\boldsymbol{N}}T)$ obeying

$$({}_{\boldsymbol{N}}S, {}_{\boldsymbol{N}}T) \prec_{\boldsymbol{N}} ({}_*S, {}_*T), \qquad (S_*, T_*) \prec_{\boldsymbol{N}} (S_{\boldsymbol{N}}, T_{\boldsymbol{N}}). \tag{4.30}$$

Using the method of Proposition 4.6.2, an interpolating metric $g_{\boldsymbol{I}}$ may now be constructed (see [48]) so that

- $\boldsymbol{I} = (\mathbb{R} \times \Sigma, g_{\boldsymbol{I}}, \mathfrak{o}, \mathfrak{t}_{\boldsymbol{I}})$ is a Loc-spacetime in standard form;
- there is a chain of Cauchy morphisms of the form (4.24) between $\boldsymbol{M}$ and $\boldsymbol{N}$, via $\boldsymbol{I}$;
- the orderings (4.29) and (4.30) hold with $\prec_{\boldsymbol{M}}, \prec_{\boldsymbol{N}}$ replaced by $\prec_{\boldsymbol{I}}$.

The last item, together with transitivity of $\prec_{\boldsymbol{I}}$, entails

$$({}_{\boldsymbol{N}}S, {}_{\boldsymbol{N}}T) \prec_{\boldsymbol{I}} (S_{\boldsymbol{M}}, T_{\boldsymbol{M}}) \prec_{\boldsymbol{I}} (S_{\boldsymbol{N}}, T_{\boldsymbol{N}}). \tag{4.31}$$

Now $\omega_{\boldsymbol{N}}$ has the Reeh–Schlieder property for $(S_{\boldsymbol{N}}, T_{\boldsymbol{N}})$ and is split for $({}_{\boldsymbol{N}}S, {}_{\boldsymbol{N}}T)$ in $\boldsymbol{N}$, and hence the same is true for $\mathscr{A}(\delta)^*\omega_{\boldsymbol{N}}$ in $\boldsymbol{F}$ and for $(\mathscr{A}(\gamma)^{-1})^*\mathscr{A}(\delta)^*\omega_{\boldsymbol{N}}$ in

$\boldsymbol{I}$, using Lemma 4.7.5 twice. By (4.31) and Remark 4.7.4 the latter state is both Reeh–Schlieder and split for (S_M, T_M), as a regular Cauchy pair in $\boldsymbol{I}$. Using Lemma 4.7.5 twice again, the same is true for $\mathscr{A}(\beta)^*(\mathscr{A}(\gamma)^{-1})^*\mathscr{A}(\delta)^*\omega_N$ in $\boldsymbol{P}$ and finally for $\nu^*\omega_N$ in $\boldsymbol{M}$, where $\nu = \mathscr{A}(\delta) \circ \mathscr{A}(\gamma)^{-1} \circ \mathscr{A}(\beta) \circ \mathscr{A}(\alpha)^{-1}$. □

Remark 4.7.7 See [48] for discussion, expanding on the following points:

1. The statement of Theorem 4.7.6 holds if modified so as to refer the split or Reeh–Schlieder properties separately.
2. We have combined the cyclic and separating aspects of the Reeh–Schlieder results for convenience. However, if the GNS vector of the state ω_N is known to be cyclic for every local von Neumann algebra corresponding to nonempty and relatively compact $O \in \mathcal{O}(N)$ with nontrivial causal complement, then it is Reeh–Schlieder for all regular Cauchy pairs in N and the conclusions of Theorem 4.7.6 apply.
3. The conclusions of Theorem 4.7.6 also apply to more general regions: if nonempty $O \in \mathcal{O}(\boldsymbol{M})$ is nonempty relatively compact with nontrivial causal complement then one may find a regular Cauchy pair (S_M, T_M) with $D_M(S_M) \subset O \subset D_M(T_M)$, whereupon Theorem 4.7.6 yields a state that is both cyclic and separating for $\pi_{\omega_M}(\mathscr{A}^{\mathrm{kin}}(\boldsymbol{M}; O))''$. Similarly, if nonempty $O_i \in \mathcal{O}(\boldsymbol{M})$ can be separated by a regular Cauchy pair (S_M, T_M), such that $O_1 \subset D_M(S_M)$, $D_M(T_M) \subset O_2$, the local von Neumann algebras corresponding to the O_i form a split inclusion in the GNS representation induced by Theorem 4.7.6.
4. Suppose that $\omega_N \in \mathscr{S}(N)$, where $\mathscr{S}$ is a state space for $\mathscr{A}$ obeying the timeslice condition. Then we also have $\omega_M \in \mathscr{S}(\boldsymbol{M})$, because the isomorphism ν is formed from a chain of Cauchy morphisms. In the case of the Klein–Gordon field, for example, if ω_N is Hadamard, then so is ω_M. If the state space $\mathscr{S}$ also obeys local quasiequivalence, then *every* state of $\mathscr{S}(\boldsymbol{M})$ has the split property for *every* regular Cauchy pair of $\boldsymbol{M}$ [48]. If, more strongly, each $\mathscr{S}(\boldsymbol{M})$ is a complete local quasiequivalence class, then there exists a full Reeh–Schlieder state in $\mathscr{S}(\boldsymbol{M})$, i.e., its GNS vector is cyclic and separating for every local von Neumann algebra of a relatively compact region with nontrivial causal complement [135].

Various applications of the partial Reeh–Schlieder result are discussed in [135]. The ability to assert both partial Reeh–Schlieder and split properties simultaneously allows one to show that local von Neumann algebras (in suitable representations) form *standard split inclusions* [43], leading to various consequences, including the local implementation of gauge transformations and to the classification of the local von Neumann algebras as the unique hyperfinite III_1 factor (up to isomorphism, and possibly tensored with an abelian centre) [48]. For the latter application, one must additionally assume the existence of a scaling limit as described in Theorem 4.5.2.

Finally, Theorem 4.7.6 would be of little utility in the absence of spacetimes N for which $\mathscr{A}(N)$ admits states that have the Reeh–Schlieder and split properties. Minkowski space provides the canonical example, but one may give reasonable physical conditions that would guarantee the existence of such states in connected ultrastatic spacetimes [48]. As every connected spacetime may be linked to a connected ultrastatic spacetime by a chain of Cauchy morphisms, one expects that Theorem 4.7.6 applies nontrivially at least in connected spacetimes $\boldsymbol{M}$ for most physically reasonable locally covariant theories.

4.8 Quantum Energy Inequalities, Passivity, μSC and All That

As already mentioned, the microlocal spectrum condition appears to be the most promising criterion for specifying physical states (and state spaces) in quantum field theory in curved spacetime. While the μSC originated in the study of linear quantum fields, it has the potential to be relevant for interacting quantum fields and has proved instrumental in the perturbative construction of locally covariant interacting quantum field theories [21, 94–96]. One of the central points is that the μSC permits the definition of renormalized Wick-ordered and time-ordered operators of the quantized linear Klein-Gordon field (and its derivatives) as operator-valued distributions, with finite fluctuations [21]. A converse of that statement has also recently been proved for ultrastatic spacetimes: In order that the Wick-products of derivatives of the quantized linear Klein-Gordon field have finite fluctuations, it is necessary that the Wick-ordering is defined with respect to a state obeying the μSC [72]. Thus, the μSC seems inevitable as the basis for any perturbative construction of interacting quantum fields on generic globally hyperbolic spacetimes.

In this section, we review various relations between the μSC and other conditions on physical states which appear reasonable to demand in locally covariant quantum field theory and which in some way express dynamical stability. We will also mention some other states which have been proposed for consideration as special states, mainly for the quantized linear Klein-Gordon field.

4.8.1 Quantum Energy Inequalities

The locally covariant setting of quantum field theory is the theoretical basis for the semiclassical Einstein equation,

$$G_{\boldsymbol{M}}(h) = 8\pi G \omega(\mathbf{T}_{\boldsymbol{M}}[h]) . \tag{4.32}$$

In this equation, $G_{\boldsymbol{M}}(h) = \int_{\boldsymbol{M}} G_{ab}(x) h^{ab}(x)\, d\mathrm{vol}_{\boldsymbol{M}}(x)$ is the Einstein tensor corresponding to the spacetime $\boldsymbol{M}$ (as usual, considered as an object of **Loc**) smeared with a C_0^∞ test-tensor field h. $\mathbf{T}_{\boldsymbol{M}}[h]$ is the stress-energy tensor of a locally covariant theory obeying the time-slice property, so it arises from the relative Cauchy evolution of the locally covariant theory as indicated in Sect. 4.4. (Actually, the relative Cauchy evolution only specifies $[\mathbf{T}_{\boldsymbol{M}}[h], A]$, i.e. the commutator of $\mathbf{T}_{\boldsymbol{M}}[h]$ with elements A in $\mathscr{A}(\boldsymbol{M})$, so fixing $\mathbf{T}_{\boldsymbol{M}}[h]$ at best up to scalar multiples of the unit operator depending linearly on h.[20] Similarly, in models where $\omega(\mathbf{T}_{\boldsymbol{M}}[h])$ is defined by a process of renormalization, there is a residual finite renormalization ambiguity. As there does

[20] In Hilbert-space representations, $\mathbf{T}_{\boldsymbol{M}}[h]$ is an unbounded operator, one also has to consider the domain of algebra elements A for which the commutator can be formed, or in which precise mathematical sense the commutator is to be understood.

not seem to be a general, locally covariant way to fix these ambiguities, (4.32) needs further input. We refer to [36, 157, 163] for further discussion.)

For linear quantum field models and states ω fulfilling the μSC, it holds that $\omega(\mathbf{T}_{\boldsymbol{M}}[h])$ is actually given by a smooth, symmetric tensor field $\omega(\mathbf{T}_{\boldsymbol{M}(ab)}(x))$ ($x \in M$) on $\boldsymbol{M}$ so that $\omega(\mathbf{T}_{\boldsymbol{M}}[h]) = \int_M \omega(\mathbf{T}_{\boldsymbol{M}(ab)}(x))h^{ab}(x)\, d\text{vol}_{\boldsymbol{M}}(x)$. Then for any smooth future-directed timelike curve γ in $\boldsymbol{M}$ parametrized by proper time τ,

$$\rho_{\omega,\gamma}(\tau) = \omega(\mathbf{T}_{\boldsymbol{M}(ab)}(\gamma(\tau)))\dot{\gamma}^a(\tau)\dot{\gamma}^b(\tau)$$

is the expectation value of the energy density along γ at $\gamma(\tau)$. It is known that

$$\inf_{\omega} \rho_{\omega,\gamma}(\tau) = -\infty$$

as ω ranges over the set of states fulfilling the μSC with γ and τ fixed. That means, at a given spacetime point x, the expectation value of the energy density, for any observer, is unbounded below as a functional of the (regular) states ω [51]. Consequently, the weak energy condition usually assumed in the macroscopic description of matter in general relativity fails to hold in general for the expectation value of the stress-energy tensor of quantized fields. (There is an argument showing such a behaviour also for stress-energy tensor expectation values of general Wightman-type quantum fields on Minkowski spacetime [45]).

This violation of the weak energy condition for quantized fields is not unconstrained, however. At least for linear quantum fields, there are quantum energy inequalities which provide restrictions on the magnitude and duration of the violation of the weak energy condition. Here, one says that a set of states $\boldsymbol{S}_{qei}(\boldsymbol{M})$ on $\mathscr{A}(\boldsymbol{M})$ fulfils a *quantum energy inequality (QEI)* if for any smooth, future-directed timelike curve γ, defined on some open proper time interval I, there is for any $f \in C_0^\infty(I, \mathbb{R})$ some constant $c_\gamma(f) > -\infty$ such that

$$\inf_{\omega \in \boldsymbol{S}_{qei}(\boldsymbol{M})} \int_I \rho_{\omega,\gamma}(\tau) f^2(\tau)\, d\tau \geq c_\gamma(f)\,. \tag{4.33}$$

In other words, when averaging the energy density expectation values with a smooth quadratic weight function along any timelike curve, one obtains a quantity which is bounded below as long as the states range over the set of states $\boldsymbol{S}_{qei}(\boldsymbol{M})$.

Quantum energy inequalities were first discussed by Ford [76], initially motivated on thermodynamic grounds, but then later derived in free models on Minkowski space [57, 75, 77, 78] and some curved spacetimes, e.g., [66, 124]. They have been established rigorously for the quantized (minimally coupled) Klein-Gordon [49], Dirac [39, 67] and free electromagnetic fields [62], in all cases for all $\boldsymbol{M}$ of **Loc**, and for $\boldsymbol{S}_{qei}(\boldsymbol{M})$ coinciding with the set of states fulfilling the μSC on $\mathscr{A}(\boldsymbol{M})$. The status of QEIs for interacting quantum field theories remains to be clarified, but it is not expected that they will hold in general without further modification [118]. Nonetheless, results are known for some interacting models in two spacetime dimensions

[16, 58] and some model-independent results are known in Minkowski space [17]. One of the main applications of QEIs is to put restrictions on the occurrence of spacetimes with unusual causal behaviour as solutions to the semiclassical Einstein equations, e.g. spacetimes with closed timelike curves. We refer to the Ref. [50] and literature cited there for considerable further discussion.

It has been shown (for the quantized, minimally coupled Klein-Gordon field) that the lower bounds $c_{\boldsymbol{M},\gamma}(f)$ in (4.33) can be chosen such that they comply with local covariance [52, 65], i.e., for any morphism $\psi : \boldsymbol{M} \to \boldsymbol{N}$, they obey

$$c_{\boldsymbol{N},\psi\circ\gamma}(f) = c_{\boldsymbol{M},\gamma}(f)\,. \tag{4.34}$$

Conversely, one can use QEIs as the basis of a selection criterion for a locally covariant state space. To be specific, suppose that for all objects $\boldsymbol{M}$ of **Loc** and timelike curves $\gamma : I \to \boldsymbol{M}$ (where I is an open interval), a map $f \mapsto c_{\boldsymbol{M},\gamma}(f) \in \mathbb{R}$ ($f \in C_0^\infty(I, \mathbb{R})$) has been selected such that the covariance condition (4.34) is fulfilled. Let $\mathscr{A}$ be a locally covariant theory and define $\boldsymbol{S}_{qei,c}(\boldsymbol{M})$ to consist of all the states ω on $\mathscr{A}(\boldsymbol{M})$ for which the expectation value of the stress-energy tensor is defined and obeys

$$\int_I \rho_{\omega,\gamma}(\tau) f^2(\tau)\, d\tau \geq c_{\boldsymbol{M},\gamma}(f)\,, \quad f \in C_0^\infty(I, \mathbb{R})\,. \tag{4.35}$$

Evidently $\boldsymbol{S}_{qei,c}(\boldsymbol{M})$ is stable under formation of convex combinations; if it is also stable under operations induced by elements in $\mathscr{A}(\boldsymbol{M})$ (and this is generally the case for quantized linear fields), then one can define a state space $\mathscr{S}$ for $\mathscr{A}$ by setting $\mathscr{S}(\boldsymbol{M}) = \boldsymbol{S}_{qei,c}(\boldsymbol{M})$. As mentioned before, this definition is consistent (up to some details not spelled out here in full) with the microlocal spectrum condition as a selection criterion for a state space for locally covariant linear quantum fields, upon appropriate choice of the $c_{\boldsymbol{M},\gamma}(f)$. In fact, the two criteria result in the same state space, as we will indicate next, with the help of yet another selection criterion.

4.8.2 Passivity

Another very natural selection criterion is that the physical states of a locally covariant quantum field theory should be locally in the folia of ground states, or thermal equilibrium states, in spacetimes which admit sufficient time-symmetry that such states exist. This is the case for ultrastatic globally hyperbolic spacetimes, i.e., those spacetimes in standard form $\boldsymbol{M} = (\mathbb{R} \times \Sigma,\, dt \otimes dt - h, \mathfrak{o}, \mathfrak{t})$ where we write spacetime points as $(t, \mathbf{x})$ with $t \in \mathbb{R}$ and $\mathbf{x} \in \Sigma$, the metric h is a (t-independent) complete Riemannian metric on Σ, and $\mathfrak{t}$ is chosen so that dt is future-directed. Then there is a global time-symmetry on that spacetime, i.e. a Killing-flow $\vartheta_t : (t_0, \mathbf{x}) \mapsto (t_0 + t, \mathbf{x})$, $t \in \mathbb{R}$. Given a locally covariant theory $\mathscr{A}$, this leads to an induced 1-parametric group $\{\alpha_t\}_{t\in\mathbb{R}}$ of unital $*$-automorphisms of $\mathscr{A}(\boldsymbol{M})$ for any ultrastatic

$\boldsymbol{M}$. An invariant state ω on $\mathscr{A}(\boldsymbol{M})$ ($\omega \circ \alpha_t = \omega$) is called a *ground state* for $\{\alpha_t\}_{t\in\mathbb{R}}$ if there is a dense unital $*$-subalgebra $\mathscr{A}_0(\boldsymbol{M})$ of $\mathscr{A}(\boldsymbol{M})$ such that

$$\frac{1}{i}\frac{d}{dt}\bigg|_{t=0} \omega(A\alpha_t(B)) \geq 0\,, \quad A, B \in \mathscr{A}_0(\boldsymbol{M})\,.$$

An invariant state ω on $\mathscr{A}(\boldsymbol{M})$ is called *passive* for $\{\alpha_t\}_{t\in\mathbb{R}}$ if there is a dense unital $*$-subalgebra $\mathscr{A}_0(\boldsymbol{M})$ of $\mathscr{A}(\boldsymbol{M})$ with the property that

$$\frac{1}{i}\frac{d}{dt}\bigg|_{t=0} \omega(U^*\alpha_t(U)) \geq 0$$

holds for all unitary elements of $\mathscr{A}_0(\boldsymbol{M})$ which are continuously connected to the unit element of $\mathscr{A}_0(\boldsymbol{M})$ (cf. [68, 126] for further details). Passive states are generalizations of KMS-states (which can be regarded as thermal equilibrium states in the setting of ultrastatic spacetimes); see [126] for further discussion on this point.

Ground states and KMS-states of the quantized linear scalar Klein-Gordon field on ultrastatic spacetimes are Hadamard states, i.e. they fulfil the microlocal spectrum condition, as was shown in [133]. The same holds for convex mixtures of KMS-states at different temperatures which are the generic examples of passive states.[21] As a consequence, ground states and KMS-states also fulfil quantum energy inequalities.

For the quantized linear Klein-Gordon field on an ultrastatic spacetime, one can show that the converse holds as well. This was established in [68] under certain additional technical assumptions on which we suppress here, contenting ourselves with a simplified statement which will now be outlined. The assumption is that the algebra $\mathscr{A}(\boldsymbol{M})$ assigned to an ultrastatic spacetime $\boldsymbol{M}$ with underlying manifold $\mathbb{R} \times \Sigma$ admits a set of states $\boldsymbol{S}_{qei}(\boldsymbol{M})$, closed under convex combinations and operations, for which the stress-energy expectation values are well-defined, and obeying a QEI of the form (4.33). The QEI then holds in particular for time-flow trajectories of the ultrastatic spacetime i.e. for all $\gamma_{\mathbf{x}}(\tau) = (\tau, \mathbf{x})$, $t \in \mathbb{R}$, $\mathbf{x} \in \Sigma$. It is then assumed that in this case, the quantum energy inequality holds in the form

$$\inf_{\omega\in \boldsymbol{S}_{qei}(\boldsymbol{M})} \int \rho_\omega(\tau, \mathbf{x}) f^2(\tau)\, d\tau \geq c_{\boldsymbol{M}}(f, \mathbf{x})\,, \quad f \in C_0^\infty(\mathbb{R}, \mathbb{R})\,,$$

using the abbreviations $\rho_\omega(\tau, \mathbf{x})$ for $\rho_{\omega,\gamma_{\mathbf{x}}}(\tau)$, and $c_{\boldsymbol{M}}(f, \mathbf{x})$ for $c_{\boldsymbol{M},\gamma_{\mathbf{x}}}(f)$. Making the assumption that $c_{\boldsymbol{M}}(f, \mathbf{x})$ is locally integrable and that Σ is compact, it has been shown in [68] that (i) there is a passive state on $\mathscr{A}(\boldsymbol{M})$, (ii) assuming a form of energy-compactness, there is a passive state which lies in the folium of some state in $\boldsymbol{S}_{qei}(\boldsymbol{M})$,

[21] In fact, the result on the Hadamard property of ground states and KMS-states on ultrastatic spacetimes holds for more general types of quantized linear fields, and more generally also on static (not necessarily ultrastatic) spacetimes.

(iii) assuming clustering properties in time, there is a ground state in $\boldsymbol{S}_{qei}(\boldsymbol{M})$.[22] This shows that—apart from some further technical details—one can generally expect that the imposition of quantum energy inequalities entails the existence of a ground (or passive) state in the folium of those obeying the QEI.

4.8.3 And All that—Relations Between the Conditions on States

Once a set of physical states has been specified on ultrastatic spacetimes, then, if a locally covariant quantum field theory $\mathscr{A}$ satisfies the time-slice property, the specification can be carried over to each spacetime $\boldsymbol{M}$ of **Loc**. For there certainly is an ultrastatic spacetime $\boldsymbol{N}$ with Cauchy surfaces oriented-diffeomorphic to those of $\boldsymbol{M}$ and therefore a chain of Cauchy morphisms linking $\boldsymbol{M}$ to $\boldsymbol{N}$, by Proposition 4.6.2. The set of physical states on $\boldsymbol{N}$ can then be pulled back along this chain to give a set of states on $\boldsymbol{M}$. This raises a question (not addressed in the literature) of whether the state space on $\boldsymbol{M}$ obtained in this way depends on the details of the construction; evidently a necessary condition is that the chosen physical states on ultrastatic spacetimes are dynamically stable in the sense that $\mathrm{rce}_{\boldsymbol{N}}[h]^* \boldsymbol{S}(\boldsymbol{N}) = \boldsymbol{S}(\boldsymbol{N})$, for arbitrary metric perturbations with time-compact support.

Alternatively, one may have a specification of the state spaces in all spacetimes, but without knowing whether any such states exist. Here, again, the deformation argument can be used, if (a) existence can be established in ultrastatic spacetimes, and (b) one has $\mathscr{A}(\psi)^*\mathscr{S}(\boldsymbol{N}) \subset \mathscr{S}(\boldsymbol{M})$ and $(\mathscr{A}(\psi)^{-1})^*\mathscr{S}(\boldsymbol{M}) \subset \mathscr{S}(\boldsymbol{N})$ for all Cauchy morphisms $\psi : \boldsymbol{M} \to \boldsymbol{N}$. Indeed this was how Fulling, Narcowich and Wald originally proved existence of Hadamard states for the quantized linear Klein-Gordon field on globally hyperbolic spacetimes: By proving that ground states on ultrastatic spacetimes have the Hadamard property, and then making use of the fact that the Hadamard property propagates throughout any globally hyperbolic spacetime once it is known to hold in the neighbourhood of a Cauchy-surface [82]. Using the equivalence of Hadamard property and microlocal spectrum condition, this propagation of the Hadamard property is equivalent to the propagation of the wavefront set along bicharacteristics via the bicharacteristic flow [44]. Thus the requirements that ground states and KMS-states for ultrastatic spacetimes should be counted among the physical states, and that all physical states should be locally quasiequivalent, are consistent with the demand that all physical states should be locally quasiequivalent to the states fulfilling the microlocal spectrum condition. But relying on the results of [68], one can even show more: The microlocal spectrum condition implies QEIs, even with a locally covariant lower bound, and this guarantees for locally covariant quantum field theories (up to some additional technical assumptions) that on ultra-

[22]While the results in [68] have only been established for compact Σ, the results could be extended to noncompact Σ upon making suitable integrability assumptions on $c_{\boldsymbol{M}}(f, \mathbf{x})$ with respect to $\mathbf{x} \in \Sigma$.

static spacetimes there will be ground states which are locally quasiequivalent to the states which fulfil locally covariant QEIs. In other words, the following three selection criteria:

- microlocal spectrum condition
- locally covariant quantum energy inequalities
- ground- or KMS-states on ultrastatic spacetimes

for the local folia of physical states are *equivalent* for the locally covariant theory of the quantized linear (minimally coupled) Klein-Gordon field. This is interesting since these selection criteria have different motivations and implications. In fact, these results can be generalized at least to a larger class of locally covariant linear quantized fields, and potentially also to certain perturbatively constructed quantum fields in a suitable version—a key result in this context may be that some perturbatively constructed interacting quantum fields have been shown to satisfy the time-slice property [33]. However, there are also examples of non-minimally coupled quantized linear scalar fields which do not fulfil quantum energy inequalities [61], and arguments for interacting models [118], indicating that quantum energy inequalities as we have stated them are a less general property than, e.g. the μSC. Therefore, as mentioned earlier, the status of quantum energy inequalities, especially in interacting quantum field theories, remains yet to be fully understood.

4.8.4 Other Special States

For quantum fields in curved spacetime, particularly for the quantized linear Klein-Gordon field, several other types of states have been proposed as physical states, or states with special interpretation, from the very beginning of the development of the theory. However, they are in some cases restricted to special spacetime geometries, or are at variance with local covariance, or fail to be locally quasiequivalent to states fulfilling the microlocal spectrum condition in general. We shall list some of them.

Adiabatic vacuum states. This class of states was originally introduced by Parker in his seminal approach to particle creation in quantum fields on expanding cosmological spacetimes [120, 121]. Adiabatic vacua for the quantized linear Klein-Gordon field have been shown to define a single local quasiequivalence class of states on Friedmann-Lemaître-Robertson-Walker spacetimes [112] and to be locally quasiequivalent to states fulfilling the μSC [103]; the latter reference extends the definition from cosmological spacetimes to general globally hyperbolic spacetimes.

Instantaneous vacuum states. This class of states is essentially defined by picking a Cauchy-surface in a globally hyperbolic spacetime and defining a two-point function for the quantized field in terms of the Cauchy-data as that two-point function which would correspond to the ultrastatic vacuum defined by the Riemannian geometry on the Cauchy-surface obtained from the ambient spacetime metric [3]. However, these states in general fail to be locally quasiequivalent to states satisfying the μSC unless the Cauchy-surface is actually part of an ultrastatic (or at least stationary) foliation [102, 148].

States of low energy. This is a class of homogeneous, isotropic states on Friedmann-Lemaître-Robertson-Walker spacetimes which minimize the averaged energy density (4.33) for given averaging function f. These states fulfil the microlocal spectrum condition [117] and have several interesting properties [40].

Local thermal equilibrium states. This is a class of states to which one can, approximately, ascribe a temperature at each point in spacetime, where the temperature together with the temperature rest frame is a function of the spacetime point. This class of states was introduced in [29] and further investigated in [27]; some applications to quantum fields in curved spacetime appear in [138, 141, 142, 157].

BMS-invariant states at conformal lightlike infinity. For a class of asymptotically flat spacetimes, conformal lightlike infinity is a boundary manifold which is invariant under the Bondi-Metzner-Sachs (BMS) group. Certain types of quantum fields induce quantum fields of "conformal characteristic data" on conformal lightlike infinity; then specifying a BMS-invariant, positive energy state on the "conformal boundary" quantum field determines a state of the quantum field on the original spacetime. This state fulfils the μSC under very general assumptions [38, 115] and has been instrumental in proving that there are solutions to the semiclassical Einstein equations for cosmological spacetimes with the non-conformally coupled massive quantized Klein-Gordon field [125].

SJ-states and FP-states. Recently, an interesting proposal for distinguished states of the quantized linear Klein-Gordon field has been made in [1]. There, the two-point function of such a state on $\mathscr{A}(\boldsymbol{M})$ is determined from $E_{\boldsymbol{M}}$, the propagator of the Klein-Gordon operator on the globally hyperbolic hyperbolic spacetime $\boldsymbol{M}$, regarded as an operator on the L^2 space of scalar functions on $\boldsymbol{M}$ induced by the volume form of $\boldsymbol{M}$. The two-point function arises from a polar decomposition of $iE_{\boldsymbol{M}}$ by taking the positive spectral part $(1/2)(|iE_{\boldsymbol{M}}| + iE_{\boldsymbol{M}})$ of $iE_{\boldsymbol{M}}$ as its L^2 kernel. The resulting (quasifree) states were named *SJ-states* in [1]. This construction can be shown to yield a well-defined pure state if there is a morphism $\psi : \boldsymbol{M} \to \boldsymbol{N}$ in Loc such that $\psi(\boldsymbol{M})$ is relatively compact in $\boldsymbol{N}$ [71]. A heuristic argument in [1] (corroborated in [71]) shows that the SJ-state of Minkowski spacetime agrees with the Minkowski vacuum state of the quantized linear Klein-Gordon field. This provided motivation in [1] to regard the SJ-states as distinguished "vacuum states" for the quantized linear Klein-Gordon field in any spacetime. However, explicit calculation for the case of "ultrastatic slab" spacetimes shows that SJ-states in general fail to fulfil the μSC, and even fail to be locally quasiequivalent to states fulfilling the μSC [71]. Furthermore, derivatives of Wick-ordered quantum fields in general fail to have finite fluctuations in SJ-states [72]. A modified construction of SJ-states, using a smoothing procedure on $E_{\boldsymbol{M}}$ and yielding states fulfilling the μSC (at least on ultrastatic slabs and similar slabs of cosmological spacetimes) has been proposed in [19] for the quantized linear Klein-Gordon field; this construction requires smoothing functions which parametrize the states. For the case of the quantized Dirac field, there is a construction method for states which is conceptually related to SJ states, they are called Fermionic projector (FP) states (and incidentally predate SJ states); see [60, 73, 74]. It has been shown in [60] that FP states also fail to fulfil the μSC in general, and that a smoothing procedure again leads to a modified FP state construction rendering states fulfilling the μSC.

4.9 Locally Covariant Fields

Conventional approaches to QFT, and the Wightman axiomatic framework, focus on quantum fields as the primary object of study. In the algebraic approach, the emphasis is rather on the local algebras of observables; quantum fields are regarded as ways of parameterising those local algebras. In the locally covariant framework, quantum fields take on a new aspect—they not only parameterise the local algebras in one given spacetime, but do so in all spacetimes in a compatible way. This idea was present in the literature on QFT in curved spacetimes for a long time in the context of the stress-energy tensor (see, e.g., [104, 163]); its use in the locally covariant context began with the treatment of the spin–statistics connection in curved spacetime [156] and the treatment of perturbation theory [95]. It was one of the motivating ideas behind [24], in which it took on a more functorial form.

4.9.1 General Considerations in **Loc**

The basic idea can be illustrated by the Klein–Gordon theory $\mathscr{A} : \mathsf{Loc} \to \mathsf{Alg}$. As described in Sect. 4.2, each algebra $\mathscr{A}(\boldsymbol{M})$ is generated by elements $\Phi_{\boldsymbol{M}}(f)$ carrying the interpretation of smeared fields; under a morphism $\psi : \boldsymbol{M} \to \boldsymbol{N}$ the smeared fields on $\boldsymbol{M}$ and $\boldsymbol{N}$ are related by equation (4.2), which can be rewritten as an equality of functions from $C_0^\infty(\boldsymbol{M})$ to $\mathscr{A}(\boldsymbol{M})$

$$\mathscr{A}(\psi) \circ \Phi_{\boldsymbol{M}} = \Phi_{\boldsymbol{N}} \circ \psi_*. \tag{4.36}$$

Indeed, as the discussion of Sect. 4.2 makes clear, much of the general theory has been structured on the basis of this observation, which can be given a more categorical form as follows. First, the assignment of test function spaces to spacetimes may be formalised as a functor

$$\mathscr{D} : \mathsf{Loc} \to \mathsf{Set}, \qquad \mathscr{D}(\boldsymbol{M}) = C_0^\infty(\boldsymbol{M}), \qquad \mathscr{D}(\boldsymbol{M} \xrightarrow{\psi} \boldsymbol{N}) = \psi_*, \tag{4.37}$$

where **Set** is the category of sets and (not necessarily injective) functions. Second, when we regard $\mathscr{A}(\psi)$ as a function, we are appealing to the existence of a forgetful functor $\mathscr{U} : \mathsf{Alg} \to \mathsf{Set}$ that maps each **Alg** object to its underlying set and each

Alg-morphism to its underlying function. Then '$\mathscr{A}(\psi)$ regarded as a function' can be represented formally by $\mathscr{U}(\mathscr{A}(\psi)) = (\mathscr{U} \circ \mathscr{A})(\psi)$. Equation (4.36) now becomes an equality of **Set**-morphisms:

$$(\mathscr{U} \circ \mathscr{A})(\psi) \circ \Phi_{\boldsymbol{M}} = \Phi_{\boldsymbol{N}} \circ \mathscr{D}(\psi) \tag{4.38}$$

which is required to hold for every **Loc**-morphism $\psi : \boldsymbol{M} \to \boldsymbol{N}$, and asserts precisely that the functions $\Phi_{\boldsymbol{M}}$ form the components of a *natural transformation* $\Phi : \mathscr{D} \stackrel{.}{\to} \mathscr{U} \circ \mathscr{A}$. The naturality condition can be represented diagrammatically as the requirement that the diagram

$$\begin{array}{ccccc} \boldsymbol{M} & \quad & \mathscr{D}(\boldsymbol{M}) & \xrightarrow{\Phi_{\boldsymbol{M}}} & (\mathscr{U} \circ \mathscr{A})(\boldsymbol{M}) \\ \Big\downarrow \psi & & \Big\downarrow \mathscr{D}(\psi) & & \Big\downarrow (\mathscr{U} \circ \mathscr{A})(\psi) \\ \boldsymbol{N} & & \mathscr{D}(\boldsymbol{N}) & \xrightarrow{\Phi_{\boldsymbol{N}}} & (\mathscr{U} \circ \mathscr{A})(\boldsymbol{N}) \end{array} \tag{4.39}$$

commutes for every morphism $\psi : \boldsymbol{M} \to \boldsymbol{N}$.

Definition 4.9.1 A locally covariant scalar field of theory $\mathscr{A}$ is a natural transformation $\Phi : \mathscr{D} \stackrel{.}{\to} \mathscr{U} \circ \mathscr{A}$. The collection of all locally covariant scalar fields is denoted $\mathrm{Fld}(\mathscr{D}, \mathscr{A})$.

This definition encompasses both linear and nonlinear fields—for example, in the Weyl formulation of the Klein–Gordon theory, the map from test functions to generators $f \mapsto W_{\boldsymbol{M}}(f_{\sim})$ defines a nonlinear locally covariant field.

Unlike many structures in locally covariant QFT, individual locally covariant fields are not stable under the evolution entailed by the timeslice property (in particular, relative Cauchy evolution). Fixing $\Phi_{\boldsymbol{M}}(f) \in \mathscr{A}(\boldsymbol{M})$, it is of course true that $\Phi_{\boldsymbol{M}}(f) \in \mathscr{A}^{\mathrm{kin}}(\boldsymbol{M}; S)$ for any subset $S \in \mathscr{O}(\boldsymbol{M})$ that contains a Cauchy surface of $\boldsymbol{M}$. In general, however, we cannot write $\Phi_{\boldsymbol{M}}(f)$ in the form $\Phi_{\boldsymbol{M}}(h)$ for h supported in S; while this can be done for linear fields such as the Klein–Gordon model, the evolution induced by the timeslice assumption becomes much more involved as soon as Wick powers are included [33].

The description of fields at the functorial level, rather than that of individual spacetimes, opens new ways of manipulating them as mathematical objects. As shown in [52], $\mathrm{Fld}(\mathscr{D}, \mathscr{A})$ can be given the structure of a unital $*$-algebra: given $\Phi, \Psi \in \mathrm{Fld}(\mathscr{D}, \mathscr{A})$, and $\lambda \in \mathbb{C}$, we may define new fields $\Phi + \lambda\Psi$, $\Phi\Psi$, Φ^* by

$$(\Phi + \lambda\Psi)_{\boldsymbol{M}}(f) = \Phi_{\boldsymbol{M}}(f) + \lambda\Psi_{\boldsymbol{M}}(f) \tag{4.40}$$

$$(\Phi\Psi)_{\boldsymbol{M}}(f) = \Phi_{\boldsymbol{M}}(f)\Psi_{\boldsymbol{M}}(f), \tag{4.41}$$

$$(\Phi^*)_{\boldsymbol{M}}(f) = \Phi_{\boldsymbol{M}}(f)^* \tag{4.42}$$

and the unit field may be defined by $\mathbf{1}_{\boldsymbol{M}}(f) = \mathbf{1}_{\mathscr{F}(\boldsymbol{M})}$, for all $f \in C_0^\infty(\boldsymbol{M})$. Furthermore, in the C^*-algebraic setting, one may even find a C^*-norm on a $*$-subalgebra of $\mathrm{Fld}(\mathscr{D}, \mathscr{A})$. The abstract algebra of fields has a number of interesting features: for example, it carries an action of the global gauge group (see Sect. 4.10).

For many quantum fields of interest the maps $\Phi_{\boldsymbol{M}}$ are linear. Such *linear fields* may be singled out as follows: regarding $\mathscr{D}$ now as a functor from **Loc** to **Vec**, the category of complex vector spaces and (not necessarily injective) linear maps, and writing $\mathscr{V}$ for the forgetful functor $\mathscr{V}$: **Alg** $\to$ **Vec**, the linear fields of the theory $\mathscr{A}$ are natural transformations $\Phi : \mathscr{D} \stackrel{.}{\to} \mathscr{V} \circ \mathscr{A}$, and form the collection $\mathrm{Fld}_{\mathrm{lin}}(\mathscr{D}, \mathscr{A})$. While $\mathrm{Fld}_{\mathrm{lin}}(\mathscr{D}, \mathscr{A})$ can be given the structure of a vector space, by (4.40), it does not in general admit a product structure, because (4.41) creates a field that is nonlinear in its argument. Similarly the $*$-operation of (4.42) creates an antilinear field and so cannot be defined as a map of $\mathrm{Fld}_{\mathrm{lin}}(\mathscr{D}, \mathscr{A})$ to itself. However, we can define an antilinear involution $\star$ on $\mathrm{Fld}_{\mathrm{lin}}(\mathscr{D}, \mathscr{A})$ by

$$\Phi^{\star}_{\boldsymbol{M}}(f) = \Phi_{\boldsymbol{M}}(\overline{f})^* \qquad (f \in \mathscr{D}(\boldsymbol{M})) \tag{4.43}$$

and it is natural to do so in this context. For example, Φ is *hermitian* if $\Phi^\star = \Phi$.[23]

The definition of a locally covariant scalar field can be generalised in various ways, by modifying the functor $\mathscr{D}$ (as well as choosing whether to work in **Set** or **Vec**). For example, $\mathscr{D}$ could be the functor assigning compactly supported smooth sections of tensor fields of some specified (but arbitrary) type; the corresponding linear fields can be regarded as tensor fields of the dual type. Again, in some applications one might allow certain subsets of compactly supported distributions (see, e.g., [52]). In all such cases, we will use the notation $\mathrm{Fld}(\mathscr{D}, \mathscr{A})$ for the natural transformations from $\mathscr{D}$ (in **Set**) to $\mathscr{U} \circ \mathscr{A}$ and $\mathrm{Fld}_{\mathrm{lin}}(\mathscr{D}, \mathscr{A})$ for those from $\mathscr{D}$ (in **Vec**) to $\mathscr{V} \circ \mathscr{A}$. Spinorial fields require additional structure beyond those of **Loc** and will be discussed briefly in Sect. 4.9.2.

We can also define multilocal fields, replacing $\mathscr{D}$ by $\mathscr{D}^{\times k}(\boldsymbol{M}) = C_0^\infty(\boldsymbol{M})^{\times k}$, $\mathscr{D}^{\times k}(\psi) = \psi_*^{\times k}$, or by $\mathscr{D}^{\otimes k}(\boldsymbol{M}) = C_0^\infty(\boldsymbol{M})^{\otimes k}$, $\mathscr{D}^{\otimes k}(\psi) = \psi_*^{\otimes k}$ in the linear case. Local fields can be combined to form multilocal fields in obvious ways. For example, given $\Phi, \Psi \in \mathrm{Fld}_{\mathrm{lin}}(\mathscr{D}, \mathscr{A})$, we may define bilocal fields $\Phi \overset{\rightarrow}{\otimes} \Psi$ and $\Phi \overset{\leftarrow}{\otimes} \Psi$, i.e., natural transformations from $\mathscr{D}^{\otimes 2}$ to $\mathscr{V} \circ \mathscr{A}$, by

$$\begin{aligned}(\Phi \overset{\rightarrow}{\otimes} \Psi)_{\boldsymbol{M}}(f \otimes h) &= \Phi_{\boldsymbol{M}}(f)\Psi_{\boldsymbol{M}}(h)\\ (\Phi \overset{\leftarrow}{\otimes} \Psi)_{\boldsymbol{M}}(f \otimes h) &= \Psi_{\boldsymbol{M}}(h)\Phi_{\boldsymbol{M}}(f)\end{aligned} \tag{4.44}$$

for $f, h \in \mathscr{D}(\boldsymbol{M})$, $\boldsymbol{M} \in$ **Loc**. These structures will be useful in Sect. 4.10.4, where it will be shown that the abstract viewpoint on fields allows the Klein–Gordon theory to be specified directly at the functorial level in terms of its generating field.

[23]The reader might wonder why (4.43) is not adopted for $\mathrm{Fld}(\mathscr{D}, \mathscr{A})$ in place of (4.42). The reason is that $\Phi^*\Phi$ is a positive element of $\mathrm{Fld}(\mathscr{D}, \mathscr{A})$ in the sense that $(\Phi^*\Phi)_{\boldsymbol{M}}(f) = \Phi_{\boldsymbol{M}}(f)^*\Phi_{\boldsymbol{M}}(f)$ is a positive element in $\mathscr{A}(\boldsymbol{M})$ for every $f \in \mathscr{D}(\boldsymbol{M})$, while $\Phi^\star\Phi$ need not be positive in this way. Order structure and functional calculus for abstract fields is discussed in [52].

4.9.2 The Inclusion of Spin

The inclusion of fields with spin requires a modification of the category of spacetimes to incorporate spin structures. For definiteness, let us work in $n = 4$ spacetime dimensions, in which there are some simplifications [99]. In particular, every globally hyperbolic manifold $\boldsymbol{M} \in \mathsf{Loc}$ admits a unique spin bundle (up to equivalence), namely the trivial bundle $S\boldsymbol{M} := \boldsymbol{M} \times \mathrm{SL}(2, \mathbb{C})$, regarded as a right principal bundle. A spin structure in this context is a smooth double covering σ from $S\boldsymbol{M}$ to the bundle $F\boldsymbol{M}$ of oriented and time-oriented orthonormal frames on $\boldsymbol{M}$ (also a right-principal bundle with structure group given by the proper orthochronous Lorentz group $\mathcal{L}_+^\uparrow$), such that $\sigma \circ R_A = R_{\Lambda(A)} \circ \sigma$, where $\Lambda : \mathrm{SL}(2, \mathbb{C}) \to \mathcal{L}_+^\uparrow$ is the standard double cover of groups and we use R for each right action.

There is always at least one spin structure in our current setting, and the distinct possibilities are classified up to equivalence by the cohomology group $H^1(\boldsymbol{M}; \mathbb{Z}_2)$. We replace **Loc** by a new category **SpinLoc**, whose objects are pairs $(\boldsymbol{M}, \sigma)$ where σ is a spin structure for $\boldsymbol{M}$; a morphism between objects $(\boldsymbol{M}, \sigma)$ and $(\boldsymbol{M}', \sigma')$ of **SpinLoc** is a bundle morphism $\Psi : S\boldsymbol{M} \to S\boldsymbol{M}'$ such that

- $\Psi(p, A) = (\psi(p), \Xi(p)A)$ for some **Loc**-morphism $\psi : \boldsymbol{M} \to \boldsymbol{M}'$ and smooth function $\Xi : \boldsymbol{M} \to \mathrm{SL}(2, \mathbb{C})$;
- $\sigma' \circ \Psi = \psi_* \circ \sigma$, where $\psi_* : F\boldsymbol{M} \to F\boldsymbol{M}'$ is induced by the tangent map of ψ.

It is convenient to write $\Psi = (\psi, \Xi)$ under these circumstances.

A locally covariant quantum field theory is now a functor $\mathscr{A} :$ **SpinLoc** $\to$ **Alg** (or **C*-Alg**, or some other category as in [156]). Note that such a functor encodes both geometric embeddings and spin rotations. The timeslice property can be defined as before, regarding $\Psi = (\psi, \Xi)$ as Cauchy in **SpinLoc** whenever ψ is Cauchy in **Loc**.

It is now possible to introduce locally covariant fields of different spin, starting with the construction of appropriate test function spaces. Let ρ be any (real or complex) representation of $\mathrm{SL}(2, \mathbb{C})$ on vector space V_ρ, and write $\mathbb{K} = \mathbb{R}$ (resp., $\mathbb{C}$) in the real (resp., complex) case. Given any object $(\boldsymbol{M}, \sigma)$ of **SpinLoc**, let $\mathscr{D}_\rho(\boldsymbol{M}, \sigma) = C_0^\infty(\boldsymbol{M}; V_\rho)$ be the space of compactly supported functions on $\boldsymbol{M}$ with values in V_ρ;[24] given any **SpinLoc** morphism $\Psi : (\boldsymbol{M}, \sigma) \to (\boldsymbol{M}', \sigma')$, define $\mathscr{D}_\rho(\Psi) : \mathscr{D}_\rho(\boldsymbol{M}, \sigma) \to \mathscr{D}_\rho(\boldsymbol{M}', \sigma')$ by

$$\mathscr{D}_\rho(\Psi) f = \psi_* (\rho(\Xi) f),$$

where $\Psi = (\psi, \Xi)$ as above, and $C_0^\infty(\boldsymbol{M}; V_\rho) \ni \rho(\Xi) f : p \mapsto \rho(\Xi(p)) f(p)$. It is easily checked that $\mathscr{D}_\rho$ is a functor from **SpinLoc** to the category of vector

[24] Somewhat more technically, $\mathscr{D}_\rho(\boldsymbol{M}, \sigma)$ may be regarded as the space of compactly supported sections of the bundle $\boldsymbol{M} \ltimes_\rho V_\rho$ associated to $S\boldsymbol{M}$ and ρ.

spaces over $\mathbb{K}$. The $\mathbb{K}$-linear locally covariant fields $\mathrm{Fld}_{\mathrm{lin}}(\mathscr{A}, \mathscr{D}_\rho)$ are now naturally regarded as fields of 'type ρ'.[25]

Particular interest attaches to the irreducible complex representations $D^{(k,l)}$ $(k, l \in \mathbb{N}_0)$ of $\mathrm{SL}(2, \mathbb{C})$ on the vector space $V^{(k,l)} = (\circledS^k \mathbb{C}^2) \otimes (\circledS^l \mathbb{C}^2)$, where $\circledS$ denotes a symmetrised tensor product, and

$$D^{(k,l)}(A) = A^{\circledS k} \otimes \overline{A}^{\circledS l} \qquad (A \in \mathrm{SL}(2, \mathbb{C})),$$

with the bar denoting complex conjugation. These representations exhaust the finite-dimensional complex irreducible representations of $\mathrm{SL}(2, \mathbb{C})$ up to equivalence, and are familiar from the Minkowski space theory [143]. Irreducible real-linear representations of interest are formed from $D^{(k,l)} \oplus D^{(l,k)}$ $(k \neq l)$ or $D^{(l,l)}$ restricted to suitable subspaces. In an obvious way, we will call locally covariant $\mathbb{K}$-linear fields associated with these representations *fields of type* (k, l). An important distinction is between the cases in which $\frac{1}{2}(k + l)$ is integer or half-integer, which (assuming the normal spin-statistics relation) correspond to bosonic or fermionic fields. We now turn to a more detailed discussion of this point.

4.9.3 The Spin–statistics Connection

It is an observed fact that particles of integer spin display bosonic statistics and those of half-integer spin obey fermionic statistics. One of the major early successes of axiomatic QFT (see, e.g., the classic presentation in [143]) was to prove that these observed facts are consequences of the basic axioms, and therefore true in a model-independent fashion. Of course, these proofs are formulated for Minkowski space QFT, and make full use of the Poincaré symmetry group, while the experimental observations take place in a curved spacetime, so it is important to understand how the spin-statistics theorem can be extended to general backgrounds.

While a number of authors had demonstrated the inconsistency of various free models with incorrect statistics in curved spacetimes [122, 160],[26] no model-independent result was available until [156], which introduced a number of ideas that now form the basis of locally covariant QFT. The framework employed in [156] differs in two important respects from that of [24]: the category of spacetimes is **SpinLoc** rather than **Loc**, and the target category is neither **Alg** nor **C*-Alg**, but rather a category whose objects are nets of von Neumann algebras indexed over relatively compact spacetime subsets. For these reasons (particularly the second) it would be a departure from our development to describe the results of [156] in detail.

In broad terms, however, the spin–statistics connection proved in [156] is as follows. We consider a theory on **SpinLoc** in which, on each spacetime, the net of local von Neumann algebras is generated by a field Φ of type (k, l). It is also assumed

[25] Terminology here is parallel to [156] but one could equally make a case for labelling the type by the dual (also known as contragredient) representation ρ^*, which was our convention in Sect. 4.9.1.

[26] An interesting variant shows what happens if negative-normed states are allowed [91].

that the instantiation of the theory in Minkowski space[27] is a Wightman theory with the corresponding component of Φ as a Wightman field. One supposes that there is a spacetime $(\boldsymbol{M}, \sigma)$ and a pair of causally disjoint and relatively compact regions $\langle O_1, O_2\rangle \in \mathscr{O}^{(2)}(\boldsymbol{M})$ so that Φ exhibits anomalous statistics

$$\Phi_{(\boldsymbol{M},\sigma)}(f_1)\Phi_{(\boldsymbol{M},\sigma)}(f_2)^* + (-1)^{k+l}\Phi_{(\boldsymbol{M},\sigma)}(f_2)^*\Phi_{(\boldsymbol{M},\sigma)}(f_1) = 0$$

for all $f_i \in C_0^\infty(O_i; V^{(k,l)})$ (or that the same holds with the adjoint removed). By a prototype of the rigidity arguments discussed in Sect. 4.6.2, it is proved that there must be a violation of the spin–statistics connection in Minkowski space, which can only happen if all smearings of Φ are trivial in Minkowski space [143, Thms. 4–10]. Thus the local algebras in Minkowski space, generated by Φ, consist only of multiples of the unit and it follows that the same is true of local algebras in all spacetimes. As these algebras are generated by Φ, one may conclude that *in every spacetime*, all smearings of Φ are multiples of the unit operator (vanishing in Minkowski space).

To close this section, we note a number of potential extensions. First, it is a slightly unsatisfactory feature that the use of spin structures invokes unobservable geometric structures from the start; similarly, the idea of spin is, to an extent, inserted by hand at the start of the construction. It would be desirable to understand more clearly why spin (which is tightly linked to rotations in Minkowski space) continues to be an appropriate notion in general curved spacetimes, and how it can be incorporated in a more operational way. Second, one would also like a spin–statistics connection that is not based on algebras generated by a single field. An account addressing these points is sketched in [56] and will appear in full shortly. The key idea is to base the framework on a category of spacetimes with global coframes (i.e., a 'rods and clocks' account of spacetime measurements). The spin–statistics theorem that emerges from this analysis is again proved by rigidity methods on the assumption that the theory obeys standard statistics in Minkowski space.

4.10 Subtheory Embeddings and the Global Gauge Group

In category theory, it is often the morphisms between functors, i.e., natural transformations, that are the main point of interest. Natural transformations appear in locally covariant QFT with important physical interpretations: they are used in the description of locally covariant fields and in order to compare theories.

The idea that equivalences of functors denote physically equivalent theories was already present in [24]; the use of general natural transformations to indicate subtheory embeddings was introduced in [69], while a systematic study of endomorphisms and automorphisms of locally covariant theories is given in [54], on which

[27] Here, the trivial spin structure $\sigma_0(A) = R_{\Lambda(A)}e$ is intended, where $e = (\partial/\partial x^\mu)_{\mu=0,\dots,3}$ is the orthonormal frame on Minkowski space associated with standard inertial coordinates x^μ.

our presentation is based. In this section we consider theories obeying Assumptions 4.3.1–4.3.3 throughout.

4.10.1 Subtheory Embeddings: Definition

Definition 4.10.1 Let $\mathscr{A}, \mathscr{B} : \mathsf{Loc} \to \mathsf{Alg}$ be locally covariant theories. Any natural transformation $\eta : \mathscr{A} \stackrel{.}{\to} \mathscr{B}$ is said to *embed $\mathscr{A}$ as a subtheory of $\mathscr{B}$.*

The requirement that η be natural means that there is a collection of morphisms $\eta_{\boldsymbol{M}} : \mathscr{A}(\boldsymbol{M}) \to \mathscr{B}(\boldsymbol{M})$ ($\boldsymbol{M} \in \mathsf{Loc}$) such that the following diagram commutes for every morphism $\psi : \boldsymbol{M} \to \boldsymbol{N}$ of Loc:

$$\begin{array}{ccc} \boldsymbol{M} & \mathscr{A}(\boldsymbol{M}) \xrightarrow{\eta_{\boldsymbol{M}}} \mathscr{B}(\boldsymbol{M}) \\ \downarrow \psi & \downarrow \mathscr{A}(\psi) \qquad\qquad \downarrow \mathscr{B}(\psi) \\ \boldsymbol{N} & \mathscr{A}(\boldsymbol{N}) \xrightarrow{\eta_{\boldsymbol{N}}} \mathscr{B}(\boldsymbol{N}) \end{array} \tag{4.45}$$

That is, the transition between theories commutes with the transitions between spacetimes. An example of a subtheory embedding is given by the even Klein–Gordon theory $\mathscr{A}^{\text{ev}}$ defined in Sect. 4.3. For each $\boldsymbol{M}$, let $\eta_{\boldsymbol{M}} : \mathscr{A}^{\text{ev}}(\boldsymbol{M}) \to \mathscr{A}(\boldsymbol{M})$ be the inclusion of the subalgebra. As $\mathscr{A}^{\text{ev}}(\boldsymbol{M})$ is generated by the unit and bilinear expressions $\Phi_{\boldsymbol{M}}(f)\Phi_{\boldsymbol{M}}(h)$ ($f, h \in \mathscr{D}(\boldsymbol{M})$) and $\mathscr{A}^{\text{ev}}(\psi)$ is a restriction of $\mathscr{A}(\psi)$, it is easily seen that $\eta_{\boldsymbol{N}} \circ \mathscr{A}^{\text{ev}}(\psi) = \mathscr{A}(\psi) \circ \eta_{\boldsymbol{M}}$ for all $\psi : \boldsymbol{M} \to \boldsymbol{N}$, and hence that $\eta : \mathscr{A}^{\text{ev}} \stackrel{.}{\to} \mathscr{A}$.

The physical interpretation of natural transformations as subtheory embeddings is supported by the following observations.

Proposition 4.10.2 *If $\eta : \mathscr{A} \stackrel{.}{\to} \mathscr{B}$, then (a)*

$$\eta_{\boldsymbol{M}} \mathscr{A}^{kin}(\boldsymbol{M}; O) \subset \mathscr{B}^{kin}(\boldsymbol{M}; O) \tag{4.46}$$

for all nonempty $O \in \mathcal{O}(\boldsymbol{M})$ and $\boldsymbol{M} \in \mathsf{Loc}$, with equality if η is a natural isomorphism (also called an equivalence*); (b) if $\psi \in \text{End}(\boldsymbol{M})$ then*

$$\eta_{\boldsymbol{M}} \circ \mathscr{A}(\psi) = \mathscr{B}(\psi) \circ \eta_{\boldsymbol{M}}; \tag{4.47}$$

(c) for all $h \in H(\boldsymbol{M})$,

$$\eta_{\boldsymbol{M}} \circ \text{rce}_{\boldsymbol{M}}^{(\mathscr{A})}[h] = \text{rce}_{\boldsymbol{M}}^{(\mathscr{B})}[h] \circ \eta_{\boldsymbol{M}}. \tag{4.48}$$

Proof Part (b) is simply a special case of the definition. Similarly, applying the definition to the embedding $\iota_{\boldsymbol{M};O}$ we have $\mathscr{B}(\iota_{\boldsymbol{M};O}) \circ \eta_{\boldsymbol{M}|_O} = \eta_{\boldsymbol{M}} \circ \mathscr{A}(\iota_{\boldsymbol{M};O})$ from which part (a) follows on taking images. Part (c) is proved in [69, Prop. 3.8] and is again simply a matter of employing the basic definitions a number of times. □

The above result shows that subtheory embeddings act locally (part (a)) and intertwine both geometric symmetries (part (b)) and the dynamics of the theory (c). In particular, if the relative Cauchy evolution can be differentiated to yield a stress-energy tensor, acting as a derivation, then (c) implies

$$[T_{\boldsymbol{M}}^{(\mathscr{B})}(f), \eta_{\boldsymbol{M}} A] = \eta_{\boldsymbol{M}}[T_{\boldsymbol{M}}^{(\mathscr{A})}(f), A] \tag{4.49}$$

which shows clearly that $\eta_{\boldsymbol{M}}$ identifies degrees of freedom of $\mathscr{A}$ with some of those of $\mathscr{B}$ in a physically meaningful way.

An equivalence of a theory $\mathscr{A}$ with itself—an automorphism of the theory—has a special significance. The automorphisms of any functor $\mathscr{A}$ form a group $\mathrm{Aut}(\mathscr{A})$ under composition, and it is a pleasing aspect of locally covariant quantum field theories that their automorphism groups can be interpreted as global gauge groups [54]: as Proposition 4.10.2(a), (b) shows, any $\zeta \in \mathrm{Aut}(\mathscr{A})$ has components $\zeta_{\boldsymbol{M}}$ that map each local algebra $\mathscr{A}^{\mathrm{kin}}(\boldsymbol{M}; O)$ isomorphically to itself and commute with the action of spacetime symmetries. These are natural generalisations of conditions set down by Doplicher et al. [42] for global gauge symmetries in Minkowski AQFT.[28] In the DHR analysis, nets of local algebras with nontrivial global gauge group are called *field algebras*. By contrast, a local algebra of observables is the subalgebra of the corresponding field algebra consisting of fixed elements under the action of the gauge group. One may make a similar construction in the locally covariant context [54]: defining $\mathscr{A}_{\mathrm{obs}}(\boldsymbol{M})$ to be the subalgebra of $\mathscr{A}(\boldsymbol{M})$ fixed under the action of all $\zeta_{\boldsymbol{M}}$ ($\zeta \in \mathrm{Aut}(\mathscr{A})$), each $\psi : \boldsymbol{M} \to \boldsymbol{N}$ induces a $\mathscr{A}_{\mathrm{obs}}(\psi) : \mathscr{A}_{\mathrm{obs}}(\boldsymbol{M}) \to \mathscr{A}_{\mathrm{obs}}(\boldsymbol{N})$ by restriction of $\mathscr{A}(\psi)$, and overall yields a new locally covariant theory $\mathscr{A}_{\mathrm{obs}}$ that can be taken as the theory of observables' relative to the 'field functor' $\mathscr{A}$. This interpretation is not entirely satisfactory (see [54, §3.3] for some cautionary remarks) but works well in a number of examples.

4.10.2 Subtheory Embeddings: Classification

The introduction of natural transformations raises the question of whether they are operationally meaningful, given the need to discuss relationships between theories on all possible spacetimes. This question is answered by a rigidity argument similar to those used in Sect. 4.6.2.

Theorem 4.10.3 *Suppose* $\eta, \zeta : \mathscr{A} \stackrel{.}{\to} \mathscr{B}$, *for theories* $\mathscr{A}, \mathscr{B}$, *with* $\mathscr{A}$ *assumed additive with respect to truncated multidiamonds. If, for some* $\boldsymbol{M} \in \mathsf{Loc}$ *and non-*

[28] DHR work in the Hilbert space representation of the Poincaré invariant vacuum state, and require that global gauge transformations should leave the vacuum vector invariant. This also has an analogue in the present setting [54].

empty $O \in \mathscr{O}(\boldsymbol{M})$, $\eta_{\boldsymbol{M}}$ *and* $\zeta_{\boldsymbol{M}}$ *agree on the local kinematic algebra* $\mathscr{A}^{\mathrm{kin}}(\boldsymbol{M}; O)$, *then* $\eta = \zeta$.

Proof This is a straightforward generalization of [54, Thm. 2.6]. We remark that the timeslice property of $\mathscr{B}$ is not used. □

This result shows that the local behaviour of a subtheory embedding in one spacetime is enough to fix it uniquely. In any individual spacetime, moreover, Proposition 4.10.2(c) gives strong constraints and facilitates the classification of subtheory embeddings.

Two examples have been worked out in full detail. For the example of finitely many independent minimally coupled Klein–Gordon fields, with ν_m denoting the number of fields of mass m, the gauge group is a direct product of factors G_m over the mass spectrum, with $G_m = \mathrm{O}(\nu_m)$ for $m > 0$ and $G_0 = \mathrm{O}(\nu_0) \ltimes \mathbb{R}^{\nu_0 *}$, where $\mathbb{R}^{k*}$ denotes the additive group of k-dimensional real row vectors, and the semidirect product is given by $(R, \ell) \cdot (R', \ell') = (RR', \ell R'_0 + \ell')$ [54]. For example, the theory $\mathscr{A}^{(\nu)}$ consisting of ν Klein–Gordon fields $\Phi^{(j)}$ $(1 \le j \le \nu)$ of common mass $m > 0$, has automorphisms ζ_R labelled by $R \in \mathrm{O}(\nu)$, acting so that

$$(\zeta_R)_{\boldsymbol{M}} \Phi^{(j)}_{\boldsymbol{M}}(f) = R_i{}^j \Phi^{(i)}_{\boldsymbol{M}}(f) \tag{4.50}$$

(summing on i), while in the massless case, there are automorphisms $\zeta_{(R,\ell)}$ labelled by $(R, \ell) \in \mathrm{O}(\nu_0) \ltimes \mathbb{R}^{\nu_0 *}$, so that

$$(\zeta_{(R,\ell)})_{\boldsymbol{M}} \Phi^{(j)}_{\boldsymbol{M}}(f) = R_i{}^j \Phi^{(i)}_{\boldsymbol{M}}(f) + \left(\int_{\boldsymbol{M}} f \,\mathrm{dvol}_{\boldsymbol{M}}\right) \ell^j \mathbf{1}_{\mathscr{A}^{(\nu)}(\boldsymbol{M})}.$$

It is not hard to verify that these formulae define automorphisms of $\mathscr{A}^{(\nu)}(\boldsymbol{M})$ that are components of natural transformations. What was shown in [54] was rather more: every *endo*morphism η of $\mathscr{A}^{(\nu)}$, at least under the additional assumption of regularity that every $\eta^*_{\boldsymbol{M}}$ maps states with distributional k-point functions to states with distributional k-point functions, is one of the *auto*morphisms described above.

The second case studied was the Klein–Gordon theory with external sources [64], which is formulated on a category of spacetimes with sources. Here, the gauge group can be determined at the purely algebraic level, without additional regularity conditions. As might be expected, the effect of the external sources is to break the $\mathrm{O}(\nu_m)$ symmetries for $m \ge 0$, leaving only a $\mathbb{R}^{\nu_0 *}$ symmetry for $m = 0$.

In both examples just mentioned, every endomorphism of the theory turns out to be an automorphism; there is no way of properly embedding the theory as a subtheory of itself. It is not hard to give examples of locally covariant theories where this is not the case: for example, the theory of countably many independent scalar fields $\mathscr{A}^{(\aleph_0)}$ of common mass and coupling constant has an endomorphism η acting on the generating fields by $\eta_{\boldsymbol{M}} \Phi^{(j)}_{\boldsymbol{M}}(f) = \Phi^{(j+1)}_{\boldsymbol{M}}(f)$ for all $f \in C_0^\infty(\boldsymbol{M})$, $\boldsymbol{M} \in \mathsf{Loc}$. However, under a condition of *energy compactness* (weaker than either of the nuclearity [31] or Haag–Swieca [88] criteria) it may be shown that proper endomorphisms are excluded and that all endomorphisms are automorphisms [54, Thm. 4.6]. The additional assumptions required are that the instantiation of the locally covariant theory in Minkowski space should comply with standard assumptions of AQFT, and also that there are no 'accidental symmetries' of the Minkowski space theory. The result also shows that, if the gauge group is given a natural topology (which requires the introduction of a state space) then it is compact.

The gauge group provides a useful invariant of locally covariant theories, because the automorphism groups of isomorphic functors are isomorphic. This allows one to read off, for example, that the theories $\mathscr{A}^{(j)}$ described above are inequivalent for distinct values of j, by virtue of their inequivalent gauge groups. In the same vein, the computation of the gauge group in [64] was used to show that an earlier quantization of the Klein–Gordon theory with sources was incorrect, because its gauge group contained unexpected symmetries. Regarding subtheory embeddings, if $\eta : \mathscr{A} \stackrel{.}{\to} \mathscr{B}$ and $\zeta : \mathscr{B} \stackrel{.}{\to} \mathscr{A}$, and (say) $\mathscr{A}$ obeys the hypotheses of [54, Thm. 4.6], then $\zeta \circ \eta$ must be an automorphism of $\mathscr{A}$, so (as ζ is monic) η and ζ are both isomorphisms [54, Cor. 4.7] (cf. the Cantor–Schröder–Bernstein theorem for sets).

A good example of *inequivalent* theories is given by Klein–Gordon theories $\mathscr{A}_1$ and $\mathscr{A}_2$ with distinct masses m_1 and m_2 [24] (we give a slightly different argument). Let ω_2 be the Poincaré-invariant vacuum state on the Minkowski space theory $\mathscr{A}_2(\boldsymbol{M}_0)$, which means that $\mathscr{A}(\psi)^*\omega_2 = \omega_2$ for every Poincaré transformation $\psi : \boldsymbol{M}_0 \to \boldsymbol{M}_0$. If there is an equivalence $\zeta : \mathscr{A}_1 \stackrel{.}{\to} \mathscr{A}_2$ then Proposition 4.10.2(b) implies that $\omega = \zeta^*_{\boldsymbol{M}_0}\omega_2$ satisfies

$$\mathscr{A}_1(\psi)^*\omega = \mathscr{A}_1(\psi)^*\zeta^*_{\boldsymbol{M}_0}\omega_2 = \zeta^*_{\boldsymbol{M}_0}\mathscr{A}_2(\psi)^*\omega_2 = \zeta^*_{\boldsymbol{M}_0}\omega_2 = \omega$$

for all Poincaré transformations ψ, so ω is a Poincaré-invariant state on $\mathscr{A}_1(\boldsymbol{M}_0)$. Indeed, the GNS representation of $\mathscr{A}_1(\boldsymbol{M}_0)$ induced by ω can be taken as $(\mathscr{H}_2, \pi_2 \circ \zeta_{\boldsymbol{M}_0}, \mathscr{D}_2, \Omega_2)$ where $(\mathscr{H}_2, \pi_2, \mathscr{D}_2, \Omega_2)$ is the GNS representation induced by ω_2. Crucially, the unitary implementation $U_2(\psi)$ of Poincaré transformations in $\mathscr{H}_2$ also implements the Poincaré transformations on $\mathscr{A}_1(\boldsymbol{M}_0)$: setting $\pi = \pi_2 \circ \zeta_{\boldsymbol{M}_0}$,

$$\begin{aligned} U_2(\psi)\pi(A)U_2(\psi)^{-1} &= U_2(\psi)\pi_2(\zeta_{\boldsymbol{M}_0}A)U_2(\psi)^{-1} = \pi_2(\mathscr{A}_2(\psi) \circ \zeta_{\boldsymbol{M}_0}A) \\ &= \pi_2(\zeta_{\boldsymbol{M}_0} \circ \mathscr{A}_1(\psi)A) = \pi(\mathscr{A}_1(\psi)A). \end{aligned}$$

Therefore ω is not only Poincaré-invariant but also obeys the spectrum condition, i.e., the momentum operators P_a corresponding to the unitary representation of the translations have joint spectrum in the forward lightcone. If ω has a distributional 2-point function, one may show $P_a P^a \pi(\Phi_1(f))\Omega_2 = \pi(-\Phi_1(\Box f))\Omega_2 = m_1^2\pi(\Phi_1(f))\Omega_2$ for all $f \in C_0^\infty(\boldsymbol{M}_0)$ (cf. e.g., the proof of [54, Prop. 5.6]). Using the Reeh–Schlieder property of Ω_2, we see that $P_a P^a$ has an eigenvalue m_1^2. But $P_a P^a$ is the mass-squared operator for the vacuum representation of $\mathscr{A}_2(\boldsymbol{M}_0)$ and so has discrete spectrum $\{0, m_2^2\}$, a contradiction. Accordingly, there is no equivalence ζ between $\mathscr{A}_1$ and $\mathscr{A}_2$ so that $\zeta^*_{\boldsymbol{M}_0}\omega_2$ has distributional 2-point function.

In a similar way, but considering e.g., de Sitter spacetime instead of Minkowski space, one can rule out the possibility of (sufficiently regular) equivalences between Klein–Gordon theories with differing curvature couplings.

4.10.3 Action on Fields

As shown in Sect. 4.9, the locally covariant (linear) fields of a theory of a given type form an abstract algebras (resp., vector spaces) $\mathrm{Fld}(\mathscr{D}, \mathscr{A})$ (resp., $\mathrm{Fld}_{\mathrm{lin}}(\mathscr{D}, \mathscr{A})$). The gauge group acts on these algebras/spaces in a natural fashion: Given any $\eta \in G$, and $\Phi \in \mathrm{Fld}(\mathscr{D}, \mathscr{A})$, define the transformed field $\eta \cdot \Phi \in \mathrm{Fld}(\mathscr{D}, \mathscr{A})$ by

$$(\eta \cdot \Phi)_{\boldsymbol{M}}(f) = \eta_{\boldsymbol{M}} \Phi_{\boldsymbol{M}}(f) \qquad (f \in C_0^\infty(\boldsymbol{M}),\ \boldsymbol{M} \in \mathsf{Loc}), \tag{4.51}$$

which clearly obeys the naturality condition

$$\begin{aligned}\mathscr{A}(\psi)(\eta \cdot \Phi)_{\boldsymbol{M}}(f) &= \mathscr{A}(\psi) \circ \eta_{\boldsymbol{M}}(\Phi_{\boldsymbol{M}}(f)) = \eta_{\boldsymbol{N}} \circ \mathscr{A}(\psi)\Phi_{\boldsymbol{M}}(f) = \eta_{\boldsymbol{N}} \Phi_{\boldsymbol{N}}(\psi_* f) \\ &= (\eta \cdot \Phi)_{\boldsymbol{N}}(\psi_* f)\end{aligned} \tag{4.52}$$

for all $\psi : \boldsymbol{M} \to \boldsymbol{N}$, $f \in C_0^\infty(\boldsymbol{M})$. Moreover, it is easily seen that $\Phi \mapsto \eta \cdot \Phi$ is a $*$-automorphism of $\mathrm{Fld}(\mathscr{D}, \mathscr{A})$, so we have defined a group homomorphism $G \mapsto \mathrm{Aut}(\mathrm{Fld}(\mathscr{D}, \mathscr{A}))$. Restricting to linear fields, (4.51) defines a representation of G on $\mathrm{Fld}_{\mathrm{lin}}(\mathscr{D}, \mathscr{A})$, which obeys $\eta \cdot \Phi^\star = (\eta \cdot \Phi)^\star$, and is thus a real linear representation. These representations are continuous with respect to a natural topology on $\mathrm{Aut}(\mathscr{A})$.

In either case, we may define a *multiplet of fields* as any subspace of $\mathrm{Fld}(\mathscr{D}, \mathscr{A})$ (or $\mathrm{Fld}_{\mathrm{lin}}(\mathscr{D}, \mathscr{A})$) transforming under an indecomposable representation of G. Every field can then be associated with an equivalence class of G-representations. Let ρ, σ be the equivalence classes corresponding to fields Φ, Ψ. Then Φ^* transforms in the complex conjugate representation $\bar{\rho}$ to ρ, while any linear combination of Φ and Ψ transforms in a subrepresentation of a quotient of $\rho \oplus \sigma$. Here, the quotient allows for algebraic relationships; for example, if Φ and Ψ belong to a common multiplet, then their linear combinations belong to the same multiplet. Similarly, $\Phi\Psi$ and $\Psi\Phi$ transform in (possibly different) subrepresentations of quotients of $\rho \otimes \sigma$.

For example, consider a locally covariant theory $\mathscr{A}^{(3)}$ consisting of three independent massive scalar fields of common mass $m > 0$ (and, for simplicity, minimal coupling), which has an O(3) of automorphisms described in (4.50). The scalar fields $\Phi^{(j)}$ $(j = 1, 2, 3)$ span a 3-dimensional multiplet associated with the defining representation σ of O(3), while the nonlinear fields $\Psi^{(S)}$ defined by

$$\Psi^{(S)}_{\boldsymbol{M}}(f) = S_{ij} \Phi^{(i)}_{\boldsymbol{M}}(f) \Phi^{(j)}_{\boldsymbol{M}}(f),$$

where S is a complex symmetric 3×3 matrix, span a 6-dimensional subspace of $\mathrm{Fld}(\mathscr{D}, \mathscr{A}^{(3)})$ (carrying a subrepresentation of $\sigma \otimes \sigma$) and decomposes into a 1-dimensional multiplet spanned by $\Psi^{(I)}$, where I is the identity matrix, and a 5-dimensional multiplet spanned by the $\Psi^{(S)}$, where S is symmetric and traceless.

4.10.4 Universal Formulation of the Free Scalar Field

Our treatment of the scalar field so far has followed the traditional route of constructing algebras in each individual spacetime and then specifying suitable morphisms between them in order to obtain a functor. One might characterise this as a bottom-up approach. We now describe an alternative top-down description, in which one specifies the theory directly at the functorial level.

First, observe that the Klein–Gordon operators $P_M\phi := (\Box_M + m^2 + \xi R_M)\phi$ form the components of a natural transformation $P : \mathscr{D} \stackrel{\cdot}{\to} \mathscr{D}$, where $\mathscr{D} : \mathsf{Loc} \to \mathsf{Vec}$ is as in (4.37) (but viewed as a functor to Vec). This follows directly from the fact that ψ is an isometry, and consequently $P_N\psi_* f = \psi_* P_M f$ for all $f \in \mathscr{D}(M)$. We may also define a bilocal natural scalar $E : \mathscr{D}^{(2)} \stackrel{\cdot}{\to} \underline{\mathbb{C}}$, whose component in each M is precisely the advanced-minus-retarded bidistribution E_M for P_M. Here, $\underline{\mathbb{C}} : \mathsf{Loc} \to \mathsf{Vec}$ is the constant functor giving $\mathbb{C}$ on all objects and $\mathrm{id}_{\mathbb{C}}$ on all morphisms. We also define $\mathbf{1}^{(\mathscr{A})} : \underline{\mathbb{C}} \stackrel{\cdot}{\to} \mathscr{V} \circ \mathscr{A}$ by $\mathbf{1}^{(\mathscr{A})}_M(z) = z\mathbf{1}_{\mathscr{A}(M)}$ $(z \in \mathbb{C})$, where $\mathscr{V} : \mathsf{Alg} \to \mathsf{Vec}$ is the forgetful functor.

Given these definitions, the Klein–Gordon theory may be given a universal form:

Definition 4.10.4 A *Klein–Gordon theory* is a pair $(\mathscr{A}, \Phi)$, where $\mathscr{A} : \mathsf{Loc} \to \mathsf{Alg}$ is a functor and $\Phi \in \mathrm{Fld}_{\mathrm{lin}}(\mathscr{D}, \mathscr{A})$ is a linear field such that

- $\Phi^\star = \Phi$, i.e., Φ is hermitian
- $\Phi \circ P = 0$, the zero field
- $\Phi \overrightarrow{\otimes} \Phi - \Phi \overleftarrow{\otimes} \Phi = i\mathbf{1}^{(\mathscr{A})} \circ E$ (see (4.44)),

and which is universal in the sense that, if $(\mathscr{B}, \Psi)$ is any other pair with these properties, then there is a unique subtheory embedding $\eta : \mathscr{A} \stackrel{\cdot}{\to} \mathscr{B}$ such that $\Psi = \eta \cdot \Phi$.

This definition specifies Klein–Gordon theories up to equivalence,[29] so it is reasonable to speak of *the* Klein–Gordon theory. The original construction of the theory is needed to show that the theory exists, but beyond that, it ought to be possible to work with Definition 4.10.4 alone. Other models of locally covariant QFT can be given similar universal formulations.

[29] Suppose $(\mathscr{A}, \Phi)$ and $(\mathscr{B}, \Psi)$ both satisfy Definition 4.10.4. Then there are naturals $\eta : \mathscr{A} \stackrel{\cdot}{\to} \mathscr{B}$ and $\zeta : \mathscr{B} \stackrel{\cdot}{\to} \mathscr{A}$ such that $\Psi = \eta \cdot \Phi$ and $\Phi = \zeta \cdot \Psi$. Hence also $\Phi = (\zeta \circ \eta) \cdot \Phi$. But by the universal property yet again, the only natural $\xi : \mathscr{A} \stackrel{\cdot}{\to} \mathscr{A}$ such that $\Phi = \xi \cdot \Phi$ is the identity, $\xi_M = \mathrm{id}_{\mathscr{A}(M)}$ for all $M \in \mathsf{Loc}$. Hence $\zeta \circ \eta = \mathrm{id}_{\mathscr{A}}$ and by similar reasoning applied to $(\mathscr{B}, \Psi)$, we also have $\eta \circ \zeta = \mathrm{id}_{\mathscr{B}}$. Hence η is an equivalence.

We now prove that $(\mathscr{A}, \Phi)$ has the universal property, where $\mathscr{A}$ is our standard Klein–Gordon functor and Φ its standard associated locally covariant field. Suppose $(\mathscr{B}, \Psi)$ satisfies the other axioms. For each $\boldsymbol{M} \in \mathsf{Loc}$, we define a unital $*$-homomorphism $\eta_{\boldsymbol{M}} : \mathscr{A}(\boldsymbol{M}) \to \mathscr{B}(\boldsymbol{M})$ by $\eta_{\boldsymbol{M}}\Phi_{\boldsymbol{M}}(f) = \Psi_{\boldsymbol{M}}(f)$ $(f \in C_0^\infty(\boldsymbol{M}))$, which is well-defined because the $\Phi_{\boldsymbol{M}}(f)$ generate $\mathscr{A}(\boldsymbol{M})$ and because both $\Phi_{\boldsymbol{M}}$ and $\Psi_{\boldsymbol{M}}$ obey the relations itemized in Definition 4.10.4. Furthermore, $\mathscr{A}(\boldsymbol{M})$ is simple, so $\eta_{\boldsymbol{M}}$ is either monic or the zero map, and the latter case is excluded because units and zeros are distinct for objects of Alg. Thus $\eta_{\boldsymbol{M}} : \mathscr{A}(\boldsymbol{M}) \to \mathscr{B}(\boldsymbol{M})$ is well-defined as an Alg-morphism. Suppose that $\psi : \boldsymbol{M} \to \boldsymbol{N}$, then

$$\begin{aligned} \eta_{\boldsymbol{N}}\mathscr{A}(\psi)\Phi_{\boldsymbol{M}}(f) &= \eta_{\boldsymbol{N}}\Phi_{\boldsymbol{N}}(\psi_* f) = \Psi_{\boldsymbol{N}}(\psi_* f) = \mathscr{B}(\psi)\Psi_{\boldsymbol{M}}(f) \\ &= \mathscr{B}(\psi)\eta_{\boldsymbol{M}}\Phi_{\boldsymbol{M}}(f) \end{aligned} \tag{4.53}$$

for all $f \in C_0^\infty(\boldsymbol{M})$. As the $\Phi_{\boldsymbol{M}}(f)$ generate $\mathscr{A}(\boldsymbol{M})$, it follows that the components $\eta_{\boldsymbol{M}}$ cohere to form a natural transformation $\eta : \mathscr{A} \stackrel{.}{\to} \mathscr{B}$. Uniqueness is clear from the foregoing argument, because η was fixed completely by requiring $\eta \cdot \Phi = \Psi$.

4.11 Dynamical Locality and SPASs

The last topic discussed in this chapter returns to the fundamental purpose of locally covariant QFT, namely, the description of common 'physical content' in all (reasonable) spacetimes. The functorial definition of a theory certainly gives a common mathematical definition across different spacetimes, but under what circumstances can this be said to represent the same physics?[30] This question was addressed in [69] and will be briefly summarised here.

We have already described how a pathological locally covariant theory $\mathscr{B}$ may be constructed from a basic theory $\mathscr{A}$ (which can be as well-behaved as one likes)—see equations (4.15) and (4.16). In spacetimes with compact Cauchy surfaces the theory corresponds to two copies of $\mathscr{A}$, while in those with noncompact Cauchy surfaces we have a single copy. Suppose we accept that $\mathscr{A}$ represents the same physics in all spacetimes (SPASs) according to some notion of what that might mean. Then $\mathscr{B}$ surely cannot also represent SPASs according to the same notion as $\mathscr{A}$—as their physical content coincides in some, but not all, spacetimes.

This may be put into a more mathematical form as follows. For each $\boldsymbol{M} \in \mathsf{Loc}$, define $\zeta_{\boldsymbol{M}} : \mathscr{A}(\boldsymbol{M}) \to \mathscr{B}(\boldsymbol{M})$ and $\eta_{\boldsymbol{M}} : \mathscr{B}(\boldsymbol{M}) \to \mathscr{A}(\boldsymbol{M}) \otimes \mathscr{A}(\boldsymbol{M})$ by

$$\zeta_{\boldsymbol{M}} A = \begin{cases} A & \text{if } \boldsymbol{M} \text{ has noncompact Cauchy surfaces} \\ A \otimes \mathbf{1} & \text{if } \boldsymbol{M} \text{ has compact Cauchy surfaces} \end{cases} \tag{4.54}$$

---

[30]In theories based on a classical Lagrangian one usually proceeds simply to use the 'same' Lagrangian (modulo some subtleties [53]) but this option is not open in a general AQFT context.

$$\eta_{\boldsymbol{M}} A = \begin{cases} A & \text{if } \boldsymbol{M} \text{ has compact Cauchy surfaces} \\ A \otimes \mathbf{1} & \text{if } \boldsymbol{M} \text{ has noncompact Cauchy surfaces.} \end{cases} \tag{4.55}$$

It is straightforward to check that these are well-defined morphisms and that, for every $\psi : \boldsymbol{M} \to \boldsymbol{N}$, the naturality conditions $\zeta_{\boldsymbol{N}} \circ \mathscr{A}(\psi) = \mathscr{B}(\psi) \circ \zeta_{\boldsymbol{M}}$ and $\eta_{\boldsymbol{N}} \circ \mathscr{B}(\psi) = (\mathscr{A} \otimes \mathscr{A})(\psi) \circ \eta_{\boldsymbol{M}}$ hold.[31] Thus we have subtheory embeddings

$$\mathscr{A} \underset{\zeta}{\stackrel{.}{\longrightarrow}} \mathscr{B} \underset{\eta}{\stackrel{.}{\longrightarrow}} \mathscr{A} \otimes \mathscr{A}, \tag{4.56}$$

which are *partial isomorphisms*—meaning that there is at least one spacetime for which the component of ζ is an isomorphism, and likewise for η. However, there are also spacetimes for which the corresponding component is not an isomorphism, so neither ζ nor η is an equivalence of theories.[32] The situation just described gives a formal expression to the idea that $\mathscr{B}$ coincides with $\mathscr{A}$ in some spacetimes and with $\mathscr{A} \otimes \mathscr{A}$ in others.

To summarize the discussion so far, let $\mathfrak{T}$ be a class of locally covariant quantum field theories. We have argued that a necessary condition for $\mathfrak{T}$ to represent theories conforming to some particular notion of SPASs is that *every partial isomorphism between theories in* $\mathfrak{T}$ *is an isomorphism.* We refer to this necessary condition as the *SPASs property.* Our example has shown immediately that the full class of locally covariant theories does not have the SPASs property.

The local structure of the pathological theory $\mathscr{B}$ is instructive. Suppose $\boldsymbol{M} \in$ **Loc** and choose $O \in \mathcal{O}(\boldsymbol{M})$ so that $\boldsymbol{M}|_O$ has noncompact Cauchy surfaces. Then $\mathscr{B}^{\text{kin}}(\boldsymbol{M}; O) \cong \mathscr{B}(\boldsymbol{M}|_O) = \mathscr{A}(\boldsymbol{M}|_O)$ whether or not $\boldsymbol{M}$ has compact Cauchy surfaces—the kinematic local algebras only 'sense' one copy of $\mathscr{A}$. However, another way to sense the local degrees of freedom, based on dynamics, was introduced in [69].

Let $\mathscr{C}$ be a locally covariant QFT and K be a compact subset of $\boldsymbol{M} \in$ **Loc**. Define

$$\mathscr{C}^\bullet(\boldsymbol{M}; K) := \{C \in \mathscr{C}(\boldsymbol{M}) : \mathrm{rce}_{\boldsymbol{M}}^{(\mathscr{C})}[h]C = C \text{ for all } h \in H(\boldsymbol{M}) \text{ with } \mathrm{supp}\, h \subset K^\perp\},$$

where $K^\perp = \boldsymbol{M} \setminus J_{\boldsymbol{M}}(K)$ is the causal complement of K. Elements of $\mathscr{C}^\bullet(\boldsymbol{M}; K)$ are precisely those unaffected by any metric perturbation supported in the causal complement of K. In Sect. 4.4 we observed that $\mathrm{rce}_{\boldsymbol{M}}^{(\mathscr{C})}[h]$ acts trivially on any $\mathscr{C}^{\text{kin}}(\boldsymbol{M}; O)$ with O causally disjoint from the support of h—here, we turn this around to give a new definition of the local content of the theory. For each $O \in \mathcal{O}(\boldsymbol{M})$, we define *dynamical algebra* $\mathscr{C}^{\text{dyn}}(\boldsymbol{M}; O)$ to be subalgebra of $\mathscr{C}(\boldsymbol{M})$ generated by the $\mathscr{A}^\bullet(\boldsymbol{M}; K)$ as K ranges over all compact subsets of O that have a multidiamond neighbourhood with base in O—see [69, §5] for details.

---

[31] Recall that if $\boldsymbol{M}$ has compact Cauchy surfaces then so does $\boldsymbol{N}$.

[32] This assumes that $\mathscr{A}$ is not isomorphic to $\mathscr{A} \otimes \mathscr{A}$. A convenient way of ruling out such isomorphisms is to check that $\mathscr{A}$ and $\mathscr{A} \otimes \mathscr{A}$ have nonisomorphic gauge groups.

In the case of our pathological theory $\mathscr{B}$, we see that $\mathrm{rce}^{\mathscr{B}}_{\boldsymbol{M}}[h] = \mathrm{rce}^{\mathscr{A}}_{\boldsymbol{M}}[h]$ if $\boldsymbol{M}$ has noncompact Cauchy surfaces and $\mathrm{rce}^{\mathscr{B}}_{\boldsymbol{M}}[h] = \mathrm{rce}^{\mathscr{A}}_{\boldsymbol{M}}[h]^{\otimes 2}$ if they are compact. Correspondingly, we see that $\mathscr{B}^{\mathrm{dyn}}(\boldsymbol{M}; O) = \mathscr{A}^{\mathrm{dyn}}(\boldsymbol{M}; O)$ in the noncompact case and $\mathscr{B}^{\mathrm{dyn}}(\boldsymbol{M}; O) = \mathscr{A}^{\mathrm{dyn}}(\boldsymbol{M}; O)^{\otimes 2}$ in the compact case. Thus the dynamical definition of locality senses degrees of freedom that are missed by the kinematical definition. This suggests focussing on theories of the following type:

Definition 4.11.1 A locally covariant QFT $\mathscr{C}$ is *dynamically local* if

$$\mathscr{C}^{\mathrm{kin}}(\boldsymbol{M}; O) = \mathscr{C}^{\mathrm{dyn}}(\boldsymbol{M}; O)$$

for all nonempty $O \in \mathcal{O}(\boldsymbol{M})$ and $\boldsymbol{M} \in \mathsf{Loc}$.

Clearly the pathological theory $\mathscr{B}$ is not dynamically local. More significantly:

Proposition 4.11.2 *[69, Thm. 6.10] The class of dynamically local and locally covariant QFTs has the SPASs property.*

Thus dynamical locality at least satisfies our necessary condition for providing a notion of SPASs (and there is no other condition known that does so and incorporates the standard free theories).

As an immediate application, we note the following. In Minkowski space AQFT there are models with a minimal localization scale, i.e., the local algebras are nontrivial only for sufficiently large regions (see, e.g., [109] for simple examples). Proposition 4.11.2 excludes the possibility that such models can be defined as locally covariant and dynamically local theories. For if there is a spacetime $\boldsymbol{M} \in \mathsf{Loc}$ and a nonempty $O \in \mathcal{O}(\boldsymbol{M})$ for which $\mathscr{A}^{\mathrm{kin}}(\boldsymbol{M}; O)$ is trivial, then $\mathscr{A}(\boldsymbol{M}|_O)$ is trivial. Now there is a *trivial theory* $\mathscr{I} : \mathsf{Loc} \to \mathsf{Alg}$ so that $\mathscr{I}(\boldsymbol{N}) = \mathbb{C}$ (regarded as a unital $*$-algebra) for all $\boldsymbol{N} \in \mathsf{Loc}$, and which maps every morphism to the identity morphism. This theory is a dynamically local subtheory of $\mathscr{A}$ in an obvious way, and we have shown that the two theories coincide on $\boldsymbol{M}|_O$. Accordingly, if $\mathscr{A}$ is dynamically local it is isomorphic to the trivial theory and hence trivial for all subregions of all spacetimes.

Every dynamically local theory $\mathscr{A}$ has a number of other nice properties: for instance, extended locality of $\mathscr{A}$ is equivalent to $\mathscr{A}^{\bullet}(\boldsymbol{M}; \emptyset) = \mathbb{C}\mathbf{1}_{\mathscr{A}(\boldsymbol{M})}$ for all $\boldsymbol{M} \in \mathsf{Loc}$, i.e., the absence of nontrivial elements in $\mathscr{A}(\boldsymbol{M})$ that are fixed under arbitrary relative Cauchy evolution [69, Thm. 6.5] and $\mathscr{A}$ is necessarily additive with respect to truncated multidiamonds [69, Thm. 6.3]. As a final application we return to one of our *leit motifs*: the nonexistence of natural states.

Theorem 4.11.3 *[69, Thm. 6.13] Suppose $\mathscr{A}$ is a dynamically local quantum field theory and has a natural state $(\omega_{\boldsymbol{M}})_{\boldsymbol{M}\in\mathsf{Loc}}$. If there is a spacetime $\boldsymbol{M}$ with noncompact Cauchy surfaces such that $\omega_{\boldsymbol{M}}$ induces a faithful GNS representation with the (full) Reeh–Schlieder property [i.e., the GNS vector corresponding to $\omega_{\boldsymbol{M}}$ is cyclic for the induced representation of $\mathscr{A}(\boldsymbol{M}|_O)$ for all relatively compact $O \in \mathcal{O}(\boldsymbol{M})$], then the relative Cauchy evolution is trivial in $\boldsymbol{M}$, and $\mathscr{A}^{\mathrm{kin}}(\boldsymbol{M}; O) = \mathscr{A}(\boldsymbol{M})$ for all nonempty $O \in \mathcal{O}(\boldsymbol{M})$. If, additionally, $\mathscr{A}$ obeys extended locality, then $\mathscr{A}$ is equivalent to the trivial theory $\mathscr{I}$.*

Proof By the natural state hypothesis, we have $\omega_M \circ \text{rce}_M[h] = \omega_M$ for each M and all $h \in H(M)$, simply because the relative Cauchy evolution is a composition of (inverses of) morphisms $\mathscr{A}(\psi)$ for Cauchy ψ. Thus the relative Cauchy evolution is unitarily implemented in the GNS representation π_M induced by ω_M:

$$\pi_M(\text{rce}_M[h]A) = U_M[h]\pi_M(A)U_M[h]^{-1},$$

where $U_M[h]$ is defined by $U_M[h]\pi_M(A)\Omega_M = \pi_M(\text{rce}_M[h]A)\Omega_M$ and leaves the GNS vector Ω_M invariant. Now let $h \in H(M)$ and choose a nonempty relatively compact connected $O \in \mathcal{O}(M)$ such that $O \subset (\text{supp}\, h)^\perp$ (such an O exists because the Cauchy surfaces are noncompact). As already mentioned, $\text{rce}_M[h]$ acts trivially on $\mathscr{A}^{\text{kin}}(M; O)$ [69, Prop. 3.7], so

$$U_M[h]\pi_M(A)\Omega_M = \pi_M(A)\Omega_M$$

for all $A \in \mathscr{A}^{\text{kin}}(M; O)$. Using the Reeh–Schlieder property of ω_M we may deduce that $U_M[h]$ agrees with the identity operator on a dense set and hence $U_M[h] = \mathbf{1}_{\mathscr{H}_M}$ for all $h \in H(M)$, so the relative Cauchy evolution is trivial on $\mathscr{A}(M)$ because π_M is faithful. Consequently, $\mathscr{A}^\bullet(M; K) = \mathscr{A}(M)$ for all compact sets K and hence by dynamical locality $\mathscr{A}^{\text{kin}}(M; O) = \mathscr{A}^{\text{dyn}}(M; O) = \mathscr{A}(M)$ for each nonempty $O \in \mathcal{O}(M)$. This proves the first part of the theorem.

For the second part, observe that there is a subtheory embedding $\eta : \mathscr{I} \dot{\to} \mathscr{A}$ of the trivial theory $\mathscr{I}$ into $\mathscr{A}$ given by $\eta_N z = z\mathbf{1}_{\mathscr{A}(N)}$ for all $N \in \textsf{Loc}$, $z \in \mathbb{C}$. Now consider two causally disjoint nonempty $O_1, O_2 \in \mathcal{O}(M)$. By the above argument together with extended locality,

$$\mathscr{A}(M) = \mathscr{A}^{\text{kin}}(M; O_1) \cap \mathscr{A}^{\text{kin}}(M; O_2) = \mathbb{C}\mathbf{1}_{\mathscr{A}(M)},$$

so η_M is an isomorphism. As $\mathscr{I}$ is obviously dynamically local, and $\mathscr{A}$ is by assumption, Proposition 4.11.2 entails that η is a natural isomorphism. □

As mentioned, the property of dynamical locality has been checked for a number of standard theories. Theories that satisfy dynamical locality include

- the Klein–Gordon scalar field in dimensions $n \geq 2$, if at least one of the mass or curvature coupling is nonzero [46, 70], and the corresponding extended algebra of Wick polynomials for nonzero mass and either minimal or conformal coupling [46] (one expects dynamical locality for general values of ξ);
- the free massless current in dimensions $n \geq 2$ (restricting to connected spacetimes) or $n \geq 3$ (allowing disconnected spacetimes) [70];
- the minimally coupled Klein–Gordon field with external sources for $m \geq 0$, $n \geq 2$—in this case relative Cauchy evolution can be induced by perturbations of both the metric and the external source and one modifies the definition of the dynamical net accordingly [64];
- the free Dirac field with mass $m \geq 0$ [47];

- the free Maxwell field in dimension $n = 4$, in a 'reduced formulation' [59].

These theories also obey the other hypotheses of Theorem 4.11.3 and so do not admit natural states. A more direct proof for the theory with sources appears in [55].

There are some cases known in which dynamical locality fails, which appears to be always related to the presence of broken global gauge symmetries or topologically stabilised charges: the free Klein–Gordon field with $m = 0$, $\xi = 0$ in dimensions $n \geq 2$, owing to the rigid gauge symmetry $\phi \mapsto \phi + \text{const}$ [70]; the free massless current in 2-dimensions allowing disconnected spacetimes [70]; and the free Maxwell field in dimension $n = 4$, in a 'universal formulation' [37, 59], owing to the presence of topological electric and magnetic charges in spacetimes with nontrivial second de Rham cohomology, which are eliminated in the reduced theory mentioned above. As already mentioned in Sect. 4.6.2 the existence of topological charges is also associated with a failure of injectivity (see [7, 137] for more discussion in related models). As suggested in [59], it would be interesting to investigate theories that are dynamically local modulo topological charges, with the aim of generalizing Proposition 4.11.2.

Acknowledgments We are grateful to Francis Wingham for comments on the text.

References

1. Afshordi, N., Aslanbeigi, S., Sorkin, R.D.: A distinguished vacuum state for a quantum field in a curved spacetime: formalism, features, and cosmology. JHEP **08**, 137 (2012)
2. Araki, H.: Mathematical Theory of Quantum Fields. Oxford University Press, Oxford (1999)
3. Ashtekar, A., Magnon, A.: Quantum fields in curved space-times. Proc. Roy. Soc. Lond. A **346**, 375–394 (1975)
4. Bär, C., Ginoux, N., Pfäffle, F.: Wave equations on Lorentzian manifolds and quantization. European Mathematical Society (EMS), Zürich (2007)
5. Barata, J., Jäkel, C., Mund, J.: The $\mathscr{P}(\varphi)_2$ model on de Sitter space (2013). arXiv:1311.2905 [math-ph]
6. Baumgärtel, H., Wollenberg, M.: Causal nets of operator algebras. Akademie-Verlag, Berlin (1992)
7. Becker, C., Schenkel, A., Szabo, R.J.: Differential cohomology and locally covariant quantum field theory (2014)
8. Beem, J.K., Ehrlich, P.E., Easley, K.L.: Global Lorentzian Geometry, Monographs and Textbooks in Pure and Applied Mathematics, vol. 202, 2nd edn. Marcel Dekker Inc., New York (1996)
9. Benini, M.: Relative cauchy evolution for spin 1 fields (2011)
10. Benini, M., Dappiaggi, C., Schenkel, A.: Quantized Abelian principal connections on Lorentzian manifolds (2013). arXiv:1303.2515
11. Bernal, A.N., Sánchez, M.: Smoothness of time functions and the metric splitting of globally hyperbolic spacetimes. Commun. Math. Phys. **257**, 43–50 (2005). Gr-qc/0401112
12. Bernal, A.N., Sánchez, M.: Further results on the smoothability of Cauchy hypersurfaces and Cauchy time functions. Lett. Math. Phys. **77**, 183–197 (2006). Gr-qc/0512095
13. Blackadar, B.: Operator Algebras, Encyclopaedia of Mathematical Sciences, vol. 122: Theory of C^*-algebras and von Neumann Algebras. Operator Algebras and Non-commutative Geometry, III. Springer, Berlin (2006)
14. Bogoliubov, N., Logunov, A., Oksak, A., Todorov, I.: General priciples of quantum field theory. Kluwer Academic Publishers, Dordrecht (1990)

15. Borchers, H.J.: On the structure of the algebra of field operators. Nuovo Cimento **24**, 214–236 (1962)
16. Bostelmann, H., Cadamuro, D., Fewster, C.J.: Quantum energy inequality for the massive ising model. Phys. Rev. D **88**(2), 025,019 (2013)
17. Bostelmann, H., Fewster, C.J.: Quantum inequalities from operator product expansions. Commun. Math. Phys. **292**, 761–795 (2009)
18. Bratteli, O., Robinson, D.W.: Operator Algebras and Quantum Statistical Mechanics 2, 2nd edn. Springer, Berlin (1996)
19. Brum, M., Fredenhagen, K.: 'Vacuum-like' Hadamard states for quantum fields on curved spacetimes. Class. Quantum Gravity **31**(2), 025024 (2014)
20. Brunetti, R., Dütsch, M., Fredenhagen, K.: Perturbative algebraic quantum field theory and the renormalization groups. Adv. Theor. Math. Phys. **13**, 1541–1599 (2009)
21. Brunetti, R., Fredenhagen, K.: Microlocal analysis and interacting quantum field theories: renormalization on physical backgrounds. Commun. Math. Phys. **208**, 623–661 (2000)
22. Brunetti, R., Fredenhagen, K., Imani, P., Rejzner, K.: The locality axiom in quantum field theory and tensor products of C^*-algebras. Rev. Math. Phys. **26**(6), 1450010, 10 (2014)
23. Brunetti, R., Fredenhagen, K., Köhler, M.: The microlocal spectrum condition and Wick polynomials of free fields on curved space-times. Commun. Math. Phys. **180**, 633–652 (1996)
24. Brunetti, R., Fredenhagen, K., Verch, R.: The generally covariant locality principle: a new paradigm for local quantum physics. Commun. Math. Phys. **237**, 31–68 (2003)
25. Brunetti, R., Ruzzi, G.: Superselection sectors and general covariance. I. Commun. Math. Phys. **270**, 69–108 (2007)
26. Brunetti, R., Ruzzi, G.: Quantum charges and spacetime topology: the emergence of new superselection sectors. Commun. Math. Phys. **287**, 523–563 (2009)
27. Buchholz, D.: On hot bangs and the arrow of time in relativistic quantum field theory. Commun. Math. Phys. **237**, 271–288 (2003)
28. Buchholz, D., Mund, J., Summers, S.A.: Transplantation of local nets and geometric modular action on Robertson-Walker spacetimes. In: Longo, R. (ed.) Mathematical physics in mathematics and physics: Quantum and operator algebraic aspects. Proceedings, Conference, Siena, Italy, 20–24 June 2000. Fields Institute Communications, vol. 30. American Mathematical Society, Providence, RI (2001)
29. Buchholz, D., Ojima, I., Roos, H.: Thermodynamic properties of nonequilibrium states in quantum field theory. Ann. Phys. **297**, 219–242 (2002)
30. Buchholz, D., Størmer, E.: Superposition, transition probability and primitive observables in infinite quantum systems. arXiv:1411.2100
31. Buchholz, D., Wichmann, E.H.: Causal independence and the energy-level density of states in local quantum field theory. Comm. Math. Phys. **106**, 321–344 (1986)
32. Budic, R., Isenberg, J., Lindblom, L., Yasskin, P.B.: On the determination of Cauchy surfaces from intrinsic properties. Commun. Math. Phys. **61**, 87–95 (1978)
33. Chilian, B., Fredenhagen, K.: The time slice axiom in perturbative quantum field theory on globally hyperbolic spacetimes. Comm. Math. Phys. **287**(2), 513–522 (2009)
34. D'Antoni, C., Hollands, S.: Nuclearity, local quasiequivalence and split property for Dirac quantum fields in curved spacetime. Comm. Math. Phys. **261**(1), 133–159 (2006)
35. Dappiaggi, C.: Remarks on the Reeh-Schlieder property for higher spin free fields on curved spacetimes. Rev. Math. Phys. **23**(10), 1035–1062 (2011)
36. Dappiaggi, C., Fredenhagen, K., Pinamonti, N.: Stable cosmological models driven by a free quantum scalar field. Phys. Rev. D **77**, 104015 (2008)
37. Dappiaggi, C., Lang, B.: Quantization of Maxwell's equations on curved backgrounds and general local covariance. Lett. Math. Phys. **101**(3), 265–287 (2012)
38. Dappiaggi, C., Moretti, V., Pinamonti, N.: Rigorous steps towards holography in asymptotically flat spacetimes. Rev. Math. Phys. **18**, 349–416 (2006)
39. Dawson, S.P., Fewster, C.J.: An explicit quantum weak energy inequality for Dirac fields in curved spacetimes. Class. Quant. Grav. **23**, 6659–6681 (2006)

40. Degner, A., Verch, R.: Cosmological particle creation in states of low energy. J. Math. Phys. **51**, 022302 (2010)
41. Dimock, J.: Algebras of local observables on a manifold. Commun. Math. Phys. **77**, 219–228 (1980)
42. Doplicher, S., Haag, R., Roberts, J.E.: Fields, observables and gauge transformations. I. Comm. Math. Phys. **13**, 1–23 (1969)
43. Doplicher, S., Longo, R.: Standard and split inclusions of von Neumann algebras. Invent. Math. **75**, 493–536 (1984)
44. Duistermaat, J., Hörmander, L.: Fourier integral operators. I. Acta Mathematica **128**, 183 (1972)
45. Epstein, H., Glaser, V., Jaffe, A.: Nonpositivity of the energy density in quantized field theories. Il Nuovo Cim. **36**, 1016–1022 (1965)
46. Ferguson, M.: Dynamical locality of the nonminimally coupled scalar field and enlarged algebra of Wick polynomials. Ann. Henri Poincaré **14**(4), 853–892 (2013)
47. Ferguson, M.T.: Aspects of dynamical locality and locally covariant canonical quantization. Ph.D. thesis, University of York (2013). http://etheses.whiterose.ac.uk/4529/
48. Fewster, C.J.: The split property for locally covariant quantum field theories in curved spacetime. To appear, Lett. Math. Phys. arXiv:1501.02682
49. Fewster, C.J.: A general worldline quantum inequality. Class. Quant. Grav. **17**, 1897–1911 (2000)
50. Fewster, C.J.: Energy inequalities in quantum field theory. In: Zambrini, J.C. (ed.) XIVth International Congress on Mathematical Physics. World Scientific, Singapore (2005). An expanded and updated version is available as math-ph/0501073
51. Fewster, C.J.: Quantum energy inequalities and stability conditions in quantum field theory. In: Boutet de Monvel, A., Buchholz, D., Iagolnitzer, D., Moschella, U. (eds.) Rigorous Quantum Field Theory: A Festschrift for Jacques Bros, Progress in Mathematics, vol. 251. Birkhäuser, Boston (2006)
52. Fewster, C.J.: Quantum energy inequalities and local covariance. II. Categorical formulation. Gen. Relativ. Gravit. **39**, 1855–1890 (2007)
53. Fewster, C.J.: On the notion of 'the same physics in all spacetimes'. In: Finster, F., Müller, O., Nardmann, M., Tolksdorf, J., Zeidler, E. (eds.) Quantum Field Theory and Gravity. Conceptual and Mathematical Advances in the Search for a Unified Framework, pp. 207–227. Birkhäuser (2012). arXiv:1105.6202
54. Fewster, C.J.: Endomorphisms and automorphisms of locally covariant quantum field theories. Rev. Math. Phys. **25**(5), 1350008, 47 (2013)
55. Fewster, C.J.: Locally covariant quantum field theory and the problem of formulating the same physics in all spacetimes. Phil. Trans. A. Roy. Soc. **373**, 20140238 (2015). arXiv:1502.04642
56. Fewster, C.J.: On the spin-statistics connection in curved spacetimes (2015). arXiv:1503.05797
57. Fewster, C.J., Eveson, S.P.: Bounds on negative energy densities in flat spacetime. Phys. Rev. D **58**, 084,010 (1998)
58. Fewster, C.J., Hollands, S.: Quantum energy inequalities in two-dimensional conformal field theory. Rev. Math. Phys. **17**, 577 (2005)
59. Fewster, C.J., Lang, B.: Dynamical locality of the free Maxwell field. In Press Ann. H. Poincaré (2014). arXiv:1403.7083
60. Fewster, C.J., Lang, B.: Pure quasifree states of the Dirac field from the fermionic projector (2014). arXiv:1408.1645
61. Fewster, C.J., Osterbrink, L.W.: Quantum energy inequalities for the non-minimally coupled scalar field. J. Phys. A **41**, 025402 (2008)
62. Fewster, C.J., Pfenning, M.J.: A quantum weak energy inequality for spin-one fields in curved spacetime. J. Math. Phys. **44**, 4480–4513 (2003)
63. Fewster, C.J., Pfenning, M.J.: Quantum energy inequalities and local covariance. I: Globally hyperbolic spacetimes. J. Math. Phys. **47**, 082303 (2006)

64. Fewster, C.J., Schenkel, A.: Locally covariant quantum field theory with external sources. Published online with Ann. H. Poincaré (2014). arXiv:1402.2436
65. Fewster, C.J., Smith, C.J.: Absolute quantum energy inequalities in curved spacetime. Ann. Henri Poincaré **9**, 425–455 (2008)
66. Fewster, C.J., Teo, E.: Bounds on negative energy densities in static space-times. Phys. Rev. D **59**, 104,016 (1999)
67. Fewster, C.J., Verch, R.: A quantum weak energy inequality for Dirac fields in curved spacetime. Commun. Math. Phys. **225**, 331–359 (2002)
68. Fewster, C.J., Verch, R.: Stability of quantum systems at three scales: passivity, quantum weak energy inequalities and the microlocal spectrum condition. Commun. Math. Phys. **240**, 329–375 (2003)
69. Fewster, C.J., Verch, R.: Dynamical locality and covariance: what makes a physical theory the same in all spacetimes? Ann. H. Poincaré **13**, 1613–1674 (2012)
70. Fewster, C.J., Verch, R.: Dynamical locality of the free scalar field. Ann. H. Poincaré **13**, 1675–1709 (2012)
71. Fewster, C.J., Verch, R.: On a recent construction of 'vacuum-like' quantum field states in curved spacetime. Class. Quant. Grav. **29**, 205017 (2012)
72. Fewster, C.J., Verch, R.: The necessity of the Hadamard condition. Class. Quant. Grav. **30**, 235027 (2013)
73. Finster, F.: Definition of the Dirac sea in the presence of external fields. Adv. Theor. Math. Phys. **2**, 963–985 (1998)
74. Finster, F., Reintjes, M.: A non-perturbative construction of the Fermionic projector on globally hyperbolic manifolds I—Space-times of finite lifetime (2013). arXiv:1301.5420
75. Flanagan, É.É.: Quantum inequalities in two-dimensional Minkowski spacetime. Phys. Rev. D **3**(56), 4922–4926 (1997)
76. Ford, L.H.: Quantum coherence effects and the second law of thermodynamics. Proc. Roy. Soc. Lond. A **364**, 227–236 (1978)
77. Ford, L.H.: Constraints on negative-energy fluxes. Phys. Rev. D **43**, 3972–3978 (1991)
78. Ford, L.H., Roman, T.A.: Restrictions on negative energy density in flat spacetime. Phys. Rev. D **55**, 2082–2089 (1997)
79. Fredenhagen, K.: On the modular structure of local algebras of observables. Comm. Math. Phys. **97**(1–2), 79–89 (1985)
80. Fredenhagen, K., Haag, R.: On the derivation of Hawking radiation associated with the formation of a black hole. Comm. Math. Phys. **127**(2), 273–284 (1990)
81. Fredenhagen, K., Rejzner, K.: Batalin-Vilkovisky formalism in perturbative algebraic quantum field theory. Commun. Math. Phys. **317**, 697–725 (2013)
82. Fulling, S.A., Narcowich, F.J., Wald, R.M.: Singularity structure of the two-point function in quantum field theory in curved spacetime. II. Ann. Phys. **136**, 243–272 (1981)
83. Geroch, R.: Domain of dependence. J. Math. Phys. **11**, 437–449 (1970)
84. Guido, D., Longo, R.: A converse Hawking-Unruh effect and dS(2)/CFT correspondence. Ann. Henri Poincaré **4**, 1169–1218 (2003)
85. Guido, D., Longo, R., Roberts, J., Verch, R.: Charged sectors, spin and statistics in quantum field theory on curved space-times. Rev. Math. Phys. **13**, 125–198 (2001)
86. Haag, R.: Local Quantum Physics: Fields, Particles. Algebras. Springer, Berlin (1992)
87. Haag, R., Narnhofer, H., Stein, U.: On quantum field theory in gravitational background. Commun. Math. Phys. **94**, 219 (1984)
88. Haag, R., Swieca, J.A.: When does a quantum field theory describe particles? Comm. Math. Phys. **1**, 308–320 (1965)
89. Halvorson, H., Clifton, R.: Generic Bell correlation between arbitrary local algebras in quantum field theory. J. Math. Phys. **41**, 1711–1717 (2000)
90. Hawking, S.W., Ellis, G.F.R.: The Large Scale Structure of Space-Time. Cambridge University Press, London (1973)
91. Higuchi, A., Parker, L., Wang, Y.: Consistency of Faddeev-Popov ghost statistics with gravitationally induced pair creation. Phys. Rev. D **42**, 4078–4081 (1990)

92. Hollands, S.: A General PCT theorem for the operator product expansion in curved space-time. Commun. Math. Phys. **244**, 209–244 (2004)
93. Hollands, S.: The Operator product expansion for perturbative quantum field theory in curved spacetime. Commun. Math. Phys. **273**, 1–36 (2007)
94. Hollands, S.: Renormalized quantum Yang-Mills fields in curved spacetime. Rev. Math. Phys. **20**, 1033–1172 (2008)
95. Hollands, S., Wald, R.M.: Local Wick polynomials and time ordered products of quantum fields in curved spacetime. Commun. Math. Phys. **223**, 289–326 (2001)
96. Hollands, S., Wald, R.M.: Existence of local covariant time ordered products of quantum fields in curved spacetime. Commun. Math. Phys. **231**, 309–345 (2002)
97. Hollands, S., Wald, R.M.: Axiomatic quantum field theory in curved spacetime. Commun. Math. Phys. **293**, 85–125 (2010)
98. Hörmander, L.: The Analysis of Partial Differential Operators. I. Springer, Berlin (1983)
99. Isham, C.J.: Spinor fields in four-dimensional space-time. Proc. Roy. Soc. Lond. Ser. A **364**(1719), 591–599 (1978)
100. Jaffe, A., Jäkel, C.D., Martinez, R.E.: Complex classical fields: a framework for reflection positivity. Commun. Math. Phys. **329**, 1–28 (2014)
101. Jaffe, A., Ritter, G.: Quantum field theory on curved backgrounds. I. The Euclidean functional integral. Commun. Math. Phys. **270**, 545–572 (2007)
102. Junker, W.: Hadamard states, adiabatic vacua and the construction of physical states for scalar quantum fields on curved space-time. Rev. Math. Phys. **8**, 1091–1159 (1996)
103. Junker, W., Schrohe, E.: Adiabatic vacuum states on general space-time manifolds: definition, construction, and physical properties. Ann. Poincaré Phys. Theor. **3**, 1113–1182 (2002)
104. Kay, B.S.: Casimir effect in quantum field theory. Phys. Rev. D **20**, 3052–3062 (1979)
105. Kay, B.S.: The principle of locality and quantum field theory on (non-globally hyperbolic) curved spacetimes. Rev. Math. Phys. (Special Issue), 167–195 (1992)
106. Kay, B.S., Radzikowski, M.J., Wald, R.M.: Quantum field theory on space-times with a compactly generated Cauchy horizon. Commun. Math. Phys. **183**, 533–556 (1997)
107. Kay, B.S., Wald, R.M.: Theorems on the uniqueness and thermal properties of stationary, nonsingular, quasifree states on space-times with a bifurcate Killing horizon. Phys. Rep. **207**, 49–136 (1991)
108. Landau, L.J.: A note on extended locality. Commun. Math. Phys. **13**, 246–253 (1969)
109. Lechner, G., Longo, R.: Localization in nets of standard spaces. ArXiv e-prints (2014)
110. Longo, R., Rehren, K.H.: Local fields in boundary conformal QFT. Rev. Math. Phys. **16**, 909 (2004)
111. Longo, R., Rehren, K.H.: Boundary quantum field theory on the interior of the lorentz hyperboloid. Commun. Math. Phys. **311**, 769–785 (2012)
112. Lueders, C., Roberts, J.: Local quasiequivalence and adiabatic vacuum states. Commun. Math. Phys. **134**, 29–63 (1990)
113. Mac Lane, S.: Categories for the Working Mathematician, 2nd edn. Springer, New York (1998)
114. Marecki, P.: Bounds on the energy densities of ground states on static spacetimes of compact objects. Phys. Rev. D **73**(12), 124009 (2006)
115. Moretti, V.: Quantum ground states holographically induced by asymptotic flatness: invariance under spacetime symmetries, energy positivity and Hadamard property. Commun. Math. Phys. **279**, 31–75 (2008)
116. Moretti, V., Pinamonti, N.: State independence for tunneling processes through black hole horizons and Hawking radiation. Commun. Math. Phys. **309**, 295–311 (2012)
117. Olbermann, H.: States of low energy on Robertson-Walker spacetimes. Class. Quant. Grav. **24**, 5011–5030 (2007)
118. Olum, K.D., Graham, N.: Static negative energies near a domain wall. Phys. Lett. B **554**, 175–179 (2003)
119. O'Neill, B.: Semi-Riemannian Geometry. Academic Press, New York (1983)
120. Parker, L.: Quantized fields and particle creation in expanding universes. 1. Phys. Rev. **183**, 1057–1068 (1969)

121. Parker, L.: Quantized fields and particle creation in expanding universes. 2. Phys. Rev. **D3**, 346–356 (1971)
122. Parker, L., Wang, Y.: Statistics from dynamics in curved spacetime. Phys. Rev. D **39**, 3596–3605 (1989)
123. Penrose, R.: Techniques of differential topology in relativity. Society for Industrial and Applied Mathematics, Philadelphia, Pa. (1972). Conference Board of the Mathematical Sciences Regional Conference Series in Applied Mathematics, No. 7
124. Pfenning, M.J., Ford, L.H.: Scalar field quantum inequalities in static spacetimes. Phys. Rev. D **57**, 3489–3502 (1998)
125. Pinamonti, N.: On the initial conditions and solutions of the semiclassical Einstein equations in a cosmological scenario. Commun. Math. Phys. **305**, 563–604 (2011)
126. Pusz, W., Woronowicz, S.: Passive states and KMS states for general quantum systems. Commun. Math. Phys. **58**, 273–290 (1978)
127. Radzikowski, M.J.: Micro-local approach to the Hadamard condition in quantum field theory on curved space-time. Commun. Math. Phys. **179**, 529–553 (1996)
128. Rehren, K.H.: Algebraic holography. Ann. Henri Poincaré **1**, 607–623 (2000)
129. Roberts, J.E.: A survey of local cohomology. In: Mathematical problems in theoretical physics (Proceedings of Internatinal Conference, University of Rome, Rome, 1977), Lecture Notes in Physics, vol. 80, pp. 81–93. Springer, Berlin (1978)
130. Roos, H.: Independence of local algebras in quantum field theory. Commun. Math. Phys. **16**, 238–246 (1970)
131. Ruzzi, G.: Homotopy of posets, net-cohomology and superselection sectors in globally hyperbolic space-times. Rev. Math. Phys. **17**, 1021–1070 (2005)
132. Ruzzi, G.: Punctured Haag duality in locally covariant quantum field theories. Comm. Math. Phys. **256**(3), 621–634 (2005)
133. Sahlmann, H., Verch, R.: Passivity and microlocal spectrum condition. Commun. Math. Phys. **214**, 705–731 (2000)
134. Sahlmann, H., Verch, R.: Microlocal spectrum condition and Hadamard form for vector valued quantum fields in curved space-time. Rev. Math. Phys. **13**, 1203–1246 (2001)
135. Sanders, K.: On the Reeh-Schlieder property in curved spacetime. Commun. Math. Phys. **288**, 271–285 (2009)
136. Sanders, K.: The locally covariant Dirac field. Rev. Math. Phys. **22**, 381–430 (2010)
137. Sanders, K., Dappiaggi, C., Hack, T.P.: Electromagnetism, local covariance, the Aharonov-Bohm effect and Gauss' law (2012). arXiv:1211.6420
138. Schlemmer, J., Verch, R.: Local thermal equilibrium states and quantum energy inequalities. Ann. Henri Poincaré **9**, 945–978 (2008)
139. Schlieder, S.: Einige Bemerkungen über Projektionsoperatoren (Konsequenzen eines Theorems von Borchers). Comm. Math. Phys. **13**, 216–225 (1969)
140. Schoch, A.: On the simplicity of Haag fields. Int. J. Theor. Phys. **1**, 107–113 (1968)
141. Solveen, C.: Local thermal equilibrium in quantum field theory on flat and curved spacetimes. Class. Quant. Grav. **27**, 235002 (2010)
142. Solveen, C.: Local thermal Equilibrium and KMS states in curved spacetime. Class. Quant. Grav. **29**, 245015 (2012)
143. Streater, R.F., Wightman, A.S.: PCT, spin and statistics, and all that. Princeton Landmarks in Physics. Princeton University Press, Princeton (2000). Corrected third printing of the 1978 edition
144. Strohmaier, A.: The Reeh-Schlieder property for quantum fields on stationary spacetimes. Comm. Math. Phys. **215**(1), 105–118 (2000)
145. Strohmaier, A., Verch, R., Wollenberg, M.: Microlocal analysis of quantum fields on curved space-times: analytic wave front sets and Reeh-Schlieder theorems. J. Math. Phys. **43**(11), 5514–5530 (2002)
146. Summers, S., Werner, R.: Maximal violation of Bell's inequalities for algebras of observables in tangent spacetime regions. Ann. Inst. Henri Poincaré **49**, 215–243 (1988)

147. Summers, S.J.: On the independence of local algebras in quantum field theory. Rev. Math. Phys. **2**(2), 201–247 (1990)
148. Torre, C.G., Varadarajan, M.: Functional evolution of free quantum fields. Class. Quant. Grav. **16**, 2651–2668 (1999)
149. Uhlmann, A.: Über die Definition der Quantenfelder nach Wightman und Haag. Wiss. Z. Karl-Marx-Univ. Leipzig Math.-Nat. Reihe **11**, 213–217 (1962)
150. Verch, R.: Antilocality and a Reeh-Schlieder theorem on manifolds. Lett. Math. Phys. **28**(2), 143–154 (1993)
151. Verch, R.: Nuclearity, split property, and duality for the Klein-Gordon field in curved spacetime. Lett. Math. Phys. **29**(4), 297–310 (1993)
152. Verch, R.: Local definiteness, primarity and quasiequivalence of quasifree Hadamard quantum states in curved spacetime. Comm. Math. Phys. **160**(3), 507–536 (1994)
153. Verch, R.: Scaling algebras, the renormalization group and the principle of local stability in algebraic quantum field theory. In: Proceedings of the Conference on Operator Algebras and Quantum Field Theory, Rome, Italy, 1–6 July 1996, International Press (1996)
154. Verch, R.: Continuity of symplectically adjoint maps and the algebraic structure of Hadamard vacuum representations for quantum fields on curved spacetime. Rev. Math. Phys. **9**, 635–674 (1997)
155. Verch, R.: Wavefront sets in algebraic quantum field theory. Commun. Math. Phys. **205**, 337–367 (1999)
156. Verch, R.: A spin-statistics theorem for quantum fields on curved spacetime manifolds in a generally covariant framework. Commun. Math. Phys. **223**, 261–288 (2001)
157. Verch, R.: Local covariance, renormalization ambiguity, and local thermal equilibrium in cosmology. In: Finster, F., Müller, O., Nardmann, M., Tolksdorf, J., Zeidler, E. (eds.) Quantum Field Theory and Gravity. Conceptual and Mathematical Advances in the Search for a Unified Framework. Birkhäuser (2012). arXiv:1105.6249
158. Verch, R., Werner, R.F.: Distillability and positivity of partial transposes in general quantum field systems. Rev. Math. Phys. **17**, 545–576 (2005)
159. Wald, R.M.: The back reaction effect in particle creation in curved space-time. Commun. Math. Phys. **54**, 1–19 (1977)
160. Wald, R.M.: Existence of the S-matrix in quantum field theory in curved space-time. Ann. Phys. **118**, 490–510 (1979)
161. Wald, R.M.: General Relativity. University of Chicago Press, Chicago (1984)
162. Wald, R.M.: Correlations beyond the horizon. Gen. Rel. Grav. **24**, 1111–1116 (1992)
163. Wald, R.M.: Quantum Field Theory in Curved Spacetime and Black Hole Thermodynamics. University of Chicago Press, Chicago (1994)
164. Wollenberg, M.: Scaling limits and type of local algebras over curved space-time. In: Arveson, W.B. et al. (eds.) Operator Algebras and Topology. Proceedings, Craiova, 1989. Pitman Research Notes in Mathematics Series, vol. 270, pp. 179–196. Longman Science and Technology, Harlow (1992)
165. Yurtsever, U.: Algebraic approach to quantum field theory on nonglobally hyperbolic spacetimes. Class. Quant. Grav. **11**, 999–1012 (1994)

Chapter 5
Algebraic QFT in Curved Spacetime and Quasifree Hadamard States: An Introduction

Igor Khavkine and Valter Moretti

Abstract Within this chapter we introduce the overall idea of the algebraic formalism of QFT on a fixed globally hyperbolic spacetime in the framework of unital $*$-algebras. We point out some general features of CCR algebras, such as simplicity and the construction of symmetry-induced homomorphisms. For simplicity, we deal only with a real scalar quantum field. We discuss some known general results in curved spacetime like the existence of quasifree states enjoying symmetries induced from the background, pointing out the relevant original references. We introduce, in particular, the notion of a Hadamard quasifree algebraic quantum state, both in the geometric and microlocal formulation, and the associated notion of Wick polynomials.

5.1 Algebraic Formalism

With this preliminary section we introduce some basic definitions and result about algebraic formulation of quantum theory reviewing some basic definitions and results about the algebraic machinery. Most literature devoted to the algebraic approach to QFT is written using C^*-algebras, in particular Weyl C^*-algebras, when dealing with free fields, nevertheless the "practical" literature mostly uses *unbounded* field operators which are encapsulated in the notion of $*$-algebra instead of C^*-algebra, whose additional feature is a multiplicatively compatible norm. Actually, at the level of free theories and quasifree (Gaussian) states the two approaches are technically equivalent. Since we think more plausible that the non-expert reader acquainted with QFT in Minkowski spacetime is, perhaps unconsciously, more familiar with $*$-algebras than C^*-algebras, in the rest of the chapter we adopt the $*$-algebra framework.

I. Khavkine · V. Moretti (✉)
Department of Mathematics, University of Trento, via Sommarive 14, 38123 Povo (Trento), Italy
e-mail: moretti@science.unitn.it

I. Khavkine
e-mail: igor.khavkine@unitn.it

R. Brunetti et al. (eds.), *Advances in Algebraic Quantum Field Theory*, Mathematical Physics Studies, DOI 10.1007/978-3-319-21353-8_5

Definition 5.1.1 (*Algebras*) An **algebra** $\mathcal{A}$ is a complex vector space which is equipped with an associative product

$$\mathcal{A} \times \mathcal{A} \ni (a, b) \mapsto ab \in \mathcal{A}$$

which is distributive with respect to the vector sum operation and satisfies

$$\alpha(ab) = (\alpha a)b = a(\alpha b) \quad \text{if } \alpha \in \mathbb{C} \text{ and } a, b \in \mathcal{A}.$$

$\mathcal{A}$ is a **$*$-algebra** if it admits an **involution**, namely an anti-linear map, $\mathcal{A} \ni a \mapsto a^*$, which is involutive, that is $(a^*)^* = a$, and such that $(ab)^* = b^*a^*$, for any $a, b \in \mathcal{A}$. $\mathcal{A}$ is **unital** if it contains a multiplicative **unit** $\mathbf{1} \in \mathcal{A}$, that is $\mathbf{1}a = a\mathbf{1} = a$ for all $a \in \mathcal{A}$.

A set $G \subset \mathcal{A}$ is said to **generate** the algebra $\mathcal{A}$, and the elements of G are said **generators** of $\mathcal{A}$, if each element of $\mathcal{A}$ is a finite complex linear combination of products (with arbitrary number of factors) of elements of G.

The **center**, $\mathcal{Z}_{\mathcal{A}}$, of the algebra $\mathcal{A}$ is the set of elements $z \in \mathcal{A}$ commuting with all elements of $\mathcal{A}$.

Regarding morphisms of algebras we shall adopt the following standard definitions

Definition 5.1.2 (*Algebra morphisms*) Consider a map $\beta : \mathcal{A}_1 \to \mathcal{A}_2$, where $\mathcal{A}_i$ are algebras.

(a) β is an **algebra homomorphism** if it is a complex linear map, preserves the product and, if the algebras are unital, preserves the unit elements.

(b) β is a **$*$-algebra homomorphism** if $\mathcal{A}_i$ are $*$-algebras, β is a **algebra homomorphism** and preserves the involution.

(c) β is an **algebra isomorphism** or a **$*$-algebra isomorphism** if it is an **algebra homomorphism** or, respectively, a **$*$-algebra homomorphism** and it is bijective.

(d) β is an **algebra automorphism** or a **$*$-algebra automorphism** if it is a **algebra isomorphism** or, respectively, a **$*$-algebra isomorphism** and $\mathcal{A}_1 = \mathcal{A}_2$.

Corresponding **anti-linear morphisms** are defined analogously replacing the linearity condition with anti-linearity.

Remark 5.1.3

(1) The unit $\mathbf{1}$, if exists, turns out to be unique. In $*$-algebras it satisfies $\mathbf{1} = \mathbf{1}^*$.

(2) Although we shall not deal with C^*-algebras, we recall the reader that a $*$-algebra $\mathcal{A}$ is a **C^*-algebra** if it is a Banach space with respect to a norm $|| \ ||$ which satisfies $||ab|| \leq ||a|| \, ||b||$ and $||a^*a|| = ||a||^2$ if $a, b \in \mathcal{A}$. It turns out that $||a^*|| = ||a||$ and, if the C^*-algebra is unital, $||\mathbf{1}|| = 1$. A unital $*$-algebra admits at most one norm making it a C^*-algebra.

Definition 5.1.4 (*Two-sided ideals*) A **two-sided ideal** of an algebra $\mathcal{A}$ is a linear complex subspace $\mathcal{I} \subset \mathcal{A}$ such that $ab \in \mathcal{I}$ and $ba \in \mathcal{I}$ if $a \in \mathcal{A}$ and $b \in \mathcal{I}$.

In a $*$-algebra, a two-sided ideal $\mathcal{I}$ is said to be a **two-sided $*$-ideal** if it is also closed with respect to the involution: $a^* \in \mathcal{I}$ if $a \in \mathcal{I}$.

An algebra $\mathcal{A}$ is **simple** if it does not admit two-sided ideals different form $\{0\}$ and $\mathcal{A}$ itself.

Remark 5.1.5 It should be evident that the intersection of a class of two-sided ideals (two-sided $*$-ideals) is a two-sided ideal (resp. two-sided $*$-ideal).

5.1.1 The General Algebraic Approach to Quantum Theories

In the algebraic formulation of a quantum theory [28], observables are viewed as abstract self-adjoint objects instead of operators in a given Hilbert space. These observable generate a *$*$-algebra* or a *C^*-algebra* depending on the context. The algebra also includes a formal identity **1** and *complex* linear combinations of observables which, consequently cannot be interpreted as observables. Nevertheless the use of complex algebras is mathematically convenient. The justification of a linear structure for the set of the observables is quite easy, the presence of an associative product is instead much more difficult to justify [61]. However, a posteriori, this approach reveals to be powerful and it is particularly convenient when the theory encompasses many unitarily inequivalent representation of the algebra of observables, as it happens in quantum field theory.

5.1.2 Defining $*$-algebras by Generators and Relations

In the algebraic approach, the $*$-algebra of observables cannot be defined simply as some concrete set of (possibly unbounded) operators on some Hilbert space. Instead, the $*$-algebra must be defined abstractly, using some more basic objects. Below we recall an elementary algebraic construction that will be of use in Sect. 5.2.1 in defining the CCR algebra of a scalar field.

We will construct a $*$-algebra from a presentation by *generators* and *relations*. As we shall see in the Sect. 5.2.1, the CCR algebra is *generated* by abstract objects, the *smeared fields*, $\phi(f)$ and the unit **1**. In other words, the elements of the algebra are finite linear combinations of products of these objects. However there also are *relations* among these objects, e.g. $[\phi(f), \phi(g)] = iE(f, g)\mathbf{1}$. We therefore need an abstract procedure to define this sort of algebras, starting form generators and imposing relations. We make each of these concepts precise in a general context.

Let us start with the notion of algebra, $\mathcal{A}_G$, *generated* by a set of generators G. Intuitively, the algebra $\mathcal{A}_G$ is the *smallest* algebra that contains the elements of the generator set G (yet without any algebraic relations between these generators). The following is an example of a definition by a *universal property* [45, Sect. I.11].

Definition 5.1.6 (*Free algebra*) Given a set G of **generators** (not necessarily finite or even countable), an algebra $\mathcal{A}_G$ is said to be **freely generated by** G (or **free on** G) if there is a map $\gamma : G \to \mathcal{A}_G$ such that, for any other algebra $\mathcal{B}$ and map $\beta : G \to \mathcal{B}$, there exists a *unique* algebra homomorphism $b : \mathcal{A}_G \to \mathcal{B}$ such that $\beta = b \circ \gamma$. We use the same terminology for $*$- and unital algebras.

Remark 5.1.7

(1) Any two algebras freely generated by G, given by say $\gamma : G \to \mathcal{A}_G$ and $\gamma' : G \to \mathcal{A}'_G$, are *naturally isomorphic*. In this sense $\mathcal{A}_G$ is uniquely determined by G. By definition, there exist unique homomorphisms $a : \mathcal{A}'_G \to \mathcal{A}_G$ and $a' : \mathcal{A}_G \to \mathcal{A}'_G$ such that $\gamma = a \circ \gamma'$ and $\gamma' = a' \circ \gamma$. Their compositions satisfy the same kind of identity as b in the above definition, namely $\gamma = \mathrm{id} \circ \gamma = (a \circ a') \circ \gamma$ and $\gamma' = \mathrm{id} \circ \gamma' = (a' \circ a) \circ \gamma'$, where we use id to denote the identity homomorphism on any algebra. Invoking once again uniqueness shows that $a \circ a' = \mathrm{id} = a' \circ a$ and hence that $\mathcal{A}_G$ and $\mathcal{A}'_G$ are naturally isomorphic. So, any representative of this isomorphism class could be called *the* algebra freely generated by G.

(2) To make the above definition useful we must prove that a pair $(\mathcal{A}_G, \gamma)$ exists for every set G. Consider the complex vector space spanned by the basis $\{e_S\}$, where S runs through all finite ordered sequences of the elements of G, say $S = (g_1, \dots, g_k)$, with $k > 0$. Define multiplication on be basis elements by concatenation, $e_S e_T = e_{ST}$, where $(g_1, \dots, g_k)(g'_1, \dots, g'_l) = (g_1, \dots, g_k, g'_1, \dots, g'_l)$ and extend it to the whole vector space by linearity. It is straight forward to see that we have defined an algebra that satisfies the property of being freely generated by G. In the case of unital $*$-algebras, we use the same construction, except that the basis is augmented by the element $\mathbf{1}$, with the extra multiplication rule $\mathbf{1} e_S = e_S \mathbf{1} = e_S$, and S now runs through finite ordered sequences of the elements of $G \sqcup G^*$, where G^* is in bijection with G, denoted by $* : G \to G^*$ and its inverse also by also $* : G^* \to G$. The $*$-involution is defined on the basis as $\mathbf{1}^* = \mathbf{1}$ and $e_S^* = e_{S^*}$, where $S^* = (*g_k, \dots, *g_1)$ for $S = (g_1, \dots, g_k)$, and extended to the whole linear space by complex anti-linearity.

Let us pass to the discussion of how to impose some algebraic relations on the algebra $\mathcal{A}_G$ freely generated by G. To be concrete, think of an algebra $\mathcal{A}_G$ freely generated by G and assume that we want to impose the relation l stating that $\mathbf{1}a - a\mathbf{1} = 0$ for all $a \in \mathcal{A}_G$ and for a preferred element $\mathbf{1} \in \mathcal{A}_G$ which will become the identity element of a new algebra $\mathcal{A}_{G,l}$. We can define $\mathcal{A}_{G,l} \cong \mathcal{A}_G / \mathcal{I}_l$, where $\mathcal{I}_l \subset \mathcal{A}_G$ is the two-sided ideal (resp. $*$-ideal, in the case of $*$-algebras) generated by l, the set of finite linear combinations of products of $(\mathbf{1}a - a\mathbf{1})$ and any other elements of $\mathcal{A}_G$. In case a set R of relations is imposed, one similarly takes the quotient with respect to the intersection $\mathcal{I}_R$ of the ideals ($*$-ideals if working with $*$-algebras) generated by each relation separately, $\mathcal{A}_{G,R} \cong \mathcal{A}_G / \mathcal{I}_R$.

The constructed algebra $\mathcal{A}_{G,R}$ satisfies the following abstract definition which again relies on a universal property.

Definition 5.1.8 (*Presentation by generators and relations*) Given an algebra $\mathcal{A}_G$ free on G and a set R whose elements are called **relations** (again, not necessarily finite or even countable), together with a map $\rho : R \to \mathcal{A}_G$, an algebra $\mathcal{A}_{G,R}$ is said to be **presented by the generators** G **and relations** R if there exists an algebra homomorphism $r\colon \mathcal{A}_G \to \mathcal{A}_{G,R}$ such that, for any other algebra $\mathcal{B}$ and map $\beta\colon G \to \mathcal{B}$ such that the composition of the relations with the canonical homomorphism $b\colon \mathcal{A}_G \to \mathcal{B}$ gives $b \circ \rho = 0$, there exists a *unique* algebra homomorphism $b_R\colon \mathcal{A}_{G,R} \to \mathcal{B}$ such that $b = b_R \circ r$. We use the same terminology for $*$- and unital algebras.

Remark 5.1.9 Analogously to the case of $\mathcal{A}_G$, this definition easily implies that any two algebras $\mathcal{A}_{G,R}, \mathcal{A}'_{G,R}$ presented by the generators G and relations R are *naturally isomorphic* as the reader can immediately prove by using the universal property of the definition. Intuitively, the algebra $\mathcal{A}_{G,R}$ is therefore *the* algebra that is generated by G satisfying *only* the relations $\rho(R) = 0$.

The presentation in terms of generators and relations works for a variety of algebraic structures, like groups, rings, module, algebras, etc. In fact, the universal property of objects defined in this way is most conveniently expressed using commutative diagrams in the corresponding category [45, Sect. I.11]. The case of groups is extensively discussed in [45, Sect. I.12]. Note that, though uniqueness of these objects is guaranteed by abstract categorical reasoning, their existence is not automatic and must be checked in each category of interest. For another example of definition by universal property in this volume, see Sect. 4.10.4.

5.1.3 The GNS Construction

When adopting the algebraic formulation, the notion of (quantum) state must be similarly generalized as follows.

Definition 5.1.10 (*States*) Given an unital $*$-algebra $\mathcal{A}$, an (algebraic) **state** ω over $\mathcal{A}$ is a $\mathbb{C}$-linear map $\omega : \mathcal{A} \to \mathbb{C}$ which is *positive* (i.e. $\omega(a^*a) \geq 0$ for all $a \in \mathcal{A}$) and *normalized* (i.e. $\omega(\mathbf{1}) = 1$).

The overall idea underlying this definition is that if, for a given observable $a = a^* \in \mathcal{A}$ we know all moments $\omega(a^n)$, and thus all expectation values of polynomials $\omega(p(a))$, we also know the probability distribution associated to every value of a when the state is ω. To give a precise meaning to this idea, we should represent observables a as self-adjoint operators $\hat{a}$ in some Hilbert space $\mathcal{H}$, where the values of a correspond to the point of spectrum $\sigma(\hat{a})$ and the mentioned probability distribution is that generated by a vector Ψ state representing ω in $\mathcal{H}$, and the spectral measure of $\hat{a}$. We therefore expect that, in this picture, $\omega(a) = \langle \Psi | \hat{a} \Psi \rangle$ for some normalized vector $\Psi \in \mathcal{H}$. This is, in fact, a consequence of the content of the celebrated GNS re-construction procedure for unital C^*-algebras [28, 49, 60]. We will discuss shortly the unital $*$-algebra version of that theorem. Note that the general problem of reconstructing even a unique classical state (a probability distribution on phase space) from the knowledge of all of its polynomial moments is much more difficult and is sometimes impossible (due to non-uniqueness). This kind of reconstruction goes under the name of the Hamburger moment problem [54, Sect. X. 6, Ex. 4]. In this case, the successful reconstruction of a representation from a state succeeds because of the special hypotheses that go into the GNS theorem, where we know not only the expectation values of a (and the polynomial $*$-algebra generated by it) but also those of all elements of the algebra of observables.

In the rest of the chapter $\mathscr{L}(V)$ will denote the linear space of linear operators $T : V \to V$ on the vector space V.

Definition 5.1.11 (*$*$-Representations*) Let $\mathcal{A}$ be a complex algebra and $\mathcal{D}$ a dense linear subspace of the Hilbert space $\mathcal{H}$.

(a) A map $\pi : \mathcal{A} \to \mathscr{L}(\mathcal{D})$ such that it is linear and product preserving is called **representation** of $\mathcal{A}$ on $\mathcal{H}$ with **domain** $\mathcal{D}$. If $\mathcal{A}$ is furthermore unital, a representation is also required to satisfy: $\pi(\mathbf{1}) = I$.

(b) If finally $\mathcal{A}$ is a $*$-algebra, a **$*$-representation** of $\mathcal{A}$ on $\mathcal{H}$ with **domain** $\mathcal{D}$ is a representation which satisfies (where † henceforth denotes the Hermitian adjoint operation in $\mathcal{H}$)

$$\pi(a)^\dagger\restriction_{\mathcal{D}} = \pi(a^*) \quad \forall a \in \mathcal{A}\,.$$

As a general result we have the following elementary proposition

Proposition 5.1.12 (On faithful representations) *If $\mathcal{A}$ is a complex algebra is simple, then every representation is either faithful—i.e., injective—or it is the zero representation.*

Proof If $\pi : \mathcal{A} \to \mathscr{L}(\mathcal{D})$ is a $*$-representation, $Ker(\pi)$ is evidently a two-sided ideal. Since $\mathcal{A}$ is simple there are only two possibilities either $Ker(\pi) = \mathcal{D}$ so that π is the zero representation, or $Ker(\pi) = \{0\}$ and thus π is injective. □

Theorem 5.1.13 (GNS construction) *If $\mathcal{A}$ is a complex unital $*$-algebra and $\omega : \mathcal{A} \to \mathbb{C}$ is a state, the following facts hold.*
(a) There is a quadruple $(\mathcal{H}_\omega, \mathcal{D}_\omega, \pi_\omega, \Psi_\omega)$, where:

(i) $\mathcal{H}_\omega$ is a (complex) Hilbert space,
(ii) $\mathcal{D}_\omega \subset \mathcal{H}_\omega$ is a dense subspace,
(iii) $\pi_\omega : \mathcal{A} \to \mathscr{L}(\mathcal{D}_\omega)$ a $*$-representation of $\mathcal{A}$ on $\mathcal{H}_\omega$ with domain $\mathcal{D}_\omega$,
(iv) $\pi_\omega(\mathcal{A})\Psi_\omega = \mathcal{H}_\omega$,
(v) $\omega(a) = \langle\Psi_\omega|\pi_\omega(a)\Psi_\omega\rangle$ for every $a \in \mathcal{A}$.

(b) If $(\mathcal{H}'_\omega, \mathcal{D}'_\omega, \pi'_\omega, \Psi'_\omega)$ satisfies (i)–(v), then there is $U : \mathcal{H}_\omega \to \mathcal{H}'_\omega$ surjective and isometric such that:

(i) $U\Psi_\omega = \Psi'_\omega$,
(ii) $U\mathcal{D}_\omega = \mathcal{D}'_\omega$,
(iii) $U\pi_\omega(a)U^{-1} = \pi'_\omega(a)$ if $a \in \mathcal{A}$.

Proof Consider $\mathcal{A}$ as complex vector space and define $N = \{a \in \mathcal{A} | \omega(a^*a) = 0\}$. N is a subspace as easily follows from sesquilinearity of $(a, b) \mapsto \omega(a^*b)$ and from the Cauchy-Schwartz inequality which holds because $(a, b) \mapsto \omega(a^*b)$ is non-negative. Define $\mathcal{D}_\omega \stackrel{\text{def}}{=} \mathcal{A}/N$ as a complex vector space and equip it with the Hermitian scalar product $\langle [a] | [b] \rangle \stackrel{\text{def}}{=} \mu(a^*b)$, which turns out to be well-defined (because $\omega(a^*b) = \omega(b^*a) = 0$ if $a \in N$ again from Cauchy-Schwartz inequality) and positive. $\mathcal{H}_\omega$ is, by definition, the completion of $\mathcal{D}_\omega$ with respect to the mentioned scalar product. Now observe that N is also a left-ideal ($\omega((ba)^*ba) = \omega((b^*(ba))^*a) = 0$ if $a \in N$) and consequently $\pi_\omega(a)[b] \stackrel{\text{def}}{=} [ab]$ is well-defined ($[ab] = [ac]$ if $c \in [b]$) and is a unital algebra representation. Defining $\Psi_\omega \stackrel{\text{def}}{=} [1]$, we have $\omega(a) = \langle \Psi_\omega | \pi_\omega(a) \Psi_\omega \rangle$. Finally:

$$\begin{aligned}\langle \pi_\omega(c)\Psi_\omega | \pi_\omega(a)\pi_\omega(b)\Psi_\omega \rangle &= \omega(c^*(a^*)^*b) = \omega((a^*c)^*b) = \langle \pi_\omega(a^*c)\Psi_\omega | \pi_\omega(b)\Psi_\omega \rangle \\ &= \langle \pi_\omega(a^*)\pi_\omega(c)\Psi_\omega | \pi_\omega(b)\Psi_\omega \rangle\end{aligned}$$

Summing up, we have:

$$\begin{aligned}\langle \pi_\omega(a)^\dagger \pi_\omega(c)\Psi_\omega | \pi_\omega(b)\Psi_\omega \rangle &= \langle \pi_\omega(c)\Psi_\omega | \pi_\omega(a)\pi_\omega(b)\Psi_\omega \rangle \\ &= \langle \pi_\omega(a^*)\pi_\omega(c)\Psi_\omega | \pi_\omega(b)\Psi_\omega \rangle\end{aligned}$$

Since c, b are arbitrary and both $\pi_\omega(b)\Psi_\omega$ and $\pi_\omega(b)\Psi_\omega$ range in $\mathcal{D}_\omega$ which is dense, we have found that $\pi(a)^\dagger|_{\mathcal{D}_\omega} = \pi(a^*)$. The proof of (b) is easy. As a matter of fact the operator U is completely defined by $U\pi_\omega(a)\Psi_\omega \stackrel{\text{def}}{=} \pi'_\omega(a)\Psi'_\omega$, we leave to the reader the proof of the fact that it is well-defined and satisfies the required properties. The proof is strictly analogous to the corresponding part of (b) in Proposition 5.1.17 below. □

There exists a stronger version of that theorem [8, 28, 49] regarding the case where $\mathcal{A}$ is a unital C^*-algebra. The quadruple $(\mathcal{H}_\omega, \mathcal{D}_\omega, \pi_\omega, \Psi_\omega)$ is called **GNS triple** (!) the name is due to the fact that for C^*-algebras $\mathcal{D}_\omega = \mathcal{H}_\omega$. In that case the representation π_ω is continuous (norm decreasing more precisely) with respect to the operator norm $|| \; ||$ in $\mathscr{B}(\mathcal{H}_\omega)$, since $\pi_\omega(a) \in \mathscr{B}(\mathcal{H}_\omega)$ if $a \in \mathcal{A}$.

As a general fact, we have that a $*$-representations π of a unital C^*-algebra $\mathcal{A}$ on a Hilbert space $\mathcal{H}$ assuming values in $\mathscr{B}(\mathcal{H})$ is automatically norm decreasing, with respect to the operator norm $|| \; ||$ in $\mathscr{B}(\mathcal{H})$. Moreover π is isometric if and only if it is injective [8, 28].

Remark 5.1.14

(1) Since $\mathcal{D}_\omega$ is dense $\pi_\omega(a)^\dagger$ is always well defined and, in turn, densely defined for (iii) in (a). Hence, $\pi_\omega(a)$ is always closable. Therefore, if $a = a^*$, $\pi(a)$ is at least symmetric. If $\pi(a)$ is self-adjoint the probability distribution of the observable a in the state ω mentioned in the comment after Definition 5.1.10 is $\mathcal{B}(\mathbb{R}) \ni E \mapsto \langle \Psi_\omega | P_E^{(\pi_\omega(a))} \Psi_\omega \rangle$, where $\mathcal{B}(\mathbb{R})$ is the class of Borel sets on $\mathbb{R}$ and $P^{(\pi_\omega(a))}$ the projection-valued measure of $\pi_\omega(a)$. The precise technical conditions,

and their physical significance, under which an operator $\pi(a)$, with $a = a^*$, might be essentially self-adjoint on $\mathcal{D}_\omega$ are poorly explored in the literature and deserve further investigation.

(2) The **weak commutant** π'_w of a $*$-representation π of $\mathcal{A}$ on $\mathcal{H}$ with domain $\mathcal{D}$, is defined as[1]

$$\pi'_w \stackrel{\text{def}}{=} \{A \in \mathscr{B}(\mathcal{H}) \mid \langle \psi | A\pi(a)\phi \rangle = \langle \pi(a)^\dagger \psi | A\phi \rangle \quad \forall a \in \mathcal{A}\,, \forall \psi, \phi \in \mathcal{D}\}\,,$$

where $\mathscr{B}(\mathcal{H})$ denotes the C^*-algebra of all bounded operators on $\mathcal{H}$. If $\mathcal{A}$ is a unital C^*-algebra, the weak commutant of π (with domain given by the whole Hilbert space) coincides to the standard commutant. We say that a $*$-representation π of $\mathcal{A}$ on $\mathcal{H}$ is **weakly irreducible** if its weak commutant is trivial, that is, it coincides with the set of operators $cI : \mathcal{H} \to \mathcal{H}$ for $c \in \mathbb{C}$.

(3) The set of states over the unital *-algebra $\mathcal{A}$ is a **convex body**. In other words *convex combinations* of states are states: $\omega = p\omega_1 + (1-p)\omega_2$ with $p \in (0, 1)$ is a state if ω_1, ω_2 are.

(4) A state ω is said to be **extremal** if $\omega = p\omega_1 + (1-p)\omega_2$, with $p \in (0, 1)$ and ω_1, ω_2 are states, is possible only if $\omega_1 = \omega_2 (= \omega)$. These states are also called **pure states**. It is possible to prove the following [28, 60]:

Proposition 5.1.15 (Pure states and irreducible representations) *Referring to the hypotheses of Theorem 5.1.13, ω is pure if and only if π_ω is weakly irreducible.*

(If $\mathcal{A}$ is a unital C^*-algebra the same statement holds but "weakly" can be omitted.) Therefore, even if ω is represented by a unit vector Ψ_ω in $\mathcal{H}_\omega$, it does not mean that ω is pure. In standard quantum mechanics it happens because $\mathcal{A}$ is implicitly assumed to coincide to the whole C^*-algebra $\mathscr{B}(\mathcal{H})$ of everywhere-defined bounded operators over $\mathcal{H}$ and π_ω is the identity when ω corresponds to a vector state of $\mathcal{H}$.

(5) When $\mathcal{A}$ is a unital C^*-algebra, the convex body of states on $\mathcal{A}$ is hugely larger that the states of the form

$$\mathcal{A} \ni a \mapsto \omega_\rho(a) \stackrel{\text{def}}{=} tr(\rho\pi_\omega(a))$$

for a fixed (algebraic) state ω and where $\rho \in \mathscr{B}(\mathcal{H}_\omega)$ is a positive trace class operator with unit trace. These trace-class operators states associated with an algebraic state ω form the **folium** of ω and are called **normal states** in $\mathcal{H}_\omega$. If $\mathcal{A}$ is not C^*, the trace $tr(\rho\pi_\omega(a))$ is not defined in general, because $\pi_\omega(a)$ is not bounded and $\rho\pi_\omega(a)$ may not be well defined nor trace class in general. Even if $\mathcal{A}$ is just a unital $*$-algebra, a unit vector $\Phi \in \mathcal{D}_\omega$ defines however a state by means of

$$\mathcal{A} \ni a \mapsto \omega_\Phi(a) \stackrel{\text{def}}{=} \langle \Phi | \pi_\omega(a)\Phi \rangle\,,$$

recovering the standard formulation of elementary quantum mechanics. *These states are pure when ω is pure.* More strongly, in this situation $(\mathcal{H}_\omega, \mathcal{D}_\omega, \pi_\omega, \Phi)$ is just a

[1] π'_w can equivalently be defined as $\{A \in \mathscr{B}(\mathcal{H}) \mid A\phi(a) = \pi(a^*)^\dagger A\,, \quad \forall a \in \mathcal{A}\}$.

GNS triple of ω_Φ, because $\pi_\omega(\mathcal{A})\Phi$ is dense in $\mathcal{H}_\omega$ (this is because the orthogonal projector onto $\pi_\omega(\mathcal{A})\Phi$ cannot vanish and belongs to the weak commutant $\pi'_{\omega w}$ which is trivial, because ω is pure). If ω is not pure, ω_Φ may not be pure also if it is represented by a unit vectors.

(6) There are unitarily non-equivalent GNS representations of the same unital $*$-algebra $\mathcal{A}$ associated with states ω, ω'. In other words there is no surjective isometric operator $U : \mathcal{H}_\omega \to \mathcal{H}_{\omega'}$ such that $U\pi_\omega(a) = \pi_{\omega'}(a)U$ for all $a \in \mathcal{A}$. (Notice that, in the notion of *unitary equivalence* it is not required that $U\Psi_\omega = \Psi_{\omega'}$). Appearance of unitarily inequivalent representations is natural when $\mathcal{A}$ has a non-trivial *center*, $\mathcal{Z}_\mathcal{A}$, i.e., it contains something more than the elements $c\mathbf{1}$ for $c \in \mathbb{C}$. Pure states ω, ω' such that $\omega(z) \neq \omega'(z)$ for some $z \in \mathcal{Z}_\mathcal{A}$ give rise to unitarily inequivalent GNS representations. This easily follows from the fact that $\pi_\omega(z)$ and $\pi_{\omega'}(z)$, by irreducibility of the representations, must be operators of the form $c_z I$ and $c'_z I$ for complex numbers c_z, c'_z in the respective Hilbert spaces $\mathcal{H}_\omega$ and $\mathcal{H}_{\omega'}$. It should be noted that such representations remain inequivalent even if the unitarity of U is relaxed. However, it can happen that some representations are unitarily inequivalent even when the algebra has a trivial center. See Sect. 5.2.6 for a relevant example.

Remark 5.1.16 The positivity requirement on states is physically meaningful when every self-adjoint element of the $*$-algebra is a physical observable. It is also a crucial ingredient in the GNS reconstruction theorem. However, in the treatment of gauge theories in the Gupta-Bleuler or BRST formalisms, in order to keep spacetime covariance, one must enlarge the $*$-algebra to include unobservable or *ghost* fields. Physically meaningful states are then allowed to fail the positivity requirement on $*$-algebra elements generated by ghost fields. The GNS reconstruction theorem is then not applicable and, in any case, the $*$-algebra is expected to be represented on an indefinite scalar product space (a *Krein space*) rather than a Hilbert space. Fortunately, several extensions of the GNS construction have been made, with the positivity requirement replaced by a different one that, instead, guarantees the reconstructed $*$-representation to be on an indefinite scalar product space. Such generalizations and their technical details are discussed in [33].

Another relevant result arising from the GNS theorem concerns *symmetries* represented by $*$-algebra (anti-linear) automorphisms.

Proposition 5.1.17 (Automorphisms induced by invariant states) *Let $\mathcal{A}$ be an unital $*$-algebra, ω a state on it and consider its GNS representation. The following facts hold.*
(a) If $\beta : \mathcal{A} \to \mathcal{A}$ is a unital $*$-algebra automorphism (resp. anti-linear automorphism) which leaves fixed ω, i.e., $\omega \circ \beta = \omega$, then there exist a unique bijective bounded operator $U^{(\beta)} : \mathcal{H}_\omega \to \mathcal{H}_\omega$ such that:
(i) $U^{(\beta)}\Psi_\omega = \Psi_\omega$ and $U^{(\beta)}(\mathcal{D}_\omega) = \mathcal{D}_\omega$,
(ii) $U^{(\beta)}\pi_\omega(a)U^{(\beta)-1}x = \pi_\omega(\beta(a))x$ if $a \in \mathcal{A}$ and $x \in \mathcal{D}_\omega$.
$U^{(\beta)}$ turns out to be unitary (resp. anti-unitary).

(b) If, varying $t \in \mathbb{R}$, $\beta_t : \mathcal{A} \to \mathcal{A}$ defines a one-parameter group of unital $*$-algebra automorphisms[2] *which leaves fixed ω, the corresponding unitary operators $U_t^{(\beta)}$ as in (a) define a one-parameter group of unitary operators in $\mathcal{H}_\omega$.*
(c) $\{U_t^{(\beta)}\}_{t\in\mathbb{R}}$ as in (b) is strongly continuous (and thus it admits a self-adjoint generator) if and only if

$$\lim_{t\to 0} \omega(a^*\beta_t(a)) = \omega(a^*a) \quad \textit{for every } a \in \mathcal{A}.$$

Proof Let us start from (a) supposing that β is a $*$-automorphism. If an operator satisfying (i) and (ii) exists it also satisfies $U^{(\beta)}\pi_\omega(a)\Psi_\omega = \pi_\omega(\beta(a))\Psi_\omega$. Since $\pi_\omega(\mathcal{A})\Psi_\omega$ is dense in $\mathcal{H}_\omega$, this identity determines $U^{(\beta)}$ on $\mathcal{D}_\omega$. Therefore we are lead to try to define $U_0^{(\beta)}\pi_\omega(a)\Psi_\omega \stackrel{\text{def}}{=} \pi_\omega(\beta(a))\Psi_\omega$ on $\mathcal{D}_\omega$. From (v) in (a) of Theorem 5.1.13 it immediately arises that $||U_0^{(\beta)}\pi_\omega(a)\Psi_\omega||^2 = ||\pi_\omega(a)\Psi_\omega||^2$. That identity on the one hand proves that $U^{(\beta)}$ is well defined because if $\pi_\omega(b)\Psi_\omega = \pi_\omega(b')\Psi_\omega$ then $U^{(\beta)}\pi_\omega(b)\Psi_\omega = U^{(\beta)}\pi_\omega(b')\Psi_\omega$, on the other hand it proves that $U^{(\beta)}$ is isometric on $\mathcal{D}_\omega$. If we analogously define the other isometric operator $V_0^{(\beta)}\pi_\omega(a)\Psi_\omega \stackrel{\text{def}}{=} \pi_\omega\left(\beta^{-1}(a)\right)\Psi_\omega$ on $\mathcal{D}_\omega$, we see that $U_0^{(\beta)}Vx = VU_0^{(\beta)}x$ for every $x \in \mathcal{D}_\omega$. Since $\mathcal{D}_\omega$ is dense in $\mathcal{H}_\omega$, these identities extend to analogous identities for the unique bounded extensions of $U_0^{(\beta)}$ and V valid over the whole Hilbert space. In particular the former operator extends into an isometric surjective operator (thus unitary) $U^{(\beta)}$ which, by construction, satisfies (i) and (ii). Notice that V, defined on $\mathcal{D}_\omega$, is the inverse of $U_0^{(\beta)}$ so that, in particular $U^{(\beta)}(\mathcal{D}_\omega) = U_0^{(\beta)}(\mathcal{D}_\omega) = \mathcal{D}_\omega$. The followed procedure also proves that $U^{(\beta)}$ is uniquely determined by (i) and (ii). The anti-linear case is proved analogously. Anti-linearity of β implies that, in $U_0^{(\beta)}\pi_\omega(a)\Psi_\omega \stackrel{\text{def}}{=} \pi_\omega(\beta(a))\Psi_\omega$, $U_0^{(\beta)}$ must be anti-linear and thus anti-unitary.

The proof of (b) immediately arises from (a). Regarding (c), we observe that, if $x = \pi_\omega(a)\Psi_\omega$ one has for $t \to 0$ by the GNS theorem,

$$\langle x|U_t^{(\beta)}x\rangle = \omega(a^*\beta_t(a)) \to \omega(a^*a) = \langle x|x\rangle$$

Since the span of the vectors x is dense in $\mathcal{H}_\omega$, $U_t^{(\beta)}$ is strongly continuous due to Proposition 9.24 in [49]. □

Remark 5.1.18

(1) Evidently, the statements (b) and (c) can immediately be generalized to the case of a representation of a generic *group* or, respectively, *connected topological group*, G. Assume that G is represented in terms of automorphisms of unital $*$-algebras $\beta_g : \mathcal{A} \to \mathcal{A}$ for $g \in G$. With the same proof of (c), it turns out that, if ω is invariant under this representation of G, the associated representation in the GNS Hilbert space of ω, $\{U_g^{(\beta)}\}_{g\in G}$ is strongly continuous if and only if

---

[2]There do not exist one-parameter group of unital $*$-algebra *anti-linear* automorphisms, this is because $\beta_t = \beta_{t/2} \circ \beta_{t/2}$ is linear both for $\beta_{t/2}$ linear or anti-linear.

$$\lim_{g \to e} \omega(a^* \beta_g(a)) = \omega(a^* a) \quad \text{for every } a \in \mathcal{A},$$

where $e \in G$ is the unit element.

(2) It could happen in physics that an *algebraic symmetry*, i.e., an automorphism (or anti-automorphism) $\beta : \mathcal{A} \to \mathcal{A}$ exists for a unital $*$-algebra with some physical interpretation, but that this symmetry cannot be completely implemented unitarily (resp. anti-unitarily) in the GNS representation of a state ω because, referring to the condition in (a) of the proved theorem, either (i) or *both* (i) and (ii) of (a) do not hold. In the first case the symmetry is broken because the cyclic vector is not invariant under a unitary representation of the symmetry, which however exists in the GNS representation of ω. Obviously, in this case, ω is not invariant under the algebraic symmetry. This situation naturally arises when one starts from a pure invariant state ω_0 and the physically relevant state is not ω_0, but another state $\omega \in \mathcal{D}_{\omega_0}$. The second, much more severe, situation is when there is no unitary map in the GNS representation of ω which fulfills (i) and (ii). In algebraic quantum theories, this second case is often called *spontaneous breaking of symmetry*.

5.2 The $*$-algebra of a Quantum Field and Its Quasifree States

This chapter mostly deals with the case of a real scalar field, we will denote by ϕ, on a given always oriented and time oriented, *globally hyperbolic spacetime* $\boldsymbol{M} = (\mathcal{M}, g, \mathfrak{o}, \mathfrak{t})$ of dimension $n \geq 2$, where g is the metric with signature $(+, -, \dots, -)$, $\mathfrak{o}$ the *orientation* and $\mathfrak{t}$ the *time orientation*. Regarding geometrical notions, we adopt throughout the definitions of Chap. 3. Minkowski spacetime will be denoted by $\mathbb{M}$ and its metric by η.

The results we discuss can be extended to charged and higher spin fields. As is well known a quantum field is a *locally covariant notion*, functorially defined in *all* globally hyperbolic spacetimes simultaneously (see Chap. 4). Nevertheless, since this chapter is devoted to discussing algebraic *states* of a QFT in a given manifold we can deal with a fixed spacetime. Moreover we shall not construct the $*$-algebras as *Borchers-Uhlmann-like* algebras (Chap. 3) nor use the *deformation approach* (see Chap. 2) to define the algebra structure, in order to simplify the technical structure and focus on the properties of the states.

5.2.1 *The Algebra of Observables of a Real Scalar Klein-Gordon Field*

In order to deal with QFT in curved spacetime, a convenient framework is the algebraic one. This is due to various reasons. Especially because, in the absence of

Poincaré symmetry, there is no preferred Hilbert space representation of the field operators, but several unitarily inequivalent representations naturally show up. Furthermore, the standard definition of the field operators based on the decomposition of field solutions in positive and negative frequency part is not allowed here, because there is no preferred notion of (Killing) time.

In the rest of the chapter $C_0^\infty(\mathcal{M})$ denotes the real vector space of compactly-supported and *real*-valued smooth function on the manifold $\mathcal{M}$.

The elementary algebraic object, i.e., a **scalar quantum field** ϕ over the globally hyperbolic spacetime $\boldsymbol{M}$ is captured by a unital $*$-algebra $\mathcal{A}(\boldsymbol{M})$ called the CCR algebra of the quantum field ϕ.

Definition 5.2.1 (*CCR algebra*) The **CCR algebra** of the quantum field ϕ over $\boldsymbol{M}$ is the unital $*$-algebra presented by the following generators and relations (cf. Sect. 5.1.2). The generators consist (**smeared abstract**) **field operators**, $\phi(f)$, labeled by functions $f \in C_0^\infty(\mathcal{M})$ (the identity $\mathbf{1}$ is of course included in the construction of the corresponding freely generated algebra). These generators satisfy the following relations:

$\mathbb{R}$-Linearity: $\phi(af+bg)-a\phi(f)-b\phi(g)=0$ if $f, g \in C_0^\infty(\mathcal{M})$ and $a, b \in \mathbb{R}$.
Hermiticity: $\phi(f)^* - \phi(f) = 0$ for $f \in C_0^\infty(\mathcal{M})$.
Klein-Gordon: $\phi\left((\Box_{\boldsymbol{M}} + m^2 + \xi R)g\right) = 0$ for $g \in C_0^\infty(\mathcal{M})$.
Commutation relations: $[\phi(f), \phi(g)] - iE(f,g)\mathbf{1} = 0$ for $f \in C_0^\infty(\mathcal{M})$.

Above E denotes the *advanced-minus-retarded fundamental solution*, also called the *causal propagator*, see Chap. 3 and (1) in Remark 5.2.2 below. The Hermitian elements of $\mathcal{A}(\boldsymbol{M})$ are the *elementary observables* of the free field theory associated with the Klein-Gordon field ϕ. The non-Hermitian elements play an auxiliary rôle. It should however be evident that $\mathcal{A}(\boldsymbol{M})$ is by no means sufficient to faithfully describe physics involved with the quantum field ϕ. For instance $\mathcal{A}(\boldsymbol{M})$ does not include any element which can be identified with the *stress energy tensor* of ϕ. Also the local interactions like ϕ^4 cannot be described as elements of this algebra either. We shall tackle this problem later.

According to the discussion in Sect. 5.1.2, the above abstract definition is sufficient to uniquely define $\mathcal{A}(\boldsymbol{M})$ up to isomorphism. An alternative, more concrete and explicit, construction using tensor products of spaces $C_0^\infty(\mathcal{M})$ is presented in Chap. 3. That construction yields a concrete representative of the isomorphism class of $\mathcal{A}(\boldsymbol{M})$.

Remark 5.2.2

(1) Let Sol indicate the real vector space of *real* smooth solutions ψ with *compact Cauchy data* of the KG equation $(\Box_{\boldsymbol{M}} + m^2 + \xi R)\psi = 0$ where $\Box_{\boldsymbol{M}} \stackrel{\text{def}}{=} g^{ab}\nabla_a\nabla_b$. Let us, as usual, use the notation $\mathcal{D}(\mathcal{M}) \stackrel{\text{def}}{=} C_0^\infty(\mathcal{M}) \oplus iC_0^\infty(\mathcal{M})$ for the space of *complex* test functions and $\mathcal{D}'(\mathcal{M})$ is the dual space of distributions. Interpreting the **advanced-minus-retarded fundamental solution** of the KG operator as a linear map

$$E : C_0^\infty(\mathcal{M}) \to \mathsf{Sol}\,,$$

we can naturally extend it by $\mathbb{C}$-linearity to the continuous linear map

$$E : \mathcal{D}(\mathcal{M}) \to \mathcal{D}'(\mathcal{M}) ,$$

which defines the bilinear functional

$$E(f_1, f_2) \stackrel{\text{def}}{=} \int_{\mathcal{M}} f_1(Ef_2)\, \text{dvol}_{\boldsymbol{M}} \quad \text{if} \quad f_1, f_2 \in C_0^\infty(\mathcal{M}) , \tag{5.1}$$

which is the one appearing in the commutation relations above. Of course, in agreement with the commutation relations,

$$\mathbb{R} \ni E(f, g) = -E(g, f) \quad \text{if } f, g \in C_0^\infty(\mathcal{M}). \tag{5.2}$$

As a map $C_0^\infty(\mathcal{M}) \to \mathsf{Sol}$, E satisfies

$$Ker(E) = \{(\Box_{\boldsymbol{M}} + m^2 + \xi R)h \mid h \in \mathcal{D}(\mathcal{M})\} . \tag{5.3}$$

Everything is a consequence of the fact that $\boldsymbol{M}$ is globally hyperbolic (see Chap. 3). Since $E(f, h) = 0$ if the support of f does not intersect $J^+_{\boldsymbol{M}}(\text{supp}h) \cup J^-_{\boldsymbol{M}}(\text{supp}h)$, we immediately have from the *commutation relations* requirement that the following important fact holds, distinguishing observable fields (Bosons) form unobservable ones (Fermions):

Proposition 5.2.3 (Causality) *Referring to $\mathcal{A}(\boldsymbol{M})$, $\phi(f)$ and $\phi(h)$ commute if the supports of f and h are causally separated.*

From standard properties of E (see Chap. 3) one also finds, if $\Sigma \subset \mathcal{M}$ is a smooth space-like Cauchy surface $f, h \in C_0^\infty(\mathcal{M})$ and $\psi_f \stackrel{\text{def}}{=} Ef$ and $\psi_h \stackrel{\text{def}}{=} Eh$ are elements of Sol,

$$E(f, g) = \int_\Sigma \left(\psi_f \nabla_{\boldsymbol{n}} \psi_h - \psi_h \nabla_{\boldsymbol{n}} \psi_f\right) d\Sigma , \tag{5.4}$$

where $d\Sigma$ is the standard measure induced by the metric g on Σ and $\boldsymbol{n}$ the future directed normal unit vector field to Σ.

(2) As $E : \mathcal{D}(\mathcal{M}) \to \mathcal{D}'(\mathcal{M})$ is continuous, due to *Schwartz kernel theorem* [39], it defines a distribution, indicated with the same symbol $E \in \mathcal{D}'(\mathcal{M} \times \mathcal{M})$, uniquely determined by

$$E(f_1, f_2) = E(f_1 \otimes f_2) \quad f_1, f_2 \in \mathcal{D}(\mathcal{M}) ,$$

and this leads to an equivalent interpretation of the left-hand side of (5.1), which is actually a bit more useful, because it permits to consider the action of E on non-factorized test functions $h \in \mathcal{D}'(\mathcal{M} \times \mathcal{M})$.

(3) The condition indicated as *Klein-Gordon* is the requirement that ϕ *distributionally* satisfies the equation of Klein-Gordon. Obviously $\Box_{\boldsymbol{M}}$ appearing in it

coincides with its formal transposed (or adjoint) operator which should appear in the distributional version of KG equation.

(4) Everything we will say holds equally for m^2 and ξ replaced by corresponding smooth real functions, also in the case where m^2 attains negative values. Also the case $m^2 < 0$ does not produce technically difficult problems.

Linearity and Commutation relations conditions together with (5.3) imply the elementary but important result which proves also the converse implication in the property *Klein-Gordon*.

Proposition 5.2.4 *Referring to $\mathcal{A}(\boldsymbol{M})$ the following facts hold.*

$$\phi(f) = \phi(g) \quad \textit{if and only if } f - g \in Ker(E), \tag{5.5}$$

so that, in particular,

$$\phi(f) = 0 \quad \textit{if and only if } f = (\Box_{\boldsymbol{M}} + m^2 + \xi R)g \ \textit{ for } \ g \in C_0^\infty(\mathcal{M}). \tag{5.6}$$

Proof $\phi(f) = \phi(g)$ is equivalent to $\phi(f - g) = 0$ and thus $iE(h, (f - g)) = [\phi(h), \phi(f-g)] = 0$ for all $h \in \mathcal{D}(\mathcal{M})$. From (5.1) one has, in turn, that $E(f-g) = 0$ that is $f - g \in Ker(E)$. Finally (5.3) implies the last statement. □

The smeared field $\phi(f)$ can be thought of as localized within the support of its argument f. However, $\phi(f)$ really depends on f only up to addition of terms from $Ker(E)$. We can use this freedom to move and shrink the support of f to be arbitrarily close to any Cauchy surface, which is a technically useful possibility.

Lemma 5.2.5 *Let $\psi \in \mathsf{Sol}$ and let Σ be a smooth space-like Cauchy surface of the globally hyperbolic spacetime $\boldsymbol{M}$. For every open neighborhood O of Σ, it is possible to pick out a function $f_\psi \in C_0^\infty(\mathcal{M})$* whose support is contained in *O, such that $\psi = Ef_\psi$.*

The proof of this elementary, but important, fact can be found in Chap. 3 and in [65] (see also the proof of our Proposition 5.3.17). This result immediately implies the validity of the so called *Time-slice axiom* for the CCR algebra (see Chap. 3).

Proposition 5.2.6 ("Time-slice axiom") *Referring to the globally hyperbolic spacetime $\boldsymbol{M}$ and the algebra $\mathcal{A}(\boldsymbol{M})$, let O be any fixed neighborhood of a Cauchy surface Σ. Then $\mathcal{A}(\boldsymbol{M})$ is generated by $\mathbf{1}$ and the elements $\phi(f)$ with $f \in C_0^\infty(\mathcal{M})$ and* $\operatorname{supp} f \subset O$.

5.2.2 States and n-point Functions

Let us focus on states. We start form the observation that the generic element of $\mathcal{A}(\boldsymbol{M})$ is always of the form

$$a = c_{(0)}\mathbf{1} + \sum_{i_1} c^{i_1}_{(1)}\phi(f^{(1)}_{i_1}) + \sum_{i_1,i_2} c^{i_1 i_2}_{(2)}\phi(f^{(2)}_{i_1})\phi(f^{(2)}_{i_2}) + \cdots + \sum_{i_1,\ldots,i_n} c^{i_1\cdots i_n}_{(n)}\phi(f^{(n)}_{i_1})\cdots\phi(f^{(n)}_{i_n}), \quad (5.7)$$

where n is arbitrarily large but finite, $c^{i_1\cdots i_k}_{(k)} \in \mathbb{C}$ and $f^{(j)}_k \in C^\infty_0(\mathcal{M})$, with all sums arbitrary but finite. Due to (5.7), if $\omega\colon \mathcal{A}(\boldsymbol{M}) \to \mathbb{C}$ is a state, its action on a generic element of $\mathcal{A}(\boldsymbol{M})$ is known as soon as the full class of the so-called **n-point functions** of ω are known. We mean the maps:

$$C^\infty_0(\mathcal{M})\times\cdots\times C^\infty_0(\mathcal{M}) \ni (f_1,\ldots,f_n) \mapsto \omega(\phi(f_1)\cdots\phi(f_n)) \stackrel{\text{def}}{=} \omega_n(f_1,\ldots,f_n)$$

At this point, the multilinear functionals $\omega_n(f_1,\ldots,f_n)$ are not yet forced to satisfy any continuity properties (in fact we have not even discussed any topologies on $\mathcal{A}(\boldsymbol{M})$ and how the states should respect it). However, in the sequel we will only be dealing with the cases where ω_n is continuous in the usual test function topology on $C^\infty_0(\mathcal{M})$. Then, by the Schwartz kernel theorem [39], we can write, as it is anyway customary, the n-point function in terms of its distributional kernel:

$$\omega_n(f_1,\ldots,f_n) = \int_{\boldsymbol{M}^n} \omega_n(x_1,\ldots,x_n) f_1(x_1)\cdots f_n(x_n)\,\mathrm{dvol}_{\boldsymbol{M}^n}\,.$$

It is worth stressing that a choice of a family of integral kernels ω_n, $n = 1, 2, \ldots$, extends by linearity and the rule $\omega(\mathbf{1}) \stackrel{\text{def}}{=} 1$ to a normalized linear functional on all of $\mathcal{A}(\boldsymbol{M})$. However, this functional generally does *not* determine a state, because the positivity requirement $\omega(a^*a) \geq 0$ may not be valid. However if two states have the same set of n-point functions they necessarily coincide in view of (5.7).

Remark 5.2.7 As defined above, the n-point functions $\omega_n(f_1,\ldots,f_n)$ need not be symmetric in their arguments. However, they do satisfy some relations upon permutation of the arguments. The reason is that the products $\phi(f_1)\cdots\phi(f_n)$ and $\phi(f_{\sigma(1)})\cdots\phi(f_{\sigma(n)})$, for any permutation σ, are not completely independent in $\mathcal{A}(\boldsymbol{M})$. It is easy to see that the CCR $*$-algebra is **filtered**, namely that $\mathcal{A}(\boldsymbol{M}) = \bigcup_{n=0}^\infty \mathcal{A}_n(\boldsymbol{M})$, where each linear subspace $\mathcal{A}_n(\boldsymbol{M})$ consists of linear combinations of $\mathbf{1}$ and products of no more than n generators $\phi(f)$, $f \in C^\infty_0(\mathcal{M})$. The product $\phi(f_1)\cdots\phi(f_n)$ belongs to $\mathcal{A}_n(\boldsymbol{M})$, as does $\phi(f_{\sigma(1)})\cdots\phi(f_{\sigma(n)})$. The commutation relation $[\phi(f),\phi(g)] = iE(f,g)\mathbf{1}$ then implies that the product $\phi(f_1)\cdots\phi(f_n)$ and the same product with any two f_i's swapped, hence also $\phi(f_{\sigma(1)})\cdots\phi(f_{\sigma(n)})$ for any permutation σ, coincide "up to lower order terms," or more precisely coincide in the quotient $\mathcal{A}_n(\boldsymbol{M})/\mathcal{A}_{n-1}(\boldsymbol{M})$. Thus, without loss of generality, the coefficients $c^{i_1\cdots i_n}_{(n)}$ in (5.7) can be taken to be, for instance, fully symmetric in their indices. So, in order to fully specify a state, it would be sufficient to specify only the fully symmetric part of each n-point function $\omega_n(f_1,\ldots,f_n)$.

Once a state ω is given, we can implement the GNS machinery obtaining a $*$-representation $\pi_\omega : \mathcal{A}(\boldsymbol{M}) \to \mathscr{L}(\mathcal{D}_\omega)$ over the Hilbert space $\mathcal{H}_\omega$ including the dense invariant linear subspace $\mathcal{D}_\omega$. The **smeared field operators** appear here as the densely defined symmetric operators:

$$\hat{\phi}_\omega(f) \stackrel{\text{def}}{=} \pi_\omega(\phi(f)) : \mathcal{D}_\omega \to \mathcal{H}_\omega\,, \quad f \in C_0^\infty(\mathcal{M})\,.$$

We stress that in general $\hat{\phi}_\omega(f)$ is not self-adjoint nor essentially self-adjoint on $\mathcal{D}_\omega$ (even if we are considering real smearing functions). That is why we introduce the following definition:

Definition 5.2.8 (*Regular states*) A state ω on $\mathcal{A}(\boldsymbol{M})$ and its GNS representation are said to be **regular** if $\hat{\phi}_\omega(f)$ is essentially self-adjoint on $\mathcal{D}_\omega$ for every $f \in C_0^\infty(\mathcal{M})$.

There are some further elementary technical properties of ω_2 and E that we list below.

Proposition 5.2.9 *Consider a state $\omega : \mathcal{A}(\boldsymbol{M}) \to \mathbb{C}$ and define $P \stackrel{def}{=} \Box_{\boldsymbol{M}} + m^2 + \xi R$. The two-point function, ω_2, satisfies the following facts for $f, g \in C_0^\infty(\mathcal{M})$:*

$$\omega_2(Pf, g) = \omega_2(f, Pg) = 0\,, \tag{5.8}$$

$$\omega_2(f, g) - \omega_2(g, f) = iE(f, g)\,, \tag{5.9}$$

$$Im(\omega_2(f, g)) = \frac{1}{2}E(f, g)\,, \tag{5.10}$$

$$\frac{1}{4}|E(f, g)|^2 \leq \omega_2(f, f)\omega_2(g, g)\,. \tag{5.11}$$

Proof The first identity trivially arises from $\omega_2(Pf, g) = \omega(\phi(Pf)\phi(g)) = 0$ and $\omega_2(f, Pg) = \omega(\phi(f)\phi(Pg)) = 0$ in view of the definition of $\phi(h)$. Next,

$$\omega_2(f, g) - \omega_2(g, f) = \omega([\phi(f), \phi(g)]) = \omega(iE(f, g)\mathbf{1}) = iE(f, g)\omega(\mathbf{1}) = iE(f, g)\,.$$

The third identity then follows immediately since $E(f, g)$ is real. Using its GNS representation and the Cauchy-Schwartz inequality we find that

$$|\omega_2(f, g)| \leq |\langle\hat{\phi}_\omega(f)\Psi_\omega|\hat{\phi}_\omega(f)\Psi_\omega\rangle|^{1/2}\, |\langle\hat{\phi}_\omega(g)\Psi_\omega|\hat{\phi}_\omega(g)\Psi_\omega\rangle|^{1/2}$$

namely

$$|\omega_2(f, g)|^2 \leq \omega_2(f, f)\omega_2(g, g)\,.$$

So that, in particular

$$|Im(\omega_2(f, g))|^2 \leq \omega_2(f, f)\omega_2(g, g)$$

and thus, due to (5.10), we end up with (5.11). □

5.2.3 Symplectic and Poisson Reformulation, Faithful Representations, Induced Isomorphisms

We recall for the reader the following elementary definitions.

Definition 5.2.10 (*Symplectic vector space*) A (**real**) **symplectic form** over the real vector space V is a *bilinear, antisymmetric* map $\tau : V \times V \to \mathbb{R}$. τ is said to be **weakly non-degenerate** if $\tau(x, y) = 0$ for all $x \in V$ implies $y = 0$. In this case (V, τ) is said to be a (**real**) **symplectic vector space**.

Next, we would like to define a Poisson vector space. In the finite dimensional case, it is simply a pair (V, Π), where V is a real vector space and $\Pi \in \Lambda^2 V$, which is the same as being a bilinear, antisymmetric form on the (algebraic) linear dual V^*. However, in our cases of interest, V is infinite dimensional and Π belongs to a larger space than $\Lambda^2 V$, that could be defined using linear duality. Constructions involving linear duality necessarily bring into play the topological structure on V (or lack thereof). We will not enter topological questions in detail, so we content ourselves with a formal notion of duality, which will be sufficient for our purposes.

Definition 5.2.11 (*Poisson vector space*) Two real vector spaces V and W, together with a bilinear pairing $\langle \cdot, \cdot \rangle : W \times V \to \mathbb{R}$, are in **formal duality** when the bilinear pairing is non-degenerate in either argument ($\langle x, y \rangle = 0$ implies $x = 0$ if it holds for all $y \in V$, and it implies $y = 0$ if it holds for all $x \in W$). Given such V and W in formal duality, we call $(V, \Pi, W, \langle \cdot, \cdot \rangle)$ a (**real**) **Poisson vector space** if $\Pi : W \times W \to \mathbb{R}$ is a bilinear, antisymmetric map, called the **Poisson bivector**. Π is said to be **weakly non-degenerate** if $\Pi(x, y) = 0$ for all $x \in W$ implies $y = 0$.

At this level, there are only subtle differences between symplectic and Poisson vector spaces. In fact, the two structures have often been confounded in the literature on QFT on curved spacetime [1, 15, 17, 18, 25, 31]. The differences become more pronounced when we consider symplectic differential forms and Poisson bivector fields on manifolds locally modeled on the vector space V. A form is a section of an antisymmetric power of the cotangent bundle, while a bivector field is a section of an antisymmetric power of the tangent bundle. In infinite dimensional settings, one has to choose a precise notion of tangent and cotangent bundle, among several inequivalent possibilities. This ambiguity is reflected in our need to introduce formal duality for the definition of a Poisson vector space.

The above abstract definitions are concretely realized in the Proposition that we present below. Let us use the formula on the right-hand side of (5.4) to define a bilinear, antisymmetric map $\tau : \mathsf{Sol} \times \mathsf{Sol} \to \mathbb{R}$ by

$$\tau(\psi, \xi) = \int_\Sigma (\psi \nabla_n \xi - \xi \nabla_n \psi) \, d\Sigma. \tag{5.12}$$

Defining the space of equivalence classes $\mathcal{E} = C_0^\infty(\mathcal{M})/(\Box_M + m^2 + \xi R) C_0^\infty(\mathcal{M})$ and recalling Eqs. (5.1) and (5.3), the advanced-minus-retarded fundamental solution defines a bilinear, antisymmetric map $E : \mathcal{E} \times \mathcal{E} \to \mathbb{R}$ by

$$E([f],[g]) = E(f,g). \tag{5.13}$$

Furthermore, there is a well-defined bilinear pairing $\langle\cdot,\cdot\rangle : \mathcal{E} \times \mathsf{Sol} \to \mathbb{R}$ given by

$$\langle [f], \psi \rangle = \int_{\mathcal{M}} f\psi \, \mathrm{dvol}_{\boldsymbol{M}}. \tag{5.14}$$

Given the above definitions for the Klein-Gordon real scalar field, we have the following.

Proposition 5.2.12 *The spaces* Sol *and* $\mathcal{E}$ *are in formal duality, with respect to the pairing* $\langle\cdot,\cdot\rangle$. *The pair* (Sol, τ) *is a symplectic vector space, while* $(\mathsf{Sol}, E, \mathcal{E}, \langle\cdot,\cdot\rangle)$ *is a Poisson vector space. Moreover, the bilinear forms* τ *and* E *respectively induce linear maps*

$$\tau \colon \mathsf{Sol} \to \mathcal{E} \quad \text{and} \quad E \colon \mathcal{E} \to \mathsf{Sol} \tag{5.15}$$

that are bijective, mutually inverse and such that $\tau(\psi,\xi) = \langle \tau\psi, \xi\rangle$ *and* $E([f],[g]) = \langle [f], E[g]\rangle$.

Proof The content of this proposition is discussed in detail in [42, Sect. 5] or [43, Sect. 3], though a basic version can be found already in [65, Sect. 3.2]. We only indicate a few salient points. The non-degeneracy of the pairing $\langle\cdot,\cdot\rangle$ implies, provided there exist linear operators τ and E such that $E([f],[g]) = \langle [f], E[g]\rangle$ and $\tau(\psi,\xi) = \langle \tau\psi, \xi\rangle$, that they are unique. These operators can be exhibited rather concretely. The linear map $E \colon \mathcal{E} \to \mathsf{Sol}$ is already defined by Remark 5.2.2(**1**), in view of Eq. (5.3). The definition of τ in (5.12) is independent of the choice of Cauchy surface $\Sigma \subset \mathcal{M}$. Let $\Sigma^+, \Sigma^- \subset \mathcal{M}$ be Cauchy surfaces, respectively to the future and to the past of the Cauchy surface Σ, and let $\chi \in C^\infty(\mathcal{M})$ be such that $\chi \equiv 0$ to the past of Σ^- and $\chi \equiv 1$ to the future of Σ^+. Then, we have the identity $\tau\psi = [(\Box_{\boldsymbol{M}} + m^2 + \xi R)(\chi\psi)]$. Finally, the non-degeneracy or bijectivity of $\langle\cdot,\cdot\rangle$, τ and E, considered either as bilinear forms or linear operators, strongly rely on the hyperbolic character and well-posedness of the Klein-Gordon equation. □

Given the isomorphism between $\mathcal{E}$ and Sol and the close relationship between E and τ, it is not surprising these two spaces and bilinear forms have often been used interchangeably in the context of the QFT of the Klein-Gordon real scalar field. However, this interchangeability may fail for more complicated field theories, as we remark next. This is another reason why it is important to keep track of the difference between the respective symplectic and Poisson vector spaces, (Sol, τ) and (Sol, E)!

Remark 5.2.13 References [42, Sect. 5] and [43, Sect. 3] also address in detail the question of whether similar statements hold for gauge theories (electrodynamics, linearized gravity, etc.) or for theories with constraints (massive vector field, etc.). Related questions were also studied in [31]. The answer turns out to be rather subtle. The bilinear forms τ and E can essentially always be defined. A reasonable choice of the spaces $\mathcal{E}$ and Sol also make sure that the linear maps $\tau \colon \mathsf{Sol} \to \mathcal{E}$ and

$E\colon \mathcal{E} \to \mathsf{Sol}$ are also well-defined and are mutually inverse. However, the pairing $\langle\cdot,\cdot\rangle$ appearing in the formulas $\tau(\psi,\xi) = \langle\tau\psi,\xi\rangle$ and $E([f],[g]) = \langle[f], E[g]\rangle$, need no longer be non-degenerate. Hence, the bilinear forms τ and E may be degenerate themselves. The conditions under which these degeneracies do or do not occur subtly depend on the geometry of the gauge transformations and the constraints of the theory.

We now turn to applying the above symplectic and Poisson structures to the study of the properties of the CCR algebra $\mathcal{A}(\boldsymbol{M})$ of a Klein-Gordon field.

Definition 5.2.14 (*CCR algebra of a Poisson vector space*) Let $(V, \Pi, W, \langle\cdot,\cdot\rangle)$ be a Poisson vector space, defined with respect to a formal duality between V and another space W. The corresponding **CCR algebra** $\mathcal{A}(V, \Pi, W, \langle\cdot,\cdot\rangle)$ is defined as the unital $*$-algebra presented by the generators $A(x)$, $x \in W$, subject to the relations $A(ax + by) - aA(x) - bA(y) = 0$, $A(x)^* - A(x) = 0$ and $[A(x), A(y)] - i\Pi(x, y)\mathbf{1} = 0$, for any $a, b \in \mathbb{R}$ and $x, y \in W$.

This generic definition allows us to state and prove the following useful result.

Proposition 5.2.15 (Simplicity and faithfulness) *Given that the spaces V and W are in formal duality and the Poisson bivector of the Poisson vector space $(V, \Pi, W, \langle\cdot,\cdot\rangle)$ is weakly non-degenerate (as a bilinear form on W), the corresponding CCR algebra $\mathcal{A}(V, \Pi, W, \langle\cdot,\cdot\rangle)$ is simple. Further, it admits only zero or faithful representations.*

Before giving the proof, we note its main consequence. It is not hard to see that the definition of the CCR algebra $\mathcal{A}(\boldsymbol{M})$, as given in Definition 5.2.1, coincides with the alternative definition $\mathcal{A}(\boldsymbol{M}) \stackrel{\text{def}}{=} \mathcal{A}(\mathsf{Sol}, E)$, using the notation of Proposition 5.2.12 and referring to the formal duality between Sol and $\mathcal{E}$. The explicit homomorphism acts on the generators as $\phi(f) \mapsto A([f])$. Thus, given Proposition 5.2.12, we have the immediate

Corollary 5.2.16 *The CCR algebra $\mathcal{A}(\boldsymbol{M})$ of a real scalar quantum field is simple and admits only either zero or faithful representations.*

Remark 5.2.17 The result established in the Corollary above is *not* valid form more complicated QFTs like *electromagnetism* [59] and *linearized gravity* [16]. The physical reason is the appearance of the *gauge invariance*. Mathematically it is related to the fact that the Poisson bivector corresponding to our E is *degenerate* on the space $\mathcal{E}$ of compactly supported observables, as discussed in [42, Sect. 5] and [43, Sect. 3].

The proof of Proposition 5.2.15 makes use of the following two lemmas.

Lemma 5.2.18 *Let Π be a bilinear form (we need not even assume it to be antisymmetric) on a vector space W. Further, let $v_i \in W$, $i = 1, \dots, N$, be a set of linearly independent vectors and $c^{i_1\cdots i_k}$ a collection of scalars, not all zero, with each index running through $i_j = 1, \dots, N$. Then, if*

$$\sum_{i_1,\ldots,i_k} c^{i_1\cdots i_k} \Pi(v_{i_1}, u_1) \cdots \Pi(v_{i_k}, u_k) = 0 \tag{5.16}$$

for each set of vectors $u_i \in W$, $i = 1, \ldots, k$. Then there exists a non-zero vector $w \in W$ such that $\Pi(w, u) = 0$ for any $u \in W$.

Proof The proof is by induction on k. Let $k = 1$, then the right-hand side of the equation in the hypothesis is $\Pi(w', u_1)$, where

$$w' = \sum_i c^i v_i. \tag{5.17}$$

Since not all c^i are zero and the v_i, $i = 1, \ldots, N$ are linearly independent, we have $w' \neq 0$. We can then set $w = w'$ and we are done, since u_1 can be arbitrary.

Now, assume that the case $k - 1$ has already been established. Note that we can write the right-hand side of the above equation as $\Pi(w', u_k)$, where

$$w' = \sum_{i_1,\ldots,i_k} c^{i_1\cdots i_k} \Pi(v_{i_1}, u_1) \cdots \Pi(v_{i_{k-1}}, u_{k-1}) v_{i_k}. \tag{5.18}$$

If $w' \neq 0$ for some choice of $u_i \in W$, $i = 1, \ldots, k - 1$, then we can set $w = w'$ and we are done, since u_k can be arbitrary.

Consider the case when $w' = 0$ for all $u_i \in W$, $i = 1, \ldots, k - 1$. Then, choose j_k such that $c^{i_1\cdots i_{k-1} j_k}$ are not all zero. Since, by linear independence, the coefficients of the v_{i_k} in w' must vanish independently, we have

$$\sum_{i_1,\ldots,i_{k-1}} c^{i_1\cdots i_{k-1} j_k} \Pi(v_{i_1}, u_1) \cdots \Pi(v_{i_{k-1}}, u_{k-1}) = 0 \tag{5.19}$$

for all $u_i \in W$, $i = 1, \ldots, k - 1$. In other words, by the inductive hypothesis, the last equality implies the existence of the desired non-zero $w \in W$, which concludes the proof. □

A bilinear form Π on W naturally defines a bilinear form $\Pi^{\otimes k}$ on the k-fold tensor product $W^{\otimes k}$. Let $S \colon W^{\otimes k} \to W^{\otimes k}$ denote the (idempotent) full symmetrization operator and denote its image, the space of fully symmetric k-tensors, by $S^k W \stackrel{\text{def}}{=} S(W^{\otimes k})$. Of course, $\Pi^{\otimes k}$ also restricts to $S^k W$. If Π is antisymmetric, then $\Pi^{\otimes k}$ is symmetric when k is even and antisymmetric when k is odd.

Lemma 5.2.19 *If the antisymmetric bilinear form Π is weakly non-degenerate on W, then the antisymmetric bilinear form $\Pi^{\otimes k}$ is weakly non-degenerate on $S^k W$.*

Proof Assume the contrary, that $\Pi^{\otimes k}$ is degenerate. By its (anti-)symmetry, we need only consider the degeneracy in its first argument. That is, there exists a vector $v = \sum_{i_1,\ldots i_k} d^{i_1\cdots i_k} v_{i_1} \otimes \cdots \otimes v_{i_k}$, where $v_i \in W$, $i = 1, \ldots, N$, constitute a linearly independent set and the $d^{i_1\cdots i_k}$ coefficients are not all zero and are symmetric under index interchange, such that

$$\Pi^{\otimes k}(v, S(u_1 \otimes \cdots \otimes u_k)) = 0. \tag{5.20}$$

for any $u_i \in W$, $i = 1, \ldots, k$. But then, the above equality is precisely of the form of the hypothesis of Lemma 5.2.18, with

$$c^{i_1\cdots i_k} = k!\, d^{i_1\cdots i_k}, \tag{5.21}$$

due to the symmetry of $d^{i_1\cdots i_k}$ under index interchanges. Therefore, by Lemma 5.2.18, there must exist a $w \in W$ such that $\Pi(w, u) = 0$ for all $u \in W$, which contradicts the weak non-degeneracy of Π on W. Therefore, $\Pi^{\otimes k}$ cannot be degenerate on $S^k W$, and hence is weakly non-degenerate. □

Proof (of Proposition 5.2.15*)* Suppose that $\mathcal{A}(V, \Pi, W, \langle\cdot,\cdot\rangle)$ is not simple, and so has a non-trivial two-sided ideal $\mathcal{I}$. If we can deduce that $\mathbf{1} \in \mathcal{I}$, then any non-trivial two-sided ideal must be all of $\mathcal{A}(V, \Pi, W, \langle\cdot,\cdot\rangle)$, implying that the algebra is simple.

Take any non-zero element $a \in \mathcal{I}$ and recall the idea behind Eq. (5.7). That is, there exists integers $k, N \geq 0$, linearly independent elements $v_i \in W$, $i = 1, \ldots, N$, and complex coefficients $c_{(l)}^{i_1\cdots i_l}$, $i_j = 1, \ldots, N$ and $l = 0, \ldots, k$, such that

$$a = c_{(0)}\mathbf{1} + \sum_{i_1} c_{(1)}^{i_1} A(v_{i_1}) + \sum_{i_1,i_2} c_{(2)}^{i_1 i_2} A(v_{i_1})A(v_{i_2}) + \cdots + \sum_{i_1,\ldots,i_k} c_{(k)}^{i_1\ldots i_k} A(v_{i_1})\cdots A(v_{i_k}), \tag{5.22}$$

where not all of the components of $c_{(k)}^{i_1\ldots i_k}$ are zero. If $k = 0$, the $\mathbf{1} \in \mathcal{I}$ and we are done. If $k > 0$, note that $\mathcal{I}$ also contains the iterated commutator $[\cdots[a, A(u_1)], \ldots, A(u_k)]$, for any $u_i \in W$, $i = 1, \ldots, k$. A straight forward calculation shows that, up to (non-zero) numerical factors, the iterated commutator is equal to

$$\Pi^{\otimes k}\left(\sum_{i_1,\ldots,i_k} c_{(k)}^{i_1,\ldots,i_k} S(v_{i_1} \otimes \cdots \otimes v_{i_k}), S(u_1 \otimes \cdots \otimes u_k)\right)\mathbf{1}. \tag{5.23}$$

By Lemma 5.2.19, since Π is weakly non-degenerate on W, $\Pi^{\otimes k}$ is weakly non-degenerate on $S^k W$. Since elements of the form $S(u_1 \otimes \cdots \otimes u_k)$ generate $S^k W$, there must exist at least one element of $S^k W$ of that form such that the coefficient in front of $\mathbf{1}$ in (5.23) is non-zero. Therefore, $\mathbf{1} \in \mathcal{I}$ and we are done. □

Automorphisms of the CCR algebra $\mathcal{A}(\boldsymbol{M})$ are important because the composition of a state with an automorphism gives a way to define more states, once at least one is known. The identity $\mathcal{A}(\boldsymbol{M}) \cong \mathcal{A}(\mathsf{Sol}, E)$ allows us to construct lots of automorphisms of $\mathcal{A}(\boldsymbol{M})$, induced by transformations of Sol or $\mathcal{E}$ that, respectively, leave τ or E invariant.

Proposition 5.2.20 (Induced homomorphism) *Let $\mathcal{A}(V, \Pi, W, \langle\cdot,\cdot\rangle)$ be as in Definition 5.2.14 and let $\sigma \colon W \to W$ be a linear map such that*

$$\Pi(\sigma x, \sigma y) = \Pi(x, y) \ \ (\textit{resp. } \Pi(\sigma x, \sigma y) = -\Pi(x, y)), \tag{5.24}$$

for all $f, g \in W$. Then, there exists a homomorphism (resp. anti-linear homomorphism) of unital $*$-algebras, $\alpha^{(\sigma)} \colon \mathcal{A}(V, \Pi, W, \langle\cdot,\cdot\rangle) \to \mathcal{A}(V, \Pi, W, \langle\cdot,\cdot\rangle)$ uniquely defined by its values

$$\alpha^{(\sigma)}(A(x)) \stackrel{def}{=} A(\sigma x), \tag{5.25}$$

for each $x \in W$, on the generators of $\mathcal{A}(V, \Pi, W, \langle\cdot,\cdot\rangle)$. Also, if σ is bijective, then $\alpha^{(\sigma)}$ is an automorphism.

Remark 5.2.21 In view of Proposition 5.2.12 and the isomorphism $\mathcal{E} \cong \mathsf{Sol}$, in the case of the CCR algebra $\mathcal{A}(\boldsymbol{M}) \cong \mathcal{A}(\mathsf{Sol}, E)$ of a real scalar quantum field, the linear endomorphisms of $\mathcal{E}$ that preserve the Poisson bivector E can be equivalently specified by linear endomorphisms of Sol that preserve the symplectic form τ.

Proof Recall the definition of an algebra presented by generators and relations by its universal property, as discussed in Sect. 5.1.2, as well as such a presentation of the algebra $\mathcal{A}(V, \Pi, W, \langle\cdot,\cdot\rangle)$ given in Definition 5.2.14.

Let us denote by $\mathcal{A}(W)$ the algebra freely generated by the elements of the vector space W. Following our notation, the map embedding the generators in this algebra can be denoted as $A \colon W \to \mathcal{A}(W)$. The composition $A \circ \sigma$ is another such map. Therefore, by the universal property, there exists a unique homomorphism $\beta \colon \mathcal{A}(W) \to \mathcal{A}(W)$ such that $\beta(A(x)) = A(\sigma x)$, for all $x \in W$, and $\beta(\mathbf{1}) = \mathbf{1}$.

We now need to check whether β leaves invariant the kernel of the projection $\mathcal{A}(W) \to \mathcal{A}(V, \Pi, W, \langle\cdot,\cdot\rangle)$. This kernel is the two-sided ideal generated by the relations $A(ax + by) - aA(x) - bA(y) = 0$, $A(x)^* - A(x) = 0$ and $[A(x), A(y)] - i\Pi(x, y)\mathbf{1} = 0$, for any $a, b \in \mathbb{R}$ and $x, y \in W$, so it is sufficient to check the invariance of these relations. The first two are obviously invariant. The last commutator identity is invariant upon invoking the hypothesis that σ preserves Π, up to sign. We deal with the two cases separately.

In the case when σ preserves Π, we have

$$[A(\sigma x), A(\sigma y)] - i\Pi(x, y)\mathbf{1} = [A(\sigma x), A(\sigma y)] - i\Pi(\sigma x, \sigma y)\mathbf{1}. \tag{5.26}$$

Hence, the homomorphism β induces a uniquely defined homomorphism on the quotiented algebra, which we call $\alpha^{(\sigma)} \colon \mathcal{A}(V, \Pi, W, \langle\cdot,\cdot\rangle) \to \mathcal{A}(V, \Pi, W, \langle\cdot,\cdot\rangle)$, which given by $\alpha^{(\sigma)}([a]) = [\beta a]$, and which has all the desired properties.

In the case when σ changes the sign of Π, we need to change perspective slightly. Recall that we defined $\mathcal{A}(\boldsymbol{M})$ as a complex algebra, which then automatically has the structure of a real algebra. Equivalently, we could have also defined it directly as a real algebra, by throwing in an extra generator i, satisfying the relations $i^2 = -\mathbf{1}$, $[i, \mathbf{1}] = [i, A(x)] = 0$ and $i^* = -i$. If the homomorphism β is extended to this generator as $\beta(i) = -i$, then it preserves the new relations that need to be satisfied by i and also the commutator identity, since

$$[A(\sigma x), A(\sigma y)] - (-i)\Pi(x, y)\mathbf{1} = [A(\sigma x), A(\sigma y)] - i\Pi(\sigma x, \sigma y)\mathbf{1}. \quad (5.27)$$

Hence, the real algebra homomorphism β induces a uniquely defined homomorphism on the quotiented algebra, which also happens to be an anti-linear homomorphism in the sense of complex algebras, which we call $\alpha^{(\sigma)}: \mathcal{A}(V, \Pi, W, \langle\cdot,\cdot\rangle) \to \mathcal{A}(V, \Pi, W, \langle\cdot,\cdot\rangle)$, and which has all the desired properties.

Finally, when σ is a bijection, we can use the universal property of $\mathcal{A}(V, \Pi, W, \langle\cdot,\cdot\rangle)$, as was done in Sect. 5.1.2, to show that $\alpha^{(\sigma^{-1})} = (\alpha^{(\sigma)})^{-1}$. Therefore, $\alpha^{(\sigma)}$ is an isomorphism and hence an automorphism of the algebra. □

We end this section by noting that there is another structure that is induced on the space $\mathcal{E} \cong C_0^\infty(\mathcal{M})/Ker(E)$ in the presence of a state ω on $\mathcal{A}(\boldsymbol{M})$, namely the symmetrized part of the 2-point function

$$\omega_2^S(f, g) \stackrel{\text{def}}{=} \frac{1}{2}(\omega_2(f, g) + \omega_2(g, f)), \quad (5.28)$$

with $f, g \in C_0^\infty(\mathcal{M})$ and $\omega_2(f, g) \stackrel{\text{def}}{=} \omega(\phi(f)\phi(g))$. By hermiticity, the symmetrized 2-point function is always real and non-negative, which was essentially already noted in (5.9) and (5.10) of Proposition 5.2.9. Also, by Proposition 5.2.4, $\phi(f)$ depends only on the equivalence class $[f] \in \mathcal{E}$. Hence, $\omega_2^S: \mathcal{E} \times \mathcal{E} \to \mathbb{R}$ defines a real symmetric bilinear form. Finally, the inequality (5.11) from Proposition 5.2.9, which we can rewrite as

$$\frac{1}{4}|E([f], [g])|^2 \leq \omega_2^S([f], [f])\omega_2^S([g], [g]), \quad (5.29)$$

shows that ω_2^S is non-degenerate on $\mathcal{E}$, since it majorizes E, which is already known to be non-degenerate by Proposition 5.2.12. Thus, a state ω on $\mathcal{A}(\boldsymbol{M})$ induces a positive scalar product ω_2^S on $\mathcal{E}$ (and also on Sol by the isomorphism of Proposition 5.2.12). We will use this scalar product structure and the inequality (5.29) to construct quasifree states in the next section.

5.2.4 Quasifree States, Also Known as Gaussian States

There is a plethora of states on $\mathcal{A}(\boldsymbol{M})$, the first class we consider is that of the *quasifree* or *Gaussian states*. They mimic the Fock representation of Minkowski vacuum and they are completely determined from the two-point function by means of a prescription generalizing the well known Wick procedure which also guarantees essential self-adjointness of the field operators $\hat{\phi}_\omega$ since they are regular (Definition 5.2.8).

Definition 5.2.22 (*Quasifree states*) An algebraic state $\omega : \mathcal{A}(\boldsymbol{M}) \to \mathbb{C}$ is said to be **quasifree** or **Gaussian** if its n-point functions agree with the so-called **Wick procedure**, in other words they satisfy the following pair of requirements for all choices of $f_k \in C_0^\infty(\mathcal{M})$,

(a) $\omega_n(f_1, \ldots, f_n) = 0, \quad \text{for } n = 1, 3, 5, \ldots$

(b) $\omega_n(f_1, \ldots, f_n) = \sum_{\text{partitions}} \omega_2(f_{i_1}, f_{i_2}) \cdots \omega_2(f_{i_{n-1}}, f_{i_n}), \quad \text{for } n = 2, 4, 6, \ldots$

For the case of n even, the *partitions* refers to the class of all possible decomposition of set $\{1, 2, \ldots, n\}$ into $n/2$ pairwise disjoint subsets of 2 elements

$$\{i_1, i_2\}, \{i_3, i_4\} \ldots \{i_{n-1}, i_n\}$$

with $i_{2k-1} < i_{2k}$ for $k = 1, 2, \ldots, n/2$.

We will prove in the next section that quasifree states exist in a generic curved spacetime for a massive scalar field and $\xi = 0$. Instead we intend to clarify here the structure of the GNS representation of quasifree states, proving that it is a Fock representation. The characterization theorem relies on the following intermediate result.

Proposition 5.2.23 (One-particle structure) *Consider the symplectic vector space* (Sol, τ), *cf. Proposition 5.2.12.*

(a) If a real scalar product $\mu : \mathsf{Sol} \times \mathsf{Sol} \to \mathbb{R}$ *satisfies*

$$\frac{1}{4}|\tau(x, y)|^2 \leq \mu(x, x)\mu(y, y) \quad \forall x, y \in \mathsf{Sol} \tag{5.30}$$

then there exists a pair (K, H), *called* **one-particle structure** *associated to* $(\mathsf{Sol}, \tau, \mu)$ *where* H *is a complex Hilbert space and* $K : \mathsf{Sol} \to H$ *is a map satisfying*

(i) K *is* $\mathbb{R}$ *linear and* $K(\mathsf{Sol}) + iK(\mathsf{Sol})$ *is dense in* H *(though* $K(\mathsf{Sol})$, *as a real subspace of* H, *need not be dense by itself),*

(ii) $\langle Kx|Ky\rangle = \mu(x, y) + \frac{i}{2}\tau(x, y)$ *for all* $x, y \in \mathsf{Sol}$.

(b) If (K', H') *satisfies (i) in (a) and* $\tau(x, y) = 2Im(\langle K'x|K'y\rangle_{H'})$, *then the scalar product* μ *on* Sol *obtained from (ii) in (a) also satisfies (5.30).*

(c) A pair (H', K') *satisfies (i) and (ii) in (a) if and only if there is an isometric surjective operator* $V : H \to H'$ *with* $VK = K'$.

Proof Barring different conventions on signs the proof is given in Proposition 3.1 in [41]. □

A characterization theorem for quasifree states can now be proved using the lemma above with the following theorem that can be obtained by Lemma A.2, Proposition 3.1 and a comment on p. 77 in [41] (again modulo different conventions on signs) where the approach based on Weyl C^*-algebras is pursued. For quasifree states the approaches relying on CCR $*$-algebras and Weyl C^*-algebras are technically equivalent. The fact that, on a Fock space, an operator as the one in (5.33) is *essentially self-adjoint* in the indicated domain [8] is well know and can be proved directly, for instance, using analytic vectors.

Theorem 5.2.24 (Characterization of quasifree states) *Consider the $*$-algebra $\mathcal{A}(\boldsymbol{M})$ associated to a real scalar KG field. Suppose that μ is a real scalar product on* Sol *which verifies* (5.30). *The following hold.*
(a) There exists a quasifree state ω on $\mathcal{A}(\boldsymbol{M})$ such that

$$\omega_2(f, g) = \mu(Ef, Eg) + \frac{i}{2}E(f, g)\,, \quad \forall f, g \in C_0^\infty(\mathcal{M})\,. \tag{5.31}$$

(b) The GNS triple $(\mathcal{H}_\omega, \mathcal{D}_\omega, \pi_\omega, \Psi_\omega)$ consists of the following:
(i) $\mathcal{H}_\omega$ is the bosonic (symmetrized) Fock space with the one-particle subspace being H, of the one structure particle (K, H) in Proposition 5.2.23;
(ii) Ψ_ω is the vacuum vector of the Fock space;
(iii) $\mathcal{D}_\omega$ is the dense subspace of the finite complex linear combinations of Ψ_ω and all of the vectors

$$a^\dagger(\psi_1)\cdots a^\dagger(\psi_n)\Psi_\omega \quad for\ n = 1, 2, \dots \ and\ \psi_k \in \mathsf{Sol} \tag{5.32}$$

where $a^*(\psi)$ is the standard creation operator[3] *corresponding to the solution $\psi \in$* Sol.
(iv) π_ω is completely determined by a and $a^\dagger$, with $a(\psi)$ being the annihilation operator corresponding to the solution $\psi \in$ Sol,

$$\hat{\phi}_\omega(f) = \pi_\omega(\phi(f)) = a(Ef) + a^\dagger(Ef) \ \ \forall f \in C_0^\infty(\mathcal{M})\,, \tag{5.33}$$

and, in particular, ω is regular, meaning that $\hat{\phi}_\omega(f)$ is essentially self-adjoint on $\mathcal{D}_\omega$.
*(d) The quasifree state ω determined by μ is pure if and only if the image K(*Sol*) is dense in the one-particle subspace H, thus strengthening (i) of Proposition 5.2.23. This condition is equivalent to:*

$$\mu(\psi, \psi) = \frac{1}{4}\sup_{\xi\neq 0}\frac{|\tau(\psi, \xi)|^2}{\mu(\xi, \xi)}\,. \tag{5.34}$$

[3]It holds that $[a(\psi), a^\dagger(\xi)] = \langle K\psi|K\xi\rangle$, $[a(\psi), a(\xi)] = 0 = [a^\dagger(\psi), a^\dagger(\xi)]$ if $\xi, \psi \in$ Sol, and $a(\xi), a(\psi)$ are defined on $\mathcal{D}_\omega$, with $a^\dagger(\psi) = a(\psi)^\dagger|_{\mathcal{D}_\omega}$.

Remark 5.2.25

(1) K is always injective because of (ii) in Proposition 5.2.23, since τ is non-degenerate.

(2) The requirement (5.30) is equivalent to saying that there is a bounded operator J everywhere defined in the real Hilbert space obtained by taking the completion $\mathcal{R}$ of Sol with respect to the real scalar product induced by μ, such that $\frac{1}{2}\tau(\psi, \xi) = \omega_2(\psi, J\xi)$, for $\psi, \xi \in \mathsf{Sol}$, and $||J|| \leq 1$. It also holds that $J^\dagger = -J$. It is not so difficult to prove that the corresponding state ω, as defined above, is pure if and only if $JJ = -I$, that is J is anti unitary. In this case $(\mathcal{R}, \mu, \frac{1}{2}\tau, J)$ defines an **almost Kähler structure** on $\mathcal{R}$.

5.2.5 Existence of Quasifree States in Globally Hyperbolic Spacetimes

In four-dimensional Minkowski spacetime $M \stackrel{\text{def}}{=} \mathbb{M}$, a distinguished real scalar product μ on $\mathsf{Sol} \cong C_0^\infty(\mathbb{M})/Ker(E)$[4] can easily be defined as follows in a Minkowski reference frame with coordinates $(t, \mathbf{x}) \in \mathbb{R} \times \mathbb{R}^3$. Consider $f \in C_0^\infty(\mathbb{M})$ and the associated solution of KG equation $\psi_f \stackrel{\text{def}}{=} Ef$:

$$\psi_f(t, \mathbf{x}) = \int_{\mathbb{R}^3} \frac{\phi_f(\mathbf{k})e^{i\mathbf{x}\cdot\mathbf{k}-itE(\mathbf{k})} + \overline{\phi_f(\mathbf{k})}e^{-(i\mathbf{x}\cdot\mathbf{k}-itE(\mathbf{k}))}}{(2\pi)^{3/2}\sqrt{2E(\mathbf{k})}} d\mathbf{k} \tag{5.35}$$

where $E(\mathbf{k}) \stackrel{\text{def}}{=} \sqrt{\mathbf{k}^2 + m^2}$ (we assume here $m > 0$) and $\phi_f \in \mathcal{S}(\mathbb{R}^3)$ (the Schwartz test function space) is obtained by the smooth compactly supported Cauchy data of ψ_f on the Cauchy surface defined by $t = 0$. If defining

$$\mu_{\mathbb{M}}([f], [f']) \stackrel{\text{def}}{=} Re \int_{\mathbb{R}^3} \overline{\phi_f(\mathbf{k})}\phi_{f'}(\mathbf{k})d\mathbf{k} \tag{5.36}$$

we obtain a well defined real scalar product on $\mathsf{Sol} \cong \mathcal{E}$ which satisfies (5.29) as can be proved by direct inspection with elementary computations. The arising quasifree state $\omega_{\mathbb{M}}$ is nothing but the **Minkowski vacuum** and we find the standard QFT free theory for a real scalar field in Minkowski spacetime. The integral kernel of $\omega_{\mathbb{M}}$ in this case is a proper distribution of $\mathcal{D}'(\mathbb{R}^4 \times \mathbb{R}^4)$ and reads

$$\omega_{\mathbb{M}2}(x, y) = w\text{-}\lim_{\varepsilon\to 0^+} \frac{m^2}{(2\pi)^2} \frac{K_1\left(m\sqrt{(|\mathbf{x}-\mathbf{y}|^2 - (t_x - t_y - i\varepsilon)^2)}\right)}{m\sqrt{|\mathbf{x}-\mathbf{y}|^2 - (t_x - t_y - i\varepsilon)^2}} \tag{5.37}$$

[4]Recall that this isomorphism was established in Proposition 5.2.12, based on the well-posedness properties of the Klein-Gordon equation. From now one, we will be making use of this isomorphism implicitly.

where the *weak* limit is understood in the standard distributional sense and the branch cut in the complex plane to uniquely define the analytic functions appearing in (5.37) is assumed to stay along the negative real axis. Another equivalent expression for $\omega_{\mathbb{M}2}$ is given in terms of Fourier transformation of distributions,

$$\omega_{\mathbb{M}2}(x, y) = \frac{1}{(2\pi)^3} \int_{\mathbb{R}^4} e^{-ip(x-y)} \theta(p^0)\delta(p^2 + m^2) d^4 p \, . \tag{5.38}$$

where $px = p^0 x^0 - \sum_{j=1}^{3} p^j x^j$ is the Minkowski scalar product. The above formula is convenient for showing the following important property of $\omega_{\mathbb{M}2}$.

Proposition 5.2.26 *If $f \in C_0^\infty(\mathcal{M})$, then $\omega_{\mathbb{M}2}(x, f)$ and $\omega_{\mathbb{M}2}(f, y)$ are smooth.*

Proof Let $\hat{f}(p) = \int_{\mathbb{R}^4} e^{ipy} f(y)\, d^4 y$. Since $f \in C_0^\infty(\mathcal{M})$, $\hat{f}$ must be a Schwartz function. Then, since $\omega_{\mathbb{M}2}(f, y) = \overline{\omega_{\mathbb{M}2}(y, f)}$, it is enough to consider

$$\begin{aligned}
\omega_{\mathbb{M}2}(x, f) &= \frac{1}{(2\pi)^3} \int_{\mathbb{R}^4} d^4 p\, e^{-ipx} \theta(p^0)\delta(p^2 + m^2) \int_{\mathbb{R}^4} d^4 y\, f(y) e^{ipy} \\
&= \frac{1}{(2\pi)^3} \int_{\mathbb{R}^3} d\mathbf{k}\, e^{-ip_{\mathbf{k}} x} \frac{\hat{f}(p_{\mathbf{k}})}{\sqrt{\mathbf{k}^2 + m^2}} \, ,
\end{aligned}$$

where $p_{\mathbf{k}} = \left(\sqrt{\mathbf{k}^2 + m^2}, \mathbf{k}\right)$. Since $\hat{f}$ is Schwartz, so is the above integrand. It is then easy to see from this integral representation that $\omega_{\mathbb{M}2}(x, f)$ is smooth. □

In view of the definition of quasifree state Definition 5.2.22, all the n-point functions of $\omega_{\mathbb{M}}$ are distributions of $\mathcal{D}'((\mathbb{R}^4)^n)$. It turns out that the associated one-particle structure $(H_{\mathbb{M}}, K_{\mathbb{M}})$ is

$$H_{\mathbb{M}} = L^2(\mathbb{R}^3, d\mathbf{k}) \, , \quad K_{\mathbb{M}} : \mathsf{Sol} \ni \psi_f \mapsto \phi_f \in L^2(\mathbb{R}^3, d\mathbf{k}))$$

The condition in part (d) of Theorem 5.2.24 is true and thus Minkowski vacuum is pure. In spite of the Poincaré non-invariant approach, the pictured procedure leads to a Poincaré invariant structure as we shall see later.

More generally, a natural *pure quasifree* state ω_ζ exists as soon as the globally hyperbolic spacetime admits a time-like Killing field ζ, i.e., in *stationary* spacetimes, provided $m > 0$, $\xi = 0$ and it when holds $g(\zeta, \zeta) \geq c > 0$ uniformly on a smooth space-like Cauchy surface, for some constant c [65, Sect. 4.3]. In that case, a ζ-invariant Hermitian scalar product can be constructed out of a certain auxiliary Hermitian scalar product $(\psi_f|\psi_g)$ induced by the stress energy tensor

$$T_{ab}(\overline{\psi_f}, \psi_g) \stackrel{\text{def}}{=} \frac{1}{2}\left(\nabla_a \overline{\psi_f} \nabla_b \psi_g + \nabla_b \overline{\psi_f} \nabla_a \psi_g\right) - \frac{1}{2} g_{ab} \left(\nabla^c \overline{\psi_f} \nabla_c \psi_g - m^2 \overline{\psi_f} \psi_g\right)$$

evaluated on solutions $\psi_f, \psi_g \in \mathsf{Sol} + i\mathsf{Sol} \stackrel{\text{def}}{=} \mathsf{Sol}_{\mathbb{C}}$ of KG equation and contracted with ζ itself.

$$(\psi_f|\psi_g) \stackrel{\text{def}}{=} \int_\Sigma T^{ab}(\overline{\psi_f}, \psi_g) n_a \zeta_b \, d\Sigma .$$

This positive Hermitian form does not depend on the Cauchy surface Σ and is ζ-invariant in view of the Killing equation for ζ and $\nabla^a T_{ab}(\overline{\psi_f}, \psi_g) = 0$ which holds as a consequence of KG equations for ψ_f and ψ_g. This Hermitian scalar product gives rise to a complex Hilbert space $\mathcal{H}_0$ obtained by taking the completion of $\mathsf{Sol}_\mathbb{C}$. It turns out that the time evolution generated by ζ in $\mathsf{Sol}_\mathbb{C}$ is implemented by a strongly continuous unitary group on $\mathcal{H}_0$, with self-adjoint generator H. The spectrum of H is bounded away from zero and thus E^{-1} exists as a bounded, everywhere defined, operator on $\mathcal{H}_0$. Let $\mathcal{H}_0^+$ be the positive spectral closed subspace of h and let $P_+ : \mathcal{H}_0 \to \mathcal{H}_0^+$ be the corresponding orthogonal projector. The distinguished scalar product defining the quasifree pure (because (5.34) holds) state ω_ζ is finally defined by means of the real scalar product

$$\mu_\zeta(\psi, \psi') \stackrel{\text{def}}{=} Re\left(P_+\psi \,\middle|\, \frac{1}{2} H^{-1} P_+\psi' \right) \quad \psi, \psi' \in \mathsf{Sol} . \tag{5.39}$$

This procedure can be viewed as a rigorous version of the popular one based of positive frequency mode decomposition with respect to the notion of time associated to ζ in particular because it can easily be proved that

$$H\psi = i\zeta^a \partial_a \psi$$

when $\psi \in \mathsf{Sol}_\mathbb{C}$. Therefore $P_+\psi$ entering the right hand side of (5.39) contains "positive frequencies" only, since P_+ project on the positive part of the spectrum of the energy H. The state ω_ζ coincides with the Minkowski vacuum in Minkowski spacetime when $\zeta = \partial_t$ with respect to any Minkowski coordinate system. This result has a well-known [65] important consequence.

Theorem 5.2.27 (Existence of quasifree states) *Consider a globally hyperbolic spacetime* $\boldsymbol{M}$ *and assume that* $\xi = 0$ *and* $m > 0$ *in the definition of* $\mathcal{A}(\boldsymbol{M})$. *There exist quasifree states on* $\mathcal{A}(\boldsymbol{M})$.

Sketch of proof Take a smooth space-like Cauchy surface $\Sigma \subset \mathcal{M}$. It is always possible to smoothly deform $\mathcal{M}$ in the past of Σ obtaining an overall globally hyperbolic spacetime still admitting Σ as a Cauchy surface and such that the open past of Σ, $\boldsymbol{M}^-$, (in the deformed spacetime) has the following property. There is a second Cauchy surface Σ_1 in $\boldsymbol{M}^-$ whose open past $\boldsymbol{M}_1^-$ includes a smooth time-like Killing field ζ satisfying the sufficient requirements for defining and associate quasifree state ω_ζ on $\mathcal{A}(\boldsymbol{M}_1^-)$. However, if $\boldsymbol{M}^+$ denotes the open future of Σ (in the original spacetime), Propositions 5.2.4 and 5.2.6 easily imply that $\mathcal{A}(\boldsymbol{M}^+) = \mathcal{A}(\boldsymbol{M}^-) = \mathcal{A}(\boldsymbol{M}_1^-)$. Therefore ω_ζ is a state on $\mathcal{A}(\boldsymbol{M}^+) = \mathcal{A}(\boldsymbol{M})$. Again Propositions 5.2.4 and 5.2.6 and the very definition of quasifree state easily prove that ω_ζ is quasifree on $\mathcal{A}(\boldsymbol{M})$ if it is quasifree on $\mathcal{A}(\boldsymbol{M}_1^-)$. □

5.2.6 Unitarily Inequivalent Quasifree States Gravitationally Produced

Coming back to what already pronounced in (6) in Remark 5.1.14, we have the following definition.

Definition 5.2.28 Two states ω_1 and ω_2 on $\mathcal{A}(\boldsymbol{M})$ and the respective GNS representations are said to be **unitarily equivalent**[5] if there is an isometric surjective operator $U : \mathcal{H}_{\omega_1} \to \mathcal{H}_{\omega_2}$ such that $U\hat{\phi}_{\omega_1}(f)U^{-1} = \hat{\phi}_{\omega_2}(f)$ for every $f \in C_0^\infty(\mathcal{M})$.

Remark 5.2.29 Notice that it is not necessary that $U\Psi_{\omega_1} = \Psi_{\omega_2}$ and it generally does not happen. As a consequence $\mathcal{H}_{\omega_2}$ includes vector states *different from the Fock vacuum* which are however quasifree.

The question if a pair of states are unitarily equivalent naturally arises in the following situation. Consider a time-oriented globally hyperbolic spacetime $\boldsymbol{M}$ such that, in the future of a Cauchy surface Σ_+, the spacetime is stationary with respect to the Killing vector field ξ_+ and it is also stationary in the past of another Cauchy surface Σ_-, in the past of Σ_+, referring to another Killing vector field ξ_-. For instance we can suppose that $\boldsymbol{M}$ coincides to (a portion of) Minkowski spacetime in the two mentioned stationary regions and a gravitational curvature bump takes place between them. This way, two preferred quasifree states ω_+ and ω_- turn out to be defined on the whole algebra $\mathcal{A}(\boldsymbol{M})$, not only in the algebras of observables localized in the two respective static regions. The natural question is whether or not the GNS representations of ω_+ and ω_- are unitarily equivalent, so that, in particular, the state ω_- can be represented as a vector state $U\Psi_{\omega_-}$ in the Hilbert space $\mathcal{H}_{\omega_+}$ of the state ω_+. Notice that, even in the case the isometric surjective operator U exists making the representations unitarily equivalent, $U\Psi_{\omega_-} \neq \Psi_{\omega_+}$ in general, so that $U\Psi_{\omega_-}$ may have non-vanishing projection in the subspace containing states with n particles in $\mathcal{H}_{\omega_+}$. This phenomenon is physically interpreted as *creation of particles due to the gravitational field* and U has the natural interpretation of an *S matrix*.

The following crucial result holds for pure quasifree states [65]. A more general result appears in [64] since it avoids the assumption that the states are pure and it deals with the notion of *quasiequivalence* of quasifree states. Quasiequivalence is weaker notion of equivalence, which essentially corresponds to unitary equivalence "up to multiplicity" [8, Sect. 2.4.4]. In particular, quasiequivalence reduces to unitary equivalence for irreducible representations, as for instance those induced by pure states.

Theorem 5.2.30 (Unitary equivalence of pure quasifree states) *If $\boldsymbol{M}$ is a globally hyperbolic spacetime, consider two pure quasifree states ω_1 and ω_2 on $\mathcal{A}(\boldsymbol{M})$ respectively induced by the scalar product μ_1 and μ_2 on* $\mathsf{Sol}(\boldsymbol{M}) \cong \mathcal{E}$ *and indicate by $\mathcal{R}_{\mu_1}$ and $\mathcal{R}_{\mu_2}$ the real Hilbert spaces obtained by respectively completing* Sol.

[5] It should be evident that the given definition does not depend on the particular GNS representation chosen for each state ω_i.

The pure states ω_1 and ω_2 may be unitarily equivalent only if they induce equivalent norms on Sol*, that is there are constants $C, C' > 0$ with*

$$C\mu_1(x,x) \leq \mu_2(x,x) \leq C'\mu_1(x,x) \quad \forall x \in \mathsf{Sol}\,.$$

When the condition is satisfied there is a unique bounded operator $Q : \mathcal{R}_{\mu_1} \to \mathcal{R}_{\mu_1}$ such that

$$\mu_1(x, Qy) = \mu_2(x,y) - \mu_1(x,y) \quad \forall x, y \in \mathsf{Sol}\,.$$

In this case ω_1 and ω_2 are unitarily equivalent if and only if Q is Hilbert-Schmidt[6] *in $\mathcal{R}_{\mu_1}$.*

In general, the said condition fails when ω_1 and ω_2 are stationary states associated with two stationary regions (in the past and in the future) of a spacetime, as discussed in the introduction of this section [65], [22, Chap. 7]. It happens in particular when the Cauchy surfaces have infinite volume. In this case the states turn out to be unitarily inequivalent. On the other hand there is no natural preferred choice between ω_+ and ω_- and this fact suggests that the algebraic formulation is more useful in QFT in curved spacetime than the, perhaps more familiar, formulation in a Hilbert space.

5.2.7 States Invariant Under the Action of Spacetime Symmetries

The quasifree state ω_ζ on $\mathcal{A}(\boldsymbol{M})$ mentioned to exist above for stationary globally hyperbolic spacetimes for massive scalar fields is *invariant* under the action of ζ (which we assume to be complete for the sake of simplicity) in the following sense. Just because ζ is a Killing field (and the subsequent construction does depend on the fact that ζ is time-like), the action of the one-parameter group of isometries $\{\chi_t^{(\zeta)}\}_{t\in\mathbb{R}}$ generated by ζ leaves Sol invariant. This is equivalent to saying that when $\{\chi_t^{(\zeta)}\}_{t\in\mathbb{R}}$ acts on $C_0^\infty(\mathcal{M})$ it preserves E (the commutation relations of quantum fields are consequently preserved in particular). In view of Proposition 5.2.20, a one-parameter group of $*$-algebra isomorphisms $\alpha_t^{(\zeta)} : \mathcal{A}(\boldsymbol{M}) \to \mathcal{A}(\boldsymbol{M})$ arises this way, completely defined by the requirement beyond the obvious $\alpha_t(\mathbf{1}) = \mathbf{1}$ if $t \in \mathbb{R}$ and

$$\alpha_t^{(\zeta)}(\phi(f)) \stackrel{\text{def}}{=} \phi\left(f \circ \chi_{-t}^{(\zeta)}\right), \quad t \in \mathbb{R}, \quad f \in C_0^\infty(\mathcal{M})\,.$$

[6]Note that this result is stated incorrectly in Theorem 4.4.1 of [65], where the condition on the operator Q is incorrectly given as *trace class* instead of *Hilbert-Schmidt*. The correct condition is actually given in Equation (4.4.21) of [65] as the Hilbert-Schmidt property of the operator $\mathcal{E}$ and the mistake appears in identifying the corresponding property of Q. We thank Rainer Verch and especially Ko Sanders for bringing this to our attention.

It turns out that, if ω_ζ is constructed by the procedure above mentioned when ζ is time-like, ω_ζ is ζ-invariant in the sense that:

$$\omega_\zeta \circ \alpha_t^{(\zeta)} = \omega_\zeta \quad \forall t \in \mathbb{R} . \tag{5.40}$$

When passing to the GNS representation, Proposition 5.1.17 implies that there is a one-parameter group of unitary operators such that

(i) $U_t^{(\zeta)} \Psi_{\omega_\zeta} = \Psi_{\omega_\zeta}$, $U_t^{(\zeta)}(\mathcal{D}_{\omega_\zeta}) = \mathcal{D}_{\omega_\zeta}$,

(ii) $U_t^{(\zeta)} \pi_{\omega_\zeta}(a) U_t^{(\zeta)*} \stackrel{\text{def}}{=} \pi_{\omega_\zeta}\left(\alpha_t^{(\zeta)}(a)\right)$ for all $t \in \mathbb{R}$ and $a \in \mathcal{A}(\boldsymbol{M})$.

Moreover we know that $\{U_t^{(\zeta)}\}_{t\in\mathbb{R}}$ is *strongly continuous* if and only if

$$\lim_{t\to 0} \omega_\zeta \left(a^* \alpha_t^{(\zeta)}(a)\right) = \omega_\zeta(a^* a) , \quad \forall a \in \mathcal{A}(\boldsymbol{M}) .$$

In this case, Stone's theorem entails that there is a unique self-adjoint operator $H^{(\zeta)}$ with $e^{-itH^{(\zeta)}} = U_t^{(\zeta)}$ for every $t \in \mathbb{R}$ and $H^{(\zeta)}\Psi_{\omega_\zeta} = 0$. If $\sigma(H^{(\zeta)}) \subset [0, +\infty)$ and Ψ_{ω_ζ} is, up to factors, the unique eigenvector of $H^{(\zeta)}$ with eigenvalue 0, ω_ζ is said to be a **ground state** (this definition generally applies to an invariant state under the action of a time-like Killing symmetry, no matter if the state is quasifree).

Remark 5.2.31 The one-parameter group $U_t^{(\zeta)}$ associated with the time-like-Killing vector field ζ has the natural interpretation of **time evolution** with respect to the *notion of time* associated with ζ and, in case the group is strongly continuous $H^{(\zeta)}$ is the natural **Hamiltonian** operator associated with that evolution. However, for a generic time-oriented globally hyperbolic spacetime, no notion of Killing time is suitable and consequently, no notion of (unitary) time evolution is possible. Time evolution *à la Schroedinger* is not a good notion to be extended to QFT in curved spacetime. Observables do not evolve, they are localized in bounded regions of spacetime by means of the smearing procedure. Causal relations are encompassed by the *Time-slice axiom* (see Chap. 3) which is a theorem for free fields (Proposition 5.2.6).

Abandoning the case of time-like Killing symmetries, it is worth stressing that, generally speaking, every isometry $\gamma : \boldsymbol{M} \to \boldsymbol{M}$, not necessarily Killing and not necessarily time-like if Killing, induces a corresponding automorphism of unital $*$-algebras, $\beta^{(\gamma)}$ of $\mathcal{A}(\boldsymbol{M})$, via Proposition 5.2.20, completely defined by the requirements $\beta^{(\gamma)}(\phi(f)) \stackrel{\text{def}}{=} \phi\left(f \circ \gamma^{-1}\right)$. If a state ω is invariant under $\beta^{(\gamma)}$, we can apply Proposition 5.1.17, in order to unitarily implement this symmetry in the GNS representation of ω. Some discrete symmetries can be represented in terms of anti-linear automorphisms, like the time reversal in Minkowski spacetime. Again $\beta^{(\gamma)}(\phi(f)) \stackrel{\text{def}}{=} \phi\left(f \circ \gamma^{-1}\right)$ completely determine the anti-linear automorphism via Proposition 5.2.20. If a state ω is invariant under $\beta^{(\gamma)}$, we can apply Proposition 5.1.17, in order to implement this symmetry anti-unitarily in the GNS representation of ω.

Remark 5.2.32

(1) It easy to prove that, if the state $\omega : \mathcal{A}(\boldsymbol{M}) \to \mathbb{C}$ is invariant under the (anti-linear) automorphism $\beta : \mathcal{A}(\boldsymbol{M}) \to \mathcal{A}(\boldsymbol{M})$ is *quasifree*, the spaces with fixed number of particles of the GNS Fock representation of ω are separately invariant under the action of the unitary (resp. anti-unitary) operator $U^{(\beta)}$ implementing β in the Fock representation of ω in view of Proposition 5.2.20.

(2) A known result [40] establishes the following remarkable uniqueness result (actually proved for Weyl algebras, but immediately adaptable to our CCR framework).

Proposition 5.2.33 (Uniqueness of pure invariant quasifree states) *Assume that a quasifree state $\omega : \mathcal{A}(\boldsymbol{M}) \to \mathbb{C}$ is pure and invariant under a one-parameter group of automorphisms $\{\beta_t\}_{t\in\mathbb{R}}$ of $\mathcal{A}(\boldsymbol{M})$, giving rise to a strongly continuous unitary group $\{U_t\}_{t\in\mathbb{R}}$ implementing $\{\beta_t\}_{t\in\mathbb{R}}$ in the GNS representation of ω. The pure quasifree state ω is uniquely determined by $\{\beta_t\}_{t\in\mathbb{R}}$ if the self-adjoint generator of $\{U_t\}_{t\in\mathbb{R}}$ restricted to the one-particle Hilbert space of ω is positive without zero eigenvalues.*

Let us focus on the *Minkowski vacuum*, that is the quasifree state $\omega_{\mathbb{M}}$ on four dimensional Minkowski spacetime $\mathbb{M}$ defined in Sect. 5.2.5 by the two-point function (5.37). As a matter of fact, $\omega_{\mathbb{M}}$ turns out to be invariant under the natural action of *orthochronous proper Poincaré group* and that the corresponding unitary representation of this connected Lie (and thus topological) group is strongly continuous. In particular the self-adjoint generator of time displacements (with respect to every timelike direction), in the one-particle Hilbert space, satisfies the hypotheses of Proposition 5.2.33. As $\omega_{\mathbb{M}}$ is pure, it *is therefore the unique pure quasifree state invariant under the orthochronous proper Poincaré group*. $\omega_{\mathbb{M}}$ is a ground state with respect to any Minkowski time evolution and, by direct inspection, one easily sees that the state it is also invariant under the remaining discrete symmetries of Poincaré group T, P and PT which are consequently (anti-)unitarily implementable in the GNS Hilbert space. Finally, it turns out that the one-particle space is irreducible under the action of the orthochronous proper Poincaré group, thus determining an *elementary particle* in the sense of the *Wigner classification*, with mass m and zero spin.

5.3 Hadamard Quasifree States in Curved Spacetime

The algebra of observables generated by the field $\phi(f)$ smeared with smooth functions is too small to describe important observables in QFT in curved spacetime. Maybe the most important is the stress energy tensor (obtained as a functional derivative of the action with respect to g^{ab}) that, for our Klein-Gordon field it reads, where G_{ab} is the standard Einstein tensor

$$T_{ab} \stackrel{\text{def}}{=} (1-2\xi)\nabla_a\phi\nabla_b\phi - 2\xi\phi\nabla_a\nabla_b\phi - \xi\phi^2 G_{\mu\nu} + g_{ab}\left\{2\xi\phi^2 + \left(2\xi - \frac{1}{2}\right)\nabla^c\phi\nabla_c\phi + \frac{1}{2}m^2\phi^2\right\} . \tag{5.41}$$

It concerns products of fields evaluated at the same point of spacetime, like $\phi^2(x)$. This observable, as usual smeared with a function $f \in C_0^\infty(\mathcal{M})$, could be formally interpreted as

$$\phi^2(f) = \int_{\mathcal{M}} \phi(x)\phi(y) f(x)\delta(x, y) \ \text{dvol}_{\boldsymbol{M}} . \tag{5.42}$$

However this object does not belong to $\mathcal{A}(\boldsymbol{M})$. Beyond the fact that T_{ab} describe the local content of energy, momentum and stress of the field, the stress-energy tensor is of direct relevance for describing the back reaction on the quantum fields on the spacetime geometry through the semi-classical Einstein equation

$$G_{ab}(x) = 8\pi\omega(T_{ab}(x)) \tag{5.43}$$

or also, introducing a smearing procedure

$$\int_{\mathcal{M}} G_{ab}(x) f(x) \ \text{dvol}_{\boldsymbol{M}} = 8\pi \int_{\mathcal{M}} \omega(T_{ab}(x)) f(x) \ \text{dvol}_{\boldsymbol{M}} ,$$

where $\omega(T_{ab}(x))$ has the interpretation of the (integral kernel of the) expectation value of the quantum observable T_{ab} with respect to some quantum state ω. Barring technicalities due to the appearance of derivatives, the overall problem is here to provide (5.42) with a precise mathematical meaning, which in fact, is equivalent to a suitable enlargement the algebra $\mathcal{A}(\boldsymbol{M})$.

5.3.1 Enlarging the Observable Algebra in Minkowski Spacetime

In flat spacetime $\boldsymbol{M} = \mathbb{M}$, for free QFT, at the level of *expectation values* and *quadratic forms* the above mentioned enlargement of the algebra is performed exploiting a *physically meaningful reference state*, the unique Poincaré invariant quasifree (pure) state introduced in Sect. 5.2.5 and discussed at the end of Sect. 5.2.7, $\omega_{\mathbb{M}}$. We call this state *Minkowski vacuum*.

Let us first focus on the elementary observable ϕ^2. We shall indicate it with $:\phi^2(x):$ and we define it as a *Hermitian quadratic form* on $\mathcal{D}_{\omega_{\mathbb{M}}}$.

We start by defining the operator on $\mathcal{D}_{\omega_{\mathbb{M}}}$ for $f, g \in C_0^\infty(\mathbb{R}^4)$

$$:\hat{\phi}(f)\hat{\phi}(g): \stackrel{\text{def}}{=} \hat{\phi}(f)\hat{\phi}(g) - \langle \Psi_{\omega_{\mathbb{M}}} | \hat{\phi}(f)\hat{\phi}(g)\Psi_{\omega_{\mathbb{M}}} \rangle I \tag{5.44}$$

(As usual $\hat{\phi}(f) \stackrel{\text{def}}{=} \hat{\phi}_{\omega_{\mathbb{M}}}(f)$ throughout this section.) Next, for $\Psi \in \mathcal{D}_{\omega_{\mathbb{M}}}$ we analyze its integral kernel, assuming that it exists, $\langle \Psi | :\hat{\phi}(x)\hat{\phi}(y): \Psi \rangle$ which is symmetric since the antisymmetric part of the right-hand side of (5.44) vanishes in view of the commutation relations of the field. The explicit form of the distribution $\omega_{\mathbb{M}2}(x, y) = \langle \Psi_{\omega_{\mathbb{M}}} | \hat{\phi}(x)\hat{\phi}(y)\Psi_{\omega_{\mathbb{M}}} \rangle$ appears in (5.37). We prove below that the mentioned formal kernel $\langle \Psi | :\hat{\phi}(x)\hat{\phi}(y): \Psi \rangle$ not only exists but it also is a jointly smooth function. Consequently we are allowed to define, for any $\Psi \in \mathcal{D}_{\omega_{\mathbb{M}}}$,

$$\langle \Psi | :\hat{\phi}^2: (f)\Psi \rangle \stackrel{\text{def}}{=} \int_{\mathbb{M}^2} \langle \Psi | :\hat{\phi}(x)\hat{\phi}(y): \Psi \rangle f(x)\delta(x, y) \, \text{dvol}_{\mathbb{M}^2}(x, y) . \quad (5.45)$$

Finally, the polarization identity uniquely defines $:\hat{\phi}^2: (f)$ as a symmetric quadratic form $\mathcal{D}_{\omega_{\mathbb{M}}} \times \mathcal{D}_{\omega_{\mathbb{M}}}$.

$$\langle \Psi' | :\phi^2: (f)\Psi \rangle \stackrel{\text{def}}{=} \frac{1}{4} \Big(\langle \Psi' + \Psi | :\phi^2: (f)(\Psi' + \Psi) \rangle - \langle \Psi' - \Psi | :\phi^2: (f)(\Psi' - \Psi) \rangle$$
$$- i\langle \Psi' + i\Psi | :\phi^2: (f)(\Psi' + i\Psi) + i\langle \Psi' - i\Psi | :\phi^2: (f)(\Psi' - i\Psi) \rangle \Big) \quad (5.46)$$

There is no guarantee that an operator $:\phi^2: (f)$ really exists on $\mathcal{D}_{\omega_{\mathbb{M}}}$ satisfying (5.45),[7] however if it exists, since $\mathcal{D}_{\omega_{\mathbb{M}}}$ is dense and (5.46) holds, it is uniquely determined by the class of the expectation values $\langle \Psi | :\hat{\phi}^2: (f)\Psi \rangle$ on the states $\Psi \in \mathcal{D}_{\omega_{\mathbb{M}}}$. As promised, let us prove that the kernel defined in (5.45) is a smooth function. First of all, notice that, as a general result arising from the GNS construction, every $\Psi \in \mathcal{D}_{\omega}$ can be written as

$$\Psi = \sum_{n \geq 0, i_1, \ldots, i_n \geq 1} C^{(n)}_{i_1 \ldots i_n} \hat{\phi}(f^{(n)}_{i_1}) \cdots \hat{\phi}(f^{(n)}_{i_n}) \Psi_{\omega_{\mathbb{M}}} \quad (5.47)$$

where only a finite number of coefficients $C^{(n)}_{i_1 \ldots i_n} \in \mathbb{C}$ is non-vanishing and the term in the sum corresponding to $n = 0$ is defined to have the form $c^0 \Psi_{\omega_{\mathbb{M}}}$. We have

$$\langle \Psi | \hat{\phi}(x)\hat{\phi}(y)\Psi \rangle = \sum_{n \geq 0, i_1, \ldots, i_n \geq 1} \sum_{m \geq 0, j_1, \ldots, j_n \geq 1} \overline{C^{(m)}_{j_1 \ldots j_n}} C^{(n)}_{i_1 \ldots i_n}$$
$$\left\langle \Psi_{\omega_{\mathbb{M}}} \middle| \hat{\phi}(f^{(m)}_{j_m}) \cdots \hat{\phi}(f^{(m)}_{j_1}) \hat{\phi}(x)\hat{\phi}(y)\hat{\phi}(f^{(n)}_{i_1}) \cdots \hat{\phi}(f^{(n)}_{i_n}) \Psi_{\omega_{\mathbb{M}}} \right\rangle . \quad (5.48)$$

Taking advantage of the quasifree property of $\omega_{\mathbb{M}}$, hence using the expansion of n-point functions in terms of the 2-point function of Definition 5.2.22, we can re-arrange the right hand side of (5.48) as (all the sums are over finite terms)

[7] By Riesz lemma, it exists if an only if the map $\mathcal{D}_{\omega_{\mathbb{M}}} \ni \Psi' \mapsto \langle \Psi' | :\phi^2: (f)\Psi \rangle$ is continuous for every $\Psi \in \mathcal{D}_{\omega_{\mathbb{M}}}$.

$$\begin{aligned}\langle\Psi|\hat{\phi}(x)\hat{\phi}(y)\Psi\rangle &= C^0_\Psi\,\omega_{\mathbb{M}2}(x,y)\\ &+ \sum_{m\geq 0, j\geq 1}\sum_{m'\geq 0, j'\geq 1} C^{(m)(m')}_{\Psi,j,j'}\,\omega_{\mathbb{M}2}(f^{(m)}_j, x)\,\omega_{\mathbb{M}2}(f^{(m')}_{j'}, y)\\ &+ \sum_{m\geq 0, j\geq 1}\sum_{n\geq 0, i\geq 1} C^{(m)(n)}_{\Psi,j,i}\,\omega_{\mathbb{M}2}(f^{(m)}_j, x)\,\omega_{\mathbb{M}2}(y, f^{(n)}_i)\\ &+ \sum_{n'\geq 0, i'\geq 1}\sum_{n\geq 0, i'\geq 1} C^{(n')(n)}_{\Psi,i',i}\,\omega_{\mathbb{M}2}(x, f^{(n')}_{i'})\,\omega_{\mathbb{M}2}(y, f^{(n)}_i)\,, \end{aligned}\tag{5.49}$$

with all sums finite and some C_Ψ-coefficients that depend on the state Ψ. We can be more specific about the first coefficient, in fact, according to the formula from Definition 5.2.22, we have $C^0_\Psi = \langle\Psi|\Psi\rangle$. Recall also, from Proposition 5.2.26, that $y \mapsto \omega_{\mathbb{M}2}(f, y)$ and $x \mapsto \omega_{\mathbb{M}2}(x, f)$ are smooth for any test function $f \in C^\infty_0(\mathcal{M})$. Hence, we can interpret Eq. (5.49) as saying that

$$\langle\Psi|\,{:}\hat{\phi}(x)\hat{\phi}(y){:}\,\Psi\rangle = \langle\Psi|\hat{\phi}(x)\hat{\phi}(y)\Psi\rangle - \langle\Psi|\Psi\rangle\omega_{\mathbb{M}}(x,y) \in C^\infty(\mathcal{M}\times\mathcal{M})\,. \tag{5.50}$$

More complicated operators, i.e. **Wick polynomials** and corresponding differentiated Wick polynomials, generated by **Wick monomials**, ${:}\hat{\phi}^n{:}\,(f)$, of arbitrary order n, can analogously be defined as quadratic forms, by means of a recursive procedure of subtraction of divergences. The stress energy operator is a differentiated Wick polynomial of order 2.

The procedure for defining ${:}\hat{\phi}^n{:}\,(f)$ as a quadratic form is as follows. First define recursively, where the tilde just means that the indicated element has to be omitted,

$$\begin{aligned} {:}\hat{\phi}(f_1){:} &\stackrel{\text{def}}{=} \hat{\phi}(f_1)\\ {:}\hat{\phi}(f_1)\cdots\hat{\phi}(f_{n+1}){:} &\stackrel{\text{def}}{=} {:}\hat{\phi}(f_1)\cdots\hat{\phi}(f_n){:}\,\hat{\phi}(f_{n+1})\\ &\qquad - \sum_{l=1}^{n} {:}\hat{\phi}(f_1)\cdots\widetilde{\hat{\phi}(f_l)}\cdots\hat{\phi}(f_n){:}\;\omega_{\mathbb{M}2}(f_l, f_{n+1})\,. \end{aligned}\tag{5.51}$$

These elements of $\mathcal{A}(\mathbb{M})$ turn out to be symmetric under interchange of $f_1, f_2, \ldots, f_n$ as it can be proved by induction.[8] By induction, it is next possible to prove that, for $n \geq 2$ and $\Psi \in \mathcal{D}_{\omega_{\mathbb{M}}}$, there is a *jointly smooth kernel*

$$\langle\Psi|\,{:}\hat{\phi}(x_1)\cdots\hat{\phi}(x_n){:}\,\Psi\rangle$$

which produces $\langle\Psi|\,{:}\hat{\phi}(f_1)\cdots\hat{\phi}(f_n){:}\,\Psi\rangle$ by integration. This result arises from (5.51) as a consequence of the fact that

(a) $\omega_{\mathbb{M}}$ is quasifree so that Definition 5.2.22 can be used to compute the said kernels,

(b) $\Psi \in \mathcal{D}_{\omega_{\mathbb{M}}}$ so that the expansion (5.47) can be used,

[8] Observe in particular that ${:}\hat{\phi}(f)\hat{\phi}(g){:} - {:}\hat{\phi}(g)\hat{\phi}(f){:} = iE(f,g)\mathbf{1} - \omega_{\mathbb{M}2}(iE(f,g)\mathbf{1})\mathbf{1} = 0$.

(c) the functions $F_k : x \mapsto \omega_{\mathbb{M}2}(x, f_k) = \overline{\omega_{\mathbb{M}2}(f_k, x)}$ are smooth when $f_k \in C_0^\infty(\mathcal{M})$ as was mentioned above.

Indeed, we have

$$\langle\Psi|\, {:}\hat{\phi}(x_1)\cdots\hat{\phi}(x_n){:}\; \Psi\rangle = \sum_{n\geq 0, i_1,\ldots,i_n\geq 1}\;\sum_{m\geq 0, j_1,\ldots,j_n\geq 1} \overline{C^{(m)}_{j_1\ldots j_n}}\, C^{(n)}_{i_1\ldots i_n}$$
$$\left\langle \Psi_{\omega_\mathbb{M}}\right| \hat{\phi}(f^{(m)}_{j_m})\cdots\hat{\phi}(f^{(m)}_{j_1})\; {:}\hat{\phi}(x_1)\cdots\hat{\phi}(x_n){:}\; \hat{\phi}(f^{(n)}_{i_1})\cdots\hat{\phi}(f^{(n)}_{i_n})\left.\Psi_{\omega_\mathbb{M}}\right\rangle . \quad (5.52)$$

after having expanded the normal product ${:}\hat{\phi}(g_1)\cdots\hat{\phi}(g_n){:}$ in the right-hand side, one can evaluate the various n-point functions arising this way by applying Definition 5.2.22. It turns out that all terms $\omega_{\mathbb{M}2}(x_i, x_j)$ always appear in a sum with corresponding terms $-\omega_{\mathbb{M}2}(x_i, x_j)$ arising by the definition (5.51) and thus give no contribution. The remaining factors are of the form $F_k(x_j)$ and thus are smooth.

We therefore are in a position to write the definition of $\langle\Psi|\; {:}\hat{\phi}^n{:}\; (f)\Psi\rangle$ if $\Psi \in \mathcal{D}_{\omega_\mathbb{M}}$

$$\langle\Psi|\, {:}\hat{\phi}^n{:}\, (f)\Psi\rangle = \int_{\mathcal{M}^n} \langle\Psi|\, {:}\hat{\phi}(x_1)\cdots\hat{\phi}(x_n){:}\; \Psi\rangle f(x_1)\delta(x_1,\ldots x_n)\mathrm{dvol}_{\boldsymbol{M}^n} \quad (5.53)$$

Exactly as before, polarization extends the definition to a quadratic form on $\mathcal{D}_{\omega_\mathbb{M}} \times \mathcal{D}_{\omega_\mathbb{M}}$. There is no guarantee that operators fitting these quadratic forms really exist.

Remark 5.3.1 The definition (5.51) can be proved to be formally equivalent to the formal definition

$${:}\hat{\phi}(x_1)\cdots\hat{\phi}(x_n){:} \stackrel{\text{def}}{=} \frac{1}{i^n}\frac{\delta^n}{\delta f(x_1)\cdots\delta f(x_n)}\bigg|_{f=0} e^{i\hat{\phi}(f)+\frac{1}{2}\omega_{\mathbb{M}2}(f,f)} \quad (5.54)$$

Though the exponential converges in the strong operator topology to a unitary operator, the **Weyl generator**, restricted to the dense domain $\mathcal{D}_{\omega_\mathbb{M}}$, $e^{i\hat{\phi}(f)}$ can be viewed here as a formal series and this series can be truncated at finite, sufficiently large, order in view of linearity of the exponent and $f = 0$.

5.3.2 Enlarging the Observable Algebra in Curved Spacetime

The discussed definition of Wick polynomials is *equivalent* in Minkowski spacetime to the more popular one based on the well known re-ordering procedure of creation and annihilation operators as can be proved by induction. Nevertheless this second approach is not natural in curved spacetime because, to be implemented, it needs the existence of a physically preferred reference state as Minkowski vacuum in flat spacetime, which in the general case it is not given. To develop a completely covariant theory another approach has been adopted, which generalises to curved spacetime the previously outlined definition of Wick polynomials based on a "divergence subtraction" instead of a re-ordering procedure. The idea is that, although it is not possible to uniquely assign each spacetime with a physically distinguishable state, it is possible to select a type of divergence in common with all physically relevant states is every spacetime. These preferred quasifree states with the same type of divergence "resembling" Minkowski vacuum in a generic spacetime are called *Hadamard states*. Minkowski vacuum belongs to this class and these states are remarkable also in view of their *microlocal features*, which revealed to be of crucial importance for the technical advancement of the theory, as we will describe later. Exploiting these distinguished states, it is possible to generalize the outlined approach in order to enlarge $\mathcal{A}(\boldsymbol{M})$, including other algebraic elements as the stress-energy tensor operator [37, 47]. Actually this is nothing but the first step to generalize the ultraviolet renormalization procedure to curved spacetime [9, 10, 35, 36, 65]. The rest of the chapter is devoted to discuss some elementary properties of Hadamard states.

Let us quickly remind some local features of (pseudo)Riemannian differential geometry [50], necessary to introduce the notion of Hadamard states from a geometric viewpoint. If $(\mathcal{M}, g)$ is a smooth Riemannian or Lorentzian manifold, an open set $C \subset M$ is said a **normal convex** neighborhood if there is a open set $W \subset TM$ with the form $W = \{(q, v) \mid q \in C, v \in S_q\}$ where $S_q \subset T_qM$ is a star-shaped open neighborhood of the origin, such that

$$exp\restriction_W : (q, v) \mapsto exp_q v$$

is a diffeomorphism onto $C \times C$. It is clear that C is connected and there is only one geodesic segment joining any pair $q, q' \in C$ if we require that it is completely contained in C. It is $[0, 1] \ni t \mapsto exp_q(t((exp_q)^{-1}q'))$. Moreover if $q \in C$ and we fix a basis $\{e_\alpha|_q\} \subset T_qM$,

$$t = t^\alpha e_\alpha|_q \mapsto exp_q(t^\alpha e_\alpha|_q)\,, \quad t \in S_q$$

defines a set of coordinates on C centered in q which is called the **normal Riemannian coordinate system** centered in q. In (M, g) as above, $\sigma(x, y)$ indicates the **squared (signed) geodesic distance** of x from y. With our signature $(+, -, \cdots, -)$, it is defined as

$$\sigma(x, y) \stackrel{\text{def}}{=} -g_x(exp_x^{-1} y, exp_x^{-1} y)\,.$$

$\sigma(x, y)$ turns out to be smoothly defined on $C \times C$ if C is a convex normal neighborhood where we also have $\sigma(x, y) = \sigma(y, x)$. The class of the convex normal neighborhoods of a point $p \in \mathcal{M}$ is a fundamental system of neighborhoods of p [3, 19].

In Euclidean manifolds σ defined as above is everywhere nonnegative with the standard Euclidean choice of the signature.

In a convex neighborhood C of a spacetime $\boldsymbol{M}$, taking in particular advantage of several properties of σ, it is possible to define a local *approximate solution* of KG equation, technically called a *parametrix*, which has essentially the same short-distance singularity of the two point function of Minkowski vacuum. Its construction uses only the local geometry and the parameters defining the equation of motion but does not refers to particular states, which are global objects. The technical idea can be traced back to Hadamard [32] (and extensively studied by Riesz [55]) and it is therefore called *Hadamard parametrix*. In the rest of the chapter we only consider a four dimensional spacetime, essentially following [30]. A quick technical discussion on the general case (details and properties of the constructions strongly depend of the dimension of the spacetime) also in relation with heath kernel expansion, can be found in [47] (see also [2, 19, 24, 27] for more extended discussions also on different types of parametrices and their use in field theory). In a convex neighborhood C of a four dimensional spacetimes the **Hadamard parametrix** of order N of the two-point function has the form

$$H_\varepsilon^{(N)}(x,y) = \frac{u(x,y)}{(2\pi)^2\sigma_\varepsilon(x,y)} + \sum_{n=0}^{N} v_n\sigma^n \log\left(\frac{\sigma_\varepsilon(x,y)}{\lambda^2}\right) \tag{5.55}$$

where $x, y \in C$, T is any local time coordinate increasing towards the future, $\lambda > 0$ a length scale and

$$\sigma_\varepsilon(x,y) \stackrel{\text{def}}{=} \sigma(x,y) + 2i\varepsilon(T(x) - T(y)) + \varepsilon^2\,, \tag{5.56}$$

finally, the cut in the complex domain of the log function is assumed along the negative axis in (5.55). Recursive differential equations (see the appendix A of [47] and also [19, 22, 24, 29, 46, 55]) determine $u = u(x, y)$ and all the **Hadamard coefficients** $v_n = v_n(x, y)$ in C as smooth functions, when assuming $u(x, x) = 1$ and $n = 0, 1, 2, \ldots$. These recurrence relations have been obtained by requiring that the sequence of the $H_0^{(N)}(x, y)$ defines a local, y-parametrized, "approximate solution" of the KG equation for $\sigma(x, y) \neq 0$ (with some further details we can say that the error with respect to a true solution is of order σ^N for each N). That solution would be exact in the $N \to \infty$ limit of the sequence provided the limit exists. The limit exists in the analytic case, but in the smooth general case the sequence diverges. However, as proved in [19, Sect. 4.3], if $\chi : \mathbb{R} \to [0, 1]$ is a smooth function with $\chi(r) = 1$ for $|r| \leq 1/2$ and $\chi(r) = 0$ for $|r| > 0$ one can always find a sequence of numbers $0 < c_1 < c_2 < \cdots < c_n \to +\infty$ for that

$$v(x,y) \stackrel{\text{def}}{=} \sum_{n=0}^{\infty} v_n(x,y)\sigma(x,y)^n \chi(c_n\sigma(x,y)) \tag{5.57}$$

uniformly converges, with all derivatives, to a C^∞ function on $C \times C$. A parametrix H_ε

$$H_\varepsilon(x,y) = \frac{u(x,y)}{(2\pi)^2\sigma_\varepsilon(x,y)} + v(x,y)\log\left(\frac{\sigma_\varepsilon(x,y)}{\lambda^2}\right) \tag{5.58}$$

arises this way. This parametrix distributionally satisfies KG equation in both arguments up to jointly smooth functions of x and y. In other words, there is a smooth function s defined in $C \times C$ such that if $f, g \in C_0^\infty(C)$ and defining $P \stackrel{\text{def}}{=} \Box_{\boldsymbol{M}} + m^2 + \xi R$,

$$\lim_{\varepsilon \to 0^+} \int_{C\times C} H_\varepsilon(x, y)(Pf)(x)g(y)\mathrm{dvol}_{\mathcal{M}\times\mathcal{M}} = \int_{C\times C} s(x, y)f(x)g(y)\mathrm{dvol}_{\mathcal{M}\times\mathcal{M}} . \tag{5.59}$$

The analog holds swapping the role of the test functions. We are in a position to state our main definition.

Definition 5.3.2 With $\boldsymbol{M}$ four dimensional, we say that a (not necessarily quasifree) state ω on $\mathcal{A}(\boldsymbol{M})$ and its two point function ω_2 are **Hadamard** if $\omega_2 \in \mathcal{D}'(\mathcal{M} \times \mathcal{M})$ and every point of $\boldsymbol{M}$ admits an open normal neighborhood C where

$$\omega_2(x, y) - H_{0^+}(x, y) = w(x, y) \quad \text{for some} \quad w \in C^\infty(C \times C) . \tag{5.60}$$

Here 0^+ indicates the standard weak distributional limit as $\varepsilon \to 0^+$ ("first integrate against test functions and next take the limit").

Remark 5.3.3

(1) The given definition does not depend either on the choice of χ or the sequence of the c_n used in (5.57) since different choices simply change w as one may easily prove. Similarly, the definition does not depend on the choice of the local time function T used in the definition of σ_ε. This fact is far from being obvious and requires a more detailed analysis [41].

(2) Using the following result arising form recurrence relations determining the Hadamard coefficients, one finds that the distribution

$$\left(v(x, y) - \sum_{k=0}^{N} v_n(x, y)\sigma(x, y)^n\right) \ln \sigma_{0^+}(x, y)$$

is a function in $C^N(O \times O)$. Exploiting this result, it is not difficult to prove that the requirement (5.60) is equivalent to the following requirement:

$$\omega_2(x, y) - H^{(N)}_{0^+}(x, y) = w_N(x, y) \quad \text{for each } N \geq 1, \text{ with } \; w_N \in C^N(C \times C) . \tag{5.61}$$

The equivalent definition of Hadamard state in [52] was, in fact, nothing but Definition 5.3.2 with (5.60) replaced by (5.61).

(3) Minkowski vacuum $\omega_{\mathbb{M}}$ defined by the two point function (5.37) is Hadamard. In particular, for $m > 0$, it holds[9]

$$\omega_{\mathbb{M}2}(x, y) = \frac{1}{4\pi^2} \frac{1}{\sigma_{0^+}(x, y)} + \frac{m^2}{2(2\pi)^2} \frac{I_1(m\sqrt{\sigma(x, y)})}{m\sqrt{\sigma(x, y)}} \ln\left(m^2 \sigma_{0^+}(x, y)\right) + w(x, y)$$

[9]The function $z \mapsto I_1(\sqrt{z})/\sqrt{z}$, initially defined for $Re(z) > 0$, admits a unique analytic extension on the whole space $\mathbb{C}$ and the formula actually refers to this extension.

where w is smooth. The result holds also for $m = 0$ and in that case, only the first term in the right-hand side does not vanish in the expansion above. Similarly, quasifree states *invariant under the symmetries generated by a timelike Killing vector field* ζ as the states considered in Sect. 5.2.5 (with all the hypotheses specified therein) are Hadamard [23, 65] if the spacetime admits spacelike Cauchy surfaces normal to ζ, that is if the spacetime is *static*. This last condition is essential because there are spacetimes admitting timelike Killing vectors but not spacelike Cauchy surfaces normal to them which do not admit *invariant Hadamard* quasifree states, like Kerr spacetime and Schwartzschild-de Sitter spacetime [41].

(4) Referring to the literature before the cornerstone results [52, 53] (we consider in Sect. 5.3.4), Definition 5.3.2 properly refers to *locally* Hadamard states. This is because there also exists a notion of *global Hadamard state* (Definition 3.4 in [52]), discussed in [41] in a completely rigorous way for the first time. This apparently more restrictive global condition essentially requires (see [41, 52] for the numerous technical details), for a certain open neighborhood $\mathcal{N}$ of a Cauchy surface of $\boldsymbol{M}$ such that $\sigma(x, y)$ is always well defined if $(x, y) \in \mathcal{N} \times \mathcal{N}$ (and this neighbourhood can always be constructed independently from the Hadamard requirement), that (5.61) is valid producing the known singularity for causally related arguments, and there are no further singularities for arbitrarily far, spacelike separated, arguments $(x, y) \in \mathcal{N} \times \mathcal{N}$. In this regard a technically important result, proved in the appendix B of [41], is that, analogous to Proposition 5.2.26 in the case of Minkowski space,

$$\mathcal{M} \ni x \mapsto \omega(\phi(f)\phi(x)) = \overline{\omega(\phi(x)\phi(f))} \in C^\infty(\mathcal{M}) \tag{5.62}$$

if $f \in C_0^\infty(\mathcal{M})$ and ω is a quasifree *globally* Hadamard state on $\mathcal{A}(\boldsymbol{M})$. We shall prove this fact later using the microlocal approach. This fact has an important consequence we shall prove later using the microlocal approach: if ω and ω' are (locally) Hadamard states, then $\mathcal{M} \times \mathcal{M} \ni (x, y) \mapsto \omega_2(x, y) - \omega'_2(x, y)$ is smooth. This fact is far from obvious, since Definition 5.3.2 guarantees only that the difference is smooth when x and y belong to the same sufficiently small neighborhood.

An important feature of the global Hadamard condition for a quasifree Hadamard state is that it *propagates* [21, 65]: If it holds in a neighborhood of a Cauchy surface it holds in a neighborhood of any other Cauchy surface. We shall come back later to this property making use of the local notion only. This fact, together with the last comment in (3) proves that quasifree Hadamard states for massive fields (and $\xi = 0$) exist in globally hyperbolic spacetimes by means of a deformation argument similar to the one exploited in Sect. 5.2.5.

We shall not insist on the distinction between the *global* and the *local* Hadamard property because, in [53], it was established that a local Hadamard state on $\mathcal{A}(\boldsymbol{M})$ is also a global one (the converse is automatic). It was done exploiting the *microlocal approach*, which we shall discuss shortly.

(5) It is possible to prove that [65] if a globally hyperbolic spacetime has one (and thus all) compact Cauchy surface, all pure quasifree Hadamard states for the massive KG field (with $\xi = 0$) are unitarily equivalent. However it is not sufficient to deal with folia of only *pure* quasifree Hadamard states as this excludes very significant

examples. Consider the massive KG field (with $\xi = 0$) on an ultrastatic spacetime with a compact Cauchy surface. Both the unique time-translation invariant pure state and any *thermal* (KMS) state with temperature $T > 0$ are Hadamard, but they are not unitarily equivalent, since the former is pure while the latter is not. There is, in fact, a more general result [64] (actually stated in terms of Weyl algebras). Consider an open region O which defines a globally hyperbolic spacetime $\boldsymbol{O}$ in its own right, in a globally hyperbolic spacetime $\boldsymbol{M}$, such that $\overline{O}$ is compact, and a pair of quasifree Hadamard states ω_1, ω_2 for the massive KG field ($\xi = 0$) on $\mathcal{A}(\boldsymbol{M})$. It is possible to prove that the restriction to $\mathcal{A}(\boldsymbol{O}) \subset \mathcal{A}(\boldsymbol{M})$ of any density matrix state associated to the GNS construction of ω_1 coincides with the restriction to $\mathcal{A}(\boldsymbol{O})$ of some density matrix state associated to the GNS construction of ω_2.

It is now possible to recast all the content of Sect. 5.3.1 in a generic globally hyperbolic spacetime $\boldsymbol{M}$ enlarging the algebra of observables $\mathcal{A}(\boldsymbol{M})$, at the level of quadratic forms, defining the expectation values of Wick monomials $:\phi^n:(f)$ with respect to Hadamard states ω or vector states $\Psi \in \mathcal{D}_\omega$ with ω Hadamard. Remarkably, all of that can be done simultaneously for all states in the said class without picking out any reference state. This is the first step for a completely *local* and *covariant* definition. First, define for smooth functions f_k supported in a convex normal neighborhood C

$$:\phi(f_1)\cdots\phi(f_n):_H \stackrel{\text{def}}{=} \int_{\mathcal{M}^n} :\phi(x_1)\cdots\phi(x_n):_H f_1(x_1)\ldots f_n(x_n)\,\mathrm{dvol}_{\boldsymbol{M}^n}(x_1,\ldots,x_n)\,, \tag{5.63}$$

where we have defined the *completely symmetrized* formal kernels,

$$:\phi(x_1)\cdots\phi(x_n):_H \stackrel{\text{def}}{=} \frac{1}{i^n}\frac{\delta^n}{\delta f(x_1)\cdots\delta f(x_n)}\bigg|_{f=0} e^{i\phi(f)+\frac{1}{2}H_{0+}(f,f)}\,. \tag{5.64}$$

Notice that H_{0+} can be replaced with its symmetric part H^S_{0+} and that, in (5.63), only the symmetric part of the product $f_1(x_1)\ldots f_n(x_n)$ produces a contribution to the left-hand side. Equivalently, these monomials regularized with respect to the Hadamard parametrix can be define recursively as

$$\begin{aligned}
:\phi(f_1):_H &\stackrel{\text{def}}{=} \phi(f_1)\\
:\phi(f_1)\cdots\phi(f_{n+1}):_H &\stackrel{\text{def}}{=} :\phi(f_1)\cdots\phi(f_n):_H\,\phi(f_{n+1})\\
&\quad -\sum_{l=1}^{n} :\phi(f_1)\cdots\widetilde{\phi(f_l)}\cdots\phi(f_n):_H\,\tilde{H}(f_l,f_{n+1})\,,
\end{aligned} \tag{5.65}$$

$$\text{where } \tilde{H} = H^S_{0+} + \frac{i}{2}E,$$

in analogy with the relation between Eqs. (5.51) and (5.54). Now consider a quasifree Hadamard state ω and indicate by ω_Ψ the generic state indexed by the normalized vector $\Psi \in \mathcal{D}_\omega$ (so that $\omega = \omega_\Psi$ when Ψ is the Fock vacuum). By induction, it is possible to prove that, for $n \geq 2$, there is a *jointly smooth kernel*

$$\omega_\Psi(:\phi(x_1)\cdots\phi(x_n):_H)$$

which produces $\omega_\Psi(:\phi(f_1)\cdots\phi(f_n):_H)$ by integration when the supports of the functions f_k belong to C.

Exactly as for the Minkowski vacuum representation, this result arises from (5.65) as a consequence of the following list of facts:

(a) ω is quasifree so that Definition 5.2.22 can be used to compute the said kernels,

(b) $\Psi \in \mathcal{D}_\omega$ so that the expansion (5.47) can be used,

(c) the functions in (5.62) are smooth (see (4) in Remark 5.3.3 and Sect. 5.3.4),

(d) the local singularity of two-point functions of quasifree Hadamard states is the same as the one of H_{0^+}.

Consider a normalized $\Psi \in \mathcal{D}_\omega$, given without loss of generality by

$$\Psi = \sum_{n\geq 0, i_1,\ldots,i_n\geq 1} C^{(n)}_{i_1\ldots i_n}\hat{\phi}_\omega(f^{(n)}_{i_1})\cdots\hat{\phi}_\omega(f^{(n)}_{i_n})\Psi_\omega\,, \tag{5.66}$$

where only a finite number of coefficients $C^{(n)}_{i_1\ldots i_n} \in \mathbb{C}$ is non-vanishing, which defines the algebraic state $\omega_\Psi(\cdot) = \langle\Psi|(\cdot)\Psi\rangle$. Then, for instance, with the same argument used to achieve (5.50) we have

$$\omega_\Psi(:\phi(x_1)\phi(x_2):_H) - \omega(\phi(x_1)\phi(x_2)) + \tilde{H}(x_1,x_2) \in C^\infty(\mathcal{M}\times\mathcal{M})\,, \tag{5.67}$$

where the smoothness is assured because the resulting expression consists of a linear combination of products like $\omega(\phi(x_1)\phi(g))\omega(\phi(f)\phi(x_2))$, with some test functions f and g. Note that the combination of the second and third terms in (5.67) can be rewritten as

$$\begin{aligned}\omega(\phi(x_1)\phi(x_2)) - \tilde{H}(x_1,x_2) &= \omega(\phi(x_1)\phi(x_2)) - H^S_{0^+}(x_1,x_2) - \frac{i}{2}E(x_1,x_2)\\ &= \frac{1}{2}\omega(\phi(x_1)\phi(x_2)) - H_{0^+}(x_1,x_2)\\ &\quad+\frac{1}{2}\omega(\phi(x_2)\phi(x_1)) - H_{0^+}(x_2,x_1),\end{aligned}$$

which is obviously smooth by the very definition of the Hadamard property of ω. Hence $\omega_\Psi(:\phi(x_1)\phi(x_2):_H)$ is also smooth. We are in a position to define the expectation values of the Wick monomials for $f \in C_0^\infty(\mathcal{M})$ such that its support is included in C,

$$\omega_\Psi(:\phi^n:_H(f)) = \int_{\mathcal{M}^n}\omega_\Psi(:\phi(x_1)\cdots\phi(x_n):_H)f(x_1)\delta(x_1,\ldots,x_n)\mathrm{dvol}_{\mathcal{M}^n} \tag{5.68}$$

Exactly as before, polarization extends the definition to a quadratic form on $\mathcal{D}_\omega \times \mathcal{D}_\omega$. There is no guarantee that operators fitting these quadratic forms really exist. The question of their existence as operators will be addressed later, in Sect. 5.3.5.

Remark 5.3.4

(1) The restriction on the support of f is not very severe. The restriction can be removed making use of a partition of unity (see for example [37, 47] referring to more generally differentiated Wick polynomials).

(2) The given definition of $\omega(:\phi^n:_H(f))$ is affected by several ambiguities due to the effective construction of H_ε. A complete classification of these ambiguities, promoting Wick polynomials to properly defined elements of a $*$-algebra, can be presented from a very general viewpoint, adopting a *locally covariant framework* [35, 44], we shall not consider in this introductory review (see Chap. 4). We only say that these ambiguities are completely described by a class of scalar polynomials in the mass and Riemann curvature tensor and their covariant derivatives. The finite order of these polynomials is fixed by scaling properties of Wick polynomials. The coefficients of the polynomials are smooth functions of the parameter ξ. We stress that this classification is the first step of the ultraviolet renormalization program which, in curved spacetime and differently from flat spacetime where all curvature vanish, starts with classifying the finite renormalization counterterms of Wick polynomials instead of only dealing with time-ordered Wick polynomials.

(3) Easily extending the said definition, using the fact that $\omega_\Psi(:\phi(x_1)\phi(x_2):_H)$ is smooth and thus can be differentiated, one can define a notion of *differentiated Wick polynomials* which include, in particular, the *stress energy tensor* as a Hermitian quadratic form evaluated on Hadamard states or vector states in the dense subspace $\mathcal{D}_\omega$ in the GNS Hilbert space of a Hadamard state ω. This would be enough to implement the computation of the back reaction of the quantum matter in a given state to the geometry of the spacetime through (5.43) especially in cosmological scenario (see Chap. 6). This program has actually been initiated much earlier than the algebraic approach was adopted in QFT in curved spacetime [7] and the notion of Hadamard state was invented, through several steps, in this context. The requirements a physically sensible object $\omega(:T_{ab}:_H(x))$ should satisfy was clearly discussed by several authors, Wald in particular (see [65] for a complete account and [30] for more recent survey). The most puzzling issue in this context perhaps concerns the interplay of the conservation requirement $\nabla_a\omega(:T^{ab}:_H(x)) = 0$ and the appearance of the *trace anomaly*. We shall come back to these issues later, at the end of Sect. 5.3.5.

5.3.3 The Notion of Wavefront Set and Its Elementary Properties

Microlocal analysis permits us to completely reformulate the theory of Hadamard states into a much more powerful formulation where, in particular, the Wick polynomials can be defined as proper operators and not only Hermitian quadratic forms.

Following [29, 62], let us start be introducing the notion of wavefront set. To motivate it, let us recall that a smooth function on $\mathbb{R}^m$ with compact support has a rapidly decreasing Fourier transform. If we take a distribution u in $\mathcal{D}'(\mathbb{R}^m)$ and multiply it by an $f \in \mathcal{D}(\mathbb{R}^m)$ with $f(x_0) \neq 0$, then uf is an element of $\mathcal{E}'(\mathbb{R}^m)$, *i.e.*, a distribution with compact support. If fu were smooth, then its Fourier transform $\widehat{fu}$ would be smooth and rapidly decreasing (with all its derivatives). The failure of fu to be smooth in a neighbourhood of x_0 can therefore be quantitatively described by the set of directions in Fourier space[10] where $\widehat{fu}$ is not rapidly decreasing. Of course it could happen that we choose f badly and therefore 'cut' some of the singularities of u at x_0. To see the full singularity structure of u at x_0, we therefore need to consider all test functions which are non-vanishing at x_0. With this in mind, one first defines the wavefront set of distributions on (open subsets of) $\mathbb{R}^m$ and then extends it to curved manifolds in a second step.

In the rest of the chapter $\mathcal{D}(\mathcal{M}) \stackrel{\text{def}}{=} C_0^\infty(\mathcal{M}, \mathbb{C})$ for every smooth manifold $\mathcal{M}$. An open neighbourhood G of $k_0 \in \mathbb{R}^m$ is called **conic** if $k \in G$ implies $\lambda k \in G$ for all $\lambda > 0$.

Definition 5.3.5 (*Wavefront set*) Let $u \in \mathcal{D}'(U)$, with open $U \subset \mathbb{R}^m$. A point $(x_0, k_0) \in U \times (\mathbb{R}^m \setminus \{0\})$ is called a **regular directed** point of u if there is $f \in \mathcal{D}(U)$ with $f(x_0) \neq 0$ such that, for every $n \in \mathbb{N}$, there is a constant $C_n \geq 0$ fulfilling

$$|\widehat{fu}(k)| \leq C_n(1 + |k|)^{-n}$$

for all k in an open conic neighbourhood of k_0. The **wavefront set** $WF(u)$, of $u \in \mathcal{D}'(U)$ is the complement in $U \times (\mathbb{R}^m \setminus \{0\})$ of the set of all regular directed points of u.

Remark 5.3.6 Obviously, if $u, v \in \mathcal{D}'(U)$ the wavefront set is not additive and, in general, one simply has $WF(u + v) \subset WF(u) \cup WF(v)$.

As, an elementary example, let us consider the wavefront set of the distribution $\delta_y(x) = \delta(x - y)$ on $\mathbb{R}^n$ [62, p. 103]:

$$WF(\delta_y) = \{(y, k_y) \in T^*\mathbb{R}^n \mid k_y \neq 0\}. \tag{5.69}$$

If $U \subset \mathbb{R}^m$ is an open and non-empty subset, T^*U is naturally identified with $U \times \mathbb{R}^m$. In the rest of the chapter $T^*U \setminus 0 \stackrel{\text{def}}{=} \{(x, p) \in T^*U \mid p \neq 0\}$.

If $U \subset \mathbb{R}^m$ is an open non-empty set, $\Gamma \subset T^*U \setminus 0$ is a **cone** when $(x, \lambda k) \in \Gamma$ if $(x, k) \in \Gamma$ and $\lambda > 0$. If the mentioned cone Γ is closed *in the topology of* $T^*U \setminus 0$, we define

$$\mathcal{D}'_\Gamma \stackrel{\text{def}}{=} \{u \in \mathcal{D}'(U) \mid WF(u) \subset \Gamma\}.$$

[10] Our convention for the Fourier transform is so that $f(x) = \frac{1}{(2\pi)^m} \int e^{-ikx} \hat{f}(k)\, d^m k$. This convention agrees with those of [29, 52, 53], but has the opposite sign in the exponential with respect to [62]. This means that our wavefont sets need to be negated to be compared to those of [62]. Fortunately, in all cases where this is done, the wavefront sets happen to be negation symmetric.

Remark 5.3.7 All these definitions can be restated for the case of U replaced with a general smooth manifold and we shall exploit this opportunity shortly.

We are in a position to define a relevant notion of convergence [39].

Definition 5.3.8 (*Convergence in Hörmander pseudotopology*) If $u_j \in \mathcal{D}'_\Gamma(U)$ is a sequence and $u \in \mathcal{D}'_\Gamma(U)$, we write $u_j \to u$ in $\mathcal{D}'_\Gamma(U)$ if both the conditions below hold.

(i) $u_j \to u$ weakly in $\mathcal{D}'(U)$ as $j \to +\infty$,

(ii) $\sup_j \sup_V |p|^N |\widehat{\phi u_j}(p)| < \infty$, $N = 1, 2, \ldots$, if $\phi \in \mathcal{D}(U)$ and $V \subset T^*U$ is any closed cone, whose projection on U is supp (ϕ), such that $\Gamma \cap V = \emptyset$.

In this case, we say that u_j converges to u in the **Hörmander pseudotopology**.

It turns out that test functions (whose wavefront set is always empty as said below) are dense even with respect to that notion of convergence [39].

Proposition 5.3.9 *If $u \in \mathcal{D}'_\Gamma(U)$, there is a sequence of smooth functions $u_j \in \mathcal{D}(U)$ such that $u_j \to u$ in $\mathcal{D}'_\Gamma(U)$.*

Let us immediately state a few elementary properties of wavefront sets [20, 38, 39, 62]. We remind the reader that $x \in U$ is a **regular point** of a distribution $u \in \mathcal{D}'(U)$ if there is an open neighborhood $O \subset U$ of x such that $\langle u, f\rangle = \langle h_u, f\rangle$ for some $h_u \in \mathcal{D}(U)$ and every $f \in \mathcal{D}(U)$ supported in O. The closure of the complement of the set of regular points is the **singular support** of u by definition.

Theorem 5.3.10 (Elementary properties of WF) *Let $u \in \mathcal{D}'(U)$, $U \subset \mathbb{R}^m$ open and non-empty.*

(a) u is smooth if and only if $WF(u)$ is empty. More precisely, the singular support *of u is the projection of $WF(u)$ on $\mathbb{R}^m$.*

(b) If P is a partial differential operator on U with smooth coefficients:

$$WF(Pu) \subset WF(u)\,.$$

(c) Let $V \subset \mathbb{R}^m$ be an open set and let $\chi : V \to U$ be a diffeomorphism. The pull-back $\chi^* u \in \mathcal{D}'(V)$ of u defined by $\chi^*u(f) = u(\chi_* f)$ for all $f \in \mathcal{D}(V)$ fulfils

$$WF(\chi^* u) = \chi^* WF(u) \stackrel{def}{=} \left\{(\chi^{-1}(x), \chi^* k) \mid (x, k) \in WF(u)\right\},$$

where $\chi^* k$ denotes the pull-back of χ in the sense of cotangent vectors.

(d) Let $V \subset \mathbb{R}^n$ be an open set and $v \in \mathcal{D}'(V)$, then $WF(u \otimes v)$ is included in

$$(WF(u) \times WF(v)) \cup ((\,supp\;\; u \times \{0\}) \times WF(v)) \cup (WF(u) \times (\,supp\;\; v \times \{0\}))\,.$$

(e) Let $V \subset \mathbb{R}^n$, $K \in \mathcal{D}'(U \times V)$ and $f \in \mathcal{D}(V)$, then

$$WF(\mathcal{K} f) \subset \{(x, p) \in TU \setminus 0 \mid (x, y, p, 0) \in WF(K) \; for\; some\;\; y \in \; supp\;\; (f)\}\,,$$

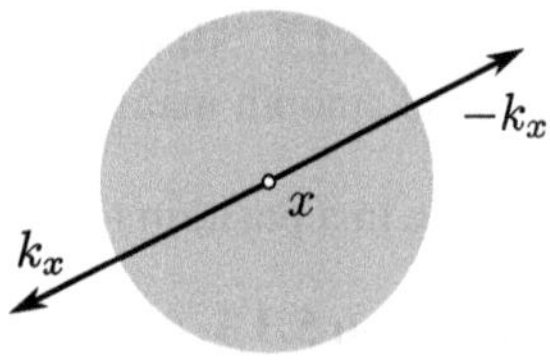

Fig. 5.1 Wavefront set of $\delta(x, y)$ on $\mathcal{M} \times \mathcal{M}$, defined in (5.70), consists of points of the form $(x, x, k_x, -k_x)$, $(x, k_x) \in T^*\mathcal{M} \setminus 0$

where $\mathcal{K} : \mathcal{D}(V) \mapsto \mathcal{D}'(U)$ is the continuous linear map associated to K in view of Schwartz kernel theorem.

The result (e), with a suitably improved statement, can be extended to the case of f replaced by a distribution [39].

From (c) we conclude that the wavefront set transforms covariantly under diffeomorphisms as a subset of T^*U, with U an open subset of $\mathbb{R}^m$. Therefore we can immediately extend the definition of WF to distributions on a manifold $\mathcal{M}$ simply by patching together wavefront sets in different coordinate patches of $\mathcal{M}$ with the help of a partition of unity. As a result, for $u \in \mathcal{D}'(\mathcal{M})$, $WF(u) \subset T^*\mathcal{M} \setminus 0$. Also the notion of convergence in the Hörmander pseudotopology easily extends to manifolds. All the statements of Theorem 5.3.10 extend to the case where U and V are smooth manifolds.

Following up on (5.69), an elementary example of a distribution on a manifold is $\delta(x, y)$ defined on $\mathcal{M} \times \mathcal{M}$. Its wavefront set is (Fig. 5.1)

$$WF(\delta) = \{(x, x, k_x, -k_x) \in T^*\mathcal{M}^2 \setminus 0 \mid (x, k_x) \in T^*\mathcal{M} \setminus 0\}. \tag{5.70}$$

The necessity of the sign reversal in the covector $-k_x$ corresponding to the second copy of $\mathcal{M}$ can be seen from the formula $\delta(x, y) = \delta(x - y)$ on $\mathbb{R}^n$.

To conclude this very short survey, we wish to stress some remarkable results of wavefront set technology respectively concerning (a) the theorem of *propagation of singularities*, (b) the *product of distributions*, (c) *composition of kernels*.

Let us start with an elementary version of the celebrated *theorem of propagation of singularities* formulated as in [62].

Remark 5.3.11

(1) Let us remind the reader that if, in local coordinates, $P = \sum_{|\alpha| \leq m} a_\alpha(x)\partial^\alpha$ is a differential operator of order $m \geq 1$ (it is assumed that $a_\alpha \neq 0$ for some α with $|\alpha| = m$) on a manifold $\mathcal{M}$, where a is a *multi-index* [39], and a_α are smooth coefficients, then the polynomial $\sigma_P(x, p) = \sum_{|\alpha|=m} a_\alpha(x)(ip)^\alpha$ is called the **principal symbol** of P. It is possible to prove that $(x, \xi) \mapsto \sigma_P(x, p)$ determines a well defined function on $T^*\mathcal{M}$ which, in general is complex valued. The **characteristic set** of P, indicated by $char(P) \subset T^*\mathcal{M} \setminus 0$, denotes the set of zeros of σ_P made of *non-vanishing* covectors. The principal symbol σ_P can be used as a *Hamiltonian function* on $T^*\mathcal{M}$ and the maximal solutions of Hamilton equations define the **local flow** of σ_P on $T^*\mathcal{M}$.

(2) The principal symbol of the Klein-Gordon operator is $-g^{ab}(x)p_a p_b$. It is an easy exercise [62] to prove that if $\boldsymbol{M}$ is a Lorentzian manifold and P is a **normally hyperbolic operator**, i.e., the principal symbol is the same as the one of Klein-Gordon operator, then the integral curves of the local flow of σ_P are nothing but the lift to $T^*\mathcal{M}$ of the *geodesics* of the metric g parametrized by an *affine parameter*. Finally, $char(P) = \{(x, p) \in T^*\mathcal{M} \setminus 0 \mid g^{ab}(x)p_a p_b = 0\}$.

Theorem 5.3.12 (Microlocal regularity and propagation of singularities) *Let P be a differential operator on a manifold $\mathcal{M}$ whose principal symbol is real valued, if $u, f \in \mathcal{D}'(\mathcal{M})$ are such that $Pu = f$ then the following facts hold.*

(a) $WF(u) \subset char(P) \cup WF(f)$,

(b) $WF(u) \setminus Wf(f)$ is invariant under the local flow of σ_P on $T^*\mathcal{M} \setminus WF(f)$.

Let us conclude with the famous Hörmander definition of *product of distributions* [38, 39]. We need a preliminary definition. If $\Gamma_1, \Gamma_2 \subset T^*\mathcal{M} \setminus 0$ are closed cones,

$$\Gamma_1 + \Gamma_2 \stackrel{\text{def}}{=} \left\{(x, k_1 + k_2) \subset T^*\mathcal{M} \mid (x, k_1) \in \Gamma_1,\ (x, k_2) \in \Gamma_2 \text{ for some } x \in \mathcal{M}\right\}.$$

Theorem 5.3.13 (Product of distributions) *Consider a pair of closed cones $\Gamma_1, \Gamma_2 \subset T^*\mathcal{M} \setminus 0$. If*

$$\Gamma_1 + \Gamma_2 \not\ni (x, 0) \quad \textit{for all } x \in \mathcal{M},$$

then there is a unique bilinear map, the **product** *of u_1 and u_2,*

$$\mathcal{D}'_{\Gamma_1} \times \mathcal{D}'_{\Gamma_2} \ni (u_1, u_2) \mapsto u_1 u_2 \in \mathcal{D}'(\mathcal{M}),$$

such that

(i) it reduces to the standard pointwise product if $u_1, u_2 \in \mathcal{D}(\mathcal{M})$,

(ii) it is jointly sequentially continuous in the Hörmander pseudotopology: If $u_j^{(n)} \to u_j$ in $D_{\Gamma_j}(\mathcal{M})$ for $j = 1, 2$ then $u_1^{(n)} u_2^{(n)} \to u_1 u_2$ in $\mathcal{D}_\Gamma(\mathcal{M})$, where Γ is a closed cone in $T^*\mathcal{M} \setminus 0$ defined as $\Gamma \stackrel{def}{=} \Gamma_1 \cup \Gamma_2 \cup (\Gamma_1 \oplus \Gamma_2)$.

In particular the following bound always holds if the above product is defined:

$$WF(u_1 u_2) \subset \Gamma_1 \cup \Gamma_2 \cup (\Gamma_1 + \Gamma_2)\ . \tag{5.71}$$

From the examples (5.69) and (5.70) and the simple observation that

$$\mathbb{R}^n \setminus \{0\} + \mathbb{R}^n \setminus \{0\} = \mathbb{R}^n \ni 0, \tag{5.72}$$

it is clear that the multiplication of two δ-functions with overlapping supports, as is to be expected, does not satisfy the above conditions.

Let us come to the last theorem concerning the composition of distributional kernels. Let X, Y be smooth manifolds. If $K \in \mathcal{D}'(X \times Y)$, the continuous map associated to K by the Schwartz kernel theorem will be denoted by $\mathcal{K} : \mathcal{D}(Y) \to \mathcal{D}'(X)$. We shall also adopt the following standard notations:

$$WF(K)_X \stackrel{\text{def}}{=} \{(x,p) \mid (x,y,p,0) \in WF(K) \quad \text{for some } y \in Y\},$$
$$WF(K)_Y \stackrel{\text{def}}{=} \{(y,q) \mid (x,y,0,q) \in WF(K) \quad \text{for some } x \in X\},$$
$$WF'(K) \stackrel{\text{def}}{=} \{(x,y,p,q) \mid (x,y,p,-q) \in WF(K)\},$$
$$WF'(K)_Y \stackrel{\text{def}}{=} \{(y,q) \mid (x,y,0,-q) \in WF(K) \quad \text{for some } x \in X\}.$$

Theorem 5.3.14 (Composition of kernels) *Consider three smooth manifolds X, Y, Z and $K_1 \in \mathcal{D}'(X \times Y)$, $K_2 \in \mathcal{D}'(Y \times Z)$. If $WF'(K_1)_Y \cap WF(K_2)_Y = \emptyset$ and the projection*

$$supp\ K_2 \ni (y,z) \mapsto z \in Z$$

is proper (that is, the inverse of a compact set is compact), then the composition $\mathcal{K}_1 \circ \mathcal{K}_2$ is well defined, giving rise to $K \in \mathcal{D}'(X,Z)$, and reduces to the standard one when the kernel are smooth. It finally holds (the symbol $\circ$ denoting the composition of relations)

$$WF'(K) \subset WF'(K_1) \circ WF'(K_2) \cup (WF(K_1)_X \times Z \times \{0\}) \cup (X \times \{0\} \times WF'(K_2)_Z). \tag{5.73}$$

Comparing with (5.70), note that $WF'(\delta)$ is the diagonal subset $\Delta \subset T^*\mathcal{M} \times T^*\mathcal{M}$. In the composition of relations, Δ acts as an identity, which is consistent with the above theorem and the fact that $\delta(x,y)$ acts as an identity for the composition of distributional kernels.

5.3.4 Microlocal Reformulation

Let us focus again on the two-point function of Minkowski quasifree vacuum state. Form (5.37) we see that the singular support of $\omega_{\mathbb{M}2}(x,y)$ is the set of couples $(x,y) \in \mathbb{M} \times \mathbb{M}$ such that $x - y$ is *light like*. From (a) in Theorem 5.3.10, we conclude that $WF(\omega_{\mathbb{M}2})$ must project onto this set. On the other hand (5.38) can be re-written as

$$\omega_{\mathbb{M}2}(x,y) = \frac{1}{(2\pi)^3} \int_{\mathbb{R}^4} e^{-i(px+qy)} \theta(p^0)\delta(p^2+m^2)\delta(p+q) d^4q d^4p\,, \tag{5.74}$$

where translational invariance is responsible for the appearance of $\delta(p+q)$ in (5.74). From this couple of facts, also noticing the presence of $\theta(p^0)$ in the integrand, one guesses that the wavefront set of the Minkowski two-point function must be

$$WF(\omega_{\mathbb{M}2}) = \left\{(x,y,p,-p) \in T^*\mathcal{M}^2 \mid p^2 = 0,\ p \,||\, (x-y),\ p^0 > 0\right\}. \tag{5.75}$$

Identity (5.75) is, in fact, correct and holds true also for $m = 0$ [54]. The condition $p^0 > 0$ encodes the *energy positivity* of the Minkowski vacuum state. Notice that the couples $(x, y) \in \mathbb{M} \times \mathbb{M}$ giving contribution to the wavefront set are always connected by a *light-like geodesic* co-tangent to p. For $x = y$ there are infinitely many such geodesics, if we allow ourselves to consider zero length curves (consisting of a single point) with a given tangent vector.

The structure (5.75) of the wavefront set of the two-point function of Minkowski vacuum is a particular case of the general notion of a Hadamard state. We re-adapt here the content of the cornerstone papers [52, 53] to our formulation. We note that we do not make use of the global Hadamard condition (see (4) in Remark 5.3.3). The following theorem collects various results of [52, 53].

Theorem 5.3.15 ("Radzikowski theorem") *For a 4-dimensional globally hyperbolic (time oriented) spacetime* $\boldsymbol{M}$ *and referring to the unital* $*$*-algebra of Klein-Gordon quantum field* $\mathcal{A}(\boldsymbol{M})$ *with* $m^2, \xi \in \mathbb{R}$ *arbitrarily fixed, let* ω *be a state on* $\mathcal{A}(\boldsymbol{M})$, *not necessarily quasifree.*
(a) The following statements are equivalent,
(i) ω *is Hadamard in the sense of Definition 5.3.2,*
(ii) the wavefront set of the two-point function ω_2 *has the* **Hadamard form** *on* $\boldsymbol{M}$ *or equivalently, it satisfies the* **microlocal spectrum condition** *on* $\boldsymbol{M}$*:*

$$WF(\omega_2) = \left\{(x, y, k_x, -k_y) \in T^*\mathcal{M}^2 \setminus 0 \mid (x, k_x) \sim (y, k_y),\ k_x \rhd 0\right\} \stackrel{def}{=} \mathcal{H}. \tag{5.76}$$

Here, $(x, k_x) \sim (y, k_y)$ *means that there exists a null geodesic* γ *connecting* x *to* y *such that* k_x *is coparallel and cotangent to* γ *at* x *and* k_y *is the parallel transport of* k_x *from* x *to* y *along* γ, *Fig. 5.2.* $k_x \rhd 0$ *means that* k_x *does not vanish and is*

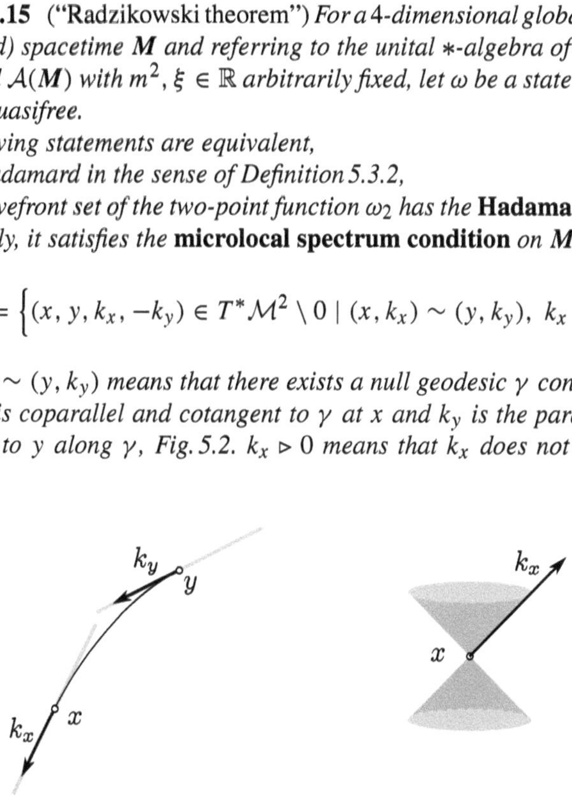

Fig. 5.2 The null geodesic relation $(x, k_x) \sim (x, k_y)$ defined in Theorem 5.3.15. The points x and y must be linked by a null geodesic, the covectors k_x and k_y must be parallel transported images of each other and both covectors must be coparallel, all with respect to the same null geodesic. Any causal ordering between x and y is admissible. Also, k_x, $-k_x$ and λk_x ($\lambda \neq 0$) are all considered coparallel to the same geodesic. In the coincident case, $x = y$, we agree that there are infinitely many (zero-length) null geodesics joining x to itself, corresponding to different non-vanishing null covectors $k_x \in T^*_x\mathcal{M}$

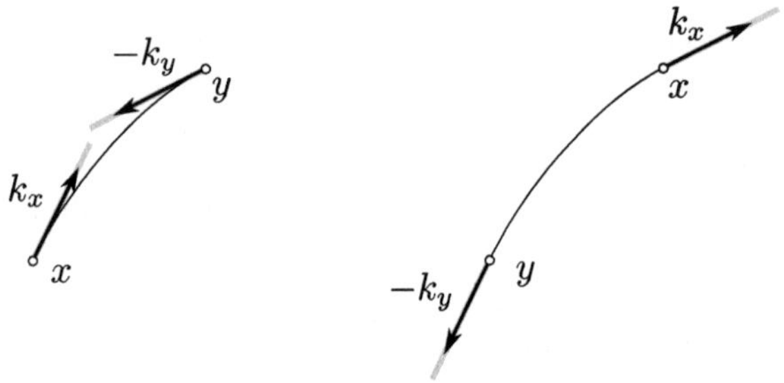

Fig. 5.3 The Hadamard form $\mathcal{H}$ of a wavefront set, as defined in Theorem 5.3.15. It consists of a subset of points $(x, y, k_x, -k_y) \in T^*\mathcal{M}^2$, where $(x, k_x) \sim (y, k_y)$ are linked but the null geodesic relation (Fig. 5.2). The restriction is that $k_x \rhd 0$, meaning that $k_x(v) \geq 0$ for any future-directed $v \in T_x\mathcal{M}$. We illustrate the two possible causal orderings $x \in J^-(y)$ and $x \in J^+(y)$

future-directed ($k_x(v) \geq 0$ for all future-directed $v \in T_x\mathcal{M}$), Fig. 5.3.
(b) If ω' is another Hadamard state on $\mathcal{A}(\boldsymbol{M})$, then $\omega_2 - \omega_2' \in C^\infty(\mathcal{M} \times \mathcal{M}, \mathbb{C})$.

Proof (a) Suppose that ω satisfies (i), then it is globally Hadamard in the sense[11] of [52] due to Theorem 9.2 in [53]. Theorem 5.1 in [52] implies that (ii) holds. Conversely, if (ii) is valid, Theorem 5.1 in [52] entails that ω is globally and thus locally Hadamard so that (i) holds true. (b) immediately arises from Theorem 4.3 in [53]. □

It is also helpful to have a characterization of the wavefront set of the retarded and advanced fundamental solutions [52, 62].

Proposition 5.3.16 *The retarded and advanced fundamental solutions of the Klein-Gordon operator $P = \Box_{\boldsymbol{M}} + m^2 + \xi R$ on $\boldsymbol{M}$, $E^+, E^- \in \mathcal{D}'(\mathcal{M} \times \mathcal{M})$ respectively, have the following wavefront sets (Fig. 5.4):*

$$WF(E^\pm) = WF(\delta) \cup \left\{(x, y, k_x, -k_y) \in T^*\mathcal{M}^2 \setminus 0 \mid (x, k_x) \sim (y, k_y), x \in J^\pm(y)\right\} \stackrel{def}{=} \mathcal{F}_\pm, \tag{5.77}$$

where $\sim$ denotes the same relation as in Theorem 5.3.15.

With this result and the microlocal technology previously introduced we can prove some remarkable properties of Hadamard states, especially in relation with what was already discussed in (4) in Remark 5.3.3. The second statement, for $n = 4$, implies that the singularity structure of Hadamard states propagates through the spacetime.

[11]Results in [52, 53] are stated for $\xi = 0$ in KG operator, however they are generally valid for m^2 replaced by a given smooth function, as specified at the beginning of p. 533 in [52].

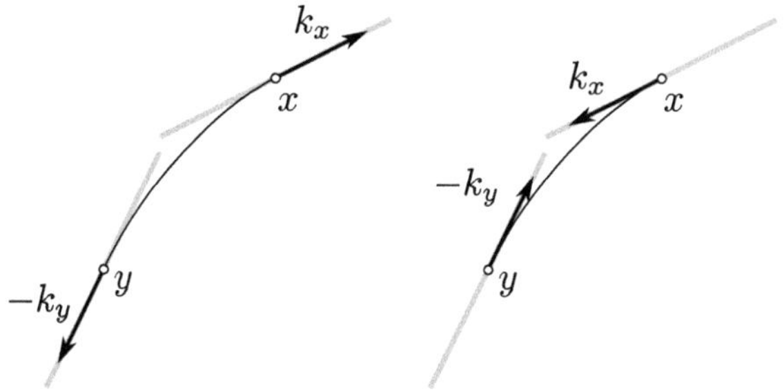

Fig. 5.4 The wavefront sets of the retarded fundamental solution E^+ of the Klein-Gordon operator, as defined in Proposition 5.3.16, consist of the union of $WF(\delta)$ (Fig. 5.1) and of the points $(x, k_x, y, -k_y) \in T^*\mathcal{M}^2$, where $(x, k_x) \sim (y, k_y)$ are linked by the geodesic relation (Fig. 5.2), with the causal precedence condition $x \in J^+(y)$. We illustrate the two cases when k_x is coparallel and anti-coparallel to the future-directed geodesic from y to x. The wavefront set of the advanced fundamental solution E^- is defined in the same way, with the exception that we require the causal precedence condition $x \in J^-(y)$ instead

Proposition 5.3.17 *Consider a state ω on $\mathcal{A}(\boldsymbol{M})$, with $\omega_2 \in \mathcal{D}'(\mathcal{M} \times \mathcal{M})$, where $\boldsymbol{M}$ is a (time oriented) globally hyperbolic spacetime with dimension $n \geq 2$. The following facts hold.*

(a) If $WF(\omega_2)$ has the Hadamard form, then $\mathcal{M} \ni x \mapsto \omega_2(x, f)$ is smooth for every $f \in C_0^\infty(\mathcal{M})$.

(b) If $WF(\omega_2\restriction_{O\times O})$ has the Hadamard form on O, where O is an open neighborhood of a smooth spacelike Cauchy surface Σ of $\boldsymbol{M}$, then $WF(\omega_2)$ has the Hadamard form on $\boldsymbol{M}$.

Proof (a) From (e) in Theorem 5.3.10 and the Hadamard form of $WF(\omega_2)$ we conclude that $WF(\omega_2(\cdot, f)) = \emptyset$. Next, (a) in Theorem 5.3.10 implies the thesis.

(b) The 2-point function $\omega_2(x, y)$ is a bisolution of the Klein-Gordon operator $P = \Box_{\boldsymbol{M}} + m^2 + \xi R$, as in (5.8). So the value of $\omega_2(f, g)$, for $f, g \in C_0^\infty(\mathcal{M})$, depends on the arguments only up to the addition of any term from $P[C_0^\infty(\mathcal{M})]$. In fact, we can choose $h, k \in C^\infty$ such that supp $(f + P[h])$ and supp $(g + P[k])$ are both contained in O. More precisely, we can define an $S \in \mathcal{D}'(O \times \mathcal{M})$ such that the corresponding operator maps $\mathcal{S}\colon C_0^\infty(\mathcal{M}) \to C_0^\infty(O)$ and we have the identity $\omega_2 = \mathcal{S}^t \circ \omega_2 \circ \mathcal{S}$. Then, using the result of Theorem 5.3.14 on the composition of kernels and the fact that ω_2 has the Hadamard form on O, we can show that ω_2 has the Hadamard form on all of $\mathcal{M}$.

Consider a smooth partition of unity $\chi_+ + \chi_- = 1$ adapted to the Cauchy surface Σ. That is, there exist two other Cauchy surfaces, Σ_+ in the future of Σ and Σ_- in the past of Σ, such that supp $\chi_+ \subset J^+(\Sigma_-)$ and supp $\chi_- \subset J^-(\Sigma_+)$. Such an adapted partition of unity always exists if O is globally hyperbolic in its own right and, if not, since $\boldsymbol{M}$ is globally hyperbolic, any open neighborhood of Σ will contain a possibly smaller neighborhood of Σ that is also globally hyperbolic [5, 6].

Let $\mathcal{S}f = f - P[\chi_+ E^- f + \chi_- E^+ f]$, with the corresponding integral kernel

$$S(x,y) = \delta(x,y) - P_x[\chi_+(x)E^-(x,y) + \chi_-(x)E^+(x,y)], \tag{5.78}$$

where the subscript on P_x means that it is acting only on the x variable. A straight forward calculation shows that S has the desired properties. Multiplication by a smooth function and the application of a differential operator does not increase the wavefront set, hence

$$WF(S) \subset WF(\delta) \cup WF(E^-) \cup WF(E^+) \tag{5.79}$$

as a subset of $T^*(O \times \mathcal{M})$. The δ-function has the wavefront set

$$WF(\delta) = \{(x,x,k_x,-k_x) \in T^*\mathcal{M}^2 \setminus 0 \mid (x,k_x) \in T^*\mathcal{M} \setminus 0\}\,. \tag{5.80}$$

The wavefront sets $\mathcal{F}_\pm = WF(E^\pm)$ of the retarded and advanced fundamental solutions was given in Proposition 5.3.16. The Hadamard form $\mathcal{H}$ of the wavefront set was defined in Theorem 5.3.15. We can now appeal to Theorem 5.3.14 on the wavefront set of the composition of kernels to how that $WF(\omega_2) = WF(\mathcal{S}^t \circ \omega_2 \circ \mathcal{S}) \subset \mathcal{H}$. The first thing to check is that $WF(S)_{\mathcal{M}_i}$, $WF(\omega_2)_{\mathcal{M}_i}$, $i = 1,2$ denoting respectively the first and the second factor in $\mathcal{M} \times \mathcal{M}$, are all empty, because they contain no element of the form $(x,y,k_x,0)$ or $(x,y,0,k_y)$. Second, due to the hypothesis $WF'(\omega_2)|_O \subset \mathcal{H}'_O$, the symmetry of the composition and the fact that composition with $\delta(x,y)$ leaves any wavefront set invariant, it is sufficient to check that the compositions of wavefront sets as relations satisfy $\mathcal{H}'_O \circ \mathcal{F}'_\pm \subset \mathcal{H}'_\mathcal{M}$.

Consider any $(x,y,k_x,k_y) \in \mathcal{H}'_O$ and $(y,z,k_y,k_z) \in \mathcal{F}'_\pm$, so that $(x,z,k_x,k_z) \in \mathcal{H}'_O \circ \mathcal{F}'_\pm$. Then $(x,k_x) \sim (y,k_y)$ and $(y,k_y) \sim (z,k_z)$ in $\mathcal{M}$ according to the relation $\sim$ defined in Theorem 5.3.15, so that $(x,k_x) \sim (z,k_z)$ by transitivity of that relation in $\mathcal{M}$. The only question is about the allowed orientations of k_x and k_y. By the Hadamard condition on O, we have $k_x \rhd 0$ and $k_y \rhd 0$. On the other hand, the condition of being a point in $\mathcal{F}_\pm$ induces the condition that either both $k_y \rhd 0$ and $k_z \rhd 0$ or both $k_y \lhd 0$ and $k_z \lhd 0$. Combining the two conditions we find that $k_z \rhd 0$, and hence that $(x,z,k_x,k_z) \in \mathcal{H}'_\mathcal{M}$. This concludes the proof. □

Remark 5.3.18 With an elementary re-adaptation, statement (b) holds true weakening the hypotheses, only requiring that $\omega_2 \in \mathcal{D}'(\mathcal{M} \times \mathcal{M})$ and that it satisfies KG equation in both arguments up to smooth functions $r, l \in C^\infty(\mathcal{M} \times \mathcal{M}, \mathbb{C})$, i.e. $P_x\omega_2(x,y) = l(x,y)$, $P_y\omega_2(x,y) = r(x,y)$. In this form, it closes a gap[12] present in the proof of the main result of [52], Theorem 5.1 (the fact that **1** implies **3**), and proves the statement on p. 548 of [52] immediately after the proof of the mentioned theorem. It should be mentioned that the same gap had been previously explicitly identified and filled in the work of Sahlmann and Verch [56]. These authors merged the partial proof of Radzikowski with the more restrictive result on the 'propagation

[12]The gap is the content of the three lines immediately before th proof **(ii) 3** $\Rightarrow$ **2** on p. 547 of [52]: The reasoning presented there cannot exclude elements of the form either $(x_1,x_2,0,p_2)$ or $(x_1,x_2,p_1,0)$ from $WF(\omega_2)$ outside $\mathcal{N}$. The idea of our proof was suggested by N. Pinamonti to the authors.

of Hadamard form' obtained previously in the works [21, 23, 41] without the methods of microlocal analysis. Somewhat later, the same gap was also filled in the thesis of Sanders [57], who relied on purely microlocal but somewhat sophisticated methods developed earlier in [63]. On the other hand, our method, though sharing some similarity in spirit with the ideas in [21, 23, 41], is both purely microlocal and rather elementary. In fact, it only takes advantage of the microlocal analysis in the guise of the theorem on the composition of wavefront sets.

The microlocal formulation gave rise to noticeable results also closing some long standing problems. In particular it was proved that the so called Unruh state describing black hole radiation is Hadamard [14] and that the analogous state, describing thermal radiation in equilibrium with a black hole, the so called Hartle-Hawking state is similarly Hadamard [58]. These results are physically important because they permit one to compute the back reaction of the quantum radiation on the geometry, since the averaged, renormalized stress-energy tensor $\omega(:T_{ab}:)$ can be defined in these states as previously discussed ((3) in Remark 5.3.4). Other recent applications concerned the definition of relevant Hadamard states in asymptotically flat spacetimes at null infinity [26, 48], and spacelike infinity [25]. Natural Hadamard states for cosmological models have been discussed [12] also in relation with the problem of the Dark Energy [13]. An improved semiclassical formulation where Einstein equations and the equation of evolution of the Hadamard quantum state and observables are solved simultaneously has been proposed in [51]. See [4, 29] for recent reviews also regarding fields with spin or helicity, in particular [15] for the vector potential field.

5.3.5 *Algebra of Wick Products*

Let us come to the proof of existence of Wick monomials $:\phi^n:(f)$ as algebraic objects, since we only have defined the expectation values $\omega_\Psi(:\phi^n:(f))$ in (5.68). We first introduce normal Wick products *defined with respect to a reference quasifree Hadamard state* ω [9, 10, 35]. Referring to the GNS triple for ω, $(\mathcal{H}_\omega, \mathcal{D}_\omega, \pi_\omega, \Psi_\omega)$ Define the elements, symmetric under interchange of $f_1, \dots, f_n \in \mathcal{D}(\mathcal{M})$,

$$\hat{W}_{\omega,0} \stackrel{\text{def}}{=} \mathbf{1}\,, \quad \hat{W}_{\omega,n}(f_1, \dots, f_n) \stackrel{\text{def}}{=} :\hat{\phi}_\omega(f_1)\cdots\hat{\phi}_\omega(f_n):_\omega \;\in \mathcal{A}(\boldsymbol{M})$$

for $n = 1, 2, \dots,$ where as before,

$$:\hat{\phi}_\omega(x_1)\cdots\hat{\phi}_\omega(x_n):_\omega \stackrel{\text{def}}{=} \frac{1}{i^n}\frac{\delta^n}{\delta f(x_1)\cdots\delta f(x_n)}\bigg|_{f=0} e^{i\hat{\phi}(f)+\frac{1}{2}\omega_2(f,f)} \tag{5.81}$$

The operators $\hat{W}_{\omega,n}(f_1, \dots, f_n)$ can be extended to (or directly defined on) [9, 35] an invariant subspace of $\mathcal{H}_\omega$, the **microlocal domain of smoothness** [9, 35], $D_\omega \supset \mathcal{D}_\omega$, which is dense, invariant under the action of $\pi_\omega(\mathcal{A}(\boldsymbol{M}))$ and the associated unitary

Weyl operators, and contains Ψ_ω and all of unit vectors of $\mathcal{H}_\omega$ which induce Hadamard quasifree states on $\mathcal{A}(\boldsymbol{M})$. The map

$$f_1 \otimes \cdots \otimes f_n \mapsto \hat{W}_{\omega,n}(f_1, \ldots, f_n)$$

uniquely extends by complexification and linearity to a map defined on

$$\mathcal{D}(\mathcal{M}) \otimes \cdots \otimes \mathcal{D}(\mathcal{M}) .$$

Finally, if $\Psi \in D_\omega$, the map $\mathcal{D}(\mathcal{M}) \otimes \cdots \otimes \mathcal{D}(\mathcal{M}) \ni h \mapsto \hat{W}_{\omega,n}(h)\Psi$ turns out to be continuous with respect to the relevant topologies: The one of $\mathcal{H}_\omega$ in the image and the one of $\mathcal{D}(\mathcal{M}^n)$ in the domain. A *vector-valued distribution* $\mathcal{D}(\mathcal{M}^n) \ni h \mapsto \hat{W}_{\omega,n}(h)$, uniquely arises this way. Actually, since $:\hat{\phi}_\omega(f_1)\cdots\hat{\phi}_\omega(f_n):_\omega$ is symmetric by construction, the above mentioned distribution is similarly symmetric and can be defined on the subspace $\mathcal{D}_n(\mathcal{M}) \subset \mathcal{D}(\mathcal{M}^n)$ of the symmetric test functions:

$$\mathcal{D}(\mathcal{M}^n) \ni h \mapsto \hat{W}_{\omega,n}(h) .$$

By Lemma 2.2 in [9], if $\Psi \in D_\omega$ the wavefront set $WF\left(\hat{W}_{\omega,n}(\cdot)\Psi\right)$ of the vector-valued distributions $t \mapsto \hat{W}_{\omega,n}(t)\Psi$, is contained in the set

$$\mathbf{F}_n(\boldsymbol{M}) \stackrel{\text{def}}{=} \{(x_1, k_1, \ldots, x_n, k_n) \in (T^*\mathcal{M})^n \setminus \{0\} | k_i \in V^-_{x_i}, i = 1, \ldots, n\}, \quad (5.82)$$

with $V_x^{+/-}$ denoting the set of all nonzero time-like and light-like co-vectors at x which are future/past directed. Theorem 5.3.13, which can be proved to hold in this case too, implies that we are allowed to define the product between a distribution t and a vector-valued distribution $\hat{W}_{\omega,n}(\cdot)\Psi$ provided $WF(t) + \mathbf{F}_n(M, \mathbf{g}) \not\ni \{(x,0) \mid x \in \mathcal{M}^n\}$. To this end, with $\mathcal{D}'_n(\mathcal{M}) \subset \mathcal{D}'(\mathcal{M}^n)$ denoting the subspace of symmetric distributions, define

$$\mathcal{E}'_n(\mathcal{M}) \stackrel{\text{def}}{=} \left\{t \in \mathcal{D}'_n(\mathcal{M}) \mid \text{ supp } t \text{ is compact, } WF(t) \subset \mathbf{G}_n(\mathcal{M})\right\}$$

where

$$\mathbf{G}_n(\mathcal{M}) \stackrel{\text{def}}{=} T^*\mathcal{M}^n \setminus \left(\bigcup_{x\in\mathcal{M}} (V_x^+)^n \cup \bigcup_{x\in\mathcal{M}} (V_x^-)^n \right).$$

It holds $WF(t) + \mathbf{F}_n(\mathcal{M}) \not\ni \{(x,0) \mid x \in \mathcal{M}^n\}$ for $t \in \mathcal{E}'_n(\mathcal{M})$. By consequence, the product

$$t \odot \hat{W}_{\omega,n}(\cdot)\Psi$$

of the distributions t and $\hat{W}_{\omega,n}(\cdot)\Psi$ can be defined for every $\Psi \in D_\omega$ and it turns out to be a well-defined vector-valued symmetric distribution, $\mathcal{D}_n(M) \ni f \mapsto t \odot \hat{W}_{\omega,n}(f)\Psi$, with values in D_ω. Thus, we have also defined an operator valued symmetric distribution, $\mathcal{D}_n(M) \ni f \mapsto t \odot \hat{W}_{\omega,n}(f)$, defined on and leaving invariant the domain D_ω, acting as $\Psi \mapsto t \odot \hat{W}_{\omega,n}(f)\Psi$. This fact permits us to smear $\hat{W}_{\omega,n}$ with $t \in \mathcal{E}'_n(\mathcal{M})$, just defining

$$\hat{W}_{\omega,n}(t) \stackrel{\text{def}}{=} \left(t \odot \hat{W}_{\omega,n}\right)(f),$$

where $f \in \mathcal{D}_n(\mathcal{M})$ is equal to 1 on supp t. It is simple to prove that the definition does not depend on f and the new smearing operation reduces to the usual one for $t \in \mathcal{D}_n(\mathcal{M}) \subset \mathcal{E}'_n(M, \mathbf{g})$. Finally, since $f\delta_n \in \mathcal{E}'_n(\mathcal{M})$ for $f \in \mathcal{D}(M)$, where δ_n is the Dirac delta supported on the diagonal of $\mathcal{M}^n = \mathcal{M} \times \cdots \times \mathcal{M}$ (n times), the following operator-valued distribution is well-defined on D_ω which, is then an invariant subspace,

$$f \mapsto :\hat{\phi}^n:_\omega (f) \stackrel{\text{def}}{=} \hat{W}_{\omega,n}(f\delta_n),$$

Definition 5.3.19 $:\hat{\phi}^n:_\omega (f)$ is the **normal ordered product of n field operators with respect to** ω. $\mathcal{W}_\omega(\boldsymbol{M})$ is the $*$-algebra generated by $\mathbf{1}$ and the operators $\hat{W}_{\omega,n}(t)$ for all $n \in \mathbb{N}$ and $t \in \mathcal{E}'_n(M, \mathbf{g})$ with involution given by $\hat{W}_{\omega,n}(t)^* \stackrel{\text{def}}{=} \hat{W}_{\omega,n}(t)^\dagger \restriction_{D_\omega}$ $(= \hat{W}_{\omega,n}(\overline{t}))$.

Remark 5.3.20

(1) As proved in [35], each product $\hat{W}_{\omega,n}(t)\hat{W}_{\omega,n'}(t')$ can be decomposed as a finite linear combination of terms $\hat{W}_{\omega,m}(s)$ extending the Wick theorem, and other natural identities, in particular related with commutation relations, hold.

(2) $\pi_\omega(\mathcal{A}(\boldsymbol{M}))$ turns out to be a sub $*$-algebra of $\mathcal{W}_\omega(\boldsymbol{M})$ since $\hat{\phi}_\omega(f) = :\hat{\phi}:_\omega (f)$ for $f \in \mathcal{D}(M)$.

If ω, ω' are two quasifree Hadamard states, $\mathcal{W}_\omega(\boldsymbol{M})$ and $\mathcal{W}_{\omega'}(\boldsymbol{M})$ are isomorphic (not unitarily in general) under a canonical $*$-isomorphism

$$\alpha_{\omega'\omega} : \mathcal{W}_\omega(\boldsymbol{M}) \to \mathcal{W}_{\omega'}(\boldsymbol{M}),$$

as shown in Lemma 2.1 in [35]. Explicitly, $\alpha_{\omega'\omega}$ is induced by linearity from the requirements

$$\alpha_{\omega'\omega}(\mathbf{1}) = \mathbf{1}, \quad \alpha_{\omega'\omega}(W_{n,\omega}(t)) = \sum_k W_{n-2k,\omega'}(\langle d^{\otimes k}, t\rangle), \tag{5.83}$$

where $d(x_1, x_2) \stackrel{\text{def}}{=} \omega(x_1, x_2) - \omega'(x_1, x_2)$ (only the symmetric part matters here) and

$$\langle d^{\otimes k}, t\rangle(x_1, \ldots, x_{n-2k}) \stackrel{\text{def}}{=} \frac{n!}{(2k)!(n-2k)!} \int_{\mathcal{M}^{2k}} t(y_1, \ldots, y_{2k}, x_1, \ldots, x_{n-2k}) \times \prod_{i=1}^{k} d(y_{2i-1}, y_{2i}) \mathrm{dvol}_{\boldsymbol{M}}(y_{2i-1}) \mathrm{dvol}_{\boldsymbol{M}}(y_{2i}) \tag{5.84}$$

for $2k \leq n$ and $\langle d^{\otimes k}, t\rangle = 0$ if $2k > n$.

These $*$-isomorphisms also satisfy

$$\alpha_{\omega''\omega'} \circ \alpha_{\omega'\omega} = \alpha_{\omega''\omega}$$

and

$$\alpha_{\omega'\omega}(\hat{\phi}_\omega(t)) = \hat{\phi}_{\omega'}(t)\,.$$

The idea behind these isomorphisms is evident: Replace everywhere ω by ω'. For instance

$$\alpha_{\omega'\omega}(:\hat{\phi}^2:_\omega (f)) = :\hat{\phi}^2:_{\omega'} (f) + \int_{\mathcal{M}} (\omega - \omega')(x,x) f(x) \mathrm{dvol}_{\boldsymbol{M}}\, \mathbf{1}$$

where $\omega - \omega'$ is smooth for (b) in Theorem 5.3.15.

One can eventually define an abstract unital $*$-algebra $\mathcal{W}(\boldsymbol{M})$, generated by elements $\mathbf{1}$ and $W_n(t)$ with $t \in \mathcal{E}'_n(\mathcal{M})$, isomorphic to each concrete unital $*$-algebra $\mathcal{W}_\omega(\boldsymbol{M})$ by $*$-isomorphisms $\alpha_\omega : \mathcal{W}(\boldsymbol{M}) \to \mathcal{W}_\omega(\boldsymbol{M})$ such that, if ω, ω' are quasifree Hadamard states, $\alpha_{\omega'} \circ \alpha_\omega^{-1} = \alpha_{\omega'\omega}$.

As above $\mathcal{A}(\boldsymbol{M})$ is isomorphic to the $*$-algebra of $\mathcal{W}(\boldsymbol{M})$ generated by $\mathbf{1}$ and $W_1(f) = :\hat{\phi}: (f) = \phi(f)$ for $f \in \mathcal{D}(\mathcal{M})$.

Remark 5.3.21 It is not evident how (Hadamard) states initially defined on $\mathcal{A}(\boldsymbol{M})$ (continuously) extend to states on $\mathcal{W}(\boldsymbol{M})$. This problem has been extensively discussed in [34] in terms of relevant topologies.

It is now possible to define a notion of local Wick monomial *which does not depend on a preferred Hadamard state.* If $t \in \mathcal{E}'_n(\mathcal{M})$ has support sufficiently concentrated around the diagonal of $\mathcal{M}^n$, realizing $\mathcal{W}(\boldsymbol{M})$ as $\mathcal{W}_\omega(\boldsymbol{M})$ for some quasifree Hadamard state ω, we define a **local covariant Wick polynomial** as

$$W_n(t)_H \stackrel{\text{def}}{=} \alpha_\omega^{-1}\left(\alpha_{H\omega}(W_{n,\omega}(t))\right)$$

where $\alpha_{H\omega}$ is defined as in (5.84) replacing ω' by the Hadamard parametrix H_{0^+}. One easily proves that this definition does not depend on the choice of the Hadamard state ω. The fact that the support of t is supposed to be concentrated around of the diagonal of $\mathcal{M}^n$ it is due to the fact that $H_\varepsilon(x,y)$ is defined only if x is sufficiently close to y. This definition is completely consistent with (5.68), where now the $:\phi(f_1)\cdots\phi(f_n):_H$ can be viewed as elements of $\mathcal{W}(\boldsymbol{M})$ and not only of $\mathcal{A}(\boldsymbol{M})$, and it makes sense to write in particular,

$$:\phi^2:_H (f) \stackrel{\text{def}}{=} W_2(f\delta_2)_H = \int_{\mathcal{M}^2} :\phi(x)\phi(y):_H \,\delta(x,y) f(x) \mathrm{dvol}_{\boldsymbol{M}^2}(x,y)\,.$$

Analogous monomials $:\phi^n:_H\ (f)$ are defined similarly as elements of $\mathcal{W}(\boldsymbol{M})$. *With the said definition (5.68) holds true literally* and not only in the sense of quadratic forms.

Remark 5.3.22 The presented definition of locally covariant Wick monomials $:\phi^n:_H\ (f)$, though satisfying general requirement of locality and covariance [11] (see also Chap. 4), remains however affected by several ambiguities. A full classification of them is the first step of ultraviolet renormalization program [35, 44]. The algebra $\mathcal{W}(\boldsymbol{M})$ also includes the so-called (locally covariant) time-ordered Wick polynomials, necessary to completely perform the renormalization procedure [36].

The constructed formalism can be extended in order to encompass differentiated Wick polynomials and it has a great deal of effect concerning the definition of the stress energy tensor operator [47]. It is defined as an element of $\mathcal{W}(\boldsymbol{M})$ by subtracting the universal Hadamard singularity from the two-point function of ω, before computing the relevant derivatives.

$$:T_{ab}:_H\ (f) = \int_{\mathcal{M}^2} D_{ab}(x,y)\ :\phi(x)\phi(y):_H\ \delta(x,y) f(x)\, \mathrm{dvol}_{\boldsymbol{M}^2}(x,y) \quad (5.85)$$

$D_{ab}(x,y)$ is a certain symmetrized second order partial differential operator obtained from (5.41) (cf. [47] Eq. (10), and [29] where some minor misprints have been corrected and the signature $(-+++)$ has been adopted),

$$D_{ab}(x,y) := D^{\mathrm{can}}_{ab}(x,y) - \frac{1}{3} g_{ab} P_x$$
$$D^{\mathrm{can}}_{ab}(x,y) := (1-2\xi) g_b^{b'} \nabla_a \nabla_{b'} - 2\xi \nabla_a \nabla_b - \xi G_{ab}$$
$$+ g_{ab} \left\{ 2\xi \Box_x + \left(2\xi - \frac{1}{2}\right) g_c^{c'} \nabla^c \nabla_{c'} + \frac{1}{2} m^2 \right\}.$$

Here, covariant derivatives with primed indices indicate covariant derivatives w.r.t. y, $g_b^{b'}$ denotes the parallel transport of vectors along the unique geodesic connecting x and y, the metric g_{ab} and the Einstein tensor G_{ab} are considered to be evaluated at x. The form of the "canonical" piece D^{can}_{ab} follows from the definition of the classical stress-energy tensor, while the last term $-\frac{1}{3} g_{ab} P_x$, giving rise to a final contribution $-\frac{g_{ab}}{3}\ :\phi(x) P\phi(x):_H$ to the stress-energy operator, has been introduced in [47]. It gives no contribution classically, just in view of the very Klein-Gordon equation satisfied by the fields, however, in the quantum realm, its presence has a very important reason. Because the Hadamard parametrix satisfies the Klein-Gordon equation only up to smooth terms, the term with P_x is non vanishing. Moreover, without this additional term, the above definition of $:T_{ab}:_H$ would not yield a conserved stress-tensor expectation value (see [47] Theorem 2.1). On the other hand the added team is responsible for the appearance of the famous *trace anomaly* [65]. An extended discussion on conservation laws in this framework appears in [37].

Acknowledgments The authors are grateful to R. Brunetti, C. Dappiaggi, C. Fewster, T. Hack, N.Pinamonti, K. Sanders, A. Strohmaier, Y. Tanimoto and R. Verch for useful discussions and suggestions.

References

1. Baez, J.C., Segal, I.E., Zhou, Z.: Introduction to Algebraic and Constructive Quantum Field Theory, Princeton Series in Physics. Princeton University Press, Princeton (1992)
2. Bär, C., Ginoux, N., Pfäffle, F.: Wave Equations on Lorentzian Manifolds and Quantization, ESI Lectures in Mathematics and Physics, vol. 2. European Mathematical Society (2007). doi:10.4171/037
3. Beem, J.K., Ehrlich, P., Easley, K.: Global Lorentzian Geometry, Pure and Applied Mathematics, vol. 202. Marcel Dekker, New York (1996)
4. Benini, M., Dappiaggi, C., Hack, T.P.: Quantum field theory on curved backgrounds—a primer. Int. J. Mod. Phys. A **28**, 1330023 (2013). doi:10.1142/s0217751x13300238
5. Bernal, A.N., Sánchez, M.: Smoothness of time functions and the metric splitting of globally hyperbolic spacetimes. Commun. Math. Phys. **257**, 43–50 (2005). doi:10.1007/s00220-005-1346-1
6. Bernal, A., Sánchez, M.: Further results on the smoothability of cauchy hypersurfaces and cauchy time functions. Lett. Math. Phys. **77**, 183–197 (2006). doi:10.1007/s11005-006-0091-5
7. Birrell, N.D., Davies, P.C.W.: Quantum Fields in Curved Space (Cambridge Monographs on Mathematical Physics). Cambridge University Press, Cambridge (1984)
8. Bratteli, O., Robinson, D.W.: Operator Algebras and Quantum Statistical Mechanics 1: C*- and W*-Algebras. Symmetry Groups. Decomposition of States, 2nd edn. Springer, New York (2002)
9. Brunetti, R., Fredenhagen, K.: Microlocal analysis and interacting quantum field theories: renormalization on physical backgrounds. Commun. Math. Phys. **208**, 623–661 (2000). doi:10.1007/s002200050004
10. Brunetti, R., Fredenhagen, K., Köhler, M.: The microlocal spectrum condition and Wick polynomials of free fields on curved spacetimes. Commun. Math. Phys. **180**, 633–652 (1996). doi:10.1007/bf02099626
11. Brunetti, R., Fredenhagen, K., Verch, R.: The generally covariant locality principle—a new paradigm for local quantum field theory. Commun. Math. Phys. **237**, 31–68 (2003)
12. Dappiaggi, C., Moretti, V., Pinamonti, N.: Distinguished quantum states in a class of cosmological spacetimes and their Hadamard property. J. Math. Phys. **50**, 062304 (2009). doi:10.1063/1.3122770
13. Dappiaggi, C., Hack, T.P., Möller, J., Pinamonti, N.: Dark energy from quantum matter (2010). http://arxiv.org/abs/1007.5009
14. Dappiaggi, C., Moretti, V., Pinamonti, N.: Rigorous construction and Hadamard property of the Unruh state in Schwarzschild spacetime. Adv. Theor. Math. Phys. **15**, 355–447 (2011). doi:10.4310/ATMP.2011.v15.n2.a4
15. Dappiaggi, C., Siemssen, D.: Hadamard states for the vector potential on asymptotically flat spacetimes. Rev. Math. Phys. **25**, 1350002 (2013). doi:10.1142/S0129055X13500025
16. Fewster, C.J., Hunt, D.S.: Quantization of linearized gravity in cosmological vacuum spacetimes. Rev. Math. Phys. **25**, 1330003 (2013). doi:10.1142/S0129055X13300033
17. Fewster, C.J., Pfenning, M.J.: A quantum weak energy inequality for spin-one fields in curved space-time. J. Math. Phys. **44**, 4480–4513 (2003). doi:10.1063/1.1602554
18. Fewster, C.J., Verch, R.: Dynamical locality of the free scalar field. Ann. Henri Poincaré **13**, 1675–1709 (2012). doi:10.1007/s00023-012-0166-z
19. Friedlander, F.G.: The Wave Equation on a Curved Space-time, Cambridge Monographs on Mathematical physics, vol. 2. Cambridge University Press, Cambridge (1975)

20. Friedlander, F.G., Joshi, M.S.: Introduction to the Theory of Distributions, 2nd edn. Cambridge University Press, Cambridge (1998)
21. Fulling, S., Sweeny, M., Wald, R.: Singularity structure of the two-point function in quantum field theory in curved spacetime. Commun. Math. Phys. **63**, 257–264 (1978). doi:10.1007/bf01196934
22. Fulling, S.A.: Aspects of Quantum Field Theory in Curved Spacetime, London Mathematical Society Student Texts, vol. 17. Cambridge University Press, Cambridge (1989)
23. Fulling, S.A., Narcowich, F.J., Wald, R.M.: Singularity structure of the two-point function in quantum field theory in curved spacetime. II. Ann. Phys. **136**, 243–272 (1981). doi:10.1016/0003-4916(81)90098-1
24. Garabedian, P.: Partial Differential Equations, Chelsea Publishing Series, vol. 325, 2nd edn. Chelsea Publishing Company, New York (1986)
25. Gérard, C., Wrochna, M.: Construction of Hadamard states by pseudo-differential calculus. Commun. Math. Phys. **325**, 713–755 (2013). doi:10.1007/s00220-013-1824-9
26. Gérard, C., Wrochna, M.: Construction of Hadamard states by characteristic Cauchy problem (2014). http://arxiv.org/abs/1409.6691
27. Günther, P.: Huygens' Principle and Hyperbolic Equations, Perspectives in Mathematics, vol. 5. Academic Press, New York (1988)
28. Haag, R.: Local Quantum Physics: Fields, Particles, Algebras, Texts and Monographs in Physics. Springer, New York (1996). doi:10.1007/978-3-642-61458-3
29. Hack, T.P.: On the backreaction of scalar and spinor quantum fields in curved spacetimes—from the basic foundations to cosmological applications. Ph.D. thesis, Hamburg (2010). http://arxiv.org/abs/1008.1776
30. Hack, T.P., Moretti, V.: On the stress-energy tensor of quantum fields in curved spacetimes—comparison of different regularization schemes and symmetry of the Hadamard/Seeley-DeWitt coefficients. J. Phys. A: Math. Theor. **45**, 374019 (2012). doi:10.1088/1751-8113/45/37/374019
31. Hack, T.P., Schenkel, A.: Linear bosonic and fermionic quantum gauge theories on curved spacetimes. Gen. Relativ. Gravit. **45**, 877–910 (2013). doi:10.1007/s10714-013-1508-y
32. Hadamard, J.: Lectures on Cauchy's Problem in Linear Partial Differential Equations, Mrs. Hepsa Ely SiLliman Memorial Lectures, vol. 15. Yale University Press, New Haven (1923)
33. Hofmann, G.: On GNS representations on inner product spaces. Commun. Math. Phys. **191**, 299–323 (1998). doi:10.1007/s002200050270
34. Hollands, S., Ruan, W.: The state space of perturbative quantum field theory in curved spacetimes. Ann. Henri Poincaré **3**, 635–657 (2002). doi:10.1007/s00023-002-8629-2
35. Hollands, S., Wald, R.M.: Local Wick polynomials and time ordered products of quantum fields in curved spacetime. Commun. Math. Phys. **223**, 289–326 (2001). doi:10.1007/s002200100540
36. Hollands, S., Wald, R.M.: Existence of local covariant time ordered products of quantum fields in curved spacetime. Commun. Math. Phys. **231**, 309–345 (2002). doi:10.1007/s00220-002-0719-y
37. Hollands, S., Wald, R.M.: Conservation of the stress tensor in perturbative interacting quantum field theory in curved spacetimes. Rev. Math. Phys. **17**, 227–311 (2005). doi:10.1142/s0129055x05002340
38. Hörmander, L.: Fourier integral operators. I. Acta Math. **127**, 79–183 (1971). doi:10.1007/bf02392052
39. Hörmander, L.: The Analysis of Linear Partial Differential Operators I: Distribution Theory and Fourier Analysis, Grundlehren Der Mathematischen Wissenschaften, vol. 256. Springer, New York (1998). doi:10.1007/978-3-642-96750-4
40. Kay, B.S.: A uniqueness result in the Segal-Weinless approach to linear bose fields. J. Math. Phys. **20**, 1712–1713 (1979). doi:10.1063/1.524253
41. Kay, B.S., Wald, R.M.: Theorems on the uniqueness and thermal properties of stationary, nonsingular, quasifree states on spacetimes with a bifurcate killing horizon. Phys. Rep. **207**, 49–136 (1991). doi:10.1016/0370-1573(91)90015-e

42. Khavkine, I.: Characteristics, conal geometry and causality in locally covariant field theory (2012). http://arxiv.org/abs/1211.1914
43. Khavkine, I.: Covariant phase space, constraints, gauge and the Peierls formula. Int. J. Mod. Phys. A **29**, 1430009 (2014) doi:10.1142/s0217751x14300099
44. Khavkine, I., Moretti, V.: Analytic dependence is an unnecessary requirement in renormalization of locally covariant QFT (2014). http://arxiv.org/abs/1411.1302v2
45. Lang, S.: Algebra, Graduate Texts in Mathematics, vol. 211, 3rd, revised edn. Springer, New York (2002). doi:10.1007/978-1-4613-0041-0
46. Moretti, V.: Proof of the symmetry of the off-diagonal heat-kernel and Hadamard's expansion coefficients in general C^∞ Riemannian manifolds. Commun. Math. Phys. **208**, 283–308 (1999). doi:10.1007/s002200050759
47. Moretti, V.: Comments on the stress-energy tensor operator in curved spacetime. Commun. Math. Phys. **232**, 189–221 (2003). doi:10.1007/s00220-002-0702-7
48. Moretti, V.: Quantum out-states holographically induced by asymptotic flatness: invariance under spacetime symmetries, energy positivity and Hadamard property. Commun. Math. Phys. **279**, 31–75 (2008). doi:10.1007/s00220-008-0415-7
49. Moretti, V.: Spectral Theory and Quantum Mechanics: With an Introduction to the Algebraic Formulation, UNITEXT, vol. 64. Springer, Berlin (2013)
50. O'Neill, B.: Semi-Riemannian Geometry With Applications to Relativity, Pure and Applied Mathematics, vol. 103. Academic Press, San Diego (1983)
51. Pinamonti, N.: On the initial conditions and solutions of the semiclassical Einstein equations in a cosmological scenario. Commun. Math. Phys. **305**, 563–604 (2011). doi:10.1007/s00220-011-1268-z
52. Radzikowski, M.J.: Micro-local approach to the Hadamard condition in quantum field theory on curved space-time. Commun. Math. Phys. **179**, 529–553 (1996). doi:10.1007/bf02100096
53. Radzikowski, M.J., Verch, R.: A local-to-global singularity theorem for quantum field theory on curved space-time. Commun. Math. Phys. **180**, 1–22 (1996). doi:10.1007/bf02101180
54. Reed, M., Simon, B.: Methods of Modern Mathematical Physics. II. Fourier Analysis, Self-Adjointness. Academic Press, New York (1975)
55. Riesz, M.: L'intégrale de Riemann-Liouville et le problème de Cauchy. Acta Math. **81**, 1–222 (1949). doi:10.1007/bf02395016
56. Sahlmann, H., Verch, R.: Microlocal spectrum condition and Hadamard form for vector-valued quantum fields in curved spacetime. Rev. Math. Phys. **13**, 1203–1246 (2001). doi:10.1142/s0129055x01001010
57. Sanders, K.: Aspects of locally covariant quantum field theory. Ph.D. thesis, University of York (2008). http://arxiv.org/abs/0809.4828
58. Sanders, K.: On the construction of Hartle-Hawking-Israel states across a static bifurcate killing horizon. Letters in Mathematical Physics (2014). Math. Phys. **105**, 575–640 (2015). http://arxiv.org/abs/1310.5537
59. Sanders, K., Dappiaggi, C., Hack, T.P.: Electromagnetism, local covariance, the Aharonov-Bohm effect and Gauss' law. Commun. Math. Phys. **328**, 625–667 (2014). doi:10.1007/s00220-014-1989-x
60. Schmüdgen, K.: Unbounded Operator Algebras and Representation Theory, Operator Theory, Advances and Applications, vol. 37. Birkhäuser, Basel (1990)
61. Strocchi, F.: An Introduction to the Mathematical Structure of Quantum Mechanics: A Short Course for Mathematicians, Advanced Series in Mathematical Physics, vol. 28. World Scientific, Singapore (2008)
62. Strohmaier, A.: Microlocal analysis. In: Bär, C., Fredenhagen, K. (eds.) Quantum Field Theory on Curved Spacetimes: Concepts and Methods, Lecture Notes in Physics, vol. 786. Springer, Berlin (2009). doi:10.1007/978-3-642-02780-2_4
63. Strohmaier, A., Verch, R., Wollenberg, M.: Microlocal analysis of quantum fields on curved space-times: analytic wave front sets and reeh-schlieder theorems. J. Math. Phys. **43**, 5514–5530 (2002). doi:10.1063/1.1506381

64. Verch, R.: Local definiteness, primarity and quasiequivalence of quasifree Hadamard quantum states in curved spacetime. Commun. Math. Phys. **160**, 507–536 (1994). doi:10.1007/bf02173427
65. Wald, R.M.: Quantum Field Theory in Curved Spacetime and Black Hole Thermodynamics, Chicago Lectures in Physics. University of Chicago Press, Chicago (1994)

Chapter 6
Cosmological Applications of Algebraic Quantum Field Theory

Thomas-Paul Hack and Nicola Pinamonti

Abstract Quantum field theory on curved spacetime is a generalisation of quantum field theory in flat spacetime which is expected to be the proper fundamental description of non–trivial physical phenomena in the presence of a spacetime curvature which is large but below Planck scale. Two prominent physical situations which fall under this characterisation are phenomena both in the vicinity of black holes and in the early universe. Focusing on the latter, we review several applications of algebraic quantum field theory on curved spacetimes to cosmology, as well as foundational results and constructions on which these applications are based. On the foundational side, we collect several proposals to construct Hadamard states on cosmological spacetimes, as this class of states is believed to encompass all physically meaningful states in quantum field theory on curved spacetimes. Afterwards we consider the solution theory of the semiclassical Einstein equation, quote a theorem of existence and uniqueness of solutions to this equation and indicate directions to go beyond the semiclassical Einstein equation. Then we highlight how the observed cosmological expansion may be understood qualitatively and quantitatively in this framework, before we discuss the quantization of perturbations in inflation in the context of algebraic quantum field theory. In the latter subject, the starting point is the assumption that the classical, rather than the semiclassical, Einstein equation is satisfied. We close this chapter briefly discussing how one may generalise the analysis of perturbations in inflation by allowing for spacetimes backgrounds which solve the semiclassical Einstein equations.

T.-P. Hack (✉) · N. Pinamonti
Department of Mathematics, University of Genova, via Dodecaneso 35, I-16146 Genova, Italy
e-mail: hack@dima.unige.it

N. Pinamonti
INFN—Sezione di Genova, via Dodecaneso 33, I-16146 Genova, Italy
e-mail: pinamont@dima.unige.it

R. Brunetti et al. (eds.), *Advances in Algebraic Quantum Field Theory*,
Mathematical Physics Studies, DOI 10.1007/978-3-319-21353-8_6

6.1 Introduction

During the last decades observational cosmology experienced a golden age in which the amount and quality of available data made it possible to test theoretical models with great accuracy. This lead to a great improvement in our understanding of the physics of our universe. A particular highlight of the very many observational sources are the anisotropies of the cosmic microwave background which have been measured to a high accuracy by the Planck satellite. It is believed that these anisotropies are sourced by primordial metric and matter perturbations which are of quantum nature. During an early phase of rapid expansion of the universe called inflation these perturbations are thought to become classical and hence visible. The assumption of rapid expansion implies that the spacetime curvature in this cosmological phase was seizable, and thus, from the theoretical side, these metric and matter perturbations should be described by quantum fields propagating over curved classical backgrounds.

The presence of a phase of rapid expansion is necessary in order to (at least partly) explain many observations such as the flatness and isotropy of the spatial slices of the universe. Moreover it is theoretically attractive because it is thought to "wash out" the distinctive features of the state of the universe prior to the expansive phase and thus it becomes possible to analyse the cosmological dynamics independently of the physics prior to inflation. Whereas presently most models attribute the origin of this accelerated expansion to the energy density of a classical scalar field, one of the first models of inflation was the Starobinsky model, which was based on the analysis of the backreaction of quantum fields propagating in curved spacetimes on the metric [44]. Later the model was considerably simplified [45] and, barring a clash with the much-debated BICEP2 data [3], it is still in perfect agreement with all observational data [2].

Due to the complexity of a fundamental analysis, many aspects of theoretical cosmology are not dealt within quantum field theory on curved spacetimes by default, but simplified frameworks are used based on the idea that in many circumstances the influence of the spacetime curvature can be taken into account in effective approximate ways. A proper legitimation of these approximations can however only be given based on a more fundamental analysis of the aforementioned aspects within quantum field theory on curved spacetimes, and it may happen that the corrections to the effective picture resulting from such an analysis have observable effects even if they are small.

For all these reasons, cosmology is increasingly becoming a field where various aspects of quantum field theory on curved spacetime play an important role. Hence, in this chapter, we shall specialise to cosmological spacetimes the general discussion of algebraic quantum field theory on curved spacetimes presented in the contributions to this book by Fredenhagen—Chap. 1, Fredenhagen and Rejzner—Chap. 2, Benini and Dappiaggi—Chap. 3, Fewster and Verch—Chap. 4 and Khavkine and Moretti—Chap. 5.

Let us recall that on large scales our universe appears isotropic. At the same time we don't have any convincing reason to assume that it contains preferred points.

Hence it should be described by a **homogeneous** and **isotropic** spacetime $\boldsymbol{M} = (\mathcal{M}, g, \mathfrak{o}, \mathfrak{t})$. The two requirements imply that $\boldsymbol{M}$ is nothing but a warped product of an oriented interval of time I and a spatial section Σ. More precisely,

$$\mathcal{M} = I \times \Sigma$$

and, fixing the **cosmological time** t and an adapted coordinate system (r, θ, φ) on Σ,

$$g = dt \otimes dt - a(t)^2 \left[\frac{1}{1 - \kappa r^2} dr \otimes dr + r^2 \left(d\theta \otimes d\theta + \sin\theta^2 d\varphi \otimes d\varphi \right) \right].$$

The above metric is nothing but a Friedmann–Lemaître–Robertson–Walker (FLRW) metric where $a(t)$ is the **scale factor**, required to be strictly positive, and the constant $\kappa \in \{-1, 0, 1\}$ is the sign of the Gaussian curvature. In particular, κ distinguishes between open, flat and closed universes.

Once the topology of the spatial section is fixed, the scale factor is the only dynamical degree of freedom in the FLRW metric. The dynamics of $a(t)$ is governed by the Einstein equations

$$G_{ab} = -8\pi T_{ab}$$

where G_{ab} is the Einstein tensor and T_{ab} is the stress tensor which describes the matter distribution. The homogeneity and isotropy of $\boldsymbol{M}$ constraint the form of T_{ab}. In fact, the admissible T_{ab} are described by two scalar quantities: the energy density ρ and the pressure P. More precisely

$$T_{ab} = \rho u_a u_b + P(u_a u_b - g_{ab})$$

where the four velocity u is related to the cosmological time introduced above by $u = \frac{\partial}{\partial t}$.

The Einstein equations imply by consistency that the stress tensor is covariantly conserved $\nabla^a T_{ab} = 0$, and thus the matter sourcing the dynamics of a cosmological spacetime is described by a stress tensor which looks like the stress tensor of a perfect fluid. However, no equation of state for this effective fluid is prescribed a priori.

We shall restrict our analysis to flat cosmological spacetimes for simplicity and because spatial flatness is favoured by observations. In this case the spatial section Σ is the manifold $\mathbb{R}^3$ equipped with the standard Euclidean metric and thus $\boldsymbol{M}$ is conformally flat. This feature can be exploited employing the **conformal time** which is related to cosmological time by the following relation

$$\tau(t) := \tau_0 + \int_{t_0}^{t} \frac{1}{a(t')} dt', \qquad (6.1)$$

where τ_0 and t_0 are necessary to fix the initial time. With respect to $(\tau, \boldsymbol{x})$, where $\boldsymbol{x} = (x_1, x_2, x_3)$ are Cartesian coordinates on $\mathbb{R}^3$, the metric g of $\boldsymbol{M}$ takes the form

$$g := a(\tau)\,(d\tau \otimes d\tau - d\boldsymbol{x} \otimes d\boldsymbol{x}) \ , \tag{6.2}$$

where, with a small abuse of notation, we have indicated $a(\tau) = a(t(\tau))$.

The set of flat cosmological spacetimes form a sub category of **Loc** introduced in the contribution by Fewster and Verch. Furthermore, all these elements are connected by a larger set of morphisms, namely, the orientation preserving conformal transformations. We indicate by **CLoc** the category where the objects are flat cosmological spacetimes and the morphisms are the above-mentioned conformal transformations. Up to an inessential redefinition of the origin of cosmological time t, elements of **CLoc** are uniquely characterised by their scale factor a as a function of t. However, since the relation (6.1) can be inverted by the assumption $a > 0$, we may equivalently consider a as a function of the conformal time τ.

Throughout this and the next section we shall for simplicity consider neutral scalar linear field theories coupled to gravity. The classical theory is described by

$$P\phi := \Box\phi + \xi R\phi + m^2\phi = 0, \tag{6.3}$$

where the real constant ξ quantifies the coupling to the Ricci scalar curvature and m is the mass of the field. The case $\xi = 1/6$ is called conformally coupled case, and we shall restrict attention to this case in the following section. The quantization of this linear model may be characterised in a first step by constructing the algebra $\mathscr{A}(\boldsymbol{M})$ generated by linear local fields, the commutation relations, the equations of motion and the $*-$ operation, cf. Sect. 3.3.1 in Benini's and Dappiaggi's contribution.

However, for our purposes we need also non–linear local fields such as $\phi^2(f)$ or $T_{ab}(f)$. Such fields can be added to $\mathscr{A}(\boldsymbol{M})$ after a deformation of this algebra. The resulting extended algebra contains observables which can be tested only on states which are sufficiently regular. Hadamard states constructed and discussed in Khavkine's and Moretti's contribution in Chap. 5 of this book provide a physically well-motivated class of admissible states. For our purposes it is important that $T_{ab}(f)$ has finite expectation value and correlations (fluctuations) in all Hadamard states. Alternatively, one may construct an algebra containing non–linear local field observables in a single step by considering arbitrary functionals on scalar field configurations whose functional derivatives satisfy a certain wave front set condition and endowing them with a $\star$-product defined by contractions with a two–point function of a Hadamard state, see the contribution of Fredenhagen and Rejzner (Chap. 2) for details. We shall briefly review the treatment of non–linear observables when discussing the backreaction of quantum fields on the background metric in the next sections.

This chapter is organised as follows. In the next section we shall recall several possible constructions of Hadamard states on cosmological spacetimes available in the literature. The Sect. 6.3 contains a discussion of the semiclassical Einstein equations for cosmological spacetimes, in particular we review existence and uniqueness of its solutions as well its relevance for understanding the cosmological expansion in quantitative terms. The analysis of metric and matter perturbations in inflation and their quantization is presented in the Sect. 6.4. The chapter ends with a few closing comments in the Sect. 6.5.

6.2 States on Cosmological Spacetimes

As argued in the introduction and in Khavkine's and Moretti's contribution in Chap. 5, Hadamard states constitute a large class of physically well motivated states in algebraic quantum field theory on curved spacetimes, in which all observables have finite expectation values; in particular, local observables have finite fluctuations in these states. Hadamard states on curved spacetimes may be thought of as being generalisations of vacuum and thermal states in flat spacetimes as well as of finite energy excitations thereof. However, in spite of the large spatial symmetry present in the class of cosmological spacetime we are considering, no preferred state can be selected as vacuum or thermal state for our theory due to the time dependence of the background encoded in the functional form of $a(\tau)$. In this section we shall collect a few constructions of Hadamard states and less regular states available in the literature.

Since the algebra of observables $\mathscr{A}(\boldsymbol{M})$ we are considering is (a suitable topological extension of an algebra) generated by linear fields $\phi(f)$, in order to fix an algebraic state ω—a positive, normalised, linear functional on $\mathscr{A}(\boldsymbol{M})$—up to physically inessential topological issues, it suffices to prescribe its n−point functions $\Lambda_n \in \mathscr{D}'(\mathcal{M}^n)$

$$\omega(\phi(f_1)\dots\phi(f_n)) := \int_{\mathcal{M}^n} \Lambda_n(x_1,\dots,x_n) f_1(x_1)\dots f_n(x_n)\, \mathrm{dvol}_{\boldsymbol{M}}(x_1)\cdots \mathrm{dvol}_{\boldsymbol{M}}(x_n)$$

In order to be as close as possible to vacuum states in flat spacetime, we shall focus on **quasi–free** (Gaussian) states which are also **pure**, **homogeneous** and **isotropic** in the following.

For quasi–free states all n−point functions with odd n vanish, while the n–point functions with even n can be written as linear combinations of products of two–point functions $\Lambda = \Lambda_2 \in \mathscr{D}'(\mathcal{M}^2)$. In particular, the "integral kernels" of these distributions are such that

$$\Lambda_n(x_1,\dots,x_n) = \sum_{\pi \in P_n} \Lambda(x_{\pi(1)}, x_{\pi(2)}) \dots \Lambda(x_{\pi(n-1)}, x_{\pi(n)})$$

where the sum is taken of all ordered partitions of n elements.

We recall that a Hadamard state is a state whose two–point function has a prescribed singular structure in position space. In particular for x, y in a normal neighbourhood it holds

$$\Lambda(x,y) := \lim_{\varepsilon\to 0^+} \frac{1}{8\pi^2}\left(\frac{U(x,y)}{\sigma_\varepsilon(x,y)} + V(x,y)\log\frac{\sigma_\varepsilon(x,y)}{\lambda^2} + W(x,y)\right) \tag{6.4}$$

where U, V are smooth functions which depend only on the local metric and the equations of motion. λ is a length-scale and W is a smooth function which characterises the state. Finally, $2\sigma_\varepsilon(x,y)$ is a suitable regularisation of the signed squared geodesic distance between the points x and y. The Hadamard coefficient V may be expanded in a formal series as $V = \sum_{i=0}^{\infty} V_i \sigma^i$.

The position-space Hadamard singular behaviour is equivalent to a condition on the wave front set of the distribution Λ, namely

$$WF(\Lambda) = \left\{(x,y;k_x,k_y) \in T^*\mathcal{M}^2 \setminus \{0\},\quad (x,k_x) \sim (y,-k_y),\quad k_x \triangleright 0\right\}, \tag{6.5}$$

where $(x,k_x) \sim (y,-k_y)$ implies that there exists an affinely parametrized light-like geodesic γ in $\boldsymbol{M}$ joining the points x and y such that $g^{-1}(k_x) = \dot\gamma(x)$ and $g^{-1}(-k_y) = \dot\gamma(y)$. Furthermore, $k_x \triangleright 0$ means that $g^{-1}(k_x)$ is future directed. We refer to Khavkine's and Moretti's contribution in Chap. 5 for further details on Hadamard states.

For explicit computations on cosmological spacetimes, it is advisable to use the high symmetry of these spacetimes and of the states sharing this property in order to expand all quantities in Fourier modes. As shown by Lüders and Roberts [32] the most general two–point function of a quasi–free, pure, homogeneous and isotropic sufficiently regular state on a flat FLRW spacetime can be written as

$$\Lambda(x_1,x_2) := \lim_{\varepsilon\to 0^+} \frac{1}{8\pi^3 a(t_1)a(t_2)}\int_{\mathbb{R}^3} \chi_k(t_1)\overline{\chi_k(t_2)}e^{i\boldsymbol{k}\cdot(\boldsymbol{x}_1-\boldsymbol{x}_2)}e^{-k\varepsilon}d\boldsymbol{k} \tag{6.6}$$

where the limit ε to 0 is considered in the weak sense and the the temporal modes χ_k have to satisfy certain regularity and integrability conditions in $k = |\boldsymbol{k}|$. Considering these modes as functions of conformal time τ and indicating by f' the derivative of f with respect to τ, the temporal modes must satisfy

$$\chi_k'' + (m^2a^2 + k^2)\chi_k + (6\xi - 1)\frac{a''}{a}\chi_k = 0, \tag{6.7}$$

and

$$\overline{\chi_k}'\chi_k - \overline{\chi_k}\chi_k' = i\ . \tag{6.8}$$

Hence, in this mode representation, a quasi–free, pure, homogeneous and isotropic state is uniquely determined by assigning to each k a temporal mode χ_k satisfying the two equations above. Hadamard states are loosely speaking characterised by a particular behaviour of χ_k for large k, which can be rigorously formulated only in special cases. However, if a set of Hadamard modes $\widetilde{\chi}_k$ is already known, then precise conditions for other modes to define a Hadamard state can always be formulated in relation to $\widetilde{\chi}_k$, see [36, 50] for details.

Apart from the already mentioned vacuum and thermal states on Minkowski spacetime as well as finite energy excitations thereof, the most prominent examples of Hadamard states in quantum field theory on curved spacetimes are the Bunch–Davies state (the unique maximally symmetric Hadamard state) on de Sitter spacetime, conformal vacuum and conformal thermal states for conformally coupled massless fields on flat FLRW spacetimes, whose two–point functions are just rescalings with suitable powers of $a(\tau)$ of the vacuum and thermal two–point functions of the same field theory on Minkowski spacetime. Examples of states who fail to satisfy the Hadamard condition are the so-called α–vacua on de Sitter spacetimes, cf. [7].

6.2.1 Adiabatic States

Before discussing further examples of Hadamard states on FLRW spacetimes in the next subsections, we consider a class of in general less regular states. The concept of **adiabatic states** was introduced by Parker [35] in order to construct states that are as close as possible to local vacua. These are quasi–free, pure and homogeneous states on cosmological spacetimes. Hence they are completely described by a two–point function of the form (6.6). In particular, in the case of minimal coupling $\xi = 0$, the considered modes $\chi_k(\tau)$ are constructed employing a WKB approximation. The construction starts with the ansatz

$$\chi_k(t) = \frac{1}{\sqrt{2a\Omega_k(t)}} \exp\left(i \int_{t_0}^{t} \Omega_k(s)ds\right)$$

where we employed modes depending on the cosmological time t. The function Ω_k must be positive and it has to satisfy

$$\Omega_k^2 = \frac{k^2}{a^2} + m^2 - \frac{3}{4}\left(\frac{\dot{a}}{a}\right)^2 - \frac{3}{2}\frac{\ddot{a}}{a} + \frac{3}{4}\left(\frac{\dot{\Omega}_k}{\Omega_k}\right)^2 - \frac{1}{2}\frac{\ddot{\Omega}_k}{\Omega_k}$$

where $\dot{}$ here stands for a derivative with respect to the cosmological time t. The previous equation is formally solved iteratively starting with $\Omega_k^0 = \omega_k = k^2/a^2 + m^2$ and computing $(\Omega_k^n)^2$ as the right side of the previous equation where Ω_k is replaced by Ω_k^{n-1}.

The iteration is in general not well–defined, because at some point, typically for small k, Ω_k^n might become imaginary for specific times thus violating the normalisation condition (6.8). For this reason and for practical purposes the iteration is stopped after a finite number n of steps. Due to this approximation, the obtained two–point function does not define a meaningful state because it does not satisfy the equation of motion. However, the related two–point function could be meaningful when considered only at a fixed time t. Starting from this observation Lüders and Roberts [32] made precise the notion of adiabatic states of order n employing the Parker construction terminated after n steps to obtain the initial values at time t for the modes χ_k which are exact solutions of (6.7). Contrary to Parker's two–point function, the such constructed two–point function solves the equation of motion exactly and defines a proper state, which is called an **adiabatic state of order** n.

The analysis of the singularity structure of these states was performed by Junker and Schrohe [28]. They characterised the degree of regularity of adiabatic states employing the notion of Sobolev wave front sets. They showed that, at least when the spatial Cauchy surfaces Σ are compact, the two–point function of adiabatic states of order n is of the Hadamard form up to functions in $C^{f(n)}(\mathcal{M}^2)$, where $f : \mathbb{N}_0 \to \mathbb{N}_0$ is a suitable monotonous function. In other words, Λ is of the form (6.4) where now $W \in C^{f(n)}(\mathcal{M}^2)$. Only adiabatic states of infinite order satisfy the Hadamard condition exactly. The results of Junker and Schrohe can be presumably generalised to a larger class of Cauchy surfaces including the flat cosmological case following the analysis of Gérard and Wrochna [20].

6.2.2 *States of Low Energy*

As seen in the previous subsection only adiabatic states of infinite order are necessarily Hadamard states, and since adiabatic states are constructed by iterating over the order, it is rather impractical to construct Hadamard states by means of the adiabatic procedure. An alternative idea which proved to be useful, was to construct states which minimise the energy density smeared over a time interval with a smooth function. These **states of low energy** have been introduced by Olbermann [34].

Starting from a generic mode function $\chi_k(t)$ which solves (6.7) and a smooth smearing function $f(t)$, Olbermann looked for modes

$$T_k = \lambda\chi_k + \mu\overline{\chi_k}$$

which satisfy the Wronsikian normalisation condition (6.8) and whose associated pure, quasi–free, homogeneous and isotropic state ω constructed via (6.6) minimises the energy density smeared in time with $f(t)$. In the minimally coupled case, the time-smeared energy density is

$$\langle T_{00}(f)\rangle_\omega = \frac{1}{2}\int_{\mathbb{R}} dt f(t)\int_{\Sigma} d\boldsymbol{k}\left(\left|\frac{d}{dt}\frac{T_k}{a}\right|^2 + \left(\frac{k^2}{a^2}+m^2\right)\left|\frac{T_k}{a}\right|^2\right)$$

where T_{00} is the t–t–component of the stress tensor and we omit the necessary regularisation of the $\boldsymbol{k}$–integral above because it is irrelevant for the construction. It turns out that μ and λ can be fixed uniquely up to an inessential global phase in terms of two constants

$$c_1 = \frac{1}{2}\int_{\mathbb{R}} dt f(t)\left(\left|\frac{d}{dt}\frac{\chi_k}{a}\right|^2 + \left(\frac{k^2}{a^2}+m^2\right)\left|\frac{\chi_k}{a}\right|^2\right),$$
$$c_2 = \frac{1}{2}\int_{\mathbb{R}} dt f(t)\left(\left(\frac{d}{dt}\frac{\chi_k}{a}\right)^2 + \left(\frac{k^2}{a^2}+m^2\right)\frac{\chi_k^2}{a^2}\right)$$

and, in particular,

$$\lambda = e^{i\alpha}\sqrt{\frac{c_1}{2\sqrt{c_1^2-|c_2|^2}}+\frac{1}{2}}, \qquad \mu = \sqrt{\frac{c_1}{2\sqrt{c_1^2-|c_2|^2}}-\frac{1}{2}},$$

where $\alpha = -\arg c_2 + \pi$.

Examples for states of low energy are the Minkowski vacuum and—in the case of conformal coupling—the conformal vacuum in flat FLRW spacetimes. Both of these are states of low energy independent of the smearing function $f(t)$. Moreover, Degner showed in his thesis [14] that on spacetimes which are asymptotically de Sitter towards $\tau \to -\infty$ (see the next section for a definition), every state of low energy converges to the Bunch–Davies state (the unique maximally symmetric Hadamard state) upon sending the support of $f(t(\tau))$ in τ to negative infinity. This is a rigorous variant of the statement that every state on de Sitter spacetimes converges to the Bunch–Davies state for positive asymptotic times. A generalisation of states of low energy was given in [31], where almost equilibrium states have been constructed. Whereas states of low energy may be thought of as generalised vacuum states, the states constructed in [31] may be interpreted as generalised thermal states. States of low energy on spacetimes with less symmetry have been constructed in [46].

6.2.3 Hadamard States on Spacetimes with Suitable Asymptotic Properties

Hadamard states are characterised by having a prescribed singularity structure for points which are lightlike related. Loosely speaking, for this reason it is difficult to provide initial conditions on the temporal modes $\chi_k(\tau)$ at fixed $\tau = \tau_0$ which lead to a Hadamard state, because the τ_0–Cauchy–surface of a FLRW spacetime is spacelike. A method to overcome this problem has been developed in [11], where the basic idea is that on spacetimes where the surfaces of fixed τ become lightlike for asymptotic τ it should be possible to construct Hadamard states by prescribing initial values for $\chi_k(\tau)$ on these asymptotic surfaces.

The particular example considered in [11] were spacetimes of **asymptotic de Sitter** type, i.e. FLRW spacetimes with a scale factor

$$a(\tau) = -\frac{1}{H\tau} + O\left(\frac{1}{\tau^2}\right), \qquad \frac{da(\tau)}{d\tau} = \frac{1}{H\tau^2} + O\left(\frac{1}{\tau^3}\right)$$

where $\overline{H}$ is the constant Hubble parameter of an exact de Sitter spacetime and $\tau \in (-\infty, 0)$. Further examples of spacetimes on which the construction of Hadamard states we shall review in the following is also possible have been discussed in [10].

In the particular asymptotically de Sitter case, the spacetime has a past boundary $\mathcal{H}$ $(\tau \to -\infty)$ which is a horizon. More precisely, this spacetime $\boldsymbol{M}$ can be smoothly extended beyond $\mathcal{H}$ in the past. We can thus meaningfully consider $\boldsymbol{M} \cup \mathcal{H}$ as a subset of a larger globally hyperbolic spacetime. In this extended spacetime, the following conditions hold

- The vector field ∂_τ, smoothly extended to $\mathcal{H}$, is tangent to $\mathcal{H}$ and on $\mathcal{H}$ it is null.
- Since ∂_τ is a conformal Killing vector field which is timelike in $\boldsymbol{M}$ and null in $\mathcal{H}$, $\mathcal{H}$ is a conformal Killing horizon.
- The induced metric on $\mathcal{H}$ is diagonal and depends only on the angular coordinates. The spatial sections of this null surface are two-dimensional spheres with constant radius.

One would now like to construct a Hadamard state for the algebra $\mathscr{A}(\boldsymbol{M})$ of the scalar field on the original spacetime $\boldsymbol{M}$ by employing geometric properties of the horizon $\mathcal{H}$ described above. For simplicity we shall here assume conformal coupling $\xi = 1/6$. The formal construction makes use of the time slice axiom, which is satisfied in the case at hand, to map the on–shell bulk algebra $\mathscr{A}(\boldsymbol{M})$ to an algebra $\mathscr{A}(\mathcal{H})$ defined intrinsically on the horizon $\mathcal{H}$

$$\iota : \mathscr{A}(\boldsymbol{M}) \mapsto \mathscr{A}(\mathcal{H})$$

and acting on the generators of $\mathscr{A}(\boldsymbol{M})$ in the following way

$$\iota(\phi(f)) = \phi_{\mathcal{H}}(E|_{\mathcal{H}}(f))$$

where, $E|_{\mathcal{H}}(f)$ is the restriction on $\mathcal{H}$ of the advanced–minus–retarded propagator E applied to an $f \in C_0^\infty(\mathcal{M})$ and $\phi_{\mathcal{H}}(\psi)$ are the generators of $\mathscr{A}(\mathcal{H})$ smeared with ψ supported on $\mathcal{H}$. The action of the restriction map preserves the $*-$operation and the commutation relations

$$[\phi(f), \phi(h)] = iE(f, h) = [\phi_{\mathcal{H}}(E|_{\mathcal{H}}(f)), \phi_{\mathcal{H}}(E|_{\mathcal{H}}(h))].$$

Hence ι is a well defined $*-$homomorphism from $\mathscr{A}(\boldsymbol{M})$ to $\mathscr{A}(\mathcal{H})$; in particular, the pullback of a state on $\mathscr{A}(\mathcal{H})$ defines a state on $\mathscr{A}(\boldsymbol{M})$. On $\mathscr{A}(\mathcal{H})$ we may introduce a pure quasi–free state $\omega_{\mathcal{H}}$ which minimizes the energy computed with respect to the conformal Killing field ∂_τ. The construction of this state is very similar to the one presented by Sewell [42] or by Kay and Wald [29]. The pullback of that state

$$\omega_{\boldsymbol{M}} := \iota^*(\omega_{\mathcal{H}})$$

defines a state on $\mathscr{A}(\boldsymbol{M})$. The obtained state is quasi–free, pure, homogeneous and isotropic and thus its two–point function $\Lambda_{\boldsymbol{M}}$ is of the form (6.6). Furthermore, it is possible to analyse the singular structure of Λ_M by interpreting it as a composition of distributions

$$\Lambda_M = (E|_{\mathcal{H}} \otimes E|_{\mathcal{H}}) \circ \Lambda_{\mathcal{H}} \,.$$

It turns out that the state constructed in this way is of Hadamard form, in fact, the wave front set condition (6.5) can be checked for Λ_M using the propagation of singularities theorem, see [12] for details.

The modes χ_k appearing in the two–point function Λ_M of the Hadamard state just constructed can be explicitly obtain by means of perturbation theory [12]. To this avail we rewrite (6.7) as

$$\chi_k'' + (V_0(k,\tau) + V(\tau))\chi_k(\tau) = 0, \tag{6.9}$$

where the unperturbed potential V_0 and the perturbation are such that

$$V_0(k,\tau) := k^2 + \left(\frac{1}{\overline{H}\tau}\right)^2 \left[m^2 + (6\xi - 1)2\overline{H}^2\right], \quad V(\tau) = O(1/\tau^3)\,.$$

The unperturbed equation coincides with the mode equation in an exact de Sitter spacetime. We indicate by ρ_k the solution of the mode equation in this spacetime, namely, the solutions of the unperturbed equation satisfying the following asymptotic initial conditions

$$\lim_{\tau\to-\infty} e^{ik\tau}\rho_k(\tau) = \frac{1}{\sqrt{2k}}, \qquad \lim_{\tau\to-\infty} e^{ik\tau}\rho_k'(\tau) = i\sqrt{\frac{k}{2}}.$$

The modes χ_k can now be constructed by means of the retarded fundamental solution of the free one–dimensional problem (6.9). To this end, let us introduce

$$r_V(f)(\tau) = i\int_{-\infty}^{\tau} \left(\overline{\rho_k(\tau)}\rho_k(s) - \overline{\rho_k(s)}\rho_k(\tau)\right) V(s) f(s) ds$$

then the Dyson series ansatz for solving (6.9) reads

$$\chi_k = \sum_{n=0}^{\infty} r_V^n(\rho_k) \tag{6.10}$$

and one can show that this series converges at least on a time interval $(-\infty, \tau_0) \subset (-\infty, 0)$. A similar construction has been employed by Anderson [5] to construct vacuum–like states.

The Hadamard states constructed as above may be interpreted as asymptotic vacuum states towards $\tau \to -\infty$. Generalising this, one may construct Hadamard states which correspond to asymptotic thermal states as demonstrated in [10].

6.2.4 Sorkin–Johnston States and Their Generalisations

Recently, a way of constructing states based on local properties of the spacetime was proposed by Sorkin–Johnston, see [4] and reference therein. The basic idea is to consider a spectral decomposition of the localised advanced–minus–retarded propagator and restricting it to the positive part of the spectrum in order to obtain a vacuum–like state. However, as argued in [18], the obtained states are not of Hadamard form and do not behave well under a change of the localisation region.

To improve on this, Brum and Fredenhagen [6] showed that a smeared version of the Sorkin–Johnston construction, which is loosely related to the construction of states of low energy, yields as a result states that are of Hadamard form. In order to illustrate this construction for cosmological spacetimes, let us recall that the integral kernel of the advanced–minus–retarded propagator has the form

$$E(x_1, x_2) = \lim_{\varepsilon \to 0^+} \frac{1}{8\pi^3 a(t_1)a(t_2)} \int_{\mathbb{R}^3} \left(\chi_k(t_1)\overline{\chi_k(t_2)} - \overline{\chi_k(t_1)}\chi_k(t_2) \right) e^{i\boldsymbol{k}\cdot(\boldsymbol{x}_1-\boldsymbol{x}_2)} e^{-k\varepsilon} d\boldsymbol{k}$$

where the temporal modes χ_k are arbitrary solutions of (6.7) and (6.8) since $E(x_1, x_2)$ does not depend on the particular choice of such χ_k. One now picks an $f \in C_0^\infty(\mathcal{M})$ such that f is equal to 1 on $\mathcal{N} \subset \mathcal{M}$, where $\boldsymbol{N} = (\mathcal{N}, g|_\mathcal{N}, \mathfrak{o}|_\mathcal{N}, \mathfrak{t}|_\mathcal{N})$, i.e. $\mathcal{N}$ endowed with metric, orientation and time orientation induced from $\boldsymbol{M}$, is required to be a globally hyperbolic spacetime in its own right.

This function is then used to construct the operator

$$A := ifEf$$

which acts on the Hilbert space $L^2(\mathcal{N}, \mathrm{dvol}_{\boldsymbol{N}})$. One may show that this operator is bounded and self-adjoint, hence, with standard functional calculus, it is possible to construct its positive part

$$A^+ := P^+ A$$

where P^+ is the spectral projection on the interval $[0, |A|]$. The Brum–Fredenhagen generalisation of the Sorkin–Johnston state is then defined by the two–point function

$$\Lambda(h_1, h_2) := \left(h_1, A^+ h_2\right)$$

for real–valued $h_1, h_2 \in C_0^\infty(\mathcal{N})$ and one may show that the resulting state is Hadamard.

The original Sorkin–Johnston construction corresponds to taking f to be the characteristic function of $\mathcal{N}$. The associated state constructed as above is not of Hadamard form due to the failure of f to be regular at the future and past boundaries of $\mathcal{N}$.

6.2.5 Further Comments

In all constructions of Hadamard states presented above, global aspects of spacetime enter in an essential way either via smearing over finite intervals of time or by demanding that the background spacetime has particular asymptotic properties. This leads to potential complications in the analysis of the back–reaction of quantum fields on the curved spacetime background via the semiclassical Einstein equation (see the next section) because one would like to solve this equation in a local way.

Finally we would like to mention two further classes of Hadamard states on cosmological spacetimes. A construction of Hadamard states via pseudodifferential calculus, which is not restricted to, but applicable to FLRW spacetimes, was developed by Gerard and Wrochna [20]. Hadamard states which possess an approximate local thermal interpretation have been constructed in [41], see [47] for a review.

6.3 Semiclassical Backreaction

In this section we shall discuss aspects of the semiclassical backreaction of matter fields on the background metric in the case of cosmological spacetimes. To this end, the equation we want to solve is the **semiclassical Einstein equation**

$$G_{ab} = -8\pi \langle T_{ab} \rangle_\omega \tag{6.11}$$

where $\langle T_{ab} \rangle_\omega$ is the expectation value of the stress tensor T_{ab} of a quantum theoretical model in some state ω. We shall accomplish this task considering only a very simple model for matter, namely, a scalar field in a homogeneous and isotropic Hadamard state. Note that the expectation values of the stress tensors of other kind of matter like Dirac fields or electromagnetic fields behave qualitatively in a very similar way, although their computation is more involved. We refer to [48] for the requirements that the expectation value of T_{ab} defined by means of a suitable renormalisation prescription needs to satisfy in order to give rise to meaningful solutions of the semiclassical Einstein equation.

In order to analyse and implement the backreaction, we need to study the expectation values of the stress tensor in a sufficiently regular state ω. As reviewed in the contribution of Fredenhagen and Rejzner in Chap. 2, if we want to consider local non–linear fields like T_{ab}, the algebra generated by linear fields needs to be deformed before being extended by these additional fields. In particular, the product in the deformed algebra is

$$\phi(f) \star_{\mathcal{H}} \phi(g) = \phi(f) \cdot \phi(g) + \mathcal{H}(f, g).$$

Here $\cdot$ is the symmetric product in the classical theory and $\mathcal{H}$ must enjoy certain properties:

- In order to preserve the commutation relations, its antisymmetric part must coincide with the advanced–minus–retarded propagator E.
- In order to be able to extend $\star_{\mathcal{H}}$ to local non–linear fields, pointwise powers of derivatives of $\mathcal{H}$ must be well–defined.

It is possible to satisfy the second condition by demanding that $\mathcal{H}$ satisfies the wave front set condition (6.5). Thanks to the work of Radzikowski [8, 39] we know that, the two-requirements implies that the singular structure of $\mathcal{H}$ is of the form (6.4).

As the state ω is originally given on the undeformed algebra, we need to keep track of its change under the discussed deformation by considering the pushforward of the original state. Notice that, the pushforward of states under this deformation evaluated on non–linear fields corresponds to **point splitting regularisation** (see also [48] for the analysis of regularisation methods in connection with the semiclassical Einstein equation). In particular we have that

$$\langle \phi(x) \cdot \phi(y) \rangle_\omega := \Lambda(x, y) - \mathcal{H}(x, y)$$

and thus, with $\phi^2(x) := \lim_{x \to y} \phi(x) \cdot \phi(y)$,

$$\langle \phi^2(x) \rangle_\omega := \lim_{x \to y} \left(\Lambda(x, y) - \mathcal{H}(x, y) \right) .$$

At this point it is important to stress that the previous two requirements do not fix $\mathcal{H}$ completely. Having defined one $\mathcal{H}$, it is always possible to obtain another reasonable $\mathcal{H}$ adding a smooth, real–valued and symmetric function. The freedom might be further restricted by requiring that $\mathcal{H}$ is constructed only based on local properties of the spacetime so that the extended algebra of observables enjoys **local covariance** [9, 24, 25]. This new requirement implies that the smooth part W in $\mathcal{H}$ has to be constructed only out of local geometrical quantities. Even then there is of course a residual freedom as we may change the value of λ in the logarithmic singular contribution present in $\mathcal{H}$, cf. (6.4), or more generally we may add a smooth, real–valued and symmetric function constructed out of local geometrical quantities.

Unfortunately local non–linear fields like the stress tensor are not invariant under changes of this residual freedom, not even when further reasonable conditions are added such as smooth/analytic dependence on ξ and m, conservation of the stress tensor. See for instance [24] for a detailed discussion or this topic. This freedom in the definition of non–linear fields is what is usually called **renormalisation freedom**.

Finally, it is important to stress that due to the requirement of local covariance, $\mathcal{H}$ satisfies the equation of motion only up to smooth terms and it is in general not possible to find a W in $\mathcal{H}$ which is local and renders $\mathcal{H}$ a weak solution of the equation of motion. In particular, choosing $W = 0$ in $\mathcal{H}$ as in (6.4), it holds that

$$8\pi^2 \langle \phi P \phi \rangle_\omega = 6[V_1], \qquad 8\pi^2 \langle (\nabla_a \phi)(P\phi) \rangle_\omega = 2\nabla_a [V_1]$$

where the fields are evaluated at the coinciding point, P is the differential operator implementing the equation of motion (6.3) and $[V_1]$ is the coinciding point limit of the second coefficient in the Hadamard expansion of V in (6.4). This observation is the source of the well–known trace anomaly present in the expectation values of the stress tensor [49].

Taking into account the present discussion, the quantum stress tensor of the theory may be explicitly constructed. For a detailed discussion we refer to the original papers of Moretti [33] and Hollands and Wald [26] and the contribution of Khavkine and Moretti in Chap. 5.

6.3.1 Analysis of the Components of T

We shall now discuss the various contributions to the expectation value of the stress tensor in a Hadamard quantum state. Since we are assuming homogeneity and isotropy, up to an initial condition, the Einstein equations (6.11) are equivalent to their trace and the conservation of the stress energy tensor, namely,

$$R = 8\pi \langle T \rangle_\omega, \qquad \nabla^a \langle T_{ab} \rangle_\omega = 0.$$

The initial condition that needs to be fixed at $\tau = \tau_0$ is $3H^2(\tau_0) = 8\pi \langle T_{00} \rangle_\omega(\tau_0) - 3\kappa/a(\tau_0)^2$. We stress that this initial condition may not be freely chosen but is uniquely determined by ω. Nonetheless, in many cases it is sufficient to consider only the expectation value of the trace of the stress tensor. The various contributions to this quantity may be classified as

$$\langle T \rangle_\omega = T_{anomaly} + T_{ren.freedom} + T_{state}$$

where the individual terms are characterised as follows.

- The term $T_{anomaly}$ quantifies the well–known trace anomaly [49]. In particular, for a conformally coupled scalar field with mass m we have

$$T_{anomaly} = \frac{[V_1]}{4\pi^2} = \frac{1}{2880\pi^2}\left(C_{abcd}C^{abcd} + R_{ab}R^{ab} - \frac{R^2}{3} + \Box R\right) + \frac{m^4}{4}$$

  where C_{abcd} is the Weyl tensor. Ignoring the $\Box R$–contribution, which we prefer to subsume in $T_{ren.freedom}$, the stress tensor of this component is a perfect fluid with an energy density and pressure given by

$$\rho = cH^4 \qquad P = -c\left(\frac{4}{3}\dot{H}H^2 + H^4\right) \qquad c := \frac{1}{960\pi^2}.$$

  Notice that it is not possible to derive this anomalous term from a local effective action, see e.g. [40], moreover it cannot be written as a mixture of radiation, dust and cosmological constant on cosmological spacetimes. If $\dot{H} > 0$, then

$P/\rho < -1$ and thus the anomalous contribution to the stress tensor behaves like so-called "phantom energy".

- The renormalisation freedom is of the form

$$T_{ren.freedom} = \alpha m^4 + \beta m^2 R + \gamma \Box R\,. \tag{6.12}$$

This contribution can be obtained from an action defined by a Lagrangean of the form

$$\alpha' m^4 + \beta' m^2 R + \gamma_1' R^2 + \gamma_2' R_{ab} R^{ab}\,.$$

Note that this implies that the anomalous contribution cannot be cancelled by the renormalisation freedom. This freedom in the definition of T_{ab} and thus T occurs because there is no intrinsic way to define the "correct" T_{ab} to serve as the right hand side of the semiclassical Einstein equation within quantum field theory on curved spacetimes as already outlined in the general discussion in the previous subsection. This problem does not play a role for non–gravitational laboratory experiments, where only differences of expectation values in two states are measured and thus this renormalisation freedom plays no role. However, gravity is sensitive to the absolute value of $\langle T_{ab}\rangle_\omega$ and thus the four constants parametrising $T_{ren.freedom}$ can (only) be fixed by a sufficient amount of observational data. Note that, by a suitable choice of γ_1' and γ_2' we can cancel from the trace the terms depending on derivatives of the metric coefficients higher than the second. This is necessary in order to control the solution of the semiclassical Einstein equation with the same amount of initial values used to control the corresponding classical problem, see e.g. [48] for an extensive discussion.

- Finally, the state–dependent part is

$$T_{state} = m^2 \langle \phi^2 \rangle_\omega\,.$$

From (6.6) and (6.7) it becomes clear that T_{state} is a complicated functional of the scale factor $a(\tau)$, thus it is non–trivial to control the state–dependent contribution in general, which complicates the solution theory of the semiclassical Einstein equation in the massive case.

6.3.2 Existence and Uniqueness of Solutions of the Semiclassical Einstein Equation in Cosmology

In this subsection we shall briefly review results about exact solutions of the semiclassical Einstein equation in cosmological spacetimes obtained in [38]. To this end, we recall that, since the stress tensor we employ is covariantly conserved, up to an initial value constraint, it suffices to consider the equation

$$6(\dot{H} + 2H^2) - 4\Lambda = 8\pi \langle T \rangle_\omega . \tag{6.13}$$

Furthermore, recalling the previous discussion, we fix the the renormalisation constants α and β in (6.12) in such a way that the renormalisation of the Newton constant and that of the cosmological constant vanish. Finally, we chose γ in such a way that $\Box R$ terms do not explicitly appear in the trace of the stress tensor. For completeness, the chosen values of the three renormalisation constants are

$$\alpha = \frac{1}{32\pi^2}, \qquad \beta = \frac{1}{288\pi^2}, \qquad \gamma = -\frac{1}{2880\pi^2}.$$

Writing (6.13) in integral form we obtain a functional Volterra equation for the function $H(\tau)$

$$H = H_0 + \int_{\tau_0}^{\tau} \frac{a}{H_c^2 - H^2} \left(H^4 - 2H_c^2 H^2 + 240\pi^2 \left(m^2 \langle \phi^2 \rangle_\omega + 4\tilde{\Lambda} \right) \right) d\eta,$$

with $H_c^2 := 1440\pi^2/(8\pi) = 180\pi/G$ and $\tilde{\Lambda} := \Lambda/(8\pi)$. Moreover, we note that both the scale factor a and the expectation value $\langle \phi^2 \rangle_\omega$ are functionals of H. For completeness we recall the explicit form of the constraint

$$H_0^2 = \frac{8\pi}{3} \left(\rho(\tau_0) + \Lambda \right).$$

The previous integral equation can be rewritten abstractly as

$$H = F(H) \quad \text{with} \quad F(H)(\tau) := H_0 + \int_{\tau_0}^{\tau} f(H)(\eta)\, d\eta. \tag{6.14}$$

This expression exhibits the structure of a Volterra integral equation, however, since f is a functional of H and not an ordinary function, its integral kernel is an integro–differential operator depending on the chosen state ω.

In order to carry on the analysis of this equation, we would like to chose a state ω in every flat cosmological spacetime based only on the properties of the metric at τ_0. Unfortunately, recalling the discussion presented in Sect. 6.2, we can only obtain states of adiabatic type from finitely many derivatives of the metric at τ_0, and, if we want to employ only derivatives of the metric up to the second order as initial values, the obtained state will be an adiabatic state of order one. In the case of conformal coupling, already adiabatic states of order zero are sufficiently regular to render the expectation values of the stress tensor finite. Hence, for simplicity we shall employ this class of states in the following.

6.3.2.1 Local Solutions

In order to study the local existence and uniqueness of solutions of (6.14) we need to control the state–dependent part $\langle\phi^2\rangle_\omega$ in an adiabatic state of order zero by $H(\tau_0)$ and $a(\tau_0)$. We recall that, up to an inessential renormalisation freedom,

$$\langle\phi^2\rangle_\omega = \frac{1}{2\pi^2 a^2}\int_0^\infty \left[\chi_k\overline{\chi}_k - \frac{1}{\sqrt{k^2+m^2a^2}}\right]k^2 dk \tag{6.15}$$

where the χ_k are adiabatic modes of order zero in τ_0 which can be constructed by perturbation theory as in (6.10) perturbing the massless conformally invariant modes (see [38] for a detailed analysis). Furthermore, perturbation theory permits to prove that the expectation value $\langle\phi^2\rangle_\omega$ computed as in (6.15) and hence $f(H)$ in (6.14) are continuously Gâteaux differentiable in the continuous function $H \in C[\tau_0, \tau_1]$ for arbitrary but fixed τ_0, τ_1 and $a_0 = a(\tau_0)$

With this result at hand it is now easy to get the following theorem taken from [38], whose proof is done constructing a contraction map out of the Gâteaux differentiable $f(H)$ and then to use the Banach fixed point theorem.

Theorem 6.3.1 *Let* (a_0, H_0), $a_0 > 0$, $|H_0| < H_c$, *be some initial conditions fixed at* τ_0 *for the functional equation* (6.14). *There is a non-empty interval* $[\tau_0, \tau_1]$ *and a closed subset* $U \subset C[\tau_0, \tau_1]$ *on which a unique solution to* (6.14) *exists.*

This theorem guarantees the existence and uniqueness of local solutions of the semi-classical Einstein equations on cosmological spacetimes.

6.3.2.2 Global Solutions

Having established local existence of solutions of (6.14) in a short interval of time $[\tau_0, \tau_1)$ we would like to show that in fact the obtained solution can always be extended further provided no singularity is encountered at τ_1. Again the perturbative construction of the adiabatic modes χ_k permits to control the regularity of $\langle\phi^2\rangle_\omega$ also far away from τ_0. In particular, again $\langle\phi^2\rangle_\omega$ is continuously Gâteaux differentiable for perturbations in every interval $[\tau_1, \tau_2]$ in the future of τ_0. We may recall a further theorem taken from [38].

Theorem 6.3.2 *Consider a solution* $H_*(\tau)$ *in* $C[\tau_0, \tau_1]$ *of the functional equation* (6.14). *If the solution is regular in* $[\tau_0, \tau_1]$, *namely, if neither* a, H_* *nor* H_*' *are singular at* τ_1, *then it is possible to find a* $\tau_2 > \tau_1$ *such that the solution* H_* *can be extended uniquely to* $C[\tau_0, \tau_2)$ *and that the solution is regular therein.*

Since the solution may be extended we may thus consider the set of all possible regular solutions $\mathcal{S}$ in the future of τ_0 with fixed initial values in τ_0. If we indicate by H_I the solution of (6.14) supported on the interval I, we may equip $\mathcal{S}$ with the a partial order relation

$$H_I \le H_J \quad \text{if} \quad I \subset J.$$

Hence, by Zorn's lemma applied to the set of all solutions with the given initial values we have that the maximal regular solution of (6.14) exists in $\mathcal{S}$. Finally, uniqueness of that solution can again be obtained by an application of Theorem 6.3.2.

Apart from this analytic approach to solving the semiclassical Einstein equation there have been efforts to solve this equation numerically. Here we would like to point out early work by Anderson [5] and also the recent work of Eltzner and Gottschalk [16] which have cast the coupled system of the Klein–Gordon equation for the modes χ_k defining the state ω and the semiclassical Einstein equation into a form of a dynamical system which is particularly suited for numerical integration.

6.3.3 Comparison with the ΛCDM–Model

In the standard model of cosmology—the ΛCDM–model—one assumes that on large scales the spacetime is a FLRW spacetime, which may be taken to have flat spatial sections as this is preferred by observations. The dynamics of the spacetime, and thus of the scale factor a, are determined by the Einstein equations, whereby it is assumed that the matter–energy content can be macroscopically subsumed into a perfect fluid stress tensor which contains various components with a particular equation of state. As we expect that quantum field theory on curved spacetimes provides an accurate fundamental microscopic description of all matter–energy in the universe, it should be possible to derive the particular perfect fluid stress tensor which is taken as an input for the ΛCDM–model within quantum field theory on curved spacetimes. While this approach may seem to be an unnecessary complication of the subject, it may provide a bridge between foundational results of quantum field theory on curved spacetimes and observations and could thus be a starting point for analysing potential quantum effects on the expansion of the universe. In this section we would like to briefly review results in this direction obtain in [21].

In the last chapter we have used the symmetry of FLRW spacetimes to replace the semiclassical Einstein equation on these spacetimes by its trace, the conservation of the stress tensor and an initial condition at an arbitrary but fixed conformal time τ_0. For the purpose of the following discussion it is more suitable to consider the conservation of the stress tensor and the time–time–component of the (semiclassical) Einstein equation as an equivalent formulation of the full equations. This component, the so-called (first) Friedmann equations reads in the classical case

$$H^2 = \frac{8\pi}{3}\rho\,.$$

According to the ΛCDM–model, our universe contains matter, radiation, and dark energy, modelled macroscopically as perfect fluids with an equation of state $P = w\rho$, $w = 0, \frac{1}{3}, -1$ for matter, radiation and Dark Energy (assuming that the latter is just due to a cosmological constant) respectively. Consequently, the Friedmann equation can be conveniently rewritten as

$$\frac{H^2}{H_0^2} = \frac{\rho_{\Lambda\text{CDM}}}{\rho_0} = \Omega_\Lambda + \frac{\Omega_m}{a^3} + \frac{\Omega_r}{a^4}, \quad \rho_0 = \frac{3H_0^2}{8\pi}, \tag{6.16}$$

where H_0 is the present Hubble rate—the Hubble constant—and the constants Ω_Λ, Ω_m, Ω_r denote the present fractions of the energy density due to dark energy, matter and radiation respectively. Observations indicate approximately

$$\Omega_m = 0.3, \quad \Omega_r = 10^{-4}, \quad \Omega_\Lambda = 1 - \Omega_m - \Omega_r \tag{6.17}$$

see [1] for the latest exact values from the Planck collaboration. In the context of cosmology the terms "matter" and "radiation" subsume all matter-energy with the respective macroscopic equation of state such that e.g. "radiation" does not encompass only electromagnetic radiation, but also the three left-handed neutrinos present in standard model of particle physics (SM) and possibly so-called dark radiation, and "matter" subsumes both the baryonic matter which is in principle well-understood in the SM and dark matter. Here, dark matter and dark radiation both quantify contributions to the macroscopic matter and radiation energy densities which exceed the ones expected from the knowledge of the SM and are believed to originate either from fields not present in the SM or from other sources, i.e. modifications of classical general relativity.

Reproducing (6.16) within quantum field theory on curved spacetimes amounts to choosing an appropriate field theoretic model, presumably (an extension of) the standard model of particle physics, and a suitable homogeneous and isotropic Hadamard state ω for this model such that the semiclassical Einstein equations are satisfied and that the energy density $\langle T_{00}\rangle_\omega$ in this state is qualitatively and quantitatively of the form (6.16) up small corrections. Clearly this is quite an ambitious task because computing $\langle T_{00}\rangle_\omega$ for the full interacting standard model of particle physics is non–trivial and, as we have seen in the previous section, it is indeed possible to construct analytic solutions of the semiclassical Einstein equations, but these solutions are not easily accessible for quantitative analyses. Thus, for a first analysis, we shall considerably simplify the problem in various ways.

- We completely disregard the field interactions and consider only free fields. To legitimate this, we introduce yet another time variable, the **redshift**

$$z := \frac{a_0}{a} - 1 \tag{6.18}$$

where $a_0 = 1$ is the scale factor of today and we note that $z = 0$ marks the present and that z is a monotonically decreasing function of either t or τ if the Hubble rate H is strictly positive which is compatible with observations and the regularity of the FLRW metric. We may now recall that it is commonly expected that field interactions become negligible on cosmological scales for redshifts smaller than 10^9, which is approximately the time when the primordial synthesis of light nuclei—the so-called Big Bang Nucleosynthesis—has ended. In the ΛCDM–model this assumption enters by demanding that the stress tensor of each perfect fluid is separately conserved. In the following we shall thus consider only the time interval $z \in [0, 10^9]$.

- We shall consider only scalar fields as a toy model for the matter content of the universe. In order to model both matter and radiation, i.e. massive and massless fields, we shall consider two scalar fields where one is massive, the second is massless and both are conformally coupled.
- We solve the semiclassical Einstein equation in the following approximate sense. We assume that the spacetime is determined by a scale factor which is an exact solution of (6.16). On this spacetime we seek to find a pair of quantum states ω^m and ω^0 for the massive and massless scalar field such that the sum of the energy densities in this states satisfies

$$\frac{\langle T_{00}\rangle_{\omega^0} + \langle T_{00}\rangle_{\omega^m}}{\rho_0} = \Omega_\Lambda + \frac{\Omega_m}{a^3} + \frac{\Omega_r}{a^4} = \frac{\rho_{\Lambda\mathrm{CDM}}}{\rho_0} \tag{6.19}$$

and (6.17) up to suitably small corrections in the time interval of interest $z \in [0, 10^9]$.

In order to pursue the strategy outlined above, we need to make a suitable ansatz for the Hadamard states ω^m and ω^0. As the radiation component in the ΛCDM–model is assumed to be of thermal nature and the same is often assumed for dark matter, which is the major part of the matter component in the ΛCDM–model, we seek to find ω^m and ω^0 among the quasi–free states defined by the two–point function

$$\Lambda(x_1, x_2) := \lim_{\varepsilon\to 0^+} \frac{1}{8\pi^3 a(t_1)a(t_2)} \tag{6.20}$$

$$\times \int_{\mathbb{R}^3} \left(\frac{\overline{\chi_k(t_1)}\chi_k(t_2)}{1 - e^{-\beta k_0}} + \frac{\chi_k(t_1)\overline{\chi_k(t_2)}}{e^{\beta k_0} - 1} \right) e^{i\boldsymbol{k}\cdot(\boldsymbol{x}_1 - \boldsymbol{x}_2)} e^{-k\varepsilon} d\boldsymbol{k}$$

with $k_0 := \sqrt{k^2 + m^2 a_F^2}$ for a fixed constant a_F and where we assume that χ_k are the modes defining a state of low energy. These states match the almost equilibrium states introduced by Küskü [31] up to the form of k_0 and their Hadamard property follows from results of [36, 50]. In the massless case, these states are independent of a_F and satisfy the conformal KMS condition with respect to the conformal Killing vector ∂_τ. In the massive case, they may be considered to describe approximately the quantum state of a field which has been in thermal equilibrium in the distant past, and has "frozen out" of equilibrium at the time $a = a_F$.

To discuss the energy density of the massless and massive conformally coupled scalar fields in this class of states, we rewrite it as follows

$$\frac{\langle T_{00}\rangle_{\omega^0} + \langle T_{00}\rangle_{\omega^m}}{\rho_0} \tag{6.21}$$

$$= \frac{\rho^m_{\text{gvac}} + \rho^0_{\text{gvac}} + \rho^m_{\text{gth}} + \rho^0_{\text{gth}}}{\rho_0} + \gamma \frac{H^4}{H_0^4} + \Omega_\Lambda + \delta \frac{H^2}{H_0^2} + \varepsilon \frac{J_{00}}{H_0^4} ,$$

where J_{00} is the time–time–component of a conserved geometric tensor J_{ab} whose trace is equal to $\Box R$.

The first terms in (6.21) denote the genuinely quantum state dependent contributions to the energy densities of the two quantum fields, whereas the last four terms denote the contribution from the trace anomaly (γ) and the renormalisation freedom (Ω_Λ, δ, ε). We have split the state–dependent contributions into parts which are already present for infinite inverse temperature parameter β in the generalised thermal states (6.20), and thus could be considered as contributions due to the states of low energy as generalised vacuum states (ρ^m_{gvac}, ρ^0_{gvac}), and into the remaining terms, which may be interpreted as purely thermal contributions (ρ^m_{gth}, ρ^0_{gth}).

We start our quantitative discussion of (6.21) by considering the geometric terms and take the point of view that δ, which effectively renormalises Newton's constant, is not a free parameter because Newton's constant has been measured already. One could also take a more conservative point of view and consider δ to be a free parameter, in this case comparison with cosmological data, e.g. from big bang nucleosynthesis, would presumably constrain δ to be very small. The trace anomaly contribution in (6.21) proportional to γ is not present in the ΛCDM-model and is thus a genuine quantum and state–independent contribution to the quantum energy momentum tensor. One has $\gamma \simeq 10^{-122}$ for two scalar fields and, as, $H < H_0 z^2$ in the ΛCDM-model for large redshifts, this term can be safely neglected for $z < 10^9$. The cosmological constant contribution Ω_Λ in (6.21) is part of the regularisation freedom and thus a free parameter just like in the ΛCDM-model. Finally, the remaining regularisation freedom contribution quantified by ε also constitutes an extension of the ΛCDM-model by the new free parameter ε. In order to fulfil the goal of this section, one may simply set $\varepsilon = 0$, but small non-zero values of ε are also compatible with observations and may have interesting physical implications, see [21] for details.

Coming to the state–dependent contributions to (6.21), it has been shown in [21] that, up to the freedom parametrised by Ω_Λ, δ and ε, $\rho^0_{\text{gvac}} = 0$ for arbitrary sampling functions f defining the state of low energy–modes χ_k, whereas $\rho^m_{\text{gth}}/\rho_{\Lambda\text{CDM}} \ll 1$ for small masses $m \simeq H_0$ and large masses $m \gg H_0$ if the sampling function f has sufficiently large support in time. This generalises results obtained by Degner on de Sitter spacetime [14] and indicates that states of low energy with broad sampling functions are reasonable generalised vacuum states on FLRW spacetimes.

As for the thermal contributions, one finds in the massless case

$$\rho^0_{\text{gth}} = \frac{\Omega_r}{a^4} \quad \text{with} \quad \Omega_r = \frac{\pi^2}{30\beta^4}.$$

Up to degree of freedom factors, this gives the ΛCDM value $\Omega_r \simeq 10^{-4}$ if the temperature parameter $1/\beta$ is in the range of the cosmic microwave background temperature $1/\beta \simeq 2.7\,\text{K}$. In the massive case, one can take typical values of β, a_F

and m from Chap. 5.2 in [30] computed by means of effective Boltzmann equations. A popular candidate for dark matter is a weakly interacting massive particle (WIMP), e.g. a heavy neutrino, for which [30] computes

$$x_F = \beta a_F m \simeq 15 + 3\log(m/\text{GeV}) \qquad a_F \simeq 10^{-12}(m/\text{GeV})^{-1}.$$

Using this one finds for large m

$$\rho^m_{\text{gth}} \simeq \frac{1}{(2\pi)^{3/2}} \frac{m}{\beta^3 a^3} x_F^{\frac{3}{2}} e^{-x_F},$$

and thus $\Omega_m \simeq 0.3$ for $m \simeq 100$ GeV.

6.3.4 Beyond the Semiclassical Einstein Equation

In the next–to–last section we have discussed the existence of solutions of the semi-classical equation

$$G_{ab} = -8\pi \langle T_{ab} \rangle_\omega .$$

However, this equation equates a classical deterministic quantity with the expectation value of the quantum stress tensor in a suitable state. The possible realisations of T_{ab} in a state ω are distributed according to a certain probability distribution, whose moments are described by $\langle T^n_{ab} \rangle_\omega$. Hence, the semiclassical equation has a meaning as an effective mean field equation only when the probability distribution is sharply peaked around the mean value $\langle T_{ab} \rangle_\omega$. A necessary condition for this to hold is that the correlations $\langle T_{ab} T_{ab} \rangle_\omega$ are negligible with respect to $\langle T_{ab} \rangle^2_\omega$. However, this is not the case, because

$$\lim_{x \to y} \langle T_{ab}(x) T_{ab}(y) \rangle_\omega$$

diverges. If the variance of $\langle T_{ab} \rangle_\omega$ is not negligible, like in the case of **Brownian motion**, the semiclassical Einstein equation could still make sense as a stochastic equation. In other words it could be understood as an **Einstein-Langevin** equation [27], namely an equation among probability distributions

$$G_{ab}(x) = -8\pi T_{ab}(x) .$$

Hence, assuming this point of view, $G_{ab}(x)$ acquires the meaning of a stochastic field and we may try to study the passive influence of matter fluctuations on curvature fluctuations. However, in spite of recent successful attempts for some special cases [19], it is not easy to compute probability distribution for T_{ab} in ω. Notwithstanding, indicating

$$\delta G_{ab} = G_{ab} - \langle G_{ab} \rangle , \qquad \delta T_{ab} = T_{ab} - \langle T_{ab} \rangle_\omega$$

we can equate their moments

$$\langle G_{ab}(x)\rangle = -8\pi\,\langle T_{ab}(x)\rangle_\omega$$
$$\langle \delta G_{ab}(x_1)\delta G_{cd}(x_2)\rangle = (-8\pi)^2\langle \delta T_{ab}(x_1)\delta T_{cd}(x_2)\rangle_\omega$$
$$\dots$$
$$\langle \delta G^n(x_1,\dots,x_n)\rangle = (-8\pi)^n\langle \delta T^n(x_1,\dots,x_n)\rangle_\omega .$$

In this way we notice that it is, at least in principle, possible to evaluate the passive influence of matter fluctuations on metric perturbations and one may apply this to the treatment of inflationary cosmology where quantized metric perturbations play an important role see [37] for details. The advantage of this point of view over the standard approach, which we review in the next section, is that the state for matter perturbations, which induces a state for the curvature perturbations as well, may not be selected freely but is strongly constrained by the semiclassical Einstein equation. Furthermore, since the correlations of $T_{ab}(x)$ are **more complicated** than in Wiener processes or Brownian motions, non–Gaussianities, which are of interest in inflationary cosmology, naturally arise.

6.4 Analysis of Perturbations in Inflation

A prominent application of quantum field theory in curved spacetime is the analysis of perturbations in inflation. In the simplest models of inflation, the characteristic exponential expansion is driven by the energy density of a classical scalar field φ. The perturbation ϕ of this field, combined in a gauge–invariant way with the perturbations γ of the classical metric g, are considered as quantum fields propagating on the classical background $\boldsymbol{M}_\varphi = (\mathcal{M}, g, \mathfrak{o}, \mathfrak{t}, \varphi)$, where $\boldsymbol{M} = (\mathcal{M}, g, \mathfrak{o}, \mathfrak{t})$ is a spatially flat FLRW spacetime and φ depends only on time in the homogeneous and isotropic FLRW coordinates. Rephrased in more abstract terms, one may say that the analysis of perturbations in inflation consists in quantizing the field theory of the tuple $\mathfrak{G} = (g, \varphi)^t$ peturbatively around a background which satisfies the classical Einstein equation and is of FLRW-type. This perturbative quantum field theory is usually truncated at linear order.

The standard textbook treatment of the quantum theory of perturbations in inflation, see e.g. [43], consists in using the FLRW-symmetry of the background $\boldsymbol{M}_\varphi$ in order to split the metric perturbation γ into components which transform as scalars, vectors and tensors under the isometry group of FLRW-backgrounds, the Euclidean group $\mathbb{E}^3$. Subsequently, gauge–invariant linear combinations of these components and the scalar field perturbation ϕ are identified and quantized in a canonical fashion. As the above-mentioned splitting is non–local and depends heavily on the FLRW-symmetry, it is a priori not clear whether it captures all local observables of the theory. A systematic analysis of this issue from the point of view of algebraic quantum field theory has been performed in [22], in this section we shall review the main steps and results of this analysis. For simplicity we consider the special and commonly assumed case where the scalar field is minimally coupled to the metric, the case of general coupling is treated in [22].

6.4.1 Quantization of the Linearised Einstein–Klein–Gordon System on Arbitrary On–Shell Backgrounds

We first consider the quantization of the linearised Einstein–Klein–Gordon system on arbitrary backgrounds $\boldsymbol{M}_\varphi = (\mathcal{M}, g, \mathfrak{o}, \mathfrak{t}, \varphi)$, such that $\mathcal{M}$ is four-dimensional, $\boldsymbol{M} = (\mathcal{M}, g, \mathfrak{o}, \mathfrak{t})$ is globally hyperbolic, and $\mathfrak{G} = (g, \varphi)^t$ satisfies the coupled Einstein–Klein–Gordon equations. Later we point our attention towards such backgrounds $\boldsymbol{M}_\varphi$ which are in addition of FLRW-type and compare the general quantization procedure with the usual approach to the quantization of perturbations in inflation.

To this avail, we introduce the vector bundles over $\mathcal{M}$ $\mathcal{V} := \bigvee^2 T^*\mathcal{M} \oplus (\mathcal{M} \times \mathbb{R})$, where $\bigvee$ denotes the symmetric tensor product, and $\mathcal{W} := T\mathcal{M}$. The space of smooth sections of a vector bundle such as $\mathcal{V}$ will be denoted by $\Gamma(\mathcal{V})$. Important subspaces of $\Gamma(\mathcal{V})$ are $\Gamma_0(\mathcal{V})$ and $\Gamma_{\mathrm{sc}}(\mathcal{V})$ the space of smooth sections of compact and space-like compact support, respectively. Both the background fields $\mathfrak{G} = (g, \varphi)^t$ and their perturbations $\Gamma = (\gamma, \phi)^t$ are elements of $\Gamma(\mathcal{V})$, whereas gauge-transformations (linearised diffeomorphisms) will be parametrised by $\varsigma \in \Gamma(\mathcal{W})$. We introduce on such sections symmetric and non–degenerate bilinear forms by

$$\langle \Gamma_1, \Gamma_2 \rangle_{\mathcal{V}} := \int\limits_{\mathcal{M}} \left(g^{ab} g^{cd} \gamma_{1,ac} \gamma_{2,bd} + \phi_1 \phi_2 \right) \mathrm{dvol}_{\boldsymbol{M}} ,$$

$$\langle \varsigma_1, \varsigma_2 \rangle_{\mathcal{W}} := \int\limits_{\mathcal{M}} g^{ab} \varsigma_{1,a} \varsigma_{2,b} \, \mathrm{dvol}_{\boldsymbol{M}} .$$

These bilinear forms are well–defined for pairs of sections with compact overlapping support.

The starting point of the analysis is the Einstein–Hilbert–Klein–Gordon action for $\mathfrak{G} = (g, \varphi)^t \in \Gamma(\mathcal{V})$

$$S(\mathfrak{G}) = \int\limits_{\mathcal{M}} \left(\frac{R}{16\pi} + \frac{(\nabla_a \varphi) \nabla^a \varphi}{2} - V(\varphi) \right) \mathrm{dvol}_{\boldsymbol{M}} ,$$

where $V(\varphi)$ is an arbitrary smooth potential. The Euler–Lagrange equations of $S(\mathfrak{G})$ are the Einstein–Klein–Gordon–equations

$$\Gamma(\mathcal{V}) \ni EL(\mathfrak{G}) = \begin{pmatrix} \frac{1}{16\pi}(G_{ab} + 8\pi T_{ab}) \\ \Box\varphi + \partial_\varphi V \end{pmatrix} = 0 .$$

In order to obtain the linearised theory, we split $\mathfrak{G}$ into a background $\mathfrak{G}$ (which, slightly abusing notation, we denote by the same symbol) and a perturbation Γ and formally expand (omitting boundary terms)

$$S(\mathfrak{G}+\Gamma) = S(\mathfrak{G}) - \langle EL(\mathfrak{G}), \Gamma\rangle_{\mathcal{V}} - \frac{1}{2}\langle \Gamma, P\Gamma\rangle_{\mathcal{V}} + \mathcal{O}(\Gamma^3).$$

Here P is a second order partial differential operator (see [22]) which is formally self-adjoint w.r.t. $\langle\cdot,\cdot\rangle_{\mathcal{V}}$.

$S(\mathfrak{G}+\Gamma)$ is invariant under diffeomorphisms of $\mathcal{M}$, in particular under those generated by an arbitrary but fixed compactly supported vector field $\varsigma \in \Gamma_0(\mathcal{W})$. Given such a diffeomorphism, $\mathfrak{G}+\Gamma$ transforms as

$$\mathfrak{G}+\Gamma \mapsto \mathfrak{G}+\Gamma+\mathcal{L}_\varsigma\mathfrak{G}+\mathcal{L}_\varsigma\Gamma+\mathcal{O}(\varsigma^2)\,,$$

where $\mathcal{L}_\varsigma$ denotes the Lie derivative w.r.t. ς. To first order in ς and Γ, the diffeomorphism-invariance of $S(\mathfrak{G}+\Gamma)$ reads

$$P\mathcal{L}_\varsigma\mathfrak{G} = \mathcal{L}_\varsigma EL(\mathfrak{G})\,,$$

where the term on the right hand side arises from the $\mathcal{L}_\varsigma\Gamma$ contribution of the transformed $\mathfrak{G}+\Gamma$. These observations imply the following: we may consistently truncate the diffeomorphism–invariant field theory for $\mathfrak{G}+\Gamma$ at joint linear order in Γ and ς if and only if we assume that the background $\mathfrak{G}$ is on–shell, i.e. $EL(\mathfrak{G})=0$.[1] In this case, one may think of the linearised Einstein–Klein–Gordon theory as originating from the quadratic action

$$S^{(2)}(\Gamma) := -\frac{1}{2}\langle \Gamma, P\Gamma\rangle_{\mathcal{V}}\,,$$

which is for all $\varsigma \in \Gamma_0(\mathcal{W})$ invariant under the affine transformation

$$\Gamma \mapsto \Gamma + \mathcal{L}_\varsigma\mathfrak{G}\,.$$

Defining $K : \Gamma(\mathcal{W}) \to \Gamma(\mathcal{V})$ by $K\varsigma := \mathcal{L}_\varsigma\mathfrak{G}$, we may express this gauge-invariance as

$$P\circ K = 0\,.$$

This automatically implies that the equation of motion $P\Gamma=0$ for the perturbations does not have a well–posed Cauchy problem as non–trivial solutions with compact support exist.

We would like to quantize the linearised Einstein–Klein–Gordon theory in a gauge–invariant manner following the strategy of [13], which deals with the electro-

[1] Strictly speaking $\mathcal{L}_\varsigma EL(\mathfrak{G})=0$ is satisfied even if $\Box\varphi+\partial_\varphi V = c$ with c constant but non–zero. However, one may absorb c by redefining $V(\varphi)$.

magnetic vector potential. This strategy has been further pursued in [17] to quantize linearised gravity on cosmological vacuum spacetimes, and axiomatised in [23]. In order to cast the linearised Einstein–Klein–Gordon theory into a form which satisfies the axioms of [23], such that the results of this work may be readily applied, we introduce a field redefinition which generalises the trace–reversal well–known from linearised gravity (the appearing numerical factors are introduced in order to homogenise the normalisation of the γ–γ and ϕ–ϕ components of the principal symbol of P).

$$\overline{\cdot} : \Gamma(\mathcal{V}) \mapsto \Gamma(\mathcal{V}) \qquad \Gamma = \begin{pmatrix} \gamma_{ab} \\ \phi \end{pmatrix} \mapsto \overline{\Gamma} = \begin{pmatrix} -\frac{1}{32\pi}\left(\gamma_{ab} - \frac{1}{2} g_{ab}\gamma_c{}^c\right) \\ \phi \end{pmatrix}$$

Using this field redefinition, we may now define

$$\overline{P} := P \circ \overline{\cdot}^{-1} \qquad \overline{K} := \overline{\cdot} \circ K \qquad \langle \cdot, \cdot \rangle_{\overline{\mathcal{V}}} := \langle \overline{\cdot}^{-1}, \cdot \rangle_{\mathcal{V}} \qquad \Theta := \begin{pmatrix} \theta_{ab} \\ \phi \end{pmatrix} := \overline{\Gamma} .$$

These definitions are tailored in such a way that the second order action for Γ may now be re-written as

$$S^{(2)}(\Gamma) = -\frac{1}{2}\langle \Gamma, P\Gamma \rangle_{\mathcal{V}} = -\frac{1}{2}\langle \Theta, \overline{P}\Theta \rangle_{\overline{\mathcal{V}}} =: \overline{S^{(2)}}(\Theta) .$$

Moreover, $\overline{S^{(2)}}(\Theta)$ is invariant under the affine transformation $\Theta \mapsto \Theta + \overline{K}\varsigma$ for all $\varsigma \in \Gamma_0(\mathcal{W})$, $\langle \cdot, \cdot \rangle_{\overline{\mathcal{V}}}$ is symmetric and non-degenerate and $\overline{P}$ is formally self-adjoint w.r.t. the redefined bilinear form.

Summing up, we may now consider the linearised Einstein–Klein–Gordon system as a linear gauge theory defined by the tuple $(\boldsymbol{M}_\varphi, \mathcal{V}, \mathcal{W}, \overline{P}, \overline{K})$, where $\boldsymbol{M}_\varphi$ plays the role of the background, the dynamical field $\Theta \in \Gamma(\mathcal{V})$ is a smooth section of $\mathcal{V}$ which satisfies the equation of motion $\overline{P}\Theta = 0$ with a formally selfadjoint $\overline{P}$ and the gauge-invariance of the theory is encoded in the relation $\overline{P} \circ \overline{K} = 0$ which implies that $\Gamma(\mathcal{W})$, via $\overline{K}$, parametrises the gauge transformations of the model. As shown in [22], $(\boldsymbol{M}_\varphi, \mathcal{V}, \mathcal{W}, \overline{P}, \overline{K})$ enjoys three further structural properties which are essential for the gauge–invariant quantization of this model:

1. There exists a "hyperbolic gauge fixing operator" $T : \Gamma(\mathcal{W}) \to \Gamma(\mathcal{V})$ such that

$$\widetilde{P} := \overline{P} + T \circ \overline{K}^\dagger = \nabla_a \nabla^a + \text{lower orders}$$

is normally hyperbolic. Here, $\overline{K}^\dagger$ is the adjoint of $\overline{K}$ defined by $\langle \overline{K}^\dagger \Gamma, \varsigma \rangle_{\mathcal{W}} := \langle \Gamma, \overline{K}\varsigma \rangle_{\overline{\mathcal{V}}}$ for all $\Gamma \in \Gamma(\mathcal{V})$, $\varsigma \in \Gamma(\mathcal{W})$ with compact overlapping support. T is not unique but we shall see that the following constructions are independent of T.
2. One may choose T above in such a way that $\overline{K}^\dagger \circ T$ is also normally hyperbolic, e.g. $T = 2\overline{K}$. This entails that the "gauge-fixing condition" $\overline{K}^\dagger\Theta = 0$ for

solutions of $\widetilde{P}\Theta = 0$, which implies that such Θ satisfy in fact $P\Theta = 0$, is compatible with $\widetilde{P}$-dynamics; in other words, it is satisfied on the full spacetime once it holds on a Cauchy surface.
3. $\overline{K}^\dagger \circ \overline{K}$ is normally hyperbolic as well and thus the condition $\overline{K}^\dagger\Theta = 0$ can always be satisfied by means of a suitable gauge transformation.

The structural properties of the linearised Einstein–Klein–Gordon system discussed so far imply that this model satisfies all axioms of linear gauge theories proposed in [23], and thus one may straightforwardly quantize this field theory in a gauge–invariant manner on arbitrary on–shell backgrounds by means of the results in [23]. The basic idea is the following. We consider the spaces

$$\overline{\mathsf{Sol}} := \{\Theta \in \Gamma(\mathcal{V}) \mid \overline{P}\Theta = 0\} \qquad \overline{\mathcal{G}} := \overline{K}\,[\Gamma(\mathcal{W})]\,.$$

The space of gauge-equivalence classes of solutions $\overline{\mathsf{Sol}}/\overline{\mathcal{G}}$ may be interpreted as the space of pure states in the classical linearised Einstein–Klein–Gordon field theory. The space of (regular) linear functionals on $\overline{\mathsf{Sol}}/\overline{\mathcal{G}}$ is parametrised by

$$\overline{\mathcal{E}} := \mathrm{Ker}_0\left(\overline{K}^\dagger\right)/\,\overline{P}\,[\Gamma_0(\mathcal{V})] \qquad \mathrm{Ker}_0\left(\overline{K}^\dagger\right) := \{h \in \Gamma_0(\mathcal{V}) \mid \overline{K}^\dagger h = 0\}.$$

Indeed, it is not difficult to check that the dual pairing

$$\overline{\mathsf{Sol}}/\overline{\mathcal{G}} \times \overline{\mathcal{E}} \ni ([\Theta],[f]) \mapsto \langle[\Theta],[f]\rangle := \langle\Theta, f\rangle_{\overline{\mathcal{V}}}$$

is independent of the representatives and thus well–defined. $\overline{\mathcal{E}}$ thus conveniently parametrises linear and gauge–invariant local obervables in the classical field theory. In order to endow $\overline{\mathcal{E}}$ with a pre–symplectic structure, we consider the advanced–minus–retarded operator $E_{\widetilde{P}}$ of the normally hyperbolic gauge–fixed equation of motion operator $\widetilde{P} = \overline{P} + T \circ \overline{K}^\dagger$ and define a bi-linear form $\overline{\sigma}$ on $\overline{\mathcal{E}}$ by

$$\overline{\sigma}([f_1],[f_2]) := \left\langle f_1, E_{\widetilde{P}} f_2\right\rangle_{\overline{\mathcal{V}}}\,.$$

One may prove that the abstract properties of $(\boldsymbol{M}_\varphi, \mathcal{V}, \mathcal{W}, \overline{P}, \overline{K})$ imply that $\overline{\sigma}$ satisfies the following properties and thus indeed defines a pre–symplectic structure on $\overline{\mathcal{E}}$ [23]:

- $\overline{\sigma}$ is independent of the representatives and thus well–defined.
- $\overline{\sigma}$ is antisymmetric.
- $\overline{\sigma}$ is independent of the gauge-fixing operator T appearing in the definition of $\widetilde{P}$.

The last property follows essentially from the fact that the gauge-fixing term $T \circ \overline{K}^\dagger$ in $\widetilde{P}$ is "invisible" to $[f] \in \overline{\mathcal{E}}$ because $\overline{K}^\dagger f = 0$; thus $\overline{\sigma}$ may be regarded as being "gauge–invariant" in that sense. One may also understand this gauge-invariance as follows. Analogously to the well-known fact that the advanced–minus–retarded operator of a hyperbolic PDE constitutes a bijection between equivalence classes of test–sections and spacelike-compact solutions, one can show that $E_{\widetilde{P}}$ induces a bijection between $\overline{\mathcal{E}}$ and suitable gauge-equivalence classes of spacelike-compact solutions of $\overline{P}\Theta = 0$. This by its own motivates to view $E_{\widetilde{P}}$ as an "effective" advanced–minus–retarded operator of $\overline{P}$. By means of this bijection on may then re–express $\overline{\sigma}$ as an equivalent bilinear form on the aforementioned suitable gauge–equivalence classes of spacelike–compact solutions, which is manifestly gauge–invariant.

The field redefinition $\overline{\cdot}$ was helpful for uncovering important structural properties of the linearised Einstein–Klein–Gordon system $(\boldsymbol{M}_\varphi, \mathcal{V}, \mathcal{W}, \overline{P}, \overline{K})$, however it is merely a computational trick and not of physical significance. To see this explicitly, we consider the spaces related to the original system $(\boldsymbol{M}_\varphi, \mathcal{V}, \mathcal{W}, P, K)$:

$$\mathsf{Sol} := \{\Gamma \in \Gamma(\mathcal{V}) \mid P\Gamma = 0\} \qquad \mathcal{G} := K\left[\Gamma(\mathcal{W})\right]$$

$$\mathcal{E} := \mathrm{Ker}_0\left(K^\dagger\right)/P\left[\Gamma_0(\mathcal{V})\right] \qquad \mathrm{Ker}_0\left(K^\dagger\right) := \{h \in \Gamma_0(\mathcal{V}) \mid K^\dagger h = 0\}.$$

It is not difficult to see that $\overline{\mathcal{E}} = \mathcal{E}$ and as before, we may now observe that $\mathcal{E}$ parametrises linear functionals on $\mathsf{Sol}/\mathcal{G}$. Due to

$$\langle \Gamma, h\rangle_{\mathcal{V}} = \langle \overline{\Gamma}, h\rangle_{\overline{\mathcal{V}}}$$

the physical interpretation of these linear functionals in terms of local observables is manifestly independent of the field redefinition. Finally, it follows from the previous discussion that

$$\widetilde{P} \circ \overline{\cdot} = P + T \circ K^\dagger \circ \overline{\cdot}$$

is not normally hyperbolic, but possesses unique advanced and retarded Green operators. The advanced–minus–retarded operator of $\widetilde{P} \circ \overline{\cdot}$ is $\overline{\cdot}^{-1} \circ E_{\widetilde{P}}$ and the induced bilinear form on $\mathcal{E}$

$$\sigma([h_1], [h_2]) := \left\langle h_1, \left[\overline{\cdot}^{-1} \circ E_{\widetilde{P}}\right] h_2\right\rangle_{\mathcal{V}}$$

manifestly equals $\overline{\sigma}$.

Given the pre–symplectic space $(\mathcal{E}, \sigma) = (\overline{\mathcal{E}}, \overline{\sigma})$ of gauge–invariant linear observables of the linearised Einstein–Klein–Gordon system, the corresponding gauge–invariant algebra of observables may be obtained by construction either the polynomial algebra or the Weyl-algebra corresponding to $(\mathcal{E}, \sigma)$, as explained in the contribution of Benini and Dappiaggi in Chap. 3. Note that, in contrast to linear theories without gauge–invariance, the resulting canonical commutation relations

$$[\Gamma(h_1), \Gamma(h_2)] = i\left\langle h_1, \left[\overline{\cdot}^{-1} \circ E_{\widetilde{P}}\right] h_2\right\rangle_{\mathcal{V}}$$

do not make sense in "unsmeared form".

6.4.2 The Special Case of Perturbations in Inflation

After discussing the gauge–invariant quantisation of the linearised Einstein–Klein–Gordon system on general on–shell backgrounds $\boldsymbol{M}_\varphi = (\mathcal{M}, g, \mathfrak{o}, \mathfrak{t}, \varphi)$, we consider the special case of FLRW-type on–shell backgrounds with flat spatial sections, which is the field–theoretic model of perturbations in inflation. We recall that these backgrounds are characterised by

$$g = a(\tau)^2(d\tau^2 - d\boldsymbol{x}^2) \qquad \varphi(\tau, \boldsymbol{x}) = \varphi(\tau)\,.$$

Two reoccurring important quantities are

$$\mathcal{H} := \frac{a'}{a} = aH \qquad z := \frac{a\varphi'}{\mathcal{H}}$$

and we shall indicate spatial indices in the homogeneous and isotropic FLRW-coordinates by Latin letters $i, j, k, \ldots$. We shall use the convention that these indices will be raised and lowered by means of the Euclidean metric δ_{ij} rather than by means of the induced metric $a(\tau)^2\delta_{ij}$.

In the following discussion, a special role is played by sections which vanish at spatial infinity (with all derivatives).

$$\Gamma_\infty(\mathcal{V}) := \{\Gamma \in \Gamma(\mathcal{V}) \mid \partial_{i_1} \cdots \partial_{i_n}\Gamma(\tau, \boldsymbol{x}) \text{ vanishes for } |\boldsymbol{x}| \to \infty \text{ for all } n \in \mathbb{N}_0\}$$

$$\Gamma_\infty(\mathcal{W}) := \{\varsigma \in \Gamma(\mathcal{W}) \mid \partial_{i_1} \cdots \partial_{i_n}\varsigma(\tau, \boldsymbol{x}) \text{ vanishes for } |\boldsymbol{x}| \to \infty \text{ for all } n \in \mathbb{N}_0\}$$

Namely, one can uniquely split $\Gamma = (\gamma_{ab}, \phi)^t \in \Gamma_\infty(\mathcal{V})$ as

$$\gamma_{ab} = a(\tau)^2 \begin{pmatrix} 2A & (\partial_i B - V_i)^t \\ \partial_i B - V_i & -2\left(\partial_i\partial_j C + \delta_{ij} D + \partial_{(i} W_{j)} + T_{ij}\right) \end{pmatrix}$$

$$A, B, C, D \in C^\infty_\infty(\mathcal{M}, \mathbb{R})\,, \quad V, W \in C^\infty_\infty(\mathcal{M}, \mathbb{R}^3)\,, \quad \partial^i V_i = \partial^i W_i = 0$$

$$T \in C^\infty_\infty(\mathcal{M}, \textstyle\bigvee^2\mathbb{R}^3)\,, \quad T^i_i = 0\,, \quad \partial^i T_{ij} = 0\,.$$

The components B, C, D, W_i are solutions of certain Poisson equations, e.g.

$$\Delta B = \frac{\partial^i \gamma_{0i}}{a^2}$$

and the uniqueness of the above splitting results from the unique solvability of such equations under the assumption that the solutions vanish at infinity. For the same reason, such a splitting for general $\Gamma \in \Gamma(\mathcal{V})$ can only be unique up to harmonic functions. In fact, this non-uniqueness is a non–trivial obstacle for proving that a splitting which is smooth in τ exists in general, though we presume that this is the case. Notwithstanding, we shall only need the existence and uniqueness of the splitting for $\Gamma_\infty(\mathcal{V})$ in the following.

Owing to their transformation properties under the Euclidean group $\mathbb{E}^3$, the components A, B, C, D, ϕ of a section in $\Gamma \in \Gamma_\infty(\mathcal{V})$ are called "scalar", V_i, W_j are called "vector" and T_{ij} are called "tensor" and similarly for the components of $\varsigma \in \Gamma_\infty(\mathcal{W})$. Following this nomenclature, we say that $\Gamma = (\gamma_{ab}, \phi)^t \in \Gamma(\mathcal{V})$ ...

...is of scalar type if γ_{ab} can be split as above with $V_i = W_i = T_{ij} = 0$.
...is of vector type if $\phi = 0$ and γ_{ab} can be split as above with $A = B = C = D = T_{ij} = 0$.
...is of tensor type if $\phi = 0$ and γ_{ab} can be split as above with $A = B = C = D = V_i = W_i = 0$.

Based on this we define the following section spaces.

$$\Gamma^{S/V/T}(\mathcal{V}) := \{\Gamma \in \Gamma(\mathcal{V}) \mid \Gamma \text{ is of scalar/vector/tensor type}\}$$

$$\Gamma^{S/V/T}_{\infty/0}(\mathcal{V}) := \Gamma_{\infty/0}(\mathcal{V}) \cap \Gamma^{S/V/T}(\mathcal{V})$$

The existence and uniqueness of the splitting of sections vanishing at spatial infinity may be cast in the following form

$$\Gamma_\infty(\mathcal{V}) = \Gamma^S_\infty(\mathcal{V}) \oplus \Gamma^V_\infty(\mathcal{V}) \oplus \Gamma^T_\infty(\mathcal{V})$$

and one may check that the splitting is orthogonal w.r.t. $\langle\cdot,\cdot\rangle_\mathcal{V}$ and $\langle\cdot,\cdot\rangle_\mathcal{W}$. However, the splitting is non–local in space, and thus one has

$$\Gamma_0{}^S(\mathcal{V}) \oplus \Gamma_0{}^V(\mathcal{V}) \oplus \Gamma_0{}^T(\mathcal{V}) \subsetneq \Gamma_0(\mathcal{V})\,!$$

Similar splitting results hold for $\varsigma \in \Gamma_\infty(\mathcal{W})$ and we extend the above nomenclature to this case in the obvious way.

By the existence and uniqueness of the splitting, the individual components induce well–defined functionals on $\Gamma_\infty(\mathcal{V})$ and $\Gamma_\infty(\mathcal{W})$, e.g. $A : \Gamma_\infty(\mathcal{V}) \to C^\infty_\infty(\mathcal{M}, \mathbb{R})$. This existence and uniqueness further implies that there exist projectors

$$\mathcal{P}^{S/V/T}_\mathcal{V} : \Gamma_\infty(\mathcal{V}) \to \Gamma^{S/V/T}_\infty(\mathcal{V}) \qquad \text{and} \qquad \mathcal{P}^{S/V}_\mathcal{W} : \Gamma_\infty(\mathcal{W}) \to \Gamma^{S/V}_\infty(\mathcal{W})$$

which are formally selfadjoint w.r.t. $\langle\cdot,\cdot\rangle_\mathcal{V}$ and $\langle\cdot,\cdot\rangle_\mathcal{W}$. It is not difficult to check that the gauge transformation operator K and the equation of motion operator P commute with these projectors, i.e.

$$\mathcal{P}^{S/V}_\mathcal{V} \circ K|_{\Gamma_\infty(\mathcal{W})} = K \circ \mathcal{P}^{S/V}_\mathcal{W} \qquad \mathcal{P}^T_\mathcal{V} \circ K|_{\Gamma_\infty(\mathcal{W})} = 0 \qquad \mathcal{P}^{S/V/T}_\mathcal{V} \circ P|_{\Gamma_\infty(\mathcal{V})} = P \circ \mathcal{P}^{S/V/T}_\mathcal{V}\,.$$

Thus the equations of motions and gauge transformations decouple for sections which vanish at spatial infinity and we may consider subspaces

$$\mathsf{Sol}_\infty^{S/V/T} \subset \mathsf{Sol}_\infty \qquad \text{and} \qquad \mathcal{G}_\infty^{S/V} \subset \mathcal{G}_\infty$$

which are defined in the obvious way.

The split equations of motion may be expressed in terms of gauge–invariant linear combinations of the split components, i.e. in terms of functionals on $\mathsf{Sol}_\infty/\mathcal{G}_\infty$.

$$\Psi := A - (\partial_\tau + \mathcal{H})(B + C') \qquad \Phi := D - \mathcal{H}(B + C')$$

$$\chi := \phi - \varphi'(B + E') \qquad \mu := -\frac{z}{a}\Phi + \chi$$

$$X_i := W_i' - V_i \qquad T_{ij}$$

Ψ and Φ are the so-called **Bardeen potentials** whereas μ is the **Mukhanov–Sasaki variable** and is of particular physical significance because it is related to the perturbation of the scalar curvature of the spatial slices. In terms of these gauge–invariant quantities, the equations of motion $P\Gamma = 0$ for $\Gamma \in \Gamma_\infty(\mathcal{V})$ read (see [22] for details):

- scalar:

$$P^\mu \mu := \left(\nabla_c \nabla^c + \frac{R}{6} - \frac{z''}{za^2}\right)\mu = 0$$

$$\Phi, \Psi, \chi = \text{ non–local functionals of } \mu$$

- vector:

$$\Delta X_i = 0 \qquad (\partial_\tau + 2\mathcal{H}) X_i = 0$$

- tensor:

$$P^T T_{ij} := \frac{1}{a^2}\left((\partial_\tau + 2\mathcal{H})\partial_\tau - \Delta\right) T_{ij} = 0$$

We see that μ is a conformally coupled scalar field with a particular time–dependent mass, whereas no non–trivial vector solutions vanishing at spatial infinity exist. Moreover, the tensor components T_{ij} satisfy a normally hyperbolic equation. The last statement may have been deduced directly from the discussion of the previous subsection by observing that the field redefinition $\bar{\cdot}$ acts trivially on $\Gamma_\infty^T(\mathcal{V})$ and that $\Gamma_\infty^T(\mathcal{V})$ lies in the kernel of $K^\dagger$. Thus P restricted to $\Gamma_\infty^T(\mathcal{V})$ coincides with the normally hyperbolic $\widetilde{P} = P \circ \bar{\cdot}^{-1} + T \circ K^\dagger$.

The standard treatment of the quantum theory of perturbations in inflation may be rephrased in the present context as follows. One considers the symplectic spaces $(\mathcal{E}^\mu, \sigma^\mu)$

$$\mathcal{E}^\mu := C_0^\infty(\mathcal{M}, \mathbb{R}) / P^\mu\left[C_0^\infty(\mathcal{M}, \mathbb{R})\right]$$

$$\mathcal{E}^\mu \times \mathcal{E}^\mu \ni ([f_1], [f_2]) \mapsto \sigma^\mu([f_1], [f_2]) := \langle f_1, E^\mu f_2 \rangle$$

$$E^\mu \text{ advanced–minus–retarded operator of } P^\mu$$

and $(\mathcal{E}^{TT}, \sigma^T)$

$$\mathcal{E}^{TT} := C_0^\infty(M,T)/P^T\left[C_0^\infty(M,T)\right]$$

$$C^\infty(\mathcal{M},T) := \{T \in C^\infty(\mathcal{M}, \textstyle\bigvee^2\mathbb{R}^3) \mid T_i^i = 0, \partial^i T_{ij} = 0\}$$

$$\mathcal{E}^{TT} \times \mathcal{E}^{TT} \ni ([f_1],[f_2]) \mapsto \sigma^T([f_1],[f_2]) := \langle f_1, E^T f_2\rangle_T$$

$$E^T \text{ advanced–minus–retarded operator of } P^T$$

$$C^\infty(M,T)^2 \ni (f_1,f_2) \mapsto \langle f_1,f_2\rangle_T := \int\limits_{\mathcal{M}} \delta^{ik}\delta^{jl} f_{1,ij} f_{2,kl}\, \mathrm{dvol}_{\boldsymbol{M}}\,.$$

P^μ and P^T are both formally selfadjoint and one may show that $\langle\cdot,\cdot\rangle_T$ is non-degenerate on $C^\infty(\mathcal{M},T)$ [22]. Thus σ^μ and σ^T are antisymmetric and non-degenerate by standard results.

In the standard treatment of perturbations in inflation one effectively assumes that all gauge–invariant (polynomial) local observables in the quantum theory are spanned by the local observables obtained from the canonical quantization of $(\mathcal{E}^\mu, \sigma^\mu)$ and $(\mathcal{E}^{TT}, \sigma^T)$. On the other hand, one may take the point of view that the construction outlined in the previous subsection, i.e. the canonical quantization of the pre–symplectic space $(\mathcal{E}, \sigma)$ should yield all (polynomial) local gauge–invariant observables. In order to compare these two approaches, we define

$$\mathrm{Ker}_0{}^{S/V/T}\left(K^\dagger\right) := \mathrm{Ker}_0\left(K^\dagger\right) \cap \Gamma^{S/V/T}(\mathcal{V})$$

$$\mathcal{E}^{S/V/T} := \mathrm{Ker}_0{}^{S/V/T}\left(K^\dagger\right) \Big/ P\left[\Gamma_0{}^{S/V/T}(\mathcal{V})\right]\,.$$

One may then prove the following results [22].

Theorem 6.4.1 *The following relations hold.*

(a) $\mathcal{E}^{S/V/T} \subset \mathcal{E}$.
(b) $\mathcal{E}^V = \{0\}$.
(c) $\left(\mathcal{E}^S, \sigma\right)$ *and* $(\mathcal{E}^\mu, \sigma^\mu)$ *are equivalent.*
(d) $\left(\mathcal{E}^T, \sigma\right)$ *and* $\left(\mathcal{E}^{TT}, \sigma^T\right)$ *are equivalent.*
(e) $\mathcal{E}^S \oplus \mathcal{E}^T \subsetneq \mathcal{E}$.
(f) $\mathcal{E}^S \oplus \mathcal{E}^T$ *is separating on* $\mathsf{Sol}_\infty/\mathcal{G}_\infty = \mathsf{Sol}^S_\infty/\mathcal{G}^S_\infty \oplus \mathsf{Sol}^T_\infty/\mathcal{G}^T_\infty$.
(g) σ *is non–degenerate on* $\mathcal{E}$.

The proofs of most of these statements are unfortunately quite cumbersome, but the results have a straightforward physical interpretation. The first statement implies that there are local observables which may be meaningfully classified as "scalar" and "tensor". This may seem surprising in view of the fact that the splitting of configurations in scalar/vector/tensor components is a priori non–local. However the second statement entails that there are indeed no non–trivial local "vector" observables. From the third and fourth statement one can infer that the standard treatment of perturbations in inflation captures the same local scalar and tensor observables that one obtains from the general gauge–invariant quantization of the linearised Einstein–Klein–Gordon system. An interesting result found in [15] implies that the scalar field μ is in fact the unique field with a second order hyperbolic PDE whose associated symplectic space is equivalent to $(\mathcal{E}^S, \sigma)$.

However, statement *e)* implies that not all local observables of the linearised Einstein–Klein–Gordon system are spanned by local observables of scalar and tensor type. In this sense, the standard approach to the quantization of perturbations in inflation "misses" some local observables. However, the sixth statement entails that the observables captured in the standard approach are still sufficient to measure configurations of the perturbations which vanish at spatial infinity. These configurations are considered to be "small" in a certain sense. Presumably this statement can be generalised by proving that local observables of scalar and tensor type separate quantum states whose correlation functions vanish at spatial infinity in each argument.

Finally, the last statement is somewhat independent of the others and may be interpreted such as to say that the quantum theory of the linearised Einstein–Klein–Gordon system on FLRW backgrounds does not contain classical observables.

6.5 Final Comments

In this chapter, we have recalled the analysis of quantum states for field theories on cosmological spacetimes and collected examples present in the literature. We have seen in Sect. 6.3 that these states may be chosen such as to solve the semiclassical Einstein equation and that the latter equation gives a strong constraint on such states. In the Sect. 6.4 we have then recalled how matter and metric perturbations of background solutions of the classical Einstein equation, which are of relevance in inflationary cosmology, may be quantized in the framework of algebraic quantum field theory in curved spacetimes. To this end, a quantum state for the matter and metric perturbations needs to be chosen, which, in the standard approach, is only restricted by comparison with observational data. On the other hand, we have seen in Sect. 6.3.3 that, in a fundamental description, our universe is not a solution of the classical Einstein equation, but of the semiclassical one. Thus one needs to generalise the standard treatment of inflationary perturbations by taking the semiclassical, rather than the classical, Einstein equations as a starting point. Apart from providing a more holistic approach to the subject, this would have the advantage that the state of matter and metric perturbations would be severely constrained by the semiclassical Einstein equation. We have seen in Sect. 6.3.4 that, going beyond the semiclassical Einstein equation, it is indeed possible to give meaning to perturbation theory around solutions of the latter. In order to obtain a complete treatment of quantized matter and metric perturbations of solutions of the semiclassical Einstein equations, it remains to work out the ansatz outlined in Sect. 6.3.4 in a way which is gauge–invariant and encompassed all degrees of freedom of the matter and metric perturbations. We hope to return to this subject in a future publication.

References

1. Ade, P.A.R., et al.: [Planck Collaboration]: Planck 2013 results. XVI. Cosmological parameters. Astron. Astrophys. **571**, A16 (2014)
2. Ade, P.A.R., et al.: [Planck Collaboration]: Planck 2013 results. XXII. Constraints on inflation. Astron. Astrophys. **571**, A22 (2014)
3. Ade, P.A.R., et al.: [BICEP2 Collaboration]: Detection of B-Mode polarization at degree angular scales by BICEP2. Phys. Rev. Lett. **112**, 241101 (2014)
4. Afshordi, N., Aslanbeigi, S., Sorkin, R.D.: A distinguished vacuum state for a quantum field in curved spacetime: formalism, features, and cosmology. JHEP **2012**, 1–12 (2012)
5. Anderson, P.R.: Effects of quantum fields on singularities and particle horizons in the early universe. 4. initially empty universes. Phys. Rev. D **33**, 1567 (1986)
6. Brum, M., Fredenhagen, K.: 'Vacuum-like' Hadamard states for quantum fields on curved spacetimes. Class. Quantum Grav. **31**, 025024 (2014)
7. Brunetti, R., Fredenhagen, K., Hollands, S.: A remark on alpha vacua for quantum field theories on de Sitter space. JHEP **0505**, 063 (2005)
8. Brunetti, R., Fredenhagen, K., Köhler, M.: The microlocal spectrum condition and wick polynomials of free fields on curved spacetimes. Commun. Math. Phys. **180**, 633–652 (1996)
9. Brunetti, R., Fredenhagen, K., Verch, R.: The generally covariant locality principle—a new paradigm for local quantum field theory. Commun. Math. Phys. **237**, 31–68 (2003)
10. Dappiaggi, C., Hack, T.-P., Pinamonti, N.: Approximate KMS states for scalar and spinor fields in Friedmann-Robertson-Walker spacetimes. Ann. Henri Poincare **12**, 1449 (2011)
11. Dappiaggi, C., Moretti, V., Pinamonti, N.: Cosmological horizons and reconstruction of quantum field theories. Commun. Math. Phys. **285**, 1129–1163 (2009)
12. Dappiaggi, C., Moretti, V., Pinamonti, N.: Distinguished quantum states in a class of cosmological spacetimes and their Hadamard property. J. Math. Phys. **50**, 062304 (2009)
13. Dimock, J.: Quantized electromagnetic field on a manifold. Rev. Math. Phys. **4**, 223–233 (1992)
14. Degner, A.: Properties of states of low energy on cosmological spacetimes. PhD thesis, University of Hamburg (2013)
15. Eltzner, B.: Quantization of perturbations in Inflation. arXiv:1302.5358 [gr-qc]
16. Eltzner, B., Gottschalk, H.: Dynamical backreaction in Robertson-Walker spacetime. Rev. Math. Phys. **23**, 531–551 (2011)
17. Fewster, C.J., Hunt, D.S.: Quantization of linearised gravity in cosmological vacuum spacetimes. Rev. Math. Phys. **25**, 1330003 (2013)
18. Fewster, C.J., Verch, R.: On a recent construction of 'Vacuum-like' quantum field states in curved spacetime. Class. Quant. Grav. **29**, 205017 (2012)
19. Fewster, C.J., Ford, L.H., Roman, T.A.: Probability distributions of smeared quantum stress tensors. Phys. Rev. D **81**, 121901 (2010)
20. Gérard, C., Wrochna, M.: Construction of Hadamard states by pseudo-differential calculus. Comm. Math. Phys. **325**, 713–755 (2014)
21. Hack, T.-P.: The Lambda CDM-model in quantum field theory on curved spacetime and dark radiation. arXiv:1306.3074 [gr-qc]
22. Hack, T.-P.: Quantization of the linearised Einstein-Klein-Gordon system on arbitrary backgrounds and the special case of perturbations in Inflation. Class. Quantum Grav. **31**, 215004 (2014)

23. Hack, T.-P., Schenkel, A.: Linear bosonic and fermionic quantum gauge theories on curved spacetimes. Gen. Rel. Grav. **45**, 877 (2013)
24. Hollands, S., Wald, R.M.: Local wick polynomials and time ordered products of quantum fields in curved spacetime. Commun. Math. Phys. **223**, 289–326 (2001)
25. Hollands, S., Wald, R.M.: Existence of local covariant time ordered products of quantum fields in curved spacetime. Commun. Math. Phys. **231**, 2309 (2002)
26. Hollands, S., Wald, R.M.: Conservation of the stress tensor in interacting quantum field theory in curved spacetimes. Rev. Math. Phys. **17**, 227 (2005)
27. Hu, B.L., Verdaguer, E.: Stochastic gravity: theory and applications. Living Rev. Rel. **11**, 3 (2008); Living Rev. Rel. **7**, 3 (2004)
28. Junker, W., Schrohe, E.: Adiabatic vacuum states on general spacetime manifolds: definition, construction, and physical properties. Ann. Henri Poincaré **3**, 1113–1181 (2002)
29. Kay, B.S., Wald, R.M.: Theorems on the uniqueness and thermal properties of stationary, nonsingular, Quasifree states on space-times with a bifurcate killing horizon. Phys. Rept. **207**, 49 (1991)
30. Kolb, E.W., Turner, M.S.: The early universe. Front. Phys. **69**, 1 (1990)
31. Küskü, M.: A class of almost equilibrium states in Robertson-Walker spacetimes. DESY-THESIS-2008-020
32. Lüders, C., Roberts, J.E.: Local quasiequivalence and adiabatic vacuum states. Commun. Math. Phys. **134**, 29–63 (1990)
33. Moretti, V.: Comments on the stress-energy tensor operator in curved spacetime. Commun. Math. Phys. **232**, 189–221 (2003)
34. Olbermann, H.: States of low energy on Robertson-Walker spacetimes. Class. Quantum Grav. **24**, 5011–5030 (2007)
35. Parker, L.: Quantized fields and particle creation in expanding universes I. Phys. Rev. **183**, 1057–1068 (1969)
36. Pinamonti, N.: On the initial conditions and solutions of the semiclassical Einstein equations in a cosmological scenario. Commun. Math. Phys. **305**, 563–604 (2011)
37. Pinamonti, N., Siemssen, D.: Scale-invariant curvature fluctuations from an extended semiclassical gravity. J. Math. Phys. **56**, 022303 (2015)
38. Pinamonti, N., Siemssen, D.: Global existence of solutions of the semiclassical Einstein equation for cosmological spacetimes. Commun. Math. Phys. **334**(1), 171–191 (2015)
39. Radzikowski, M.J.: Micro-local approach to the hadamard condition in quantum field theory on curved space-time. Commun. Math. Phys. **179**, 529–553 (1996)
40. Riegert, R.J.: A nonlocal action for the trace anomaly. Phys. Lett. B **134**, 56 (1984)
41. Schlemmer, J.: Ph.D. thesis, Faculty of Physics, University of Leipzig (2010)
42. Sewell, G. L.: Quantum fields on manifolds: PCT and gravitationally induced thermal states, Ann. Phys. (NY) **141**, 201 (1982)
43. Straumann, N.: From primordial quantum fluctuations to the anisotropies of the cosmic microwave background radiation. Annalen Phys. **15**, 70 (2006)
44. Starobinsky, A.A.: A new type of isotropic cosmological models without singularity. Phys. Lett. B **91**, 99 (1980)
45. Starobinsky, A.A.: The perturbation spectrum evolving from a nonsingular initially de-sitter cosmology and the microwave background anisotropy. Sov. Astron. Lett. **9**, 302 (1983)
46. Them, K., Brum, M.: States of low energy in homogeneous and inhomogeneous. Expanding Spacetimes. Class. Quant. Grav. **30**, 235035 (2013)
47. Verch, R.: Local covariance, renormalization ambiguity, and local thermal equilibrium in cosmology. In: Finster, F., Müller, O., Nardmann, M., Tolksdorf, J., Zeidler, E. (eds.) Quantum field theory and gravity. Conceptual and mathematical advances in the search for a unified framework, Birkhäuser (2012)
48. Wald, R.M.: Axiomatic renormalization of stress tensor of a conformally invariant field in conformally flat spacetimes. Ann. Phys. **110**, 472 (1978)
49. Wald, R.M.: Trace anomaly of a conformally invariant quantum field in curved space-time. Phys. Rev. D **17**, 1477 (1978)
50. Zschoche, J.: The Chaplygin gas equation of state for the quantized free scalar field on cosmological spacetimes. Ann. Henri Poincare **15**, 1285 (2014)

Chapter 7
Quantum Spacetime and Algebraic Quantum Field Theory

Dorothea Bahns, Sergio Doplicher, Gerardo Morsella and Gherardo Piacitelli

Abstract We review the investigations on the quantum structure of spacetime, to be found at the Planck scale if one takes into account the operational limitations to the localization of events which result from the concurrence of Quantum Mechanics and General Relativity. We also discuss the different approaches to (perturbative) Quantum Field Theory on Quantum Spacetime, and some of the possible cosmological consequences.

7.1 Quantum Nature of Spacetime at the Planck Scale: Why and How

According to Classical General Relativity, at large scales spacetime is a pseudo Riemanniann manifold locally modelled on Minkowski space. But the concurrence with the principles of Quantum Mechanics renders this picture untenable *in the small.*

Those theories are often reported as hardly reconcilable, but they do meet at least in a single *partial* principle, the *Principle of Gravitational Stability against localization of events* formulated in [17, 18]:

D. Bahns
Mathematisches Institut and Courant Research Centre "Higher Order Structures in Mathematics", Universität Göttingen,
Bunsenstr. 3-5, D-37073 Göttingen, Germany
e-mail: bahns@uni-math.gwdg.de

S. Doplicher
Dipartimento di Matematica, Università di Roma "La Sapienza",
P.le Aldo Moro, 5, I-00185 Roma, Italy
e-mail: dopliche@mat.uniroma1.it

G. Morsella (✉)
Dipartimento di Matematica, Università di Roma Tor Vergata,
V.le Della Ricerca Scientifica, 1, I-00133 Roma, Italy
e-mail: morsella@mat.uniroma2.it

G. Piacitelli
SISSA, Via Bonomea, 265, I-34136 Trieste, Italy
e-mail: piacitel@sissa.it

R. Brunetti et al. (eds.), *Advances in Algebraic Quantum Field Theory*,
Mathematical Physics Studies, DOI 10.1007/978-3-319-21353-8_7

> The gravitational field generated by the concentration of energy required by the Heisenberg Uncertainty Principle to localize an event in spacetime should not be so strong to hide the event itself to any distant observer-distant compared to the Planck scale.

The effect of this principle is best seen considering first the effect of an observation which locates an event, say, in a spherically symmetric way around the origin in space with accuracy a; according to Heisenberg principle an uncontrollable energy E of order $1/a$ has to be transferred, which will generate a gravitational field with Schwarzschild radius $R \simeq E$ (in universal units where $\hbar = c = G = 1$). Hence we *must* have that $a \gtrsim R \simeq 1/a$; so that $a \gtrsim 1$, i.e. in CGS units

$$a \gtrsim \lambda_P \simeq 1.6 \times 10^{-33}\text{cm}. \tag{7.1}$$

This folklore argument is certainly very old, but its elaborations in two significant directions are surprisingly recent.

First, if we consider generic uncertainties, the argument above suggests that they ought to be limited by uncertainty relations.

Indeed, if we measure one of the space coordinates of our event with great precision a, but allow large uncertainties L in the knowledge of the other coordinates, the energy $1/a$ may spread over a thin disk of radius L and thus generate a gravitational potential that would vanish everywhere as $L \to \infty$ (provided a, as small as we like but non zero, remains constant).

This is shown by trivial computation of the Newtonian potential generated by the corresponding *mass* distribution; whenever such a potential is nearly vanishing, nobody would expect large General Relativistic or Quantum Gravitational corrections; so we can rely on that estimate.

An equally elementary computation would show that the same conclusion holds if *two* space coordinates are measured with small but fixed precision a and the third one with an uncertainty L, and $L \to \infty$.

Second, if we consider the energy content of a generic quantum state where the location measurement is performed, the bounds on the uncertainties should depend also upon that energy content [16, 20] .

To see this point, just suppose that our background state describes the spherically symmetric distribution of the total energy E within a sphere of radius R, with $E < R$. If we localize, in a spherically symmetric way, an event at the origin with space accuracy a, due to the Heisenberg Principle the total energy will be of the order $1/a + E$. We must then have

$$\frac{1}{a} + E < R,$$

otherwise our event will be hidden to an observer located far away, out of the sphere of radius R around the origin. Thus, if $R - E$ is much smaller than 1, the "minimal distance" will be much larger than 1. But if a is anyway larger than R the condition implies rather

$$\frac{1}{a} + E < a.$$

Thus, if $R - E$ is very small compared to 1 and R is much larger than 1, a cannot be essentially smaller than R.

Now the causal relations between events should also break down at scales which are so small that events cannot be localized that sharply; hence we have to expect that scale to express the range of propagation of acausal effects.

This naive picture suggests that, due to the principle of Gravitational Stability, initially all points of the Universe should have been causally connected.

Thus we can expect that Quantum Spacetime (QST) *solves* the horizon problem (cf. [16] for hints in that direction, [20] or Sect. 7.4.3 below for an indication that a Quantum Spacetime with a constant Planck length should generate dynamically a range of propagation of acausal effects which solves the horizon problem).

We come back to the general discussion. If we aim at a merge of Quantum Mechanics and General Relativity we should reason in terms of concepts which are physically legitimate from the general relativistic point of view as well. One might doubt from the start about concepts like *local energy and coordinates* to which the Heisenberg Principle refers.

Concerning the use of coordinates, one should better talk of measurements conditioned to the measurement of a finite number of auxiliary local quantities; in some appropriate limit, in Minkowski space, that auxiliary measurement should become the specification of a frame. Thus the use of coordinates should be legitimate at a semiclassical level.

Another important reason to work with coordinates is that we are interested in the tangent space at a point equipped with normal coordinates, describing a free falling system in Einstein's lift. Or a system in a constant gravitational field; for the outside distribution of matter on the large scale, such as the structure of the Virgo supercluster of galaxies to which we belong, ought to have no influence on a high energy collision in the CERN collider; even if we were so clever to detect (quantum) effects of the gravitational forces *between* the colliding particles.

Thus in a first stage it is legitimate, and physically reasonable, to study the small scale structure of Minkowski space. The spacetime symmetries of our space ought to be described by the *classical* Poincaré group: for the *global motions* of our space should look the same in the large as they do in the small, and, in the large, they should be precisely the *classical symmetries.*

One other remark in order here concerns the very nature of the coordinates. In the Quantum Mechanics of systems with finitely many degrees of freedom, they are *observables* describing the particle positions.

In Quantum Field Theory, the observables are *local* quantities associated each with a finite region in spacetime. They can never describe *exactly* a property of one particle or n—particle states, which are global (asymptotic) constructs. If that region reduces to a point, we find only the multiples of the identity. We ought to consider open regions. We might consider such a region as a neighbourhood of a spacetime point, defining it with some uncertainty, and the measurement of associated local quantities as leading to information on that location.

Thus Spacetime appears as a space of parameters, which, in absence of gravitational forces, can be specified with arbitrarily high (but finite!) precision, with higher and higher energy cost for higher and higher precision. The consideration of the gravitational effects of that energy cost will cause, as we will see, that space of parameters to become *noncommutative*.

The semiclassical level of a first analysis justifies also the use of concepts like energy; but a more careful analysis shows, as briefly mentioned here in the sequel, that in essence the conclusions remain true without any reference to the concept of energy.

At a semiclassical level, the main consequence of the Principle stated above is the validity of *Spacetime Uncertainty Relations*; furthermore, they have been shown to be implemented by Commutation Relations between coordinates, thus turning Spacetime into *Quantum Spacetime* [17, 18].

The word "Quantum" is very appropriate here, to stress that noncommutativity does not enter just as a formal generalization, but is strongly suggested by a compelling *physical* reason, unlike the very first discussions of possible noncommutativity of coordinates in the pre-renormalization era, by Heisenberg, Snyder and Yang, where noncommutativity was regarded as a curious, in itself physically doubtful, possible *regularisation device*, without any reference to General Relativity and Gravitational forces; the qualitative fact that the quantum structure of gravitational forces ought to have consequences on the nature of spacetime in the small was anticipated by Bronstein [8], where, however, the focus was on the extension of the Bohr-Rosenfeld argument to the Christoffel symbols, and on the proposal of a Quantum Theory of *linearized* Gravity, without any mention of spacetime uncertainty relations.

The analysis based on the Principle of Gravitational Stability against localization of events leads to the following conclusions:

(i) There is no a priori lower limit on the precision in the measurement of any *single* coordinate (it is worthwhile to stress once more that the apparently opposite conclusions, still often reported in the literature in connection with the ACV variant of the Heisenberg principle [1], are drawn under the *implicit* assumption that *all* the space coordinates of the event are simultaneously sharply measured). Every alerted reader will note that nobody knows an operational prescription to measure, say, *only one spacetime coordinate* of the location of an event with a terrific (ultra Planckian) precision. But of course we cannot say that such a measurement is *impossible* just because we are not capable of inventing a device; we could say that only if we could show that it is *forbidden* by the presently known physical principles. Which at present does *not* seem to be the case.

(ii) The uncertainties Δq_μ in the measurement of the coordinates of an event in Minkowski space should be at least bounded by the following Spacetime Uncertainty Relations:

$$\Delta q_0 \cdot \sum_{j=1}^{3} \Delta q_j \gtrsim 1; \tag{7.2a}$$

$$\sum_{1 \leq j < k \leq 3} \Delta q_j \Delta q_k \gtrsim 1. \tag{7.2b}$$

> Thus points become fuzzy and *locality looses any precise meaning*. We believe it should be replaced at the Planck scale by an equally sharp and compelling principle, which reduces to locality at larger distances. Such a principle is nowadays totally unknown, and unaccessible by operational reasoning.

Some comments on the derivation of these relations are in order. In the analysis of 1994–95, they were justified *in special cases* by their consistency with the *exact* solutions of Einstein Equations (EE), as Schwarzschild and Kerr's solutions. But in general they were derived using the *linearized* approximation to EE.

Furthermore the concept of energy was central: in a *semiclassical* approach, the expectation value in a state describing an *ansatz* for the outcome of a localization experiment (a coherent state in a free field theory) of the *energy-momentum tensor* for that field, was used as a source for the linearized EE.

Then, the requirement of non-formation of trapped surfaces hiding the observed event was formulated as the condition of non negativity of the time-time component of the metric tensor. The relations above follow as a weaker simplified necessary condition.

Both the use of the linearized approximation and of the notion of energy are doubtful.

But in recent works [30, 31] Tomassini and Viaggiu have shown that (a stronger form of) the above relations do follow from an *exact* treatment, if one adopts the *Hoop Conjecture*, which limits the energy content of a space volume in terms of the area of the boundary, as a condition for the non-formation of bounded trapped surfaces. Moreover, their analysis applies to a curved background as well.

The treatment is again semiclassical, and involves the notion of energy, but the conflict about the use of the linearized approximations to derive bounds, and imposing those bounds in situations close to singularities, disappears.

Eventually, in [20] the special case of spherically symmetric experiments, with all spacetime uncertainties taking the same value, was treated with use of the *exact* semiclassical EE, without any reference to the energy observables. The state describing the outcome of the localization experiment was taken *not* as a strictly localized state, but as the state, with weaker localization properties, obtained acting on the vacuum state with the field operators themselves, smeared with test functions having the appropriate symmetry, in a theory of a single scalar massless field coupled semiclassically to gravity. The solution of the Raychaudhuri equation yields to the universal lower bound for the common value of the uncertainties, of the order of Planck length (see also Sect. 7.4.2 below for more details). We stress that this result gives a possibly weaker condition than the condition which could be derived by a choice of better localized *ansätze* for the probe state.

We can conclude that the above Spacetime Uncertainty Relations are reasonably well grounded for Minkowski space; they are to be expected to hold in similar variant in curved spacetimes, by the Tomassini-Viaggiu argument; a basic consequence of those relations, when implemented by the Quantum Conditions we will now discuss, namely that the Planck scale is a universal minimal length, is well grounded on the basis of the most general assumptions, in the spherically symmetric case.

The Spacetime Uncertainty Relations strongly suggest that spacetime has a *Quantum Structure* at small scales, expressed, in generic units, by

$$[q_\mu, q_\nu] = i\lambda_P^2 Q_{\mu\nu}, \tag{7.3}$$

where Q has to be chosen not as a random toy mathematical model, but in such a way that (7.2) follows from (7.3).

To achieve this in the simplest way, it suffices to select the model where the $Q_{\mu\nu}$ are *central*, and impose the "Quantum Conditions" on the two invariants

$$Q_{\mu\nu}Q^{\mu\nu}; \tag{7.4}$$

$$\begin{aligned}[q_0, \ldots, q_3] &\equiv \det \begin{pmatrix} q_0 & \cdots & q_3 \\ \vdots & \ddots & \vdots \\ q_0 & \cdots & q_3 \end{pmatrix} \\ &\equiv \varepsilon^{\mu\nu\lambda\rho} q_\mu q_\nu q_\lambda q_\rho = \\ &= -(1/2) Q_{\mu\nu}(*Q)^{\mu\nu}; \end{aligned} \tag{7.5}$$

whereby the first one must be zero and the square of the half of the second is I (in Planck units; we must take the square since it is a pseudoscalar and not a scalar).

One obtains in this way [17, 18] a model of Quantum Spacetime which implements exactly our Spacetime Uncertainty Relations and is fully Poincaré covariant.

As anticipated, here the *classical* Poincaré group acts as symmetries; translations, in particular, act adding to each q_μ a real multiple of the identity.

Thus "coordinates" and "translation parameters", classically described by the same objects, hear split into different entities; but this happens already in non relativistic Quantum Mechanics: rotations apart, the Galilei group acts by adding numerical multiples of the identity to the non commuting position and momentum operators.

In view of the Gel'fand–Naimark Theorem, the classical Minkowski Space M is described by the commutative C*-algebra of continuous functions vanishing at infinity on M; the classical coordinates can be viewed as commuting selfadjoint operators affiliated to that C*-algebras.

Similarly a *noncommutative* C*-algebra $\mathscr{E}$ of Quantum Spacetime can be associated to the above relations. It was proposed in [17, 18] by a procedure which applies to more general cases (see also Sects. 7.2.1 and 7.2.3).

Assuming that the q_λ, $Q_{\mu\nu}$ are selfadjoint operators and that the $Q_{\mu\nu}$ commute *strongly* with one another and with the q_λ, the relations above can be seen as a bundle of Lie algebra relations based on the joint spectrum of the $Q_{\mu\nu}$.

We are interested only in representations which are regular in the sense that in their central decomposition only integrable representations of the corresponding Lie algebras appear.

Such representations are described by representations of the group C*-algebra of the unique simply connected Lie group associated to the corresponding Lie algebra.

Hence the C*-algebra of Quantum Spacetime $\mathscr{E}$ is the C*-algebra of a continuous field of group C*-algebras based on the spectrum of a commutative C*-algebra.

In our case, that spectrum—the joint spectrum of the $Q_{\mu\nu}$—is the manifold Σ of the real valued antisymmetric 2-tensors fulfilling the same relations as the $Q_{\mu\nu}$ do: a homogeneous space of the proper orthochronous Lorentz group, identified with the coset space of $SL(2, \mathbb{C})$ mod the subgroup of diagonal matrices. Each of those tensors can be taken to its rest frame, where the electric and magnetic part are parallel unit vectors, by a boost specified by a third vector, orthogonal to those unit vectors; thus Σ can be viewed as the tangent bundle to two copies of the unit sphere in 3-space—its base Σ_1.

The fibers, with the condition that I is not an independent generator but is represented by I, are the C*-algebras of the Heisenberg relations in 2 degrees of freedom—the algebra of all compact operators on a fixed infinite dimensional separable Hilbert space.

The continuous field can be shown to be trivial, since it must contain a continuous field of one dimensional projectors—those corresponding to the orthogonal projection on the one dimensional subspace of multiples of the ground state vector for the harmonic oscillator (see [18]).

The states whose central decomposition is supported by the base Σ_1, and for each point of the base correspond to the ground state for the harmonic oscillator, are precisely the states of *optimal localization*, where the *sum* of the four squared uncertainties of the coordinates is minimal, and equal to 2 (see Sect. 7.2.2).

Thus the C*-algebra of Quantum Spacetime $\mathscr{E}$ is identified with the tensor product of the continuous functions vanishing at infinity on Σ and the algebra of compact operators.

In the *classical limit* $\lambda_P \rightarrow 0$ the second factor deforms to the commutative C*-algebra of Minkowski space, but the first factor survives. When Quantum Spacetime is probed with optimally localized states its classical limit is $M \times \Sigma_1$, i.e. M acquires *compact extra dimensions*.

Note that the mathematical generalization of points are pure states, but only optimally localized pure states are physically appropriate.

But to explore more thoroughly the Quantum Geometry of Quantum Spacetime we must consider *independent events*.

Quantum mechanically n independent events ought to be described by the n-fold tensor product of $\mathscr{E}$ with itself; considering arbitrary values on n we are led to use the direct sum over all n.

If A is the C*-algebra with unit over $\mathbb{C}$, obtained adding the unit to $\mathscr{E}$, we will view the $(n+1)$ tensor power $\Lambda_n(A)$ of A over $\mathbb{C}$ as an A-bimodule with the product in A, and the direct sum

$$\Lambda(A) = \bigoplus_{n=0}^{\infty} \Lambda_n(A)$$

as the A-bimodule tensor algebra, where

$$(a_1 \otimes a_2 \otimes \cdots \otimes a_n)(b_1 \otimes b_2 \otimes \cdots \otimes b_m) = a_1 \otimes a_2 \otimes \cdots \otimes (a_n b_1) \otimes b_2 \otimes \cdots \otimes b_m.$$

This is the natural ambient for the *universal differential calculus*, where the differential is given by

$$d(a_0 \otimes \cdots \otimes a_n) = \sum_{k=0}^{n} (-1)^k a_0 \otimes \cdots \otimes a_{k-1} \otimes I \otimes a_k \otimes \cdots \otimes a_n.$$

As usual d is a graded differential, i.e., if $\phi \in \Lambda(A)$, $\psi \in \Lambda_n(A)$, we have

$$d^2 = 0;$$
$$d(\phi \cdot \psi) = (d\phi) \cdot \psi + (-1)^n \phi \cdot d\psi.$$

Note that $A = \Lambda_0(A) \subset \Lambda(A)$, and the d-stable subalgebra $\Omega(A)$ of $\Lambda(A)$ generated by A is the *universal differential algebra.* In other words, it is the subalgebra generated by A and

$$da = I \otimes a - a \otimes I$$

as a varies in A.

In the case of n independent events one is led to describe the spacetime coordinates of the jth event by $q_j = I \otimes \cdots I \otimes \otimes q \otimes I \cdots \otimes I$ (q in the jth place); in this way, the commutator between the different spacetime components of the q_j would depend on j.

A better choice is to require that it does not; this is achieved as follows. The centre Z of the multiplier algebra of $\mathscr{E}$ is the algebra of all bounded continuous functions on Σ with values in the complex numbers; so that $\mathscr{E}$, and hence A, is in an obvious way a Z-bimodule.

Therefore we can, and will, replace, in the definition of $\Lambda(A)$, the $\mathbb{C}$-tensor product by the Z-bimodule-tensor product, so that

$$dQ = 0.$$

As a consequence, the q_j and the $2^{-1/2}(q_j - q_k)$, j different from k, and $2^{-1/2}dq$, obey the same spacetime commutation relations, as does the normalised barycenter coordinates, $n^{-1/2}(q_1 + q_2 + \cdots q_n)$; and the latter commutes with the difference coordinates.

These facts allow us to define a *quantum diagonal map* from $\Lambda_n(A)$ to A, which leaves the functions of the barycenter coordinates alone, and evaluates on functions of the difference variables the *universal optimally localized map* which, when composed with a probability measure on Σ_1, would give the generic optimally localized state (see Sect. 7.2.4).

Replacing the classical diagonal evaluation of a function of n arguments on Minkowski space by the quantum diagonal map allows us to define the *Quantum Wick Product* [5].

But working in $\Omega(A)$ as a subspace of $\Lambda(A)$ allows us to use two structures [7]:

- the tensor algebra structure described above, where both the A bimodule and the Z bimodule structures enter, essential for our reduced universal differential calculus;
- the pre-C*-algebra structure of $\Lambda(A)$, which allows us to consider, for each element a of $\Lambda_n(A)$, its modulus $(a^*a)^{1/2}$, its spectrum, and so on.

In particular we can study the geometric operators: separation between two independent events, area, 3-volume, 4-volume, given by

$$
\begin{aligned}
&dq,\\
&dq \wedge dq,\\
&dq \wedge dq \wedge dq,\\
&dq \wedge dq \wedge dq \wedge dq,
\end{aligned}
$$

where, for instance, the latter is given by

$$
\begin{aligned}
V &= dq \wedge dq \wedge dq \wedge dq\\
&= \varepsilon_{\mu\nu\rho\sigma} dq^\mu dq^\nu dq^\rho dq^\sigma .
\end{aligned}
$$

Each of these forms has a number of spacetime components: e.g. 4 the first one (a vector), 1 the last one (a pseudoscalar).

It is found that, for each of those forms, each component is a normal operator, and that the sum of the square moduli of all spacetime components is bounded below by a multiple of the identity of unit order of magnitude. Although that sum is (except for the 4-volume!) not Lorentz invariant, the bound holds in any Lorentz frame (see Sect. 7.2.5).

In particular, the *Euclidean* distance between two independent events can be shown to have a lower bound of order one in Planck units. Two distinct points *can never merge to a point*. However, of course, the state where the minimum is achieved will depend upon the reference frame where the requirement is formulated. (The structure of length, area and volume operators on QST has been studied in full detail [7].)

Thus the existence of a minimal length is not at all in contradiction with the Lorentz covariance of the model; note that models where the commutators of the coordinates are just numbers θ, which appear so often in the literature, arise as irreducible representations of our model; such models, taken for a fixed choice of θ rather than for its full Lorentz orbit, necessarily break Lorentz covariance. To restore it as a twisted symmetry is essentially equivalent to going back to the model where the commutators are operators. This point has been recently clarified in great depth [26].

On the other side, a theory with a fixed, numerical commutator (a θ in the sky; it could be hardly believed, but at least, in case, it ought to be thought in the CMB reference system, with respect to which we fly at a speed of 600 km/s) can hardly be realistic.

The geometry of Quantum Spacetime and the free field theories on it are *fully Poincaré covariant*. The various formulation of interaction between fields, all equivalent on ordinary Minkowski space, provide inequivalent approaches on QST; but all of them, sooner or later, meet problems with *Lorentz covariance*, apparently due to the nontrivial action of the Lorentz group on the *centre* of the algebra of Quantum Spacetime. On this point in our opinion a deeper understanding is needed.

One can however introduce interactions in different ways, all preserving spacetime translation and space rotation covariance, that we discuss in Sect. 7.3; among these it is just worth mentioning here one of them, where one takes into account, in the very definition of Wick products, the fact that in our Quantum Spacetime n (larger or equal to two) distinct points can never merge to a point. But we can use the canonical quantum diagonal map mentioned above, which associates to functions of n independent points a function of a single point, evaluating a conditional expectation which on functions of the differences takes a numerical value, associated with the minimum of the Euclidean distance (in a given Lorentz frame!).

The "Quantum Wick Product" obtained by this procedure leads to a perturbative Gell-Mann and Low formula free of ultraviolet divergences at each term of the perturbation expansion [5] . However, those terms have a meaning only after a sort of adiabatic cutoff: the coupling constant should be changed to a function of time, rapidly vanishing at infinity, say depending upon a cutoff time T. But the limit $T \to \infty$ is difficult problem, and there are indications it does not exist.

A major open problems is the following. Suppose we apply this construction to the normalised Lagrangean of a theory which is renormalizable on the ordinary Minkowski space, with the counter terms defined by that ordinary theory, and with finite renormalization constants depending upon both the Planck length λ_P and the cutoff time T. Can we find a natural dependence such that in the limit $\lambda_P \to 0$ and $T \to \infty$ we get back the ordinary renormalized Gell-Mann Low expansion on Minkowski space? This should depend upon a suitable way of performing a joint limit, which hopefully yields, for the physical value of λ_P, to a result which is essentially independent of T within wide margins of variation, and can be taken as source of predictions to be compared with observations.

The common feature of all approaches is that, due to the quantum nature of spacetime at the Planck scale, *locality is broken* (even at the level of free fields, for explicit estimates see [18]); in perturbation theory, its breakdown manifests itself in a non local kernel, which spreads the interaction vertices [4, 5, 18] ; this forces on us the appropriate modifications of Feynman rules [25].

However, it is worth noting that in Quantum Field Theory on the Minkowski space (and similarly on curved classical backgrounds) there are two aspects of locality.

First, the theory is defined by the *assignment* to bounded (nice) open regions in spacetime of algebras generated by the observables which can be measured within those regions.

Covariance is expressed by the fact that that assignment intertwines the actions of the spacetime symmetries on the regions and on the observables.

Second, that assignment should reflect Einstein causality: observables that are measured in regions between which no signal can be transmitted, ought to commute.

As we mentioned, the second assertion is bound to be lost if the gravitational forces *between* the elementary particles are taken into account.

But the first assertion, at least partially, can well be maintained.

Indeed, if we describe Minkowski space by the algebra of continuous functions vanishing at infinity, we can describe open sets through their characteristic functions, which are special selfadjoint idempotents in the Borel completion.

Similarly, a "region" in Quantum Spacetime can be described by a selfadjoint idempotent E in the Borel completion of the C*-algebra of Quantum Spacetime.

To associate algebras of observables to such projections assume first that we wish to define on the basic model of Quantum Spacetime the ordinary free field ϕ over Minkowski space.

The analogue of the von Neumann functional calculus on the q_μ's with functions whose Fourier transform is L^1 can be extended to operator valued distributions as Wightman fields (cf. [18] and Sect. 7.2). This applies in particular to free fields.

The evaluation of ϕ on the non commuting operators q can be given by

$$\phi(q) = \frac{1}{(2\pi)^{3/2}} \int (e^{iq_\mu k^\mu} \otimes a(\mathbf{k}) + e^{-iq_\mu k^\mu} \otimes a(\mathbf{k})^*) d\Omega_m^+(\mathbf{k}) \qquad (7.6)$$

where $d\Omega_m^+(\mathbf{k}) = \frac{d^3\mathbf{k}}{2\sqrt{\mathbf{k}^2+m^2}}$ is the usual invariant measure over the positive energy hyperboloid of mass m:

$$\Omega_m^+ = \{k \in \mathbb{R}^4 \ / \ k_\mu k^\mu = m^2 \ , k_0 > 0\}.$$

This is an unbounded operator affiliated to the C*-tensor product $\mathscr{E} \otimes \mathcal{B}(\mathcal{H})$, where $\mathcal{H}$ is the Fock space.

Similarly, using the full Fourier transform of the field, any Wightman field on Minkowski space could be evaluated on $\mathscr{E}$.

The free field defines a map from states $\omega \in \mathcal{S}(\mathscr{E})$ to operators on $\mathcal{H}$ by

$$\phi(\omega) \equiv \langle \omega \otimes id, \phi(q) \rangle \ , \quad \omega \in \mathcal{S}(\mathscr{E}).$$

The von Neumann algebra generated by bounded functions of these operators, as ω varies in the set of states supported by E, will be the *local algebra* $\mathfrak{A}(E)$ associated to E.

This map preserves inclusions and intertwines the actions of the Poincaré group, since the free field is covariant. The same would apply to any covariant field.

However, the local commutativity is lost, as well as the notion "E is spacelike to F".

The local algebras $\mathfrak{A}(E)$ might show many unexpected behaviours. In the case of a free scalar neutral field, to a minimal E given by the product of the characteristic function of a point in Σ_1 with the spectral projection of the sum of squares of the coordinates associate to the interval $[0, 2]$, we would get a commutative algebra; in the case of a free Dirac Field, a finite dimensional algebra. But spreading those algebras with spacetime translations in any tiny neighbourhood would lead to an irreducible algebra [15]. These results partly survive even for the scale invariant model of Quantum Spacetime with $\lambda_P = 0$ [24].

Can we formulate an analogue of Locality as a sharp, physically compelling, principle, which reduces to ordinary locality at large scales?

The only way we can figure out to address this question relates to the Principle of local gauge invariance and of minimal form of the interactions.

In ordinary Field Theory these principles select *local point* interactions, and thus can be viewed as the root of locality.

We could speculate on the extension of those principles to Quantum Field Theory on Quantum Spacetime as the way to extend Locality.

But, unfortunately, already on Minkowski space those principle seem to have a crystal clear form only in classical field theory, and to be not amenable to any formulation in terms of local observables. And they seem to require anyway a formulation in terms of non observable quantities.

Hence at the moment we cannot say more than the fact that locality must break down on Quantum Spacetime.

But nonlocal effects should be visible only at Planck scales, and vanish fast for larger separations. If Lorentz invariance can be maintained by interactions, a point quite open at present, then we ought to expect that the analysis of the superselection structure, the notion of Statistics, conjugate sectors, the emergence of a compact group of global gauge symmetries, and even the Spin and Statistics Theorem, all deduced on the basis of the Principle of Locality, ought to remain true [19].

That argument might, however, raise the objection that, in a theory which accounts for gravitational interactions as well, there might be no reasonable scattering theory at all, due to the well known paradox of loss of information, if black holes are created in a scattering process, destroying the unitarity of the S matrix.

Of course, this is an open problem; but one might well take the attitude that a final answer to it will come only from a complete theory, while at the moment we are rather relying on semiclassical arguments. Which might be quite a reasonable guide in order to get indications of local behaviours; but scattering theory involves the limit to infinite past/future times; and it might well be that interchanging these limits with those in which the semiclassical approximations are valid, or with the infinite volume limit in which the thermal behaviour of the vacuum for a uniformly accelerated observer becomes an exact mathematical statement, is dangerous, if not misleading. And whatever theory will account for Quantum Gravity, it should also describe the world of Local Quantum Field Theory as an appropriate approximation.

One might expect that a complete theory ought to be covariant under general coordinate transformations as well. This principle, however, is grounded on the conceptual experiment of the falling lift, which, in the classical theory, can be thought of as occupying an infinitesimal neighbourhood of a point. In a quantum theory the size of a "laboratory" must be large compared with the Planck length, and this might pose limitations on general covariance.

On the other side elementary particle theory deals with collisions which take place in narrow space regions, studied irrespectively of the surrounding large scale mass distributions, which we might well think of as described by the vacuum, and worry only about the short scale effects of gravitational forces.

We are thus lead to consider Quantum Minkowski Space as a more realistic geometric background for Elementary Particle Physics. But, as we briefly mentioned at the beginning, the energy distribution in a generic quantum state will affect the Spacetime Uncertainty Relations, suggesting that the commutator between the coordinates ought to depend in turn on the metric field.

Thus the spacetime commutation relations would become part of the *equations of motion*.

While in Classical General Relativity Geometry is part of the Dynamics, in this scenario also *Algebra* would be part of the Dynamics.

This might well be the clue to restore Lorentz covariance in the theory of interactions between fields on Quantum Spacetime.

On the other side, we mentioned how heuristic arguments suggest that the distance of acausal propagation of effects could increase near singularities.

This scenario could be related to the large scale thermal equilibrium of the cosmic microwave background (horizon problem). Actually, taking into account only of the Planck length as a universal lower bound for that distance of propagation, and assuming the simple model of a scalar massless field semiclassically interacting with the gravitational field (but treating EE exactly) shows that the *effect* of the divergence of the minimal distance of acausal propagation shows up, solving the horizon problem without any inflationary hypothesis.

Similarly one could wonder whether the non vanishing of the Cosmological Constant is related to the dependence of the commutators of the coordinates upon the metric [16]. And to the fact that noncommutativity at the Planck scale might manifest itself as an effective repulsion; in which case it might well be an explanation of an inflationary potential.

7.2 The Basic Model: An Example of Quantum Geometry

7.2.1 The Basic Model and Its Covariant Representations

The basic model arises from the simplifying *ansatz* that the commutators $Q^{\mu\nu} = -i[q^\mu, q^\nu]$ are central, namely they strongly commute with the coordinates q^μ. To fix domain ambiguities and select reasonably regular representations, we understand the formal definition of the antisymmetric 2-tensor $Q^{\mu\nu}$ as a reminder of the Weyl relations

$$e^{ih_\mu q^\mu} e^{ik_\nu q^\nu} = e^{-\frac{i}{2}h^\mu Q^{\mu\nu} h_\nu} e^{i(h+k)_\mu q^\mu}, \quad h, k \in \mathbb{R}^3, \tag{7.7}$$

where we took care of using the Lorentz metric to parametrise the 4-parameters group $k \mapsto e^{ikq} = e^{ik_\mu q^\mu}$. In what follows, formal commutation rules will always be understood as shorthands for the regular Weyl form.

As described in Sect. 7.1, covariant quantitative conditions on the commutators amount to make a choice of the quantities a, b of the two independent "scalars" which can be formed out of an antisymmetric tensor:

$$Q^{\mu\nu} Q_{\mu\nu} = aI, \quad \left(\frac{1}{4} Q^{\mu\nu}(\star Q)_{\mu\nu}\right)^2 = bI.$$

The choice $a = 0, b = 1$—which in a sense is the most symmetric, see [18]—results in Heisenberg-like uncertainty relations which have the same form as the desired heuristically motivated relations (7.2).

A first, a priori only partial classification of the irreducible representations is provided by the remark that, by the Schur lemma, the 2 tensor of the commutators must be of the form $i\sigma^{\mu\nu}$ for some constant real antisymmetric 2-tensor $\sigma = (\sigma^{\mu\nu})$. It follows from the quantisation conditions that such a σ should fulfil

$$\sigma^{\mu\nu}\sigma_{\mu\nu} = 0, \quad \left(\frac{1}{4}\sigma^{\mu\nu}(\star\sigma)_{\mu\nu}\right)^2 = 1.$$

Let Σ be the manifold of all antisymmetric 2-tensors fulfilling the above conditions; it is by construction a homogeneous space under the natural action $\sigma \mapsto {}_\Lambda\sigma = \Lambda\sigma\Lambda^t = (\Lambda^\mu{}_{\mu'}\sigma^{\mu'\nu'}\Lambda^\nu{}_{\nu'})$ of the full Lorentz group.

Therefore, in order to classify all irreducible representations, it is sufficient to classify all the equivalence classes of irreducible regular representations with commutators which are multiples of the identity.

We next observe that there is a natural choice for σ_0: the standard symplectic matrix

$$\sigma_0 = \begin{pmatrix} 0 & 0 & 1 & 0 \\ 0 & 0 & 0 & 1 \\ -1 & 0 & 0 & 0 \\ 0 & -1 & 0 & 0 \end{pmatrix} \in \Sigma.$$

Upon renaming

$$(P_1, P_2, Q_1, Q_2) := (q^0_{\sigma_0}, q^1_{\sigma_0}, q^2_{\sigma_0}, q^3_{\sigma_0}),$$

the relations

$$[q^\mu_0, q^\nu_0] = i\sigma^{\mu\nu}_0$$

take the form of the Canonical Commutation Relations

$$[P_j, Q_k] = -i\delta_{jk}, \quad [P_j, P_k] = [Q_j, Q_k] = 0,$$

for two canonical pairs (P_1, Q_1) and (P_2, Q_2).

This fact—which of course must be regarded solely as a mathematical identification without any *direct* physical interpretation—is very lucky, as it completely solves the classification problem for irreducible representations of our spacetime relations, by reducing it to von Neumann uniqueness: there is only one irreducible representation

$$q_0 := q_{\sigma_0} = (P_1, P_2, Q_1, Q_2) \tag{7.8}$$

with commutators $i\sigma_0 I$, up to equivalence; where P_j, Q_j are canonical Schrödinger operators.

According to the previous remark, it follows that for every $\sigma \in \Sigma$ there is one and one only regular irreducible representation $q_\sigma = \Lambda_\sigma q_0$ with commutators $i\sigma = i\Lambda_\sigma \sigma_0 \Lambda^t_\sigma$, up to equivalence.

The manifold Σ may be identified with the quotient $\mathscr{L}/G_0$ of the full Lorentz group by the stabiliser of σ_0, which provides the possibility of building $\sigma \mapsto \Lambda_\sigma$ as a Borel section. Hence we have a complete classification of the representation theory of the spacetime commutation relations.

Not only the occurrence of the standard symplectic matrix σ_0 in Σ is lucky; it also is fascinating, for two quantum models with quite distant underlying physical motivations and interpretation—the non relativistic quantum mechanics of a material point on the plane and the gravity-induced (semiclassical) quantisation of the Minkowski spacetime—both rely on the very same basic building blocks: canonical pairs of Schrödinger operators. We also observe that the whole argument would have failed if the dimension of spacetime were odd, precisely because the canonical operators come in pairs.

Next we address the question whether there is a representation q^μ which is Lorentz covariant, in the precise sense that there is a strongly continuous unitary representation U of the Lorentz group on the representation Hilbert space $\mathscr{H}_q$ such that

$$U(\Lambda)^* q^\mu U(\Lambda) = \Lambda^\mu{}_{\mu'} q^{\mu'},$$

where the closure of the operator on the right is implicitly understood, or equivalently we regard the above as a shorthand of the corresponding transformation of the Weyl operators.

Correspondingly,

$$U(\Lambda)^* Q^{\mu\nu} U(\Lambda) = \Lambda^\mu{}_{\mu'} \Lambda^\nu{}_{\nu'} Q^{\mu'\nu'},$$

which prevents the possibility for a covariant representation to be irreducible; on the contrary it will have to be highly reducible.

For every representation q of the relations (7.7), the joint spectrum jSp(Q) of the 16 operators $Q^{\mu\nu}$ may be regarded as a manifold of antisymmetric real tensors $\sigma = (\sigma^{\mu\nu})$, namely a submanifold of Σ. If q^μ is a covariant representation in the sense of above, necessarily jSp(Q) is a homogeneous space under the Lorentz action; hence it must coincide with the whole Σ:

$$q^\mu \text{ covariant} \quad \Rightarrow \text{jSp}(Q) = \Sigma.$$

As a consequence, a covariant representation must weakly contain at least one representative q_σ for every $\sigma \in \Sigma$.

To construct a covariant representation, it would be sufficient to use a quasi-invariant regular positive measure. However, such a measure can be chosen to be even invariant: we may use the projection map $\mathscr{L} \mapsto \mathscr{L}/G_0 = \Sigma$ and the Haar measure on $\mathscr{L}$. Hence we take the Hilbert space

$$\mathscr{H}_q = L^2(\mathscr{L}, \mathfrak{H})$$

of square summable, $\mathfrak{H}$-valued functions of $\mathscr{L}$, where $\mathfrak{H}$ is the Hilbert space on which the Schrödinger operators P_1, P_2, Q_1, Q_2 act. Using the basic representation (7.8), we may set

$$(q^\mu \Psi)(M) = M^\mu{}_\nu q_0^\nu \Psi(M), \quad \Psi \in \mathscr{D}(q^\mu), \tag{7.9}$$

$$(U(\Lambda)\Psi)(M) = \Psi(\Lambda^{-1}M), \quad \Psi \in \mathscr{H}_q. \tag{7.10}$$

If we choose the Schrödinger representation $P_j = -i\partial/\partial s_j$, $Q_j = s_j\cdot$ on $\mathfrak{H} = L^2(\mathbb{R}^2, d^2s)$, then $\mathscr{H}_q \simeq L^2(\mathscr{L} \times \mathbb{R}^2, d\Lambda\, d^2s)$, and the operators q^μ are essentially selfadjoint e.g. on the smooth, compactly supported functions of $\mathscr{L} \times \mathbb{R}^2$. Every other covariant representation is quasi-equivalent to the above.

The problem of obtaining a Poincaré covariant representation is easily solved by doubling the underlying Schrödinger pairs, see [18].

7.2.2 Uncertainty Relations and Optimal Localization

It is convenient to identify the antisymmetric 4×4 matrices with the pairs $(\mathbf{e}, \mathbf{m})$ of their "electric" and "magnetic" parts of

$$\begin{pmatrix} 0 & e_1 & e_2 & e_3 \\ -e_1 & 0 & m_3 & -m_2 \\ -e_2 & -m_3 & 0 & m_1 \\ -e_3 & m_2 & -m_1 & 0 \end{pmatrix}$$

One easily checks that, if $\sigma = (\sigma^{\mu\nu}) = (\mathbf{e}, \mathbf{m})$, then $(\sigma_{\mu\nu}) = (-\mathbf{e}, \mathbf{m})$ and $(\star\sigma_{\mu\nu}) = (-\mathbf{m}, \mathbf{e})$. Moreover,

$$\sigma = (\mathbf{e}, \mathbf{m}), \tau = (\mathbf{f}, \mathbf{n}) \quad \Longrightarrow \quad \sigma^{\mu\nu}\tau_{\mu\nu} = 2(\mathbf{m} \cdot \mathbf{n} - \mathbf{e} \cdot \mathbf{f}).$$

With these notations,

$$\Sigma = \{(\mathbf{e}, \mathbf{m}) : \mathbf{e} \cdot \mathbf{m} = \pm 1, |\mathbf{e}| = |\mathbf{m}|\},$$

where $|\cdot|$ is the Euclidean length; moreover,

$$(\mathbf{e}, \mathbf{m}) \in \Sigma \quad \Rightarrow \quad |\mathbf{e}| \geqslant 1, |\mathbf{m}| \geqslant 1;$$

this fact will be important in the derivation of the uncertainty relations. Note also that the standard symplectic matrix corresponds to the second vector of the canonical bases $\{\mathbf{n}_1, \mathbf{n}_2, \mathbf{n}_3\}$ of $\mathbb{R}^3$: $\sigma_0 = (\mathbf{n}_2, -\mathbf{n}_2)$.

If $\Lambda = \left(\begin{smallmatrix} 1 & 0 \\ 0 & R \end{smallmatrix}\right)$ for $R \in O(3, \mathbb{R})$, then $\Lambda(\mathbf{e}, \mathbf{m})\Lambda^t = (R\mathbf{e}, \pm R\mathbf{m})$ where $\pm \det R = 1$. The only subset of Σ which is invariant under orthogonal transformations is

$$\Sigma^{(1)} = \{(\mathbf{e}, \pm\mathbf{e}) : |\mathbf{e}| = 1\}$$

which has two connected components $\Sigma^{(1)}_{\pm}$, both evidently isomorphic to the 2-sphere S^2. It follows (cf introduction) that Σ itself has two connected components $\Sigma_{\pm}$, each of which is isomorphic with the tangent space of S^2.

We may now sketch the argument by which the uncertainty relations (7.2) follow from the quantisation conditions; it is sufficient to prove (7.2) for every irreducible q_σ, and for every vector state $\omega(\cdot) = (\psi, \cdot\psi)$, $\psi \in \mathscr{D}(q_\sigma)$). With $\sigma = (\mathbf{e}, \mathbf{m}) \in \Sigma$, the (generalized) Heisenberg uncertainty theorem gives

$$\Delta_\omega(q_\sigma^0)\Delta_\omega(q_\sigma^j) \geqslant \frac{1}{2}\omega(|[q_\sigma^0, q_\sigma^j]|) = \frac{1}{2}|e_j|;$$

(7.2a) then follows from $1 \leqslant |\mathbf{e}| = \left(\sum_j |e_j|^2\right)^{1/2} \leqslant \sum_j |e_j|$. A similar argument (using $|\mathbf{m}| \geqslant 1$) gives (7.2b).

The non-invariant quantity $\sum_\mu \Delta(q_\mu)^2$ provides information about the localization properties of a state according to a given observer. Given a state ω on an irreducible representation q_σ, we have

$$\sum_\mu \Delta_\omega(q_\sigma^\mu)^2 \geqslant \sqrt{2+|\mathbf{e}|^2+|\mathbf{m}|^2},$$

where $\sigma = (\mathbf{e},\mathbf{m}) \in \Sigma$, and provided ω is in the domain of the involved operators (see [18, Proposition 3.4] for more details).

Two questions arise:

1. given any $\sigma \in \Sigma$, do states ω on q_σ exist, such that the above bound is attained?
2. the bound itself is minimal when $\sigma \in \Sigma^{(1)}$, in which case it becomes

$$\sum_\mu \Delta_\omega(q_\sigma^\mu)^2 \geqslant 2, \quad \sigma \in \Sigma^{(1)};$$

do states ω on q_σ for $\sigma \in \Sigma^{(1)}$ exist, such that the above bound is attained?

While the answer to the general question (1) is unknown, question (2) is easy to deal with. If $\sigma \in \Sigma^{(1)}$, then $\sigma = \Lambda\sigma_0\Lambda^t$ for some $\Lambda = \left(\begin{smallmatrix}1&0\\0&R\end{smallmatrix}\right)$, where $R = (R_{jk}) \in O(3,\mathbb{R})$. Then $q_\sigma^j = \sum_{k=1}^3 R_{jk}q_0^k$ and

$$\begin{aligned}\sum_\mu {q_\sigma^\mu}^2 &= {q_\sigma^0}^2 + \sum_{j=1}^3 {q_\sigma^j}^2 = \sum_\mu {q_0^\mu}^2 \\ &= P_1^2 + P_2^2 + Q_1^2 + Q_2^2,\end{aligned}$$

namely twice the Hamiltonian of the harmonic oscillator on the plane; the optimal localization states are precisely the translates of the ground state of the harmonic oscillator (the canonical coherent states) and $\sum_\mu \Delta(q_\sigma^\mu)^2 \geqslant 2$.

If instead we work with a state ω on the fully covariant representation q, define the probability measure μ_ω on Σ by $f \mapsto \omega(f(Q)) = \int_\Sigma f(\sigma)d\mu_\omega(\sigma)$, where $f(Q)$ is the joint bounded continuous functional calculus of the $Q^{\mu\nu}$'s. If ω is in the domain of all $q^\mu, {q^\mu}^2$, then

$$\sum_\mu \Delta_\omega(q^\mu)^2 \geqslant \int_\Sigma d\mu_\omega(\sigma)\sqrt{2+|\mathbf{e}_\sigma|^2+|\mathbf{m}_\sigma|^2},$$

where $(\mathbf{e}_\sigma,\mathbf{m}_\sigma) = \sigma$. Hence the lower bound becomes

$$\sum_\mu \Delta_\omega(q^\mu)^2 \geqslant 2$$

which is attained if μ_ω has support in Σ_1 and ω acts as a superposition of canonical coherent states on each q_σ contained in q, with $\sigma \in \Sigma_1$; we shall make this more transparent in the next section.

7.2.3 The C-Algebra of the Basic Model*

It is intuitively clear that we face a trivial bundle structure over Σ: over each $\sigma \in \Sigma$ there is a CCR-Weyl algebra, so that the universal C*-algebra to which every regular representation of the Weyl relations is affiliated is

$$\mathscr{E} = \mathcal{C}_0(\Sigma, \mathcal{K}) \simeq \mathcal{C}_0(\Sigma) \otimes \mathcal{K},$$

namely the trivial continuous field of C*-algebras over Σ with standard fibre $\mathcal{K}$, the compact operators over the separable, infinite dimensional Hilbert space $\mathfrak{H}$. The multipliers C*-algebra $M(\mathscr{E})$ is easily identified with $\mathcal{C}_b(\Sigma, B(\mathfrak{H}))$.

While we refer to [18] for the details of the proof why that bundle is trivial, we shall describe here how to work with this algebra.

We follow Weyl's prescription for quantisation:

$$f(q) = \int dk \hat{f}(k) e^{ik_\mu q^\mu}, \quad f \in L^1(\mathbb{R}^4) \cap \widehat{L^1(\mathbb{R}^4)},$$

where $\hat{f}(k) = (4\pi)^{-4} \int dx\, f(x) e^{ik_\mu x^\mu}$ is the usual Fourier transform; in practice, the idea is to replace the usual plane waves which build f up with their quantised counterpart, the Weyl operators.

Since the commutators are not multiples of the identity, a product $f(q)g(q)$ is not of the form $h(q)$; the Weyl-quantised functions do not close to an algebra of operators.

To circumvent this, we enlarge the class of functions to be quantised. We consider functions $f(\sigma; x)$ of both σ, x as elements of $\mathcal{C}_0(\Sigma, L^1(\mathbb{R}^4))$, the space of continuous $L^1(\mathbb{R}^4)$ valued functions of Σ, vanishing at infinity. For each σ define $\hat{f}(\sigma; k) = (4\pi)^{-4} \int dx\, f(\sigma; x) e^{ik_\mu x^\mu}$.

Whenever both $f, \hat{f}$ are in $\mathcal{C}_0(\Sigma, L^1(\mathbb{R}^4))$—in which case we call f a *symbol*—we may construct the operator $f(Q; q)$, where the Q dependence is understood in the sense of joint functional calculus, and the q dependence in the sense of Weyl quantisation. In more detail, if $Q^{\mu\nu} = \int_\Sigma \sigma^{\mu\nu} dE(\sigma)$ is the joint spectral resolution of the $Q^{\mu\nu}$'s,

$$f(Q, q) = \int_\Sigma dE(\sigma) \int_{\mathbb{R}^4} dk \hat{f}(\sigma, k) e^{ikq},$$

which is unambiguous, since the Weyl operators e^{ikq} and the joint spectral projections of the $Q^{\mu\nu}$'s commute.

A short computation with the Weyl relations gives the generalized symbolic calculus, defined as the pull-back of the operator product to symbols:

$$f(Q;q)g(Q;q) = (f \star g)(Q;q),$$

where the $\star$-product

$$(f \star g)(\sigma;x) = \frac{1}{(2\pi)^4} \int da \int db\, f(\sigma;a)g(\sigma;b)e^{2i(a-x)_\mu \sigma^{\mu\nu}(b-x)_\nu}. \tag{7.11}$$

may be regarded as a field of $\star_\sigma$-products over Σ:

$$(f \star g)(\sigma;\cdot) = f(\sigma,\cdot) \star_\sigma g(\sigma;\cdot)$$

Moreover, $f(Q,q)^* = \bar{f}(Q;q)$. We thus equipped the space $\mathscr{S}(\Sigma)$ of symbols with a product and an involution which make it a *-algebra, since they inherit all the relevant properties (associativity, involutivity,...) from being the pull-back of the operator product and involution; it may be turned into a Banach *-algebra taking its completion under the norm $\|f\| = \sup_\sigma \|\hat{f}(\sigma,\cdot)\|_{L^1}$, with universal enveloping C*-algebra $\mathscr{E}$. The algebra $\mathscr{S}(\Sigma)$ of symbols may be regarded as an algebra of continuous sections for $\mathscr{E}$. Note that, if q is the fully covariant representation, $f \mapsto f(Q;q)$ defines a faithful, covariant representation of $\mathscr{S}(\Sigma)$:

$$U(a,\Lambda)f(Q;q)U(a,\Lambda) = f(\Lambda^{-1}Q\Lambda^{-1^t}, \Lambda^{-1}(q-aI)), \quad (a,\Lambda) \in \mathscr{P},$$

which extends to a faithful covariant representation of $(\mathscr{E},\alpha)$, where the action α of the Poincaré group is the normal extension of the natural action on symbols. Hence we have an essentially unique covariant representation of the C*-dynamical system $(\mathscr{E},\alpha)$. We thus feel free to understand $f(Q;q)$ as indicating equivalently an operator (a represented element of the algebra), a symbol, or an abstract element of the algebra.

Let ω is a state on q with optimal localization and expectations $\omega(q^\mu) = a^\mu$. If μ_ω is the associated measure on Σ—supported by Σ_1—as described at the end of Sect. 7.2.2, we have

$$\omega(f(Q;q)) = \int_{\Sigma_1} d\mu_\omega(\sigma)(\eta_a f)(\sigma)$$

where for each localization centre a the localization map $\eta_a : M(\mathscr{E}) \to \mathcal{C}(\Sigma_1)$ is defined by

$$(\eta_a f)(\sigma) = \int_{\mathbb{R}^4} dk\, \hat{f}(\sigma;k)e^{ika-|k^2|/2}$$

and normal extension. It may be convenient to define $\mathscr{E}_1 = \mathcal{C}(\Sigma_1, \mathcal{K})$, $Z_1 = \mathcal{C}(\Sigma_1)$ the centre of $M(\mathscr{E}_1)$ and $\eta_{a,1} = \eta_a \restriction_{\Sigma_1}$ as the restriction of η_a to $\mathscr{E}_1$. Hence $\eta_{a,1}$ is a morphism of Z_1-modules, and η_a is a conditional expectation, in a natural way.

7.2.4 Many Events and the Diagonal Map

In order to develop a Quantum Geometry, we must identify the coordinates of multi-events. Since we want them to be independent, the usual prescription is to take tensor products: we regard each set

$$q_j^\mu = I^{\otimes(j-1)} \otimes q^\mu \otimes I^{\otimes(n-j-1)}, \qquad \mu = 0, \dots, 3, \tag{7.12}$$

as the coordinates of the jth event.

Then a segment may be identified by its two independent endpoints q_j, q_k, or even better with the separation operator $q_j - q_k$.

Since the theory is covariant under translations, we should expect the separations $q_j - q_k$ of two events to be statistically independent from the average position

$$\bar{q} = \frac{1}{n}\sum_{j=1}^{n} q_j$$

of all the n events. We immediately check that

$$[\bar{q}^\mu, (q_j - q_k)^\nu] = \frac{1}{n}(Q_j^{\mu\nu} - Q_k^{\mu\nu}),$$

where $Q_j^{\mu\nu} = -i[q_j^\mu, q_j^\nu]$; which does not vanish if $\otimes$ is understood as the tensor product of complex spaces. This forces us to understand $\otimes$ as the tensor product $\otimes_Z$ of Z-modules, where $Z \simeq C_b(\Sigma)$ is the centre of the multiplier algebra $M(\mathscr{E})$; intuitively, this amounts to take the usual tensor product fibrewise on Σ . Since Q is affiliated to Z, with this position

$$Q_j = Q, \quad j = 1, \dots, n,$$

and

$$[q_j^\mu, q_k^\nu] = iQ^{\mu\nu}, \tag{7.13}$$

$$[\bar{q}^\mu, (q_j - q_k)^\nu] = 0, \quad j, k = 1, \dots, n. \tag{7.14}$$

The coordinates q_j^μ are then affiliated with the C*-algebra[1]

[1]Of course $\mathcal{K}^{\otimes n} \simeq \mathcal{K}$, so that $\mathscr{E}^{(n)} \simeq \mathscr{E}$.

$$\mathscr{E}^{(n)} = \underbrace{\mathscr{E} \otimes_Z \cdots \otimes_Z \mathscr{E}}_{n \text{ factors}} \simeq \mathcal{C}_0(\Sigma, \underbrace{\mathcal{K} \otimes \cdots \otimes \mathcal{K}}_{n \text{ factors}}).$$

We now are in condition to construct a natural (non surjective) *-monomorphism from $M(\mathscr{E}^{(n)})$ to $M(\mathscr{E}^{(n+1)})$.

By construction

$$\bar{Q}^{\mu\nu} := -i[\bar{q}^\mu, \bar{q}^\nu] = -\frac{i}{n} Q^{\mu\nu};$$

namely the same commutation relations of the basic model, with the Planck length 1 replaced by $\sqrt{1/n}$ (in natural units where $\lambda_P = 1$). It follows that $\bar{q}$ is an amplification of $q/\sqrt{n}$. Moreover we have the identity

$$q^j = \bar{q} + \frac{1}{n} \sum_k (q_k - q_j).$$

The commutation relations (7.13) may be equivalently realised by taking

$$\boldsymbol{q}_j^\mu = \frac{1}{\sqrt{n}} q^\mu \otimes_Z I^{\otimes n} + \frac{1}{n} I \otimes_Z \sum_k (q_k - q_j),$$

so that

$$[\boldsymbol{q}_j^\mu, \boldsymbol{q}_k^\nu] = i\delta_{jk} Q^{\mu\nu}$$

and we recognise that $\bar{\boldsymbol{q}} = \frac{1}{\sqrt{n}} q \otimes_Z I^{\otimes n}$ and $\boldsymbol{q}_j - \boldsymbol{q}_k = I \otimes_Z (q_j - q_k)$ live in different tensor factors.

It follows that, with

$$f(Q; q_1, \ldots, q_n) = \int dk_1 \cdots dk_n \hat{f}(Q; k_1, \ldots, k_n) e^{i(k_1 q_1 + \cdots + k_n q_n)},$$

the map

$$\begin{aligned} \beta : f(Q; q_1, \ldots, q_n) &\mapsto f(Q; \boldsymbol{q}_1, \ldots, \boldsymbol{q}_n) \\ &= \int dk_1 \cdots dk_n \hat{f}(Q; k_1, \ldots, k_n) e^{\frac{i}{\sqrt{n}} \sum_j k_j q} \otimes_Z e^{\frac{i}{n} \sum_{jh} k_j (q_h - q_j)} \end{aligned}$$

extends to the announced *-monomorphism. This is interesting because it provides a tensor separation between the average position of a family of n independent events, and the algebra of the relative positions. This suggests to set the relative positions as close to zero as possible, compatibly with positivity in the algebra, leaving a function of the average position (and the centre) alone, to be understood as a noncommutative analogue of the classical evaluation of a function $f(x_1, \ldots, x_n)$ at $x_1 = x_2 = \cdots = x_2$.

Now, let ω_a be an optimally localized state with localization centre a and associated measure μ_ω on Σ_1; the idea we have in mind is to compose the above *-monomorphism with "id $\otimes\, \omega_a \otimes \cdots \otimes \omega_a$", so to set the separations $q_j - q_k$ to their minima, while leaving a function of $\bar{q}$ alone. However ω_a is not a Z-module map, hence such a tensor product is not well defined. Taking seriously that the centre should be regarded as a point independent background, and recalling from the end of Sect. 7.2.3 that $\omega_a = \int d\mu_\omega(\sigma) \circ \eta_a$ and $\eta_{a,1} = \eta_a \circ \restriction_{\Sigma_1}$, we may define the desired quantum diagonal map $E^{(n)}$ as

$$\mathcal{C}_b(\Sigma, \mathcal{K}^{\otimes n}) \xrightarrow{\beta} \mathcal{C}_b(\Sigma, \mathcal{K}^{\otimes(n+1)}) \xrightarrow{\restriction_{\Sigma_1}} \mathcal{C}(\Sigma_1, \mathcal{K}^{\otimes(n+1)}) \xrightarrow{\Phi} \mathcal{C}(\Sigma_1, \mathcal{K})$$

where

$$\Phi = \text{id} \otimes_{Z_1} \underbrace{\eta_{a,1} \otimes_{Z_1} \cdots \otimes_{Z_1} \eta_{a,1}}_{n \text{ factors}} .$$

It is an obvious consequence of translation covariance that the resulting map does not depend on the choice of a. We find

$$\begin{aligned} &E^{(n)} f(Q; q_1, \ldots, q_n) \\ &\quad = \int da_1 \cdots da_n e^{-\frac{1}{2}\sum_j |a_j|^2} \delta^{(4)}\Big(\frac{1}{n}\sum_j a_j\Big) f(Q; \bar{q} + a_1, \ldots, \bar{q} + a_n), \end{aligned} \tag{7.15}$$

where $\bar{q}$ now are the coordinates with characteristic length $\sqrt{1/n}$ and affiliated to $\mathcal{C}(\Sigma_1, \mathcal{K})$, while $|a|^2 = \sum_\mu (a_\mu)^2$.

The map so constructed is naturally covariant under orthogonal transformations, but not under Lorentz boosts.

7.2.5 Planckian Bounds on Geometric Operators

The choice of the Z-module tensor product to form coordinates of many events, discussed in the preceding section, was motivated by the necessity that $[q^\mu, q^\nu] \otimes_Z I - I \otimes_Z [q^{\mu'}, q^{\nu'}] = 0$ which, in the universal differential calculus, reads

$$dQ = 0.$$

With $q_j = I^{\otimes_Z (j-1)} \otimes_Z q \otimes I^{\otimes_Z (n-j-1)}$ the coordinates of the jth of n events,

$$dq_j = q_{j+1} - q_j, \quad j = 1, \ldots, n-1$$

is the separation between 2 of n events.

The operator $\sum_\mu q^{\mu 2}$ may be regarded as the square Euclidean distance between the event and the (classical) origin, and thus has no direct physical interpretation; we already observed that it is bounded below by 2. More interesting is the Euclidean distance $\sum_\mu dq^{\mu 2}$ between two events. We easily compute

$$[dq^\mu, dq^\nu] = 2i\, Q^{\mu\nu},$$

namely the same commutation relations as the basic coordinates, with characteristic length $\sqrt{2}$ (or $\sqrt{2}\lambda_P$, in generic units). It follows that the same bound on the square Euclidean length of q—appropriately scaled—holds true for the square Euclidean length of dq:

$$\sum_\mu dq^{\mu 2} \geqslant 4.$$

While observers connected by a Lorentz boost will disagree in general about the localization states where this bound can be attained, they agree on the bound itself, which thus is a quantity with an invariant meaning and a physical interpretation, and may be experimentally tested (at least in principle). This shows that a fully covariant theory may well be characterised by two distinct physically meaningful invariant quantities—the light speed and the Planck length—*without any contradiction with the Lorentz-Fitzgerald contraction.* In a sense, Special Relativity already is "Doubly-Special" in the sense of [12], *without any modification (deformation) of the Lorentz action.*

This is already an interesting geometric bound, though very elementary; by the way, it provides a clear example why a minimal length needs not being realised as a limitation on the precision which can be attained when measuring a single coordinate,[2] nor by requiring a discrete spectrum (in this model—as well as in any translation-invariant model—the spectrum of the coordinates is continuous).

In [7], the spectra of 2, 3 and 4-volume operators mentioned in Sect. 7.1 are discussed in some detail, for the case of the coordinates of the basic model. Note that, in the definition of such "quantum form-operators", the order of products does matter, so that they are not, and cannot be, (essentially) selfadjoint. The presence of a non trivial polar decomposition may be regarded as a quantum generalization of the classical notion of orientation. However—quite surprisingly—it is possible to show that they all are normal, so that they have a well defined spectral theory.

The findings of [7] are

1. the square Euclidean length of the separation dq between two independent events is bounded below by 4; its square Lorentzian length has continuous spectrum, pure Lebesgue, including the whole real line;

[2] Such a limitation could not be obtained in any case if coordinates have to be represented by selfadjoint operators, unless the availability of (generalized) eigenstates is restricted.

2. the sum of the squares of the components of both the space-time and space-space area operators $dq^0 \wedge dq^j$ and $dq^j \wedge dq^k$ have spectral values with absolute value bounded below by 1;
3. the 4-vector $V^\mu = \bigwedge_{\nu \neq \mu} dq^\nu$ whose components are the 3-volume operators has Euclidean length bounded below by 8; its time component alone has spectrum $\mathbb{C}$;
4. the 4-volume operator V has spectrum

$$\sigma(V) = \pm 2 + \sqrt{5}\mathbb{Z} + i\mathbb{R},$$

whose distance from 0 is $\sqrt{5} - 2$.

Apart the numeric factors, all bounds on n-volume operators (above expressed in natural units where $\lambda_P = 1$) are of order λ_P^n, consistently with their physical dimensions.

7.3 Quantum Field Theory on Quantum Spacetime: The Various Approaches and Their Problems

The problem of a Quantum Field Theory of Gravitation, eighty years after the pioneering paper by M. P. Bronstein on the quantum *linearized* Einstein theory [8], is still open.

It is therefore not entirely surprising if, twenty years after the publication of [18], the study of the interactions between quantum fields on Quantum Spacetime remains somewhat unsatisfactory. For, even if very simple forms of interactions are studied, the underlying geometry keeps into account some quantum aspects of gravitation near singular regimes.

While a large number of calculations have been performed and some conceptual issues have been raised, leading to a better insight, some fundamental issues still remain unsolved, such as, typically, the apparently unavoidable break down of Lorentz invariance as a result of the presence of nontrivial interactions. The expectation that ultraviolet (short distance) divergences would be removed or lessened, has been partly and in some case fully fulfilled, but generally, the models investigated exhibit a strange mixing of ultraviolet and infrared divergences. In the case when UV divergences disappear completely, the prize to pay for this positive feature lies in serious difficulties in taking an adiabatic limit in time.

7.3.1 Free Fields and "Local Algebras" on QST

In the approaches to QFT on Quantum Spacetime investigated by a number of the present authors, the free field equation remains unchanged, and therefore, the free massive bosonic quantum field on QST can be understood as follows: after evaluation in a (suitable) state on QST, one obtains an operator on the ordinary Fock space $\mathcal{H}$ by the assignment

$$\phi(\omega) := \frac{1}{(2\pi)^{3/2}} \int (\psi_\omega(k)\, a(\mathbf{k}) + \psi_\omega(-k)\, a(\mathbf{k})^*) d\Omega_m^+(\mathbf{k}) = \varphi(\widehat{\psi_\omega}) \qquad (7.16)$$

where $\psi_\omega(k) = \omega(e^{iq_\mu k^\mu})$ is the corresponding (inverse Fourier transformed) Wigner function, and where φ is the quantum field on classical spacetime, $\widehat{}$ denotes the Fourier transform, and $d\Omega_m^+(\mathbf{k})$ is the Lorentz-invariant measure on the positive mass shell as usual. For definiteness, the set of states might be chosen to be such that the resulting Wigner functions are Schwartz functions.[3] Short-hand notation for the above construction is the formula (7.6) for $\phi(q)$ found in the Introduction.

One obtains a fully Poincaré covariant field, which gives rise to a Poincaré covariant net of local algebras, as a map

$$E \mapsto \mathfrak{A}(E) \subset B(\mathcal{H})$$

which assigns to selfadjoint idempotents E in the Borel completion of the C*-algebra $\mathcal{E}$ of Quantum Spacetime the von Neumann algebra generated by the (bounded functions of the appropriate self adjoint extensions of the real and imaginary parts of the) field operators in (7.16), when ω has support in E, i.e. $\omega(E) = 1$.

This map would be covariant: if $\mathcal{P}$ is the covering group of the Poincaré group, and τ, α respectively denote its action on the C*-algebra of Quantum Spacetime $\mathscr{E}$, extended by normality to the Borel completion, and on the C*-algebra $\mathfrak{F}$ of field operators, then

$$\alpha_L \mathfrak{A}(E) = \mathfrak{A}(\tau_L(E)) \qquad (7.17)$$

But Locality breaks down: if ω is translated by a in a spacelike direction, even if ω is optimally localized, the commutator between $\phi(\omega)$ and $\phi(\omega_a)$ is never zero. But, as explicitly computed for the typical case of free massless fields, it vanishes as a Gaussian of Planckian width as a goes to spacelike infinity.

Therefore the fields are no longer local—which is perhaps to be expected on QST—but only at Planckian separations, for the free fields. It is not clear, however, if this results in a violation of causality at large scales in presence of interactions.

Moreover, as we do not know how to deal with interactions in a Lorentz covariant way, we cannot be sure that a covariant net as in (7.17) can still be a picture of an interacting theory.

But even if it were, the formalism would still miss an essential ingredient to be significant: a clear cut algebraic property which replaces Locality and reduces to it in the limit where the Planck length is neglected. As Locality does in the classical Minkowski case [19], this axiom ought to imply most conceptual features of QFT on Quantum Spacetime, independently of the specific form of the interactions.

[3] This set is nonempty, as the Gauss function is the Wigner function of the best localized states.

An indication of how radically new ideas are needed here is given by the dependence of the local algebras from our choices, already in the case of free fields.

If, as an example, we let $E(\lambda)$ denote the spectral family of $q_0^2 + q_1^2 + q_2^2 + q_3^2$, whose spectrum is the half line with minimum 2, and consider the local algebra $\mathfrak{A}(E(2))$, there will be only one function $k \mapsto \omega(e^{iqk})$ for all states ω on $\mathscr{E}$ such that $\omega(E(2)) = 1$. This function is a Gaussian. Hence, in the case of a single scalar and neutral field, the local algebra $\mathfrak{A}(E(2))$ will be generated by a single self adjoint operator (with spectrum the real line, pure Lebesgue), and hence isomorphic to the commutative von Neumann algebra of the Lebesgue L^∞ complex functions on the unit circle (on the other hand, in the case of finitely many generating fields, all of Fermi type, it would be finite dimensional).

But if we generate a von Neumann algebra with the translates of that algebras over any tiny neighbourhood of the origin in the translation group, we find all bounded operators [15] (see also [24]).

7.3.2 Perturbation Theory

When putting a quantum field theoretic model on Quantum Spacetime, several choices have to be made. In the absence of a good notion of locality, most publications have focused on perturbative approaches. Even so, the ordinary setup allows for a number of different generalizations. While on Minkowski space a number of approaches turn out to be equivalent (inductive construction of time-ordered products in the sense of Epstein and Glaser, Yang-Feldman approach, Dyson series, even Feynman graphs calculated via the Wick rotation), this ceases to be true on Quantum Spacetime.

For one thing, only the Dyson series and the Yang-Feldman approach seem to be even definable on Quantum Spacetime (where time does not commute with the space coordinates). And it then seems that, even on the simplest model of Quantum Spacetime, they yield theories which are inequivalent. Both approaches, however, share the feature that that they were based on the introduction of a commutative time parameter t—in the Hamiltonian approach this was caused by taking a partial trace on the algebra,

$$\mathcal{H}_I(t) = \int_{q^0=t} d^3\boldsymbol{q}\ H(q) \tag{7.18}$$

to define the interaction Hamiltonian $\mathcal{H}_I(t)$, and in the Yang-Feldman approach, such a time t was introduced in order to define the incoming field and to even formulate the initial value problem. Here, the interacting field is calculated recursively, as a formal power series in the coupling constant, formally written (for the massive Klein-Gordon field), for the simplest choice of an interaction term,

$$(\Box + m^2)\phi = -g\phi^{n-1}, \quad \phi = \sum_{k=0}^{\infty} g^k \phi_k.$$

Here, the Klein-Gordon operator is defined via $\partial_\mu f(q) := \frac{d}{dt} f(q+te_\mu I)|_{t=0}$. Fixing the initial condition by assuming that for $t \to -\infty$, the field $\phi(q + te_0 I)$ is the free field, the power series starts with the free field ϕ_0, while higher orders are calculated as convolutions with the retarded propagator G_{ret} of the ordinary Klein-Gordon equation, e.g.,

$$\phi_1(q) = \int g(x) G_{ret}(x) \phi_0^{n-1}(q + xI)\, dx,$$

with an infrared cutoff given by an x-dependent coupling constant g. Of course, the need for renormalization occurs here, since products of (even free) fields are ill-defined. Different methods of defining the interaction term have been investigated in e.g. [6, 14].

In the Hamiltonian formalism, on the other hand, it is important to note that in the expression for $\mathcal{H}_I(t)$ products of field operators appear which are spread in space and in time with a non local kernel, which is produced by the quantum nature of spacetime, see Sect. 7.3.3. Thus the time argument for the fields is not the parameter in $\mathcal{H}_I(t)$, but in the Dyson expansion for the S-matrix

$$\begin{aligned} S &= T \exp\left[i\lambda \int_{-\infty}^{\infty} dt\, \mathcal{H}_I(t) \right] \qquad (7.19) \\ &= I + \sum_{n=1}^{+\infty} \frac{(i\lambda)^n}{n!} \int_{-\infty}^{+\infty} dt_1 \cdots \int_{-\infty}^{+\infty} dt_n\, T\left[\mathcal{H}_I(t_1) \cdots \mathcal{H}_I(t_n)\right] \end{aligned}$$

the *time ordering* T *has to be performed in terms of the arguments of the factors* $\mathcal{H}_I(t)$, and not in terms of the time arguments in the fields [18]. Otherwise an unjustified violation of unitarity is introduced [4]. This prescription can be summarised in modified Feynman rules [25].

7.3.3 Interaction Terms

The next choice, which turns out to be just as delicate, is the generalization of even so simple an interaction term as ϕ^n. We do not comment on gauge theories here, see however, e.g. [33] and the comments on covariant coordinates and gauge invariant quantities in [7].

The first possibility that comes to mind is to use the product in the Quantum Spacetime C*-algebra $\mathcal{E}$ to define $\phi(q)^n$. If this prescription is used, in the Dyson series approach, to define the Hamiltonian density $H(q)$ appearing in (7.18), it turns out that the resulting $\mathcal{H}_I(t)$ is still a function of the commutators Q.

In terms of interpretation, this means that besides the localization, an experimentalist would also have to specify which measure on the spectrum Σ of the centre he prepared. This problem would equally show up in the Yang-Feldman approach.

This cannot be solved by evaluating a Lorentz invariant state on the centre, for the Lorenz group is not amenable. Already in [18] it was proposed to use a distinguished state on the centre in order to lessen this problem. More specifically, if $H(q, \sigma)$ denotes the evaluation of the Hamiltonian density at the point $\sigma \in \Sigma$, and $d\mu$ the rotation invariant regular probability measure on Σ carried by the base Σ_1, in (7.19) the following expression was used

$$\mathcal{H}_I(t) = \int d\mu(\sigma) \int_{q^0=t} d^3\boldsymbol{q}\ H(q, \sigma), \tag{7.20}$$

but of course the *ad hoc* choice of $d\mu$ breaks Lorentz invariance.

Note that, by power counting arguments, the resulting ϕ^3 theory was shown to be finite in this frame [3].

It must be mentioned that the twisting of the product of functions of q caused by non commutativity suggested a very interesting approach initiated in [11, 21]. This framework of so-called warped products still holds potential to be an effective tool in the construction of two dimensional models, with non trivial S matrix; which can even be preassigned as a phase function in two particles elastic scattering, solving the inverse scattering problem in terms of wedge local algebras. In these approaches, locality is replaced by the weaker notion of wedge locality.

Coming back to QFT on Minkowski QST, in order to specify the quantum Hamiltonian density $H(q)$ in (7.18), apart from using the product of $\mathscr{E}$, one sees that products of fields on QST, which generalize the ordinary interaction term, can now be defined in various ways, of which we mention two.

The first one, originally adopted in the above mentioned works, relies on the interpretation that an interaction is produced by bringing fields close to each other—in the end to bring them to coinciding points (at the cost, of course, of having to renormalize the corresponding term). This is the classical Wick procedure. But on QST it is not allowed to bring independent events at a coinciding point.

Thus, in our framework, it is natural to redefine this limit of coinciding points using the quantum diagonal map introduced in Sect. 7.2.4.

A (classical) interaction term $\phi^n(x)$ is then replaced by

$$:\phi(\bar{q})^n:_Q = E^{(n)}\big(:\phi(q_1)\phi(q_2)\ldots\phi(q_n):\big),$$

with $E^{(n)}$ as in Eq. (7.15) and with the actual dependence on the quantum coordinate $\bar{q}$ of characteristic length $1/\sqrt{n}$ (the mean coordinate) already spelled out explicitly.

The interaction Hamiltonian on the Quantum Spacetime is then given by

$$\mathcal{H}_I(t) = \lambda \int_{q^0=t} d^3\bar{\boldsymbol{q}}\ :\phi(\bar{q})^n:_Q$$

This expression is independent of the commutators $Q^{\mu,\nu}$, hence no *ad hoc* integration on Σ is needed. But the definition of the quantum diagonal map chooses a particular Lorentz frame, hence Lorentz covariance is broken *ab initio*.

The above choice leads to a unique prescription for the interaction Hamiltonian on Quantum Spacetime. When used in the Dyson perturbative expansion for the S matrix, this gives the same result as the ***effective non local Hamiltonian*** determined by the kernel

$$\exp\left\{-\frac{1}{2}\sum_{j,\mu} a_j^{\mu 2}\right\}\delta^{(4)}\left(\frac{1}{n}\sum_{j=1}^{n} a_j\right).$$

The corresponding perturbative Gell-Mann and Low formula is then **free of ultraviolet divergences** at each term of the perturbation expansion [5].

However those terms have a meaning only after a sort of adiabatic cutoff: the coupling constant should be changed to a function of time λ_τ, rapidly vanishing at infinity, say depending upon a cutoff time τ, i.e., the Gell-Mann and Low formula for the time-ordered products of the interacting field of the effective non local theory should read

$$\frac{T\left(\phi(x_1)\ldots\phi(x_n)\exp\left[i\int_{-\infty}^{+\infty} dt\,\lambda_\tau(t)\mathcal{H}_I(t)\right]\right)}{\left\langle T\exp\left[i\int_{-\infty}^{+\infty} dt\,\lambda_\tau(t)\mathcal{H}_I(t)\right]\right\rangle_0}$$

where the vacuum-vacuum contributions have to be divided out as usual, and where T indicates the time ordering, of course with respect to the t-values in the expansion of the exponential, not the time values in the arguments of the field operators, as already remarked above.

Thus this prescription leads to an ultraviolet finite theory, thereby finally fulfilling one of the original hopes of the whole approach. However, it remains to be shown that the adiabatic limit in time can be performed; otherwise, ultraviolet-infrared mixing problems cannot be excluded. This is an open problem, and there are indications that the limit might not exist.

Moreover, of course, a λ_P-dependent *finite renormalization* would be needed anyhow, otherwise the results would not have any physical meaning, for they would include meaningless large contribution, divergent in the limit of classical Minkowski space.

From this perspective, a major open problem, anticipated in the Introduction, is the following. Suppose we apply this construction to the renormalized Lagrangean of a theory which is renormalizable on the ordinary Minkowski space, with the counterterms defined by that ordinary theory, and with finite renormalization constants depending upon both the Planck length λ_P and the cutoff time τ. Can we find a natural dependence such that in the limit $\lambda_P \to 0$ and $\tau \to \infty$ we get back the ordinary renormalized Gell-Mann and Low expansion on Minkowski space?

This should depend upon a suitable way of performing a joint limit, which hopefully yields, for the physical value of λ_P, to a result which is essentially independent of τ within wide margins of variation, and can be taken as source of predictions to be compared with observations.

The other possibility for obtaining an interaction term is to consider auxiliary variables $x_i \in \mathbb{R}^4$ to define fields at separate points $q + x_i I$ in Quantum Spacetime, and to define the limit of coinciding points by letting $x_i \rightarrow x$, i.e., using this set of commutative extra parameters. This was motivated mostly by the fact that, in the Yang-Feldman approach, such commutative separations occur anyhow. Also, it makes mathematically precise the idea that after evaluation in a state on QST, one gets an ordinary operator valued tempered distribution, similar to what one has in the Wightman formalism. In fact, after the choice of a localization state ω, one considers as the n-fold tensor product of a quantum field on QST the tempered distribution

$$\mathcal{S}(\mathbb{R}^{4n}) \ni g \mapsto \phi^{\otimes n}(\Psi_\omega \times g)$$

where $\times$ is the convolution, and $\Psi_\omega(k_1, \ldots, k_n) = \omega(e^{i(\sum_j k_j^\mu)q_\mu})$. Formally, this corresponds to considering products

$$\phi(q + x_1) \cdots \phi(q + x_n)$$

The crucial point is that one can now give a precise notion of what a local counterterm should be. The resulting Wick products which are defined by subtracting only such local counterterms were conjectured to be well-defined in the limit of coinciding points—the proof which was sketched in [6] has been superseded by general considerations on twisted products of tempered distributions, which are currently being applied. But unfortunately this cannot be the end of the story. Some of the unsubtracted terms, even if finite for non zero values of the Planck length, are bound to diverge as that length is allowed to tend to zero. This means that they would contribute with possibly very large unphysical values, which ought to be removed by a finite renormalization.

Moreover, it turned out that this approach leads (in the Yang-Feldman approach) to a strange dispersion relation (modified in the infrared), which cannot be absorbed by local counterterms. Furthermore, it was shown later [2], that in the Hamiltonian formalism at least, the approach also exhibits a mixing of ultraviolet and infrared divergences.[4]

At the heart of these problems seems to be the fact that we cannot control the effects of noncommutativity at large scales. In particular, we cannot control how would those effects cumulate at higher and higher orders of the perturbation expansion, and decide whether they would keep being sensible only at Planckian distances.

[4]Note however, that in a Euclidean realm at least, there is hope that an infrared-cutoff model, the so-called Grosse-Wulkenhaar model might have a chance to be resummable and thus give way even to a constructible theory.

To understand these issues better, seems to be one of the essential points to better understand quantum field theory in QST.

But, at a more fundamental level, the difficulties with Lorentz covariance posed by the non triviality both of the centre of the algebra of QST and of the action on it of the Lorentz group, might be a spy of the need of a more dynamical meaning of the commutators. As mentioned in the Introduction (cf. also next section), Physics suggest that those commutators should depend on the fields, hence they should be acted upon by the Lorentz group in a more essential way. This might be the key to solve the problems with the correct definition of covariant interacting theories; however, in a scenario of which the only thing which is clear is that it would be extremely difficult to treat.

7.4 Quantum Spacetime and Cosmology

7.4.1 Beyond Minkowski: A Dynamical Quantum Spacetime Scenario

The model of Quantum Minkowski Spacetime presented in the previous sections should be thought of as a geometric background for Quantum Field Theory, which is more realistic than standard Minkowski Spacetime, as it implements in the noncommutative nature of the underlying geometry some of the limitations to localizability of events dictated by our present understanding of the basic principles of Quantum Mechanics and General Relativity. As shown above, the development of Quantum Field Theory on it allows us to avoid at least some of the problems and contradictions which we are otherwise bound to meet on commutative Minkowski.

As already mentioned, such a model seems to be sufficient for describing the typical regime of Particle Physics, in which the large scale spacetime structure is expected to have essentially no effect on particle collisions in an accelerator, even at very high energies.

On the other hand, it is widely believed that a quantum description of Gravity becomes of relevance near classical gravitational singularities, e.g., at cosmological times smaller than the Planck time $t_P \simeq 10^{-43}s$, where it could provide a better understanding of the initial state of the universe. Moreover, it can be foreseen that gravitational effects that demand for such a quantum description can have observational consequences, for instance in the structure of the Cosmic Microwave Background. It seems therefore compelling to extend the analysis of the quantum structure of spacetime to the curved case, also in view of the fact that it is conceivable that Quantum Spacetime may serve as a more suitable background for Quantum Gravity too.

If we turn then to the consideration of a generally curved (commutative) spacetime and of quantum fields propagating on it, it is to be expected that the energy density of the prevailing quantum state affects the Spacetime Uncertainty Relations, as shown, e.g., by the argument presented in Sect. 7.1. Since this energy density determines the dynamics of spacetime itself, through Einstein's Equations, we are led to the conclusion that, on an arbitrary spacetime, the Spacetime Uncertainty Relations, and therefore the commutator between the coordinates of a generic event, should depend on the underlying metric tensor. This leads us to a scenario where the equations of motion of the system should then become (in natural units) [16]:

$$[q_\mu, q_\nu] = i Q_{\mu\nu}(g), \tag{7.21}$$

$$R_{\mu\nu} - \frac{1}{2} g_{\mu\nu} R = 8\pi T_{\mu\nu}(\phi), \tag{7.22}$$

$$F(\phi) = 0, \tag{7.23}$$

where ϕ denotes the collection of the quantum fields under consideration, which should be thought as functions of the q_μ's, $T_{\mu\nu}(\phi)$ is their stress-energy tensor, and the last equation is symbolic for the fields' equations of motion (where the metric g also appears via the covariant derivatives).

In order to turn this general picture into a model apt to perform actual calculations, it is of course necessary, among other things, to investigate more closely the possible form of the right hand side of (7.21). This is obviously a hard problem. Which maybe ought to be tackled without forgetting that keeping the familiar form of (7.22) all the way down to the Planck scale is a terrific extrapolation: the experimental verification of Newton's law is not available for distances shorter than few millimetres.

Having said that, and lacking any clue on the possible modifications of Gravity at small scales, the simplest thing to do in order to study the possible form of the right hand side of (7.21) is to try to generalize the original derivation of the Spacetime Uncertainty Relations in [18] to a generic curved spacetime treating the gravitational field in the semiclassical approximation. This means that one should estimate the backreaction of spacetime to the localization of the state of a quantum field propagating on it, in order to detect the formation of trapped surfaces enclosing the localization region. The first problem which arises is that the concept of energy, which enters the argument of [18] through Heisenberg's Uncertainty Principle, is in general ill-defined on a curved background. Moreover it seems advisable to avoid also the other sharp simplifications made there, as, e.g., the use of the linearized form of the Einstein Equations to derive limitations which are relevant precisely in the extremely relativistic regime, where the linear approximation cannot be expected to be a good one. Or the use of a crude criterion, such as $g_{00} > 0$, for the non-formation of trapped surfaces.

7.4.2 Localization on a Spherically Symmetric Spacetime

The problems pointed out above have been solved in [20] in the case of a spherically symmetric background and a spherically symmetric localization region. The result, not surprisingly, is that in order to prevent the formation of trapped surfaces, the spatial sphere of localization should have a radius whose proper length is bounded below by a constant of the order of the Planck length.

More specifically, consider a globally hyperbolic spacetime M which is spherically symmetric. This means that M is diffeomorphic to $I \times \mathbb{R}_+ \times \mathbb{S}^2$, with $I \subset \mathbb{R}$ an open interval, and that the metric on M takes the form

$$ds^2 = -A(u,s)du^2 - 2ds\,du + r(u,s)^2 d\mathbb{S}^2. \tag{7.24}$$

The coordinates $(u, s) \in I \times \mathbb{R}_+$ are the so called *retarded coordinates*, and they have the following geometrical meaning. The coordinate u is the proper time along the worldline γ spanned by the centre of the spherical symmetry, while s is the affine parameter along the future pointing null geodesics which emanate from the point $\gamma(u)$, normalised in such a way that the scalar product between the tangent vector to the considered geodesics and to γ is one. The collection of all the lightlike geodesics emanating from $\gamma(u)$ forms a cone in M which we will denote by C_u.

The surface of the spatial 2-sphere described by the points of M at fixed $(u, s) \in I \times \mathbb{R}_+$ is given by $4\pi r(u,s)^2$, and, as intuitively clear, a trapped surface occurs when this quantity is decreasing with increasing s at fixed u, as this means that the geodesics spanning C_u are focusing. Thus, in order to detect the emergence of trapped surfaces, it is necessary to study the rate of change of this quantity. The latter is measured, along a fixed cone C_u, by the *expansion parameter of null geodesics* $s \mapsto \theta(s)$, whose evolution is governed by the *Raychaudhuri equation* (see, e.g., [32]), which, under the present symmetry assumptions, reads

$$\dot{\theta} = -\frac{\theta^2}{2} - R_{ss}, \qquad \theta \sim \frac{2}{s} \quad \text{for } s \to 0^+. \tag{7.25}$$

Here, R_{ss} is the s-s component of the Ricci tensor, which, due to spherical symmetry, is only dependent on (u, s).

Consider now a scalar, massless, conformally coupled quantum field ϕ propagating on M, a background metric $g^{(0)}$ of the form (7.24), and an initial Hadamard state ω on the *-algebra $\mathcal{A}(M, g^{(0)})$ generated by the Wick monomials of ϕ, in equilibrium with such background, namely a triple $(\phi, \omega, g^{(0)})$ satisfying the Klein-Gordon and the semiclassical Einstein Equations coupled together:

$$\Box_{g^{(0)}}\phi = 0, \tag{7.26}$$

$$R^{(0)}_{\mu\nu} - \frac{1}{2}g^{(0)}_{\mu\nu}R^{(0)} = 8\pi\omega(T_{\mu\nu}), \tag{7.27}$$

being $T_{\mu\nu}$ the stress-energy tensor of the field ϕ (we refer the reader to [9] and references therein for a detailed discussion of a free scalar quantum field on a general globally hyperbolic background from the algebraic point of view). We emphasise that solutions to (7.27) exist at least for spacetimes of cosmological interest [27, 28].

The field ϕ is used to model an experiment of spherically symmetric localization of an event on the background spacetime $(M, g^{(0)})$, and the state in which ϕ is prepared to perform such an experiment will be represented by the following simple perturbation of ω:

$$\omega_f(A) = \frac{\omega(\phi(f)A\phi(f))}{\omega(\phi(f)^2)}, \qquad A \in \mathcal{A}(M, g^{(0)}), \tag{7.28}$$

with f a spherically symmetric real smooth function whose support describes the localization region of the event under consideration. Such a state is obviously not strictly localized in supp f, and this entails that the limitations obtained on the size of the localization region will be weaker that those deriving from a strictly localized one, whose energy density, at fixed total energy, will be larger.

These limitations arise in principle by considering the backreaction of the underlying metric to the localization, i.e., the solution $g_{\mu\nu}$ to the semiclassical Einstein Equations with source the stress-energy tensor of ϕ (on the fixed background $g^{(0)}$) in the perturbed state ω_f,

$$R_{\mu\nu} - \frac{1}{2} g_{\mu\nu} R = 8\pi \omega_f(T_{\mu\nu}), \tag{7.29}$$

and imposing that, in accordance to the Principle of Gravitational Stability, no trapped surface appears preventing signals from supp f to reach a distant observer.

In practice, this is accomplished in [20] by first evaluating the change in the expectation value of the stress-energy tensor $\langle T_{ss}\rangle_{f,0} := \omega_f(T_{ss}) - \omega(T_{ss})$; then by fixing a cone C_0 containing supp f in its causal future and considering (7.25) on it, where, in the right hand side,

$$R_{ss} = 8\pi\omega_f(T_{ss}) = R^{(0)}_{ss} + \langle T_{ss}\rangle_{f,0}$$

(remember that $g_{ss} = 0$); and finally by requiring that its solution remains positive for all $s > 0$. This, according to the above discussion, entails that no trapped surface appears in the future of C_0. We notice explicitly that in this procedure no use of ill-defined concepts like energy is made, as the estimate of $\langle T_{ss}\rangle_{f,0}$ is solely a consequence of the free field properties, namely of the CCR.

The outcome of this discussion is summarised in the following theorem.

Theorem 7.4.1 ([20]) *Under the above hypotheses and notations, assume moreover that:*

(i) $R^{(0)}_{ss} \geq 0$ *on* C_0*;*
(ii) there exists a constant $C > 0$ *such that*

$$|\omega(\phi(f)\phi(f))| \leq C\|s\psi_f\|_{L^2(C_0)}\|\partial_s(s\psi_f)\|_{L^2(C_0)},$$

Fig. 7.1 Past null shadow of supp f

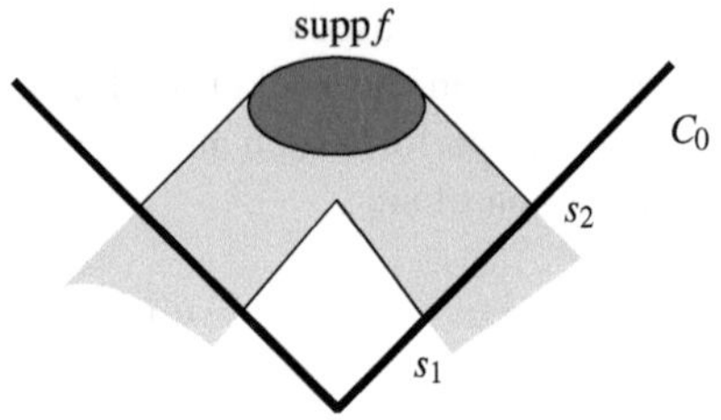

where $\psi_f := \Delta(f)|_{C_0}$ *is the restriction to* C_0 *of the image of* f *under the causal propagator* Δ *of Eq.* (7.26)*;*

(*iii*) *defining* $s_2 > 0$ *to be the value of the affine parameter such that the points of* C_0 *in the past causal shadow of* supp f *satisfy* $s < s_2$, *there exists an* $s_1 < s_2 < \frac{3}{2}s_1$ *such that*

$$\|\partial_s \psi_f\|^2_{L^2(C_0)} \leq 8\pi \int_{s_1}^{s_2} |\partial_s \psi_f|^2 \, ds.$$

Then, for the expansion parameter θ *to be positive on* C_0, *it is necessary that* $s_2 \geq \bar{s} := 1/\sqrt{12C}$.

We remark that assumption (i) is verified at least in all reasonable cosmological spacetimes, and that assumption (ii) is satisfied (with $C = 1$) by the massless Minkowski vacuum [20, Appendix], and that a similar property holds for a large class of Hadamard states on curved backgrounds [13]. Finally, assumption (iii) appears to be reasonable if s_1, s_2 are related to the past null shadow of supp f like in Fig. 7.1, due to the fact that the dominant contribution to $\|\partial_s \psi_f\|^2_{L^2(C_0)}$ comes from the singularities of the causal propagator $\Delta(x, y)$ for lightlike separations of the arguments.

Thus, for the localization experiment to be physically realisable, the size of the localization sphere, as measured in terms of the affine parameter, has to be bounded below by some constant $\bar{s}$ of order 1. We obtain in this way a generalization to a curved spherically symmetric space-time of the particular case of the Spacetime Uncertainty Relations in which all the uncertainties are of the same order of magnitude. In order to get a full set of Spacetime Uncertainty Relations, it would be of course necessary to treat the case in which supp f is not spherically symmetric.

The achieved result, anyway, means in particular that in a flat Friedmann-Robertson-Walker (FRW) background (which is spherically symmetric with respect to every point), with metric, in spatial spherical coordinates, $ds^2 = -dt^2 + a(t)^2[dr^2 + r^2 d\mathbb{S}^2]$, the size of a localization region centred around an event at cosmological time t, measured by the radial coordinate r, must be at least of order $1/a(t)$ (and therefore of order 1 in terms of proper length). Thus this gives further support to the expectation that the Spacetime Uncertainty Relations are affected by the background metric, and therefore to (7.21).

7.4.3 Backreaction on Quantum Spacetime and the Horizon Problem

The above derived behaviour of the effective Planck length on a flat FRW spacetime suggests of course that the acausal effects induced by the quantum structure of spacetime should become more important near the Big Bang, when $a(t) \to 0$, in agreement with previous remarks. In particular, as seen also in Sect. 7.3, it can be expected that this results in a high-energy modification of the product of quantum fields at the same spacetime point, and in particular of the stress energy-tensor. In the spirit of the scenario outlined in Sect. 7.4.1, this should entail, in turn, a modified cosmological evolution.

In [20] this issue has been analysed in more detail in the simplified situation of a universe only filled with radiation (modeled by a massless scalar field). In order to circumvent the problem of not knowing the explicit form of the commutation relations (7.21) implementing the Spacetime Uncertainty Relations on a generic background, and therefore without a full-fledged QFT on the resulting Quantum Spacetime algebra, the following strategy was adopted: first evaluate the modification of the stress-energy tensor on ordinary Minkowski Quantum Spacetime, then use the conformal isometry of (commutative) flat FRW with Minkowski to propose an *ansatz* for the stress-energy tensor on (the yet unknown) Quantum FRW Spacetime, and finally solve the semiclassical Einstein Equation with source given by the expectation value of such modified stress-energy tensor in a thermal state, used as a simple approximation of the initial hot state of the universe, whose relics we see today as the Cosmic Microwave Background (CMB). The interesting result of this analysis is that, although the Big Bang singularity is still present in this model, the scaling behaviour of radiation density near the singularity in significantly modified, in a way such that the resulting cosmological evolution avoids the horizon problem of standard cosmology.

More in detail, given a free massless scalar field ϕ on Minkowski Quantum Spacetime $\mathcal{E}$, its energy density is defined by replacing the coinciding point limit with the quantum diagonal map of Sect. 7.2.4:

$$:\rho:_Q(\bar{q}) := E^{(2)}\left(:\partial_0\phi(q_1)\partial_0\phi(q_2): -\frac{1}{2}\eta_{\mu\nu}:\partial^\mu\phi(q_1)\partial^\nu\phi(q_2):\right).$$

Consequently, the expectation value of $:\rho:_Q$ in the unique KMS state ω_β at inverse temperature $\beta > 0$ is easily calculated to be, in generic units,

$$\omega_\beta(:\rho:_Q(\bar{q})) = \frac{1}{2\pi^2}\int_0^{+\infty} dk\, k^3 \frac{e^{-\lambda_P^2 k^2}}{1 - e^{\beta k}}, \tag{7.30}$$

where the $\bar{q}$ dependence disappears in the right hand side because of translation invariance of ω_β. The result differs from the analogous quantity on commutative Minkowski spacetime by the Gaussian damping at high energies in the integrand.

Consider now a free, massless, conformally coupled field ϕ on flat FRW spacetime M. Introducing the conformal time $\tau = \int_{t_0}^{t} \frac{dt'}{a(t')}$, the metric of M becomes $ds^2 = a(\tau)^2[-d\tau^2 + d\boldsymbol{x}^2]$ and therefore M is conformally isometric to a subset of Minkowski spacetime. This entails that the state ω_β induces a state ω_β^M on (the algebra generated by the Wick powers of ϕ on) M, by simply replacing, in its two-point function, β with $\beta a(t)$. Accordingly, its physical interpretation (e.g., in the framework of [10]) can be seen to be that of a state describing local thermal equilibrium at inverse temperature $\beta(t) = \beta a(t)$.

This fact, together with the observation, made at the end of Sect. 7.4.2, that the effective Planck *proper* length is constant in time, i.e., it does not scale with $a(t)$, is at the basis of the following *ansatz* [20]: in passing from Quantum Minkowski Spacetime to Quantum Spacetime modeled on flat FRW, the only effective change on the expectation value of the energy density of ϕ is given by replacing β in (7.30) with $\beta(t) = \beta a(t)$.

The resulting expression for the energy density is therefore

$$\rho_\beta(t) := \omega_t \otimes \omega_\beta^M(:\rho:_Q(\bar{q})) = \frac{1}{2\pi^2}\int_0^{+\infty} dk\, k^3 \frac{e^{-\lambda_P^2 k^2}}{1 - e^{\beta a(t) k}}, \tag{7.31}$$

where ω_t is a state in which the cosmological time coordinate of M is sharply localized, and the dependence on the other components of $\bar{q}$ disappears again due to the spatial translation invariance of ω_β^M. It is easy to see that, while this expression is a negligibly small correction of the standard one on commutative flat FRW for $\frac{\lambda_P}{\beta a(t)} \to 0$, its asymptotic behaviour for $\frac{\lambda_P}{\beta a(t)} \to +\infty$ is given by

$$\rho_\beta(t) \sim \frac{C}{\beta a(t) \lambda_P^3},$$

significantly different from the standard one $\sim 1/a^4$.

Thanks to the assumed symmetry of the metric, the semiclassical Einstein Equations reduce to the first Friedmann equation, which, for $a(t) \to 0$ (i.e., near the Big Bang), takes then the simple form

$$\left(\frac{\dot{a}}{a}\right)^2 = \frac{c}{a},$$

and has therefore solutions, in terms of the conformal time τ, of the form $a(\tau) = (\alpha\tau + \beta)^{-2}$, from which one sees that the Big Bang occurs for conformal time $\tau \to -\infty$. This means that the singularity is the lightlike past boundary of the conformally related Minkowski spacetime, and thus in this spacetime every couple of points have been in causal contact at some time in the past after the Big Bang. This is to be compared with the standard cosmological evolution driven by a radiation field, where the Big Bang corresponds to some spacelike surface at finite conformal time $\tau = \tau_0$, which produces the horizon problem, illustrated by Fig. 7.2: on any

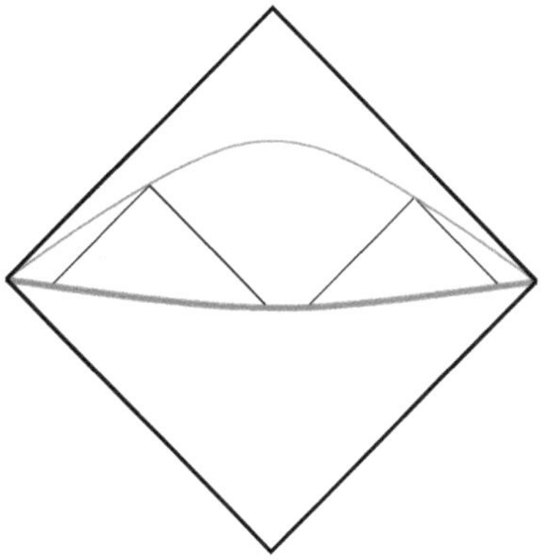

Fig. 7.2 The horizon problem

spacelike surface, there exists events which were never in causal contact since the Big Bang, which conflicts with the high degree of homogeneity of the CMB over the entire sky.

We recall that the commonly accepted solution to this problem, the inflationary scenario (see, e.g., [23]), typically postulates the existence of an *ad hoc* field, the inflaton, with a specific interaction, which has the role of driving a cosmological evolution without the horizon problem, and then decouples. In the model presented in [20], on the contrary, the field ϕ is just a free field, and the inflationary expansion is produced by the high-energy modification of its energy density caused by the quantum structure of spacetime, so that it can be expected that this is a generic feature, occurring also in more realistic situations of Standard Model fields interacting with a background "Quantum FRW Spacetime".

Finally, we stress that these results give further support to the discussion in Sect. 7.1 motivating the scenario (7.21)–(7.23), and also agree with the heuristic argument [16] which suggests to modify the Planck length on a curved background by, as a rough approximation, the factor $g_{00}^{-1/2}$. Such a rough argument points too to an infinite extension of non local effects near a singularity, so that, near the Big Bang, thermal equilibrium would have been established globally.

7.4.4 Further Possible Cosmological Applications

The findings reported above, although obtained in a semiclassical, oversimplified version of the scenario presented in Sect. 7.4.1 resulting from extrapolations of properties of QFT on commutative curved spacetime, provide a strong motivation for its further analysis.

Among the issues to be considered in a more refined framework, a prominent position is certainly taken by the analysis of the possible solutions given to the other basic problems of standard cosmology (as, e.g., the flatness problem), usually solved by inflationary models. Moreover, a study of the structure of CMB anisotropies induced by QFT on Quantum Spacetime promises to be a very stringent test, in view of the many experimental data which have become available in recent times [29], which already put rather severe constraints on inflation [22].

It is also worth mentioning that the apparent increase of the range of acausal effects near classical curvature singularities, together with the characteristic property of QFT on Quantum Spacetime of mixing short and long range effects, could also be expected to be at the root of the emergence of a non-zero Cosmological Constant, as a form of effective repulsion at very short distances [16]. In another direction, the same feature points to a minimal size for black holes, where Hawking evaporation would stop. The minimal black holes would be stable, and fill the universe with a gas which would contribute to the dark matter. Due to the spherical symmetry involved, some indications on this issue could come from an adaptation of the arguments presented in Sect. 7.4.2. The problem, however, would be displaced from the nature of dark matter to that of the possible formation of such black holes.

In an equally speculative attitude, it is conceivable that a dynamical Quantum Spacetime could solve the problem of maintaining Lorentz covariance in interacting QFT on it, and also that it could serve as a suitable geometrical background for the formulation of a consistent Quantum Gravity theory.

To conclude this section, we mention that a full set of Spacetime Uncertainty Relations for a flat FRW background has been obtained in [31], starting from an *ansatz* on the formation of trapped surfaces which generalizes exact results for spherically symmetric or equipotential surfaces, but still using Heisenberg Principle to evaluate the energy content of the localized quantum state. Moreover, an implementation of these Spacetime Uncertainty Relations is proposed in terms of concrete Hilbert space operators satisfying specific commutation relations. This could then be taken as a starting point for the formulation of a symmetry-reduced, semiclassical version of (7.21)–(7.23), in which some of the above mentioned problems could be addressed.

References

1. Amati, D., Ciafaloni, M., Veneziano, G.: Higher-order gravitational deflection and soft bremsstrahlung in Planckian energy superstring collisions. Nucl. Phys. B **347**, 550 (1990)
2. Bahns D.: On the ultraviolet-infrared mixing problem on the noncommutative Minkowski space. Ann. Henri Poincar/'e. To appear
3. Bahns D.: Ultraviolet finiteness of the averaged Hamiltonian on the noncommutative Minkowski space. arXiv:hep-th/0405224
4. Bahns, D., Doplicher, S., Fredenhagen, K., Piacitelli, G.: On the Unitarity problem in space-time noncommutative theories. Phys. Lett. B **533**, 178–181 (2002)
5. Bahns, D., Doplicher, S., Fredenhagen, K., Piacitelli, G.: Ultraviolet finite quantum field theory on quantum space-time. Commun. Math. Phys. **237**, 221–241 (2003)
6. Bahns, D., Doplicher, S., Fredenhagen, K., Piacitelli, G.: Field theory on noncommutative spacetimes: quasiplanar wick products. Phys. Rev. D **71**, 1–12 (2005)
7. Bahns, D., Doplicher, S., Fredenhagen, K., Piacitelli, G.: Quantum geometry on quantum spacetime: distance, area and volume operators. Commun. Math. Phys. **308**, 567–589 (2011)

8. Bronstein M.P.: Quantum theory of weak gravitational fields. (Republication from Physikalische Zeitschrift der Sowjetunion, Band 9, Heft 23, pp. 140–157 (1936), English translation by M.A. Kurkov, edited by S. Deser) Gen. Relativ. Gravit. **44**, 267–283 (2012)
9. Brunetti, R., Fredenhagen, K., Verch, R.: The generally covariant locality principle: a new paradigm for local quantum physics. Commun. Math. Phys. **237**, 31–68 (2003)
10. Buchholz, D., Ojima, I., Roos, H.: Thermodynamic properties of non-equilibrium states in quantum field theory. Ann. Phys. **297**, 219–242 (2002)
11. Buchholz, D., Lechner, G., Summers, S.: Warped convolutions, rieffel deformations and the construction of quantum field theories. Commun. Math. Phys. **304**, 95–123 (2011)
12. Camelia G.A.: Testable scenario for relativity with minimum length. Phys. Lett. B **510**, 255–263 (2001)
13. Dappiaggi, C., Pinamonti, N., Porrmann, M.: Local causal structures, Hadamard states and the principle of local covariance in quantum field theory. Commun. Math. Phys. **304**, 459–498 (2011)
14. Doescher, C., Zahn, J.: Infrared cutoffs and the adiabatic limit in noncommutative spacetime. Phys. Rev. D **73**, 045024 (2006)
15. Doplicher, S., Fredenhagen, K.: Unpublished
16. Doplicher S.: Space-time and fields: a quantum texture. In: Karpacz, New developments in fundamental interaction theories, pp. 204–213. arXiv:hep-th/0105251 (2001)
17. Doplicher, S., Fredenhagen, K., Roberts, J.E.: Spacetime quantization induced by classical gravity. Phys. Lett. B **331**, 39–44 (1994)
18. Doplicher, S., Fredenhagen, K., Roberts, J.E.: The quantum structure of spacetime at the planck scale and quantum fields. Commun. Math. Phys. **172**, 187–220 (1995)
19. Doplicher, S.: The principle of locality: effectiveness, fate, and challenges. J. Math. Phys. **51**, 015218 (2010)
20. Doplicher, S., Morsella, G., Pinamonti, N.: On quantum spacetime and the horizon problem. J. Geom. Phys. **74**, 196–210 (2013)
21. Grosse, H., Lechner, G.: Wedge-local quantum fields and noncommutative minkowski space. JHEP **0711**, 012 (2007)
22. Ijjas, A., Steinhardt, P.J., Loeb, A.: Inflationary paradigm in trouble after Planck 2013. Phys. Lett. B **723**, 261–266 (2013). arXiv:1304.2785
23. Peebles P.J.E.: Principles of Physical Cosmology, Princeton University Press, Princeton (1993)
24. Perini, C., Tornetta G.N.: A scale covariant quantum spacetime. Rev. Math. Phys. To appear. arXiv:1211.7050
25. Piacitelli, G.: Nonlocal theories: new rules for old diagrams. JHEP **0408**, 031 (2004). arXiv:hep-th/0403055
26. Piacitelli, G.: Twisted covariance as a non invariant restriction of the fully covariant dfr model. arXiv:0902.0575[hep-th]
27. Pinamonti, N., Siemssen, D.: Global existence of solutions of the semiclassical Einstein equation for cosmological spacetimes. Commun. Math. Phys. **334**, 171–191 (2015). arXiv:1309.6303
28. Pinamonti, N.: On the initial conditions and solutions of the semiclassical Einstein equations in a cosmological scenario. Commun. Math. Phys. **305**, 563–604 (2011)
29. Planck Collaboration. Planck 2013 results. I. Overview of products and scientific results. A&A. To appear. arXiv:1303.5602
30. Tomassini, L., Viaggiu, S.: Physically motivated uncertainty relations at the Planck length for an emergent non commutative spacetime. Class. Quantum Grav. **28**, 075001 (2011)
31. Tomassini, L., Viaggiu, S.: Building non commutative spacetimes at the Planck length for Friedmann flat cosmologies. Class. Quantum Grav. **31**, 185001 (2014)
32. Wald, R.M.: General Relativity, Chicago University Press, Chicago (1984)
33. Zahn, J.: Noncommutative (supersymmetric) electrodynamics in the Yang-Feldman formalism. Phys. Rev. D **82**, 105033 (2010). arxiv:1008.2309

Chapter 8
Algebraic Conformal Quantum Field Theory in Perspective

Karl-Henning Rehren

Abstract Conformal quantum field theory is reviewed in the perspective of Axiomatic, notably Algebraic QFT. This theory is particularly developed in two spacetime dimensions, where many rigorous constructions are possible, as well as some complete classifications. The structural insights, analytical methods and constructive tools are expected to be useful also for four-dimensional QFT.

8.1 Introduction

We give an overview of the methods and results of conformal quantum field theory (CFT), accumulated in the last three decades, in the perspective of axiomatic approaches. In particular, we advocate the point of view that a CFT is just a relativistic quantum field theory (QFT) which is invariant under the group of conformal spacetime symmetries. Thus, there are no independent "CFT axioms", but the usual QFT axioms apply with an enlarged symmetry.

Starting from the Wightman axiomatic setting in Sect. 8.2, we emphasize the crucial importance of inequivalent representations (superselection sectors) for several aspects of QFT. This motivates the formulation in the Haag-Kastler axiomatic setting ("algebraic quantum field theory", AQFT), which is particularly powerful to address superselection sectors. We give a brief review of this setting in Sect. 8.3, and turn to its specific application to chiral CFT in Sect. 8.4.

The Sects. 8.5 and 8.6 give an overview of various constructive methods to produce models of CFT (and QFT), and related classification results.

K.-H. Rehren (✉)
Institut für Theoretische Physik, Universität Göttingen,
Friedrich-Hund-Platz 1, D-37077 Göttingen, Germany
e-mail: rehren@theorie.physik.uni-goettingen.de

R. Brunetti et al. (eds.), *Advances in Algebraic Quantum Field Theory*,
Mathematical Physics Studies, DOI 10.1007/978-3-319-21353-8_8

8.2 CFT in the Context of Relativistic QFT

Since the conformal group contains the proper orthochronous Poincaré group, a conformal QFT is in particular a relativistic QFT in the usual sense. The simplest examples are the massless Klein-Gordon and Dirac fields and the Maxwell field in four spacetime dimensions (4D). Their conformal symmetry arises as a consequence of the massless field equations, and not because it was "postulated" as an extra feature.

The Wightman axioms describe quantum fields ϕ as operator-valued distributions on (a common invariant domain in) a Hilbert space $\mathcal{H}$, subject to the principle of Locality (Einstein Causality = commutation at spacelike distance). The Hilbert space carries a unitary representation U of the Poincaré group $P_+^\uparrow = SO(1, D-1)_0 \ltimes \mathbb{R}^{1,D-1}$ in D dimensions, which extends to a representation of the conformal group $\mathcal{C}_D$ (see below). Conformally covariant fields transform with a transformation law

$$U_g\phi(x)U_g^* = \alpha_g(\phi(x)) = D(g,x)^{-1}\phi(gx), \tag{8.1}$$

where the fields may be multiplets, and accordingly $D(g, x)$ is a suitable matrix-valued cocycle. Finally, the vacuum state is a unique vector $\Omega \in \mathcal{H}$ which is a zero-energy ground state for the Hamiltonian (the generator of the subgroup of time-translations) in every Lorentz frame, which implies that it is invariant under U_g for all $g \in \mathcal{C}_D$ (in $D > 2$).

For Poincaré and scale transformations $g_\lambda : x \mapsto \lambda x$ ($\lambda \in \mathbb{R}_+$), the cocycle $D(g, x)$ is independent of x and is just a matrix representation of $(SO(1, D-1)_0 \times \mathbb{R}_+) \ltimes \mathbb{R}^{1,D-1}$. For scale transformations, one has $D(g_\lambda)^{-1} = \lambda^d$ where the parameter $d \geq 0$ is the scaling dimension of the field ϕ.

8.2.1 *Conformal Symmetry*

We take the conformal group $\mathcal{C}_D$ to be the connected component of the group of transformations of D-dimensional Minkowski spacetime that preserve the Minkowski metric $ds^2 = \eta_{\mu\nu}dx^\mu dx^\nu$ up to a positive factor $\omega^2(x)$ that may depend on x. This group is generated (in $D > 2$) by the translations

$$T_a: \quad x^\mu \mapsto x^\mu + a^\mu \qquad (a \in \mathbb{R}^{1,D-1}) \tag{8.2}$$

and the involutive "conformal inversion"

$$I: \quad x^\mu \mapsto \frac{x_\mu}{(x \cdot x)}, \tag{8.3}$$

such that $S_b = I \circ T_b \circ I$ are the special conformal transformations

$$S_b : \quad x^\mu \mapsto \frac{x^\mu - (x \cdot x)b^\mu}{1 - 2(x \cdot b) + (x \cdot x)(b \cdot b)} \qquad (b \in \mathbb{R}^{1,D-1}). \tag{8.4}$$

Proper orthochronous Lorentz transformations $x \mapsto \Lambda x$ and scale transformations $x \mapsto \lambda x$, as well as the inversion I are generated by translations and special conformal transformations.

The singularity of the conformal inversion and the special conformal transformations can be dealt with as follows (see, e.g., [13, 55]). One introduces the "Dirac manifold" which is the projective null cone $\{\xi \in \mathbb{R}^{2,D} : \xi \cdot \xi = 0\}/_{\xi=\lambda\xi,\lambda\neq 0} \simeq (S^1 \times S^{D-1})/\mathbb{Z}_2$ in an auxiliary space with metric $(+, - \ldots -, +)$ S^1 is timelike, S^{D-1} spacelike. The conformal group acts perfectly regular as $\mathscr{C}_D \sim \mathrm{SO}(2, D)_0/\mathbb{Z}_2$ through the linear action of $\mathrm{SO}(2, D)_0$ on the cone.

Minkowski spacetime is just a (dense) chart with coordinates

$$x^\mu = \frac{\xi^\mu}{\xi^D + \xi^{D+1}} \qquad (\mu = 0, \ldots D-1) \tag{8.5}$$

of the Dirac manifold, so that the singularity of the inversion and the special conformal transformations is just a coordinate effect. The Poincaré group is the subgroup of $\mathscr{C}_D$ preserving the auxiliary coordinate $\xi^D + \xi^{D+1}$; its generators are $m^{\mu\nu}$ and $p^\mu = \frac{1}{2}(m^{\mu D} + m^{\mu D+1}) \in \mathrm{so}(2, D)$ $(\mu, \nu = 0, \ldots D-1)$. Other Minkowski charts are obtained by acting with $\mathscr{C}_D$ on the Dirac manifold.

Denoting p_0 the generator of the time translations in the Lie algebra of $\mathscr{C}_D$, and $k_0 = Ip_0I$ the generator of the timelike special conformal transformations, $h_{\mathrm{conf}} := \frac{1}{2}(p_0 + k_0)$ is the generator of the compact subgroup $\sim \mathrm{SO}(2) \subset \mathscr{C}_D$ in the "time" plane (0-$D+1$-plane) of the auxiliary space, in which I is the rotation by π.

Thus, quantum fields of a QFT with conformal symmetry must be defined as operator-valued distributions on the Dirac manifold. Their restriction to any Minkowski chart are Poincaré covariant Wightman fields in the usual sense.

The discrete spacetime symmetries (parity P and time reflection T) are not part of the axioms; instead, the CPT theorem applies, stating that while P and T may not be separate symmetries, there is an antiunitary operator Θ preserving the vacuum vector and acting on the fields like the combination of P, T, and a charge conjugation C.

Conformal transformations may take pairs of points which are timelike separated in a Minkowski chart to pairs of points at spacelike separation. In other words: the distinction between "spacelike" and "timelike" is Poincaré invariant, but not conformally invariant. As a consequence, causal commutativity of fields at spacelike distance implies causal commutativity also at timelike distance, and the support of the commutator is constrained to the null cone.

This conclusion is avoided if one admits projective representations of the conformal group, such as they occur, e.g., for the massless Klein-Gordon field in odd spacetime dimensions D which has half-integer scaling dimension $d = \frac{D-2}{2}$. In these cases, fields are not defined on the Dirac manifold, but on a suitable covering space thereof [55].

8.2.2 Two Dimensions

Conformal QFT in two dimensions is in our focus of interest because it admits a multitude of models which can be rigorously constructed. Most of these models possess (finitely or infinitely) many positive-energy representations, which admit, e.g., case studies of the general theory of superselection sectors as originally formulated for Poincaré covariant QFT in four dimensions (generalizing the univalence superselection rule or the electromagnetic charge superselection rule). This theory in particular defines the statistics (a representation of the permutation group) as an intrinsic feature of a positive-energy representation. In two dimensions, a new feature occurs, due to the disconnectedness of the causal complement of a point in two-dimensional Minkowski spacetime: the statistics in general is a representation of the braid group, related (via the Spin-Statistics theorem) to a much wider range of helicities than the Fermi-Bose alternative in $D = 4$.

In contrast to $D > 2$ dimensions, the conformal group in two spacetime dimensions (2D) turns out to be infinite-dimensional. In fact, the Dirac manifold in two dimensions can be identified with the product of two "chiral" circles $S^1 \times S^1$ on which the conformal group $\mathscr{C}_2$ acts as a product of two infinite-dimensional Lie groups $\mathrm{Diff}_+(S^1)$ (the orientation-preserving diffeomorphisms).

Namely, one can parameterize the solutions to $\xi \cdot \xi = 0$ for $\xi \in \mathbb{R}^{2,D}$ as $\xi = \lambda \cdot (\sin\alpha, \sin\beta, \cos\beta, \cos\alpha)$, so that $[\xi] \leftrightarrow (z^+ = e^{i(\alpha+\beta)}, z^- = e^{i(\alpha-\beta)})$ is a bijection between the quotient manifold and the product $S^1 \times S^1$ of two "chiral circles". Moreover, the chiral Minkowski coordinates are

$$x^{\pm} \equiv x^0 \pm x^1 = \frac{\xi^0 \pm \xi^1}{\xi^2 + \xi^3} = \frac{\sin\alpha \pm \sin\beta}{\cos\beta + \cos\alpha}, \tag{8.6}$$

such that

$$z^{\pm} = \frac{1 + ix^{\pm}}{1 - ix^{\pm}}. \tag{8.7}$$

This is the Cayley transform $\mathbb{R} \to S^1$ mapping the light ray into the circle (excepting the point $z = -1$), whose inverse is the stereographic projection $S^1 \backslash \{-1\} \to \mathbb{R}$. Because the Minkowski metric factorizes as $ds^2 = dx^+ dx^-$, the independent diffeomorphisms of the chiral circles preserve the metric up to a factor.

As it turns out, nontrivial unitary positive-energy representations of the group $\mathrm{Diff}_+(S^1)$ are necessarily projective representations, and a state invariant under U_g for all diffeomorphisms g does not exist. As a consequence, the vacuum vector of a conformal QFT in two dimensions is invariant only under the product of the two Möbius groups $\mathrm{SL}(2,\mathbb{R})/\mathbb{Z}_2 \sim \mathrm{SU}(1,1)/\mathbb{Z}_2 \subset \mathrm{Diff}_+(S^1)$. This unbroken subgroup of $\mathscr{C}_2 = \mathrm{Diff}_+ \times \mathrm{Diff}_+$ coincides with the group $\mathrm{SO}(2,2)_0/\mathbb{Z}_2$ (which is the conformal group $\mathscr{C}_D$ in $D > 2$ dimensions if one puts $D = 2$).

An important feature of 2D conformal QFT (also related to the factorization of the metric) is the presence of chiral observables, which depend only on x^+, or on x^-. These fields are therefore defined on the circle S^1, into which $\mathbb{R}$ is embedded via the Cayley map. The Möbius group acts by fractional linear transformations $\mathrm{PSL}(2,\mathbb{R})$ on $\mathbb{R}\cup\{\infty\}$ or by fractional linear transformations $\mathrm{PSU}(1,1)$ on S^1:

$$x \mapsto \frac{ax+b}{cx+d}, \quad \begin{pmatrix} a & b \\ c & d \end{pmatrix} \in \mathrm{SL}(2,\mathbb{R}), \qquad z \mapsto \frac{\alpha z+\beta}{\overline{\beta} z+\overline{\alpha}}, \quad \begin{pmatrix} \alpha & \beta \\ \overline{\beta} & \overline{\alpha} \end{pmatrix} \in \mathrm{SU}(1,1).$$

The presence of chiral fields is strongly connected with conservation laws. Because a symmetric tensor field that transforms irreducibly under the conformal group is traceless, it has only two independent tensor components $T_{+\dots+}$ and $T_{-\dots-}$ (where $\pm$ stands for Lorentz indices 0 ± 1). If a symmetric tensor current is conserved, the continuity equation implies that these components are chiral fields, i.e., $\partial_- T_{+\dots+} = 0$ and $\partial_+ T_{-\dots-} = 0$. (This simple argument fails for rank 1, because the current conservation law $\partial_\mu J^\mu = 0$ gives only one equation. But one notes by inspection that the unique conformally covariant two-point function of a conserved current is also dually conserved, hence by the Reeh-Schlieder theorem also the dual current is conserved: $\partial_\mu\,\varepsilon^{\mu\nu} J_\nu = 0$. Then again, J_+ and J_- are chiral fields.) Notice also that the conservation of the conformally invariant two-point function fixes the scaling dimension d of a conserved tensor current, to coincide with the tensor rank r.

Conversely, every pair of chiral fields of equal dimension constitutes a conserved traceless symmetric tensor current of rank $r = d$ as a two-dimensional tensor field. Thus, chiral fields naturally occur in 2D conformal QFT models whenever this theory has local conservation laws or, by Noether's theorem, continuous symmetries, and they actually represent the local generators of these symmetries.

The most important conserved tensor current is the stress-energy tensor (SET; dimension = rank = 2), which is (by definition) the local generator of the conformal symmetry $\mathrm{Diff}_+(S^1)\times\mathrm{Diff}_+(S^1)$ itself. The Lüscher-Mack theorem [56] fixes the self-commutator of its chiral components up to the "central charge" c, which can be regarded as the coefficient of the unique central extension of the Lie algebra of the diffeomorphism group of the circle. This quantity is a distinctive invariant of the chiral CFT at hand (where a priori, unless the 2D theory is parity symmetric, the two chiral central charges need not to coincide).

By local commutativity in the two-dimensional Minkowski spacetime, + and − chiral fields ("left- and right-movers") commute with each other, and chiral fields of the same chirality commute at non-coinciding points, so that their commutators are linear combinations of other fields multiplied with (derivatives of) δ-distributions in the chiral variables $x^\pm$.

Any algebraic relations among chiral fields, including their local commutators, are strongly restricted by conformal covariance. Early attempts at classification of chiral CFTs tried to find consistent commutation and operator product relations ("W-algebras"); but in this approach it is in general not known how to assess the existence of representations on a Hilbert space. The most prominent cases where this is possible, are the quantizations of the central charge c below 1 [29], and of the "level" k of chiral current algebras [37], which arise precisely due to the unitarity of the vacuum representation. Both these classification results are made possible by the fact that local commutation relations of the stress-energy tensor and of current algebras can be regarded as central extensions of infinite-dimensional Lie algebras (the Virasoro algebra, and affine Kac-Moody algebras, respectively), which are obtained by extending the fields as operator-valued distributions on S^1 and Fourier-decomposing the latter. Then, methods of highest-weight representations of Lie algebras can be applied to obtain the mentioned results.

The most basic examples of chiral fields are free fields related to the 2D massless Dirac and Klein-Gordon fields, which can be constructed in a standard way in terms of creation and annihilation operators on a Fock space:

• The massless 2D Dirac Fermi field decouples into its two chiral components, where "chiral" stands for the projections onto the eigenvalues of the Dirac matrix $\gamma^5 = \gamma^0\gamma^1$ which fix the sign of the helicity relative to the momentum of the massless particles states; by virtue of the massless Dirac equation, the chiral components depend only on x^+ or x^-, respectively. (While this original meaning of the term "chiral" refers to the helicity, we shall below understand it in the sense of "dependence on x^+ or x^- only".)

• The canonical massless Klein-Gordon field is ill-defined as an operator-valued distribution on all test functions in $\mathcal{S}(\mathbb{R}^2)$. It is, however, well-defined on test functions that are derivatives of test functions—which is tantamount to considering the "gradient of the Klein-Gordon field" as an operator-valued distribution on all test functions; this gradient field is of course a conserved current. Its chiral components are chiral fields of scaling dimension $d = r = 1$.

The stress-energy tensors of these free fields are not free fields themselves, but can be written as Wick products of the free fields. The SET of a real chiral Fermi field ψ has $c = \frac{1}{2}$, that of the chiral current j has $c = 1$. A remarkable feature arises here, known as "fermionization": The chiral current, which is a free Bose field, is at the same time a neutral Wick product $:\psi^*\psi:$ of a complex free Fermi field (= two real free Fermi fields), acting as the local generator of its $U(1)$ gauge symmetry $\psi(x) \mapsto e^{i\alpha(x)}\psi(x)$. Likewise, the SET of the Bose current $T = \pi :j^2:$ coincides with the SET of the complex Fermi field, $T = \frac{i}{2}:\psi^*\partial\psi - \partial\psi^*\psi:$, so that the same field has two representations in terms of free Bose or free Fermi fields.

(For the more remarkable converse, called "bosonization", i.e., the representation of fermionic fields in terms of bosonic fields, see Sect. 8.5.3.)

Writing "the same field" in the previous exposition means, that all vacuum correlation functions coincide in both representations. But the vacuum Hilbert space of the SET is just a subspace of the Fock space of the current, which is in turn a subspace of the Fock space of the complex chiral Fermi field. Thus, the latter Fock space carries reducible representations of the current and of the SET.

This simple example leads us to the issue of *representations*.

8.2.3 Representations

The specific algebraic relations defining a QFT in general admit many inequivalent Hilbert space representations. Covariant representations π of the field algebra come with a (projective) unitary representation U_π of $\mathscr{C}_D$ whose adjoint action on the covariant fields implements the transformation law:

$$U_\pi(g)\pi(\phi(x))U_\pi(g)^* = \pi(\alpha_g(\phi(x))) = D(g,x)^{-1}\pi(\phi(gx)). \quad (8.8)$$

A positive-energy representation is a covariant representation of the field algebra in which the generator of the unitary one-parameter group of time translations (the Hamiltonian) has positive spectrum: $P_0 \geq 0$. This implies that the commuting generators P_μ of the subgroup of translations have joint spectrum in the closed forward lightcone ($P \cdot P \geq 0$, $P_0 \geq 0$). A vacuum representation has in addition a unique ground state of zero energy, $P_0\Omega = 0$.

In conformal QFT, positivity of the Hamiltonian P_0 is equivalent to positivity of the generator $K_0 = U(I)P_0U(I)^*$ of the timelike special conformal transformations, and of the "conformal Hamiltonian" $H_{\rm conf} = \frac{1}{2}(P_0 + K_0)$. In 2D, positivity of the generators of chiral Möbius transformations $P^\pm = P_0 \pm P_1$, $K^\pm = K_0 \mp K_1$ and $L_0^\pm = \frac{1}{2}(P^\pm + K^\pm)$ follows. $L_0^\pm$ are the generators of the "rotations" of S^1.

In the free field example above, the Fock space of the complex Fermi field splits into an infinite direct sum of charged positive-energy representations of the current (the neutral representation being the vacuum representation); in turn, as a representation of the SET, the vacuum representation of the current decomposes into an infinite direct sum of representations of the $c = 1$ Virasoro algebra. The precise decompositions can be read off the "chiral characters": these are the power series $\chi(t) = Tr\exp(-\beta L_0)$ or $\chi(t,q) = Tr\exp(-\beta L_0 - \mu Q)$ in the variables $t = e^{-\beta}$ and $q = e^{-\mu}$, which yield the multiplicities of the (integer or half-integer) eigenvalues of the conformal Hamiltonian L_0 and charge operator Q in the respective representation spaces.

The classification of the positive-energy representations of the Virasoro algebra and of affine Kac-Moody algebras [29, 37] is a breakthrough for QFT without precedent, and without analog in $D = 4$: not only are the algebras generated by the stress-energy tensor, resp. by currents of dimension 1, universally fixed by conformal invariance (up to the central charge in the first case, and up to the structure constants of a Lie algebra and the level in the latter case), but also all their positive-energy representations are explicitly known without a perturbative construction. These models stand at the beginning of many remarkable findings in the representation theory of more general chiral CFTs, including modular tensor categories and the Verlinde formula.

Many issues in QFT are of a basically representation theoretic nature; e.g., the vacuum representation of an extension of a given QFT is in general a reducible representation of the latter, and certain data pertaining to this representation can be used to classify extensions. (An "extension" is, broadly speaking, a QFT containing a given

QFT with the same stress-energy tensor as generator of the covariance; see Sect. 8.6.2 for more details.) Another issue are QFTs with boundaries, whose boundary conditions can be understood as different representations of a suitable quotient algebra [7], see Sect. 8.6.5. These topics can be nicely addressed in two-dimensional CFT and give rise to several nontrivial classifications, because the structure of superselection sectors of chiral CFT is well understood.

In particular, a two-dimensional CFT is an extension of the tensor product of a pair of chiral CFTs; but also many chiral CFTs can be constructed as extensions or subtheories of other chiral CFTs (see Sect. 8.6.2).

8.2.4 Different Axiomatizations

Representation theoretic issues are best captured in the algebraic formulation of QFT (AQFT), emphasizing QFT as a given algebraic structure that admits many inequivalent Hilbert space representations. The Haag-Kastler axioms of AQFT therefore do not presume the existence of a vacuum vector—its presence is rather a feature of the representation considered.

Apart from this, the main difference to the Wightman axiomatics is that the same physical principles are reformulated in terms of local observables rather than quantum fields, thus offering a somewhat broader generality. We shall briefly review this approach in Sect. 8.3, and be more detailed for chiral conformal QFT in Sect. 8.4, where we also present its most important general results.

It follows from these general axioms that the structure of superselection sectors (=positive-energy representations) of completely rational (see Sect. 8.4) models of chiral CFT is captured by a modular C* tensor category. The latter is therefore the basic tool to describe the algebraic structure of two-dimensional CFT models (as extensions of chiral ones). There is a wealth of abstract mathematical results about braided and modular tensor categories. We shall point out in Sects. 8.6.2 and 8.6.5 that many of these abstract results have natural algebraic and representation-theoretic counterparts in the setting of chiral and two-dimensional CFT.

This nearly perfect match is the reason why certain popular axiomations of CFT actually "start from the other end", taking a modular tensor category as the initial axiomatic data. This point of view is justified by the above line of argument, but it tends to create the impression that conformal QFT, rather than a special case of relativistic QFT, were a "world of its own", with all its peculiar features—notably the presence of a braided or modular tensor category—being not only *admitted* by the general axiomatic frameworks, but being actually *consequences* of the usual axioms, specialized to 2D and augmented by conformal symmetry.

Especially, modularity of the representation category should not be regarded as an axiom reflecting some fundamental physical principle, since important models, like the $u(1)$ current algebra, do not share this property. As the characterization in [42] shows, it follows from complete rationality which is rather a regularity property than a fundamental feature.

Yet different axiomatisations, e.g., Euclidean CFT or vertex operator algebras (VOA), also capture essential features of rational CFT, without requiring all the features of relativistic QFT. Notably, the Hilbert space axiom is not essential in some approaches, which therefore admit even more classes of models. Euclidean CFT has a direct physical interpretation independent of its possible correspondence to a real-time relativistic CFT, as critical limits of classical lattice systems in two space dimensions (even with experimental verification). The physical interpretation of "the most general VOA" is not known, but with additional assumptions a close tie with relativistic conformal QFT can be established [23].

8.3 Algebraic QFT

The algebraic approach to quantum field theory is formulated by the Haag-Kastler axioms [35]. The localization of local observables is captured not by the use of quantum fields, but rather by specifying "local subalgebras" of a suitable global C* algebra. To each open spacetime region is associated a local algebra, whose self-adjoint elements are supposed to represent the physical observables that can be measured (or operations that can be performed) in that region:

$$O \mapsto A(O).$$

Unitary exponentials or spectral projections of self-adjoint smeared field operators would generate local algebras, but this kind of construction is not necessarily assumed in the AQFT approach. It is a nontrivial challenge to find criteria to decide whether local algebras come from Wightman fields, and to extract the latter from the former [16].

The principles of covariance and causality are easily formulated: The group of spacetime symmetries acts by automorphisms α_g of the global algebra properly transforming its local subalgebras into each other, and the local algebras associated with two spacelike separated regions are mutually commuting subalgebras.

Appropriate spacetime regions in Minkowski spacetime are doublecones O (intersections of past and future lightcones). With respect to the order relation among doublecones by inclusion, the assignment $O \mapsto A(O)$ is a "net" of local algebras. (Since the set of doublecones in the Dirac manifold is only partially ordered, the local algebras constitute only a pre-cosheaf of algebras.)

By postulating the local algebras to be C* algebras, and the "global" algebra to be the C*-inductive limit of the local algebras on Minkowski spacetime (usually called the quasilocal algebra), it is ensured that there exist Hilbert space representations. Among all Hilbert space representations, one should select those which describe states of physical interest. This accounts for the fact, that quantum field theory—unlike quantum mechanics—admits in general many inequivalent representations, among which positive-energy representations (distinguished by the implementation of time translations by a unitary one-parameter group with positive generator) and thermal equilibrium representations (distinguished by the presence of a KMS state ensuring the appropriate thermodynamic stability and passivity properties) are the most important ones [16]. Our focus in the sequel will be only on positive-energy representations.

We shall describe various ways of specifying local algebras in Sect. 8.5.

One may object that a "net of local algebras" is a structure too abstract for specifying a particular model—in particular, it does not involve an explicit specification of a Lagrangian. But recall that a model of Quantum Mechanics is fixed by specifying the set of observables on the Hilbert space (typically all selfadjoint operators) and the Hamiltonian, i.e., the time evolution automorphisms $\alpha_t = \mathrm{Ad}_{U(t)}$, $U(t) = e^{iHt}$. The same is true in QFT: one has to specify the algebras of observables and the relativistic covariance.

It is a crucial fact in this respect that scattering theory can be carried out (at least in positive-energy representations with a mass gap), by constructing multi-particle states and the scattering matrix, using only the net of local algebras and its covariance. Thus, the local algebras "contain" all the information that is needed to provide the interpretation of a model in terms of particles and their interactions. This is in accord with the fact that, in collider experiments, one usually does not measure a particular field strength but rather deposits of "something" in the detector arrays, and the physical interpretation of this "something" as a particle of either kind is imposed by the correlations of signals in different detector cells (naturally interpreted as "particle tracks"). In Haag-Ruelle scattering theory, the asymptotic dynamics of any local observable applied to the vacuum state (as long as it is not orthogonal to the desired particle state) is sufficient to identify the asymptotic particle states [35].

The specific dynamics of course enters through the specification of the time evolution automorphisms as part of the covariance. Thus, the Lagrangian (if there is any) is hidden as an "inverse scattering problem" in the scattering theory of the net of local algebras.

There is a marked difference to the standard approach (quantization of a classical Lagrangean theory): it is not a priori required that the generators of spacetime symmetries (in particular the Hamiltonian) are integrals over densities (components of the stress-energy tensor) which are some local "functions" of the field observables. In the classical theory, such relations would imply, through the canonical Poisson brackets, the correct infinitesimal transformation laws on the fields. In quantum field theory, on the other hand, canonical commutation relations cannot hold (in general) in a strict sense. Retaining only the "correct infinitesimal transformation laws" is therefore an appropriate substitute for the quantization of a classical Lagrangean formulation. One should keep in mind that classical physics is only a limit of the "true" quantum physics, and there is no reason to believe in a fundamental 1:1 correspondence between the two realms.

The "lack of fields" in the AQFT framework is in fact another strength: while different sets of quantum fields (relatively local w.r.t. each other) may generate the same local algebras, and hence have the same scattering states and the same scattering matrix (thus they belong to the same "Borchers class"), the local algebras they generate are the same, and should be regarded as the invariant content of the theory. The actual choice of fields may rather be regarded as an auxiliary device to describe the algebras (analogous to the choice of coordinates for a manifold), which may be very convenient but is not intrinsic to the physical interpretation.

In the application of the AQFT framework to conformal QFT in two dimensions, nets of local algebras are not just *assumed* to be given (as often in axiomatic approaches), but can be explicitly constructed by a large variety of methods. Some of these models actually do use fields (e.g., free Fermi fields which are bounded operators after smearing, or currents whose unitary Weyl operators are taken as the generators of the local algebras), but manipulations on such elementary constructions give rise to new nets that cannot always be easily described in the language of fields. We shall contrast the usual "field-theoretical" construction methods with the algebraic methods in Sect. 8.5.

Finally, it is a mathematical benefit that one may work with bounded operators, with norm given by the C* structure, and has not to worry about domains of definition. In this setup, it is easy to say what a "representation" is, and the issue of superselection sectors as unitary equivalence classes of positive-energy representations can be addressed. Algebraic quantum field theory is therefore the ideal setup to approach representation-theoretic issues.

Most of the seminal breakthrough achievements in conformal quantum field theory in two spacetime dimensions are of a representation theoretic nature (classification of the central charge and of conformal dimensions at $c < 1$ [29], fusion rules [4], coset models and branching rules [32], ...). Indeed, AQFT provides a unifying framework for these insights, that has proven one of the places where the AQFT formulation is most powerful, by the general theory of superselection sectors initiated by Doplicher et al. [26], see Sect. 8.4.2.

8.4 Algebraic CFT on the Circle

8.4.1 Axioms

We give here the AQFT axioms for a Möbius covariant chiral QFT (chiral CFT) directly in its vacuum representation, as in [13].

As emphasized earlier, in order to be prepared for the study of more general representations, one should rather axiomatize a pre-cosheaf of abstract local algebras on which the Möbius group acts by automorphisms; but one may as well read off this pre-cosheaf from its vacuum representation, by regarding the latter as the defining representation.

A chiral CFT is thus given by a family of local algebras $A(I)$ on a Hilbert space H where I runs over the proper open intervals of the circle S^1 into which the real axis $\mathbb{R}$ is embedded via the Cayley transform Eq. (8.7). One may take $A(I)$ to be von Neumann algebras. The fundamental axioms are:

(1) **Isotony**. If $I_1 \subset I_2$, then

$$A(I_1) \subset A(I_2). \tag{8.9}$$

(2) **Locality**. If $I_1 \cap I_2 = \emptyset$, then

$$[A(I_1), A(I_2)] = 0. \tag{8.10}$$

(3) **Möbius covariance**. There is a unitary representation U of the Möbius group $\mathrm{PSU}(1,1) = \mathrm{PSL}(2,\mathbb{R})$ on H such that for any interval I and $g \in \mathrm{PSU}(1,1)$,

$$\alpha_g\big(A(I)\big) := U_g \pi_0(A(I)) U_g^* = A(gI). \tag{8.11}$$

(4) **Positive energy**. The generator P of the one-parameter subgroup of translations in the representation U has positive spectrum.

(5) **Vacuum**. There is unique (up to a phase) unit vector $\Omega \in H$ which is invariant under the action of U, and cyclic for $\bigvee_I A(I)$.

As mentioned above, positivity of the chiral Hamiltonian P is equivalent to positivity of the generator L_0 of the one-parameter subgroup of rotations (the conformal Hamiltonian). By (4) and (5), the spectrum of L_0 in the vacuum representation is a subset of $\mathbb{N}_0$.

The axioms (1)–(5) imply the following properties [13]:

(6) **Reeh-Schlieder property**. The vector Ω is cyclic and separating for each local algebra $A(I)$.

(7) **Additivity**. If $I = \bigcup_i I_i$, then $A(I) = \bigvee_i A(I_i)$.

(8) **Haag duality on S^1**. For any proper interval I one has

$$A(I) = A(I')', \tag{8.12}$$

where I' is the interior of the complement of I in S^1.

(9) **Bisognano-Wichmann property**. The Tomita modular group [68] Δ^{it} of $A(\mathbb{R}_+)$ with respect to the vector Ω coincides with $U(\delta_{-2\pi t})$, where $\delta_t \in \mathrm{PSL}(2,\mathbb{R})$ is the one-parameter group of dilations.

(Here and below we are freely using the Cayley identification of $\mathbb{R}$ as a subset of S^1, and $\mathrm{PSU}(1,1) = \mathrm{PSL}(2,\mathbb{R})$. By virtue of Möbius covariance, the analog of (9) is true for every local algebra $A(I)$ and the associated subgroup of "dilations" of I.)

An interesting consequence of (9) is that the modular groups of the local algebras of any three intervals generate the representation U of the Möbius group $\mathrm{PSU}(1,1)$, in particular, spacetime symmetries are of modular origin. Conversely, it was shown in [34] that if M_i are three commuting von Neumann algebras with a joint cyclic and separating vector Ω, and $M_i \subset M'_{i+1 \bmod 3}$ are halfsided modular inclusions, then the three modular groups generate a positive-energy representation U of $\mathrm{PSU}(1,1)$. From this, one can construct a conformal net in its vacuum representation satisfying the axioms (1)–(5) (and strong additivity (12) below) by identifying M_i with the local algebras of three intervals arising by subdividing the circle by removing three points, and their modular groups with the corresponding dilation subgroups of $\mathrm{PSL}(2,\mathbb{R})$, and using the action of the resulting representation U to consistently define $A(I)$ for general I.

(This seems to be an interesting way to construct new chiral CFT models, but ideas to provide triples of commuting von Neumann algebras with the stated properties are scarce if one does not *start* from a CFT.)

Beyond the basic axioms (1)–(5), one may require further properties that are satisfied for many models.

The presence of a stress-energy tensor is axiomatized as a stronger version of (3):

(10) **Diffeomorphism covariance**. The representation U of $\mathrm{PSU}(1,1)$ extends to a projective unitary representation of $\mathrm{Diff}_+(S^1)$ such that for any interval I, one has

$$\mathrm{Ad}_{U(g)}A(I) = A(gI), \quad \text{for} \quad g \in \mathrm{Diff}_+(S^1), \tag{8.13}$$

and

$$\mathrm{Ad}_{U(g)}a = a, \quad \text{if} \quad a \in A(I), \ \mathrm{supp}\, g \subset I'. \tag{8.14}$$

By Haag duality (8), the latter property implies that $U(g) \in A(I)$ if $\mathrm{supp}\, g \subset I$, and the subnet generated by such $U(g)$ is called a Virasoro net.

The next two properties express "maximal decoupling" of local observables in intervals at a finite distance, and "maximal interaction" of local observables in touching intervals [54]:

(11) **Split property**. If I_1 and I_2 are two intervals with disjoint closure, then the map $a_1 \otimes a_2 \mapsto a_1 a_2$ is an isomorphism of von Neumann algebras

$$A(I_1) \otimes A(I_2) \simeq A(I_1) \vee A(I_2). \tag{8.15}$$

An equivalent assertion is that states can be independently prepared on the subalgebras $A(I_1)$ and on $A(I_2)$, such that a joint state restricting to the given states on $A(I_i)$ always exists. This feature depends on the energy level density, and is known to be true [18] if $e^{-\beta L_0}$ is a trace-class operator (in the vacuum representation) for every $\beta > 0$. (Such traces, regarded as power series in $t = e^{-\beta}$ whose coefficients give the multiplicities of the spectrum of L_0, are usually referred to as "characters" and are a very useful tool for the decomposition of reducible representations.)

(12) **Strong additivity**. Whenever two intervals I_1 and I_2 are obtained by removing an interior point from a proper interval I, then

$$A(I) = A(I_1) \vee A(I_2). \tag{8.16}$$

Thinking in terms of quantum fields, (12) may be regarded as a regularity property to the effect that the smearing can be approximated by test functions that vanish at a given point.

In view of (8), strong additivity (12) is equivalent to Haag duality on $\mathbb{R}$, namely for any proper interval $I \subset \mathbb{R}$ one has

$$A(I) = A(I^c)', \tag{8.17}$$

where I^c is the interior of the complement of I in $\mathbb{R}$.

For the algebra of two intervals $I, J \subset S^1$ with disjoint closure, Haag duality will generally fail. The index of the inclusion (in the vacuum representation)

$$A(I \cup J) \equiv A(I) \vee A(J) \subset A\big((I \cup J)'\big)' \tag{8.18}$$

is called the μ-index (which may be infinite).

A chiral CFT is called "completely rational" [42, 49, 54] if it is split (11), strongly additive (12), and has finite μ-index. A chiral CFT is called "rational" if it possesses only finitely many irreducible superselection sectors, see Sect. 8.4.2. It was shown in [54] that a split chiral CFT is completely rational if and only if it is rational; in particular, rationality together with split implies strong additivity. Moreover, in a completely rational chiral CFT the braiding is completely non-degenerate, turning the C* category of superselection sectors (see Sect. 8.4.2) into a modular category [42].

It should be stressed, however, that complete rationality is by no means an obvious feature. One of the most elementary models, the chiral $u(1)$ current algebra, satisfies the split property and strong additivity, but possesses a continuum of charged sectors, and thereby fails to be rational, hence completely rational.

8.4.2 Superselection Sectors

A general theory of superselection sectors was originally developed by Doplicher et al. [26] to describe sectors in massive QFT in four spacetime dimensions. With minor modifications in the setup, but big differences in the outcome (see below), it has proven to be applicable in two-dimensional and chiral CFT [27].

The crucial assumption on the representations π that can be treated by the DHR theory of superselection sectors is that, upon restriction to the causal complement of any doublecone region, π is indistinguishable from the vacuum representation up to unitary equivalence:

$$\pi\big|_{A(O')} \simeq \pi_0\big|_{A(O')}. \tag{8.19}$$

The heuristics behind this criterion is the argument that a charge distinguishing two representations can always sit inside the doublecone O, which is inaccessible to measurements in its causal complement O'.

Even for massive theories, this heuristic idea may fail for charges that can only be localized in "topological strings" (narrow cones extending to spacelike infinity) [15], requiring some mild adaptation of the theory. More dramatically, the criterium equation (8.19) excludes theories with long-range forces, notably QED, because an electric charge can be detected by measurements in the causal complement due to Gauß' law. An adaptation of the theory to this case has recently been formulated by Buchholz and Roberts [17].

In contrast, the chiral analogue of Eq. (8.19) is automatically satisfied in chiral CFT [20].

Assuming the validity of the criterium equation (8.19), positive-energy representations π can be described in terms of DHR endomorphisms ρ of the quasi-local algebra [26]. (In the case of CFT on S^1, where the set of intervals is only partially ordered, one may restrict the CFT to a net on $\mathbb{R} \subset S^1$. Otherwise, the definition of DHR endomorphisms is slightly more involved, invoking the pre-cosheaf structure.) Up to unitary equivalence, one has

$$\pi = \pi_0 \circ \rho, \tag{8.20}$$

where π_0 is the (defining) vacuum representation.

The DHR endomorphisms are "localized" in some region, meaning that their action is trivial on the subalgebras of local observables at spacelike distance of that region, and "transportable", meaning that that region can be chosen arbitrarily. The first property is consistent with the idea that ρ arises by conjugation with some localized charged field operator. The second is (a consequence of) covariance, and is consistent with the idea that the total charge does not depend on the localization of the charged field operator. But it should be stressed that these properties are derived *without assuming* the existence of such a field operator, and are entirely intrinsic in terms of the observables.

Equivalence, direct sums and subrepresentations of positive-energy *representations* can be formulated directly for the corresponding DHR *endomorphisms*, in terms of intertwiners, which are local observables satisfying the intertwiner relation $t\rho_1(a) = \rho_2(a)t$. The crucial insight of [26] is that locality, Haag duality and the localization of DHR endomorphisms imply algebraic properties of the intertwiners, which turn the representation theory into a C* tensor category (the "DHR category").

The composition of DHR endomorphisms describes a product of representations ("fusion") with the vacuum representation (the identical endomorphism) as the "neutral" element. The fusion product is commutative up to unitary equivalence, implemented by distinguished unitary intertwiners ("statistics operators") [26, 27]. The statistics operators turn the DHR category into a braided tensor category. In particular, each irreducible sector can be assigned a "statistics phase" κ_ρ and a "statistical dimension" $\dim(\rho)$.

In four dimensions, the braiding is in fact a permutation symmetry. As a consequence, $\kappa_\rho = \pm 1$ and $\dim(\rho) \in \mathbb{N}$, and these quantum numbers are related to the statistics (and hence the spin) of particles in the associated charge sector [26, II]. In chiral CFT, the conformal spin-statistics theorem [33] relates the statistics phase with the conformal spin, namely the value of the unitary representative $U(2\pi)$ of the full rotation of S^1, and hence the spectrum of the generator L_0 (mod $\mathbb{Z}$).

In a large class of chiral CFT models (the completely rational ones, see Sect. 8.4.1), the braiding is in fact maximally non-degenerate, so that the DHR category is even a modular C* tensor category [42], and the sum $\sum_\rho \dim(\rho)^2$ over all irreducible sectors equals the μ-index, see Sect. 8.4.1.

The latter is a very interesting result, quantitatively relating the existence of nontrivial superselection sectors to a quantity that can be "measured" in the vacuum representation.

The large variety of available models opens the door to model studies of general concepts, exploring the range of possibilities admitted by the general principles of local quantum field theory. Although these methods are presently limited to two dimensional conformal QFT, viewing them in the context of general QFT may be instructive for the construction of more realistic models in four-dimensional spacetime. We shall discuss some of these issues in Sect. 8.6.

8.5 Chiral Model Constructions

8.5.1 Free Fields

The most elementary constructions of chiral models are, as always, free fields. As the conformal scaling dimension h specifies the Möbius invariant two-point function $\propto \big(-i/(x-y-i\varepsilon)\big)^{-2h}$, and the two-point function completely determines a free field, the only "choice" is a scaling dimension h which must be positive in order that the two-point function is a positive-definite scalar product. The two-point function is local iff $h \in \mathbb{N}$, and it is anti-local iff $h \in \frac{1}{2} + \mathbb{N}_0$.

The case $h = \frac{1}{2}$ is identical with the chiral component of a massless Majorana resp. Dirac field, which are real resp. complex chiral Fermi fields. The case $h = 1$ is identical with the "chiral derivative of the massless Klein-Gordon field", called the free current, see Sect. 8.2.2.

In the AQFT framework, one would rather define the chiral free Fermi field as a CAR algebra over $L^2(S^1)$ with the vacuum state $\omega\big(\psi(f)\psi(g)\big) = (\overline{f}, \Pi_+ g)$ specified by the projection Π_+ onto the positive-frequency part; and the chiral free current can be defined by the CCR algebra over $C(S^1)$ with symplectic form $\sim i\int(f'g - fg')$ with the vacuum state given by a Gaussian $\omega(W(f)) = e^{-\frac{1}{4}\|f\|^2_C}$ with a suitable inner product of the complexified symplectic space to guarantee positive energy. Local subalgebras are specified by specifying the support of the functions f.

8.5.2 Wick Products and Fermionization

Quadratic Wick products of free fields are well-defined by standard quantum field theory methods.

In particular, the stress-energy tensor of the free Fermi field is $T = \frac{i}{2}{:}\psi\partial\psi{:}$, and the stress-energy tensor of a free current is $T = \pi :j^2:$, giving the central charge $c = \frac{1}{2}$ and $c = 1$, respectively.

"Fermionization" is the remarkable feature that the bosonic free current can be represented as the neutral Wick product $:\psi^*\psi:$ of a complex free Fermi field. The bosonic current is therefore defined on the fermionic Fock space, which as a representation splits into an infinite direct sum of charged representations with integer charge. The bosonic stress-energy tensor $\pi :j^2:$ coincides with the fermionic stress-energy tensor $\frac{i}{2}{:}\psi^*\partial\psi - \partial\psi^*\psi{:}$.

By way of generalization, one obtains "nonabelian currents" associated with a (semi-simple) Lie algebra $\mathfrak{g}$ by the "quark model construction": choosing an n-dimensional real or complex matrix representation $\tau^a = \pi(X^a)$ of the generators X^a of g, one defines the currents as quadratic Wick products of n real or complex free Fermi fields: $j^a \sim \tau^a_{ij}{:}\psi_i\psi_j{:}$ resp. $j^a \sim \tau^a_{ij}{:}\psi^*_i\psi_j{:}$.

Their commutation relations can be viewed as central extensions of the Lie algebra $L\mathfrak{g}$ of the loop group LG, i.e., the $\mathfrak{g}$-valued functions on S^1; the central term is universal up to a factor, called the level k, and the level in the given construction is a function of the Lie algebra $\mathfrak{g}$ and its representation π.

The resulting currents $j^a(f_a)$ act as infinitesimal generators of gauge transformations $\underline{\psi}(x) \mapsto e^{-if_a(x)X^a}\underline{\psi}(x)$.

Given a nonabelian current algebra, its stress-energy tensor is given by the Sugawara construction, $T_S \sim h_{ab}{:}j^a j^b{:}$ where h_{ab} is the invariant Killing metric, and the normalization factor is determined by the Lie algebra $\mathfrak{g}$. It should be noted that if the currents are obtained by the "quark model construction" on a fermionic Fock space, the Sugawara stress-energy tensor for the currents will in general not coincide with the fermionic stress-energy tensor (see Sect. 8.5.6).

This construction has an analogue in the AQFT framework [60, 72]. Let G be a compact Lie group with a unitary representation π on $\mathbb{C}^N$. One notes that the gauge transformations $\alpha_g : \underline{\psi}(x) \to \pi(g(x)^{-1})\underline{\psi}(x)$ for G-valued functions g on S^1, i.e., $g \in LG$, are automorphisms of the CAR algebra of N Fermi fields. The criterium for implementability by unitary operators $W(g)$ in the GNS-Hilbert space of the vacuum state (i.e., the Fock space) can be verified to be fulfilled. It follows that the unitaries $W(g)$ define a projective representation of the loop group LG. The cohomology class of the cocycle of this representation can be identified with the level, such that, up to a coboundary, $W(g) = \exp ij^a(f_a)$ if $g(x) = \exp if_a(x)X^a$.

Similarly, the orientation-preserving diffeomorphisms of S^1 are automorphisms of the CAR algebra via $\psi(f) \mapsto \psi(\gamma'^{\frac{1}{2}} f \circ \gamma)$. Again, these are implemented by unitaries $V(\gamma)$, giving rise to a projective representation of $\mathrm{Diff}_+(S^1)$ on the Fock space. This representation is related to the fermionic stress-energy tensor by $V(\gamma) = \exp iT_F(\delta)$ (again, up to a cocycle) if δ is an infinitesimal diffeomorphism and $\gamma = \exp(\delta)$.

On the other hand, for $\gamma \in \mathrm{Diff}_+(S^1)$, the map $g \mapsto g \circ \gamma$ is an automorphism α_γ of the loop group LG. Hence, for any projective representation π of LG, one obtains another projective representation $\pi \circ \alpha_\gamma$. It turns out that this representation is unitarily equivalent to π, i.e., α_γ are unitarily implemented by operators $V(\gamma)$ giving rise to a projective representation of $\mathrm{Diff}_+(S^1)$ on the representation space of π. This is the AQFT analogue of the Sugawara construction.

One might object that these theories are just subtheories of free QFT. However, the Sugawara stress-energy tensor of these theories is different from the free-field stress-energy tensor, indicating that the dynamics is different. Accordingly, the current algebras possess many positive-energy representations that do not arise by restriction of free-field representations. The study of the representation theory of chiral CFT models has been in the focus of interest for three decades, and has revealed a host of fascinating connections, including modular invariance of characters [71] and A-D-E classifications [21].

8.5.3 Bosonization

A much less-to-be-expected "converse" of fermionization is bosonization. It goes back to Mandelstam's vertex operator construction of free Fermi fields as the exponential of the (non-existing) massless scalar field, where the latter is to be written as an integral over the chiral current. Clearly, this integral makes the construction highly nonlocal. Frenkel and Kac [28] rediscovered this (formal) construction in a clean mathematical setup in terms of well-defined infinite normal ordered exponentials of the Fourier modes of the current on S^1 on the vacuum Fock space of the current, times a quantum-mechanical factor for the zero mode. The latter factor requires to extend the Hilbert space by a factor $L^2(S^1)$ on which the total charge operator acts like a rotation with discrete spectrum. The construction therefore selects a discrete one-dimensional lattice from the continuum of charges (superselection sectors) of the $u(1)$ current.

By exponentiating several $u(1)$ currents with coefficients that take values in an even higher-dimensional lattice, one can construct new *local* fields in a similar way. Some of these theories coincide with nonabelian current algebras, as in Sect. 8.5.2, at level $k = 1$, where the original $u(1)$ currents are the currents for the Cartan subalgebra.

In the AQFT framework, the analogous construction is understood as a crossed product of the Weyl algebra corresponding to the abelian current algebra by a lattice subgroup of the continuous group of DHR automorphisms [20, 67]. This amounts to an extension of the quasilocal algebra by charged intertwiners ("fields") whose charges take value in the lattice. The commutation relations of the fields are determined by the statistics operators of the DHR automorphisms, which turn out to be just complex phases. The extension by the new fields is a local extension if and only if all phases are $=1$, which is precisely the condition that the lattice is even.

A lattice extension corresponding to the "moonshine" vertex operator algebra has been constructed using the 24-dimensional Leech lattice [41]. Its central charge is $c = 24$, its μ-index is 1 (i.e., it has no nontrivial superselection sectors), its vacuum character is the modular invariant J-function and its automorphism group is the Monster group.

8.5.4 Orbifold Constructions

One can always descend from a chiral CFT with a compact gauge group of inner symmetries to the gauge-invariant subtheory. There are many examples with finite gauge groups; but one may as well consider the global symmetry of the Lie group G on a current algebra associated with its Lie algebra $\mathfrak{g}$. The fixed point subalgebra contains the Sugawara stress-energy tensor, but is in general larger, with very few exceptions.

The formulation as fixed points under the action of a gauge group by automorphisms of the net of local algebras is the same in the AQFT framework.

8.5.5 Simple Current Extensions

The positive-energy representations (sectors) do in general not form a group under the fusion product; in particular, there is no inverse but only a conjugate such that the product with the conjugate contains the trivial (=vacuum) sector. Sectors which have an inverse (i.e., the fusion product is the trivial sector), are called simple sectors (or "simple currents" in some communities). In the DHR theory, simple sectors are given by DHR automorphisms, rather than endomorphisms.

Simple current extensions are extensions of a chiral CFT by local fields that carry the charge of simple sectors. By the Spin-Statistics Theorem, such fields can only be local if they carry integer spin, hence the statistics phases must be $=1$. In order to define the extensions, one has to specify consistent algebraic relations between the new fields and the old fields, and among the new fields. The problem can often be reduced to a control of the representation category of the extended theory, in terms of that of the original theory.

The simple sectors with trivial statistics phases form a group under the fusion product. In the AQFT framework, simple current extensions can be defined as crossed products of the original net by the action of this group, in much the same way as the lattice extension of Weyl algebras in Sect. 8.5.3.

8.5.6 The Coset Construction

If a local QFT contains a sub-theory, the "coset QFT" is generated by all local fields of the larger theory that commute with the sub-theory. Thinking of the latter as generators of a symmetry (currents as generators of gauge transformations, the SET as generator of diffeomorphisms), the coset fields are invariant under that symmetry. Specifically, an inclusion $\mathfrak{h} \subset \mathfrak{g}$ of Lie algebras induces an inclusion of the current algebra chiral CFTs. Both these CFTs have their own Sugawara stress-energy tensor. Both stress-energy tensors are generators of the same (universal) diffeomorphism transformations of the h-currents, which means that their difference commutes with these currents. One obtains the "coset stress-energy tensor" $T_{\mathfrak{g}} - T_{\mathfrak{h}}$ with central charge $c_{\mathfrak{g}}(k) - c_{\mathfrak{h}}(k')$ [32].

Since the coset stress-energy tensor is different from the pair of given stress-energy tensors, the coset theory has its own dynamics. In particular, in this way all stress-energy tensors with central charge $c < 1$ have been constructed [32], thus rounding off the classification of admissible values of $c < 1$ "by exclusion of the continuum" [29] by an existence result for the remaining discrete set.

The coset construction allows to construct "new models from old models", and is by no means restricted to current algebras associated with Lie subalgebras $\mathfrak{h} \subset \mathfrak{g}$.

In the AQFT framework, the coset construction is given by a relative commutant of local algebras, namely, if $A(I) \subset B(I)$ are the local algebras of a chiral CFT and a subtheory, then the local algebras of the coset theory are

$$C(I) := A(I)' \cap B(I). \tag{8.21}$$

Some of the Virasoro algebra theories with $c < 1$ admit extensions by further local fields, without increasing the central charge. A complete classification has been obtained by AQFT methods (building on the earlier classification of modular invariant matrices), see Sect. 8.6.2.

8.6 Algebraic Constructions (not only Conformal)

To construct a model in the AQFT framework, one has to specify the local algebras along with the automorphic action of the spacetime symmetries, and one has to provide a vacuum representation with the appropriate spectral properties.

Starting from a given model, one possibility is to extend it by enlarging the local algebras. In order to exclude trivial "extensions by tensor products", one would require the extension to be irreducible, i.e., to have trivial relative commutant. The extended model will in general require a larger Hilbert space.

Another possibility is to deform the local algebras on the same Hilbert space, while preserving the local commutativity and covariance. For a more extensive review of this approach, see Chap. 10.

A third idea is "holographic" in the sense that the spacetime association of observables in a given net of local algebras is radically reorganized, such that, e.g., the quantum observables of a chiral CFT are re-arranged to become the observables of a two-dimensional model.

We shall present examples of these new construction ideas in the sequel.

8.6.1 Superselection Sectors and Symmetry

Unbroken inner symmetries give rise to superselection sectors.

Let G be a compact global gauge group with an action by automorphisms γ_g ($g \in G$) on a local QFT B such that γ_g preserve the local algebras and commute with the spacetime covariance. Let furthermore the vacuum state ω_B of B be invariant under γ_g, i.e., the vacuum does not break the symmetry, so that the gauge transformations are implemented by a unitary representation of G on the vacuum Hilbert space H_B.

The net of fixed-point subalgebras $A(O) = B(O)^G$ inherits the local structure and the spacetime covariance, and H_B carries a reducible positive-energy representation of A. It decomposes into subspaces H_π generated from the vacuum vector by elements of B transforming in some representation π of G, and each subspace H_π is invariant under the invariants A. In particular, the cyclic subspace H_A generated by A from the vacuum vector is a proper subspace of H_B.

One can reconstruct the full vacuum representation of B from the vacuum state ω_A of A, via the GNS construction of the state $\omega_B = \omega_A \circ \mu$, where μ is the conditional expectation from B onto A given by the Haar average over the group action.

A breakthrough result by [25], valid in QFT in four-dimensional spacetime, shows that the scenario just described is the generic origin of superselection sectors. It relies on the fact that in spacetime dimension >3 (and often also =3), the category of DHR superselection sectors [26] is a *permutation symmetric* C* tensor category, and such categories can be identified with the dual of a compact group. It proceeds by reconstructing from the given local net A and its DHR category a (unique up to isomorphism) universal "field algebra" B (which may be graded local) together with an action of a compact gauge group G such that $A = B^G$ is the fixed-point subalgebra, and the DHR sectors ρ of A are in 1:1 correspondence with the unitary representations π of G with statistical dimension $\dim(\rho) =$ matrix dimension $\dim(\pi)$.

As a consequence of this result, every irreducible extension of A is the intermediate algebra of invariants of B under a subgroup $H \subset G$.

It may appear natural to expect some inner symmetry also to be at the origin of superselection sectors in low-dimensional spacetime. However, there are obstructions due to the fact that the braiding is not a permutation symmetry in low dimensions (which is in turn a consequence of the geometric property that the causal complement of a finite connected region has two connected components). As a consequence, the dimensions $\dim(\rho)$ fail to be integer in general, and a 1:1 association with finite-dimensional representations π of some inner symmetry such that $\dim(\rho) = \dim(\pi)$ as before cannot be expected.

Yet, it is possible to associate a (non-unique) weak C* Hopf algebra [61] with the DHR category of a local QFT A, where the non-integrality of dimensions enforces the failure of the property $\Delta(1) = 1 \otimes 1$ of the coproduct. The "reduced field bundle" F of [27, II] can be interpreted as a nonlocal sector-generating field algebra with an action of a weak C* Hopf algebra such that the invariants are the observables A. An undesired but unavoidable feature of this construction is that the embedding $A \subset F$ is not irreducible, and F contains elements which belong to each of its local algebras.

8.6.2 Extensions by Q-Systems

An alternative approach was initiated in [50], by characterizing irreducible covariant extensions $A \subset B$ of a given local QFT A. It is assumed that B is relatively local w.r.t. A, i.e., observables of B commute with observables of A at spacelike distance, and that there is a conditional expectation $\mu : B \to A$ taking B onto A and preserving the vacuum state. One does not assume any specific symmetry concept (like group or weak Hopf algebra), but retains only the conditional expectation as a substitute for the Haar average over the action of the gauge group.

It is also assumed that the index of the local subfactors $A(O) \subset B(O)$ (which is independent of O) is finite, which is automatic if A is completely rational.

This scenario includes simple current extensions as well as orbifold constructions (regarding the full algebra as an extension of the fixed points), and conformal embeddings (local CFTs that share the same stress-energy tensor [64]) as well as coset constructions (regarding B as an extension of $A \otimes C$ if C is the coset model of $A \subset B$). But it is more general since it is not required that the extension B of A is itself local.

The main result is that every such extension is characterized by a "Q-system" in the DHR category (or "DHR triple"), and every Q-system allows to reconstruct the extension B in terms of data pertaining solely to A and its DHR category.

A Q-system is a triple, consisting of a DHR endomorphism θ of A and a pair of intertwiners $w \in \mathrm{Hom}(\mathrm{id}_A, \theta)$, $x \in \mathrm{Hom}(\theta, \theta^2)$ satisfying the relations

$$w^*w = x^*x = d \cdot \mathbf{1}, \quad w^*x = \theta(w^*)x = \mathbf{1}, \quad xx = \theta(x)x, \tag{8.22}$$

where $d^2 = \dim(\theta)$. In a more abstract category setting, a Q-system is the same thing as a (standard) Frobenius algebra in a C* tensor category [8], where the category at hand is the DHR category of superselection sectors.

From the data of the Q-system, the net B is reconstructed as an extension of A. It comes equipped with a local structure $A(O) \subset B(O)$ and covariance, and with a conditional expectation μ respecting the local structure and commuting with the covariance. The GNS representation of the state

$$\omega_B = \omega_A \circ \mu \tag{8.23}$$

as in Sect. 8.6.1 gives the vacuum representation of B, which—as a representation of A—is equivalent to the DHR representation θ. To every subsector $\rho \prec \theta$ corresponds a generator ψ_ρ of B that interpolates the vacuum subspace to the subspace carrying the representation ρ, and that implements ρ "in the average", namely

$$\rho(a) = \mu(\psi_\rho \, a \, \psi_\rho^*). \tag{8.24}$$

The algebraic relations among the "charged fields" ψ_ρ are encoded in the intertwiners w and x that specify the Q-system.

The extension is by construction relatively local w.r.t. A, and it is local if and only if the intertwiner x satisfies the condition $\varepsilon_{\theta,\theta}\,x = x$, where $\varepsilon_{\theta,\theta}$ is the statistics operator.

Thus, exhibiting Q-systems by solving their defining algebraic relations within the DHR category of the given subtheory, amounts to a construction of (relatively local or local) extensions. This is a finite-dimensional problem in rational theories, since for irreducible extensions, one can show that the multiplicity of every sector $\rho \prec \theta$ is bounded by its dimension; hence there are only finitely many "a priori candidates" for θ, and the intertwiner spaces, where w and x take values, are also finite-dimensional. Thus, even lacking more inspired methods (see below), Q-systems in a given rational C* tensor category can in principle be classified "by brute force".

A very useful fact is that possibly nonlocal chiral extensions $A \subset B$ induce local two-dimensional extensions $A \otimes A \subset B_2$, that are "CFT realizations of modular matrices". The original construction [62] uses "α-induction", and was recently recognized [6] to coincide, in terms of the corresponding Q-systems, with the "full centre", which is a most interesting concept in braided tensor categories [31, 44].

Complete classifications of local extensions have been achieved [39, 40] for the chiral and two-dimensional Virasoro nets with $c < 1$. The authors have exploited the fact that the Virasoro nets with $c < 1$ are completely rational, hence their DHR categories are modular (the braiding is non-degenerate). In this case, one can associate a modular invariant matrix with every chiral Q-system [10, 11], and these matrices have been classified before [21]. Since the DHR representation θ of the underlying chiral Q-system can be read off the modular invariant matrix, the number of candidates for θ is drastically reduced. The authors then show existence and uniqueness of the Q-system for each θ (in the chiral case) by using more abstract existence and uniqueness results of [43], and they classify the Q-systems for a given θ by a certain second cohomology [36] in the two-dimensional case.

It turns out that all local chiral extensions are: an infinite series of $\mathbb{Z}_2$ simple current extensions, and four exceptional extensions labelled (A_{10}, E_6), (E_6, A_{12}), (A_{28}, E_8), and (E_8, A_{30}) according to the A-D-E classification [21] of their modular invariants. Of these, three have been identified with coset extensions using current algebras [9], except (A_{28}, E_8) which occurs at $c = \frac{144}{145}$. This one was later identified with the "mirror" of a coset extension, where the mirror construction [73] is an operation on Q-systems relating a Q-system in A to a Q-system in C if A and C are each other's relative commutants (coset models) within some common extension B.

8.6.3 Borchers Triples and Deformation Methods

A net of local algebras $A(O)$ in any dimension can be constructed from a "Borchers triple" (or "causal triple"). A Borchers triple consists of a von Neumann algebra $M \subset B(H)$ with a cyclic and separating vector $\Omega \in H$, and a unitary positive-energy representation U of the proper orthochronous Poincaré group $P_+^\uparrow$ on H for which Ω is the unique invariant ground state. It is required that

$$U(\lambda)MU(\lambda)^* \subset M \text{ whenever } \lambda W_0 \subset W_0$$
$$U(\lambda)MU(\lambda)^* \subset M' \text{ whenever } \lambda W_0 \subset W_0', \tag{8.25}$$

where λ stands for $(\underline{a}, \Lambda) \in P_+^\uparrow$, $W_0 = \{\underline{x} \in \mathbb{R}^{1,D-1} : x^1 > |x^0|\}$ is the standard wedge region of Minkowski spacetime, and $W_0' = \{\underline{x} : x^1 < -|x^0|\}$ its causal complement.

Clearly, in every QFT, the algebra $M = A(W_0)$ of the standard wedge in the vacuum representation gives a Borchers triple, by virtue of covariance and locality.

Conversely, every Borchers triple defines a net by a simple construction. The construction proceeds by defining $A(W_0) := M$ and $A(\lambda W_0) := U(\lambda)MU(\lambda)^*$ for $\lambda \in P_+^\uparrow$. Then one defines

$$A(O) := \bigcap_{W \supset O} A(W), \tag{8.26}$$

where the intersection runs over all Poincaré transforms $W = \lambda W_0$ of the standard wedge which contain O. The assumptions Eq. (8.25) ensure that the algebras $A(W)$ are well-defined, and that the net $A(O)$ is local and covariant. (Unfortunately, the algebras $A(O)$ may fail to satisfy the Reeh-Schlieder property ($A(O)\Omega$ dense in H), and may be as small as $\mathbb{C} \cdot \mathbf{1}$.)

For Borchers triples in two dimensions, the second condition of Eq. (8.25) is obsolete (because there are no such $\lambda \in P_+^\uparrow$). Moreover, it is sufficient to have a positive-energy representation of the translations only: one can then use Tomita's Modular Theory to reconstruct also the representation of the Lorentz group including the CPT conjugation. Namely, Borchers [12] has discovered that the inclusions $U(a)MU(a)^*M \subset M$ for $a \in W_0$ are half-sided modular, and the modular group Δ^{it} and the modular conjugation J of (M, Ω) satisfy the same commutation relations with the translations $U(a)$ as the Lorentz group $U(\Lambda_{-2\pi t})$ and the CPT conjugation $U(\Theta)$, so one can define $U(\Lambda_t) := \Delta^{-it/2\pi}$ and $U(\Theta) := J$, and the second of Eq. (8.25) is automatic for λ involving the conjugation.

(In an effort to generalize this powerful result to $D = 4$, Kähler and Wiesbrock [38] have given a characterisation of the "relative modular position" of several von Neumann algebras with a joint cyclic and separating vector, such that their modular groups generate the four-dimensional Poincaré group.)

Exhibiting Borchers triples amounts to the construction of a QFT, with any prescribed particle content specified by the representation U.

The difficulty is, of course, to find algebras M satisfying the assumptions, and to find criteria such that the intersections $A(O)$ are large enough. This is easy for free theories, where M is generated by the Weyl operators of the free field smeared within W_0. Using Modular Theory, one can also define the wedge algebras for free fields associated with Wigner's massless "infinite-spin" representation [14], but the last step defining $A(O)$ for doublecones fails [45, 48]: the intersections of algebras turn out to be trivial unless O contains an (arbitrarily narrow) infinite spacelike cone. Indeed, "string-local" fields associated with the infinite-spin representation have been constructed in [57].

The prevailing ideas for finding more examples proceed by deformations $\widetilde{M}$ of free-field (or any other given) algebras $M = A(W_0)$, so as to produce deformed nets $\widetilde{A}(O)$. They are particularly successful in two dimensions:

Lechner [46] has constructed integrable massive models with factorizing scattering matrix by deformations of the canonical commutation relations (Zamolodchikov-Faddeev algebra). The input in this approach is a scattering function (in the rapidity variable), subject to restrictive analyticity conditions. The question whether the local algebras $A(O)$ are sufficiently large can be answered by a regularity criterium on the scattering function [46]. In the affirmative case, the scattering matrix of the resulting deformed local quantum field theory factorizes into two-particle scattering matrices given by the input scattering function.

Another deformation method uses "warped convolutions" [1, 2, 19]. These can be regarded as "momentum dependent translations" of the elements of wedge algebras where the spectrum of the momentum ensures that wedge-locality is preserved. This deformation violates Lorentz invariance in more than two dimensions.

Chiral conformal QFT can be used as a starting point to construct both massive and massless new two-dimensional QFT models through "Longo-Witten endomorphisms", as follows.

If A is a chiral CFT net, then $(M = A(\mathbb{R}_+), \Omega, T = U|_{\mathbb{R}})$ is a chiral Borchers triple, namely, the translations $T(a)$ with $a > 0$ satisfy $T(a)MT(a)^* \subset M$ for $a > 0$. In order to get a two-dimensional Borchers triple, one has to extend T to a unitary representation T_2 of the two-dimensional translations $\mathbb{R}^2$ with positive energy, such that $T(\underline{a})MT(\underline{a})^* \subset M$ for $\underline{a} \subset W_0$.

A Longo-Witten endomorphism [53] is an endomorphism of M of the form Ad_V, where V is a unitary operator commuting with T and preserving Ω. Therefore, every one-parameter semigroup of Longo-Witten endomorphisms $V(b) = e^{-ib\widetilde{P}}$ $(b > 0)$ with positive generator $\widetilde{P}$ gives rise to a two-dimensional Borchers triple by putting $T_2(t, x) = T(t + x)V(t - x)$.

For the net A generated by the free chiral current, semigroups of Longo-Witten endomorphisms that arise by second quantization of a unitary semigroup $V_1(b)$ on the one-particle subspace, have been classified [53], namely $V_1(b)$ turn out to be "singular symmetric inner functions" $\varphi_b(P_1)$ of the chiral one-particle momentum operator P_1. (A symmetric inner function is the boundary limit of an analytic function on the upper complex halfplane with $|\varphi(p)| = 1$ and $\varphi(-p) = \overline{\varphi(p)}$ for almost all $p \in \mathbb{R}$. These conditions precisely ensure that the unitary $V = \varphi(P_1)$ implements a Longo-Witten endomorphism at the one-particle level. Symmetric inner functions are closely related to the admissible scattering functions in Lechner's massive deformations [46], but the physical significance of this relation remains to be explored.) To get a one-parameter semigroup, $\varphi_b(p) = e^{ibf(p)}$ must be singular, i.e., it must not have zeros in $\mathbb{C}_+$.

The corresponding generator $\widetilde{P}_1 = -f(P_1)$ is positive iff $f(p) = -m^2p^{-1}$ for some $m^2 > 0$. The resulting two-dimensional QFT with chiral translations P_1 and $\widetilde{P}_1$

is just the free massive scalar field (defined on the Hilbert space of the free bosonic current), because $\underline{P}_1^2 = P_1^+ P_1^- = P_1 \widetilde{P}_1 = m^2 \cdot \mathbf{1}$. This is the converse of the well-known fact that the restriction of the massive free field to a light ray is the conformal free current.

The interesting challenge is to find other one-parameter semigroups of Longo-Witten endomorphisms with positive generator that are not of this simple (second-quantized) form.

Pursuing this idea, Tanimoto [69] constructed a large class two-dimensional (massless) Borchers triples from a chiral Borchers triple ($M = A(\mathbb{R}_+), \Omega, T = U|_{\mathbb{R}}$). These constructions proceed by deformations of the two-dimensional tensor product theory $A \otimes A$ whose Borchers triple is $(A(W_0) = M \otimes M', \Omega \otimes \Omega, T_2)$ with $T_2(\underline{a}) = T(a^+) \otimes T(a^-)$. The deformations act by conjugations on the subalgebras $M \otimes \mathbf{1}$ and $\mathbf{1} \otimes M'$ with different unitary operators:

$$M_V = V(M \otimes \mathbf{1})V^* \vee V^*(\mathbf{1} \otimes M')V.$$

With suitable conditions on the unitary operator V, the triple $(M_V, \Omega \otimes \Omega, T_2)$ is a deformed two-dimensional Borchers triple, whose energy-momentum spectrum is unchanged. With an appropriate adaptation of scattering theory to the massless situation, the nontrivial scattering matrix of this deformed QFT coincides with the square V^2 of the deformation unitary.

Depending on the choice of V as a function of the chiral momentum operators, one obtains models that are equivalent, respectively [47, 69], to a massless version of the integrable deformations by a scattering function as in [46], or to the deformations by warped convolutions [19]. Yet different choices of V of the form $e^{i\kappa Q \otimes Q}$, where Q is the generator of an inner symmetry of the chiral theory and κ a real deformation parameter, give new classes of deformed models [5, 69].

Starting instead from a *two-dimensional* (massive) Borchers triple ($M = A(W_0)$, $\Omega, T = U|_{\mathbb{R}^2}$), the tensor product $(M \otimes M, \Omega \otimes \Omega, T^{(2)})$ with $T^{(2)}(\underline{a}) = T(\underline{a}) \otimes T(\underline{a})$ is the Borchers triple of two uncoupled identical QFT models. A deformation interaction is introduced by deforming the wedge algebra:

$$M_V^{(2)} := V(M \otimes \mathbf{1})V^* \vee V^*(\mathbf{1} \otimes M)V,$$

where V is again of the form $e^{i\kappa Q \otimes Q}$ with a suitable self-adjoint generator Q. Depending on this choice, the Reeh-Schlieder property of the local algebras can be established [70].

8.6.4 Holographic Models

Let B be a chiral QFT net on the real line. For any pair of intervals $K \subset L$ with non-touching end points, let the intervals I and J be the two connected components of $L\backslash\overline{K} = I \cup J$ such that $I > J$ (elementwise). Let

$$O = I \times J := \{(t, x) \in \mathbb{R}^{1,1} : t + x \in I, t - x \in J\}. \tag{8.27}$$

Then O is a doublecone contained in the Minkowski halfspace $\mathbb{R}^{1,1}_+ = \{(t, x) \in \mathbb{R}^{1,1} : x > 0\}$.

Defining

$$B_+(O) := B(K)' \cap B(L), \tag{8.28}$$

one obtains a local net of local algebras on the halfspace, covariant under the diagonal of the direct product of two diffeomorphism groups, acting on $t + x \in \mathbb{R}$ and $t - x \in \mathbb{R}$, respectively.

This is the prototype of a "holographic" construction, since every local operator of B (a one-dimensional net on $\mathbb{R}$) is a local observable of B_+ (a two-dimensional net on $\mathbb{R}^{1,1}_+$), but the localization assigned to it is very different. In the conformal case, this is precisely the algebraic AdS-CFT correspondence [63], where $\mathbb{R}^{1,1}_+$ appears as a chart of the two-dimensional Anti-deSitter spacetime.

In order to ensure locality of B_+, it is actually not necessary that B is local: it is sufficient that B is relatively local w.r.t. a local subnet A on $\mathbb{R}$. In this case, $B_+(O)$ contains at least the subalgebra $A_+(O) := A(I) \vee A(J)$, but already for $B = A$, $B_+(O)$ is strictly larger than $A_+(O)$ whenever A possesses nontrivial DHR sectors. E.g., if A is a Virasoro net, then the halfspace net with local algebras $A_+(O) = A(I) \vee A(J)$ has the obvious physical interpretation as the algebra generated by a two-dimensional stress-energy tensor localized in O, whose chiral components $T_+ = T_-$ are identified by the boundary condition $T^{01}(t, x)|_{x=0} = 0$. This condition is just the conservation of energy at the boundary $x = 0$.

More generally speaking, the holographic halfspace models Eq. (8.28) are extensions of chiral halfspace CFTs with local algebras $A_+(O)$, which arise by means of a boundary condition on two-dimensional chiral fields; and every such extension is of the form Eq. (8.28) (or intermediate between $A_+(O)$ and $B_+(O)$) [51].

There also exist algebraic prescriptions to "remove the boundary" [52] and to "add a boundary" [22], which allow to pass between extensions $A_+ \subset B_+$ on the halfspace and extensions $A_2 \subset B_2$ of CFTs on two-dimensional Minkowski spacetime, where $A_2(O) = A(I) \otimes A(J)$ is the local algebra of a pair of independent (although isomorphic) chiral algebras. Under the local isomorphism (in completely rational models) $A(I) \vee A(J) \sim A(I) \otimes A(J)$, these pairs of extensions are locally (but of course not globally) isomorphic.

As discussed in Sect. 8.6.2, we can think of $A \subset B$ as a relatively local chiral extension, described by a Q-system of A. Via the holographic construction, this produces a local extension $A_+ \subset B_+$ on the halfspace, and by "removing the boundary", one arrives at a local two-dimensional extension $A_2 \subset B_2$. The Q-system for $A_2 \subset B_2$ as a "functional" of the underlying chiral Q-system for $A \subset B$ turns out to be precisely the α-induction construction [62] which was discovered without knowing the steps just described. More recently [6], it was also identified with the "full centre" of the chiral Q-system, as defined in [31, 44] in the abstract tensor category setting, providing yet a different line of construction for the same extension $A_2 \subset B_2$.

8.6.5 Phase Boundaries

In contrast to the "hard" boundaries as encountered in the previous subsection (physics is defined only in a halfspace), phase boundaries may separate "different physics" on both sides of the boundary. In two-dimensional conformal QFT, imposing the conservation of energy *and* momentum at the boundary implies that the two local QFTs on both sides share the same $2D$ stress-energy tensor, and further boundary conditions may imply more common chiral observables.

The issue is thus to have two possibly different local quantum field theories B^L and B^R, each defined in halfspaces $\mathbb{R}^{1,1}_L, \mathbb{R}^{1,1}_R$, to be represented on the same Hilbert space such that two requirements are respected: B^L and B^R share a common subtheory A, defined on full Minkowski spacetime $\mathbb{R}^{1,1}$, and all observables satisfy Einstein causality, i.e., any two observables at spacelike distance commute.

Because the stress-energy tensor (contained in A) generates the full spacetime covariance, it can be used to extend both halfspace nets to the full Minkowski spacetime (with their local operators "on the wrong side of the boundary" not being considered as observables). Thus, one has two full local QFTs with a common subtheory defined on the same Hilbert space, where Einstein causality for the observables is equivalent to B^L being "left local" w.r.t. B^R, i.e., a pair of observables of B^L and B^R commutes if the former is localized at the spacelike left of the latter.

This algebraic situation has been analysed in [7], see also [8]. It is found that, if the local extensions $A \subset B^L$ and $A \subset B^R$ are given by their Q-systems, there is a universal extension $A \subset C$ given by the "braided product" of the Q-systems. This extensions contains both B^L and B^R as intermediate extensions, and B^L is left local w.r.t. B^R. It is universal in the sense that every irreducible joint representation of B^L left local w.r.t. B^R is a quotient of C. Thus, to classify such representations, one has to compute the centre of the universal extension C.

As a linear space, the centre of the algebra C is generated by neutral products $\Psi^{L*}_\rho \Psi^R_\rho$ of charged fields from B^L and B^R (where $\rho = \rho_1 \otimes \rho_2$ is a DHR sector of $A \otimes A$ common to both Q-systems). It is more ambitious to compute the centre as an algebra, in order to determine its minimal projections. This can be achieved [7, 8] if the underlying chiral CFT is completely rational and both extensions are given as full centres (α-induction construction from chiral Q-systems, see Sect. 8.6.2): In this situation, the minimal central projections are in 1:1 correspondence with the irreducible bimodules between the underlying chiral Q-systems. Each minimal projection assigns numerical values to the operators $\Psi^{L*}_\rho \Psi^R_\rho$, and thereby specifies the boundary conditions valid among the charged fields. In some cases, but not always, they become linear relations of the form $\Psi^L_\rho = \omega \Psi^R_\rho$ with phase factors ω.

As an example, we give the classification for the two-dimensional Ising model, which is originally defined as the continuum limit of a lattice model of two-dimensional Statistical Mechanics at the critical point, but can (via an Osterwalder-Schrader "Wick rotation") be regarded as a relativistic quantum field theory. Its chiral net is given by the Virasoro net A with central charge $c = \frac{1}{2}$. The Ising model is then the unique maximal local two-dimensional extension $B \supset A \otimes A$, which has two charged fields $\Psi_{\sigma\otimes\sigma}$ (the "order parameter") and $\Psi_{\tau\otimes\tau}$ (the product of two chiral Fermi fields).

One finds three bimodules, hence three boundary conditions, expressed in terms of relations between the charged fields:

$$\begin{aligned} \text{(i)}\ \ \Psi^L_{\tau\otimes\tau} &= \Psi^R_{\tau\otimes\tau}, & \Psi^L_{\sigma\otimes\sigma} &= \Psi^R_{\sigma\otimes\sigma}; \\ \text{(ii)}\ \Psi^L_{\tau\otimes\tau} &= \Psi^R_{\tau\otimes\tau}, & \Psi^L_{\sigma\otimes\sigma} &= -\Psi^R_{\sigma\otimes\sigma}; \\ \text{(iii)}\ \Psi^L_{\tau\otimes\tau} &= -\Psi^R_{\tau\otimes\tau}, & \Psi^{L*}_{\sigma\otimes\sigma}\Psi^R_{\sigma\otimes\sigma} &= 0. \end{aligned}$$

The first case is the trivial boundary, and the second is the "fermionic" boundary where the field $\Psi_{\sigma\otimes\sigma}$ changes sign. The third is the "dual" boundary, which allows the coexistence of the order and disorder parameter fields σ and μ (in the original Statistical Mechanics terminology, see [30, 66]) on either side: these are the two isomorphic but independent fields $\Psi^R_{\sigma\otimes\sigma}$ and $\Psi^L_{\sigma\otimes\sigma}$.

8.7 Final Remarks

8.7.1 What is Special About CFT in Two Dimensions?

Quantum field theory in two dimensions offers a wealth of algebraic methods for the construction of models, especially conformal models. Why are these methods so efficient in two dimensions, and can one put them into perspective with QFT in four dimensions?

1. The prominent reason is the kinematical simplicity of CFT in 2D, especially the existence of chiral fields related to conserved tensors fields. Chiral fields live in a "one-dimensional spacetime", namely the light ray, which shares crucial features of space *and* time: local commutativity and spectral positivity of the generator of translations.

As a consequence, chiral commutators are ultralocal (δ functions), supported only in coinciding points rather than in or on a lightcone). Based on this feature, the Lüscher-Mack theorem provides an explicit form of the possible commutators of the stress-energy tensor, with the central charge as the only free parameter. A similar parametrization of the commutator in higher dimensions, or without conformal symmetry, is not known.

Moreover, the algebra of the stress-energy tensor field (and also of chiral fields of scaling dimension 1) is that of an infinite-dimensional Lie algebra, permitting the application of highest-weight representation methods for the efficient study of its positive-energy representation theory. In contrast, Lie fields in 4D (whose commutators are linear in the field) don't exist [3]. (The argument, based on geometric properties of lightcones and the spectrum condition, was worked out only for scalar fields, but is presumably true more generally.)

2. Another consequence of the ultralocal commutation relations of chiral free Fermi fields is the feature that chiral gauge transformations $\psi(x) \mapsto e^{i\alpha(x)}\psi(x)$ are *automorphisms* of the free Fermi (CAR) algebra. Hence, current fields are their infinitesimal generators, and can be constructed algebraically by exploiting this property.

In contrast, in $D \geq 2$, the free Fermi algebra is not gauge invariant, and gauge invariance requires the coupling to a gauge field. Currents are generators of gauge transformations at a fixed time only. Notice, however, that in two dimensions the massless free Dirac field ψ is nothing but a pair of two chiral Fermi fields $\psi_\pm(t \pm x) = P_\pm\psi(t,x)$, where $P_\pm = \frac{1}{2}(1 \pm \gamma_0\gamma_1)$. While local gauge transformations of the general form $\psi(t,x) \mapsto e^{i(\alpha(t,x)+\beta(t,x)\gamma_0\gamma_1}\psi(t,x)$ do not preserve the equation of motion, the chiral gauge transformations are of the more special form $\psi(t,x) \mapsto e^{i(\alpha_+(x^+)P^+ + \alpha_-(x^-)P^-}\psi(t,x)$. The latter commute with $\gamma^0\gamma^\mu\partial_\mu = P_+\partial_- + P_-\partial_+$, hence preserve the free equation of motion and the commutation relations at all times. In fact, they may be regarded as gauge transformations of the Cauchy data. Thus, the chiral gauge transformations are the subgroup of the gauge group which is a symmetry without the interaction with a gauge field.

3. For general conformal fields in two dimensions, the chiral factorization manifests itself (in rational theories) in the form of the conformal block decomposition of their correlation functions [4]. The analytic behaviour of conformal blocks under field exchange can be formulated algebraically as an "exchange algebra" of chiral components, which are in turn most naturally understood in terms of charged intertwiners among the chiral sectors subject to braid group statistics [27, II].

It is worth a remark that also in four dimensions, conformal partial waves—which are akin to but not exactly the same as conformal blocks—exhibit a factorization into "chiral" factors [24]. This feature has been exploited for the study of Wightman positivity (positivity of the Hilbert space inner product defined by $2n$-point correlation functions) [58, 59]; but the algebraic counterpart of an underlying exchange algebra, as suggested in [65], has not been identified.

4. Many classification results obtained in two-dimensional conformal QFT have been obtained thanks to rationality (finitely many positive-energy representations), or related properties (strong additivity, split property, finite μ-index) in the AQFT framework. These results are then owing to the ensuing rigidity of the DHR category with finitely many irreducible sectors.

QFT models with finitely many sectors exist also in four dimensions—e.g., whenever the global gauge group as in Sect. 8.6.1 is a finite group. The corresponding classification results, also in the case of compact gauge groups, are all in terms of groups and subgroups, and do not exhibit a comparably rich structure as in the low-dimensional case with sectors with braid group statistics.

5. Recall that every local QFT can be encoded in a Borchers triple (M, Ω, U), see Sect. 8.6.3. The defining properties of Borchers triples include the proper adjoint

action of the representation U of the Poincaré group on the von Neumann algebra M. In four dimensions, these are quite difficult to satisfy, and compelling ideas how to construct such triples (if one does not want to *start* from a QFT) are lacking. In contrast, in two dimensions, only a positive-energy representation of the translation group is required, and the representation of the Poincaré group can be constructed from the former with Modular Theory (Borchers' Theorem), and its required properties are automatic. Therefore, Borchers triples are much easier to obtain in two dimensions, and their data can be subjected to algebraic deformations while preserving their defining properties. This appears to be a promising non-perturbative approach to obtain new QFT models by deformation of given ones. This idea does not require conformal symmetry.

8.7.2 *What can We Learn for QFT in Four Dimensions?*

Algebraic QFT is a powerful approach that conceptually clarifies many features known to be true in QFT models, or expected to be true in QFT in general. Its value is not least that it allows to sharply exhibit the (in other approaches often tacit) assumptions that are responsible for these features. At the same time, it allows to investigate the consequences when some of these assumptions are not fulfilled—be it systematically due to the structure of the two-dimensional spacetime and its symmetry group, or model-dependently due to specific properties of the dynamics.

The "bifurcation" between four dimensions and two dimensions has particularly strong consequences in the theory of superselection sectors, where the braided tensor category is completely degenerate in four dimensions, and maximally non-degenerate (modular) in large classes of two-dimensional conformal QFT models. Yet, it is rewarding to view either extreme in the light of the other, or in the context of the general structure, as this opens the mind to the many options QFT has in store that might be missed by model studies.

Once the underlying abstract structure has been identified (and separated from the dynamical details of specific models), the road is open to classifications. Many classifications have been obtained, mostly of representation theoretic nature.

Let us return to the kinematic simplicity of QFT in two dimensions, with the Poincaré group being a subgroup of the product of two translation-dilation groups. Representations of the translation-dilation group with positive generator of the translations can be constructed by Modular Theory [68]. This feature is exploited in new algebraic deformation approaches. The passage to four dimensions is presently not yet very satisfactory; e.g., the "warped convolutions" deformations [19] break parts of the Lorentz symmetry. In contrast, the modular approach indicated in [38] seems not very practical, but it points out a direction: The problem is to control the relations between the many translation-dilation subgroups that generate the Poincaré group. This is reminiscent of the classification of semisimple Lie algebras by controlling the relations between the many $su(2)$ subalgebras that generate them. Gaining experience with two-dimensional models, one may expect progress also in four dimensions.

Acknowledgments I thank Yoh Tanimoto and Jakob Yngvason for a critical reading of the manuscript. Supported by the German Research Foundation (Deutsche Forschungsgemeinschaft (DFG)) through the Institutional Strategy of the University of Göttingen. The hospitality and support of the Erwin Schrödinger International Institute for Mathematical Physics, Vienna, is gratefully acknowledged.

References

1. Alazzawi, S.: Deformation of fermionic quantum field theories and integrable models. Lett. Math. Phys. **103**, 37–58 (2012)
2. Alazzawi, S.: Deformation of quantum field theories and the construction of interacting models. Ph.D. thesis, University of Vienna (2015)
3. Baumann, K.: There are no scalar Lie fields in three or more dimensional space-time. Commun. Math. Phys. **47**, 69–74 (1976)
4. Belavin, A.A., Polyakov, A.M., Zamolodchikov, A.B.: Infinite conformal symmetry in two-dimensional quantum field theory. Nucl. Phys. B **241**, 333–380 (1984)
5. Bischoff, M., Tanimoto, Y.: Construction of wedge-local nets of observables through Longo-Witten endomorphisms. II. Commun. Math. Phys. **317**, 667–695 (2013)
6. Bischoff, M., Kawahigashi, Y., Longo, R.: Characterization of 2D rational local conformal nets and its boundary conditions: the maximal case. arXiv:1410.8848
7. Bischoff, M., Kawahigashi, Y., Longo, R., Rehren, K.-H.: Phase boundaries in algebraic conformal QFT. arXiv:1405.7863
8. Bischoff, M., Kawahigashi, Y., Longo, R., Rehren, K.-H.: Tensor categories and endomorphisms of von Neumann algebras, with applications to quantum field theory. SpringerBriefs in Mathematical Physics, vol. 3 (2015). arXiv:1407.4793v3
9. Böckenhauer, J., Evans, D.: Modular invariants, graphs and α-induction for nets of subfactors. II. Commun. Math. Phys. **200**, 57–103 (1999)
10. Böckenhauer, J., Evans, D., Kawahigashi, Y.: On α-induction, chiral projectors and modular invariants for subfactors. Commun. Math. Phys. **208**, 429–487 (1999)
11. Böckenhauer, J., Evans, D., Kawahigashi, Y.: Chiral structure of modular invariants for subfactors. Commun. Math. Phys. **210**, 733–784 (2000)
12. Borchers, H.-J.: The CPT-theorem in two-dimensional theories of local observables. Commun. Math. Phys. **143**, 315–332 (1992)
13. Brunetti, R., Guido, D., Longo, R.: Modular structure and duality in conformal quantum field theory. Commun. Math. Phys. **156**, 201–219 (1993)
14. Brunetti, R., Guido, D., Longo, R.: Modular localization and Wigner particles. Rev. Math. Phys. **14**, 759–785 (2002)
15. Buchholz, D., Fredenhagen, K.: Locality and the structure of particle states. Commun. Math. Phys. **84**, 1–54 (1982)
16. Buchholz, D., Haag, R.: The quest for understanding in relativistic quantum physics. J. Math. Phys. **41**, 3674–3697 (2000)
17. Buchholz, D., Roberts, J.E.: New light on infrared problems: sectors, statistics, symmetries and spectrum. Commun. Math. Phys. **330**, 935–972 (2014)
18. Buchholz, D., Wichmann, E.: Causal independence and energy-level density in quantum field theory. Commun. Math. Phys. **106**, 321–344 (1986)
19. Buchholz, D., Lechner, G., Summers, S.J.: Warped convolutions, Rieffel deformations and the construction of quantum field theories. Commun. Math. Phys. **304**, 95–123 (2011)
20. Buchholz, D., Mack, G., Todorov, I.: The current algebra on the circle as a germ of local field theories. Nucl. Phys. B (Proc. Suppl.) **5B**, 20–56 (1988)
21. Cappelli, A., Itzykson, C., Zuber, J.-B.: The A-D-E classification of minimal and $A_1^{(1)}$ conformal invariant theories. Commun. Math. Phys. **113**, 1–26 (1987)

22. Carpi, S., Kawahigashi, Y., Longo, R.: How to add a boundary condition. Commun. Math. Phys. **322**, 149–166 (2013)
23. Carpi, S., Kawahigashi, Y., Longo, R., Weiner, M.: From vertex operator algebras to conformal nets and back. arXiv:1503.01260
24. Dolan, F.A., Osborn, H.: Conformal four point functions and the operator product expansion. Nucl. Phys. B **599**, 459–496 (2001)
25. Doplicher, S., Roberts, J.E.: Why there is a field algebra with a compact gauge group describing the superselection structure in particle physics. Commun. Math. Phys. **131**, 51–207 (1990)
26. Doplicher, S., Haag, R., Roberts, J.E.: Local observables and particle statistics. I, Commun. Math. Phys. **23**, 199–230 (1971), and II, Commun. Math. Phys. **35**, 49–85 (1974)
27. Fredenhagen, K., Rehren, K.-H., Schroer, B.: Superselection sectors with braid group statistics and exchange algebras. I, Commun. Math. Phys. **125**, 201–226 (1989), and II, Rev. Math. Phys. **SI1** (Special Issue) 113–157 (1992)
28. Frenkel, J.B., Kac, V.G.: Basic representations of affine Lie algebras and dual resonance models. Invent. Math. **62**, 23–66 (1980)
29. Friedan, D., Qiu, Z., Shenker, S.: Conformal invariance, unitarity, and critical exponents in two dimensions. Phys. Rev. Lett. **52**, 1575–1578 (1984)
30. Fröhlich, J., Fuchs, J., Runkel, I., Schweigert, C.: Kramers-Wannier duality from conformal defects. Phys. Rev. Lett. **93**, 070601 (2004)
31. Fröhlich, J., Fuchs, J., Runkel, I., Schweigert, C.: Correspondences of ribbon categories. Ann. Math. **199**, 192–329 (2006)
32. Goddard, P., Kent, A., Olive, D.: Unitary representations of the Virasoro and super-Virasoro algebras. Commun. Math. Phys. **103**, 105–119 (1986)
33. Guido, D., Longo, R.: The conformal spin and statistics theorem. Commun. Math. Phys. **181**, 11–35 (1996)
34. Guido, D., Longo, R., Wiesbrock, H.-W.: Extensions of conformal nets and superselection structures. Commun. Math. Phys. **192**, 217–244 (1998)
35. Haag, R.: Local Quantum Physics. Springer, Berlin (1996)
36. Izumi, M., Kosaki, H.: On a subfactor analogue of the second cohomology. Rev. Math. Phys. **14**, 733–737 (2002)
37. Kac, V.G.: Infinite Dimensional Lie Algebras. Cambridge University Press, Cambridge (1985)
38. Kähler, R., Wiesbrock, H.-W.: Modular theory and the reconstruction of four-dimensional quantum field theories. J. Math. Phys. **42**, 74–86 (2001)
39. Kawahigashi, Y., Longo, R.: Classification of local conformal nets. Case $c < 1$. Ann. Math. **160**, 493–522 (2004)
40. Kawahigashi, Y., Longo, R.: Classification of two-dimensional local conformal nets with $c < 1$ and 2-cohomology vanishing for tensor categories. Commun. Math. Phys. **244**, 63–97 (2004)
41. Kawahigashi, Y., Longo, R.: Local conformal nets arising from framed vertex operator algebras. Adv. Math. **206**, 729–751 (2006)
42. Kawahigashi, Y., Longo, R., Müger, M.: Multi-interval subfactors and modularity of representations in conformal field theory. Commun. Math. Phys. **219**, 631–669 (2001)
43. Kirillov Jr, A., Ostrik, V.: On q-analog of McKay correspondence and ADE classification of sl(2) conformal field theories. Adv. Math. **171**, 183–227 (2002)
44. Kong, L., Runkel, I.: Morita classes of algebras in modular tensor categories. Adv. Math. **219**, 1548–1576 (2008)
45. Köhler, Ch.: On the localization properties of quantum fields with zero mass and infinite spin, Ph.D. thesis, Vienna (2015)
46. Lechner, G.: Construction of quantum field theories with factorizing S-matrices. Commun. Math. Phys. **277**, 821–860 (2008)
47. Lechner, G., Schlemmer, J., Tanimoto, Y.: On the equivalence of two deformation schemes in quantum field theory. Lett. Math. Phys. **103**, 421–437 (2013)
48. Longo, R., Morinelli, V., Rehren, K.-H.: Where infinite-spin particles are localizable. Commun. Math. Phys. arXiv:1505.01759 (to appear)

49. Longo, R.: Conformal subnets and intermediate subfactors. Commun. Math. Phys. **237**, 7–30 (2003)
50. Longo, R., Rehren, K.-H.: Nets of subfactors. Rev. Math. Phys. **7**, 567–597 (1995)
51. Longo, R., Rehren, K.-H.: Local fields in boundary CFT. Rev. Math. Phys. **16**, 909–960 (2004)
52. Longo, R., Rehren, K.-H.: How to remove the boundary in CFT—an operator algebraic procedure. Commun. Math. Phys. **285**, 1165–1182 (2009)
53. Longo, R., Witten, E.: An algebraic construction of boundary quantum field theory. Commun. Math. Phys. **303**, 213–232 (2011)
54. Longo, R., Xu, F.: Topological sectors and a dichotomy in conformal field theory. Commun. Math. Phys. **251**, 321–364 (2004)
55. Lüscher, M., Mack, G.: Global conformal invariance in quantum field theory. Commun. Math. Phys. **41**, 203–234 (1975)
56. Mack, G.: Introduction to conformal invariant quantum field theory in two and more dimensions. In: 't Hooft, G., et al. (eds.) Nonperturbative QFT, pp. 353–383. Plenum Press, New York (1988)
57. Mund, J., Schroer, B., Yngvason, J.: String-localized quantum fields and modular localization. Commun. Math. Phys. **268**, 621–672 (2006)
58. Neumann, C., Rehren, K.-H., Wallenhorst, L.: New methods in conformal partial wave analysis. In: Dobrev, V. (ed.) Lie Theory and Its Applications in Physics: IX International Workshop, Springer Proceedings in Mathematics and Statistics, vol. 36, pp. 109–126 (2013)
59. Nikolov, N.M., Rehren, K.-H., Todorov, I.T.: Partial wave expansion and Wightman positivity in conformal field theory. Nucl. Phys. B **722 [PM]**, 266–296 (2005)
60. Pressley, A., Segal, G.: Loop Groups. Oxford University Press, Oxford (1986)
61. Rehren, K.-H.: Weak C* Hopf symmetry. In: Doebner, H.-D. et al. (eds.) Quantum Groups Symposium at "Group21", Goslar 1996 Proceedings. Heron Press, Sofia, pp. 62–69. arXiv:q-alg/9611007 (1997)
62. Rehren, K.-H.: Canonical tensor product subfactors. Commun. Math. Phys. **211**, 395–406 (2000)
63. Rehren, K.-H.: Algebraic holography. Ann. H. Poinc. **1**, 607–623 (2000)
64. Schellekens, A., Warner, N.: Conformal subalgebras of Kac-Moody algebras. Phys. Rev. D **34**, 3092–3096 (1986)
65. Schroer, B.: Braided structure in 4-dimensional conformal quantum field theory. Phys. Lett. B **506**, 337–343 (2001)
66. Schroer, B., Truong, T.T.: The order/disorder quantum field operators associated with the two-dimensional Ising model in the continuum limit. Nucl. Phys. B **144**, 80–122 (1978)
67. Staszkiewicz, C.-P.: Die lokale Struktur abelscher Stromalgebren auf dem Kreis. Ph.D. thesis (in German), Freie Universität Berlin (1995)
68. Takesaki, M.: Theory of Operator Algebras II. Springer Encyclopedia of Mathematical Sciences, vol. 125 (2003)
69. Tanimoto, Y.: Construction of wedge-local nets of observables through Longo-Witten endomorphisms. Commun. Math. Phys. **314**, 443–469 (2012)
70. Tanimoto, Y.: Construction of two-dimensional quantum field theory models through Longo-Witten endomorphisms. Forum Math. Sigma **2**, e7 (2014). arXiv:1301.6090
71. Verlinde, E.: Fusion rules and modular transformations in 2D conformal field theory. Nucl. Phys. B **300**, 360–376 (1988)
72. Wassermann, A.: Kac-Moody and Virasoro Algebras. arXiv:1004.1287
73. Xu, F.: Mirror extensions of local nets. Commun. Math. Phys. **270**, 835–847 (2007)

Chapter 9
Kitaev's Quantum Double Model from a Local Quantum Physics Point of View

Pieter Naaijkens

Abstract A prominent example of a topologically ordered system is Kitaev's quantum double model $\mathcal{D}(G)$ for finite groups G (which in particular includes $G = \mathbb{Z}_2$, the toric code). We will look at these models from the point of view of local quantum physics. In particular, we will review how in the abelian case, one can do a Doplicher-Haag-Roberts analysis to study the different superselection sectors of the model. In this way one finds that the charges are in one-to-one correspondence with the representations of $\mathcal{D}(G)$, and that they are in fact *anyons*. Interchanging two of such anyons gives a non-trivial phase, not just a possible sign change. The case of non-abelian groups G is more complicated. We outline how one could use amplimorphisms, that is, morphisms $\mathfrak{A} \to M_n(\mathfrak{A})$ to study the superselection structure in that case. Finally, we give a brief overview of applications of topologically ordered systems to the field of quantum computation.

9.1 Introduction

A fundamental result in local quantum physics is that in high enough dimensional space-times (or, for well enough localized particles), charged particles are either (para-)bosons or (para-)fermions [19]. This is no longer true in lower dimensional space-times [23]. Instead of a representation of the symmetric group, representations of the braid group may be obtained from interchanging identical particles. Even though our world appears to have three spatial dimensions, many systems effectively behave like two or one dimensional systems, opening the possibility that they may be described by a low dimensional effective theory, with excitations with braided statistics.

Interest in such systems has sparked in recent years, getting attention from the theoretical and mathematical physics communities, condensed matter physicists, quantum information theorists, and mathematicians. Part of the reason for this is

P. Naaijkens (✉)
Institut für Theoretische Physik, Leibniz Universität Hannover, Appelstraße 2, 30167 Hannover, Germany
e-mail: pieter.naaijkens@itp.uni-hannover.de

R. Brunetti et al. (eds.), *Advances in Algebraic Quantum Field Theory*,
Mathematical Physics Studies, DOI 10.1007/978-3-319-21353-8_9

the discovery of topological quantum computation (first proposed independently by Kitaev [35] and Freedman [25]), a field at the intersection of quantum theory, quantum computation and quantum topology [58]. One of the attracting features is that the use of topological properties of the system can lead to much better fault-tolerance with respect to (local) perturbations. An overview and candidates for systems that can be used for topological quantum computation can be found in the review [48].

On the mathematical level the behaviour of the anyons, the quasi-particle excitations with braided statistics, can be captured by the concept of a *braided fusion category*. This category encodes all information on how two anyons can combine ("fusion"), what happens if we interchange them ("braiding"), which types can exist in the system, and so on. This structure is in fact very well known in local quantum physics. One of the highlights of algebraic quantum field theory is the study of superselection sectors initiated by Doplicher et al. [19, 20], see also this volume, Chaps. 1 and 8. This leads in a natural way to a fusion category as above. Also in the local quantum physics approach to conformal field theories, such category appear. One can show that for *rational* conformal field theories on the circle, this category is in fact *modular* [32], see also Chap. 8. A modular tensor category is a special type of fusion category that is highly non-degenerate. It is precisely such anyon models that are of interest to topological quantum computing.

Many of the models that have such anyonic excitations are what is called *topologically ordered*. Topological order is a new type of order that does not fall into the Landau theory of spontaneous symmetry breaking, and there is no local order parameter distinguishing different phases. The examples known are not relativistic theories, and most of them are not described by a conformal field theory (at least, not directly). For example, the model that we will study here, Kitaev's quantum double model, is a quantum spin model, defined on a lattice. Nevertheless, one could try to apply the ideas of Doplicher, Haag and Roberts to these systems, to study the superselection sectors and the properties of the anyons. This is indeed possible [22, 43].

This approach gives a strong mathematical founding to the study of anyons in topologically ordered systems. Not only it allows us to borrow techniques from local quantum physics, it also opens up the way to the use of, for example, operator algebraic methods (see [46] for such an application). A rigorous mathematical framework may also be of use in this fast-moving field, where no consensus on the right definitions for fundamental concepts is reached. A precise setting where operator algebraic methods can be used may be the right setting to study important questions such as about the stability of such systems: suppose that the Hamiltonian of the system is perturbed by a suitable perturbation, how much of the structure remains? One expects that the topological nature of the system will preserve the interesting properties (as long as, for example, the Hamiltonian remains gapped), but this is something that one wants to prove rigorously (however, see e.g. [11, 12] for results in this direction).

The remainder part of this contribution is outlined as follows. First, we discuss the basic idea behind topological order, and introduce the quantum double model in the setting of local quantum physics. Section 9.3 discusses the technical property of Haag duality in this context. We then come to the main part: an overview of the

DHR-type analysis of the superselection sectors in the quantum double model. There we also relate it to the theory of (modular) tensor categories. This construction only works for the *abelian* quantum double models. Therefore, in the following section (which contains new results) we outline how one could use so-called amplimorphisms to study the non-abelian case. Finally, in the last section we briefly comment on applications to quantum computing.

9.2 Topological Order

Around the late '80s, it was found that there are states of matter that do not fall into the Landau theory of spontaneous symmetry breaking. One of the first papers discussing this is [30]. Such states were called *topologically ordered*, because some of their properties depend on the topology of the manifold on which the system is defined [59]. As a particular example one can think of two dimensional systems defined on a surface. For topologically ordered systems, the ground state degeneracy typically depends on the genus of the surface.

Around the turn of the millennium, quantum information theorists started to take an increased interest in topologically ordered systems. One of the main reasons is that it was realised that the topological properties of such systems might be useful for quantum computation. This was first suggested by Kitaev [35], who introduced a simple quantum system, the *toric code*, that could act as a quantum memory, storing a pair of qubits. A qubit is the quantum analogue of a bit in a classical computer. Mathematically it is nothing but a copy of the Hilbert space $\mathbb{C}^2$, whose basis is usually denoted by $|0\rangle$ and $|1\rangle$. Although the basics of quantum computing are not difficult to explain, we refer to the standard textbook [49] for more information. In any case, the idea is to embed the two qubit Hilbert space $\mathbb{C}^2 \otimes \mathbb{C}^2$ into the physical Hilbert space $\mathcal{H}$. We can then initialize our system in one of the states. The point of a good quantum memory is then that after some time, we should be able to recover the state (at least through the gathering of measurement statistics). In practice, however, quantum systems are not completely isolated, and influences from the environment might drive the system into a different state.

This is where topologically ordered systems come into play. A nice feature of topologically ordered systems is that there is no local[1] order parameter. Hence, local observables cannot distinguish between different ground states, and on the other hand local operations cannot put the system from one ground state into another. Hence the idea is to encode information in the ground space, since this will be robust, at least against local perturbations. Of course, nature does not make it easy for us, and there are some more subtle points to being a good quantum memory. For example, because the information is stored non-locally, accessing it or acting on it also necessarily requires non-local operations, which may be difficult to implement. In addition, although local perturbations of the system do not destroy the information,

[1] For finite systems, "local" means small compared to the system size.

this requires the use of an error correcting protocol. One should be able to do this fast enough, to prevent the errors from spreading out over the system to a *non-local* error, which *does* corrupt the information in the memory. At least for a wide class of 2D systems, it turns out that these systems by themselves are not good memories, so one indeed has to do some error correction [2, 13, 33, 36].

Another interesting aspect of topologically ordered systems is that they generally have anyonic excitations. That is, excitations of such systems behave like quasi-particles, with anyonic or braided statistics: the global state of the system changes non-trivially (that is, differently from just a sign change) under the interchange of two such quasi-particles. It is this property that we will focus on here. In particular, we will outline how the Doplicher-Haag-Roberts analysis of superselection sectors in algebraic quantum field theory can be translated to the setting of topologically ordered systems, and how this can be used to recover the statistics of the anyons. To illustrate how this works we consider Kitaev's *toric code*, and more generally his quantum double model [35], which is the prototypical example of a topologically ordered system.

Since the toric code is relatively simple and nevertheless has many interesting features, one can use it as a testbed to see if ideas from algebraic quantum field theory can be applied to this class of models. This indeed turns out to work very well, and allows one to explicitly realize many of the fundamental concepts in the theory of superselection sectors. This shows that the concept can be transferred from relativistic theories to non-relativistic condensed matter systems. Although the explicit constructions we present here depend on the specific knowledge of the toric code, these features are common to a range of topologically ordered systems, and the theory can in principle be applied to them as well.

To make the connection between the toric code and the theory of superselection sectors, it is convenient to depart from the usual setting of Kitaev's model and define it on an *infinite* 2D lattice, instead of on some compact surface of genus g. This is most conveniently done in the operator algebraic framework, were one assigns algebras of local observables to finite regions of space [9, 10]. In particular, there is a clear distinction between local and global observables, just like in local quantum physics. That is, one does not have to keep track of the system size. Concretely, we consider the lattice $\mathbb{Z}^2$, and write Γ for the set of *edges* or *bonds* between them. We give each edge an orientation. For simplicity, all vertical edges are assumed to point upwards, the horizontal edges will point to the right (see Fig. 9.1). Now let G be a finite dimensional group. Then to each edge we assign a "G-spin". This means that there is a quantum system at each edge, with corresponding Hilbert space $\mathcal{H}_e := \mathbb{C}[G]$, where the right hand side is the group algebra of G, seen as a Hilbert space in the natural way.

The local operators that act on an edge e are $\mathfrak{B}(\mathcal{H}_e) \cong M_{|G|}(\mathbb{C})$. If $\Lambda \in \mathcal{P}_f(\Gamma)$, the set of finite subsets of Γ, then the local observables associated to Λ are defined as $\mathfrak{A}(\Lambda) := \bigotimes_{e\in\Lambda}$. If $\Lambda_1 \subset \Lambda_2$ there is a natural inclusion of the corresponding algebras, by tensoring with the identity operator at the sites of Λ_2 that are not in Λ_1. Hence we are in the familiar setting of local quantum physics, where we have a net $\Lambda \mapsto \mathfrak{A}(\Lambda)$ of observables associated to bounded regions (cf. Chap. 1). It is not difficult to see that this net is local, that is, $[\mathfrak{A}(\Lambda_1), \mathfrak{A}(\Lambda_2)] = \{0\}$ if $\Lambda_1 \cap \Lambda_2 = \emptyset$.

We then proceed in the same way as for relativistic systems: the local operators are defined as $\mathfrak{A}_{loc} = \bigcup_{\Lambda \in \mathcal{P}_f(\Gamma)} \mathfrak{A}(\Lambda)$. Note that $\mathfrak{A}_{loc}$ is a $*$-algebra in a natural way, and that the standard operator norm on matrix algebras induces a C^*-norm on $\mathfrak{A}_{loc}$. The quasilocal algebra $\mathfrak{A}$ is the completion of $\mathfrak{A}_{loc}$ with respect to this norm. If $\Lambda \subset \Gamma$ is an arbitrary subset (not necessarily finite), then we define

$$\mathfrak{A}(\Lambda) = \overline{\bigcup_{\Lambda_f \in \mathcal{P}_f(\Gamma \cap \Lambda_f)} \mathfrak{A}(\Lambda_f)}^{\|\cdot\|}$$

as the algebra of all observables that can be measured inside the region Λ.

Finally, note that there is a natural translation symmetry on the system, although we do not need this for our purposes, except for picking out a translationally invariant ground states. This touches upon one of the fundamental differences between relativistic quantum field theory and discrete systems. For the discrete systems there is no Lorentz group or Poincaré group, which play an important role in relativistic theories. Nevertheless, one can mimic some of these concepts using the translation symmetries mentioned, together with locality estimates for local observables evolved under the time evolution of the system [47]. For our purposes these aspects play no role.

9.2.1 The Quantum Double Model

So far the description has been completely general. To consider a specific system one has to specify the dynamics of the model. Note that because the model is infinite, the corresponding Hamiltonian generally is unbounded, and hence not an element of $\mathfrak{A}$. Fortunately, in practice there is a lot more structure, and the dynamics are local in a suitable sense. More concretely, for each $\Lambda \in \mathcal{P}_f(\Gamma)$ one can define a self-adjoint H_Λ, describing the interactions within that region. Heuristically, the dynamic evolution of an observable A is then obtained as $\alpha_t(A) := \lim_{\Lambda \to \infty} e^{itH_\Lambda} A e^{-itH_\Lambda}$, where $\Lambda \to \infty$ means that we take an increasing sequence of finite sets Λ that exhaust Γ. If the strength of the interaction decays fast enough, this expression converges and one obtains a strongly continuous one-parameter group $t \mapsto \alpha_t$ of automorphisms [10].

Once the dynamics are defined one can talk about ground states. The most convenient way to do this is in terms of (generally unbounded) $*$-derivations δ, which are obtained as the generator of an automorphism group $t \mapsto \alpha_t$. In the cases of interest to us these are simply obtained as (the closure of) $\delta : \mathfrak{A}_{loc} \to \mathfrak{A}_{loc}$, with

$$\delta(A) := i \lim_{\Lambda \to \infty} [H_\Lambda, A].$$

Because the interactions are local and of finite range in the cases of interest, this converges, and $\alpha_t(A) = e^{t\delta}(A)$. A state ω_0 on $\mathfrak{A}$ is then called a *ground state* if $-i\omega_0(A^*\delta(A)) \geq 0$ for all $A \in D(\delta)$. This (at first sight) perhaps strange looking condition can be interpreted as a positivity of the energy condition. Such a state is automatically invariant, $\omega_0 \circ \alpha_t = \omega_0$, so that in the GNS representation the dynamics are implemented by a strongly continuous group of unitaries $t \mapsto U(t)$. By Stone's theorem one obtains a Hamiltonian H, which can be shown to be positive using the condition mentioned above [10].

There is some extra notation that has to be introduced to define the dynamics in Kitaev's quantum double model. A combination $s = (v, f)$ of a vertex v and a choice of an adjacent face f is called a *site*. To each site we associate the following operators. Let $g \in G$. Note that for a vertex v, there are four edges that start or end in v. A basis for the Hilbert space of this four sites is given by $|g_1\rangle \otimes \cdots \otimes |g_4\rangle$, with $g_i \in G$. The operator A_s^g acts on this basis vector by multiplying g_i from the *left* with g if the corresponding edge points away from the vertex v, and g_i gets sent to $g_i g^{-1}$ if the edge points inwards. If $h \in G$, a projection B_s^h is defined as follows. First, list the edges around the face f in counter-clockwise order, starting in v. On the basis labelled by group elements of the Hilbert space corresponding to these four edges, B_s^h acts as the identity if $\sigma(g_1) \cdots \sigma(g_4) = h$, and zero otherwise. Here $\sigma(g) = g$ if the corresponding edge is in the same direction as the counter-clockwise labelling, and $\sigma(g) = g^{-1}$ otherwise. Pictorially the operators can be depicted as follows:

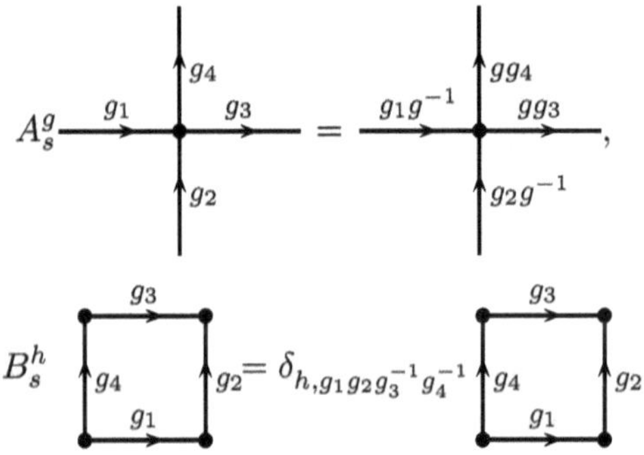

Sometimes one says that B_s^h projects on the states with *flux* h through the face f. Indeed, the structure is very similar to that of lattice gauge theory [51], and the quantum double model can in fact also be interpreted in such terms [35].

The operators A_s and B_s are also called *star* and *plaquette* operators respectively, and satisfy the following algebraic relations, which can easily be verified:

$$A_s^g A_s^h = A_s^{gh}, \quad B_s^g B_s^h = \delta_{g,h} B_s^h, \quad A_s^g B_s^h = B_s^{ghg^{-1}} A_s^g, \quad (A_s^g)^* = A_s^{g^{-1}}. \quad (9.1)$$

Star and plaquette operators acting on different sites s and s' always commute. At this point the name "quantum double" can be explained: the algebraic relations above are exactly those of the quantum double $\mathcal{D}(G)$ of the Hopf algebra $\mathbb{C}[G]$ (see e.g. [31] for an introduction). That is, at each site there is an action of the Hopf algebra which acts via the star and plaquette operators. The representation theory of $\mathcal{D}(G)$ plays an important role in the analysis of the superselection sectors, as we will see below.

We are now in a position to define the dynamics of the system. First, write $A_s = \frac{1}{|G|}\sum_{g\in G} A_s^g$ for the average of the star operators, and $B_s := B_s^e$. Then $[A_s, B_s] = 0$. The local Hamiltonians are then defined as

$$H_\Lambda = -\sum_{s\in\Lambda} A_s - \sum_{s\in\Lambda} B_s, \tag{9.2}$$

where in the first sum the summation is over all sites $s = (v, f)$, such that the edges starting or ending at v are contained in Λ. Similarly, the second summation is over all sites such that the edges of the face f are all contained in Λ. The case $G = \mathbb{Z}_2$ is called the *toric code*. These local Hamiltonians generate a one-parameter group of automorphisms α_t as described above. It turns out that there is a unique *translationally invariant* ground state [1, 44].[2]

Proposition 9.2.1 *There is a unique translationally invariant ground state ω_0. This state is pure and is the unique state on $\mathfrak{A}$ which satisfies $\omega_0(A_s) = \omega_0(B_s) = 1$ for all star and plaquette operators A_s and B_s.*

This ground state will be the starting point from the analysis: the different superselection sectors will be realized, roughly speaking, by creating single excitations of this ground state. The GNS representation corresponding to the ground state representation from the proposition will be denoted by $(\pi_0, \Omega, \mathcal{H}_0)$. Since $\mathfrak{A}$ is a UHF (and hence simple) algebra, π_0 is automatically injective, and we will often identify $\pi_0(A)$ with A.

To understand how this works it is necessary to understand what we mean by an excitation, and how these can be obtained. From the proposition above it follows that in the ground state representation, $A_s\Omega = B_s\Omega = \Omega$. We can interpret this as constraints, and violating some of these constraints carries an energy penalty, according to the local Hamiltonians (9.2). We will interpret the violation of such a constraint as a (quasi-)particle sitting at the site s (if the state is an eigenstate of H_Λ).[3] So the excitations live at sites of the lattice.

The excitations can be obtained by acting with so-called *ribbon operators* on the ground state. We only list their main properties here. A more complete treatment and proofs can be found in [5]. A *ribbon* is, roughly speaking, a continuous path of triangles (see Fig. 9.1 for an example). To each such ribbon ξ and pair g, h of group elements one can assign a ribbon operator $F_\xi^{h,g}$. These operators act only on the edges that are part of the ribbon (or cross any of the triangles). They satisfy the following algebraic relations:

---

[2]Bruno Nachtergaele pointed out that the remark in [43] is in fact false, and there are additional (non-translationally invariant) ground states. Indeed, the charged states constructed in [43] have dynamics implemented by a positive Hamiltonian.

[3]Note that we disregard any momentum variables. The pairs of excitations form bound states, so the excitations mentioned here are not quite the single-particle excitations one encounters in scattering theory.

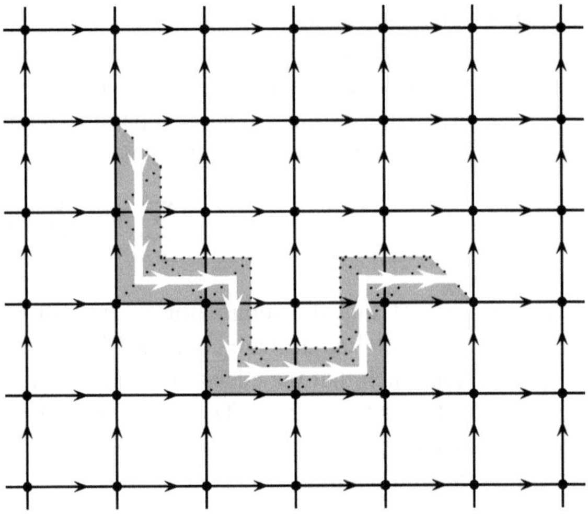

Fig. 9.1 The lattice describing the system, together with a ribbon between two sites. Note that the ribbon has an orientation, pointing from start to end

$$F_\xi^{h_1,g_1} F_\xi^{h_2,g_2} = \delta_{g_1,g_2} F_\xi^{h_1 h_2,g_1}, \quad (F_\xi^{h,g})^* = F_\xi^{h^{-1},g}, \quad \sum_{g \in G} F_\xi^{e,g} = I.$$

Note that the ribbons carry an orientation. This allows us to talk about the *starting* and *ending* sites of the ribbon. Let us denote them by s_1 and s_2 for the moment. An important property of the ribbon operators is that they commute with any star or plaquette operator, except for those at the ending sites of the ribbon. There we have the following commutation relations:

$$A_{s_1}^k F_\xi^{h,g} = F_\xi^{khk^{-1},kg} A_{s_1}^k, \quad B_{s_1}^k F_\xi^{h,g} = F_\xi^{h,g} B_{s_1}^{kh}. \tag{9.3}$$

Similarly, for the ending site

$$A_{s_2}^k F_\xi^{h,g} = F_\xi^{h,gk^{-1}} A_{s_2}^k, \quad B_{s_1}^k F_\xi^{h,g} = F_\xi^{h,g} B_{s_2}^{g^{-1}h^{-1}gk}. \tag{9.4}$$

From these commutation relations it follows that the vector state $F_\xi^{h,g}\Omega$ violates some of the constraints the Hamiltonian gives for the ground state, namely precisely those at the start- and ending sites. Hence this vector state can be thought of as an excited state of the system. These excited states satisfy the following important property, whose proof we will omit (see e.g. [5, 22] for the proof).

Lemma 9.2.2 *Let ξ_1 and ξ_2 be two ribbons with the same starting and ending sites. Then $F_{\xi_1}^{h,g}\Omega = F_{\xi_2}^{h,g}\Omega$.*

In other words, the state does not depend on the ribbon itself, only on the endpoints. This already shows the topological nature of the system.

Now suppose that ξ is a non-trivial ribbon. It can then be written as the concatenation of ribbons ξ_1 and ξ_2. The ribbon operator ξ is related to the smaller ribbons by means of the following recursion relation (which indeed can also be used to recursively define the ribbon operators):

$$F_\xi^{h,g} = \sum_{k \in G} F_{\xi_1}^{h,k} F_{\xi_2}^{\overline{k}hk,\overline{k}g}. \tag{9.5}$$

This feature will be important later.

The basis $F_\xi^{h,g}$ of ribbon operators is not always the most convenient one. Recall that for each site s there is an action of $\mathcal{D}(G)$ on the Hilbert space. One can now consider the vector space $V = \operatorname{span}\{F_\xi^{h,g}\Omega : h, g \in G\}$ for a fixed ribbon ξ. By Eq. (9.3) this space has a natural interpretation as a left $\mathcal{D}(G)$-module, because $A_s^g \Omega = \Omega$ and $B_s^h \Omega = \delta_{h,e}\Omega$. Similarly, Eq. (9.4) also give it the structure of a $\mathcal{D}(G)$-module, induced by the action of the star and plaquette operators at the ending site.[4] This suggests that it may be useful to decompose this space into irreducible subspaces, and find a new basis of the ribbon operators accordingly. This indeed turns out to be a good idea.

The representation theory of $\mathcal{D}(G)$ is well understood: constructions of all irreducible representations can be found in [3, 17]. They are in one-to-one correspondence with pairs (C, ρ), where C is a conjugacy class of G and ρ an irreducible representation of $Z_G(c)$, where $Z_G(c)$ is the centralizer in G of a representative c of the conjugacy class. It turns out to be convenient to label the elements of the conjugacy class in a particular way (compare [5]). First of all, let C be a conjugacy class of G. Choose a representative $r \in C$, and let $Z_G(r)$ be the centraliser of r in G. We label the elements of C by $c_1, \ldots, c_n$, where $n = |C|$. Then there are q_i such that $c_i = q_i r \overline{q}_i$, where we used the notation $\overline{g}$ for the inverse of g, to improve readability. The set $\{q_i\}$ is denoted by Q_C. Note that each $g \in G$ can be uniquely written as $g = q_i n$ for some $q_i \in Q_C$ and $n \in Z_G(r)$.

With this notation we can choose a new basis of the ribbon operators associated to a ribbon ξ. Suppose that ρ is a unitary representation of $Z_G(r)$. We regard each $\rho(g)$ as a unitary matrix, and write $\rho(g)_{jj'}$ for the corresponding matrix elements in the standard basis. Let $i, i' = 1, \ldots, n$ and $j, j' = 1, \ldots, \dim(\rho)$. We then define

$$F_\xi^{C\rho;i,i',j,j'} = \sum_{g \in Z_G(r)} \overline{\rho}_{jj'}(g) F_\xi^{\overline{c}_i, q_i g \overline{q}_{i'}}.$$

As C runs over all conjugacy classes of G, and ρ runs over the corresponding irreducible representations of the centralisers, these operators form a basis of the space spanned by $F_\xi^{h,g}$. We refer to [5] for a proof. In essence, the point is that the space of operators is decomposed into subspaces transforming according to some irreducible representation of $\mathcal{D}(G)$. For convenience we sometimes drop the notation for C and ρ when these are implied by the context, and write $I = (i, j)$, $J = (i', j')$ for the pairs of indices: F_ξ^{IJ}.

[4] This is in fact the contragradient module of the representation at the starting site.

If the group G is abelian the notation can be simplified, since each conjugacy class consists of exactly one element, and clearly the centralizer $Z_G(r)$ is always equal to G. Hence we can label the basis by a pair (ω, c), where ω is a character of G and $c \in G$. It can be checked that in that case the corresponding ribbon operators $F_\xi^{\omega,c}$ are unitaries and that $F_\xi^{\omega,c} F_\xi^{\chi,d} = F_\xi^{\omega\chi,cd}$. There are many more useful relations, describing for example the commutation relations between two ribbon operators acting on crossing ribbons, these can be found in [5].

9.3 Haag Duality

Just as in the Doplicher-Haag-Roberts theory of superselection sectors, Haag duality plays a very useful role. Let us first recall what Haag duality actually is: it is a commutation property of von Neumann algebras generated by algebras of observables in the ground state representation. More precisely, let Λ be a cone-like region.[5] Then $\pi(\mathfrak{A}(\Lambda))''$ is the von Neumann algebra generated by all (quasi-) local observables localised in Λ. Note that, by locality, these observables commute with any local observable localised *outside* the cone, that is, in Λ^c. In other words, $\pi_0(\mathfrak{A}(\Lambda))'' \subset \pi_0(\mathfrak{A}(\Lambda^c)'$, where Λ^c is the complement of Λ in Γ. Haag duality says that these sets are actually equal. It means that we cannot add operators to the cone algebra $\pi_0(\mathfrak{A}(\Lambda))''$ without violating locality. This property is fulfilled for Kitaev's model for abelian groups G:

Theorem 9.3.1 *Let G be a finite abelian group and let Λ be a cone. Write π_0 for the translational invariant ground state representation of the quantum double model for G. Then Haag duality holds:*

$$\pi_0(\mathfrak{A}(\Lambda))'' = \pi_0(\mathfrak{A}(\Lambda^c))',$$

where the prime denotes the commutant in the set of all bounded operators.

This theorem was first proven in [45] for the case of $G = \mathbb{Z}_2$ (the toric code) and later extended to all finite abelian groups in [22].

The proof of the theorem depends on a good knowledge of the pairs of excitations and the operators that generate them. This makes it possible to find a convenient description of the Hilbert space of the system, and the Hilbert space $\mathcal{H}_\Lambda$ describing pairs of excitations localised in a fixed cone Λ. With this description it is possible to demonstrate that the self-adjoint parts of certain algebras (the restriction to $\mathcal{H}_\Lambda$ of the algebras of observables localised in Λ, and the algebra of observables localised in Λ^c), when acting on the cyclic GNS vector generate a dense subset of $\mathcal{H}_\Lambda$. It is

[5] The precise shape is not so important, but see [22] for a precise definition. Heuristically, one can take a point in the lattice, draw two semi-infinite lines from this point, and consider all edges that either intersect these lines, or lie in the convex set bounded by the lines.

known that this is related to commutation properties of algebras [54], and this allows one to conclude Haag duality.

Much of the proof boils down to a good understanding of the representation theory of $\mathcal{D}(G)$, the quantum double of G, since the operators that create pairs of excitations correspond to irreducible representations of the Hopf algebra. Their behaviour under the interchange of two such operators (or on the representation theory side, the braiding of the category of representations) also plays a role. These properties are well-known and studied for a wider class of quantum doubles, particularly those associated to Hopf-$*$ algebras that are quasi-triangular (which means a braiding can be defined). As Kitaev already remarked in his paper [35], the quantum double model can also be defined for such Hopf algebras. The following conjecture therefore seems very natural:

Conjecture 9.3.2 Consider a quasi-triangular Hopf-$*$ algebra H. Then the corresponding Kitaev model on the plane satisfies Haag duality for cones.

The main difficulty in proving this is that the combinatorics get much more involved. In particular, in the case of abelian algebras all irreducible representations are one dimensional. For non-abelian algebras one has to consider multiplets of operators that create pairs of excitations, transforming according to some irreducible representation of the Hopf algebra $\mathcal{D}(H)$. We will see some of the consequences later on, when we discuss non-abelian theories.

It should be noted that many of the constructions we present below do not depend on Haag duality. The only place where it is used is in going from a representation that satisfies the selection criterion to a localised endomorphism. Representatives of these endomorphisms (and intertwiners between them that change the localisation region) can be constructed without an appeal to Haag duality. However, it is then not possible to conclude that each representative of the equivalence class is indeed given by a localised endomorphism.

9.4 Superselection Theory for Abelian Models

It is well-known that the superposition principle of quantum mechanics does not hold unrestrictedly. A familiar example is that of bosons and fermions [60]. Consider the vector $\psi = \frac{1}{\sqrt{2}}(\psi_f + \psi_b)$, where ψ_f is a single-particle fermionic state, while ψ_f is bosonic. Under a rotation of the system by 360°, the fermionic state will acquire a minus sign, while the bosonic part is unchanged. Physically, however, the states are indistinguishable, and will lead to the same expectation values. In general, one can argue that the physical Hilbert space can be decomposed into different sectors corresponding to the different types of charges in the system. This is called a *superselection rule*.

On a mathematical level, superselection sectors arise because there are inequivalent representations of the observable algebra $\mathfrak{A}$. A C^*-algebra has many inequivalent representations in general, hence one should somehow select the physically relevant representations. There are several conditions one might impose. For example, in relativistic theories a natural condition is to look at representations that are covariant with respect translations, such that the spectrum of the generators (that is, the momentum) is contained in the forward light cone. Doplicher, Haag and Roberts proposed to look at those representations that, roughly speaking, look like the vacuum representation in the spacelike complement of a double cone, the intersection of a forward and backward light-cone [19]. Although this certainly does not cover all physical systems of interests (it excludes, for example, electromagnetic charges [14]), it is nevertheless a useful criterion, and can for example also be applied to conformal theories on the circle [27].

What Doplicher, Haag and Roberts (DHR) showed [19, 20] (see also Chaps. 1 and 8) is that these representations can be studied in a systematic way. In this way, one can learn something about the statistics of the different charges (i.e., if they are bosons or fermions in space-times of dimensions $\geq 2+1$), or how they behave under composition of two charges (so-called *fusion rules*). Mathematically, it amounts to studying the structure of the equivalence classes satisfying the selection criterion as a *braided fusion category*, which is a type of *tensor category*. Some of the essential steps will be outlined below, but a more thorough introduction can be found in, e.g., [28, 29]. The study of tensor categories is a whole field on its own (an introduction can be found in [40]), but familiarity with the main terminology is not strictly necessary for our purposes here.

In the remainder of this section G will be a finite abelian group. We will see how one can construct different superselection sectors for Kitaev's quantum double model, and do a Doplicher-Haag-Roberts type of analysis to recover the properties of the different charges. The results outlined here have been obtained in [22, 43], to which the reader is referred for details. It is perhaps somewhat surprising to see that this theory, which was originally developed for relativistic quantum systems, can be applied so successfully to discrete lattice quantum spin systems, even though there are fundamental technical differences.

9.4.1 Localized Representations

As mentioned, the different superselection sectors are identified as equivalence classes of representations of the observable algebra, satisfying additional selection criteria. Here we take the ground state representation π_0, corresponding to the translationally invariant ground state, as a reference representation, and look at all representations that look like π_0 when considering observables outside a *cone-like region* Λ. The reason for this will become clear later, but at this point we mention that this is similar to the work of Buchholz and Fredenhagen, who show that in relativistic theories, massive particles can be localised in *spacelike* cones [14, 15], leading to a similar criterion. We will write $\mathcal{L}$ for the set of cones.

Definition 9.4.1 A representation π of $\mathfrak{A}$ is called *localizable* if for each $\Lambda \in \mathcal{L}$ we have

$$\pi_0 \upharpoonright \mathfrak{A}(\Lambda^c) \cong \pi \upharpoonright \mathfrak{A}(\Lambda^c), \tag{9.6}$$

where $\cong$ denotes unitary equivalence of representations and $\upharpoonright$ means that we restrict the representation to the subalgebra $\mathfrak{A}(\Lambda^c)$.

An equivalence class of representations satisfying the criterion is called a *(superselection) sector* or simply a *charge*.

To proceed in the analysis of these representations, we have to pass from representations to endomorphisms of the observable algebra. This can be done with the help of Haag duality. Indeed, let $\Lambda \in \mathcal{L}$. Then from Eq. (9.6) there is a unitary V_Λ setting up the equivalence with π_0 for the subalgebra $\mathfrak{A}(\Lambda^c)$. Define $\rho(A) = V_\Lambda \pi(A) V_\Lambda^*$ and let $\Lambda_2 \in \mathcal{L}$ contain Λ. Let $A \in \mathfrak{A}(\Lambda_2)$ and $B \in \mathfrak{A}(\Lambda_2^c)$. Note that $\rho(B) = \pi_0(B)$. Moreover, by locality,

$$\rho(AB) = \rho(A)\pi_0(B) = \rho(BA) = \pi_0(B)\rho(A),$$

hence by Haag duality $\rho(A) \in \pi_0(\mathfrak{A}(\Lambda_2^c))' = \pi_0(\mathfrak{A}(\Lambda_2))''$. Since π_0 is faithful, we can identify $\mathfrak{A}$ with $\pi_0(\mathfrak{A})$ and regard ρ as an endomorphism of $\mathfrak{A}$.[6] The endomorphism is called *localized* in Λ. It is also *transportable*: if $\Lambda' \in \mathcal{L}$, there is a unitary V and endomorphism ρ' (localized in Λ') such that $V\rho(A) = \rho'(A)V$ for all $A \in \mathfrak{A}$. Such an operator V (not necessarily unitary) is called an *intertwiner* from ρ to ρ'. If V is unitary we will also call it a *charge transporter*, since it moves a charge from one cone to another. Using Haag duality one can show that in fact $V \in \mathfrak{A}(\widehat{\Lambda})$ for suitable $\widehat{\Lambda} \in \mathcal{L}$ (where $\widehat{\Lambda}$ should contain the localization regions of both ρ and ρ'). There is in fact a 1-1 correspondence between superselection sectors and equivalence classes of these localized and transportable endomorphisms. From now on we will work with the latter.

9.4.2 *Localized Sectors in the Quantum Double Model*

Recall that the ribbon operators $F_\xi^{\omega,c}$ create a *pair* of excitations (or charges). We are however interested in the properties of a *single* charge. Since we are in an infinite system, it is possible to create a pair of excitations, and move one of them to infinity. As may already be anticipated from the discussion in Sect. 9.2.1, the different charges are in 1-1 correspondence with pairs (ω, c) of a character ω and group element c. The ribbon operator $F_\xi^{\omega,c}$ then creates the corresponding charge at the beginning of ξ, and a conjugate charge at the other end. On the level of the observables, this

[6]Strictly speaking, this is only true if $\mathfrak{A}(\Lambda_2)'' \subset \mathfrak{A}$, which in general is not the case for unbounded regions Λ_2. One can however solve this by passing to a larger algebra $\mathfrak{A}^{\Lambda_a} \supset \mathfrak{A}$ and extend ρ to a proper endomorphism of $\mathfrak{A}^{\Lambda_a}$ [14, Sect. 4].

means that we map $A \mapsto F_\xi^{\omega,c} A (F_\xi^{\omega,c})^*$.[7] Hence to implement the idea of moving one charge to infinity, one can choose a semi-infinite ribbon ξ, which for simplicity we assume to be completely inside some $\Lambda \in \mathcal{L}$, and write ξ_n for the (finite!) ribbon consisting of the first n parts. Then we define

$$\alpha(A) = \lim_{n\to\infty} F_{\xi_n}^{\omega,c} A (F_{\xi_n}^{\omega,c})^*. \tag{9.7}$$

Considering the dense set of local observables, and using the decomposition rule (9.5), it is not so difficult to show that this expression converges, and defines an automorphism of $\mathfrak{A}$. It describes how the observables change in the presence of a *single* charge (ω, c) in the background.

The map α defined above leads us to an example of a representation that satisfies the selection criterion, by defining $\pi = \pi_0 \circ \alpha$. Clearly, if Λ is any cone containing the ribbon, $\alpha(A) = A$ for all $A \in \mathfrak{A}(\Lambda^c)$ by locality, so α is localised. It is also transportable. There is a neat way of seeing this: because the vector state $F_{\xi_n}^{\omega,c}\Omega$ depends only on the endpoints of the path, it follows that two for automorphisms α_1 and α_2 defined in terms of the same charge, but different ribbons (with the same fixed endpoints), the states $\omega_0 \circ \alpha_1$ and $\omega_0 \circ \alpha_2$ coincide. In addition, note that both triples $(\pi_0 \circ \alpha_i, \Omega, \mathcal{H}_0)$ are GNS triples for this state, so that by the uniqueness of the GNS construction, $\pi_0 \circ \alpha_1$ must be unitarily equivalent to $\pi_0 \circ \alpha_2$. The argument can be extended if the endpoints of the ribbon do not coincide, by conjugating with a suitable ribbon operator. It follows that the automorphisms are transportable. The corresponding intertwiners can be shown to be in $\pi_0(\mathfrak{A}(\Lambda))''$ for a suitable cone Λ using Haag duality, but for the quantum double model an explicit construction of a net converging in the weak operator topology to an intertwiner is also possible.

The construction above gives for each pair (ω, c) an equivalence class of localized and transportable automorphisms, but it is not clear yet that these classes are distinct. This does turn out to be the case. The key idea in showing this is by considering an analogue of Wilson loops, which measure the charge in the region enclosed by the loop. Such operators can be obtained by considering ribbons with the same start and ending sites. In this way one can construct a charge measurement operator in an arbitrarily large region, that have expectation value 1 in the state $\omega_0 \circ \alpha$, where α has charge (ω, c), and zero expectation value in states obtained from a different pair (ω', c'). It follows that the corresponding representations $\pi_0 \circ \alpha$ must be inequivalent. The discussion can be summarized in the following theorem:

Theorem 9.4.2 *Let G be an abelian group. For each pair (ω, c) of a character of G and an element $c \in G$, there is an equivalence class of localised and transportable automorphisms. If α_1 and α_2 are such automorphisms, then $\pi_0 \circ \alpha_1 \cong \pi_0 \circ \alpha_2$ if and only if they belong to the same charge class (ω, c).*

[7] This actually corresponds to creating charges with $(F_\xi^{\omega,c})^*$, which turns out to be slightly more convenient.

This result can also be understood as an instance of *charge conservation*. The total charge of the state can be obtained by finding all excitations ω (resp. c) in the state, and multiply all of them together. Note that this is possible because the dual of an abelian group is a group again. The ribbon operators span a dense subset of the quasilocal algebra $\mathfrak{A}$. Because the two distinct charges at the end of the ribbons transform according to *conjugate* representations, it follows that one cannot change the total charge of the system just by local operations. In particular, one cannot go from a charged sector with a certain total charge to another sector with a different total charge by acting with local operators, so that indeed we have inequivalent representations of $\mathfrak{A}$.

9.4.3 Braiding and Fusion

So far we have constructed endomorphisms that describe single charges localized in cones, and charge transporters or intertwiners that can move the charges around. There is however much more structure, and this is where it becomes essential that we have endomorphisms rather than representations: in contrast to representations, endomorphisms can be composed. If ρ_1 and ρ_2 are two localized endomorphisms, we define $\rho_1 \otimes \rho_2(A) = \rho_1 \circ \rho_2(A)$. The interpretation is that we first create a charge ρ_2, and then add a charge ρ_1 to the system. This operation is called *fusion*. Note that $\rho_1 \circ \rho_2$ is localized again, more particularly it is localized in any cone that contains the localization regions of ρ_1 and ρ_2. It is also transportable again. This can be seen by defining a product operation for intertwiners as well: if $S\rho_1(A) = \rho_2(A)S$ and $T\sigma_1(A) = \sigma_2(A)T$ for all $A \in \mathfrak{A}$, we define $S \otimes T := S\rho_1(T)$. Then an easy calculation shows that $S \otimes T$ intertwines $\rho_1 \otimes \sigma_1$ and $\rho_2 \otimes \sigma_2$. There is however a slight technical issue. The intertwiners S are in general only elements of weak closures of the form $\pi_0(\mathfrak{A}(\Lambda))''$ for some cone Λ (this follows from Haag duality), while the endomorphisms are *a priori* only defined on $\mathfrak{A}$. It is however possible to extend the endomorphisms to a slightly larger algebra $\mathfrak{A}^{\Lambda_a}$, where Λ_a is some fixed auxiliary cone, that does contain the intertwiners. Since this is a minor technical point, we ignore the issue here, and refer to [14, 43] for technical details.

Even if ρ_1 and ρ_2 are irreducible, their composition need not be irreducible any more. A natural question therefore is if it is possible to decompose the composition into irreducibles again. The *fusion rules* give this decomposition. That is, if ρ_i and ρ_j are irreducible endomorphisms, there are integers N_{ij}^k, where k runs over a set of representatives of all irreducible localized and transportable endomorphisms, such that

$$\rho_i \circ \rho_j \cong \sum_k N_{ij}^k \rho_k. \tag{9.8}$$

The sum operation is a direct sum, which can be described in terms of intertwiners. This is very similar to representation theory of finite (or compact) groups: the tensor product of two irreducible representations can be decomposed again as the direct sum of finitely many irreducible representations.[8] Note that the identical endomorphism $\iota(A) = A$ acts as a unit for the fusion operation. Finally, a charge ρ has a *dual* or *conjugate* $\overline{\rho}$ if there is such a localized and transportable $\overline{\rho}$ with $N^{\iota}_{\rho\overline{\rho}} = 1$. The existence of conjugates is not automatic, and has to be either proven or taken as an assumption. Physically, it is related to the existence of (charge) anti-particles.

In algebraic quantum field theory, it is usually shown that such a decomposition exists with the help of a technical property, *Property B* [8], that essentially allows any projection in the local algebra to be written in the form $P = WW^*$ for some isometry W, localized in a slightly bigger region. In the present situation of spin systems, one could go about by showing that direct sums can be constructed, and by explicitly verifying the fusion rules (9.8) [43]. This is particularly easy in abelian models, where there always is a unique fusion outcome. That is, N^k_{ij} is equal to one for exactly one value of k, and zero otherwise. Symbolically, for the abelian quantum double model the fusion rules are

$$(\omega_1, c_1) \otimes (\omega_2, c_2) = (\omega_1\omega_2, c_1c_2).$$

This can be seen by considering, without loss of generality, a single ribbon ξ, and define the automorphisms α_1 and α_2 corresponding to the choices of charges. Using the multiplication rule for two ribbon operators in abelian models, mentioned at the end of Sect. 9.2.1, the result follows.

The final piece of structure that we discuss here is that of *braiding*. This is related to the statistics of identical particles, that is, their behaviour under interchange. This amounts to the study of the relation between $\rho_1 \otimes \rho_2$ and $\rho_2 \otimes \rho_1$ for general localized endomorphisms ρ_1 and ρ_2. An intertwiner $\varepsilon_{\rho_1,\rho_2}$ relating the two can be constructed explicitly by, quite literally, moving ρ_2 around (or, ρ_1 of course, in similar way). More precisely, one can choose a localization region $\widehat{\Lambda}$ that is disjoint from the localization regions of both ρ_1 and ρ_2. By transportability there is a unitary U and a endomorphism $\widehat{\rho}$ localized in $\widehat{\Lambda}$ such that $U\rho_2(A) = \widehat{\rho}(A)U$. Because $\rho_1 \otimes \widehat{\rho} = \widehat{\rho} \otimes \rho_1$ since their localization regions are disjoint, it follows that $\varepsilon_{\rho_1,\rho_2} := (U^* \otimes I)(I \otimes U) = U^*\rho_1(U)$ is an intertwiner from $\rho_1 \circ \rho_2$ to $\rho_2 \circ \rho_1$.

In three or more spatial dimensions, the definition of $\varepsilon_{\rho_1,\rho_2}$ is completely independent of the choices made, and one can show that $\varepsilon_{\rho_1,\rho_2}\varepsilon_{\rho_2,\rho_1} = I$. Note that this is the situation of ordinary bosons and fermions: moving one particle around the other doesn't change the system. One can show that the operators $\varepsilon_{\rho,\rho}$ indeed induce a representation of the permutation group, which interchanges charged excitations [19]. In lower dimensional space times, things get more interesting. It is no longer true that there is a unique choice of $\varepsilon_{\rho_1,\rho_2}$ [23, 24, 26]. The reason is that one cannot continuously move the cone $\widehat{\Lambda}$ around, but instead has to choose between either "left" or "right". This can be defined unambiguously by choosing an auxiliary

[8] Indeed, if the theory contains only bosonic sectors, the localized and transportable endomorphisms are in one-to-one correspondence with the representations of a compact global symmetry group G [18].

cone Λ_a as above. To get a consistent definition for the operators $\varepsilon_{\rho_1,\rho_2}$ (for example, to make sure that the different ways to go from ($\rho \otimes \sigma \otimes \tau$ to $\tau \otimes \rho \otimes \sigma$ coincide), one has to choose one of the two alternatives, and stick to that. A consequence of this ambiguity is that it is no longer true that $\varepsilon_{\rho_1,\rho_2}\varepsilon_{\rho_2,\rho_1} = I$. In addition, one does not obtain a representation of the permutation group any more, but rather one of the braid group. Consequently, such charges are said to have *braid statistics*. If this representation is abelian, one calls the charges (abelian) anyons (because they can pick up "any" phase under interchange). If the representation is non-abelian, one usually speaks of non-abelian anyons.[9]

The charges of the quantum double model are abelian anyons. This can be verified directly, by calculating the operators $\varepsilon_{\rho_1,\rho_2}$ and $\varepsilon_{\rho_2,\rho_1}$, together with the explicit construction of charge intertwiners, mentioned above. In the end the calculation boils down to commutation relations of the ribbon operators: $F^{\omega,c}_{\xi_1} F^{\sigma,d}_{\xi_2} = \omega(d)\sigma(c) F^{\sigma,d}_{\xi_2} F^{\omega,c}_{\xi_1}$ for two crossing ribbons ξ_1 and ξ_2. In the end one finds that

$$\varepsilon_{(\omega,c),(\sigma,d)}\varepsilon_{(\sigma,d),(\omega,c)} = \overline{\omega}(d)\overline{\sigma}(c),$$

showing that the charges are indeed abelian anyons.

9.4.4 The Category of Localized Endomorphisms

The set of localized endomorphisms has a very rich structure, of which we have given some examples above. Mathematically, it has the structure of a *braided unitary fusion category*.[10] The theory of such categories is quite rich, and an active field of study, in particular considering the classification of such categories. There are by now a few texts that provide an accessible entry to the literature. We mention for example [29, 31, 40, 58], each of which has a different focus.

We will denote this category by Δ. Its objects are the cone localized and transportable endomorphisms of $\mathfrak{A}$. The morphisms are intertwiners: $T \in \mathrm{Hom}(\rho, \sigma)$ if $T\rho(A) = \sigma(A)T$ for all $A \in \mathfrak{A}$. Note that if $T \in \mathrm{Hom}(\rho, \sigma)$, then $T^* \in \mathrm{Hom}(\sigma, \rho)$, where T^* is the adjoint of T. The isomorphisms in this category are unitary operators, so that the we indeed obtain equivalence classes of localized and transportable endomorphisms. An object is irreducible, by definition, if $\mathrm{Hom}(\rho, \rho) \cong \mathbb{C}$. Note that this is just another way of stating that the commutant of $\rho(\mathfrak{A})$ is trivial, hence this coincides with the usual notion of an irreducible representation.

[9] In the local quantum physics literature the name *plektons* is used. This name does not seem to have caught on outside of that community: the name non-abelian anyons prevails for example in the field of topological quantum computing, and this is why we adhere to that name here.

[10] Strictly speaking, "fusion" implies that there are only finitely many equivalence classes of irreducible objects. This is the case for the quantum double models, but need not be true in general. Dropping that condition does not make a difference in many cases.

In the previous section we defined a tensor product. This is actually a tensor product in the category, where the trivial endomorphism is the tensor unit. It is in fact a *strict* tensor category: associativity of the tensor product holds on the nose, not only up to isomorphism as is often the case in category theory. For each pair of objects ρ_1 and ρ_2, the definition of the braid operator $\varepsilon_{\rho_1,\rho_2}$ gives an isomorphism $\varepsilon_{\rho_1,\rho_2} \in \operatorname{Hom}(\rho_1, \rho_2)$. This assignment is in fact natural in both variables, and satisfies the so-called "braid equations", which give consistency conditions. This makes Δ into a *braided* tensor category. It is *symmetric* if $\varepsilon_{\rho_1,\rho_2}^{-1} = \varepsilon_{\rho_2,\rho_1}$ for all objects ρ_1, ρ_2.

Direct sums can be described in the category theory language as well. If ρ_1 and ρ_2 are objects in Δ, an object $\rho_1 \oplus \rho_2$ is a direct sum if there are $V_i \in \operatorname{Hom}(\rho_i, \rho_1 \oplus \rho_2)$ with $\sum_i V_i V_i^* = I$ and $V_i^* V_j = \delta_{i,j} I$. In algebraic quantum field theory the existence of such direct sums follows from Property B discussed above. In the case of the quantum double model we can explicitly construct the direct sums. Let Λ be a cone and suppose that ρ_1 and ρ_2 are localized in Λ. Adapting the arguments of [34, 37], it follows that $\mathcal{R}_\Lambda := \pi_0(\mathfrak{A}(\Lambda))''$ is an infinite factor, and hence there are isometries $V_i \in \mathcal{R}_\Lambda$ such that $\sum_i V_i V_i^* = I$ and $V_i V_j^* = \delta_{i,j} I$. We can then define

$$\rho_1 \oplus \rho_2(A) := V_1 \rho_1(A) V_1^* + V_2 \rho_2(A) V_2^*.$$

Because of locality it follows that $\rho_1 \oplus \rho_2$ is localized in Λ, and a straightforward calculation shows that this indeed is a direct sum in the categorical sense. Note that the direct sum is only defined up to unitary equivalence. This also explains how the summation in the fusion rules has to be understood, namely as a direct sum in this category. Note that in the abelian models we consider here, there is only one non-zero fusion coefficient, so the direct sum construction is not necessary.

Finally, there is the notion of *duals*. Physically, these can be interpreted as anti-charges. If ρ is an irreducible object of Δ, a dual (or conjugate) is a $\overline{\rho} \in \Delta$ such that the trivial endomorphism ι of $\mathfrak{A}$ appears exactly once in the direct sum composition of $\rho \otimes \overline{\rho}$ and of $\overline{\rho} \otimes \rho$. Note that this coincides with the notion of a conjugate in terms of the fusion coefficients N_{ij}^k mentioned before. This implies that there is an isometry $R \in \operatorname{Hom}(\iota, \rho \otimes \overline{\rho})$. We assume all objects have conjugates. In this case, it follows that all Hom-sets are *finite-dimensional* vector spaces (over $\mathbb{C}$). It follows from the fusion rules that for the abelian quantum double model, the conjugate of (ω, c) is $(\overline{\omega}, c^{-1})$.

So far we have constructed different superselection sectors of the abelian quantum double model, and studied their properties. The question remains if there are perhaps additional sectors, that we have so far overlooked. In other words, are there perhaps representations that satisfy Definition 9.4.1, but are not equivalent to one of the representations constructed so far? This question can be answered by adapting techniques from rational conformal nets [32]. In particular, one can consider two disjoint cones Λ_1 and Λ_2, and look at the von Neumann algebra $\pi_0(\mathfrak{A}(\Lambda_1 \cup \Lambda_2))''$. While Haag duality holds for a single cone, for a *pair* of cones it generally does not hold anymore. This means that $\pi_0(\mathfrak{A}(\Lambda_1 \cup \Lambda_2))'' \subset \pi_0(\mathfrak{A}((\Lambda_1 \cup \Lambda_2)^c)''$ in general is a proper inclusion (it is an irreducible subfactor, in fact). Essentially, the bigger algebra also contains the intertwiners or charge transporters that move a charge from one cone to the other. Hence by studying how much bigger the algebra is, one can learn something about the number of sectors in the theory. This relative size can be quantified by the Jones-Kosaki-Longo index $[\pi_0(\mathfrak{A}((\Lambda_1 \cup \Lambda_2)^c)'' : \pi_0(\mathfrak{A}(\Lambda_1 \cup \Lambda_2))'']$, which in turn gives an upper bound on the number of sectors [46]. For the toric code this number is four, and hence it follows that we have indeed construct all sectors: the ground state sector, and the ones corresponding to the pairs (ι, g), (σ, e), (σ, g), where σ is the sign representation of $\mathbb{Z}_2$ and g is the only non-trivial element in the group. This leads to the following conclusion:

Theorem 9.4.3 *The category of localised and transportable endomorphisms of the toric code is equivalent as a braided fusion category to category* $\text{Rep}_f\,\mathcal{D}(G)$ *of finite dimensional representations of the quantum double* $\mathcal{D}(\mathbb{Z}_2)$.

It should be rather straightforward (but tedious) to extend this result to all abelian quantum double models.

There is one aspect that we have not mentioned so far. The category of the quantum double model is *modular*. Modular tensor categories are fusion categories with an additional non-degeneracy property: it is called modular if it has trivial braided centre. That is, suppose that ρ is irreducible, and $\varepsilon_{\rho,\sigma}\varepsilon_{\sigma,\rho} = I$ for all irreducible σ. If this implies that ρ is the trivial object, then the category is modular [39, 53]. In this sense, modular categories are "as far away" as possible from being symmetric, where $\varepsilon_{\rho,\sigma}\varepsilon_{\sigma,\rho}$ is always the identity.

There is an equivalent condition that is related to Verlinde's matrix S in conformal field theory [57]. In the categorical setting it can be defined as follows. One can show that the duality allows one to define a *trace* on morphisms in the category. A matrix S is than defined by having entries $S_{ij} = \text{tr}(\varepsilon_{\rho_i,\rho_j}\varepsilon_{\rho_j,\rho_i})$, where ρ_i is a set of representatives of the irreducible objects. A category is modular if and only if this matrix S is invertible. It also allows us to explain the name modular: together with a matrix T that can also be canonically defined, the matrices S and T induce a projective representation of the modular group $\text{SL}_2(\mathbb{Z})$. In modular categories it turns out that there is in fact a relation between the matrix S and the fusion coefficients, given by the Verlinde rule [57]:

$$N_{ij}^k = \sum_r \frac{S_{ij} S_{jr} \overline{S}_{kr}}{S_{1r}^2},$$

where the label 1 stands for the trivial object. This shows that the structure of a modular tensor category is quite rigid, and not any given fusion rule can be realized in some modular category.

Modular tensor categories can be realized as *topological quantum field theories* (TQFTs), see for example [56]. One can therefore think, in a sense, of the type of topologically ordered systems that we have considered here as systems that in the low energy limit can be described by a TQFT. It should be noted than in general these are really effective theories, giving an effective description of oft-times very complex and poorly understood condensed matter systems. They also play a fundamental role in the field of topological quantum computing [58], on which we comment briefly in the last section.

9.5 Extension to Non-abelian Models

So far we have only considered abelian models. A natural question is if the methods can be extended to *non*-abelian quantum spin models. We again consider the quantum double model, but now for a non-abelian (but still finite) group G. In that case, there still is a unique translational invariant ground state ω_0. Many of the proofs, however, do not directly carry over from the abelian case. The reason for this will become clear below, but the underlying difficulty is that the irreducible representations of $\mathcal{D}(G)$ are no longer all one-dimensional. In particular, it is not clear how one could construct endomorphisms describing these "non-abelian" charges.

A similar problem appeared in one-dimensional spin chains, with compactly localized charges. There the problem is that the algebra $\mathfrak{A}(\Lambda)$ of such a localization region is finite dimensional, namely the tensor product of finitely many matrix algebras. All endomorphisms of this algebras are in fact automorphisms, and one can show that these cannot have non-abelian statistics. This problem can be circumvented by the methods of Szlachányi and Vecsernyés [55], and Nill and Szlachányi [50]. Instead of looking at endomorphisms, they look at *amplimorphisms*, that is, morphisms $\chi : \mathfrak{A} \to M_n(\mathfrak{A})$ for some integer n. One can then do a study of these endomorphisms in the spirit of the DHR theory.

In the present case we are interested in two dimensional systems with localization in *cones*. The cone algebras are certainly not finite dimensional, so the obstruction of the only endomorphisms being automorphisms does not play a role here.[11] Nevertheless, as mentioned it is not easy to explicitly construct examples of the endomorphisms. This is where the amplimorphisms come in. The strategy is to mimic the amplimorphism construction in [55] in the context of cone-localized charges, to construct representatives of the different charge classes, and will then show how we can go back to the usual setting of cone-localized endomorphisms. The amplimorphism description is much more explicit, making it possible to explicitly calculate intertwiners, fusion rules, *et cetera*. Here we will mainly restrict to the construction of the different sectors. Combining the techniques developed in the abelian case with the amplimorphism results in [55] should enable one to completely solve the model.

The ribbon operators again play a fundamental role. We will use the same notation as in Sect. 9.2.1, e.g. r will always be a fixed representative in a conjugacy class. Let ξ be a fixed ribbon, and recall that one can choose a basis of the ribbon operators acting on ξ in terms of the irreducible representations of $\mathcal{D}(G)$. For a fixed representation

[11] To be a bit more precise, one actually needs that $\pi_0(\mathfrak{A}(\Lambda))''$ is not a factor of Type I, that is, not of the form $\mathfrak{B}(\mathcal{H})$ for some Hilbert space $\mathcal{H}$. For the quantum double models that is the case.

there is a multiplet of ribbon operators F_ξ^{IJ}, where $I = (i_1, i_2)$ and $J = (j_1, j_2)$. The first indices run over the elements of the corresponding conjugacy class C, while the second runs over the dimension of the group. These multiplets satisfy certain completeness relations, namely

$$\sum_I (F_\xi^{IJ})^* F_\xi^{IK} = \delta_{JK} I, \quad \sum_J F_\xi^{IJ} (F_\xi^{KJ})^* = \delta_{IK} I, \tag{9.9}$$

where the summations are over all pairs (i_1, i_2) and the I on the right hand side is the unit of $\mathfrak{A}$. These equations can be verified by a simple calculation using the definitions and the algebraic relations for $F_\xi^{h,g}$. In [55] these F^{IJ} are called *irreducible* and *complete* multiplets, with the difference that here we have not defined an action γ_a of $\mathcal{D}(G)$ acting on these multiplets. Nevertheless, they do transform under an irreducible representation of $\mathcal{D}(G)$, cf. equation (B69) of [5].

Recall that for the abelian model, a ribbon operator acting on a large ribbon is just the product of two ribbon operators acting on smaller ribbons, but related to the same irreducible representation. If the irreducible representation is not one-dimensional any more, this is no longer true. Nevertheless, a suitable analogue of Eq. (9.5) still holds:

Lemma 9.5.1 *Choose a pair (C, ρ) of a conjugacy class and an irreducible representation of the centralizer $Z_G(r)$, where r is as explained in Sect. 9.2.1. Let $\xi = \xi_1\xi_2$ be a ribbon that is decomposed into two ribbons. The corresponding multiplets will be denoted by F_ξ^{IJ} and $F_{\xi_i}^{IJ}$ respectively. Then we have the following relation*

$$F_\xi^{IJ} = \sum_K F_{\xi_1}^{IK} F_{\xi_2}^{KJ}, \tag{9.10}$$

where the sum is over all pairs $K = (k_1, k_2)$

Proof First write out the right hand side of Eq. (9.10) in terms of the elementary ribbon operators $F_{\xi_i}^{h,g}$, where we set $I = (i_1, i_2)$, and similarly for J and K:

$$\sum_{g,h \in Z_G(r)} \sum_{k_1=1}^{|C|} \sum_{k_2=1}^{\dim(\rho)} \overline{\rho}_{i_2 k_2}(g) \overline{\rho}_{k_2 j_2}(h) F_{\xi_1}^{\overline{c}_{i_1}, q_{i_1} g \overline{q}_{k_1}} F_{\xi_2}^{\overline{c}_{k_1}, q_{k_1} h \overline{q}_{j_1}}.$$

Since ρ is a representation, the summation over k_2 yields a term $\overline{\rho}(gh)$. After a substitution $h \mapsto \overline{g}h$ we obtain

$$\sum_{g,h \in Z_G(r)} \sum_{k_1=1}^{|C|} \overline{\rho}_{i_2 j_2}(h) F_{\xi_1}^{\overline{c}_{i_1}, q_{i_1} g \overline{q}_{k_1}} F_{\xi_2}^{\overline{c}_{k_1}, q_{k_1} \overline{g} h \overline{q}_{j_1}}.$$

As remarked before, every element $s \in G$ can be written uniquely in the form $s = nq_i$ for some i and $n \in Z_G(r)$. Hence the summation over k_1 and g can be replaced by a summation over $s \in G$. More precisely, we set $s = g\overline{q}_{k_1}$. Note that $c_{k_1} = q_{k_1} r \overline{q}_{k_1} = sr\overline{s}$. With this observation the expression above reduces to

$$\sum_{h \in Z_G(r)} \sum_{s \in G} \overline{\rho}_{i_2 j_2}(h) F_{\xi_1}^{\overline{c}_{i_1}, q_{i_1} s} F_{\xi_2}^{\overline{s}\overline{r}s, \overline{s}h\overline{q}_{j_1}} = F_{\xi}^{IJ},$$

where the equality follows after a substitution $s \mapsto \overline{q}_i s$ and with the help of Eq. (9.5). This completes the proof.

Analogously to the automorphisms for the abelian models, we now define linear maps $\chi_{IJ}(A)$ of $\mathfrak{A}$. Choose again a semi-infinite ribbon ξ and let $F_{\xi_n}^{IJ}$ denote the corresponding multiplets, where ξ_n is the first part of the ribbon, consisting of n triangles. Then we set for any local observable A:

$$\chi_{IJ}(A) := \lim_{n \to \infty} \sum_K F_{\xi_n}^{IK} A \left(F_{\xi_n}^{JK} \right)^*. \tag{9.11}$$

Note that (assuming for the moment that the expression converges) this defines a linear map defined on a dense subset of $\mathfrak{A}$. Since it is bounded, it can be extended to $\mathfrak{A}$. This extension will also be denoted by χ_{IJ}. So it remains to be shown that the expression indeed converges (in norm, even). The main idea is similar as in the abelian case, only here we have to use Lemma 9.5.1.

Lemma 9.5.2 *Let χ_{IJ} be as above. Then the limit on the right hand side of Eq.* (9.11) *converges. We have the following properties:*

1. *for $A \in \mathfrak{A}_{loc}$, $\chi_{IJ}(A) = \sum_K F_{\xi_N}^{IK} A \left(F_{\xi_N}^{JK} \right)^*$ for N big enough;*
2. *$\chi_{IJ}(I) = \delta_{IJ} I$;*
3. *$\chi_{IJ}(A) = \delta_{IJ} A$ if A is localised away from the ribbon;*
4. *$\chi_{IJ}(AB) = \sum_K \chi_{IK}(A) \chi_{KJ}(B)$;*
5. *$\chi_{IJ}(A)^* = \chi_{JI}(A^*)$.*

Proof We show the first property. The others then follow straightforwardly using orthogonality and completeness for the multiplets, as well as the definition of χ_{IJ}. Consider $A \in \mathfrak{A}_{loc}$. Let N be such that $\text{supp}(A) \cap (\xi_n \setminus \xi_N) = \emptyset$ for all $n \geq N$. The idea is to decompose the ribbon ξ_n as $\xi_n = \xi_N \widehat{\xi}$, where $\widehat{\xi} = \xi_n \setminus \xi_N$. Let us now write $\chi_{IJ}^n(A)$ for $\sum_K F_{\xi_n}^{IK} A \left(F_{\xi_n}^{JK} \right)^*$, and set $\xi_1 = \xi_N$, $\xi_2 = (\xi_n \setminus \xi_N)$. By Lemma 9.5.1, locality, and Eq. (9.9) we get

$$\begin{aligned} \chi_{IJ}^n(A) &= \sum_K \sum_L \sum_M F_{\xi_1}^{IL} F_{\xi_2}^{LK} A (F_{\xi_1}^{JM} F_{\xi_2}^{MK})^* \\ &= \sum_K \sum_L \sum_M F_{\xi_1}^{IL} A (F_{\xi_1}^{JM})^* F_{\xi_2}^{LK} (F_{\xi_2}^{MK})^* \end{aligned}$$

$$= \sum_L \sum_M F_{\xi_1}^{IL} A (F_{\xi_1}^{JM})^* \delta_{LM}$$
$$= \chi_{IJ}^N(A).$$

From this it is clear that the limit in Eq. (9.11) converges for operators $A \in \mathfrak{A}_{loc}$. As mentioned the other properties are easy to verify.

Note that the properties stated in the Lemma are precisely those that one needs to define an amplimorphism. Note that there are $n = |C| \dim(\rho)$ pairs $I = (i, j)$. We then define a map $\chi : \mathfrak{A} \to M_n(\mathfrak{A})$ by setting $[\chi(A)]_{IJ} = \chi_{IJ}(A)$. It follows that χ is an amplimorphism. In addition, it is localized in any cone Λ that contains the ribbon ξ used in the definition of χ_{IJ}, in the sense that $\chi(A)$ is the matrix with entries A on the diagonal and otherwise zeros, if $A \in \mathfrak{A}(\Lambda^c)$. It remains to show that the amplimorphisms are transportable. Here we show that the charges can be transported over a *finite* region.

Lemma 9.5.3 *The amplimorphisms χ constructed above are transportable over finite distances.*

Proof Fix a semi-infinite ribbon ξ. We demonstrate how we can move the charge at the endpoint of the ribbon around. That is, let $\widehat{\xi}$ be a ribbon, such that $\widehat{\xi}\xi$ is again a semi-infinite ribbon. Define $V \in M_n(\mathfrak{B}(\mathcal{H}))$, where $n = |C| \dim(\rho)$, by having entries $V_{IJ} = F_{\widehat{\xi}}^{IJ}$. From Eq. (9.9) it follows that V is unitary. The claim is that $V\chi(A)V^* = \widehat{\chi}(A)$, where the amplimorphism in the right hand side is defined with respect to the ribbon $\widehat{\xi}$. This can be verified for local operators A, by carrying out the matrix multiplication, and using Lemma 9.5.1 together with 1. of Lemma 9.5.2.

The case of cone transportability is more complicated. Because of the lemma it is enough to consider two semi-infinite ribbons ξ_1 and ξ_2 starting at the same site. Let $\xi_{i,n}$ denote the ribbon consisting of the first n triangles. For each n, choose a ribbon $\widehat{\xi}_n$ that goes from the endpoint of $\xi_{2,n}$ to the endpoint of $\xi_{1,n}$ in such a way that as n goes to infinity, so does the distance of the ribbon to the endpoint of ξ_1. Now define a unitary operator $V_n = V_{\xi_{2,n}\widehat{\xi}_n} V_{\xi_{1,n}}^*$, where $V_{\xi_{i,n}}$ is the unitary obtained from the multiplet $F_{\xi_{i,n}}^{IJ}$. Now if A is local, it follows that $V_n \chi_1(A) = \chi_2(A) V_n$ for all n large enough. This can be seen by the argument in the proof of the above lemma.

This gives a uniformly bounded sequence of operators, since each of them is unitary. By the compactness of the unit ball in the weak operator topology, there is a subnet that converges to some operator V. Since multiplication on the right with a fixed $\chi(A)$ is weakly continuous, it follows that V intertwines $\chi(A)$ and $\widehat{\chi}(A)$. The problem remains to show that V is unitary. There are different ways that one might achieve this. For example, one could first try to show that χ and $\widehat{\chi}$ are irreducible, so that $\mathrm{Hom}(\chi, \widehat{\chi})$ must be either zero or one-dimensional. If V is non-zero, it then follows that one can choose V to be unitary. The other option is to realize both representations as the GNS representation of the same state. The vector state with an Ω in the first component (and otherwise zero) is a good candidate. By the independence of the ribbon operators on the exact choice of ribbon, this leads to the same state in both representations. If one can show that this vector is in fact cyclic, the proof is complete. We will leave this issue open for now.

It is natural to look at the amplimorphisms in the ground state representation, that is, look at $(\pi_0 \otimes \mathrm{id}) \circ \chi$. This amounts to applying the ground state representation to each matrix element. In fact, that is what we have been doing implicitly above. Note that by localization of the amplimorphisms, for observables outside the localization region, this representation looks like n copies of the ground state representation. Hence it is natural to adapt the selection criterion (9.6) a bit to allow for this case. It turns out that in the end this does not really matter, and one can go back from the amplimorphism picture to cone-localized endomorphisms (or representations). This is the content of the next theorem.

Theorem 9.5.4 *Suppose that π_0 satisfies Haag duality for cones. Let π be a representation of $\mathfrak{A}$ and n be a positive integer such that for some cone Λ, we have*

$$n \cdot \pi_0 \upharpoonright \mathfrak{A}(\Lambda^c) \cong \pi \upharpoonright \mathfrak{A}(\Lambda^c), \tag{9.12}$$

where $n \cdot \pi_0$ is the direct sum of n copies of the representation. Then the following hold:

1. *There is an amplimorphism $\chi : \mathfrak{A} \to M_n(\mathfrak{A}^{\Lambda_a})$, localised in Λ, such that we have $(\pi_0 \otimes \mathrm{id}) \circ \chi \cong \pi$.*
2. *There is a morphism $\rho : \mathfrak{A} \to \mathfrak{A}^{\Lambda_a}$ such that $\pi_0 \circ \rho \cong \pi$ when restricted to $\mathfrak{A}$.*

Note that the unitary equivalence in the second point can be used to map intertwiners between amplimorphisms to intertwiners between morphisms. In particular, if χ is transportable, so is the corresponding morphism, and ρ can be extended to an endomorphism of $\mathfrak{A}^{\Lambda_a}$.

Proof 1. Let $U : \mathcal{H}_\pi \to \bigoplus_{i=1}^n \mathcal{H}_0$ be the unitary setting up the equivalence and suppose that $A \in \mathfrak{A}(\Lambda)$. For $A \in \mathfrak{A}$, define $\chi(A) := U\pi(A)U^*$. Note that $\chi(A) \in \mathfrak{B}(\bigoplus_{i=1}^n \mathcal{H}_0) \cong M_n(\mathfrak{B}(\mathcal{H}_0))$, so that the matrix elements $\chi_{ij}(A) \in \mathfrak{B}(\mathcal{H}_0)$. Now let $B \in \mathfrak{A}(\Lambda^c)$ and $A \in \mathfrak{A}(\Lambda)$. Note that from Eq. (9.12) it follows that $\chi(B) = \mathrm{diag}(\pi_0(B), \ldots, \pi_0(B))$.

Remark that A and B commute by locality, so that $\chi(A)\chi(B) = \chi(AB) = \chi(B)\chi(A)$. Writing out the definitions and comparing matrices element wise, it follows that $\pi_0(B)\chi_{ij}(A) = \chi_{ij}(A)\pi_0(B)$. Hence $\chi_{ij}(A) \in \pi_0(\mathfrak{A}(\Lambda^c))' = \pi_0(\mathfrak{A}(\Lambda))''$, by Haag duality. Since the algebra on the right hand side is contained in the auxiliary algebra $\mathfrak{A}^{\Lambda_a}$, and χ acts trivially on $\mathfrak{A}(\Lambda^c)$, it follows that $\chi : \mathfrak{A} \to M_n\left(\mathfrak{A}^{\Lambda_a}\right)$ is an amplimorphism.

2. Since $\pi_0(\mathfrak{A}(\Lambda))''$ is an infinite factor, one can find isometries V_i, $i = 1, \ldots, n$ generating a Cuntz algebra [16]. That is, they satisfy $V_i^* V_j = \delta_{ij} I$ and $\sum_{i=1}^n V_i V_i^* = I$. Write χ for the amplimorphism obtained in part (1), and $\chi_{ij}(A)$ for its matrix elements when evaluated in $\mathfrak{A}$. Define a map $\rho : \mathfrak{A} \to \mathfrak{A}^{\Lambda_a}$ by

$$\rho(A) = \sum_{i,j=1}^{n} V_i \chi_{ij}(A) V_j^*.$$

Suppose that $A, B \in \mathfrak{A}$. A straightforward calculation then shows

$$\begin{aligned}\rho(A)\rho(B) &= \sum_{i,j,k,l} V_i \chi_{ij}(A) V_j^* V_k \chi_{kl}(B) V_l^* \\ &= \sum_{i,j,l} V_i \chi_{ij}(A) \chi_{jl}(B) V_l^* \\ &= \rho(AB).\end{aligned}$$

Moreover, $\rho(A^*) = \rho(A)^*$ since $\chi_{ij}(A^*) = \chi_{ji}(A)^*$. If $B \in \mathfrak{A}(\Lambda^c)$, then $\chi_{ij}(B) = \delta_{ij} B$ and $B \in \mathfrak{A}(\Lambda)'$, hence $\rho(B) = B$ and ρ is localised in Λ.

Next we show that $\pi_0 \circ \rho$ is unitarily equivalent to $\pi_0 \otimes \chi$. To this end, identify $\mathcal{H}_0 \otimes \mathbb{C}^n$ with $\mathcal{H} = \bigoplus_{i=1}^n \mathcal{H}_0$. Define a map $U : \mathcal{H} \to \mathcal{H}_0$ by setting

$$U(\psi_1 \oplus \cdots \oplus \psi_n) = \sum_{i=1}^{n} V_i \psi_i.$$

Using the properties of the V_i it is easy to check that U preserves the inner product (and hence is an isometry and well-defined). Since $\sum_{i=1}^n V_i V_i^* = I$ it also has dense range, hence U is unitary.

A few remarks are in order at this point. First of all, the condition of Haag duality is only used to obtain an amplimorphism from a representation satisfying Eq. (9.12). Without Haag duality, one can still obtain an morphism ρ such that $\pi_0 \circ \rho$ and $\pi_0 \otimes \chi$ are unitarily equivalent. However, we then have little control over the range of the morphism, and we cannot extend it to an endomorphism of the auxiliary algebra without any additional information.

The theorem shows that in principle one can restrict to the study of localised and transportable endomorphisms. Nevertheless, it can be very helpful to look at amplimorphisms as well. One reason is that it may be easier to construct such amplimorphisms explicitly in concrete models, as we have done above. They also provide more information on the symmetries of the model. In particular, the vector space $\mathbb{C}^n$ in $\mathcal{H}_0 \otimes \mathbb{C}^n$ carries a representation of the symmetry algebra, through the symmetry transformations of the multiplets F_ξ^{IJ}. This plays an important role in the analysis of the superselection sectors in [50, 55]. We expect that the methods used there to study the category of amplimorphisms (and hence, by the theorem above, the category of localized endomorphisms). In particular, to define fusion and braiding. This leads to the following conjecture:

Conjecture 9.5.5 Let G be a finite group. Then the category Δ of localized and transportable endomorphisms of the quantum double model is equivalent to the representation category $\mathrm{Rep}_f\, \mathcal{D}(G)$ of the quantum double $\mathcal{D}(G)$.

In [44] we obtained through computer algebra the fusion rules of $\mathcal{D}(S_3)$ by taking the composition of certain positive maps. These results are consistent with the approach here, since the positive maps there are the traces of the amplimorphisms we have defined here. The composition of these positive maps yields the trace of the amplimorphism $\chi_1 \otimes \chi_2$, where the tensor product is defined as in [55].

To study this conjecture it would be helpful to go from endomorphisms to amplimorphisms. An important question is also if there is a "canonical" way to obtain an amplimorphism from an endomorphism. Of course, there is always the "trivial" way, since and endomorphism ρ gives rise to an amplimorphism $\widehat{\rho} : \mathfrak{A} \to \mathfrak{A} \otimes \mathbb{C}$. In general, let $T_1, \dots, T_n \in \mathfrak{B}(\mathcal{H})$ be such that $T_i^* T_j = \delta_{i,j} I$ and $\sum_{k=1}^n T_k T_k^* = I$. Then it is easy to check that $[\chi(A)]_{ij} := T_i^* \rho(A) T_j$ defines an amplimorphism. Hence to have a meaningful equivalence between the category of amplimorphisms and of localised morphisms, we would have to impose some additional conditions on the amplimorphisms. One of them is that they transform in the right way, as explained above. To find this symmetry one could look for an analogue of the Doplicher-Roberts theorem [18], which gives a group symmetry for bosonic/fermion sectors. In general, one could only expect a so-called weak Hopf-algebra symmetry [52].

We have outlined here how one could proceed with an analysis of the superselection structure of the non-abelian quantum double model. Although the analysis is not complete, we hope that it is a helpful starting point. It should be remarked that here we have provided an explicit model whose ground space representation should lead to non-abelian charges. This should be contrasted with [50, 55], where the existence of such a representation is taken as an assumption.

9.6 An Application: Topological Quantum Computing

There are different reasons why there has been a huge interest in topologically ordered systems in recent years. One of the reasons is that they provide examples of new phases of matter, that go beyond the Landau paradigm of symmetry breaking. This is not just of theoretical interest—these phases really exist in nature. There is also experimental evidence for the existence of quasi-particle excitations with anyonic excitations, see for example [38]. A good theoretical understanding is therefore very welcome.

Here we focus on another aspect that has sparked the interest of the quantum computation community. One of the goals of quantum computation is to use the full power of quantum mechanics to solve computationally hard problems, for which a computation on a usual, classical computer is infeasible. To illustrate this one can think of a simple spin-1/2 quantum system, with Hilbert space $\mathbb{C}^2$. The dimension of n copies of such a system scales as 2^n, so if one wants to simulate the whole Hilbert space of a n-particle system, one quickly runs out of memory in a classical computer. The idea behind quantum computing, which goes back to Feynman [21], is to use the laws of quantum mechanics to solve computationally complex problems, or even simulate other quantum systems.

This is not the place for a full-fledged introduction to quantum computing (for this Nielsen and Chuang's [49] is a good start), but let us summarise the main points. The "quantum memory" of a quantum computer is modelled by a (generally finite dimensional) Hilbert space, ususally of the form $\mathcal{H} = (\mathbb{C}^2)^{\otimes n}$ for some n. A computation then consists of three steps:

1. Initialise the system in a known state;
2. Perform a unitary operation on the system to implement the algorithm;
3. Measure the result (and if necessary, repeat to get statistics).

This encompasses classical computing. To see this, we can reformulate each computational problem in the calculation of a function $f : \{0, 1\}^n \to \{0, 1\}^n$, where for simplicity we take f to be injective. Now let $|0\rangle, |1\rangle$ be a basis of $\mathbb{C}^2$. Then we can identify each $(x_i) \in \{0, 1\}^n$ with a basis vector $|x\rangle := |x_1\rangle \otimes \cdots \otimes |x_n\rangle$. We can then define a map U_f by $U_f |x\rangle = U_f |f(x)\rangle$, which is unitary because f is injective. Hence we can initialize the system in a known state $|x\rangle$, apply U_f, and measure to learn something about $|f(x)\rangle$. This is where quantum mechanics comes in: it allows us to act with U_f on a *superposition*, say of the form $\frac{1}{\sqrt{2^n}} \sum_x |x\rangle$, so that we can learn something about *all* values $f(x)$ in a *single* operation. This is not possible on a classical computer.

Although the main idea is quite simple, it turns out to be very difficult to implement this in physical systems in a reliable and scalable way. This is were topological quantum computing comes in [35, 48]. We have already briefly discussed the problem of storing quantum information over an extended period of time. Here we focus on the computation part, that is, processing this quantum information to implement an algorithm. As mentioned above, this amounts to acting with a certain unitary operator on the system. In practice this could work, for example, by coupling the system to some external magnetic field for a period of time, and let it evolve. The problem is that it generally is very difficult to exactly perform the unitary that you want, and not something slightly different (because the magnetic field is left on too long, for example). The idea is therefore to use "topological" operations to implement the necessary unitary. A small disturbance should not change the topological property, and hence have no effect on the computation. The braiding operation of two anyons for example is independent of the path that the anyons take (as long as the paths don't cross).

To use this idea, the quantum information on which we want to operate has to be encoded using anyons. Let us consider a non-abelian anyon ρ, which for the sake of simplicity we assume to be self-dual: $\rho = \overline{\rho}$. We can then take n copies of it, that is, consider $\rho^{\otimes n}$. The idea is to create these by pulling pairs of them from the vacuum, which is possible because ρ is its own anti-charge. Because the anyon is non-abelian, this can actually be done in different ways. More precisely, the possible states are described by $\mathrm{Hom}(\iota, \rho^{\otimes n})$, where ι is again the trivial charge. This vector space can be given the structure of a Hilbert space. It is this space that we use to encode the qubits in. It should be noted that in general $\mathrm{Hom}(\iota, \rho^{\otimes})$ does *not* have a nice decomposition as the tensor product of n copies of some Hilbert space. Nevertheless, one can embed the state space of a number of qubits into this space. The dimension of $\mathrm{Hom}(\iota, \rho^{\otimes n})$ grows exponentially in n.

Step 1 in the quantum computation scheme can now be accomplished by pulling charges from the vacuum in a suitable way. To implement unitary operations on the qubits, we note that there is a natural representation of the braid group B_n on $\mathrm{Hom}(\iota, \rho^{\otimes n})$: if $T \in \mathrm{Hom}(\iota, \rho^{\otimes n})$, we can consider

$$\pi(b_i)T := (I \otimes \cdots \otimes \varepsilon_{\rho,\rho} \otimes \cdots \otimes I) \circ T,$$

where b_i is the generator of the braid group that swaps the i-th and $(i+1)$-th strand, while on the right hand side, the $\varepsilon_{\rho,\rho}$ term acts on the i-th and $(i+1)$-th tensor factors. Note that the result is again in $\mathrm{Hom}(\iota, \rho^{\otimes n})$, hence this gives a unitary rotation on the encoded qubits, and hence allows us to implement (a part of) a quantum computation.

This is not the end of the story, because it is not clear if every unitary on the encoded qubits can be obtained in this way. It is enough for $\pi(B_n)$ to generate a dense subset of the unitaries. If this is the case for a certain anyon model, it is said to be *universal*, because it means that each quantum algorithm can in principle be implemented on it. Kitaev's quantum double model is universal for a wide range of non-abelian groups [41, 42], but in general this is a rather special feature of a model. If a model is not universal, one can supplement the braiding operations with other, non-topological operations to be able to implement each unitary operation. Even if the model is universal, one still has to find out which combination of braidings one has to do to obtain a certain unitary. Luckily, this can be found efficiently using the Solovay-Kitaev theorem [49]. This is worked out explicitly for a simple anyon model, the so called Fibonacci model, in [7].

Finally, after the appropriate braidings have performed, it is time to measure the outcome. That is, we have to find out which state in $\mathrm{Hom}(\iota, \rho^{\otimes n})$ the system is in. This is done by fusing some of the anyons again. Recall that the fusion rules are of the form $\rho \otimes \rho = \sum_k N^k_{\rho\rho}\rho_k$. The integers $N^k_{\rho\rho}$ essentially says in how many ways the fusion of two ρ-anyons can lead to an anyon ρ_k. It are precisely these different ways to fuse n anyons to the vacuum that label the states. Hence, by fusion some anyons and observing the outcome (which amounts to doing a charge measurement), we can obtain statistics on which state the system was in. The charge measurement is where it becomes important to have modularity of the category: this allows us to distinguish the charges by pulling pairs of charges from the vacuum, move one of the anyons around the region for which we want to determine the charge, and fuse to the vacuum again and observe if there is anything left or not.

It should be noted that while topological quantum computing has advantages with respect to the stability of the operations, there are also drawbacks. For example, in practice it may not be so easy to physically move the anyons around, especially over larger distances. This may be circumvented by measurement based quantum computation. There, the braiding operations are mimicked by doing a series of measurements [6]. One could also restrict to more local operations, although this will mean that the anyon models are no longer universal [4].

To conclude, we have seen that there is a rich class of so-called topologically ordered states, for which methods of local quantum physics provide useful tools to study such systems. Besides the possible applications to quantum computing that we have mentioned, there are also very interesting condensed matter and mathematical aspect related to such phases. Finally, also on the experimental side the field is very active. It would be good to see if the tools of local quantum physics can be further employed to advance progress in this multi-disciplinary field.

Acknowledgments The author wishes to thank Courtney Brell for helpful comments and discussions and Leander Fiedler for collaboration on [22]. This work is supported by the Dutch Organisation for Scientific Research (NWO) through a Rubicon grant and partly through the EU project QFTCMPS and the cluster of excellence EXC 201 Quantum Engineering and Space-Time Research

References

1. Alicki, R., Fannes, M., Horodecki, M.: A statistical mechanics on Kitaev's proposal for quantum memories. J. Phys. A **40**, 6451–6467 (2007)
2. Alicki, R., Fannes, M., Horodecki, M.: On thermalization in Kitaev's 2D model. J. Phys. A **42**, 065303 (2009)
3. Bakalov, B., Kirillov Jr, A.: Lectures on Tensor Categories and Modular Functors. American Mathematical Society, Providence (2001)
4. Beverland, M.E., König, R., Pastawski, F., Preskill, J., Sijher, S.: Protected gates for topological quantum field theories. Preprint, arXiv:1409.3898 (2014)
5. Bombin, J., Martin-Delgado, M.A.: Family of non-Abelian Kitaev models on a lattice: topological condensation and confinement. Phys. Rev. B. **78**, 115421 (2008)
6. Bonderson, P., Freedman, M., Nayak, C.: Measurement-only topological quantum computation via anyonic interferometry. Ann. Phys. **324**, 787–826 (2009)
7. Bonesteel, N.E., Hormozo, L., Zikos, G., Simon, S.H.: Braid topologies for quantum computation. Phys. Rev. Lett. **95**, 140503 (2005)
8. Borchers, H.-J.: A remark on a theorem of B. Misra. Commun. Math. Phys. **4**, 315–323 (1967)
9. Bratteli, O., Robinson, D.: Operator Algebras and Quantum Statistical Mechanics 1, 2nd edn. Springer, New York (1987)
10. Bratteli, O., Robinson, D.: Operator Algebras and Quantum Statistical Mechanics 2, 2nd edn. Springer, New York (1997)
11. Bravyi, S., Hastings, M.B.: A short proof of stability of topological order under local perturbations. Commun. Math. Phys. **307**, 609–627 (2011)
12. Bravyi, S., Hastings, M.B., Michalakis, S.: Topological quantum order: stability under local perturbations. J. Math. Phys. **51**, 093512 (2010)
13. Bravyi, S., Terhal, B.: A no-go theorem for a two-dimensional self-correcting quantum memory based on stabilizer codes. New J. Phys. **11**, 043029 (2009)
14. Buchholz, D., Fredenhagen, K.: Locality and the structure of particle states. Commun. Math. Phys. **84**, 1–54 (1982)
15. Buchholz, D., Fredenhagen, K.: Locality and the structure of particle states in gauge field theories. In: Schrader, R., Seiler, R., Uhlenbrock, D. (eds.) Mathematical Problems in Theoretical Physics, pp. 368–371. Springer, Berlin (1982)
16. Cuntz, J.: Simple C^*-algebras generated by isometries. Commun. Math. Phys. **57**, 173–185 (1977)
17. Dijkgraaf, R., Pasquier, V., Roche, P.: Quasi Hopf algebras, group cohomology and orbifold models. Nucl. Phys. B (Proc. Suppl.) **18B**, 60–72 (1990)

18. Doplicher, S., Roberts, J.E.: Why there is a field algebra with a compact gauge group describing the superselection structure in particle physics. Commun. Math. Phys. **131**, 51–107 (1990)
19. Doplicher, S., Haag, R., Roberts, J.E.: Local observables and particle statistics. I. Commun. Math. Phys. **23**, 199–230 (1971)
20. Doplicher, S., Haag, R., Roberts, J.E.: Local observables and particle statistics. II. Commun. Math. Phys. **35**, 49–85 (1974)
21. Feynman, R.P.: Simulating physics with computers. Int. J. Theor. Phys. **21**, 467–488 (1982)
22. Fiedler, L., Naaijkens, P.: Haag duality for Kitaev's quantum double model for abelian groups. Preprint, arXiv:1406.1084 (2014)
23. Fredenhagen, K., Rehren, K.-H., Schroer, B.: Superselection sectors with braid group statistics and exchange algebras. I. General theory. Commun. Math. Phys. **125**, 201–226 (1989)
24. Fredenhagen, K., Rehren, K.-H., Schroer, B.: Superselection sectors with braid group statistics and exchange algebras. II. Geometric aspects and conformal covariance. Rev. Math. Phys. **4**(Special Issue), 113–157 (1992)
25. Freedman, M.H.: P/NP, and the quantum field computer. Proc. Natl. Acad. Sci. USA **95**, 98–101 (1998)
26. Fröhlich, J., Gabbiani, F.: Braid statistics in local quantum theory. Rev. Math. Phys. **2**, 251–353 (1990)
27. Gabbiani, F., Fröhlich, J.: Operator algebras and Conformal Field Theory. Commun. Math. Phys. **155**, 569–640 (1993)
28. Haag, R.: Local Quantum Physics: Fields, Particles, Algebras, 2nd edn. Springer, Berlin (1996)
29. Halvorson, H.: Algebraic quantum field theory (with an appendix by M. Müger). In: Butterfield, J., Earman, J. (eds) Philosophy of Physics, pp. 731–922, Elsevier, Amsterdam (2006)
30. Kalmeyer, V., Laughlin, R.B.: Equivalence of the resonating-valence-bond and fractional quantum Hall states. Phys. Rev. Lett. **59**, 2095–2098 (1987)
31. Kassel, C.: Quantum Groups. Springer, New York (1995)
32. Kawahigashi, Y., Longo, R., Müger, M.: Multi-interval subfactors and modularity of representations in conformal field theory. Commun. Math. Phys. **219**, 631–669 (2001)
33. Kay, A., Colbeck, R.: Quantum self-correcting stabilizer codes. Preprint arXiv:0810.3557 (2008)
34. Keyl, M., Matsui, T., Schlingemann, D., Werner, R.F.: Entanglement, Haag-duality and type properties of infinite quantum chains. Rev. Math. Phys. **18**, 935–970 (2006)
35. Kitaev, A.: Fault-tolerant quantum computation by anyons. Ann. Phys. **303**, 2–30 (2003)
36. Landon-Cardinal, O., Poulin, D.: Local topological order inhibits thermal stability in 2D. Phys. Rev. Lett. **110**, 090502 (2013)
37. Matsui, T.: The split property and the symemtry breaking of the quantum spin chain. Commun. Math. Phys. **218**, 393–416 (2001)
38. Mourik, V., Zuo, K., Frolov, S.M., Plissard, S.R., Bakkers, E.P.A.M., Kouwenhoven, L.P.: Signatures of Majorana fermions in hybrid superconductor-semiconductor nanowire devices. Science **336**, 1003–1007 (2012)
39. Müger, M.: On the structure of modular categories. Proc. London Math. Soc. **87**, 291–308 (2003)
40. Müger, M.: Tensor categories: a selective guided tour. Rev. Unión Mat. Argentina **51**, 95–163 (2010)
41. Mochon, C.: Anyons from nonsolvable finite groups are sufficient for universal quantum computation. Phys. Rev. A **67**, 022315 (2003)
42. Mochon, C.: Anyon computers with smaller groups. Phys. Rev. A **69**, 032306 (2004)
43. Naaijkens, P.: Localized endomorphisms in Kitaev's toric code on the plane. Rev. Math. Phys. **23**, 347–373 (2011)
44. Naaijkens, P.: Anyons in infinite quantum systems: QFT in $d = 2 + 1$ and the Toric Code. Ph.D. thesis, Radboud Universiteit Nijmegen (2012)
45. Naaijkens, P.: Haag duality and the distal split property for cones in the toric code. Lett. Math. Phys. **101**, 341–354 (2012)

46. Naaijkens, P.: Kosaki-Longo index and classification of charges in 2D quantum spin models. J. Math. Phys. **54**, 081901 (2013)
47. Nachtergaele, B., Sims, R.: Lieb-Robinson bounds and the exponential clustering theorem. Commun. Math. Phys. **265**, 119–130 (2006)
48. Nayak, C., Simon, S.H., Stern, A., Freedman, M., Das Sarma, S.: Non-abelian anyons and topological quantum computation. Rev. Modern Phys. **80**, 1083–1159 (2008)
49. Nielsen, M.A., Chuang, I.L.: Quantum Computation and Quantum Information. Cambridge University Press, Cambridge (2000)
50. Nill, F., Szlachányi, K.: Quantum chains of Hopf algebras with quantum double cosymmetry. Commun. Math. Phys. **187**, 159–200 (1997)
51. Oeckl, R.: Discrete Gauge Theory: From Lattices to TQFT. Imperial College Press, London (2005)
52. Ostrik, V.: Module categories, weak Hopf algebras and modular invariants. Transform. Groups **8**, 177–206 (2003)
53. Rehren, K.-H.: Braid group statistics and their superselection rules. In: Kastler, D. (ed.) The Algebraic Theory of Superselection Sectors, pp. 333–355. World Scientific Publishing, River Edge (1990)
54. Rieffel, M., Van Daele, A.: The commutation theorem for tensor products of von Neumann algebras. Bull. London Math. Soc. **7**, 257–260 (1975)
55. Szlachányi, K., Vecsernyés, P.: Quantum symmetry and braid group statistics in G-spin models. Comm. Math. Phys. **156**, 127–168 (1993)
56. Turaev, V.G.: Quantum Invariants of Knots and 3-manifolds. Walter de Gruyter & Co., Berlin (1994)
57. Verlinde, E.: Fusion rules and modular transformations in 2D conformal field theory. Nucl. Phys. B. **300**, 360–376 (1988)
58. Wang, Z.: Topological quantum computation. Conference Board of the Mathematical Sciences, Washington, DC (2010)
59. Wen, X.-G.: Vacuum degeneracy of chiral spin states in compactified space. Phys. Rev. B **40**, 7387–7390 (1989)
60. Wick, G.C., Wightman, A.S., Wigner, E.P.: The intrinsic parity of elementary particles. Phys. Rev. Lett. **88**, 101–105 (1952)

Chapter 10
Algebraic Constructive Quantum Field Theory: Integrable Models and Deformation Techniques

Gandalf Lechner

Abstract Several related operator-algebraic constructions for quantum field theory models on Minkowski spacetime are reviewed. The common theme of these constructions is that of a *Borchers triple*, capturing the structure of observables localized in a Rindler wedge. After reviewing the abstract setting, we discuss in this framework (i) the construction of free field theories from standard pairs, (ii) the inverse scattering construction of integrable QFT models on two-dimensional Minkowski space, and (iii) the warped convolution deformation of QFT models in arbitrary dimension, inspired from non-commutative Minkowski space.

10.1 Algebraic Constructions of Quantum Field Theories

Models in quantum field theory (QFT) are usually constructed with the help of a classical analogue: One starts from a classical relativistic field theory, and then uses some quantization procedure (typically involving renormalization) to arrive at a corresponding QFT model. Such constructions have led to theoretical predictions that in some cases match experimental data to a remarkable degree of accuracy.

This success must however be contrasted with the difficulties of rigorously defining any interacting QFT model. Although at the level of perturbation theory, many QFT models are nowadays well understood (see Chap. 2 for discussions of this subject), one typically has no control over the convergence of the perturbation series, and no control over the error made by truncating it.

Constructive QFT, on the other hand, aims at non-perturbative constructions of models of interacting quantum fields. This program was very successful in two and three spacetime dimensions, where a large family of interacting QFTs with polynomial self-interaction ("$P(\phi)_2$ models" and the "ϕ_3^4 model") was constructed, chiefly by Glimm and Jaffe [76]. We refer to the recent review [133] for a detailed account of these constructions and the relevant literature. In four dimensions, however, no comparable results are known.

G. Lechner (✉)
School of Mathematics, Cardiff University, Senghennydd Road, Cardiff CF24 4AG, Wales, UK
e-mail: LechnerG@cardiff.ac.uk

R. Brunetti et al. (eds.), *Advances in Algebraic Quantum Field Theory*,
Mathematical Physics Studies, DOI 10.1007/978-3-319-21353-8_10

The models obtained in constructive QFT were shown to fit [76] into both, the operator-algebraic Haag-Kastler framework [84] as well as the field-theoretic Wightman setting [130]. The tools used in their construction, in particular the Euclidean formulation, are however more closely linked to the field-theoretic picture.

This review article focuses on a different approach to the rigorous construction of interacting models, namely on those which are constructed by operator-algebraic methods. In comparison to "constructive QFT", this "algebraic constructive QFT"[1] is a much more recent topic. Due to page constraints, this review is far from exhaustive. However, several of the topics not treated here are the subject of other chapters in this book: For example, we will not discuss the recent progress in perturbative algebraic QFT [65] (see Chap. 2), or conformal algebraic QFT [91] (see Chap. 8), or the recent algebraic construction of models on two-dimensional de Sitter space by Barata et al. [13].

To motivate the discussion of the topics that *are* treated in this review, we first mention that from a philosophical point of view, there is no good reason to construct quantum field theories on the basis of classical field theories (that is, by quantization). Since quantum theories are supposed to provide the more fundamental description of reality, classical theories should rather appear only as a limiting case. One would like to take the quantum theories as the fundamental data, and consider their classical limits only for comparison with classical notions.

Desirable as it is, a purely quantum description of relativistic physics poses several major challenges. The first question is what the relevant mathematical structures are that one tries to construct. As we will focus on the algebraic setting of QFT [84] here, the data comprising a model theory in our case are a family of operator algebras $\mathcal{A}(O)$, labeled by localization regions O in spacetime, subject to a number of conditions (Chap. 1).

In model-independent investigations of QFT, the algebras of observables localized in certain spacelike wedge-shaped regions of Minkowski space (wedges), play a prominent role, a point emphasized in particular by D. Buchholz. We will recall their definition and relevant properties in Sect. 10.2.1. By the results of Bisognano and Wichmann [21, 22], and later Borchers [23], algebras localized in wedges provide a link between the geometric properties of Minkowski space, encoded in its Poincaré symmetry, and certain algebraic properties of the net, encoded in its modular data.[2] Via the idea of *modular localization* (see Sect. 10.2.3), this link also connects Wigner's classification of elementary particles by positive energy representations of the Poincaré group to the modular structure of wedge algebras.

Moreover, because wedges are unbounded regions, observables localized in them can have much milder momentum space properties than point-like localized quantum fields, which typically fluctuate enormously in energy and momentum. As argued

[1]A term coined by S.J. Summers, see also his online article http://people.clas.ufl.edu/sjs/constructive-quantum-field-theory/ for a review.

[2]For large parts of this review, we will rely on Tomita-Takesaki modular theory, see for example [35] for an introduction and [24] for an overview of applications to QFT.

by B. Schroer, one can in particular consider wedge-localized observables that are free of vacuum polarization, i.e. just create single particle states from the vacuum, also in interacting theories [126]. Such *polarization-free generators* do not exist for smaller localization regions in general, but can be used to generate algebras localized in wedges, and are directly related to the two-particle S-matrix [26, 115].

Finally, the family of wedges on Minkowski space forms a causally separating set (see Sect. 10.2.1), so that it is possible to construct a complete net of local algebras in terms of a *single* algebra and a suitable representation of the Poincaré group. Making use of this observation, the construction of a QFT model is reduced to the construction of a so-called Borchers triple [53], consisting of an algebra localized in a wedge, together with a suitable representation of the Poincaré group and a vacuum vector. This general construction scheme is reviewed in the Sects. 10.2.2–10.2.4.

While model-independent investigations did lead to the idea of constructing local nets from wedge algebras, they did not (yet) shed much light onto the question *how* this single algebra, on which the whole construction rests, should be realized. This question is closely related to the question of how to model interaction without making use of classical concepts, and as of now, has found no general answer.

Thinking of quantum descriptions of interactions, the S-matrix is an object of central importance. Unfortunately, in theories with particle production, the S-matrix is also of such a complicated form that it is not a manageable quantity for describing interactions. There is, however, an exception to this rule: For certain *integrable* models on two-dimensional Minkowski space, the S-matrix is of a simple factorizing form, and in particular does not allow for production processes. In that setting, it is therefore possible to use it as suitable description of the quantum dynamics, and generate wedge algebras based on such an S-matrix.

This approach was initiated by Schroer [125], who introduced certain wedge-local fields in this context (see Sect. 10.3.2). This idea was then thoroughly investigated and generalized, in particular with regard to the analysis of *local* observables, by several authors. We will in Sect. 10.3 review the construction of integrable models on two-dimensional Minkowski space by these methods, which led to the solution of the corresponding inverse scattering problem [97].

To complement the concrete construction of integrable models on the basis of a factorizing S-matrix, we will also review a different construction scheme. As in the case of integrable models, the central object is that of a Borchers triple. However, here the input does not consist of an S-matrix, but rather amounts to a *deformation procedure*: Starting from the Borchers triple of some arbitrary QFT (in arbitrary dimension), one modifies/deforms it to a new, inequivalent one. The method to be used here is inspired [78] from non-commutative Minkowski space (cf. Chap. 7), and now goes under the name of *warped convolution* [46, 53]. We review this deformation procedure in Sect. 10.4, where it is also compared to the approach taken in Sect. 10.3.

10.2 Operator-Algebraic Constructions Based on Wedge Algebras

Most operator-algebraic approaches to constructing quantum field theory models on Minkowski space split the construction problem into two steps: First one constructs a *single* von Neumann algebra $\mathcal{M}$ and a representation of the Poincaré group with specific properties, and then these data are used to generate a full local net. The algebra $\mathcal{M}$ considered in the first step contains all observables localized in a special wedge-shaped region of Minkowski space, *wedge* for short. Before going into the quantum field theoretic constructions, we define these regions and discuss their geometric properties.

10.2.1 Wedges

In this section we will be working in Minkowski spacetime $\mathbb{R}^d$ of general dimension $d \geq 2$, equipped with proper coordinates $x = (x_0, x_1, \ldots, x_{d-1})$, with x_0 being the time coordinate.[3] The following regions will play a special role.

Definition 10.2.1 (*Wedges*) The *right wedge* is the set

$$W_R := \{x \in \mathbb{R}^d \; : \; x_1 > |x_0|\}, \tag{10.1}$$

Any set $W \subset \mathbb{R}^d$ which is a proper Poincaré transform of W_R, i.e. $W = \Lambda W_R + x$ for some $\Lambda \in \mathcal{L}_+$, $x \in \mathbb{R}^d$, is called a *wedge*. The set of all wedges is denoted $\mathcal{W}$.

Wedges can equivalently be defined as regions that are bounded by two non-parallel characteristic hyperplanes [135], thereby avoiding reference to the particular wedge W_R. However, for our purposes the above definition will be convenient.

One might wonder why wedge regions deserve particular attention, and as a first answer to this question, we note that wedges have the special property that their causal complements are of the same form. In fact, one directly checks that the causal complement of W_R is $W_R' = -W_R$, that is, a proper Poincaré image of W_R. Thus $W_R' \in \mathcal{W}$; it is customary to call W_R' the *left wedge* and denote it by W_L. By covariance, one then finds that for any wedge W, also W' is a wedge. In the later

[3] Although we will mostly be working with Minkowski space here, it should be noted that similar families of regions can also be defined in other situations: On the one-dimensional line, the half lines (a, ∞) and $(-\infty, a)$, $a \in \mathbb{R}$, have the same properties as the wedges in Minkowski space (see also the discussion in Sect. 10.3.5). Also the family of all intervals on a circle, of prominent importance in chiral conformal field theory (Chap. 8), shares many properties with the family of wedges on Minkowski space, as it is the orbit of a reference region (e.g., the upper semi circle) under a symmetry group (the Möbius group PSL(2, $\mathbb{R}$)), see for example [108].

Furthermore, on certain curved spacetimes, such as de Sitter space [25, 36], anti de Sitter space [45, 93], and more general curved spacetimes [51, 58], families of regions with properties analogous to Minkowski space wedges exist.

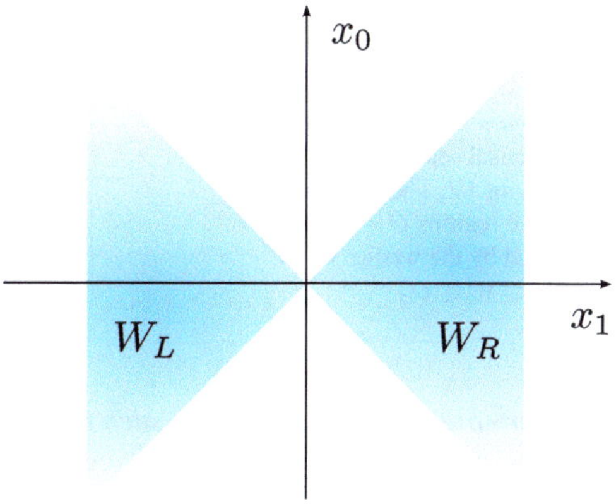

Fig. 10.1 The *right* and *left* wedge in two-dimensional Minkowski space. Both regions extend to (*right*, respectively *left*) spacelike infinity

constructions, this symmetry between wedges and their causal complements will be parallel to that of von Neumann (wedge) algebras and their commutants (Fig. 10.1).

As a consequence of $W_R' = -W_R$, we also have $W_R'' = (-W_R)' = -W_R' = W_R$, so W_R is *causally complete*.[4] By covariance, this implies $W'' = W$ for any wedge W.

In the quantum field theory setting, we will be interested in mappings from $\mathcal{W}$ into the family of all von Neumann algebras on a fixed Hilbert space, complying with the usual assumptions of isotony, locality, and covariance (cf. Chap. 1). As a preparation for this, we here consider the inclusion, causal separation, and covariance properties of wedges.

Beginning with inclusions, let $(x, \Lambda) \in \mathcal{P}_+$ denote a proper Poincaré transformation. It is clear that if $\Lambda W_R = W_R$ and $x \in \overline{W_R}$, then $\Lambda W_R + x \subset W_R$ (observe that wedges are in particular convex cones). In fact, also the converse is true, namely $\Lambda W_R + x \subset W_R$ implies $\Lambda W_R = W_R$ and $x \in \overline{W_R}$ [135].

We thus see that there are only relatively few pairs of wedges W_1, W_2 that form inclusions. Namely, $W_1 \subset W_2$ if and only if $W_1 = W_2 + x$ with $x \in \overline{W_2}$. Since causal complements of wedges are also wedges, the same applies to pairs of spacelike separated wedges: $W_1 \subset W_2'$ if and only if $W_1 = W_2' + x'$ with $x' \in \overline{W_2'}$. This simple structure of the family of causal configurations of wedges is used in constructions based on wedge algebras in a crucial manner.

Finally, we point out that the set $\mathcal{W}$ is *causally separating* in the following sense: Given any two bounded, convex, causally complete sets $O_1, O_2 \subset \mathbb{R}^d$ (such as double cones), that are spacelike separated, $O_1 \subset O_2'$, there exists $W \in \mathcal{W}$ such that $O_1 \subset W \subset O_2'$ [135, Proposition 3.7] (see Fig. 10.2).

We collect these properties in a proposition.

Proposition 10.2.2 *Any wedge $W \in \mathcal{W}$ is an open, convex, unbounded, causally complete set. The set $\mathcal{W}$ of all wedges in Minkowski space $\mathbb{R}^d$ is causally separating, and invariant under the action of the Poincaré group and causal complementation.*

[4] This implies that W_R is globally hyperbolic, and can be regarded as a spacetime in its own right.

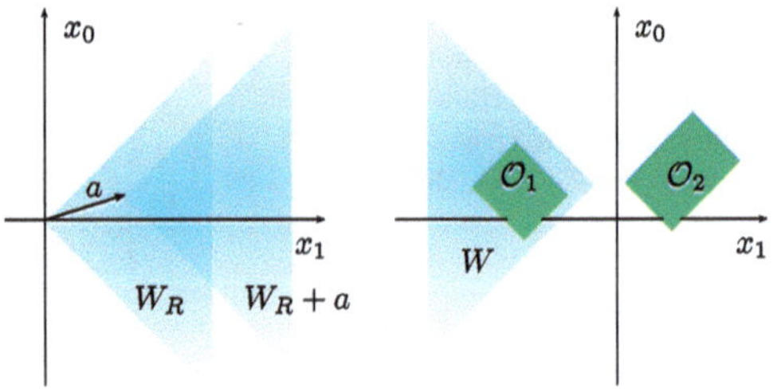

Fig. 10.2 *Left figure* An inclusion of two *right* wedges $W_R + a \subset W_R$. *Right figure* An illustration of the causal separation property of $\mathcal{W}$: The two spacelike regions O_1, O_2 are separated by the wedge W, i.e. $O_1 \subset W \subset O_2'$

Furthermore, given $W_1, W_2 \in \mathcal{W}$ *such that* $W_1 \subset W_2$*, then* $W_1 = W_2 + x$ *for some* $x \in \overline{W_2}$.

By Definition 10.2.1, arbitrary Poincaré transformations leave $\mathcal{W}$ invariant as a set. For any given wedge $W \in \mathcal{W}$, there also exist specific Lorentz transformations which preserve W or map it to its causal complement, respectively. Consider the proper Lorentz transformations, $t \in \mathbb{R}$,

$$j_{W_R}(x) := (-x_0, -x_1, x_2, \ldots, x_{d-1}), \tag{10.2}$$
$$\Lambda_{W_R}(t)x := (\cosh(2\pi t)x_0 + \sinh(2\pi t)x_1, \sinh(2\pi t)x_0 + \cosh(2\pi t)x_1, x_2, \ldots, x_{d-1}).$$

The map j_{W_R} in the first line is the reflection about the $(d-2)$-dimensional *edge* $E(W_R) := \{x : x_0 = x_1 = 0\}$ of W_R, and maps W_R onto $-W_R = W_R'$. The second line defines the one parameter group of Lorentz boosts $\Lambda_{W_R}(t)$ in the x_1-direction. By computing the eigenvectors and eigenvalues of $\Lambda_{W_R}(t)$, $t \in \mathbb{R}$, one finds that these belong to the group of all Lorentz transformations that leave W_R invariant as a set.

More generally, to any wedge $W = \Lambda W_R + x$ we can assign its edge $E(W) := \Lambda E(W_R) + x$, the reflection $j_W := (x, \Lambda) j_{W_R}(x, \Lambda)^{-1}$ about $E(W)$, satisfying $j_W(W) = W'$, and a one-parameter group of boosts, $\Lambda_W(t) := (x, \Lambda)\Lambda_{W_R}(t)(x, \Lambda)^{-1}$, which preserve W.

Wedges in $d = 2$ **dimensions**. While so far the spacetime dimension $d \geq 2$ was arbitrary, we now specialize to the two-dimensional situation, where wedges have a number of additional properties.

To begin with, the causal complement of any one point set $\{x\}$ in $\mathbb{R}^2$ consists precisely of the disjoint union of the two wedges $W_R + x$ and $W_L + x$ (cf. Fig. 10.1). These are in fact all wedges in this setting: In $d = 2$ dimensions, the proper Lorentz group is generated by the one parameter group $\{\Lambda_{W_R}(t)\}_{t \in \mathbb{R}}$ and the spacetime reflection $j_{W_R}(x) = -x$ (which maps W_R onto W_L, as just observed). Thus in this case,

$$\mathcal{W} = \{W_R + x,\ W_L + x \ :\ x \in \mathbb{R}^2\} \qquad (d = 2). \tag{10.3}$$

In two dimensions, wedges can also be most easily visualized.

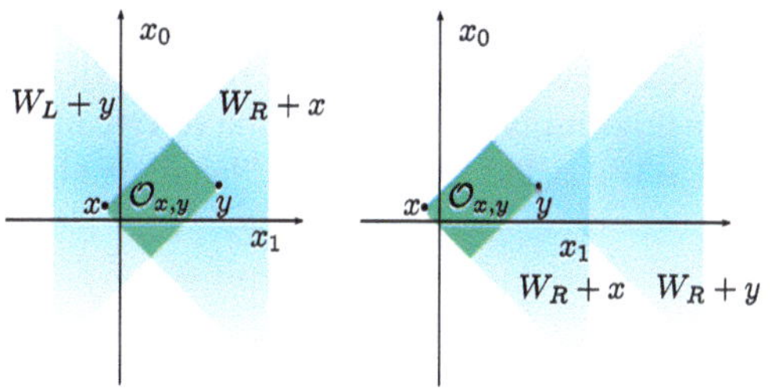

Fig. 10.3 *Left figure* The double cone $O_{x,y}$ as the wedge intersection $(W_R + x) \cap (W_L + y)$. *Right figure* The double cone $O_{x,y}$ and the associated inclusion of *right* wedges $W_R + y \subset W_R + x$

Double cones, typically the most important localization regions in quantum field theory, are defined as intersections of forward and backward light cones (Chap. 1). However, in two dimensions, this is the same as taking intersections of left and right wedges,

$$O_{x,y} := (W_R + x) \cap (W_L + y) = (W_R + x) \cap (W_R + y)' ; \tag{10.4}$$

and this set is nonempty if and only if $(y - x) \in W_R$. Any double cone arises in this way, and from the second equality in (10.4), we see that the double cone $O_{x,y}$ is the *relative causal complement* of the inclusion $W_R + y \subset W_R + x$ (see Fig. 10.3).

In $d > 2$ dimensions, nonempty intersections of two wedges are unbounded regions, here one needs an intersection of several wedges to arrive at a bounded region. In particular, double cones are not given as relative causal complements of inclusions of wedges in $d > 2$. Quite generally, inclusions are easier to analyze than intersections. This is true on the geometric level of wedges,[5] but even more on the level of von Neumann algebras, and provides one of the many reasons why certain construction procedures are easier in $d = 2$.

10.2.2 Wedge Algebras and Borchers Triples

Having clarified the geometrical preliminaries, we now turn to studying algebras of observables localized in wedges. This discussion will take place in the setting of a vacuum representation of a quantum field theory on Minkowski space $\mathbb{R}^d$, $d \geq 2$ (Chap. 1). We therefore consider a Hilbert space $\mathcal{H}$, carrying a strongly continuous (anti-)unitary positive energy representation U of the proper Poincaré group with a U-invariant unit vector Ω, implementing the vacuum state. The observables of a quantum field theory are represented as operators on $\mathcal{H}$: Associated with any localization region $O \subset \mathbb{R}^d$, we have the C^*-algebra $\mathfrak{A}(O) \subset \mathcal{B}(\mathcal{H})$ of all observables localized in O, and the usual assumptions of isotony, locality and covariance under U are assumed to hold for the net $O \mapsto \mathfrak{A}(O)$. As we are in a vacuum representation,

[5]Note, for example, that for any $n \in \mathbb{N}$, there exists a family of n wedges $W_1, \ldots, W_n \subset \mathbb{R}^4$ such that $W_i \cap W_j = \emptyset$ for $i \neq j$ [135].

we also assume that Ω is cyclic and separating for $\mathfrak{A}(O)''$, for each double cone O (Reeh-Schlieder property, Theorem 1.1.1).

In this setting, we introduce the von Neumann algebra $\mathcal{M}$ associated with the right wedge W_R as the smallest von Neumann algebra containing all $\mathfrak{A}(O)$, $O \subset W_R$,

$$\mathcal{M} := \bigvee_{O \subset W_R} \mathfrak{A}(O)''. \tag{10.5}$$

This algebra has a number of properties which reflect the geometric properties of W_R, and which follow directly from the definition of $\mathcal{M}$ and the properties of the net $O \mapsto \mathfrak{A}(O)$: For any Poincaré transformation g with $gW_R \subset W_R$, we have $U(g)\mathcal{M}U(g)^{-1} \subset \mathcal{M}$, and for any Poincaré transformation $\tilde{g}$ with $\tilde{g}W_R \subset W_R'$, we have $U(\tilde{g})\mathcal{M}U(\tilde{g})^{-1} \subset \mathcal{M}'$. Furthermore, the vacuum vector Ω is cyclic for $\mathcal{M}$ (because W_R contains a double cone) and separating for $\mathcal{M}$ (because W_R' also contains a double cone).

As these properties will be essential in the following, we isolate them in a definition.[6] The term "Borchers triple", in honor of H.-J. Borchers who studied such systems (see, e.g., [23]), was suggested in [53].

Definition 10.2.3 A d-dimensional *Borchers triple* $(\mathcal{M}, U, \Omega)$ relative to $W \in \mathcal{W}$ consists of

(a) a strongly continuous (anti-)unitary positive energy representation U of the proper Poincaré group $\mathcal{P}_+$ of $\mathbb{R}^d$ on some Hilbert space $\mathcal{H}$,
(b) a unit vector $\Omega \in \mathcal{H}$ that is invariant under U, and
(c) a von Neumann algebra $\mathcal{M} \subset \mathcal{B}(\mathcal{H})$ which has Ω as a cyclic and separating vector and which satisfies

$$U(g)\mathcal{M}U(g)^{-1} \subset \mathcal{M} \quad \text{for all } g \in \mathcal{P}_+ \text{ with } gW \subset W\,, \tag{10.6}$$

$$U(g)\mathcal{M}U(g)^{-1} \subset \mathcal{M}' \quad \text{for all } g \in \mathcal{P}_+ \text{ with } gW \subset W'\,. \tag{10.7}$$

A von Neumann algebra $\mathcal{M}$ in a Borchers triple relative to W describes quantum observables that are localized (in the specified sense) in the wedge W, and will also be referred to as *wedge algebra* when U and Ω are fixed and clear from the context. For the sake of concise formulations, we agree to drop the specification "relative to W" for Borchers triples relative to our standard reference wedge $W = W_R$, or if W is clear from the context.

In comparison to a full quantum field theory, described by an infinite collection of algebras in specific relative positions, the data $(\mathcal{M}, U, \Omega)$ of a Borchers triple are much simpler. However—and this observation is central for all that follows—one can reconstruct a full net of local algebras from a Borchers triple, essentially by Poincaré symmetry (cf. in particular [21–23], [14, Sect. 7.3.6]), as shall be explained below.

---

[6] Note that we deviate here slightly from the definition in [53, Definition 4.1], where the term "causal Borchers triple" has been used. Also note that in [82], there is a related but different definition of the term "Borchers triple". We will always stick to the definition given here.

Setting ourselves the task to define a local net $O \mapsto \mathcal{A}(O)$ of von Neumann algebras corresponding to a given Borchers triple $(\mathcal{M}, U, \Omega)$, we have to give definitions of the algebras $\mathcal{A}(O)$, $O \subset \mathbb{R}^d$. Assuming for the sake of concreteness that the Borchers triple is relative to the right wedge W_R, one first sets

$$\mathcal{A}(\Lambda W_R + x) := U(x, \Lambda)\mathcal{M}\, U(x, \Lambda)^{-1}, \qquad (x, \Lambda) \in \mathcal{P}_+ . \tag{10.8}$$

This defines for any wedge $W \in \mathcal{W}$ a von Neumann algebra $\mathcal{A}(W) \subset \mathcal{B}(\mathcal{H})$ (Note that (10.8) is well-defined because of (10.6)). Making use of the properties of the Borchers triple, one checks that (10.8) yields a map $\mathcal{W} \ni W \mapsto \mathcal{A}(W) \subset \mathcal{B}(\mathcal{H})$ from wedges on $\mathbb{R}^d$ to von Neumann algebras in $\mathcal{B}(\mathcal{H})$ that (a) is inclusion preserving (isotony), (b) maps spacelike separated wedges to commuting algebras (locality), and (c) transforms covariantly under the adjoint action of U by its very definition. Moreover, Ω is cyclic and separating for any $\mathcal{A}(W)$, $W \in \mathcal{W}$.

To proceed from the algebras $\mathcal{A}(W)$ to algebras associated with smaller regions, we first consider double cones. Any double cone O is an intersection of wedges, $O = \bigcap_i W_i$, where i runs over some index set and $W_i \in \mathcal{W}$. We associate with it the von Neumann algebra

$$\mathcal{A}(\bigcap_i W_i) := \bigcap_i \mathcal{A}(W_i) . \tag{10.9}$$

Making use of the properties listed in Proposition 10.2.2, one can show that this yields a map from double cones to von Neumann algebras which inherits the isotony, locality, and covariance properties from the wedge net $\mathcal{W} \ni W \mapsto \mathcal{A}(W) \subset \mathcal{B}(\mathcal{H})$ [21].

Finally, given an arbitrary bounded region O, we define $\mathcal{A}(O)$ as the smallest von Neumann algebra containing all $\mathcal{A}(D)$, where $D \subset O$ is a double cone. Also this step preserves isotony, locality, and covariance. We therefore note:

Proposition 10.2.4 *Any d-dimensional Borchers triple $(\mathcal{M}, U, \Omega)$ defines a local net $O \mapsto \mathcal{A}(O)$ of von Neumann algebras on $\mathbb{R}^d$ such that*

(a) $O \mapsto \mathcal{A}(O)$ is isotonous and local, and transforms covariantly under U.
(b) $\mathcal{A}(W_R) = \mathcal{M}$ is the algebra associated with the right wedge W_R.
(c) For any $\tilde{W} \in \mathcal{W}$, the vacuum vector Ω is cyclic and separating for $\mathcal{A}(\tilde{W})$.

In view of this observation, the construction of local nets of von Neumann algebras, i.e. models of quantum field theories, is closely related to the construction of Borchers triples $(\mathcal{M}, U, \Omega)$. We will see below (p. 407) that in $d = 2$ dimensions, a slight variation of this construction is available.

The representation U (and the vector Ω) describe the Poincaré symmetry of the model theory constructed from $(\mathcal{M}, U, \Omega)$. By decomposition into irreducible components, U yields a list of all species of stable particles in this theory, and can thus be thought of as data implementing kinematic properties. From the point of view of constructing examples, the representation U and vacuum vector Ω pose no problems—In fact, it follows from Haag-Ruelle scattering theory (cf. Sect. 1.3) that in any theory, one may use a standard Fock space construction for realizing these data in terms of a single particle representation U_1 of the Poincaré group, cf. Sect. 10.2.3.

The dynamics and interaction of the model corresponding to $(\mathcal{M}, U, \Omega)$ is encoded in an indirect manner in the choice of von Neumann algebra $\mathcal{M}$, completing U, Ω to a Borchers triple. In the present general context, where in particular no link to a classical Lagrangian or equation of motion is made, there is currently no general principle known to select $\mathcal{M}$ in such a way that the model constructed along the lines described above exhibits features of a particular type of interaction. As we will see later in Sect. 10.3, it is however well possible to realize $\mathcal{M}$ in terms of a two-particle S-matrix in the setting of integrable models on two-dimensional Minkowski space, or to modify a given $\mathcal{M}$ to a new one by a deformation procedure (Sect. 10.4).

Independent of concrete construction ideas for $\mathcal{M}$, it should be noted that the conditions in Definition 10.2.3 impose strong restrictions on $\mathcal{M}$ in general. This is illustrated by a theorem of Longo [107, Theorem 3] (see also Driessler [64]), which in the context of Borchers triples reads as follows.

Theorem 10.2.5 *Let $(\mathcal{M}, U, \Omega)$ be a Borchers triple. If $\mathcal{M} \neq \mathbb{C}$, then $\mathcal{M}$ is a factor of type III_1.*

Since the hyperfinite type III_1 factor is known to be unique [85], this implies that the internal algebraic structure of $\mathcal{M}$ is almost uniquely fixed by Definition 10.2.3.

A result of a similar nature is a famous theorem of Borchers, stating that there exist also strong restrictions on the modular data of a wedge algebra [23]. Since Ω is cyclic and separating for the wedge algebra $\mathcal{M}$ of a Borchers triple $(\mathcal{M}, U, \Omega)$, we can consider the corresponding modular data[7] J_W, Δ_W (see Sect. 1.5). In the present setting, Borchers' result can be formulated in the following way. (See [23] for the original work, [72] for a simplified proof, and [53] for a discussion in the context of Borchers triples.)

Theorem 10.2.6 *Let $(\mathcal{M}, U, \Omega)$ be a d-dimensional Borchers triple. Then the modular conjugation J_W and modular unitaries Δ_W^{it} act on the translations $U(x) := U(x, 1)$ according to, $t \in \mathbb{R}$, $x \in \mathbb{R}^d$,*

$$\Delta_W^{it} U(x) \Delta_W^{-it} = U(\Lambda_W(t)x)\,, \tag{10.10}$$

$$J_W U(x) J_W = U(j_W x)\,, \tag{10.11}$$

where j_W, $\Lambda_W(t)$ denote the Lorentz transformations associated with W (10.2).

According to this theorem, the modular data can not be distinguished from the represented reflections $U(j_W)$ and boosts $U(\Lambda_W(t))$ via their action on the translations. This is in line with theorems of Bisognano and Wichmann [21, 22], who

[7] Here and in many places in the following text, we will make use of the Tomita-Takesaki modular theory of von Neumann algebras with cyclic separating vector, see for example [35] for an introduction.

showed that for wedge algebras generated by *Wightman fields* [130], one has[8]

$$\Delta_W^{it} = U(\Lambda_W(t)), \qquad J_W = U(j_W) \qquad \text{(Bisognano-Wichmann property).} \tag{10.12}$$

In comparison to these strong results on the inner structure of wedge algebras, and on their modular data, very little is known about the double cone algebras $\mathcal{A}(O)$, which are given rather indirectly as intersections of wedge algebras (10.9). In the general situation described here, it is in particular not known whether the vacuum vector Ω is cyclic for the double cone algebras, and not even if these algebras are non-trivial in the sense that $\mathcal{A}(O) \neq \mathbb{C} \cdot 1$.

Physically interesting models complying with the principle of causality have many local observables. One therefore has to add extra conditions on a Borchers triple, implying a sufficiently rich local structure, and in particular non-triviality of the intersections (10.9). This question will be discussed in Sect. 10.2.4.

Constructing local nets from Borchers triples in $d = 2$. Borchers' theorem also applies in situations where there is no Lorentz symmetry present a priori. Namely, if $(\mathcal{M}, T, \Omega)$ is a Borchers triple with only translational symmetry—that is, T is a positive energy representation of the translation group $\mathbb{R}^d$ instead of the Poincaré group, Eq. (10.6) in Definition 10.2.3(c) is required to hold only for translations x with $W + x \subset W$, and condition (10.7) drops out because there is no translation mapping a wedge W into its causal complement W'—the conclusion of Theorem 10.2.6 still holds. In $d = 2$ dimensions, this circumstance can then be used to extend T to a representation U of the proper Poincaré group by taking (10.12) as the *definition* of $U(\Lambda_W(t))$, $U(j_W)$. Since $\Delta^{it}\mathcal{M}\Delta^{-it} = \mathcal{M}$, $t \in \mathbb{R}$, and $J\mathcal{M}J = \mathcal{M}'$ by Tomita's Theorem, it then follows that $(\mathcal{M}, U, \Omega)$ is a Borchers triple in the usual sense.

This observation also brings us to the previously mentioned variation of constructing a local net from a Borchers triple in $d = 2$. Here it can be advantageous to start from a Borchers triple $(\mathcal{M}, T, \Omega)$ with only translational symmetry, as just described, and define the net $O \mapsto \mathcal{A}(O)$ via $\mathcal{M}$ and the Poincaré representation U generated by T and the modular data Δ^{it}, J of $(\mathcal{M}, \Omega)$. In that case, one observes that the definition of the wedge algebras (10.8) is Haag-dual, i.e. satisfies $\mathcal{A}(W') = \mathcal{A}(W)'$ for all $W \in \mathcal{W}$, and the definition of the double cone algebras (10.9) amounts to

$$\mathcal{A}(O_{x,y}) = \mathcal{M}(x) \cap \mathcal{M}'(y), \tag{10.13}$$

where $\mathcal{M}(x) = U(x)\mathcal{M}U(x)^{-1}$, $\mathcal{M}'(y) = U(y)\mathcal{M}'U(y)^{-1}$, $(y - x) \in W_R$. This intersection is the relative commutant of the inclusion $\mathcal{M}(y) \subset \mathcal{M}(x)$, and closely resembles the geometric situation, where the double cone $O_{x,y}$ coincides with the relative causal complement of the inclusion $W_R + y \subset W_R + x$ (10.4). In our

[8]This statement is stronger than the one of Theorem 10.2.6, which does not yield equality of the modular data J_W, Δ_W^{it} with the Lorentz transformations $U(j_W)$, $U(\Lambda_W(t))$. However, in the context of a local net satisfying further assumptions, including asymptotic completeness, Mund proved that the Bisognano-Wichmann property *does* follow from Borchers' theorem [114].

subsequent analysis, this will be of advantage in comparison to the general construction, where $\mathcal{A}(O_{x,y})$ is only a subalgebra of the relative causal complement.

10.2.3 Standard Pairs and Free Field Theories

The simplest quantum field theories are those describing particles without any interaction ("free" theories). Such models are very thoroughly studied, and can be presented in many different ways. Whereas the usual approach is to present them as quantized versions of free field theories (cf. Chap. 3), we would like to stress here that free field theories can be constructed without any reference to their classical counterparts, and perfectly fit into the setting of Borchers triples. The construction of such "free" Borchers triples will not only give us first examples of Borchers triples and the construction procedure based on them, but it will also serve as the starting point for the construction of interacting theories, considered later.

As one might expect in a free theory, the only required input is a description of the single particle spectrum. Such single particle data can, together with a suitable notion of localization, be described conveniently in terms of a so-called *standard pair*. Recall for the following definition that a real-linear subspace $\mathcal{K}_1 \subset \mathcal{H}_1$ of a complex Hilbert space $\mathcal{H}_1$ is called *standard* if it is cyclic in the sense that $\mathcal{K}_1 + i\,\mathcal{K}_1 \subset \mathcal{H}_1$ is dense, and separating in the sense that $\mathcal{K}_1 \cap i\,\mathcal{K}_1 = \{0\}$ [108].

Definition 10.2.7 (*Standard pairs*) A d-dimensional standard pair $(\mathcal{K}_1, U_1)$ (with Poincaré symmetry) relative to a wedge $W \in \mathcal{W}$ consists of a closed real standard subspace $\mathcal{K}_1 \subset \mathcal{H}_1$ of some complex Hilbert space $\mathcal{H}_1$, which carries a unitary strongly continuous positive energy representation U_1 of $\mathcal{P}_+$ such that

$$U_1(g)\mathcal{K}_1 \subset \mathcal{K}_1 \quad \text{for all } g \in \mathcal{P}_+ \text{ with } gW \subset W\,, \tag{10.14}$$

$$U_1(g)\mathcal{K}_1 \subset \mathcal{K}_1' \quad \text{for all } g \in \mathcal{P}_+ \text{ with } gW \subset W'\,, \tag{10.15}$$

where $\mathcal{K}_1' = \{\psi \in \mathcal{H}_1 \,:\, \mathrm{Im}\langle\psi, \xi\rangle = 0 \text{ for all } \xi \in \mathcal{K}_1\}$ is the symplectic complement of $\mathcal{K}_1$ in $\mathcal{H}_1$ w.r.t. the symplectic form $\mathrm{Im}\langle\,\cdot\,,\,\cdot\,\rangle$.

We added the term "with Poincaré symmetry" here because standard pairs are often considered with translational symmetry only [19, 100, 109]. In this text, we will however always consider standard pairs with Poincaré symmetry, and therefore suppress this term from now on. Just as for Borchers triples, we will also drop the specification "relative to W" in case the reference region is the right wedge W_R, or clear from the context.

The relation between standard pairs and Borchers triples is two-fold. We first consider the step from a Borchers triple to a standard pair.

Lemma 10.2.8 *Let $(\mathcal{M}, U, \Omega)$ be a Borchers triple on a Hilbert space $\mathcal{H}$, and*

$$\mathcal{K} := \{A\Omega \,:\, A = A^* \in \mathcal{M}\}^{\|\cdot\|}\,. \tag{10.16}$$

Let furthermore $Q \in \mathcal{B}(\mathcal{H})$ be an orthogonal projection which commutes with the representation U. Then $(Q\mathcal{K}, U|_{Q\mathcal{H}})$ is a standard pair on $Q\mathcal{H}$.

The proof of this lemma uses the modular theory of standard subspaces, where to any real standard subspace $\mathcal{K}_1$ one associates a Tomita operator $S_{\mathcal{K}_1} : \mathcal{K}_1 + i\,\mathcal{K}_1 \mapsto \mathcal{K}_1 + i\,\mathcal{K}_1$, $S_{\mathcal{K}_1}(k + i\ell) := k - i\ell$, which in turn defines $\mathcal{K}_1$ by $\mathcal{K}_1 = \ker(1 - S_{\mathcal{K}_1})$. These objects satisfy properties closely analogous to the von Neumann algebra case, see [108] for more details and an account of the literature. In the situation of the above lemma, with $Q = 1$, the modular data J, Δ of $(\mathcal{M}, \Omega)$ coincide with those of the standard subspace, $J = J_{\mathcal{K}_1}$, $\Delta = \Delta_{\mathcal{K}_1}$. Since $J_{\mathcal{K}_1}$ is known to map $\mathcal{K}_1$ onto its symplectic complement $\mathcal{K}_1'$ by the subspace version of Tomita's Theorem, the conclusion follows in this case. The generalization to $Q \neq 1$ is straightforward; one uses that Q commutes with the modular data by virtue of Theorem 10.2.6.

Lemma 10.2.8 can be applied to extract single particle information from Borchers triples. To illustrate this, consider the case that in the representation U of the Borchers triple, there exists an isolated eigenvalue $m > 0$ of the mass operator, and take the projection $Q := E_{\{m\}}$ as the corresponding spectral projection. Then $\mathcal{H}_1 := Q\mathcal{H}$ describes single particle vectors of mass m, and $\mathcal{K}_1$ the "single particle vectors localized in W". In this projection process, a lot of information is lost,[9] and only single particle data remain. Hence many QFTs give rise to the same standard pairs by projecting their Borchers triples to the single particle level.

However, for each standard pair one can without further input construct a *specific* Borchers triple, representing an interaction-free theory. This brings us to the link in the other direction, namely from a standard pair to a Borchers triple. This step can be carried out by second quantization. In this context, we denote by $\Gamma(\mathcal{H}_1)$ the Bose Fock space over a Hilbert space $\mathcal{H}_1$, and by $V(\xi)$, $\xi \in \mathcal{H}_1$, the Weyl operators on $\Gamma(\mathcal{H}_1)$, characterized by the familiar Weyl relation $V(\xi)V(\psi) = e^{-\frac{i}{2}\mathrm{Im}\langle\xi,\psi\rangle}\,V(\xi + \psi)$ and $V(\xi)\Omega = e^{-\frac{1}{4}\|\xi\|^2}\,e_\otimes^\xi$, with $e_\otimes^\xi = \bigoplus_{n=0}^\infty \xi^{\otimes n}/\sqrt{n!}$, and Ω the Fock vacuum (cf. Sect. 4.2).

Proposition 10.2.9 *Let $(\mathcal{K}_1, U_1)$ be a standard pair on a Hilbert space $\mathcal{H}_1$. On the Fock space $\Gamma(\mathcal{H}_1)$, consider the second quantization $\Gamma(U_1)$ of U_1, the Fock vacuum Ω, and the von Neumann algebra*

$$\mathcal{M} := \{V(\xi) \,:\, \xi \in \mathcal{K}_1\}''. \tag{10.17}$$

Then $(\mathcal{M}, U, \Omega)$ is a Borchers triple, and projecting it with $Q = P_1$ (the projection onto $\mathcal{H}_1$) returns the standard pair $(\mathcal{K}_1, U_1)$.

This relation between real standard spaces and the algebras of a free field are known from the work of Araki [6, 7]. Their modular data were shown to be of second quantized form by Eckmann and Osterwalder [68], see also [105].

Whereas the version presented here is suitable for Bosonic systems with commuting fields at spacelike separation, there is also a version adapted to the Fermionic

[9] This is even the case for the projection $Q = 1$.

case, where fields anticommute [16, 73]. This formulation makes use of so-called "twisted duality" [61], and requires only minor modifications. We will not discuss it any further here.

In view of these relations between standard pairs and Borchers triples, all that is required for the construction of a free (second quantization) Borchers triple is a corresponding (single particle) standard pair. This requires in particular a (single particle) representation U_1 of the proper Poincaré group and a standard subspace $\mathcal{K}_1$, which, as mentioned above, is determined by its modular data according to $\mathcal{K}_1 = \ker(1 - J_{\mathcal{K}_1}\Delta_{\mathcal{K}_1}^{1/2})$.

For a concrete construction of this space, one can therefore anticipate the Bisognano-Wichmann relation (10.12) between geometric data and modular data, and use it as a *definition* for the modular data, and hence the standard subspace. This idea is known as *modular localization* [36, 70, 116], see also [15].

In more detail, Brunetti et al. consider a (anti-)unitary strongly continuous positive energy representation U_1 of the proper Poincaré group on a Hilbert space $\mathcal{H}_1$, and the one parameter groups $\Lambda_W(t)$ and reflections j_W, $W \in \mathcal{W}$, in this representation [36]. By Stone's theorem, there exists a selfadjoint generator R_W such that $U_1(\Lambda_W(t)) = e^{itR_W}$, and one defines for the right wedge $W = W_R$

$$\Delta_1 := e^{-\pi R_{W_R}}\,, \qquad J_1 := U(j_{W_R})\,, \qquad S_1 := J_1\Delta_1^{1/2}\,. \tag{10.18}$$

By comparison with (10.12), we see that this assignment mimics the Bisognano-Wichmann relation, and one defines further

$$\mathcal{K}_1 := \{\psi \in \mathrm{dom}\,\Delta_1^{1/2} \,:\, S_1\psi = \psi\}\,. \tag{10.19}$$

Theorem 10.2.10 ([36]) *$(\mathcal{K}_1, U_1)$ is a standard pair, with Tomita operator $S_{\mathcal{K}_1} = S_1$.*

The main point of this theorem is to demonstrate the inclusion property (10.14), which is linked to the positive energy condition of U_1. For the proof of this, and further results, see [36].

In view of this theorem, we have, for any considered representation U_1 of $\mathcal{P}_+$, an associated standard pair and thus also an associated second quantization Borchers triple. These triples can now be used in the general construction outlined in Sect. 10.2.2 to generate local nets of von Neumann algebras, corresponding to free QFT models.

At the end of Sect. 10.2.2, we mentioned the problem that the algebras corresponding to smaller regions, defined as intersections of wedge algebras (10.4), are not guaranteed to be non-trivial in the setting of a general Borchers triple. In the present context of free field constructions, this problem can however be resolved. To begin with, Brunetti et al. have shown that for any of the representations U_1

considered here, the von Neumann algebras corresponding to spacelike cones have the Fock vacuum as a cyclic vector[10] [36].

Moreover, in the case of "usual" representations U_1, i.e. direct sums of mass $m \geq 0$ finite spin s representations, also cyclicity of the Fock vacuum for algebras associated with double cones is known. In fact, the net resulting from the Borchers triple by application of the procedure in Sect. 10.2.2 (or its twisted version in the Fermionic case) is then a known free field net, which in particular has Ω as a cyclic vector for each double cone algebra.

As an aside, we mention that there also exist "continuous spin" representations of the Poincaré group, for which Wightman fields do not exist [138]. The algebraic construction presented here applies also to such representations, and in fact there exist models of free quantum fields which are localizable only in spacelike cones [116]. The double cone algebras in such models are currently under investigation, and it seems that they might be trivial in such models [92].

10.2.4 Relative Commutants of Wedge Algebras

After this excursion to free field models and modular localization, we return to the setting of a general Borchers triple $(\mathcal{M}, U, \Omega)$, and the question how to ensure large double cone algebras in the construction in Sect. 10.2.2. As mentioned earlier, there are no efficient tools available for analyzing intersections of general families of von Neumann algebras, and therefore we focus on the more particular situation of a relative commutant $\mathcal{M}_1' \cap \mathcal{M}_2$ of an inclusion $\mathcal{M}_1 \subset \mathcal{M}_2$. Since double cones are relative causal complements of wedges in two dimensions (10.4), such an analysis directly applies to double cone algebras in $d = 2$. In higher dimensions $d > 2$, relative commutants of wedge algebras correspond to cylinder like regions which are unbounded in $(d-2)$ perpendicular directions.

The best studied type of inclusions of von Neumann algebras are so-called split inclusions [59], and, as we shall see, they will also play a prominent role in our present context. We first recall the definition of split inclusions and some of their most important properties before we discuss applications of these concepts to relative commutants of wedge algebras.

Definition 10.2.11 Let $\mathcal{M}_1 \subset \mathcal{M}_2 \subset \mathcal{B}(\mathcal{H})$ be an inclusion of von Neumann algebras on some Hilbert space $\mathcal{H}$.

(a) $\mathcal{M}_1 \subset \mathcal{M}_2$ is called split if there exists a type I factor[11] $\mathcal{N}$ such that

$$\mathcal{M}_1 \subset \mathcal{N} \subset \mathcal{M}_2 . \tag{10.20}$$

[10] In case U_1 does not contain the trivial representation, as is adequate for a single particle representation, the algebras corresponding to spacelike cones are also known to be factors of type III_1 [36, 71].

[11] That is, a von Neumann algebra isomorphic to $\mathcal{B}(\tilde{\mathcal{H}})$ for some Hilbert space $\tilde{\mathcal{H}}$.

(b) $\mathcal{M}_1 \subset \mathcal{M}_2$ is called standard if there exists a vector which is cyclic and separating for $\mathcal{M}_1$, $\mathcal{M}_2$, and the relative commutant $\mathcal{M}_1' \cap \mathcal{M}_2$.

In the standard case, split inclusions can be characterized as follows [56, 59].

Lemma 10.2.12 *Let $\mathcal{M}_1 \subset \mathcal{M}_2$ be a standard inclusion of von Neumann algebras on the Hilbert space $\mathcal{H}$. Then $\mathcal{M}_1 \subset \mathcal{M}_2$ is split if and only if there exists a unitary $V : \mathcal{H} \to \mathcal{H} \otimes \mathcal{H}$ such that*

$$V A_1 A_2' V^* = A_1 \otimes A_2', \qquad A_1 \in \mathcal{M}_1, \quad A_2' \in \mathcal{M}_2' . \tag{10.21}$$

Remark: Note that if the assumptions of this lemma are satisfied, the inclusion under consideration has a large relative commutant, namely $\mathcal{M}_1 \cap \mathcal{M}_2' \cong \mathcal{M}_1 \otimes \mathcal{M}_2'$.

In view of Lemma 10.2.12, the split property of an inclusion $\mathcal{M}_1 \subset \mathcal{M}_2$ can be understood as a form of statistical independence between the subsystems described by the commuting algebras $\mathcal{M}_1$ and $\mathcal{M}_2'$ of the larger system identified with $\mathcal{M}_1 \vee \mathcal{M}_2'$ (see the review [132] for a detailed discussion of these matters, and references to the original literature). Namely, it implies that for any pair of normal states φ_1 on $\mathcal{M}_1$ and φ_2 on $\mathcal{M}_2'$, there exists a normal state φ on $\mathcal{M}_1 \vee \mathcal{M}_2'$ such that $\varphi|_{\mathcal{M}_1} = \varphi_1$, $\varphi|_{\mathcal{M}_2'} = \varphi_2$, expressing the fact that states in the subsystems $\mathcal{M}_1$ and $\mathcal{M}_2'$ can be prepared independently of each other. Moreover, φ can be chosen in such a way that there are no correlations between "measurements" in $\mathcal{M}_1$ and $\mathcal{M}_2'$, i.e. as a product state

$$\varphi(A_1 A_2') = \varphi_1(A_1) \cdot \varphi_2(A_2') , \qquad A_1 \in \mathcal{M}_1 , \; A_2' \in \mathcal{M}_2' .$$

Taking $\mathcal{M}_1 = \mathcal{A}(O_1)$ and $\mathcal{M}_2' = \mathcal{A}(O_2)$ as the observable algebras of two spacelike separated regions $O_1 \subset O_2'$ in a quantum field theory given by a net $\mathcal{A}$, some form of statistical independence between $\mathcal{M}_1$ and $\mathcal{M}_2'$ can be expected on physical grounds. For the massive free field, the existence of normal product states for such pairs of local algebras was shown by Buchholz [37]. A corresponding analysis for algebras of free Fermi fields, and for the Yukawa$_2$+$P(\varphi)_2$ model has been carried by Summers [131].

Examples of theories violating the split property can be obtained by considering models with a non-compact global symmetry group, or certain models with infinitely many different species of particles [59]. Such theories have an immense number of local degrees of freedom, and according to the analysis of Buchholz and Wichmann [48], it is precisely this feature which is responsible for the breakdown of the split property.

One can therefore expect that the split property (for proper inclusions of double cones) holds in theories which do not exhibit pathologically large numbers of local degrees of freedom. Such theories, in turn, can be expected to have a reasonable thermodynamical behavior. In the literature, there exist several "nuclearity" conditions [44, 48, 50, 52], reminiscent of the trace class condition $\mathrm{Tr}(e^{-\beta H}) < \infty$ for Gibbs states in quantum mechanics, which are related to the split property and thermodynamical properties.

For applications to relative commutants of wedge algebras, the relevant condition is the so-called "modular nuclearity condition" [49, 50]. Given a Borchers triple $(\mathcal{M}, U, \Omega)$, one considers the inclusions

$$\mathcal{M}(x) := U(x, 1)\mathcal{M}U(x, 1)^{-1} \subset \mathcal{M}, \qquad x \in W_R, \tag{10.22}$$

and defines the maps

$$\Xi(x) : \mathcal{M} \to \mathcal{H}, \qquad \Xi(x)A := \Delta^{1/4}U(x)A\Omega. \tag{10.23}$$

Here Δ is the modular operator of $(\mathcal{M}, \Omega)$. Using elementary properties of modular theory, it is easy to see that $\Xi(x)$ is a bounded operator between the Banach spaces $\mathcal{M}$ (equipped with the operator norm of $\mathcal{B}(\mathcal{H})$) and $\mathcal{H}$.

If $\Xi(x)$ is even compact, and more particularly, *nuclear* (i.e. $\Xi(x)$ can be written as a norm convergent sum of rank one operators), then one has the following result.

Theorem 10.2.13 ([50]) *Let $(\mathcal{M}, U, \Omega)$ be a Borchers triple and assume that for some $x \in W_R$, the map $\Xi(x)$ (10.23) is nuclear. Then the inclusion $\mathcal{M}(x) \subset \mathcal{M}$ is split. Conversely, if $\mathcal{M}(x) \subset \mathcal{M}$ is split, then $\Xi(x)$ is compact.*

This theorem provides a sufficient condition for an inclusion to be split. However, it must be noticed that the split property is a very strong condition. It is a reasonable assumption for inclusions of *local* algebras in theories which satisfy some rough bound on the number of their local degrees of freedom, but some care is needed when dealing with unbounded regions like wedges, even in such theories. In fact, there is an argument by Araki [37, p. 292] to the effect that inclusions of wedge algebras cannot be split if the spacetime dimension is larger than two. Araki's argument exploits the translation invariance of wedges along their edges and does not apply in two dimensions, where these edges are zero-dimensional points.

In two dimensions, the split property for wedges is known to hold in the theory of a free, scalar, massive field [43, 113]. It is, however, not fulfilled for arbitrary mass spectra. For example, the split property for wedges does not hold in massless theories, and is also violated in the model of a generalized free field with continuous mass spectrum [59]. But for models describing finitely many species of massive particles, there is no a priori reason for the split property for wedges not to hold. We can therefore take it as a tentative assumption (to be verified in concrete models), and now discuss its consequences.

Proposition 10.2.14 [43] *Let $(\mathcal{M}, U, \Omega)$ be a two-dimensional Borchers triple, and $x \in W_R$. If the inclusion $\mathcal{M}(x) \subset \mathcal{M}$ is split, then $\mathcal{M}$, $\mathcal{M}(x)$ and the relative commutant $\mathcal{M}(x)' \cap \mathcal{M}$ are all isomorphic to the unique hyperfinite type III_1 factor. In particular, the relative commutant has cyclic vectors, and $\mathcal{M}(x) \subset \mathcal{M}$ is standard.*

In the light of this result, we can view the split property as a sufficient condition for non-trivial relative commutants of inclusions of wedge algebras. Whereas non-triviality of local algebras is a minimal requirement in a local theory, also the Reeh-Schlieder property (Theorem 1.1.1), i.e. cyclicity of the vacuum vector for algebras of observables localized in arbitrarily small regions, is of importance in quantum field theory.

To arrive at such a statement from the split property, we first recall that on the basis of Lemma 10.2.12 one can easily show that a standard split inclusion is *normal*, i.e. $(\mathcal{M}_1' \cap \mathcal{M}_2)' \cap \mathcal{M}_2 = \mathcal{M}_1$ [59]. By similar arguments in the setting of a Borchers triple $(\mathcal{M}, U, \Omega)$ for which $\mathcal{M}(x) \subset \mathcal{M}$ is split for some $x \in W$, it follows that $\mathcal{M}$ is *locally generated*, i.e. it coincides with the smallest von Neumann algebra containing all relative commutants $\mathcal{M} \cap \mathcal{M}'(x)$, $x \in W$ [97]. In combination with a result of Müger [113], stating that algebras corresponding to double cones of different sizes are closely related, this provides sufficient information for application of the usual Reeh-Schlieder arguments, making use of positivity of the energy. One arrives at the following statement [97].

Proposition 10.2.15 *Let $(\mathcal{M}, U, \Omega)$ be a Borchers triple, and $x \in W_R$. If the inclusion $\mathcal{M}(x) \subset \mathcal{M}$ is split, then Ω is cyclic for the relative commutant $\mathcal{M} \cap \mathcal{M}'(x)$.*

Thus the modular nuclearity and split conditions yield nets satisfying all the basic assumptions of algebraic quantum field theory (see Chap. 1). Below we summarize these and additional results in a theorem, which strengthens Proposition 10.2.4 under the assumption of the split property for wedges. In its formulation, we make use of the *diameter* of a two-dimensional double cone $O_{x,y} = (W_R + x) \cap (W_L + y)$, defined as $d(O_{x,y}) := \sqrt{-(x-y)^2} \geq 0$.

Theorem 10.2.16 *Let $(\mathcal{M}, U, \Omega)$ be a two-dimensional Borchers triple, such that the inclusion $\mathcal{M}(x) \subset \mathcal{M}$ is split for some $x \in W_R$ (this is in particular the case if the map $\Xi(x)$ is nuclear). Let $s := \sqrt{-x^2} > 0$ be the "splitting distance". Then the net $\mathcal{A}$ constructed from $(\mathcal{M}, U, \Omega)$ has (in addition to what is stated in Proposition 10.2.4) the following properties: For any double cone O with $d(O) > s$,*

(a) $\mathcal{A}(O)$ is isomorphic to the hyperfinite type III_1 factor.
(b) The vacuum vector Ω is cyclic and separating for $\mathcal{A}(O)$.
(c) Haag duality holds, i.e. $\mathcal{A}(O)' = \mathcal{A}(O')$.
(d) Weak additivity holds, i.e.

$$\bigvee_{x \in \mathbb{R}^2} \mathcal{A}(O + x) = \mathcal{B}(\mathcal{H}). \tag{10.24}$$

(e). The time slice property (in its von Neumann version) holds above the splitting distance: If $t_0, t_1 \in \mathbb{R}$ with $t_1 - t_0 > s$, then the algebra $\mathcal{A}(S(t_0, t_1))$ associated to the time slice $S = \{x \in \mathbb{R}^2 : t_0 < x_0 < t_1\}$ is $\mathcal{A}(S(t_0, t_1)) = \mathcal{B}(\mathcal{H})$.

In this Theorem, (a), (b) follow from Propositions 10.2.14 and 10.2.15. For (c) and (d), see [97], and for (e), [113]. Note that the above theorem gives slightly generalized statements over the ones found in, say, [97, 113]: In these works, the split property was assumed to hold for arbitrarily small splitting distances $s > 0$. However, the corresponding results for finite splitting distance are straightforward to obtain by the same arguments. Also see the work of Müger for further implications of the split property for wedges, in particular with regard to superselection theory [113].

Although it is possible to construct models of algebraic quantum field theory in which there exists a minimal length in the sense that $\mathcal{A}(O) = \mathbb{C} \cdot 1$ for double cones below a minimal diameter [100], this is not an expected feature in typical QFT models. We mostly stated the theorem in the above form because in certain models, discussed in Sect. 10.3, the modular nuclearity condition can so far only be proven for large enough splitting distance. Of course, Theorem 10.2.16 also applies to the case where the split property holds for *all* the inclusions $\mathcal{M}(x) \subset \mathcal{M}$, $x \in W_R$, and then gives the usual unrestricted forms of cyclicity, additivity, duality, and the time slice property.

Theorem 10.2.16 can be seen as the abstract form of a general construction scheme, which we can summarize as "first construct a Borchers triple, then check modular nuclearity for its wedge inclusions". It provides physically reasonable properties of the emerging net under assumptions which are both natural (for massive theories, in $d = 2$) and manageable in concrete models. Its main drawback is that it does not apply to more than two spacetime dimensions. Finding conditions that suitably weaken the split property for wedges and apply to dimension $d > 2$ is currently an open question and subject of ongoing research.

10.3 Integrable Models and Inverse Scattering Theory

In this section we discuss a concrete implementation of the general construction scheme of QFT models via Borchers triples, presented in Theorem 10.2.16. As we saw in Sect. 10.2.3, free theories can be completely described in terms of their (stable) single particle content, formalized as a specific representation U_1 of the Poincaré group and its associated net $W \mapsto \mathcal{K}_1(W)$ of real standard subspaces of the single particle space $\mathcal{H}_1$. Here we will describe how to proceed in a similar manner for certain *interacting* theories in two dimensions.

From a mathematical perspective, we will start from a specific single particle representation U_1 of $\mathcal{P}_+$ as before, and then "deform" the second quantization step (cf. Proposition 10.2.9) from the subspace net $W \mapsto \mathcal{K}_1(W)$ to the net of von Neumann algebras $W \mapsto \mathcal{A}(W)$. The "deformation parameter" will take the form of a unitary S on $\mathcal{H}_1 \otimes \mathcal{H}_1$ with specific properties, which enters into both, the definition of the multi particle Hilbert space, and the definition of the wedge algebras.

Physically speaking, S describes the two-particle S-matrix, and we are considering theories in which this two-particle scattering operator completely fixes the full (multi particle) S-matrix. These are theories in which no particle production occurs in collision processes of arbitrary energy. Examples of such models are well known as *integrable quantum field theories*, referring to the existence of an infinite number of conversation laws which constrain the dynamics in such a way that each collision process factorizes into two-particle processes ("factorizing S-matrix", see [87]).

Specific examples are field theories like the Sinh-Gordon model, the Ising model, the Sine-Gordon model, the Thirring model, and many more [1, 127].

Since we are starting our construction from the (two particle) S-matrix S, we consider the *inverse scattering problem*, in contrast to the canonical approach, where the interaction is specified in terms of a classical Lagrangian or Hamiltonian density, which is then quantized. In fact, in the case of integrable models, the S-matrix is typically much simpler than the Lagrangian (which is not of polynomial form), and therefore S suggests itself as a suitable quantity for characterizing the interaction.

This inverse scattering point of view also lies at the heart of the form factor program, see the monograph [127] and review article [9]. In that approach, one characterizes local fields/observables by their expectation values in scattering states ("form factors"), which are severely restricted by factorizability and analyticity of the S-matrix. In many cases, it is possible to obtain explicit expressions for form factors, see [11, 12, 74, 128] for just some sample articles. After the determination of the form factors, the next step in the form factor program is to compute Wightman n-point functions, which are given by series of integrals over form factors. Controlling these complicated series, as required for a complete construction of a model, has so far only been possible in very few examples [10].

In comparison, the algebraic approach presented here circumvents the explicit construction of local field operators (see however the end of Sect. 10.3.4 for results in that direction), and analyzes the local observable content via the modular nuclearity property from Tomita-Takesaki theory.

In Sect. 10.3.1, we introduce the class of S-matrices we consider, and then use them to construct Borchers triples in Sect. 10.3.2. The known results pertaining to the modular nuclearity condition are reviewed in Sect. 10.3.3. In Sect. 10.3.4, it is then shown that this construction solves the inverse scattering problem and yields asymptotically complete theories. Finally, in Sect. 10.3.5 we discuss certain massless versions of these models, and compare to related constructions in the literature.

10.3.1 Factorizing S-Matrices

Just as in Sect. 10.2.3, the construction of the models we are interested in begins with the specification a single particle representation of the proper Poincaré group, fixing the single particle spectrum. Recall that in two dimensions, to any mass $m > 0$, there exists a unique (up to unitary equivalence) irreducible, unitary, strongly continuous, positive energy representation $U_{1,m}$ of the proper orthochronous Poincaré group $\mathcal{P}_+^\uparrow$. It can be realized on the representation space $\mathcal{H}_{1,m} := L^2(\mathbb{R}, d\theta)$ as

$$(U_{1,m}(x,\lambda)\psi)(\theta) := e^{ip_m(\theta)\cdot x} \cdot \psi(\theta - \lambda)\,, \tag{10.25}$$

where $(x, \lambda) \in \mathcal{P}_+^\uparrow$ denotes the Poincaré transformation consisting of a boost with rapidity $\lambda \in \mathbb{R}$ and a subsequent space-time translation by $x \in \mathbb{R}^2$. The variable θ is the *rapidity*, which is connected to the on-shell momentum and mass via

$$p_m(\theta) := m \begin{pmatrix} \cosh\theta \\ \sinh\theta \end{pmatrix} . \tag{10.26}$$

We will allow for several species of particles, and therefore consider a direct sum of several representations $U_{1,m}$ with different masses.

We can also include the single particle charges in our description, identified with equivalence classes q of unitary irreducible representations of a compact Lie group G (the global gauge group) as usual. As charges carried by single particles, we consider a set $\mathcal{Q}$ of finitely many charges, and to account for antiparticles, we assume that with each class $q \in \mathcal{Q}$, also the conjugate class $\overline{q}$ is contained in $\mathcal{Q}$. We are interested in constructing massive stable quantum field theories and must therefore guarantee that in each sector, the masses are positive isolated eigenvalues of the mass operator. This will in particular be the case when to each charge q, there corresponds a single mass $m(q) > 0$ (with $m(\overline{q}) = m(q)$), and for simplicity, we restrict ourselves to this setting.

The single particle Hilbert space then has the form

$$\mathcal{H}_1 = \bigoplus_{q\in\mathcal{Q}} \mathcal{H}_{1,q} = \bigoplus_{q\in\mathcal{Q}} L^2(\mathbb{R}, d\theta) \otimes \mathcal{V}_q = L^2(\mathbb{R}, d\theta) \otimes \mathcal{V} , \tag{10.27}$$

where $\mathcal{V} = \bigoplus_{q\in\mathcal{Q}} \mathcal{V}_q$ is a finite-dimensional Hilbert space, $D := \dim \mathcal{V}$. Picking unitary irreducible representations $V_{1,q}$ of G in the class $q \in \mathcal{Q}$, the representations of the Poincaré group $\mathcal{P}_+^\uparrow$ and the gauge group G on $\mathcal{H}_1$ are

$$U_1 := \bigoplus_{q\in\mathcal{Q}} \left(U_{1,m(q)} \otimes \mathrm{id}_{\mathcal{V}_q}\right) , \qquad V_1 := \bigoplus_{q\in\mathcal{Q}} \left(\mathrm{id}_{L^2(\mathbb{R},d\theta)} \otimes V_{1,q}\right) . \tag{10.28}$$

In the following, we will always tacitly refer to a fixed particle spectrum given by the data G, $\mathcal{Q}$, $\{V_{1,q}\}_{q\in\mathcal{Q}}$, $\{m_q\}_{q\in\mathcal{Q}}$ and complying with the above assumptions. It will be convenient to use a particular orthonormal basis for $\mathcal{V}$ (10.27): For each subspace $\mathcal{V}_q$ of fixed charge, we choose an orthonormal basis, and denote their direct sum by $\{e^\alpha : \alpha = 1, \ldots, D\}$. We can thus associate with each index α a definite charge $q_{[\alpha]}$ and mass $m_{[\alpha]} := m(q_{[\alpha]})$. The corresponding components of vectors $\Psi_1 \in \mathcal{H}_1$ will be denoted by $\theta \mapsto \Psi_1^\alpha(\theta)$. Using standard multi index notation, this notation is extended to tensor products: We write $\xi^{\boldsymbol{\alpha}}$, $\boldsymbol{\alpha} = (\alpha_1, \ldots, \alpha_n)$, for the components of vectors $\xi \in \mathcal{V}^{\otimes n}$, $T^{\boldsymbol{\alpha}}_{\boldsymbol{\beta}}$, $\boldsymbol{\alpha} = (\alpha_1, \ldots, \alpha_n)$, $\boldsymbol{\beta} = (\beta_1, \ldots, \beta_n)$, for the components of tensors $T \in \mathcal{B}(\mathcal{V}^{\otimes n})$, and $\mathbb{R}^n \ni \boldsymbol{\theta} \mapsto \Psi_n^{\boldsymbol{\alpha}}(\boldsymbol{\theta})$ for the component functions of $\Psi_n \in \mathcal{H}_1^{\otimes n}$, $n \in \mathbb{N}$.

To proceed as in Sect. 10.2.3, we need a representation of the *proper* Poincaré group, i.e. we still need to define a single particle TCP operator J_1, implementing the spacetime reflection $j(x) := -x$. In view of our above assumption regarding conjugate charges $q, \overline{q} \in \mathcal{Q}$, there exists such an operator on $\mathcal{H}_1$ (see for example [41, 62, 81, 114]). It is the product of a charge conjugation operator exchanging the representation spaces $\mathcal{V}_q$ and $\mathcal{V}_{\overline{q}}$, and a space-time reflection, acting by complex conjugation on $L^2(\mathbb{R}, d\theta)$. When working in the basis chosen above, this simply means that we have an involution $\alpha \mapsto \overline{\alpha}$ of $\{1, \dots, D\}$ (that is, a permutation of D elements with $\overline{\overline{\alpha}} = \alpha$) such that $m_{[\overline{\alpha}]} = m_{[\alpha]}$ and $q_{[\overline{\alpha}]} = \overline{q_{[\alpha]}}$, and the TCP operator reads

$$(J_1\Psi_1)^\alpha(\theta) := \overline{\Psi_1^{\overline{\alpha}}(\theta)}\,. \tag{10.29}$$

By straightforward calculation, one checks that J_1 is an antiunitary involution which commutes with V_1 and extends the representation U_1 to the proper Poincaré group $\mathcal{P}_+$ via $U_1(j) := J_1$.

As in (10.18), we introduce the "geometric modular group" $\Delta_1^{it} := U_1(0, -2\pi t)$ and the real standard subspace (10.19)

$$\mathcal{K}_1 := \ker(1 - J_1\Delta_1^{1/2}) \subset \mathcal{H}_1\,. \tag{10.30}$$

For later reference, we recall that $\mathcal{K}_1 + i\,\mathcal{K}_1$ coincides with the (vector-valued) Hardy space $\mathbb{H}^2(-\pi, 0) \otimes \mathcal{V} \subset \mathcal{H}_1$, consisting of all $\Psi_1 \in \mathcal{H}_1$ such that $\theta \mapsto \Psi_1^\alpha(\theta)$ is the boundary value of a function analytic in $\mathrm{S}(-\pi, 0) := \{\zeta \in \mathbb{C} : -\pi < \mathrm{Im}\zeta < 0\}$ with

$$\sup_{-\pi<\lambda<0} \sum_\alpha \int_{\mathbb{R}} d\theta\, |\Psi_1^\alpha(\theta + i\lambda)|^2 < \infty,$$

and the real standard subspace is

$$\mathcal{K}_1 = \{\Psi_1 \in \mathbb{H}^2(-\pi, 0) \otimes \mathcal{V} : \Psi_1^\alpha(\theta + i\pi) = \overline{\Psi_1^{\overline{\alpha}}(\theta)} \text{ for all } \theta, \alpha\}, \tag{10.31}$$

see, for example, [100].

We now come to specifying the two-particle S-matrix of the models to be constructed. This S-matrix can be formulated as a unitary $\mathbf{S}$ on $\mathcal{H}_1 \otimes \mathcal{H}_1$ which has to satisfy a number of compatibility conditions with the single particle data U_1, V_1, and additional properties. This unitary will have the form

$$(\mathbf{S}\Psi_2)^{\alpha_1\alpha_2}(\theta_1, \theta_2) = S^{\alpha_1\alpha_2}_{\beta_1\beta_2}(\theta_1 - \theta_2)\Psi_2^{\beta_1\beta_2}(\theta_1, \theta_2)$$

(here and in the following, we use Einstein's summation convention) with some function $S : \mathbb{R} \to \mathcal{U}(\mathcal{V} \otimes \mathcal{V})$, expressing the Poincaré invariance of $\mathbf{S}$. The properties of $\mathbf{S}$ can therefore be most explicitly formulated in terms of the function S, as we shall do below. For a manifestly basis-independent formulation in terms of $\mathbf{S}$, see [18, 20]. All the listed properties are standard in the context of integrable models, and for example discussed in [1, 63, 117].

Definition 10.3.1 A (two-particle) S-matrix is a continuous bounded function[12] $S : \overline{S(0, \pi)} \to \mathcal{B}(\mathcal{V} \otimes \mathcal{V})$ which is analytic in the interior of this strip and satisfies for arbitrary $\theta, \theta' \in \mathbb{R}$,

(a) Unitarity:

$$S(\theta)^* = S(\theta)^{-1} \tag{10.32}$$

(b) Hermitian analyticity:

$$S(\theta)^{-1} = S(-\theta) \tag{10.33}$$

(c) Yang-Baxter equation:

$$\begin{aligned}(S(\theta) \otimes 1_1)(1_1 \otimes S(\theta + \theta'))(S(\theta') \otimes 1_1) \\ = (1_1 \otimes S(\theta'))(S(\theta + \theta') \otimes 1_1)(1_1 \otimes S(\theta))\end{aligned} \tag{10.34}$$

(d) Poincaré symmetry:

$$S^{\alpha\beta}_{\gamma\delta}(\theta) = 0 \text{ if } m_{[\alpha]} \neq m_{[\delta]} \text{ or } m_{[\beta]} \neq m_{[\gamma]}\,, \tag{10.35}$$

$$S^{\alpha\beta}_{\gamma\delta}(\theta) = S^{\overline{\delta}\overline{\gamma}}_{\overline{\beta}\overline{\alpha}}(\theta)\,. \tag{10.36}$$

(e) Gauge invariance:

$$[S(\theta),\ V_1(g) \otimes V_1(g)] = 0\,, \qquad g \in G\,. \tag{10.37}$$

(f) Crossing symmetry: For all $\alpha, \beta, \gamma, \delta \in \{1, \ldots, D\}$,

$$S^{\alpha\beta}_{\gamma\delta}(i\pi - \theta) = S^{\overline{\gamma}\alpha}_{\delta\overline{\beta}}(\theta) \tag{10.38}$$

The family of all S-matrices will be denoted $\mathcal{S}$.

Many examples of such S-matrices are known. In the *scalar case*, pertaining to the irreducible representation $U_1 = U_{1,m}$ with $\mathcal{V} = \mathbb{C}$, a complete characterization of $\mathcal{S}$ can be given [96]. Some typical examples are listed in Table 10.1.

In the tensor case, where $D = \dim \mathcal{V} > 1$, the general solution to the constraints summarized in Definition 10.3.1 is not known, mostly because of the complicated Yang-Baxter equation (c). However, special S-matrices *are* known, for example the S-matrix of the $O(N)$ Sigma models. In this case, $D = N > 2$, $G = O(N)$, and $\mathcal{Q}$ consists of the defining representation of $O(N)$ on $\mathbb{C}^N$, with S-matrix

$$S(\theta)^{\alpha_1\alpha_2}_{\beta_1\beta_2} := \sigma_1(\theta)\delta^{\alpha_1\alpha_2}\delta^{\beta_1\beta_2} + \sigma_2(\theta)\delta^{\alpha_1\beta_2}\delta^{\alpha_2\beta_1} + \sigma_3(\theta)\delta^{\alpha_1\beta_1}\delta^{\alpha_2\beta_2}\,. \tag{10.39}$$

[12] The continuity assumption can be relaxed, cf. [104].

Table 10.1 Examples of scalar S-matrices

$S(\theta)$	Name of associated QFT model	References
$+1$	Free field theory	
-1	Ising model	[17, 95]
$\frac{\sinh\theta - i\sin b(g)}{\sinh\theta + i\sin b(g)}$	Sinh-Gordon model with coupling constant g, and $b(g) = \frac{\pi g^2}{4\pi + g^2}$	[1, 8]
$e^{i\kappa m^2 \sinh\theta}$	S-matrix of non-commutative Minkowski space with noncommutativity parameter $\kappa > 0$	[78]

Here $\sigma_1, \sigma_2, \sigma_3$ are specific combinations of rational functions and Gamma functions, see [1, Chap. 8.3.2] for details.

The assumption that S is analytic in the physical strip is not satisfied for all integrable models. In general S is only expected to be *meromorphic* in that strip, with poles signifying the presence of bound states [1]. The known results are strongest for S-matrices without such bound state poles, and we will therefore restrict to this case in this review. However, some steps of the construction program to follow have also been accomplished in the meromorphic case by Cadamuro and Tanimoto [55]; we will comment on this point later.

10.3.2 Construction of Borchers Triples from Two-Particle S-Matrices

Given a single particle spectrum (with representations U_1, V_1), and a two-particle S-matrix $S \in \mathcal{S}$, we now set out to construct a corresponding Borchers triple $(\mathcal{M}_S, U_S, \Omega_S)$. We first introduce a convenient Hilbert representation space $\mathcal{H}_S$, the *S-symmetric Fock space*[13] over our single particle space $\mathcal{H}_1$. In a different context, the construction of $\mathcal{H}_S$ was first carried out by Liguori and Mintchev [106], then for the S-matrix case in [94, 102]. For a more abstract formulation, emphasizing the functorial properties of the construction, see [20].

Starting from the single particle space (10.27), we consider the n-fold tensor products $\mathcal{H}_1^{\otimes n} = L^2(\mathbb{R}^n, d^n\boldsymbol{\theta}) \otimes \mathcal{V}^{\otimes n}$, and the subspace $\mathcal{H}_{S,n} \subset \mathcal{H}_1^{\otimes n}$ of *S-symmetric wave functions*, i.e. those $\Psi_n \in \mathcal{H}_1^{\otimes n}$ which satisfy

$$S(\theta_{k+1} - \theta_k)_{k,k+1}\Psi_n(\theta_1, \ldots, \theta_{k+1}, \theta_k, \ldots, \theta_n) = \Psi_n(\theta_1, \ldots, \theta_k, \theta_{k+1}, \ldots, \theta_n)\,, \tag{10.40}$$

for all $\theta_1, \ldots, \theta_n \in \mathbb{R}$, $k \in \{1, \ldots, n-1\}$. Here the subscript $k, k+1$ on S signifies that this tensor acts on the tensor factors k and $k+1$ of $\mathcal{V}^{\otimes n}$.

---

[13] Picking this particular Hilbert space is a matter of choice, see [3, 98] for other possibilities.

For the constant "flip" S-matrix $S(\theta) = \pm F$ (with $F(v \otimes w) = w \otimes v$, $v, w \in \mathcal{V}$), the space $\mathcal{H}_{S,n}$ then coincides with the totally symmetric (+) respectively totally antisymmetric (−) n-fold tensor product of $\mathcal{H}_1$ with itself. For general $S \in \mathcal{S}$, one can describe $\mathcal{H}_{S,n}$ as the invariant subspace of $\mathcal{H}_1^{\otimes n}$ of an S-dependent representation of the permutation group [102, 106].

We now define the *S-symmetric Fock space* as the direct sum of these "n-particle Hilbert spaces",

$$\mathcal{H}_S := \bigoplus_{n=0}^{\infty} \mathcal{H}_{S,n}\,, \tag{10.41}$$

where we understand $\mathcal{H}_{S,1} := \mathcal{H}_1$ and $\mathcal{H}_{S,0} := \mathbb{C}$. The n-particle component of a vector $\Psi \in \mathcal{H}_S$ will be denoted $\Psi_n \in \mathcal{H}_{S,n}$. Occasionally we will also use the "particle number operator" $(N\Psi)_n := n\Psi_n$, and the dense subspace $\mathcal{D}_S \subset \mathcal{H}_S$ of "finite particle number". For the time being, these are just names for certain subspaces, their physical interpretation in terms of particle states will be justified later in scattering theory.

Each $\mathcal{H}_S$, $S \in \mathcal{S}$, is a closed subspace of the "Boltzmann Fock space" $\widehat{\mathcal{H}} := \bigoplus_{n=0}^{\infty} \mathcal{H}_1^{\otimes n}$, and we denote by $P_S : \widehat{\mathcal{H}} \to \mathcal{H}_S$ the corresponding orthogonal projection. The Fock vacuum $\Omega_S \in \mathcal{H}_S$, given by $(\Omega_S)_0 = 1$, $(\Omega_S)_n = 0$ for $n \geq 1$, will be the vacuum vector of the Borchers triple to be constructed.

The second ingredient of the Borchers triple, the representation of the Poincaré group, is given by a variant of the standard second quantization procedure.

Lemma 10.3.2 *(a) For $(x, \lambda) \in \mathcal{P}_+^{\uparrow}$, let*

$$U_S(x, \lambda) := P_S \bigoplus_{n=0}^{\infty} \left(U_1(x, \lambda)^{\otimes n} \otimes \mathrm{id}_{\mathcal{V}^{\otimes n}}\right) P_S\,. \tag{10.42}$$

Then U_S is a strongly continuous, unitary, positive energy representation of $\mathcal{P}_+^{\uparrow}$ on $\mathcal{H}_S$, with (unique) invariant vector Ω_S.

(b) Let $J : \mathcal{H}_S \to \mathcal{H}_S$ be defined as

$$(J\Psi)_n^{\boldsymbol{\alpha}}(\boldsymbol{\theta}) := \overline{\Psi_n^{\overline{\alpha_n}\ldots\overline{\alpha_1}}(\theta_n, \ldots, \theta_1)}\,. \tag{10.43}$$

Then J is an antiunitary involution satisfying $JU_S(x, \lambda)J = U_S(-x, \lambda)$.

(c) For $g \in G$, let

$$V_S(g) := P_S \bigoplus_{n=0}^{\infty} \left(\mathrm{id}_{L^2(\mathbb{R}^n)} \otimes V_1(g)^{\otimes n}\right) P_S\,. \tag{10.44}$$

Then V_S is a unitary representation of the gauge group G on $\mathcal{H}_S$, commuting with U_S and J.

The proof of this lemma essentially amounts to showing that the various operators considered here respect the S-symmetry, i.e. restrict from $\widehat{\mathcal{H}}$ to the subspace $\mathcal{H}_S$. This is the case because of property (d) and (e) of S in Definition 10.3.1. Explicitly, we have

$$[U(x,\lambda)\Psi]_n^{\boldsymbol{\alpha}}(\boldsymbol{\theta}) := \exp(i\sum_{l=1}^{n} p_{\alpha_l}(\theta_l)\cdot x)\, \Psi_n^{\boldsymbol{\alpha}}(\theta_1-\lambda,\dots,\theta_n-\lambda)\,, \tag{10.45}$$

where p_{α_l} is shorthand for $p_{m_{[\alpha_l]}}$. Following our earlier convention also on the multi particle level, we write $\Delta^{it} := U(0,-2\pi t)$ for the rescaled boosts.

Having fixed the representation U_S and its vacuum vector Ω_S, we now turn to the construction of the most important ingredient, the von Neumann algebra $\mathcal{M}_S$ of wedge-local observables, completing U_S, Ω_S to a Borchers triple. As a prerequisite for this, we first introduce creation and annihilation operators on the S-symmetric Fock space $\mathcal{H}$.

On $\widehat{\mathcal{H}}$, we have the usual unsymmetrized operators $a(\varphi)$, $a^\dagger(\varphi)$, $\varphi\in\mathcal{H}_1$. They are defined by linear and continuous extension from

$$a^\dagger(\varphi)\psi_1\otimes\cdots\otimes\psi_n := \sqrt{n+1}\,\varphi\otimes\psi_1\otimes\cdots\otimes\psi_n\,,\qquad \psi_1,\dots,\psi_n\in\mathcal{H}_1\,, \tag{10.46}$$

$$a(\varphi)\psi_1\otimes\cdots\otimes\psi_n := \sqrt{n}\,\langle\varphi,\psi_1\rangle\,\psi_2\otimes\cdots\otimes\psi_n\,,\qquad a(\varphi)\widehat{\Omega} := 0\,, \tag{10.47}$$

to $\mathcal{H}_1^{\otimes n}$, and then to the subspace of finite particle number. We introduce their projections onto $\mathcal{H}_S$ as

$$z_S^\dagger(\varphi) := P_S\, a^\dagger(\varphi)P_S\,,\qquad z_S(\varphi) := P_S\, a(\varphi)P_S\,,\qquad \varphi\in\mathcal{H}_1, \tag{10.48}$$

and the distributional kernels $z_{S,\alpha}^\#(\theta)$ of these operators by

$$z_S^\dagger(\varphi) = \sum_\alpha\int d\theta\,\varphi_\alpha(\theta)z_{S,\alpha}^\dagger(\theta)\,,\qquad z_S(\varphi) = \sum_\alpha\int d\theta\,\overline{\varphi_\alpha(\theta)}z_{S,\alpha}(\theta)\,. \tag{10.49}$$

Their essential properties are listed next.

Proposition 10.3.3 *Let $\varphi\in\mathcal{H}_1$ and $\Psi\in\mathcal{D}_S$ be arbitrary.*

(a) The operators (10.48) *are in general unbounded, but well-defined on $\mathcal{D}_S$.*
(b) We have

$$z_S(\varphi)^* \supset z_S^\dagger(\varphi)\,. \tag{10.50}$$

(c) For $(x,\lambda)\in\mathcal{P}_+^\uparrow$ and $g\in G$, we have

$$U_S(x,\lambda)z_S^\#(\varphi)U_S(x,\lambda)^{-1} = z_S^\#(U_1(x,\lambda)\varphi)\,,$$

$$V_S(g)z_S^{\#}(\varphi)V_S(g)^{-1} = z_S^{\#}(V_1(g)\varphi),$$

where $z_S^{\#}$ *stands for either* z_S *or* $z_S^{\dagger}$.

(d) *With respect to the particle number operator* N, *there hold the bounds*

$$\|z_S(\varphi)\Psi\| \leq \|\varphi\|\|N^{1/2}\Psi\|, \qquad \|z_S^{\dagger}(\varphi)\Psi\| \leq \|\varphi\|\|(N+1)^{1/2}\Psi\|. \quad (10.51)$$

(e) $z_S, z_S^{\dagger}$ *form a representation of the Zamolodchikov-Faddeev algebra with* S*-matrix* S*: The distributional kernels* $z_{S,\alpha}^{\#}(\theta)$ *satisfy*

$$z_{S,\alpha}(\theta)z_{S,\beta}(\theta') - S_{\delta\gamma}^{\beta\alpha}(\theta-\theta')z_{S,\gamma}(\theta')z_{S,\delta}(\theta) = 0, \quad (10.52)$$

$$z_{S,\alpha}^{\dagger}(\theta)z_{S,\beta}^{\dagger}(\theta') - S_{\alpha\beta}^{\gamma\delta}(\theta-\theta')z_{S,\gamma}^{\dagger}(\theta')z_{S,\delta}^{\dagger}(\theta) = 0, \quad (10.53)$$

$$z_{S,\alpha}(\theta)z_{S,\beta}^{\dagger}(\theta') - S_{\beta\delta}^{\alpha\gamma}(\theta'-\theta)z_{S,\gamma}^{\dagger}(\theta')z_{S,\delta}(\theta) = \delta^{\alpha\beta}\delta(\theta-\theta')\cdot 1. \quad (10.54)$$

The algebraic relations in item (e) are known as the *Zamolodchikov–Faddeev* algebra [69, 139], and are frequently used in the context of integrable quantum field theories (see for example [10, 127], and references cited therein). Note in particular that for the constant S-matrices $S_{\gamma\delta}^{\alpha\beta}(\theta) = \pm\delta_\delta^\alpha\delta_\gamma^\beta$, they coincide with the familiar CCR/CAR relations.

The exchange relations in Proposition 10.3.3(e) were proposed by the Zamolodchikov brothers [139]. Their heuristic basis is that a product $z_{S,\alpha_1}^{\dagger}(\theta_1)\cdots z_{S,\alpha_n}^{\dagger}(\theta_n)$ of n creation operators (acting on a vacuum vector) represents a configuration of n particles with rapidities θ_j, such that the order of factors in the product corresponds to the ordering of the particles on the line (from left to right). An exchange of two particles should correspond to a two-particle scattering process and thus produce an S factor. Faddeev completed this structure [69] by adding a corresponding annihilation operator and the exchange relation between z_S and $z_S^{\dagger}$.

In our subsequent analysis, we will not rely on this motivation. Rather, we take the operators $z_S^{\#}$ as a tool for defining (generators of) a wedge algebra, and will then *derive* the intuitive picture drawn by the Zamolodchikov's from Haag-Ruelle scattering theory (Sect. 10.3.4).

In the CCR situation $S = F$ (tensor flip), we can form the usual free field as the (selfadjoint closure of) the sum of a creation and an annihilation operator, and use the unitaries $\exp(i(z_F^{\dagger}(\xi) + z_F(\xi)))$, $\xi \in \mathcal{K}_1$, to generate the wedge algebra $\mathcal{M}_F$ as in (10.17). In this case, the same field operator can be used to generate the observables in left and right wedges, because the covariance statement in Proposition 10.3.3(c) also holds for the TCP operator J.

For other S-matrices S, however, this is not the case: Proposition 10.3.3(c) does *not* hold if $U(x,\lambda)$ is replaced by J, and different generators are needed for left and right wedges. We therefore introduce

$$z_S^{\dagger}(\varphi)' := Jz_S^{\dagger}(J\varphi)J, \qquad z_S(\varphi)' := Jz_S(J\varphi)J. \quad (10.55)$$

Taking into account Lemma 10.3.2(b), it becomes apparent that items (b)–(d) of Proposition 10.3.3 apply to the $z_S^{\#}(\varphi)'$ without any changes.

The TCP-reflected creation/annihilation operators satisfy commutation relations analogous to the ones listed in Proposition 10.3.3(e) with the only difference that $S_{\gamma\delta}^{\alpha\beta}(\theta)$ has to be replaced by $S_{\delta\gamma}^{\beta\alpha}(-\theta)$. For controlling commutators between observables in left and right wedges, it will be important to know the *relative* commutation relations of the $z_S^{\#}(\varphi)$ and $z_S^{\#}(\varphi)'$. They are stated next, see also [119] for a related analysis.

Proposition 10.3.4 ([102]) *Let $\varphi, \psi \in \mathcal{H}_1$. Then, on the domain $\mathcal{D}_S$,*

$$[z_S(\varphi)', z_S(\psi)] = 0, \tag{10.56}$$

$$[z_S^{\dagger}(\varphi)', z_S^{\dagger}(\psi)] = 0, \tag{10.57}$$

$$[z_S(\varphi)', z_S^{\dagger}(\psi)] = K_S(\varphi, \psi), \tag{10.58}$$

$$[z_S^{\dagger}(\varphi)', z_S(\psi)] = L_S(\varphi, \psi), \tag{10.59}$$

where $K_S(\varphi, \psi)$ and $L_S(\varphi, \psi)$ are bounded operators on $\mathcal{H}_S$ which commute with N. If $\varphi \in \mathcal{K}_1$ and $\psi' \in \mathcal{K}_1'$, then

$$L_S(\varphi, \psi') = -K_S(\varphi, \psi'). \tag{10.60}$$

The proof of the first part of this proposition follows by explicit calculation of the commutators, which also gives the explicit form of the operators $K_S(\varphi, \psi)$ and $L_S(\varphi, \psi)$ as multiplication operators with certain tensor-valued bounded functions $K_{S,n}(\varphi, \psi), L_{S,n}(\varphi, \psi)$ on each n-particle space $\mathcal{H}_{S,n}$. These functions are given by integrating the components of φ and ψ against a kernel which consists of a product of S-factors. To establish (10.60), one relies on the analyticity properties of both, $\varphi, J\psi' \in \mathcal{K}_1$ (10.31) and S, as well as their boundary conditions $J\Delta^{1/2}\varphi = \varphi$, $J\Delta^{-1/2}\psi' = \psi'$ and the crossing symmetry in Definition 10.3.1(f). These features allow for a contour shift in the integrals defining K_S, L_S, and lead to (10.60).

When S has first order poles in the physical strip, (10.60) fails, which is the reason why we restrict ourselves to analytic S. However, for certain S-matrices with poles, Cadamuro and Tanimoto found a way to modify the generators Φ_S (see below) to cancel these residue contributions. This modification, and the emerging subtle domain questions, are explained in [55].

Sticking to the case of analytic S and following Schroer [126], we now define

$$\Phi_S(\xi) := z_S^{\dagger}(\xi)' + z_S(J\Delta^{1/2}\xi)', \tag{10.61}$$

$$\Phi_S'(\xi) := z_S^{\dagger}(\xi) + z_S(J\Delta^{-1/2}\xi). \tag{10.62}$$

The counter intuitive placement of the primes is chosen here to have Φ_S generate $\mathcal{M}_S$ (instead of $\mathcal{M}_S'$), but also be consistent with the literature.

The above defined operators depend complex-linearly on their arguments, and it is clear that $\Phi_S(\xi)$ is a well-defined operator on $\mathcal{D}_S$ for $\xi \in \mathrm{dom}\, \Delta_1^{1/2}$, whereas $\Phi'_S(\xi)$ is well-defined for $\xi \in \mathrm{dom}\, \Delta_1^{-1/2}$. It also follows quickly from the definitions that

$$J\Phi_S(\xi)J = \Phi'_S(J\xi), \qquad \xi \in \mathrm{dom}\, \Delta^{1/2}. \tag{10.63}$$

Proposition 10.3.5 *Let* $\xi \in \mathrm{dom}\, \Delta^{1/2}$.

(a) $\Phi_S(\xi)^* \supset \Phi_S(J\Delta^{1/2}\xi)$.
(b) All vectors in $\mathcal{D}_S$ *are entire analytic for* $\Phi_S(\xi)$. *For* $\xi \in \mathcal{K}_1$, *the field operator* $\Phi_S(\xi)$ *is essentially selfadjoint.*
(c) Φ_S *transforms covariantly under the proper orthochronous Poincaré group* $\mathcal{P}_+^\uparrow$ *and the gauge group* G, *i.e. on* $\mathcal{D}_S$, *there holds*

$$U_S(x,\lambda)\Phi_S(\xi)U_S(x,\lambda)^{-1} = \Phi_S(U_1(x,\lambda)\xi), \qquad (x,\lambda) \in \mathcal{P}_+^\uparrow, \tag{10.64}$$

$$V_S(g)\Phi_S(\xi)V_S(g)^{-1} = \Phi_S(V_1(g)\xi), \qquad g \in G. \tag{10.65}$$

(d) The vacuum vector Ω_S *is cyclic for* Φ_S, *i.e. the subspace*

$$\mathbb{C}\text{-span}\{\Phi_S(\xi_1)\cdots\Phi_S(\xi_n)\Omega_S \;:\; \xi_1,\ldots,\xi_n \in \mathcal{K}_1,\; n \in \mathbb{N}_0\}$$

is dense in $\mathcal{H}_S$.

All these statements also hold if Φ_S *is exchanged with* Φ'_S, *and* $\mathcal{K}_1$ *with* $\mathcal{K}'_1$.

All these properties follow from straightforward calculations, except part (b) which uses the bound from Proposition 10.3.3(d) and an application of Nelson's Theorem.

Part (a) of this proposition explains the factors $J\Delta^{\pm 1/2}$ appearing in the definition of Φ_S and Φ'_S, anticipating the Bisognano-Wichmann property of a wedge algebra $\mathcal{M}_S$ generated by $\Phi_S(\xi)$. In fact, we have, $\xi \in \mathrm{dom}\, \Delta^{1/2}$,

$$J\Delta^{1/2}\Phi_S(\xi)\Omega_S = J\Delta^{1/2}\xi = \Phi_S(J\Delta^{1/2}\xi)\Omega_S = \Phi_S(\xi)^*\Omega_S,$$

and $J\Delta^{-1/2}\Phi'_S(\xi')\Omega_S = \Phi'_S(\xi')^*\Omega_S$ for $\xi' \in \mathrm{dom}\, \Delta^{-1/2}$. This is consistent with $\Phi_S(\xi)$, $\xi \in \mathrm{dom}\, \Delta^{1/2}$, being affiliated to a von Neumann algebra which has Ω_S as a cyclic separating vector, and modular data J, Δ, as well as $\Phi'_S(\xi')$, $\xi' \in \mathrm{dom}\, \Delta^{-1/2}$, being affiliated with the commutant of that algebra.

We therefore define the von Neumann algebra

$$\mathcal{M}_S := \{e^{i\Phi_S(\xi)} \;:\; \xi \in \mathcal{K}_1\}'', \tag{10.66}$$

and want to convince ourselves that $(\mathcal{M}_S, U_S, \Omega_S)$ is a Borchers triple. To this end, we first note that in view of $U_1(x,\lambda)\mathcal{K}_1 \subset \mathcal{K}_1$ for $x \in \overline{W_R}$, $\lambda \in \mathbb{R}$, and Proposition 10.3.5(c), we have $U(x,\lambda)\mathcal{M}_S U(x,\lambda)^{-1} \subset \mathcal{M}_S$ for $x \in \overline{W_R}$, $\lambda \in \mathbb{R}$.

Next, by using Proposition 10.3.5(d) and standard techniques, one can show that Ω_S is cyclic for $\mathcal{M}_S$.

The crucial point is to show that Ω_S is also separating, which amounts to showing that $\mathcal{M}_S$ has a large commutant in $\mathcal{B}(\mathcal{H}_S)$. At this point, the second field Φ'_S comes into play: For $\varphi \in \mathcal{K}_1$, $\psi' \in \mathcal{K}'_1$, we find the following equality on $\mathcal{D}_S$:

$$[\Phi_S(\varphi), \Phi'_S(\psi')] = [z_S^\dagger(\varphi)' + z_S(\varphi)', z_S^\dagger(\psi') + z_S(\psi')] = L_S(\varphi, \psi') + K_S(\varphi, \psi') = 0\,.$$

Here we have first used that $J\Delta^{1/2}\varphi = \varphi$, $J\Delta^{-1/2}\psi' = \psi'$, and then the commutation relations of Proposition 10.3.4. By linearity, it then follows that $\Phi_S(\varphi)$ and $\Phi'_S(\psi')$ commute on $\mathcal{D}_S$ for any $\varphi \in \mathrm{dom}\,\Delta_1^{1/2}$, $\psi' \in \mathrm{dom}\,\Delta_1^{-1/2}$, so that bounded functions of $\Phi'_S(\psi')$ are good candidates for operators commuting with $\mathcal{M}_S$. Indeed, by a calculation on analytic vectors [4, 98], one finds

$$\left[e^{i\Phi_S(\varphi)}, e^{i\Phi'_S(\psi')}\right] = 0\,, \qquad \varphi \in \mathcal{K}_1,\ \psi' \in \mathcal{K}'_1\,. \tag{10.67}$$

This implies that the von Neumann algebra

$$\widetilde{\mathcal{M}_S} := \{e^{i\Phi'_S(\xi')} \,:\, \xi' \in \mathcal{K}'_1\}''\,, \tag{10.68}$$

commutes with $\mathcal{M}_S$. As Ω_S is cyclic for $\widetilde{\mathcal{M}_S}$ by the same arguments as for $\mathcal{M}_S$, we arrive at the following result.

Theorem 10.3.6 *Let $S \in \mathcal{S}$. Then the triple $(\mathcal{M}_S, U_S, \Omega_S)$ is a Borchers triple.*

This result was proven in [94] for the scalar case, and in [102] for the general tensor case. It completes the first step in the construction program presented in Sect. 10.2, and we therefore obtain a local covariant net $O \mapsto \mathcal{A}_S(O)$ of von Neumann algebras on $\mathcal{H}_S$ for each S-matrix $S \in \mathcal{S}$.

As discussed in Sect. 10.2.4, the next step is to analyze the relative commutants of the inclusions $\mathcal{M}_S(x) \subset \mathcal{M}_S$, $x \in W_R$, by means of the split property and modular nuclearity. In this context, it is important to know the modular data of $(\mathcal{M}_S, \Omega_S)$.

Theorem 10.3.7 *Let $S \in \mathcal{S}$.*

(a) The Bisognano-Wichmann property holds, i.e. the modular data $\tilde{J}, \tilde{\Delta}$ of $(\mathcal{M}_S, \Omega_S)$ are given by

$$\tilde{\Delta} = \Delta\,, \qquad \tilde{J} = J\,. \tag{10.69}$$

(b) The commutant of $\mathcal{M}_S$ in $\mathcal{B}(\mathcal{H}_S)$ is $\mathcal{M}_S{}' = \widetilde{\mathcal{M}_S}$.

This theorem was proven in the scalar case in [43], and in the tensor case in [4].

10.3.3 Regular S-matrices and the Modular Nuclearity Condition

As discussed in the abstract setting in Sect. 10.2, the operator-algebraic construction program of a QFT model proceeds in two main steps. The first step is to find a suitable Borchers triple, and has been accomplished in Theorem 10.3.6 for any S-matrix $S \in \mathcal{S}$. The second step consists of analyzing the local observable content which is (indirectly) defined by the Borchers triple. In this section, we will follow the general strategy explained in Sect. 10.2.4, and summarize the known results on modular nuclearity and the split property in the model theories with Borchers triple $(\mathcal{M}_S, U_S, \Omega_S)$, $S \in \mathcal{S}$.

We first recall that for split distance $s > 0$, the map in question is (10.23)

$$\Xi_S(s) : \mathcal{M}_S \to \mathcal{H}_S\,, \qquad \Xi_S(s)A := \Delta^{1/4} U_S(s) A \Omega_S\,, \tag{10.70}$$

where $U_S(s)$ is shorthand for the purely spatial translation by $(0, s) \in W_R$. Making use of the explicit form (10.45) of the representation U_S, one finds

$$(\Xi_S(s)A)_n^{\boldsymbol{\alpha}}(\boldsymbol{\theta}) = \prod_{k=1}^{n} e^{-m_{\alpha} s \cosh \theta_k} \cdot (A\Omega)_n^{\boldsymbol{\alpha}}(\boldsymbol{\theta} + i\boldsymbol{\lambda}_0), \quad \boldsymbol{\lambda}_0 := \left(-\tfrac{\pi}{2}, \ldots, -\tfrac{\pi}{2}\right). \tag{10.71}$$

Here the complex argument $\boldsymbol{\theta} + i\boldsymbol{\lambda}_0$ has to be understood in terms of analytic continuation, and suggests that for understanding nuclearity properties of the map $\Xi_S(s)$, complex analysis will be essential.

For general S, one can not expect $\Xi_S(s)$ to be nuclear. We therefore make the following definition of "regular" S [97], demanding stronger analyticity of S than can be expected from first principles [110].

Definition 10.3.8 An S-matrix $S \in \mathcal{S}$ is called regular if there exists $\varepsilon > 0$ such that $\theta \mapsto S(\theta)$ extends to a bounded analytic function on the enlarged strip $\mathrm{S}(-\varepsilon, \pi + \varepsilon)$. The family of all regular S-matrices will be denoted $\mathcal{S}_0 \subset \mathcal{S}$.

This condition demands in particular that all singularities of S lie a minimal distance $\varepsilon > 0$ off the physical strip. Poles in $\mathrm{S}(-\pi, 0)$ are usually interpreted as evidence for unstable particles with a finite lifetime [137], with lifetime of such a resonance becoming arbitrarily long if the corresponding pole lies sufficiently close to the real axis. The regularity condition rules out S-matrices with infinitely many resonances with arbitrarily long lifetimes and "masses" such that the thermodynamical partition function diverges, a situation in which we cannot expect modular nuclearity to hold [42, 49]. The additional condition of a *bounded* analytic extension of S to $\mathrm{S}(-\varepsilon, \pi + \varepsilon)$, $\varepsilon > 0$, is a condition on the phase shift that can also be found in other approaches [89]. In both the scalar and tensor case, there exist many regular S-matrices.

The detailed analysis of the nuclearity properties of the map (10.71) requires the discussion of many technical points and will not be presented here. Rather, we will give an outline of the strategy which was so far used for proving that $\Xi_S(s)$ is nuclear for regular S. It consists of the following three steps.

Step (1) *Show that for $A \in \mathcal{M}_S$, the functions $(A\Omega)_n : \mathbb{R}^n \to \mathcal{V}^{\otimes n}$ have an analytic continuation to a tube $\mathcal{T}_n = \mathbb{R}^n + i\mathcal{B}_n \subset \mathbb{C}^n$ which contains the point $\boldsymbol{\lambda}_0$ in the interior of its base $\mathcal{B}_n \subset \mathbb{R}^n$.*

For a heuristic argument why such analyticity can be expected, take A to be a polynomial in the generators $\Phi_S(\xi_1), \ldots, \Phi_S(\xi_N)$, $\xi_1, \ldots, \xi_N \in \mathcal{K}_1$. (This unbounded operator is not an element of $\mathcal{M}_S$, but affiliated with this algebra.) Considering only the term with only creation operators for concreteness, $\widetilde{A} = z_S^\dagger(\xi_1) \cdots z_S^\dagger(\xi_n)$, we find the n-particle wave functions

$$(\widetilde{A}\Omega)_n(\boldsymbol{\theta}) = \frac{1}{n!} \sum_{\pi \in \mathfrak{S}_n} S^\pi(\boldsymbol{\theta}) \left(\xi_1(\theta_{\pi(1)}) \otimes \cdots \otimes \xi_n(\theta_{\pi(n)})\right) , \qquad (10.72)$$

where the $S^\pi(\boldsymbol{\theta}) \in \mathcal{B}(\mathcal{V}^{\otimes n})$ are tensors consisting of several S-factors, depending on differences $\theta_k - \theta_l$ of rapidities. As each ξ_k is analytic in $S(-\pi, 0)$ (cf. (10.31)), we see that $(\widetilde{A}\Omega)_n$ is analytic in a tube containing an open neighborhood of $\boldsymbol{\lambda}_0$ in its base, provided S is regular.

For general observables $A \in \mathcal{M}_S$, such analyticity properties were shown to hold for regular S in [97] in the scalar case, and by S. Alazzawi in the general tensor case [4]. A generalization of the detailed analytic structure of these wave functions has later also proven to be important in the work of Bostelmann and Cadamuro on characterizations of local observables [30].

Step (2) *Show that there exists $0 < C_n(2s) < \infty$ such that for all $A \in \mathcal{M}$*

$$\sup_{\boldsymbol{\lambda} \in \mathcal{B}_n} \left(\int_{\mathbb{R}^n} d^n\boldsymbol{\theta}\, \|(U_S(s)A\Omega)_n(\boldsymbol{\theta} + i\boldsymbol{\lambda})\|^2 \right)^{1/2} \leq C_n(2s) \cdot \|A\| . \qquad (10.73)$$

Such Hardy type bounds were established in [4, 97].

It then follows that the linear map $\Sigma_n(s) : \mathcal{M}_S \to \mathbb{H}^2(\mathcal{T}_n)$, $\Sigma_n(s)A := (U(\frac{s}{2})A\Omega)_n$, from the wedge algebra $\mathcal{M}_S$ into the (vector-valued) Hardy space $\mathbb{H}^2(\mathcal{T}_n)$ on the tube $\mathcal{T}_n$, is bounded with $\|\Sigma_n(s)\| \leq C_n(s) < \infty$.

To formulate the last step, we first note that by employing tools from complex analysis, one can show that the map $E_n(s) : \mathbb{H}^2(\mathcal{T}_n) \to L^2(\mathbb{R}^n, d^n\boldsymbol{\theta})$ given by $E_n(s)(F)(\boldsymbol{\theta}) := \prod_{k=1}^n e^{ism\sinh\theta_k} F(\boldsymbol{\theta} + i\boldsymbol{\lambda}_0)$ is nuclear.[14] As $\Xi_S(s) = \sum_n E_n(s) \circ \Sigma_S(s)$, this gives the nuclearity bound

$$\|\Xi_S(s)\|_1 \leq \sum_{n=0}^{\infty} \|E_n(s)\Sigma_n(s)\|_1 \leq \sum_{n=0}^{\infty} C_n(s)\|E_n(s)\|_1 . \qquad (10.74)$$

[14] It is in fact p-nuclear for any $p > 0$, i.e. its singular values decay faster than any inverse power.

To prove nuclearity of $\Xi_S(s)$, one has to estimate the norm bound $C_n(s)$ and nuclear norm $\|E_n(s)\|_1$ sharp enough to have this series converging. This is only possible if one properly takes into account the statistics (S-symmetry). Therefore the last step is:

Step (3) *Exploit the S-symmetry to obtain bounds $C_n(s)$ and $\|E_n(s)\|_1$ sharp enough such that $\sum_{n=0}^{\infty} C_n(s)\|E_n(s)\|_1 < \infty$, at least for sufficiently large split distance $s > 0$.*

In the scalar case, Step 3 has been accomplished in case $S(0) = -1$, where an effective Pauli principle becomes available[15] [97].

Theorem 10.3.9 *Let $S \in \mathcal{S}_0$ be a scalar S-matrix with $S(0) = -1$. Then there exists a splitting distance $s_0 > 0$ such that $\Xi_S(s)$ is nuclear at least for $s > s_0$.*

In the tensor case, a similar situation occurs when demanding S to be regular and $S(0) = -F$ (with F the flip on $\mathcal{V} \otimes \mathcal{V}$ as before), these assumptions are in particular met by the S-matrices of the $O(N)$ sigma models. There is good evidence that this will give rise to a proof of modular nuclearity, as in the scalar case [4].

The additional assumption $S(0) = -1$ (only the two possibilities $S(0) = \pm 1$ exist for regular scalar S) amounts to a kind of "hard core" condition, and is satisfied in most of the interacting models known from Lagrangian formulations.

10.3.4 Discussion of the Constructed Models

Having constructed a large class of QFT models, one would next like to analyze their properties, a minimum requirement being that these actually do describe non-trivial interaction.[16]

In the case at hand, where an S-matrix is the input into the construction, it is most natural to consider scattering theory. Whenever there exist non-trivial observables interpolating between the vacuum and single particle states and localized in some double cone (of arbitrary size), one is in the position to apply Haag-Ruelle scattering theory (Sect. 1.3). In the models at hand, such observables exist, but are given only indirectly as elements of intersections of wedge algebras. However, it turns out that explicit knowledge of local operators is not necessary, one can express all collision states in terms of the explicit wedge-local fields by making use of the adaptation of Haag-Ruelle theory to the case of wedge-local observables [26]. The following result holds.

Theorem 10.3.10 *Let $S \in \mathcal{S}$ be a two-particle S-matrix such that the vacuum vector is cyclic for some double cone (This is in particular the case for S satisfying the*

---

[15] Note that the stronger results claimed in [97] contain an incorrect estimate [99], spotted by S. Alazzawi. At the time of writing, the result stated here is the strongest one with a complete proof.

[16] That this is a non-trivial issue can be seen (in the setting of Wightman QFT) at the example of a family of complicated Wightman functions [121] which were only later realized to be equivalent to generalized free fields [122].

assumptions of Theorem 10.3.9). Then the collision operator of the QFT model generated from the Borchers triple $(\mathcal{M}_S, U_S, \Omega_S)$ is the factorizing S-matrix with two-particle S-matrix S. Furthermore, this theory is then asymptotically complete.

This result shows that the presented construction provides a solution to the inverse scattering problem for two-particle S-matrix $S \in \mathcal{S}$. Under the assumption that Ω_S is cyclic for some double cone (of arbitrary size), one can also explicitly compute incoming and outgoing n-particle scattering states. Restricting to the scalar case for simplicity, one finds the (improper) asymptotic collision states of n particles of rapidities $\theta_1, \ldots, \theta_n \in \mathbb{R}$, $n \in \mathbb{N}_0$,

$$|\theta_1, \ldots, \theta_n\rangle_{S,\text{out}} = z_S^\dagger(\theta_1) \cdots z_S^\dagger(\theta_n)\Omega_S\,, \qquad \text{if } \theta_1 < \cdots < \theta_n\,, \tag{10.75}$$

$$|\theta_1, \ldots, \theta_n\rangle_{S,\text{in}} = z_S^\dagger(\theta_1) \cdots z_S^\dagger(\theta_n)\Omega_S\,, \qquad \text{if } \theta_1 > \cdots > \theta_n\,. \tag{10.76}$$

These identifications can be proven with the tools of Haag-Hepp-Ruelle scattering theory [83, 86, 124]. They are in perfect agreement with the intuition behind the Zamolodchikov-Faddeev algebra, identifying products of n creation operators in order of increasing (decreasing) rapidities as creating outgoing (incoming) collision states, and rearrangements of these operators as two-particle scattering processes, producing S-factors [69, 139].

We conclude this section with a comparison with other approaches to solving the inverse scattering problem for integrable QFTs, notably the form factor program [10, 127]. In that approach, one aims at constructing a *Wightman* QFT [88, 130] by specifying the n-point functions of local field operators. These are *not* the field operators Φ_S appearing in the approach presented here, which can also be formulated as operator-valued distributions over Minkowski space, but are localized only in wedges $W_{R/L} + x$ rather than points $\{x\}$ [102]. The role of the wedge-local fields Φ_S is to serve as generators of Borchers triples. It might well happen that the emerging net of von Neumann algebras can equivalently be generated by certain point-local Wightman fields, see [27, 28] for the right tools for analyzing the local observable content. However, there is no straightforward connection between the wedge-local fields and point like fields.

Hence the approach presented here is complementary to the form factor program, and also to conventional constructive QFT [76]: It does give a rigorous solution of the inverse scattering problem by operator-algebraic methods, but does not provide explicit formulae for strictly local quantities like Wightman n-point functions. Comparison with the conventional approach to constructive QFT is therefore indirect: Only if a Lagrangian model is conjectured to have a certain S-matrix (such as the Sinh-Gordon model, the $O(N)$-Sigma models, etc.) then the approach starting from the Lagrangian can be compared to the one starting from the S-matrix. For most of the S-matrices $S \in \mathcal{S}$, however, no corresponding Lagrangian is known, and the classical counterparts of these theories are therefore unknown.

The problem of characterizing the local observables $A \in \mathcal{A}_S(O)$ more explicitly has been taken up by Bostelmann and Cadamuro. Generalizing the well-known fact that on the totally symmetric Bose Fock space, any bounded operator can be expanded into a series of normal ordered creation and annihilation (CCR) operators [6], they showed that for any quadratic form A on $\mathcal{H}_S$, where $S \in \mathcal{S}$ is a scalar S-matrix and A is subject to certain regularity assumptions, one has [29]

$$A = \sum_{m,n=0}^{\infty} \int \frac{d\boldsymbol{\theta}\, d\boldsymbol{\eta}}{m!n!}\, f_{m,n}^{[A]}(\boldsymbol{\theta}, \boldsymbol{\eta})\, z_S^{\dagger}(\theta_1)\cdots z_S^{\dagger}(\theta_m)\, z_S(\eta_1)\cdots z_S(\eta_n)\,. \qquad (10.77)$$

The expansion coefficients $f_{m,n}^{[A]}$ are sums of contracted matrix elements of A. It is interesting to note that such distributions also appear in the context of proving the modular nuclearity condition [97] (for wedge-local bounded A). The expansion (10.77) is essentially the form factor expansion. We refer to [29] for details, also regarding the convergence properties of this series.

The expansion (10.77) holds for any (sufficiently regular) A, independent of its localization properties. If the expanded operator A is an element of a local algebra, one has more information on the expansion coefficients[17]—for example, compact localization in spacetime leads to strong analyticity properties of the $f_{m,n}^{[A]}$. In [30], Bostelmann and Cadamuro give complete, necessary and sufficient, conditions on the coefficients $f_{n,m}^{[A]}$ for A to be localized in a double cone O_r of radius $r > 0$ around the origin. These conditions are rather involved, and solutions are currently known only for the constant scalar S-matrices $S = 1$ (the free field) and $S = -1$ (the Ising model) [54]. The Ising model S-matrix $S = -1$, although looking almost trivial as an S-matrix, already gives rise to rather complicated local operators, see also [17] for a related analysis in the form factor program.

This state of affairs is typical for the comparison of the algebraic approach and the form factor program: Whereas the operator-algebraic tools are more efficient for questions like proving existence of local observables, the form factor methods (and also the expansion (10.77)) give more explicit information about these local quantities, see for example [31, 34] for applications to quantum energy inequalities and energy densities in the models considered here.

10.3.5 Massless Models

Up to this point we have considered theories with purely massive particle spectrum. While many integrable models are of this type, and scattering theory is best understood in the massive case,[18] there are also good reasons to consider massless theories. On the one hand, for an analysis of phenomena like confinement or asymptotic freedom, one has to consider short distance (scaling) limits, in which the masses go to zero (see [32, 47, 57] for discussions within the algebraic framework). Since certain

[17] However, boundedness of A is not directly reflected in its coefficients $f_{n,m}^{[A]}$ because the expansion (10.77) involves the unbounded operators $z_S^{\#}$.

[18] See, however, [38, 40, 66] for results on scattering involving massless particles.

integrable QFTs are believed to be asymptotically free (see, for example, [1]), this point of view is also relevant here. On the other hand, massless integrable models often exhibit conformal symmetry, providing a link to conformal QFT (Chap. 8) with all its specialized tools, which can give additional insight into their structure. In this section, we therefore consider various massless versions of the constructions presented so far. Due to page constraints, we will be rather brief, and often refer to the literature for details.

As for massive models, we first consider the interaction-free case, and recall the Borchers triple of the free massless current. Since the definition of the rapidity (10.26) depends on the mass, we work here instead in the momentum picture, and consider the Hilbert space $\mathcal{H}_1 = \mathcal{H}_1^+ \oplus \mathcal{H}_1^-$, with $\mathcal{H}_1^\pm := L^2(\mathbb{R}_\pm, dp/|p|)$ and the mass zero Poincaré representation $U_1 = U_1^+ \oplus U_1^-$, given by $(U_1^\pm(x,\lambda)\psi)(p) = e^{i(\pm x_0 - x_1)p}\,\psi(e^{\mp\lambda}p)$. The massless free current decomposes into a direct sum of chiral operators $j_\pm$, each depending only on one of the light cone coordinates $x_0 \pm x_1$. Since the right wedge W_R is in these coordinates the set $x_0 + x_1 > 0$, $x_0 - x_1 < 0$, one considers the von Neumann algebras $\mathcal{M}_\pm = \{\exp(ij_\pm(f)) \,:\, f \in C_{c,\mathbb{R}}^\infty(\mathbb{R}_\pm)\}''$ on the Bose Fock spaces $\mathcal{H}^\pm$ over $\mathcal{H}_1^\pm$. Denoting the respective Fock vacua by $\Omega^\pm \in \mathcal{H}^\pm$, one obtains on $\mathcal{H} := \mathcal{H}^+ \otimes \mathcal{H}^-$ the Borchers triple

$$(\mathcal{M}_0^+ \otimes \mathcal{M}_0^-, U^+ \otimes U^-, \Omega^+ \otimes \Omega^-)\,. \tag{10.78}$$

This construction is equivalent to the one presented in the preceding sections, with the trivial scalar S-matrix $S = 1$, and the modification that instead of the field Φ_S, one has to consider its directional derivatives (current) in order to avoid the infrared singularity of the measure $dp/|p|$.

Since all data decompose into light cone coordinates, one extends in this setting the definition of wedges, Borchers triples and standard pairs to the case of dimension $d = 1$. The space $\mathbb{R}^1$ is thought of as a light ray, the affine group of $\mathbb{R}$ ("$ax + b$ group") plays the role of the Poincaré group, and "positive energy representation" means that the generator of the translation subgroup is positive. Replacing the right wedge by the right half line $\mathbb{R}_+$, and double cones by intervals $I \subset \mathbb{R}$, the framework presented in Sects. 10.2.1 and 10.2.2 also applies to the one-dimensional case.

To motivate the following construction, we first recall a result by Longo and Witten on endomorphisms of standard pairs [109]. In the setting of a one-dimensional standard pair $(\mathcal{K}, U)$ on a Hilbert space $\mathcal{H}$, they defined an endomorphism to be a unitary $V \in \mathcal{B}(\mathcal{H})$ such that $V\mathcal{K} \subset \mathcal{K}$ and $[V, U(x)] = 0$ for all translations $x \in \mathbb{R}$. In their result, they make use of symmetric inner functions on the upper half plane, that is, analytic bounded functions φ on the upper half plane such that

$$\overline{\varphi(p)} = \varphi(p)^{-1} = \varphi(-p) \quad \text{for almost all } p \in \mathbb{R}\,. \tag{10.79}$$

Theorem 10.3.11 ([109]) *Let $(\mathcal{K}, U)$ be a one-dimensional standard pair with U non-degenerate and irreducible. Then a unitary V is an endomorphism of $(\mathcal{K}, U)$ if* *and only if $V = \varphi(P)$, where $P > 0$ is the generator of the translations, and φ is a symmetric inner function on the upper half plane.*

For generalizations of this theorem, see [100].

The Eq. (10.79) appearing in the definition of a symmetric inner function are reminiscent of the constraints imposed on a (scalar) S-matrix. In fact, when changing variables from p to $\theta = \log p$, the upper half plane is transformed to the strip $S(0, \pi)$ appearing in Definition 10.3.1, and (10.79) translates to $\overline{\psi(\theta)} = \psi(\theta)^{-1} = \psi(\theta + i\pi), \theta \in \mathbb{R}$. In this form, the similarities to scalar S-matrices are most striking, and reveal a remarkable match between the properties of endomorphisms of standard pairs, and the completely independently defined S-matrices. At the time of writing, no deep reason for this match was known, and we refer to [100] for further discussions of this point.

In the context of deformations of Borchers triples, the result by Longo and Witten provides the background to a construction by Tanimoto [134], which amounts to twisting one tensor factor in the chiral Borchers triple (10.78). He defines

$$\mathcal{M}_{0,\varphi} := (\mathcal{M}_0^+ \otimes 1) \vee S_\varphi (1 \otimes \mathcal{M}_0^-) S_\varphi^* , \tag{10.80}$$

where $S_\varphi \in \mathcal{B}(\mathcal{H}^+ \otimes \mathcal{H}^-)$ is a suitably chosen unitary depending on a symmetric inner function φ, namely

$$[S_\varphi \Psi]_{n,n'}(p_1, \ldots, p_n; q_1, \ldots, q_{n'}) := \prod_{\substack{l=1,\ldots,n \\ l'=1,\ldots,n'}} \varphi(-p_l q_{l'}) \cdot \Psi_{n,n'}(p_1, \ldots, p_n; q_1, \ldots, q_{n'}) . \tag{10.81}$$

Here the indices n, n' refer to the Fock structure of $\Psi \in \mathcal{H}^+ \otimes \mathcal{H}^-$, and the momenta to the spectra of the second quantized generators of translations in the light like directions, see [134].

Theorem 10.3.12 ([134]) *Let φ be a symmetric inner function on the upper half plane, and S_φ defined as* (10.81). *Then the triple $(\mathcal{M}_{0,\varphi}, U^+ \otimes U^-, \Omega^+ \otimes \Omega^-)$ is a two-dimensional Borchers triple.*

The physical significance of the operator S_φ appearing here is that of a "wave S-matrix" [39], as has been shown by Dybalski and Tanimoto [67]. In the sense of scattering of waves, one even has asymptotic completeness in this situation, and can recover the Borchers triple from its S-matrix and an asymptotic algebra [67, 134].

Tanimoto's deformation formula (10.80) is a result of the chiral (tensor product) structure present in the massless case, and has no direct analogue in the massive situation. Nonetheless, the twisted chiral triple is equivalent to a massless version of the inverse scattering construction discussed in the preceding sections. We refrain from giving the details of the massless version, which can be found in [33]. The equivalence proof is contained in [104].

The twist (10.80) is not the only possibility to obtain deformations by symmetric inner functions in the massless case. In fact, since the mass shell decomposes into two half light rays which are both invariant under Lorentz boosts, one can use *three* such functions $\varphi, \varphi_+, \varphi_-$ [20, 104]. Here φ corresponds to a twist "between the two light rays" which mixes the two tensor factors as in (10.80), with the interpretation of wave scattering. The functions $\varphi_\pm$, on the other hand, are used to perform a construction of a "deformed field theory" on a line in close analogy to the (scalar) construction presented in Sect. 10.3.2 [33]. This amounts to deformations of the individual half line algebras $\mathcal{M}_0^\pm \rightsquigarrow (\mathcal{M}_0^\pm)_{\varphi_\pm}$, leaving the tensor product structure between them unchanged.

A combination of these effects appears in the short distance scaling limits of the massive two-dimensional integrable models. As argued in [33], if for a regular, scalar S-matrix S the limits $\lim_{\theta\to\pm\infty} S(\theta)$ exist, then the integrable model with mass $m > 0$ and S-matrix S has a well-defined short distance scaling limit (performed at the level of the generating fields Φ_S, Φ'_S). The resulting limit theory has zero mass and decomposes into a (possibly twisted) chiral tensor product, which differs from the free current triple (10.78) by the deformations $\mathcal{M}_0^\pm \rightsquigarrow (\mathcal{M}_0^\pm)_{\varphi_\pm}$, with $\varphi_\pm = S$, and a constant twist $\varphi = \pm 1$ (this sign depends on the limit of S). At the time of writing, these limit theories were completely analyzed only in the simple cases $S = \pm 1$ [33]. The main challenge is to analyze the relative commutants of the half line inclusions $(\mathcal{M}_0^\pm)_{\varphi_\pm}(x_\pm) \subset (\mathcal{M}_0^\pm)_{\varphi_\pm}$, $x_\pm > 0$, which, in contrast to the massive two-dimensional case, are not split.

A common feature of all the constructions discussed in this chapter is the fact that they are based on data which respect the particle number: Factorizing S-matrices correspond to scattering processes without particle production, and the deformations based on Longo-Witten endomorphisms of standard pairs have a structure akin to second quantization. From this point of view, it is not surprising that the constructed models have the typical features of integrable models, and their higher-dimensional generalizations [98] exhibit non-local features (see also the following section for such effects).

Extensions of this construction program to interactions with particle productions, as they are typical for relativistic quantum field theories in more than two dimensions [2], require new ideas. In this context, it is promising to note that in [19], Bischoff and Tanimoto already found a wave S-matrix which does not preserve the Fock structure of its representation space. Although this can not be seen as particle creation due to the considered theory being massless, it is an indication that substantial generalizations of the program presented here might well be possible.

10.4 Deformations of Quantum Field Theories by Warped Convolution

This section is devoted to another class of examples of operator-algebraic constructions of QFT models. As before, we follow the general approach of Sect. 10.2 and proceed by identifying certain Borchers triples. However, in contrast to the inverse scattering approach in Sect. 10.3, the dimension $d \geq 2$ of Minkowski space is arbitrary here, and the input into the construction does not consist of an S-matrix, but rather of a representation of the translation group.

The point of view taken here is that of *deformation theory*, i.e. we will start with a given (arbitrary) Borchers triple $(\mathcal{M}, U, \Omega)$, and then construct new triples by "deforming" the algebra[19] $\mathcal{M}$. In the situations we consider, U and Ω will be held fixed.[20] Recently, several related deformation methods have been investigated (see, for example, [3, 98, 120]), of which the approach considered here is a particularly general and representative example.

The mathematical method to be used for this deformation, the warped convolution, has its roots in non-commutative geometry, and will be reviewed in Sect. 10.4.1. The application to Borchers triples is then presented in Sect. 10.4.2.

10.4.1 Warped Convolutions and Rieffel deformations

In this section we concentrate on the mathematical backbone of the deformation procedure, and begin by recalling Rieffel's deformations of C^*-algebras (see [123] for Rieffel's original work, and [90, 103] for generalizations in several different directions).

The setting is a C^*-algebra $\mathfrak{A}$ with a strongly continuous automorphic $\mathbb{R}^d$-action α. One chooses a non-degenerate inner product on $\mathbb{R}^d$ which is normalized to determinant ± 1 and will be denoted px for $p, x \in \mathbb{R}^d$. As a deformation parameter, we pick a real matrix $Q \in \mathbb{R}_-^{d\times d}$ which is skew-symmetric w.r.t. this inner product. In this setting, one wants to deform $\mathfrak{A}$ by introducing a new product, while keeping the linear structure and *-involution unchanged. Motivated by the desire to formulate an abstract, C^*-algebraic version of quantization, one considers the integral expression

$$A \times_Q B = (2\pi)^{-d} \int_{\mathbb{R}^d} \int_{\mathbb{R}^d} dp\, dx\, e^{-ipx}\, \alpha_{Qp}(A)\alpha_x(B)\,, \tag{10.82}$$

which is strongly reminiscent of the Weyl-Moyal star product of deformation quantization.[21] For A, B in the dense subalgebra $\mathfrak{A}^\infty \subset \mathfrak{A}$ of all elements for which $x \mapsto \alpha_x(A)$ is smooth, the above integral exists in a precise oscillatory sense, defined

---

[19] As we saw in Theorem 10.2.5, wedge algebras are always type III_1 factors, and are in fact expected to be isomorphic to the unique hyperfinite type III_1 factor in generic cases. Thus "deforming $\mathcal{M}$" does not mean deforming the (fixed) internal algebraic structure of $\mathcal{M}$, as in other algebraic deformation procedures [75]. Rather, "deforming $\mathcal{M}$" means deforming/changing the position of $\mathcal{M}$ inside $\mathcal{B}(\mathcal{H})$, i.e. deforming the subfactor $\mathcal{M} \subset \mathcal{B}(\mathcal{H})$.

[20] However, it is entirely possible to "deform" Borchers triples by keeping the wedge algebra fixed while changing U and Ω.

[21] The Weyl-Moyal product (see, for example [77]) features prominently in deformation quantization [136], where it describes the transition from classical mechanics to quantum mechanics. In that setting, one considers suitable number-valued functions f, g on the simple classical phase space $\mathbb{R}^d$ (d even) and equips them with the non-commutative product

as a limit of removing a smooth cutoff. This limit exists in the (Fréchet) topology of $\mathfrak{A}^\infty$, i.e. in particular $A \times_Q B \in \mathfrak{A}^\infty$. Furthermore, the new product $\times_Q$ is compatible with the star involution and identity element (if it exists) of $\mathfrak{A}$, and there exists a C^*-norm $\|\cdot\|_Q$ on $(\mathfrak{A}^\infty, \times_Q)$, completing in which yields the Rieffel-deformed C^*-algebra $(\mathfrak{A}_Q, \times_Q)$. In this procedure, Q plays the role of a deformation parameter: Setting $Q = 0$ results in the old "undeformed" product $A \times_0 B = AB$, and the C^*-algebras $\mathfrak{A}_Q$ depend on Q in a continuous manner (see [123] for details).

The *warped convolution* was introduced in [46] as a generalization of a deformation procedure in [78]. It was then thoroughly studied in [53], and we refer to this article for proofs of all the claims made in this section. Because of its origin in quantum field theory on non-commutative, "quantized" Minkowski spacetime [78], it is closely related to the abstract quantization procedure of Rieffel.[22]

For defining the warped convolution, one uses a concrete setting instead of an abstract C^*-algebra, namely a Hilbert space $\mathcal{H}$ with a unitary strongly continuous representation U of the translation group $\mathbb{R}^d$, $d \geq 2$. As in the Rieffel setting, one fixes a non-degenerate inner product on $\mathbb{R}^d$ (in the applications to QFT, typically the Minkowski inner product), and a skew-symmetric real matrix $Q \in \mathbb{R}_-^{d\times d}$.

It is then the aim to deform operators A on $\mathcal{H}$ to new operators A_Q on the same Hilbert space. That is, the action of A, $\Psi \to A\Psi$ on vectors $\Psi \in \mathcal{H}$ is changed to $A, \Psi \to A_Q\Psi$, but the product of operators is unchanged. As in the Rieffel setting, this is accomplished by an integral formula, namely

$$A_Q\Psi := (2\pi)^{-d} \int_{\mathbb{R}^d} dp \int_{\mathbb{R}^d} dx\, e^{-ipx}\, U(Qp)AU(Qp)^{-1}U(x)\Psi\,. \qquad (10.83)$$

If A is smooth in the sense that $x \mapsto U(x)AU(x)^{-1}$ is smooth in norm (that is, $A \in \mathcal{C}^\infty$, where $\mathcal{C} \subset \mathcal{B}(\mathcal{H})$ denotes the C^*-algebra on which the adjoint action of U acts strongly continuously), and Ψ is smooth in the sense that $x \mapsto U(x)\Psi$ is smooth ($\Psi \in \mathcal{H}^\infty$), then this integral exists in an oscillatory sense, and defines a smooth vector $A_Q\Psi \in \mathcal{H}^\infty$. Thus (10.83) yields a densely defined (smooth) operator A_Q. It is easy to see that setting $Q = 0$ results in the old operator, $A_0 = A$, so that one can think of the mapping $A \mapsto A_Q$ as a deformation of operators.

The mathematical status of these operators is settled in the following theorem.

Theorem 10.4.1 *Let $A \in \mathcal{C}^\infty$. Then the map $\mathcal{H}^\infty \ni \Psi \mapsto A_Q\Psi$ (10.83) extends to bounded (smooth) operator $A_Q \in \mathcal{C}^\infty$. More precisely, the map $A \mapsto A_Q$ is a linear bijection of $\mathcal{C}^\infty$ onto itself, such that, $A \in \mathcal{C}^\infty$,*

$$\|A_Q\| = \|A\|_Q < \infty\,, \qquad (10.84)$$

(Footnote 21 continued)

$$(f \star g)(y) = (2\pi)^{-d} \int_{\mathbb{R}^d} dp \int_{\mathbb{R}^d} dx\, e^{-ipx} f(y + \hbar\theta p)g(y + x),$$

where the antisymmetric matrix θ is given by the Poisson bracket.

[22]See also [118] for another recent closely related approach, drawing from [90].

where $\|A\|_Q$ *denotes the norm of the Rieffel-deformed* C^**-algebra* $\mathcal{C}_Q$*, corresponding to the undeformed* C^**-algebra* $\mathcal{C}$ *and action* $\alpha_x(A) = U(x)AU(x)^{-1}$.

Definition 10.4.2 Let $A \in \mathcal{C}^\infty$ and $Q \in \mathbb{R}_-^{d\times d}$. The warped convolution of A is the operator $A_Q \in \mathcal{C}^\infty$ defined by extending (10.83) to all of $\mathcal{H}$.

Having defined the warped convolution (or "warping", for short), we now summarize its properties. We begin with the algebraic aspects.

Theorem 10.4.3 *The warped convolution extends to a representation of the Rieffel-deformed* C^**-algebra* $\mathcal{C}_Q$*. In particular,*

(a) Warping carries the operator product into the Rieffel product,

$$A_Q B_Q = (A \times_Q B)_Q\,, \qquad A, B \in \mathcal{C}^\infty\,, \tag{10.85}$$

(b) Warping commutes with taking adjoints,

$$A_Q{}^* = A^*{}_Q\,, \qquad A \in \mathcal{C}^\infty\,. \tag{10.86}$$

(c) Furthermore, any U*-invariant operator* $A = U(x)AU(x)^{-1}$ *is invariant under warping,* $A_Q = A$*. In particular,*

$$1_Q = 1\,. \tag{10.87}$$

In the light of these results, the warped convolution seems very similar to Rieffel deformations, and in fact, it can be viewed as a module version of the Rieffel deformation [103]. But despite these similarities, warping has some advantages over the abstract Rieffel setting in application to QFT, in particular when it comes to comparing different deformation parameters Q_1, Q_2, or when spectral data of the translation representation are needed.

The following proposition, listing covariance properties of the warped convolution, shows that in typical QFT situations (in contrast to the situation in deformation quantization), several different deformation parameters must appear.

Proposition 10.4.4 *Let* $A \in \mathcal{C}^\infty$ *be smooth and* $Q \in \mathbb{R}_-^{d\times d}$.

(a) The warping procedure is U*-covariant,*

$$U(x)A_Q U(x)^{-1} = \left(U(x)AU(x)^{-1}\right)_Q\,. \tag{10.88}$$

(b) If $\Omega \in \mathcal{H}$ *is a* U*-invariant vector, then*

$$A_Q\Omega = A\Omega\,. \tag{10.89}$$

(c) *If* V *is a unitary or antiunitary operator on* $\mathcal{H}$ *such that* $VU(x)V^{-1} = U(Mx)$ *for some invertible matrix* $M \in \mathrm{GL}(d, \mathbb{R})$*, then*

$$VA_QV^{-1} = \left(VAV^{-1}\right)_{\pm MQM^T}, \tag{10.90}$$

where M^T *is the transpose of* M *w.r.t. the inner product used in the oscillatory factor in* (10.83)*, and the sign is "+" for unitary* V*, and "−" for antiunitary* V*.*

The last statement of this proposition applies in particular to the situation where we have a representation of the Poincaré group, and carry out the warped convolution with the representation of the translation subgroup. In that case, one sees that by introducing a warping w.r.t. some matrix Q, and insisting on Lorentz symmetry, one automatically ends up with all Lorentz transformed matrices $\Lambda Q \Lambda^T$, $\Lambda \in \mathcal{L}_+^\uparrow$ as well.

It will therefore be important to also consider situations where (at least) two different deformation parameters $Q_1, Q_2 \in \mathbb{R}_-^{d\times d}$ enter. We first mention the following simple lemma.

Lemma 10.4.5 *Let* $A \in \mathcal{C}^\infty$ *and* $Q_1, Q_2 \in \mathbb{R}_-^{d\times d}$*. Then*

$$(A_{Q_1})_{Q_2} = A_{Q_1+Q_2}. \tag{10.91}$$

In particular, the inverse of the warping map $A \mapsto A_Q$ *is given by the negative parameter, i.e.* $A \mapsto A_{-Q}$*.*

In the context of a Borchers triple $(\mathcal{M}, U, \Omega)$, we will consider operators $A \in \mathcal{M}$ and generate a new wedge algebra $\mathcal{M}_Q$ by the warped convolutions A_Q. Thinking of locality and commutators between observables in spacelike separated wedges (see Definition 10.2.3(c)), it is clear that also commutators $[A_{Q_1}, B_{Q_2}]$ between operators with different warping parameters will be relevant.

In general, commuting operators $[A, B] = 0$ do not give rise to commuting warped convolutions, i.e. in general $[A_{Q_1}, B_{Q_2}] \neq 0$. This holds in particular for $Q_1 = Q_2$, as is clear from the origin of warping in quantization, where a commutative algebra is deformed into a non-commutative one. However, there does exist an interesting commutation theorem for warped convolutions with *opposite* deformation parameters Q and $-Q$. In this theorem, due to Buchholz and Summers [46], a spectral condition enters. To understand the relevance of this spectral information, let us first mention that the warped convolution can also be expressed as

$$A_Q = \int dE(x)\, \alpha_{Qx}(A) = \int \alpha_{Qx}(A) dE(x), \tag{10.92}$$

where E is the joint spectral measure of the generators $P_0, \ldots, P_{d-1}$ of the translations $U(x) = \exp i P_\mu x^\mu$ (with support the joint spectrum, denoted $\mathrm{Sp}\, U$), and $\alpha_x(A) = U(x)AU(x)^{-1}$ as before. Both the above integrals can be defined as strong limits of suitably cut off expressions, and then coincide with each other and the warped convolution A_Q.

Whereas in most calculations, the oscillatory form (10.83) is more convenient to work with, the spectral integral form (10.92) is better suited for establishing the following compatibility result on the warped convolution $A \mapsto A_Q$ and its inverse $A \mapsto A_{-Q}$.

Proposition 10.4.6 *Let $A, B \in C^\infty$ be operators such that $[\alpha_{Qp}(A), \alpha_{-Qq}(B)] = 0$ for all $p, q \in \operatorname{Sp} U$. Then*

$$[A_Q, B_{-Q}] = 0\,. \tag{10.93}$$

This result completes the list of properties of the warped convolution that we will need in the next subsection, where application to deformations of Borchers triples are discussed. Besides that application, warped convolutions have by now also been used in a variety of other situations, such as Wightman QFT [79], QFT on curved spacetimes [111], Wick rotation on non-commutative spacetime [80], deformations of quantum mechanical Hamiltonians [112], and quantum measurement theory [5].

10.4.2 Borchers Triples and Warped Convolutions

We now apply the warped convolution to deformations of Borchers triples. We therefore start with a fixed but arbitrary d-dimensional Borchers triple $(\mathcal{M}, U, \Omega)$, $d \geq 2$, according to Definition 10.2.3. The warped convolution will always be carried out w.r.t. the translations $U(x, 1)$, $x \in \mathbb{R}^d$, from the representation U of the triple, the inner product in the oscillatory integrals will be the Minkowski inner product, and the deformation parameter Q skew-symmetric w.r.t. this inner product.

The basic idea for deforming $(\mathcal{M}, U, \Omega)$ is to keep U and Ω unchanged (as in Sect. 10.3), and replace $\mathcal{M}$ by an algebra containing "all A_Q, $A \in \mathcal{M}$". Some remarks are in order here: First, one checks that as a consequence of its half-sided translational invariance, the wedge algebra $\mathcal{M}$ contains a strongly dense subalgebra $\mathcal{M}^\infty = \mathcal{M} \cap C^\infty$ of smooth elements. Thus the warped convolutions A_Q are well-defined for sufficiently many $A \in \mathcal{M}$. However, products of such operators are typically not of the form C_Q for some $C \in \mathcal{M}$. This is so because $A_Q B_Q = (A \times_Q B)_Q$ (Theorem 10.4.3(a)), and $A \times_Q B$ involves integration over translations in all directions (10.82). Thus the set $\{A_Q : A \in \mathcal{M}^\infty\}$ is typically not an algebra. One therefore passes to the algebra generated, and writes (with a slight abuse of notation)

$$\mathcal{M}_Q := \{A_Q : A \in \mathcal{M}^\infty\}''\,. \tag{10.94}$$

Note that in view of Theorem 10.4.3(b), $\mathcal{M}_Q$ is a von Neumann algebra, and using Proposition 10.4.4(b), one can show that it has Ω as a cyclic vector.

In order for the so defined triple $(\mathcal{M}_Q, U, \Omega)$ to be a Borchers triple, the deformation parameter Q has to be chosen in a suitable manner, essentially by adapting it to the geometry of the right wedge W_R. Looking at Definition 10.2.3, it is clear that we only have to check the conditions in part (c). Since (orthochronous) Poincaré transformations $U(x, \Lambda)$ act on warped convolutions according to (cf. Proposition 10.4.4(c))

$$U(x, \Lambda)A_Q U(x, \Lambda)^{-1} = \Big(U(x, \Lambda)AU(x, \Lambda)^{-1}\Big)_{\Lambda Q \Lambda^{-1}}, \tag{10.95}$$

and this is required to be an element of $\mathcal{M}_Q$ for $\Lambda W_R + x \subset W_R$, we should choose Q in such a way that $\Lambda Q \Lambda^{-1} = Q$ for any Lorentz transformation Λ with $\Lambda W_R = W_R$. Furthermore, since the only commutativity result we have for warped convolutions is Proposition 10.4.6, we ought to choose Q in such a way that also $\Lambda Q \Lambda^{-1} = -Q$ for any Lorentz transformation Λ mapping W_R onto its causal complement, $\Lambda W_R = -W_R$. Finally, we have to take into account the energy momentum spectrum, lying in the forward light cone, which appears in Proposition 10.4.6. These considerations suggest to consider only *admissible* matrices Q, defined by the following three conditions:

(i) $\Lambda Q \Lambda^{-1} = Q$ for any $\Lambda \in \mathcal{L}_+^\uparrow$ such that $\Lambda W_R = W_R$,
(ii) $\Lambda Q \Lambda^{-1} = -Q$ for any $\Lambda \in \mathcal{L}_+^\uparrow$ such that $\Lambda W_R = -W_R$, and $jQj = Q$ (with j the reflection at the edge of W_R),
(iii) $Q\overline{V^+} \subset \overline{W_R}$.

These conditions drastically constrain the form of Q. In fact, Q is admissible if and only if [78]

$$Q = \begin{pmatrix} 0 & \kappa & 0 & \cdots & 0 \\ \kappa & 0 & 0 & \cdots & 0 \\ 0 & 0 & 0 & \cdots & 0 \\ \vdots & \vdots & \vdots & \ddots & \vdots \\ 0 & 0 & 0 & \cdots & 0 \end{pmatrix} \tag{10.96}$$

for some $\kappa \geq 0$, in case the spacetime dimension is $d \neq 4$. In the physically most interesting case $d = 4$, there is a little more freedom[23] for choosing Q. In this case, Q is admissible if and only if

$$Q = \begin{pmatrix} 0 & \kappa & 0 & 0 \\ \kappa & 0 & 0 & 0 \\ 0 & 0 & 0 & \kappa' \\ 0 & 0 & -\kappa' & 0 \end{pmatrix},$$

with parameters $\kappa \geq 0$, $\kappa' \in \mathbb{R}$.

[23] This is related to the fact that Q is skew-symmetric and for $d > 4$, the edge of W_R is fixed by the subgroup $SO(d-2)$ of rotations in the edge.

The condition (iii) on admissible Q's implies that for $A \in \mathcal{M}$ (localized in W_R) and $B \in \mathcal{M}'$ (localized in $-W_R$), one does not only have $[A, B] = 0$, but even $[\alpha_{Qp}(A), \alpha_{-Qq}(B)] = 0$ for all $p, q \in \overline{V^+}$. Since the energy-momentum spectrum is contained in $\overline{V^+}$, we are in the situation to apply Proposition 10.4.6, which is the key to the following result.

Theorem 10.4.7 *Let $(\mathcal{M}, U, \Omega)$ be a d-dimensional Borchers triple, $d \geq 2$, and let Q be admissible (see above).*

(a) The warped triple $(\mathcal{M}_Q, U, \Omega)$ is again a d-dimensional Borchers triple.
(b) The modular data Δ_Q, J_Q of $(\mathcal{M}_Q, \Omega)$ coincide with those of the original triple, i.e.

$$\Delta_Q = \Delta\,, \qquad J_Q = J\,. \tag{10.97}$$

(c) The commutant of the deformed wedge algebra is the inverse deformation of the original commutant,

$$\mathcal{M}_Q{}' = \mathcal{M}'_{-Q}\,.$$

In view of this theorem, warping yields a one-parameter family (two-parameter family in dimension $d = 4$) of "new" Borchers triples, which can again be taken to generate QFT models. In particular, when applying this method to a Borchers triple which is explicitly known (such as that of a free field theory, or of any other completely constructed QFT), it yields new QFT models in which Q plays the role of a coupling constant.

In comparison to the construction in Sect. 10.3, where we started from an S-matrix, the input into the construction by warped convolution is of a more abstract nature. It is therefore necessary to investigate the physical properties of these models. To begin with, there is the question how much the theory given by the deformed triple $(\mathcal{M}_Q, U, \Omega)$ depends on Q. Since the algebras $\mathcal{M}_Q$ are generically expected to be isomorphic to each other (generically being isomorphic to the unique hyperfinite type III_1 factor), any argument for showing that the Borchers triples $(\mathcal{M}, U, \Omega)$ and $(\mathcal{M}_Q, U, \Omega)$, $Q \neq 0$, are not equivalent, must also take into account the representation U. Indirect arguments to this effect exist.

On the one hand, one can consider massive theories, in which scattering states can be constructed. Then, using methods developed in [26], one can show that the S-matrix elements of collision processes with two incoming and two outgoing particles depend on Q via a factor e^{ipQq} (with p, q the momenta of the incoming particles) [46, 78]. Hence at least in this case, one does have a true dependence on Q, and inequivalence of the models with different values of Q. In particular, one sees in the scattering data that the warping procedure changes the interaction of the model under consideration.[24]

[24]There is also a sharp difference between the undeformed and deformed case in the thermal equilibrium states. In the deformed ($Q \neq 0$) situation, the thermal representation leads to a decoupling of the noncommutativity parameters $\Lambda Q \Lambda^{-1}$ related to different wedges [101].

For comparison with the approach taken in Sect. 10.3, it is also instructive to consider the two-dimensional case. In this case, the deformation parameter must be of the form

$$Q = \begin{pmatrix} 0 & \kappa \\ \kappa & 0 \end{pmatrix}, \qquad \kappa \geq 0 .$$

In the case of the representation $U = U_m$ given by the second quantization of the irreducible $U_{1,m}$ with fixed mass $m > 0$ (10.25), and using the rapidity parametrization (10.26), one notes that

$$e^{ip_m(\theta')Qp_m(\theta)} = e^{i\kappa m^2 \sinh(\theta-\theta')} =: S(\theta - \theta') \tag{10.98}$$

is a (scalar) S-matrix in the sense of Definition 10.3.1. In fact, one can show that for the Borchers triple $(\mathcal{M}, U, \Omega)$ of the massive scalar field in two dimensions, its deformation $(\mathcal{M}_Q, U, \Omega)$ gives rise to a QFT which is unitarily equivalent to the integrable model constructed in Sect. 10.3, with the S-matrix (10.98) [78].

Note, however, that this S-matrix is *not* regular in the sense of Definition 10.3.8, because it is unbounded on any strip $S(-\varepsilon, \pi + \varepsilon)$, $\varepsilon > 0$. Hence the modular nuclearity results of Sect. 10.3.3 do not apply, so that we cannot conclude existence of local observables for this model.

The models obtained by warped convolution have certain non-local features in general, which can also be seen by other arguments: In application to Wightman QFT, one observes that warping modifies the n-point functions in a non-local manner [79, 129]. In the general case, one can show (in dimension $d \geq 3$, and under a mild assumption on the energy-momentum spectrum, see [53]) that if one starts from a QFT which has the vacuum Ω as a cyclic vector for its double cone algebras, then Ω will *not* be cyclic for the double cone algebras of the QFT generated by the deformed Borchers triple $(\mathcal{M}_Q, U, \Omega)$, $Q \neq 0$.

These non-local aspects of the warped models show that the deformation procedure employed here is still too simple to yield realistic quantum field theory models in physical spacetime. To find deformation methods which are also compatible with strict localization in four dimensions is the subject of ongoing research.

In conclusion, we also mention how the inherent non-locality can be understood by considering the origin of the warped convolution in quantum field theory on noncommutative Minkowski space [78]: The initial motivation to consider the warped convolution was to define quantum field operators on a space in which the coordinates $X_0, \ldots, X_{d-1}$ do not commute, but rather satisfy a relation of the form

$$[X_\mu, X_\nu] = i\, Q_{\mu\nu}\,, \qquad [Q_{\mu\nu}, X_\kappa] = 0\,, \tag{10.99}$$

where Q is either a (non-zero) matrix times an identity element in the algebra generated by the X_μ, or an operator with spectrum consisting of a Lorentz orbit of such matrices [60] (Chap. 7). On a space with non-commuting coordinates, localization in bounded regions is impossible. This fact is reflected in the weaker than usual localization properties of the algebras constructed by warped convolution, and shows that more refined deformation methods are needed to obtain strictly local quantum field theories in higher dimensions.

Acknowledgments The publications reviewed in this article include joint work with my colleagues S. Alazzawi, H. Bostelmann, D. Buchholz, C. Dappiaggi, H. Grosse, R. Longo, T. Ludwig, E. Morfa-Morales, G. Morsella, J. Schlemmer, C. Schützenhofer, S.J. Summers, Y. Tanimoto, R. Verch, and S. Waldmann. I wish to thank them all for fruitful and enjoyable collaborations.

References

1. Abdalla, E., Abdalla, C., Rothe, K.D.: Non-perturbative Methods in 2-Dimensional Quantum Field Theory. World Scientific, Singapore (1991)
2. Åks, S.: Proof that scattering implies production in quantum field theory. J. Math. Phys. **6**, 516–532 (1965)
3. Alazzawi, S.: Deformations of fermionic quantum field theories and integrable models. Lett. Math. Phys. **103**(1), 37–58 (2013)
4. Alazzawi, S.: Deformations of quantum field theories and the construction of interacting models. Ph.D. thesis, University of Vienna, 2014. arXiv:1503.00897
5. Andersson, A.: Operator deformations in quantum measurement theory. Lett. Math. Phys. **104**(4), 415–430 (2013)
6. Araki, H.: A lattice of von Neumann algebras associated with the quantum theory of a free bose field. J. Math. Phys. **4**(11), 1343–1362 (1963)
7. Araki, H.: Von Neumann algebras of local observables for free scalar field. J. Math. Phys. **5**(1), 1–13 (1964)
8. Arinshtein, A.E., Fateev, V.A., Zamolodchikov, A.B.: Quantum S-matrix of the (1+1)-dimensional Toda chain. Phys. Lett. B **87**, 389–392 (1979)
9. Babujian, H.M., Karowski, M.: Towards the construction of Wightman functions of integrable quantum field theories. Int. J. Mod. Phys. **A19S2**, 34–49 (2004)
10. Babujian, H.M., Foerster, A., Karowski, M.: The form factor program: a review and new results—the nested SU(N) off-shell Bethe ansatz. SIGMA **2**, 082 (2006)
11. Babujian, H.M., Foerster, A., Karowski, M.: SU(N) and O(N) off-shell nested Bethe ansatz and exact form factors. J. Phys. A: Math. Theor. **41**(27), 275202 (2008)
12. Balog, J., Hauer, T.: Polynomial form-factors in the O(3) nonlinear sigma model. Phys. Lett. B **337**, 115–121 (1994)
13. Barata, J., Jäkel, C., Mund, J.: The $\mathcal{P}(\varphi)_2$ Model on the de Sitter Space. Preprint, arXiv:1311.2905 (2013)
14. Baumgärtel, H., Wollenberg, M.: Causal Nets of Operator Algebras. Akademie Verlag, Berlin (1992)
15. Baumgärtel, H., Jurke, M., Lledó, F.: On free nets over Minkowski space. Rept. Math. Phys. **35**(1), 101–127 (1995)
16. Baumgärtel, H., Jurke, M., Lledo, F.: Twisted duality of the CAR algebra. J. Math. Phys. **43**, 4158–4179 (2002)
17. Berg, B., Karowski, M., Weisz, P.: Construction of Green's functions from an exact S-matrix. Phys. Rev. D **19**(8), 2477–2479 (1979)
18. Bischoff, M.: Construction of models in low-dimensional quantum field theory using operator algebraic methods. Ph.D. thesis, Rome (2012)
19. Bischoff, M., Tanimoto, Y.: Construction of wedge-local nets of observables through Longo-Witten endomorphisms. II. Comm. Math. Phys. **317**(3), 667–695 (2013)
20. Bischoff, M., Tanimoto, Y.: Integrable QFT and Longo-Witten endomorphisms. Ann. Henri Poincaré **16**(2), 569–608 (2015)

21. Bisognano, J.J., Wichmann, E.H.: On the duality condition for a Hermitian scalar field. J. Math. Phys. **16**, 985–1007 (1975)
22. Bisognano, J.J., Wichmann, E.H.: On the duality condition for quantum fields. J. Math. Phys. **17**, 303–321 (1976)
23. Borchers, H.-J.: The CPT theorem in two-dimensional theories of local observables. Comm. Math. Phys. **143**, 315–332 (1992)
24. Borchers, H.-J.: On revolutionizing quantum field theory with Tomita's modular theory. J. Math. Phys. **41**, 3604–3673 (2000)
25. Borchers, H.-J., Buchholz, D.: Global properties of vacuum states in de Sitter space. Ann. Poincare Phys. Theor. **A70**, 23–40 (1999)
26. Borchers, H.-J., Buchholz, D., Schroer, B.: Polarization-free generators and the S-matrix. Comm. Math. Phys. **219**, 125–140 (2001)
27. Bostelmann, H.: Operator product expansions as a consequence of phase space properties. J. Math. Phys. **46**, 082304 (2005)
28. Bostelmann, H.: Phase space properties and the short distance structure in quantum field theory. J. Math. Phys. **46**, 052301 (2005)
29. Bostelmann, H., Cadamuro, D.: An operator expansion for integrable quantum field theories. J. Phys. A: Math. Theor. **46**, 095401 (2012)
30. Bostelmann, H., Cadamuro, D.: Characterization of local observables in integrable quantum field theories. Comm. Math. Phys. (2014) (To appear)
31. Bostelmann, H., Cadamuro, D.: Negative energy densities in integrable quantum field theories at one-particle level. Preprint, arXiv:1502.01714 (2015)
32. Bostelmann, H., D'Antoni, C., Morsella, G.: Scaling algebras and pointlike fields: a nonperturbative approach to renormalization. Comm. Math. Phys. **285**, 763–798 (2009)
33. Bostelmann, H., Lechner, G., Morsella, G.: Scaling limits of integrable quantum field theories. Rev. Math. Phys. **23**(10), 1115–1156 (2011)
34. Bostelmann, H., Cadamuro, D., Fewster, C.: Quantum energy inequality for the massive Ising model. Phys. Rev. D **88**(2), 025019 (2013)
35. Bratteli, O., Robinson, D.W.: Operator Algebras and Quantum Statistical Mechanics I. Springer, New York (1987)
36. Brunetti, R., Guido, D., Longo, R.: Modular localization and Wigner particles. Rev. Math. Phys. **14**, 759–786 (2002)
37. Buchholz, D.: Product states for local algebras. Comm. Math. Phys. **36**, 287–304 (1974)
38. Buchholz, D.: Collision theory for massless fermions. Comm. Math. Phys. **42**, 269 (1975)
39. Buchholz, D.: Collision theory for waves in two dimensions and a characterization of models with trivial S-matrix. Comm. Math. Phys. **45**, 1–8 (1975)
40. Buchholz, D.: Collision theory for massless bosons. Comm. Math. Phys. **52**, 147 (1977)
41. Buchholz, D., Fredenhagen, K.: Locality and the structure of particle states. Comm. Math. Phys. **84**, 1 (1982)
42. Buchholz, D., Junglas, P.: On the existence of equilibrium states in local quantum field theory. Comm. Math. Phys. **121**, 255–270 (1989)
43. Buchholz, D., Lechner, G.: Modular nuclearity and localization. Ann. Henri Poincaré **5**, 1065–1080 (2004)
44. Buchholz, D., Porrmann, M.: How small is the phase space in quantum field theory? Ann. Poincare Phys. Theor. **52**, 237 (1990)
45. Buchholz, D., Summers, S.J.: Stable quantum systems in anti-de Sitter space: causality, independence and spectral properties. J. Math. Phys. **45**, 4810–4831 (2004)
46. Buchholz, D., Summers, S.J.: Warped convolutions: a novel tool in the construction of quantum field theories. In: Seiler, E., Sibold, K. (eds.) Quantum Field Theory and Beyond: Essays in Honor of Wolfhart Zimmermann, pp. 107–121. World Scientific, Singapore (2008)
47. Buchholz, D., Verch, R.: Scaling algebras and renormalization group in algebraic quantum field theory. II: instructive examples. Rev. Math. Phys. **10**, 775–800 (1998)
48. Buchholz, D., Wichmann, E.H.: Causal independence and the energy level density of states in local quantum field theory. Comm. Math. Phys. **106**, 321 (1986)

49. Buchholz, D., D'Antoni, C., Longo, R.: Nuclear maps and modular structures 2: applications to quantum field theory. Comm. Math. Phys. **129**, 115 (1990)
50. Buchholz, D., D'Antoni, C., Longo, R.: Nuclear maps and modular structures. I. General properties. J. Funct. Anal. **88**, 233–250 (1990)
51. Buchholz, D., Mund, J., Summers, S.J.: Transplantation of local nets and geometric modular action on Robertson-walker space-times. Fields Inst. Commun. **30**, 65–81 (2001)
52. Buchholz, D., D'Antoni, C., Longo, R.: Nuclearity and thermal states in conformal field theory. Comm. Math. Phys. **270**, 267–293 (2007)
53. Buchholz, D., Lechner, G., Summers, S.J.: Warped convolutions, Rieffel deformations and the construction of quantum field theories. Comm. Math. Phys. **304**, 95–123 (2011)
54. Cadamuro, D.: A characterization theorem for local operators in factorizing scattering models. Ph.D. thesis, Göttingen University, 2012
55. Cadamuro, D., Tanimoto, Y.: Wedge-local fields in integrable models with bound states. Preprint, arXiv:1502.01313 (2015)
56. D'Antoni, C., Longo, R.: Interpolation by type I factors and the flip automorphism. J. Funct. Anal. **51**, 361 (1983)
57. D'Antoni, C., Morsella, G., Verch, R.: Scaling algebras for charged fields and short-distance analysis for localizable and topological charges. Ann. Henri Poincare **5**, 809–870 (2004)
58. Dappiaggi, C., Lechner, G., Morfa-Morales, E.: Deformations of quantum field theories on spacetimes with Killing fields. Comm. Math. Phys. **305**(1), 99–130 (2011)
59. Doplicher, S., Longo, R.: Standard and split inclusions of von Neumann algebras. Invent. Math. **75**, 493–536 (1984)
60. Doplicher, S., Fredenhagen, K., Roberts, J.E.: The Quantum structure of space-time at the Planck scale and quantum fields. Comm. Math. Phys. **172**, 187–220 (1995)
61. Doplicher, S., Haag, R., Roberts, J.E.: Fields, observables and gauge transformations. I. Comm. Math. Phys. **13**, 1–23 (1969)
62. Doplicher, S., Haag, R., Roberts, J.E.: Local observables and particle statistics. II. Comm. Math. Phys. **35**, 49–85 (1974)
63. Dorey, P.: Exact S-matrices. In: Horváth, Z., Palla, L. (eds.) Conformal Field Theories and Integrable Models. Lecture Notes in Physics, vol. 498, pp. 85–125. Springer, New York (1997)
64. Driessler, W.: Comments on lightlike translations and applications in relativistic quantum field theory. Comm. Math. Phys. **44**, 133–141 (1975)
65. Dütsch, M., Fredenhagen, K.: Algebraic quantum field theory, perturbation theory, and the loop expansion. Comm. Math. Phys. **219**, 5–30 (2001)
66. Dybalski, W.: Haag-Ruelle scattering theory in presence of massless particles. Lett. Math. Phys. **72**, 27–38 (2005)
67. Dybalski, W., Tanimoto, Y.: Asymptotic completeness in a class of massless relativistic quantum field theories. Comm. Math. Phys. **305**, 427–440 (2011)
68. Eckmann, J.-P., Osterwalder, K.: An application of Tomita's theory of modular Hilbert algebras: duality for free Bose fields. J. Funct. Anal. **13**(1), 1–12 (1973)
69. Faddeev, L.D.: Quantum completely integrable models in field theory, volume 1 of Mathematical Physics Reviews, pp. 107–155. In: Novikov, S.P. (ed.) Mathematical Physics Reviews, vol. 1, pp. 107–155 (1984)
70. Fassarella, L., Schroer, B.: Wigner particle theory and local quantum physics. J. Phys. A **35**, 9123–9164 (2002)
71. Figliolini, F., Guido, D.: On the type of second quantization factors. J. Oper. Theory **31**(2), 229–252 (1994)
72. Florig, M.: On Borchers' theorem. Lett. Math. Phys. **46**, 289–293 (1998)
73. Foit, J.J.: Abstract twisted duality for quantum free Fermi fields. Publ. Res. Inst. Math. Sci. Kyoto **19**, 729–741 (1983)
74. Fring, A., Mussardo, G., Simonetti, P.: Form-factors of the elementary field in the Bullough-Dodd model. Phys. Lett. B **307**, 83–90 (1993)
75. Gerstenhaber, M.: On the deformation of rings and algebras. Ann. Math. **79**, 59–103 (1964)

76. Glimm, J., Jaffe, A.: Quantum Physics. A Functional Integral Point of View, 535p. Springer, New York, (1987)
77. Gracia-Bondia, J.M., Varilly, J.C.: Algebras of distributions suitable for phase space quantum mechanics. I. J. Math. Phys. **29**, 869–879 (1988)
78. Grosse, H., Lechner, G.: Wedge-local quantum fields and noncommutative Minkowski space. JHEP **11**, 012 (2007)
79. Grosse, H., Lechner, G.: Noncommutative deformations of wightman quantum field theories. JHEP **09**, 131 (2008)
80. Grosse, H., Lechner, G., Ludwig, T., Verch, R.: Wick rotation for quantum field theories on degenerate Moyal space (-time). J. Math. Phys. **54**, 022307 (2013)
81. Guido, D., Longo, R.: An algebraic spin and statistics theorem. Comm. Math. Phys. **172**(3), 517 (1995)
82. Guido, D., Longo, R., Wiesbrock, H.W.: Extensions of conformal nets and superselection structures. Comm. Math. Phys. **192**, 217–244 (1998)
83. Haag, R.: Quantum field theories with composite particles and asymptotic conditions. Phys. Rev. **112**, 669–73 (1958)
84. Haag, R.: Local Quantum Physics—Fields, Particles, Algebras, 2nd edn. Springer, Berlin (1996)
85. Haagerup, U.: Conne's bicentralizer problem and uniqueness of the injective factor of type III1. Acta Math. **158**(1), 95 (1987)
86. Hepp, K.: On the connection between Wightman and LSZ quantum field theory. Comm. Math. Phys. **1**, 95–111 (1965)
87. Iagolnitzer, D.: Scattering in Quantum Field Theories. Princeton University Press, Princeton (1993)
88. Jost, R.: The General Theory of Quantized Fields. American Mathematical Society, Providence (1965)
89. Karowski, M., Thun, H.J., Truong, T.T., Weisz, P.H.: On the uniqueness of a purely elastic S-matrix in (1+1) dimensions. Phys. Lett. B **67**(3), 321–322 (1977)
90. Kasprzak, P.: Rieffel Deformation via crossed products. J. Funct. Anal. **257**(5), 1288–1332 (2009)
91. Kawahigashi, Y., Longo, R.: Classification of local conformal nets: case $c < 1$. Ann. Math. **160**, 493–522 (2004)
92. Köhler, C.: Ph.D. thesis, University of Vienna (2015) (To appear)
93. Lauridsen-Ribeiro, P.: Structural and dynamical aspects of the AdS/CFT correspondence: a rigorous approach. Ph.D. thesis, Sao Paulo, 2007
94. Lechner, G.: Polarization-free quantum fields and interaction. Lett. Math. Phys. **64**, 137–154 (2003)
95. Lechner, G.: On the existence of local observables in theories with a factorizing S-matrix. J. Phys. A **38**, 3045–3056 (2005)
96. Lechner, G.: On the construction of quantum field theories with factorizing S-matrices. Ph.D. thesis, University of Göttingen, 2006
97. Lechner, G.: Construction of quantum field theories with factorizing S-matrices. Comm. Math. Phys. **277**, 821–860 (2008)
98. Lechner, G.: Deformations of quantum field theories and integrable models. Comm. Math. Phys. **312**(1), 265–302 (2012)
99. Lechner, G.: Erratum (2015)
100. Lechner, G., Longo, R.: Localization in nets of standard spaces. Comm. Math. Phys. **336**, 27–61 (2015)
101. Lechner, G., Schlemmer, J.: Thermal equilibrium states for quantum fields on noncommutative spacetimes. arXiv:1503.01639 (2015)
102. Lechner, G., Schützenhofer, C.: Towards an operator-algebraic construction of integrable global gauge theories. Ann. Henri Poincaré **15**(4), 645–678 (2014)
103. Lechner, G., Waldmann, S.: Strict deformation quantization of locally convex algebras and modules. Preprint, arXiv:1109.5950 (2011)

104. Lechner, G., Schlemmer, J., Tanimoto, Y.: On the equivalence of two deformation schemes in quantum field theory. Lett. Math. Phys. **103**(4), 421–437 (2013)
105. Leylands, P., Roberts, J.E., Testard, D.: Duality for Quantum Free Fields. Preprint, 1978
106. Liguori, A., Mintchev, M.: Fock representations of quantum fields with generalized statistics. Comm. Math. Phys. **169**, 635–652 (1995)
107. Longo, R.: Notes on algebraic invariants for noncommutative dynamical systems. Comm. Math. Phys. **69**, 195–207 (1979)
108. Longo, R.: Lectures on conformal Nets—Part 1. In: Von Neumann algebras in Sibiu, Theta (2008)
109. Longo, R., Witten, E.: An algebraic construction of boundary quantum field theory. Comm. Math. Phys. **303**(1), 213–232 (2011)
110. Martin, A.: Can one continue the scattering amplitude through the elastic cut? (1969)
111. Morfa-Morales, E.: Deformations of quantum field theories on curved spacetimes. Ph.D. thesis, University of Vienna, 2012
112. Much, A.: Quantum mechanical effects from deformation theory. J. Math. Phys. **55**, 022302 (2014)
113. Müger, M.: Superselection structure of massive quantum field theories in 1+1 dimensions. Rev. Math. Phys. **10**, 1147–1170 (1998)
114. Mund, J.: The Bisognano-Wichmann theorem for massive theories. Ann. Henri Poincare **2**, 907–926 (2001)
115. Mund, J.: An algebraic Jost-Schroer theorem for massive theories. Comm. Math. Phys. **315**, 445–464 (2012)
116. Mund, J., Schroer, B., Yngvason, J.: String-localized quantum fields and modular localization. Comm. Math. Phys. **268**, 621–672 (2006)
117. Mussardo, G.: Off critical statistical models: factorized scattering theories and bootstrap program. Phys. Rept. **218**, 215–379 (1992)
118. Neshveyev, S.: Smooth crossed products of Rieffel's deformations. Lett. Math. Phys. **104**(3), 361–371 (2014)
119. Niedermaier, M.R.: A derivation of the cyclic form factor equation. Comm. Math. Phys. **196**, 411–428 (1998)
120. Plaschke, M.: Wedge local deformations of charged fields leading to anyonic commutation relations. Lett. Math. Phys. **103**(5), 507–532 (2013)
121. Read, C.J.: Quantum field theories in all dimensions. Comm. Math. Phys. **177**(3), 631 (1996)
122. Rehren, K.-H.: Comments on a recent solution to Wightman's axioms. Comm. Math. Phys. **178**, 453–466 (1996)
123. Rieffel, M.A.: Deformation Quantization for Actions of R^d. Memoirs of the American Mathematical Society, vol. 106. American Mathematical Society, Providence (1992)
124. Ruelle, D.: On the asymptotic condition in quantum field theory. Helv. Phys. Acta **35**, 147–163 (1962)
125. Schroer, B.: Modular localization and the bootstrap-formfactor program. Nucl. Phys. B **499**, 547–568 (1997)
126. Schroer, B.: Modular wedge localization and the d=1+1 formfactor program. Ann. Phys. **275**, 190–223 (1999)
127. Smirnov, F.A.: Form Factors in Completely Integrable Models of Quantum Field Theory. World Scientific, Singapore (1992)
128. Smirnov, F.A.: A New set of exact form-factors. Int. J. Mod. Phys. A **9**, 5121–5144 (1994)
129. Soloviev, M.A.: On the failure of microcausality in noncommutative field theories. Phys. Rev. D **77**, 125013 (2008)
130. Streater, R.F., Wightman, A.: PCT, Spin and Statistics, and All That. Benjamin-Cummings, Reading (1964)
131. Summers, S.J.: Normal product states for fermions and twisted duality for CCR- and CAR-type algebras with application to the Yukawa2 quantum field model. Comm. Math. Phys. **86**, 111–141 (1982)

132. Summers, S.J.: On the independence of local algebras in quantum field theory. Rev. Math. Phys. **2**, 201–247 (1990)
133. Summers, S.J.: A Perspective on Constructive Quantum Field Theory. arXiv:1203.3991 (2012)
134. Tanimoto, Y.: Construction of wedge-local nets of observables through Longo-Witten endomorphisms. Comm. Math. Phys. **314**(2), 443–469 (2012)
135. Thomas, L.J., Wichmann, E.H.: On the causal structure of Minkowski space-time. J. Math. Phys. **38**, 5044–5086 (1997)
136. Waldmann, S.: Poisson-Geometrie und Deformationsquantisierung. Springer, New York (2007)
137. Weinberg, S.: The Quantum Theory of Fields I—Foundations. Cambridge University Press, Cambridge (1995)
138. Yngvason, J.: Zero-mass infinite spin representations of the Poincaré group and quantum field theory. Comm. Math. Phys. **18**, 195–203 (1970)
139. Zamolodchikov, A.B.: Factorized S-matrices in two dimensions as the exact solutions of certain relativistic quantum field models. Ann. Phys. **120**, 253–291 (1979)

Index

R. Brunetti et al. (eds.), *Advances in Algebraic Quantum Field Theory*,
Mathematical Physics Studies, DOI 10.1007/978-3-319-21353-8

MIX
Papier aus verantwortungsvollen Quellen
Paper from responsible sources
FSC® C105338

If you have any concerns about our products,
you can contact us on
ProductSafety@springernature.com

In case Publisher is established outside the EU,
the EU authorized representative is:
Springer Nature Customer Service Center GmbH
Europaplatz 3, 69115 Heidelberg, Germany

Printed by Libri Plureos GmbH
in Hamburg, Germany